Relative Acidities of Various Compounds

pK_a	Conjugate acid			Conjugate base
~ -9.5	HI	$\longrightarrow$	$H^+ +$	I^-
~ -9	HBr	$\longrightarrow$	$H^+ +$	Br^-
~ -7	HCl	$\longrightarrow$	$H^+ +$	Cl^-
~ -5	H_2SO_4	$\longrightarrow$	$H^+ +$	HSO_4^-
-1.7	H_3O^+	$\rightleftharpoons$	$H^+ +$	H_2O
2.1	H_3PO_4	$\rightleftharpoons$	$H^+ +$	$H_2PO_4^-$
3.45	HF	$\rightleftharpoons$	$H^+ +$	F^-
4.75	CH_3CO_2H	$\rightleftharpoons$	$H^+ +$	$CH_3CO_2^-$
$3-6$	RCO_2H	$\rightleftharpoons$	$H^+ +$	RCO_2^-
6.4	H_2CO_3	$\rightleftharpoons$	$H^+ +$	HCO_3^-
9	$CH_3\overset{O}{\overset{\|}{C}}CH_2\overset{O}{\overset{\|}{C}}CH_3$	$\rightleftharpoons$	$H^+ +$	$CH_3\overset{O}{\overset{\|}{C}}\overset{-}{C}HCCH_3$
9.3	HCN	$\rightleftharpoons$	$H^+ +$	^-CN
9.4	NH_4^+	$\rightleftharpoons$	$H^+ +$	NH_3
10.00	⟨O⟩—OH	$\rightleftharpoons$	$H^+ +$	⟨O⟩—O$^-$
11	$CH_3\overset{O}{\overset{\|}{C}}CH_2\overset{O}{\overset{\|}{C}}OC_2H_5$	$\rightleftharpoons$	$H^+ +$	$CH_3\overset{O}{\overset{\|}{C}}\overset{-}{C}H\overset{O}{\overset{\|}{C}}OC_2H_5$
15.74	H_2O	$\rightleftharpoons$	$H^+ +$	^-OH
15.9	CH_3CH_2OH	$\rightleftharpoons$	$H^+ +$	$CH_3CH_2O^-$
$15-19$	ROH	$\rightleftharpoons$	$H^+ +$	RO^-
20	$CH_3\overset{O}{\overset{\|}{C}}CH_3$	$\rightleftharpoons$	$H^+ +$	$^-CH_2\overset{O}{\overset{\|}{C}}CH_3$
25	$CH_3\overset{O}{\overset{\|}{C}}OC_2H_5$	$\rightleftharpoons$	$H^+ +$	$^-CH_2\overset{O}{\overset{\|}{C}}OC_2H_5$
~ 26	$CH \equiv CH$	$\rightleftharpoons$	$H^+ +$	$^-C\equiv CH$
~ 35	NH_3	$\rightleftharpoons$	$H^+ +$	$^-NH_2$
~ 45	$CH_2 = CH_2$	$\rightleftharpoons$	$H^+ +$	$^-CH=CH_2$
~ 50	CH_3CH_3	$\rightleftharpoons$	$H^+ +$	$^-CH_2CH_3$

**Fundamentals of
Organic Chemistry**

Fundamentals of Organic Chemistry

Ralph J. Fessenden

Joan S. Fessenden

University of Montana

1817

HARPER & ROW, PUBLISHERS, New York

Grand Rapids, Philadelphia, St. Louis, San Francisco,
London, Singapore, Sydney, Tokyo

Sponsoring Editor: Glyn Davies
Cover Design: Lucy Krikorian
Cover Photo: © Photo Researchers, Inc., Sidney Moulds
Text Art: Accurate Art, Inc.
Production Manager: Kewal Sharma
Production Assistant: Linda Murray
Compositor: Science Typographers, Inc.
Printer and Binder: R. R. Donnelley & Sons Co.
Cover Printer: Phoenix Color Corp.

Fundamentals of Organic Chemistry

Library of Congress Cataloging in Publication Data

Fessenden, Ralph J., 1932–
    Fundamentals of organic chemistry/Ralph J. Fessenden, Joan S.
Fessenden.
      p.   cm.
    Includes index.
    ISBN 0-06-042035-9
      1. Chemsitry, Organic.   I. Fessenden, Joan S.   II. Title.
QD253.F45   1990
                                              547--dc20   89-34074
                                                                  CIP

92   9   8   7   6   5   4   3   2

Contents

Preface

This text is a one-semester (or one to two quarters) introduction to organic chemistry intended for majors in science and science-related fields, including agriculture, biological sciences, home economics, nursing, and pharmacy.

Because students using this text may not take additional courses in chemistry, we have attempted, whenever feasible, to relate organic chemistry to everyday life and to other disciplines. For additional interest value, we have included a brief "Point of Interest" at the end of each chapter. The varied topics covered in these include biographical asides, applications of organic chemistry in consumer products, and organic chemistry in nature.

ORGANIZATION

The organization of this text is relatively classical. Chapter 1 contains a brief review of structure and bonding. In Chapter 1 we also introduce acid–base chemistry because we use these concepts throughout the book. More detail concerning acids and bases is found in the chapters on carboxylic acids and amines.

Chapter 2 (alkanes) includes an introduction to nomenclature. This chapter is followed by chapters on alkenes and alkynes (Chapter 3), dienes (Chapter 4), and aromatic compounds (Chapter 5).

Chapter 6 covers stereochemistry. This material is covered early so that it can be used in the discussions of mechanisms in the coming chapters.

The next three chapters (Chapters 7–9) include compound classes with single-bond functional groups—alkyl halides, alcohols and phenols, and ethers and epoxides. These chapters are followed by Chapters 10–12 on carbonyl compounds —aldehydes and ketones, carboxylic acids, and derivatives of carboxylic acids.

Chapter 13 covers additions and condensations (enolates and active methylene chemistry). We have found it convenient and efficient to combine aldol additions, ester condensations, and related reactions in one chapter because of their similarities. Also, this approach allows an instructor to emphasize synthesis reactions if this is desired.

Chapter 14 (amines) is placed this late in the text because many of the important reactions of amines are those with carbonyl compounds. Amines are used throughout the text as bases and nucleophiles; therefore, many of the reactions in this chapter have already been covered in earlier chapters.

Chapter 15 (spectroscopy) is an introduction to infrared, ultraviolet/visible, and nuclear magnetic resonance (^{1}H and ^{13}C) spectroscopy. We include simplified explanations of the theoretical basis of each form of spectroscopy, the principal effects of structure on the positions and intensities of absorption, and the general techniques for interpreting spectra.

The final chapters are devoted to the structures and organic chemistry of the major classes of biological compounds—carbohydrates, lipids, proteins, and nucleic acids.

NOMENCLATURE

In this text we emphasize nomenclature of the International Union of Pure and Applied Chemistry (IUPAC); however, we do include the trivial and semitrivial names, such as acetone and isopropyl alcohol, that are commonly used in chemistry and other fields. To prevent confusion, we generally indicate which name is IUPAC and which is trivial when we present both.

We have included the (E) and (Z) system for the naming of alkenes; however, we present this system in Chapter 6 (stereochemistry) instead of in Chapter 3 (alkenes). The reason for this placement is that the Cahn-Ingold-Prelog rules are not presented until Chapter 6.

MECHANISMS

The major reactions for each class of compounds are explained with stepwise "electron-pushing" mechanisms where possible. Mechanisms such as these help a student see *how* and *why* reactants go to products and thus help the student interrelate and remember the various reactions.

PROBLEMS

Several types of problems appear in the text.

Examples. Numerous carefully solved examples show the student how to work out problems in organic chemistry.

Study problems within each chapter. These expand on the examples or check the student's grasp of newly presented material. The solutions to these problems are at the end of the text.

Chapter-end study problems. These generally follow the chapter order until the last few, which are more challenging problems. The solutions to the chapter-end problems are contained in the *Student Study Guide* that accompanies this text.

STUDY AIDS

We strongly urge any student of organic chemistry to purchase an inexpensive set of molecular models. These are usually available at campus bookstores.

The Study Guide that accompanies this text contains hints on how to study organic chemistry, reviews the most important features in each chapter, shows step-by-step techniques for solving problems, and provides the solutions to the chapter-end study problems.

ACKNOWLEDGMENTS

We are indebted to a number of our colleagues for their advice and suggestions during the development of this text. We are especially grateful to those who reviewed the manuscript during various stages:

Peter R. Adams, Pennsylvania State University, York;

Merle Battiste, University of Florida, Gainesville;

Otis C. Dermer, Aklahoma State University;

Guy Mattson, University of Central Florida;

Daniel H. O'Brien, Texas A & M University;

John H. Penn, West Virginia University;

Iris Kaye Stovall, University of Illinois, Urbana–Champaign;

and Kay G. Turner, Rochester Institute of Technology.

Their help has been invaluable

We also thank the editors and staff at Harper & Row. It has been a pleasure to work with them.

Ralph J. Fessenden
Joan S. Fessenden

To the Student

Organic chemistry, the chemistry of carbon compounds, is an integral part of everyday life. Life exists because of organic reactions. Gasoline and lubricating oils, plastics, synthetic fibers, cotton, wool, medicines, and soaps—all are organic compounds. Organic chemistry is thus an important part of many areas of study—agriculture, medicine, nutrition, and many others.

Although organic chemistry is not difficult to learn, much of the terminology is new and the formulas may seem complex. You may find it helpful to read the *summary* of a chapter first to familiarize yourself with new terms. Then skim the chapter to see what these terms mean and how they are used. If you do these two steps *before* attending each class, you will find the lectures to be more meaningful. Finally, study the chapter for learning and understanding. *Outlining the material* is one of the best ways to approach this final step.

Many students find that the difficult part of learning organic chemistry is its cumulative nature. The instructor might say, "The material in today's lecture is based on the subjects presented last week." The only way to master a cumulative subject is to set aside time during each study session for review of material presented earlier. Review the material "backward," spending most of your review time on recently presented topics and less time on more distant subjects.

Work all problems *on paper*, not in your head. Writing down your answers will give you practice in the important art of writing formulas. Work the in-chapter problems as best you can before checking the answers at the end of the book.

Use *molecular models* right from the start. Molecular models are indispensable in discussions involving stereochemistry, such as Newman projections or determining (R) or (S) configuration.

Ralph J. Fessenden
Joan S. Fessenden

Chapter 1

Structure and Bonding in Organic Compounds

The name "organic" chemistry means the chemistry of compounds that come from living things. In the early years of organic chemistry, chemists believed that a living plant or animal was necessary for the synthesis of organic compounds. An experiment performed by the German chemist Friedrich Wöhler in 1828 showed this belief to be false. He prepared urea (a component of urine) in the laboratory from an inorganic compound. Afterward, he wrote to a colleague, "I can make urea without the necessity of a kidney."

Wöhler's experiment is generally cited as the dawn of modern organic chemistry. We now know that neither life force nor magic is needed to prepare organic compounds. Our synthetic fabrics, plastics, pharmaceuticals, and many other organic products are routinely synthesized in laboratories and factories. Organic compounds are diverse, but have one feature in common: they all contain carbon. Today, organic chemistry is defined as the *chemistry of compounds that contain carbon*.

All organic molecules have atoms joined together by covalent bonds—bonds in which the atoms share the bonding electrons. To present organic chemistry, we must first discuss covalent bonding. Therefore, in this chapter we begin our study of organic chemistry with chemical bonds. We also discuss how the bonding in organic compounds affects some of the properties of these compounds.

1.1 ATOMIC STRUCTURE

The smallest particle of an element that retains the unique properties of that element is the **atom**. On the earth, only a few elements, the noble gases (helium, argon, etc.), are commonly composed of individual atoms. Most elements, such as oxygen (O_2), and all compounds, such as sucrose (table sugar), are composed of **molecules** or **ions**, which are discussed in Sections 1.2 and 1.3.

An atom is composed of a small, but relatively heavy, positively charged **nucleus** surrounded by lightweight, negatively charged **electrons**. The major nuclear components are the positively charged **protons** and the neutral **neutrons**.

We can characterize an atom by its **atomic number** and the **atomic weight** or **atomic mass**. The atomic number of an element is the number of protons in the nucleus. The atomic weight of an element is the sum of the number of protons and the number of neutrons in the nucleus. Atomic weight is reported as a weighted average to account for both the mass and the abundance of naturally occurring isotopes. **Formula weight** is the sum of atomic weights of atoms in a molecule.

A. Electron Structure of the Atom

The electrons of an atom are restricted to specific regions in space called **electron shells**. The first electron shell is the one closest to the nucleus, the second shell is the next one out, and so forth up to the seventh and final shell.

Within each shell, electrons are further restricted to special regions called **atomic orbitals**. The number of orbitals a shell contains depends on its distance from the nucleus. The first shell contains only a single orbital, the 1s orbital. The number

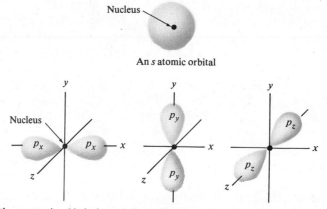

The three p atomic orbitals shown individually along with their mutually perpendicular axes

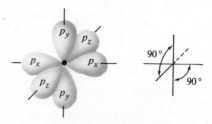

The three p orbitals shown together

Figure 1.1 Geometry of the s and p atomic orbitals.

1 refers to the first shell, and the letter s refers to the shape of the orbital. An s orbital is a spherical orbital with the nucleus at the center (see Figure 1.1). The second shell contains two types of orbitals, one $2s$ orbital and three $2p$ orbitals. The $2s$ orbital, like the $1s$ orbital, is spherical. A p orbital is not spherical but is shaped somewhat like a dumbbell with the nucleus at the center. The three $2p$ orbitals are positioned at right angles to one another, as shown in Figure 1.1.

As the orbitals become progressively farther from the nucleus, the energy of any electrons they contain becomes greater. The lowest-energy orbital, the one in which an electron is most stable, is the $1s$ orbital. The next lowest-energy orbital is the $2s$ orbital, followed by the three $2p$ orbitals. All orbitals in the third shell are of higher energy than those in the second shell.

Although negatively charged particles repel one another, electrons of opposite spin are allowed to occupy the same orbital, or become *paired* in that orbital. Each orbital in any shell can hold a maximum of two electrons. As atomic number increases in a series of elements, the additional electrons that must be added seek the lowest-energy orbitals possible. Thus a neutral atom of hydrogen (H, atomic number 1), which contains one proton, contains its single electron in the $1s$ orbital. A neutral atom of helium (He, atomic number 2) contains *two* electrons in the $1s$ orbital. The $1s$ orbital and the first electron shell are now filled. Helium, with a filled first electron shell, is the first noble gas. The element lithium (Li, atomic number 3) contains two electrons in the $1s$ orbital and one electron in the next highest orbital, the $2s$ orbital. The electron structures of the first 18 elements are shown in Table 1.1.

TABLE 1.1   ELECTRON CONFIGURATIONS OF THE FIRST 18 ELEMENTS

Element	Atomic number	Orbital								
		$1s$	$2s$	$2p_x$	$2p_y$	$2p_z$	$3s$	$3p_x$	$3p_y$	$3p_z$
H	1	1								
He	2	2								
Li	3	2	1							
Be	4	2	2							
B	5	2	2	1						
C	6	2	2	1	1					
N	7	2	2	1	1	1				
O	8	2	2	2	1	1				
F	9	2	2	2	2	1				
Ne	10	2	2	2	2	2				
Na	11	2	2	2	2	2	1			
Mg	12	2	2	2	2	2	2			
Al	13	2	2	2	2	2	2	1		
Si	14	2	2	2	2	2	2	1	1	
P	15	2	2	2	2	2	2	1	1	1
S	16	2	2	2	2	2	2	2	1	1
Cl	17	2	2	2	2	2	2	2	2	1
Ar	18	2	2	2	2	2	2	2	2	2

B. Atomic Radii

The **atomic radius** of an atom is the distance from the center of its nucleus to its outermost electrons. This value is calculated by dividing the distance between the two nuclei of a symmetrical diatomic molecule, such as H_2 or O_2, by two. The atomic radius of an element does not vary significantly from compound to compound. By simply adding the atomic radii of two elements, we can approximate the length of a covalent bond between the two.

Values for atomic radius are generally reported in angstroms (Å), where $1 \text{ Å} = 10^{-8}$ cm, or in nanometers (nm), where $1 \text{ nm} = 10^{-7}$ cm. The atomic radius of hydrogen, for example, is 0.37 Å, or 0.037 nm.

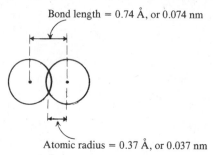

Bond length = 0.74 Å, or 0.074 nm

Atomic radius = 0.37 Å, or 0.037 nm

The atomic radius of an element depends on two factors: the number of electron shells and the number of protons in the nucleus. When comparing two atoms that have different numbers of electron shells, we find that the atom with the larger number of electron shells has the larger atomic radius. For example, fluorine with two shells has an atomic radius of 0.72 Å, while chlorine with three shells has an atomic radius of 0.99 Å.

When comparing two atoms that have the same number of electron shells, we find that atomic radius decreases with increasing number of protons in the nucleus. For example, oxygen contains eight protons, while fluorine contains nine; the atomic radius of fluorine is smaller. The reason for this trend is that an increasing number of protons in the nucleus attracts the electrons more strongly and thus decreases the atomic radius.

O   F

Atomic number 8   Atomic number 9
Atomic radius = 0.74 Å   Atomic radius = 0.72 Å

C. Electronegativity

Electronegativity is a measure of the attraction of an atom's nucleus for its outermost electrons, in particular the bonding electrons in covalent molecules. Atoms that tend to attract or to gain electrons, such as O or Cl, are more electronegative than atoms that tend to lose electrons, such as Na or other metals. Elements whose atoms tend to attract electrons are called **electronegative elements**, and elements whose atoms tend to lose electrons are called **electropositive elements**. These terms are relative; oxygen is more electronegative than carbon but more electropositive than fluorine.

H 2.1						
Li 1.0	Be 1.5	B 2.0	C 2.5	N 3.0	O 3.5	F 4.0
Na 0.9	Mg 1.2	Al 1.5	Si 1.8	P 2.1	S 2.5	Cl 3.0
						Br 2.8

Figure 1.2 Electronegativities (Pauling scale) of selected elements.

The elements have been ranked numerically according to their electronegativities. Figure 1.2 shows a few of the more common elements as they appear in the periodic table along with their electronegativity values. This scale of values is called the **Pauling scale of electronegativities**, after the American chemist Linus Pauling, who received the Nobel Prize in Chemistry in 1954 for his work in bonding and structure. The scale of electronegativities for the elements shown in Figure 1.2 ranges from 0.9 for Na (low electronegativity) to 4.0 for F, the most electronegative element. Carbon has an intermediate electronegativity of 2.5.

Note the inverse relationship between atomic radius and electronegativity. Atomic radii decrease as we move from left to right in the periodic table, and electronegativity increases. Atomic radii increase as we proceed from top to bottom in the periodic table and add electron shells. Electronegativity decreases as we add electron shells because the outer electrons are more shielded from the nucleus.

Increasing electronegativity →

Li	Be	B	C	N	O	F
Na	Mg	Al	Si	P	S	Cl

↑ Increasing electronegativity

PROBLEM 1.1.   Referring to the periodic table inside the cover of this book, predict the relative electronegativities of the following two series:

(a) N   P   As   **(b)** As   Se   Br

1.2 IONIC COMPOUNDS

An ion is an atom or a group of covalently bonded atoms with an electrical charge. If the charge is positive, the ion is called a **cation**; if the charge is negative, the ion is called an **anion**. An ionic bond is the electrostatic attraction between oppositely charged ions.

Typical cations: Na^+, K^+, Ca^{2+}, $^+NH_4$

Typical anions: Cl^-, Br^-, ^-OH

TABLE 1.2 LEWIS STRUCTURES FOR THE FIRST 18 ELEMENTS

IA	IIA	IIIA	IVA	VA	VIA	VIIA	VIIIA
H·							He:
Li·	Be:	Ḃ:	·Ċ:	·N̈:	:Ö:	:F̈:	:N̈e:
Na·	Mg:	Ȧl:	·Ṡi:	·P̈:	:S̈:	:C̈l:	:Är:

A. Lewis Formulas

The electrons in the outermost shell, or **valence electrons**, of an atom are used for bonding. Formulas showing only valence electrons and ignoring electrons in underlying shells are called **Lewis formulas** after the chemist G. N. Lewis (1875–1946), a professor at the University of California, Berkeley, who developed them. Ionic formulas, and also sometimes covalent formulas, can be represented by Lewis formulas. Table 1.2 shows the Lewis formulas for the first 18 elements in the periodic table. The Lewis formulas for the noble gases helium and neon show their valence shells to be filled with two electrons and eight electrons, respectively. The Lewis formula of argon also shows eight electrons; the third shell acts like a full shell when it contains an octet.

B. Ionic Bond Formation

An ionic bond is formed in a chemical reaction by the complete transfer of one or more electrons from one atom to another. Sodium chloride ($Na^+ Cl^-$), common table salt, is a typical ionic compound.

Formation of an Ionic Bond:

$$Na· \ + \ :\overset{..}{\underset{..}{C}}l: \longrightarrow Na^+ \ + \ :\overset{..}{\underset{..}{C}}l:^-$$

When an ionic compound dissolves in water, the ions become surrounded by water molecules because of the strong attractions between charged ions and water molecules. On dissolving, the ions separate, or dissociate, from one another.

PROBLEM 1.2.   Write equations, similar to the preceding one, showing **(a)** the formation of calcium chloride ($Ca^{2+} + 2Cl^-$) from atoms of calcium and chlorine and **(b)** the formation of magnesium oxide ($Mg^{2+} + O^{2-}$) from atoms of magnesium and oxygen.

1.3 COVALENT COMPOUNDS

Ionic compounds are formed when the difference in electronegativity between two reacting elements is relatively large, about 1.7 or more. For example, sodium (0.9) and chlorine (3.0) have an electronegativity difference of 2.1; therefore, on reaction they form the ionic compound sodium chloride. Atoms with electronegativity differences smaller than 1.7 do not form ionic bonds on reaction. They may, however, react to form a covalent bond. A **covalent bond** is the sharing of a pair of electrons between two atoms. The two electrons are commonly represented as a single line (—) joining the atomic symbols. This type of formula is called a **line-bond formula**.

$$\text{H:H} \quad \text{or} \quad \text{H—H} \qquad\qquad \text{H:}\overset{\overset{\displaystyle H}{\cdot\cdot}}{\underset{\overset{\cdot\cdot}{\displaystyle H}}{C}}\text{:H} \quad \text{or} \quad \text{H—}\overset{\displaystyle H}{\underset{\displaystyle H}{C}}\text{—H}$$

Hydrogen   Methane

The ions of ionic compounds, such as $Na^+ Cl^-$, separate from one another, or dissociate, when dissolved in water. This behavior is in direct contrast to that of covalent compounds. When dissolved, the molecules of most covalently bonded compounds do not separate to any extent. For example, when methanol dissolves in water, the hydrogen atoms and oxygen atoms remain bonded to the carbon just as they are in the pure state.

$$\text{H—}\overset{\displaystyle H}{\underset{\displaystyle H}{C}}\text{—O—H}$$

Methanol

a covalently bonded compound

The amount of energy required to separate bonded atoms into neutral particles is called the **bond dissociation energy**. Bond dissociation energies are used to compare the stabilities of different bonds: the greater the bond dissociation energy, the more stable the bond. For example, the bond dissociation energy for the Cl—Cl bond is 58 kcal/mol, while that for the CH_3—H bond is 104 kcal/mol. The CH_3—H bond is stronger and more stable than the Cl—Cl bond.

A. Why Atoms Share Electrons

In 1916 G. N. Lewis and others proposed that atoms tend to lose, gain, or share electrons in order to attain a **noble-gas electron configuration**. For example, when hydrogen and lithium form covalent bonds to other elements, they form an electron configuration containing two valence electrons, similar to that of helium. Most common elements, however, have stable electron configurations when their valence shell contains eight electrons—the same electron configuration as that of neon or argon. Consequently, Lewis's proposal is often called the **octet rule**.

$$H\cdot \; + \; \cdot \ddot{\underset{\cdot\cdot}{Cl}}: \; \longrightarrow \; \left(H \left(\overset{\cdot\cdot}{\cdot}\right) \ddot{\underset{\cdot\cdot}{Cl}} :\right)$$

H has two electrons
in its outer shell

Cl has eight electrons
in its outer shell

Carbon has an intermediate electronegativity, and the electronegativity differences between carbon and the elements commonly found in organic compounds are small. For example, the electronegativity difference between C and H is only 0.4, and the difference between C and O is 1.0. Larger differences in electronegativity are required for complete electron transfer and ionic bond formation. Therefore, carbon rarely forms ionic bonds but almost always forms covalent bonds.

A carbon atom has four valence electrons and, to attain an octet, it must share four additional electrons with other atoms. Carbon forms four covalent bonds.

$$\cdot \overset{\cdot}{\underset{\cdot}{C}} \cdot \; + \; 4H\cdot \; \longrightarrow \; H : \overset{\overset{\displaystyle H}{\cdot\cdot}}{\underset{\underset{\displaystyle H}{\cdot\cdot}}{C}} : H$$

C forms four bonds

B. The Number of Covalent Bonds an Atom Forms

Table 1.3 lists some common elements found in organic compounds along with the numbers of covalent bonds they usually form and their positions in the periodic table. Hydrogen always forms one bond and carbon always forms four bonds in stable covalent compounds.

TABLE 1.3   NUMBER OF COVALENT BONDS SOME ELEMENTS FORM
IN NEUTRAL COVALENT COMPOUNDS

Element	Number of bonds	Number of group in periodic table
Hydrogen (H)	1	IA
Carbon (C)	4	IVA
Nitrogen (N)	3	VA
Oxygen (O)	2	VIA
Halogens (F, Cl, Br, I)	1	VIIA

One unique feature of carbon is its ability to share electrons with other carbon atoms, an ability that allows the formation of an almost infinite variety of chains and rings.

In stable compounds, oxygen generally forms two covalent bonds and nitrogen, three. Both these elements, however, also form negative and positive ions, as shown in the following formulas.

The sharing of two electrons by two atoms is called a **single bond**. Many elements that form more than one covalent bond, such as C, N, and O, can also form multiple bonds. A **double bond** is the sharing of two pairs of electrons between two atoms. Carbon can form double bonds with other carbon atoms, with oxygen, or with nitrogen.

Carbon can also form triple bonds with other carbon atoms or with nitrogen. A **triple bond** is the sharing of three pairs of electrons between two atoms.

$$H-C\equiv C-H \qquad H-C\equiv N$$

Regardless of whether carbon is bonded by single bonds, double bonds, or a triple bond, carbon almost always has four bonds.

C. Polar Covalent Bonds

When two identical atoms (such as two carbon atoms of a C—C bond) share a pair of electrons, they share the electrons equally. The reason is that the two identical atoms have identical electronegativities and thus equal attractions for the bonding electrons. A bond in which the electrons are shared equally is called a **nonpolar covalent bond**.

Typical Molecules with Nonpolar Covalent Bonds:

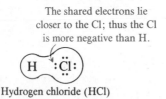

H—H   Cl—Cl   N≡N

Electrons equally shared

Covalent bonds formed from dissimilar elements, such as the C—H bond, are also considered nonpolar provided that the difference in electronegativity is small.

But what of covalent bonds between elements that have relatively large differences in electronegativity, such as the covalent bond in HCl? In the gaseous state, hydrogen chloride is a covalent molecule. (If it were ionic, it would be a crystalline solid like sodium chloride and not gaseous.) The chlorine has a stronger attraction for the bonding electrons than does the hydrogen, and the electrons are *not* shared equally by the two bonded atoms. A bond in which the shared electrons are pulled closer to one of the bonded atoms is called a **polar covalent bond**.

The shared electrons lie
closer to the Cl; thus the Cl
is more negative than H.

H   :C̈l:

Hydrogen chloride (HCl)

In HCl, the Cl is more negative than is H. Because Cl and H are still joined by a covalent bond, the Cl does not carry a complete negative ionic charge, but it does have a slight charge. We say that the chlorine has a **partial negative charge**, which we represent as $\delta-$ ("delta minus"). The H carries a **partial positive charge**, represented as $\delta+$ ("delta plus"). Because of these partial charges, a molecule of H—Cl is polar.

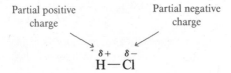

Partial positive   Partial negative
charge   charge

$\delta+$ $\delta-$
H—Cl

Water is another example of a polar compound. In a molecule of water, the more electronegative oxygen carries a partial negative charge and each hydrogen carries a partial positive charge. The polarity of water molecules makes water a good solvent for ionic compounds and for polar organic compounds.

The Polar Bonds in Water:

$$\overset{\delta-}{\underset{\delta+\,H}{\diagup}}\overset{O}{\diagdown}\underset{H\,\delta+}{}$$

Examples of Polar Organic Molecules:

$$
\begin{array}{ccc}
\overset{\displaystyle H}{\underset{\displaystyle H}{H-C-O-H}} & \overset{\displaystyle O}{H-C-O-H} & \overset{\displaystyle H}{\underset{\displaystyle H}{H-C-N\diagup^{H}_{\diagdown H}}}
\end{array}
$$

Examples of Some Relatively Nonpolar Organic Molecules:

$$
\begin{array}{ccc}
\overset{\displaystyle H}{\underset{\displaystyle H}{H-C-H}} & \overset{H\diagdown\;\diagup H}{\underset{H\diagup\;\diagdown H}{C=C}} & \overset{\displaystyle H\;\;H}{\underset{\displaystyle H\;\;H}{H-C-C-H}}
\end{array}
$$

D. Formal Charge

In the formulas of some compounds, we find it is not possible to draw the usual number of line bonds to each atom. For example, we might find that an oxygen in a formula has eight valence electrons but only one covalent bond, instead of two. In such structures, one or more atoms must carry an ionic charge, called a **formal charge**.

Consider the line-bond formula for nitric acid, HNO_3.

An oxygen with eight valence electrons
and one covalent bond

$$
\overset{\displaystyle :\ddot{O}:}{\underset{}{H-O-N=\ddot{O}:}}
$$

N has an octet
but has four bonds

In the line-bond formula of nitric acid, the electron pairs cannot be distributed so that the nitrogen has three bonds as well as an octet while each oxygen has two bonds and an octet. To make this structure fit these rules of valence, we must assign an ionic charge to one of the two oxygen atoms and another ionic charge to the nitrogen atom.

The formal charge for an atom is calculated by the following equation:

Formal charge = (number of valence e^- in the neutral atom)

$$-\tfrac{1}{2}\text{ (number of shared } e^-)$$

$$-\text{ (number of unshared valence } e^-)$$

Nitrogen as a neutral atom has five valence electrons. In nitric acid, the nitrogen has eight shared electrons and no unshared valence electrons. Substituting these values into the equation, we obtain

$$\text{Formal charge for nitrogen in } HNO_3 = (5) - \tfrac{1}{2}(8) - 0$$
$$= +1$$

We can carry out an identical calculation for the single-bonded oxygen:

$$\text{Formal charge for single-bonded oxygen in } HNO_3 = (6) - \tfrac{1}{2}(2) - 6$$
$$= -1$$

We can now insert the charges in the formula.

Lewis Formula for Nitric Acid Showing Formal Charges:

$$H:\ddot{\underset{..}{O}}:\underset{+1}{\ddot{N}}::\ddot{O}\overset{:\ddot{\underset{..}{O}}:^{-1}}{}$$

In nitric acid, the sum of the two formal charges is zero; therefore, the molecule as a whole is electrically neutral. If the formal charges do not sum to zero, the formula represents an ion.

EXAMPLE

Calculate the formal charge on N in $CH_3NH_3{}^{+}$.

$$\left[CH_3 : \underset{\underset{H}{..}}{\overset{\overset{H}{..}}{N}} : H \right]^{+}$$

Solution: A neutral N atom contains five valence electrons. The above N has eight shared electrons and no unshared valence electrons. Substituting,

$$\text{Formal charge} = (5) - \tfrac{1}{2}(8) - 0$$
$$= +1$$

With practice, you will be able to assign formal charges by inspection. Any N with four bonds has a formal charge of $+1$. Any O with an octet, but with only one bond, has a formal charge of -1. In covalent compounds hydrogen never has a formal charge, and carbon has a formal charge only rarely.

Formal charge, 0   Formal charge, $+1$

$$\overset{|}{-}\underset{\cdot\cdot}{\overset{\cdot}{O}}:\quad\text{or}\quad=\underset{\cdot\cdot}{\overset{\cdot}{O}}:$$

$$:\underset{|}{\overset{\cdot\cdot}{O}}:^-$$

Formal charge, 0   Formal charge, −1

PROBLEM 1.3. Calculate the formal charge on each N and each O.

(a) $H:\underset{\cdot\cdot}{\overset{\cdot\cdot}{O}}:\overset{\cdot\cdot}{N}::\underset{\cdot\cdot}{\overset{\cdot}{O}}:$ (b) $H:\overset{\overset{\textstyle H}{|}}{\underset{\cdot\cdot}{C}}: \ :\overset{\cdot\cdot}{N}::\overset{\cdot\cdot}{N}:$ (c) $H:\overset{\overset{\textstyle H}{|}}{C}:N\overset{:\overset{\cdot\cdot}{O}:}{\underset{\overset{\cdot\cdot}{O}\cdot}{}}$

E. Orbital Theory of Covalent Bonding

The Molecular Orbital of Hydrogen   The simplest molecule is the hydrogen molecule (H_2). In the formation of the hydrogen molecule, two hydrogen atoms (each containing a single electron in a $1s$ atomic orbital) approach each other. When the two orbitals are close, they merge into a new orbital, called a **molecular orbital**. This new orbital encompasses both hydrogen atoms and is the covalent bond in the H_2 molecule. For H_2, this one pair of electrons in one molecular orbital fills the outer shell of each hydrogen atom, allowing each to attain the helium configuration. Figure 1.3 depicts the formation of a hydrogen molecule.

A molecular orbital, like an atomic orbital, can contain a maximum of two electrons, with opposite spins. This is why a covalent bond is formed from only one pair of shared electrons. Although a double bond appears to involve the sharing of

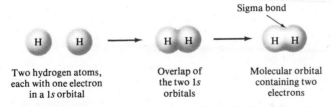

Two hydrogen atoms,   Overlap of   Molecular orbital
each with one electron   the two $1s$ containing two
in a $1s$ orbital   orbitals   electrons

Figure 1.3 Formation of the molecular orbital of H_2 from the atomic orbitals of two hydrogen atoms.

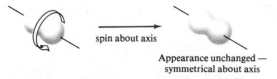

spin about axis

Appearance unchanged —
symmetrical about axis

Figure 1.4 A sigma orbital is symmetrical about the axis between nuclei.

four electrons, the double bond is actually composed of two different molecular orbitals, each of which can have a maximum of two electrons. A triple bond consists of three separate molecular orbitals.

An orbital that is symmetrical around an axis passing through two nuclei is called a **sigma (σ) orbital**. The symmetry is illustrated in Figure 1.4. A covalent bond whose molecular orbital is a sigma orbital is called a **sigma bond**.

The Molecular Orbitals of Carbon   Carbon has four electrons in the second energy level, which is the valence shell. This second energy level contains a $2s$ orbital and three $2p$ orbitals. Carbon, however, does not use these atomic orbitals for bonding. When carbon forms compounds it *hybridizes*, or blends, these second-level atomic orbitals into hybrid orbitals so that it can form stronger bonds with maximal overlap.

In methane, carbon forms four single bonds, each to a hydrogen atom. These four bonds are identical because they are formed from hybrid orbitals that are a complete blend of carbon's second-level atomic orbitals. Because these four hybrid orbitals are formed from one s orbital and three p orbitals, they are called sp^3**-hybrid orbitals**. Figure 1.5 shows the energy changes in the hybridization.

In methane, the four sp^3-hybrid orbitals of the carbon atom form four single bonds to four hydrogen atoms. These four hybrid orbitals place the hydrogen atoms as far apart as possible. Each bond forms an angle of approximately 109.5° with each of the other bonds in the structure. This geometry places the sp^3-hybrid orbitals pointing toward the corners of a regular tetrahedron (a triangular-based pyramid with equivalent sides), as shown in Figure 1.6. A carbon with four single bonds to other atoms is called a **tetrahedral carbon**.

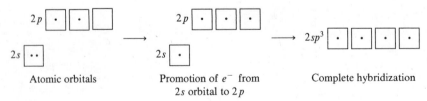

Atomic orbitals	Promotion of e^- from $2s$ orbital to $2p$	Complete hybridization

Figure 1.5 Carbon blends its atomic orbitals to form four identical sp^3-hybrid orbitals, which can form four identical single bonds.

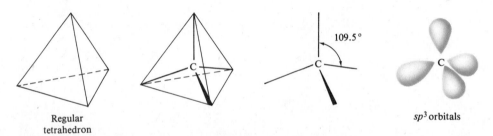

Regular
tetrahedron

sp^3 orbitals

Figure 1.6 The bonds of an sp^3-hybridized carbon point toward the corners of a regular tetrahedron.

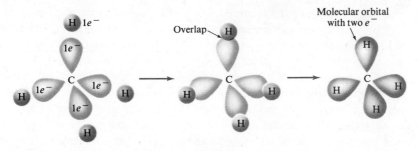

Figure 1.7 The bonding in methane, CH_4.

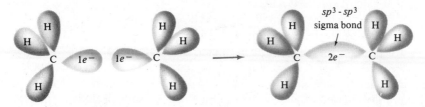

Figure 1.8 The bonding in ethane, CH_3CH_3.

Figure 1.7 shows the molecular orbitals in methane. Each sp^3-hybrid orbital of carbon overlaps with the $1s$ orbital of hydrogen. The carbon and the hydrogen contribute one electron each for bond formation. The resulting sp^3-$1s$ molecular orbital contains two electrons. This orbital, like the molecular orbital in H_2, is symmetrical around an axis passing through both nuclei and is also a sigma bond.

Any carbon atom bonded by four single bonds (not double bonds or triple bonds) is in the sp^3-hybrid state. Its orbitals point toward the corners of a tetrahedron just as they do in methane. The bond angles around most single-bonded carbons are approximately 109.5°. (These angles can be slightly compressed or expanded because of attractions and repulsions of groups within a molecule.) Figure 1.8 shows the bonding in ethane (CH_3CH_3). Other compounds are similar.

When carbon forms double bonds and triple bonds, its atomic orbitals are hybridized differently from those of the single-bonded carbon. We will discuss these orbitals in Chapter 3 when we introduce alkenes, compounds containing carbon–carbon double bonds, and alkynes, compounds containing carbon–carbon triple bonds.

1.4 FORMULAS FOR ORGANIC COMPOUNDS

A. Empirical, Molecular, and Structural Formulas

Several different types of formulas are commonly used in chemistry. An **empirical formula** shows the simplest ratio of elements in a compound. For example, both acetylene and benzene have the same empirical formula, CH, which tells us that the molecules of both of these compounds contain one hydrogen for each carbon atom. Empirical formulas can be calculated from experimental data, but they are rarely used in organic chemistry because they tell us nothing about the structure of a molecule—the order in which the atoms are bonded together.

A **molecular formula** is more useful than an empirical formula because it tells us the number of atoms in a molecule as well as their ratio. The molecular formulas for acetylene and benzene are C_2H_2 and C_6H_6, respectively. These formulas tell us that an acetylene molecule has a total of two carbons and two hydrogens and a benzene molecule has a total of six carbons and six hydrogens. Molecular formulas, such as H_2SO_4 and H_3PO_4, are commonly used in inorganic chemistry; however, these formulas are rarely used in organic chemistry because they tell us little about the structure of the molecule.

The structure of a molecule must be evident from its formula if the formula is to be of value in organic chemistry. Therefore, when working with organic compounds, we prefer to use **structural formulas**. Lewis formulas, with dots representing the valence electrons, are one type of structural formula. Lewis formulas are not used extensively because they are inconvenient to draw. Closely related to Lewis formulas are **line-bond formulas**, in which each pair of shared electrons is represented by a line. In line-bond formulas, unshared pairs of valence electrons may or may not be shown.

Structural Formulas for Water:

$$H:\overset{..}{\underset{..}{O}}: \qquad \overset{\displaystyle H}{\underset{}{\overset{|}{H-O}}} \quad \text{or} \quad \overset{\displaystyle H}{\underset{}{\overset{|}{H-\underset{..}{O}}}}:$$

Lewis formula   Line-bond formulas

Structural Formulas for Chloroethane:

$$\begin{matrix} H & H \\ H:\overset{..}{\underset{..}{C}}:\overset{..}{\underset{..}{C}}:\overset{..}{\underset{..}{Cl}}: \\ H & H \end{matrix} \qquad \overset{H \quad H}{\underset{H \quad H}{\overset{|\quad|}{H-C-C-Cl}}}\overset{}{\underset{|\quad|}{}} \quad \text{or} \quad \overset{H \quad H}{\underset{H \quad H}{\overset{|\quad|}{H-C-C-\overset{..}{\underset{..}{Cl}}}}}:$$

Lewis formula   Line-bond formulas

By inspecting the line-bond formulas of acetylene and benzene, we can see that their atoms are bonded together in completely different ways.

Structural Formulas for Acetylene and Benzene, Each with the Empirical Formula CH:

$$H-C\equiv C-H$$

Acetylene

Benzene

TABLE 1.4 EXAMPLES OF TYPES OF FORMULAS

Lewis	Line bond	Condensed structural	Molecular	Empirical
H H:C:H H	H | H—C—H | H	CH_4	CH_4	CH_4
H :Ö· H:C: C :Ö:H H	H O | ‖ H—C—C—O—H | H	O ‖ CH_3COH	$C_2H_4O_2$	CH_2O
H H H:Ö:C:C:Ö:H H H	H H | | H—O—C—C—O—H | | H H	$HOCH_2CH_2OH$	$C_2H_6O_2$	CH_3O

For convenience, line-bond formulas may be simplified to **condensed structural formulas**, formulas in which a few, if any, bonds are shown but from which we can still deduce the structure. For clarity, multiple bonds are generally shown in condensed formulas.

Structural Formulas for Ethane:

$$H—\overset{\overset{\displaystyle H}{|}}{C}—\overset{\overset{\displaystyle H}{|}}{C}—H \qquad H_3C—CH_3 \quad \text{or} \quad CH_3CH_3$$

Line-bond formula   Condensed structural formulas

Structural Formulas for Propene:

$$H_3C—CH=CH_2 \quad \text{or} \quad CH_3CH=CH_2$$

Line-bond formula   Condensed structural formulas

The type of formulas we have been discussing are summarized in Table 1.4.

EXAMPLE

Write line-bond formulas for the following condensed formulas, showing unshared pairs of valence electrons.

(a) $CH_3CH_2CH_3$ **(b)** $(CH_3)_2C=O$

Solution:

(a)

$$H-\overset{\underset{|}{H}}{\underset{\underset{H}{|}}{C}}-\overset{\underset{|}{H}}{\underset{\underset{H}{|}}{C}}-\overset{\underset{|}{H}}{\underset{\underset{H}{|}}{C}}-H$$

(b)

(structure) or

$$H-\overset{\underset{|}{H}}{\underset{\underset{H}{|}}{C}}-\overset{\underset{|}{\ddot{O}}}{\underset{\underset{H}{|}}{C}}-\overset{\underset{|}{H}}{\underset{\underset{H}{|}}{C}}-H$$

PROBLEM 1.4. Write condensed structural formulas for the following line-bond formulas.

(a) $H-\overset{O}{\overset{\|}{C}}-H$

(b) (structure)

(c) (structure)

PROBLEM 1.5. Write line-bond formulas for the following condensed structural formulas.

(a) CH_3CH_2Cl (b) $(CH_3CH_2)_2CHCH_2CH_3$ (c) $CH_2(\overset{O}{\overset{\|}{C}}OCH_2CH_3)_2$

B. Resonance Formulas

Line-bond formulas are convenient, but they do not always correctly represent the structure of a compound. For example, the nitrate ion (NO_3^-) cannot be adequately represented by one line-bond formula. The usual line-bond formula for the nitrate ion follows.

(structure of nitrate ion)

In this formula, the nitrogen has a formal charge of $+1$ and two of the three oxygens have formal charges of -1. Overall, the nitrate ion has an ionic charge of -1. Why is this formula inadequate? Experimental studies show that the nitrate ion does *not* contain two nitrogen-oxygen single bonds and one nitrogen-oxygen double bond. Instead, all three nitrogen-oxygen bonds are identical. To communicate this fact, we need *three* different line-bond formulas to represent the nitrate ion.

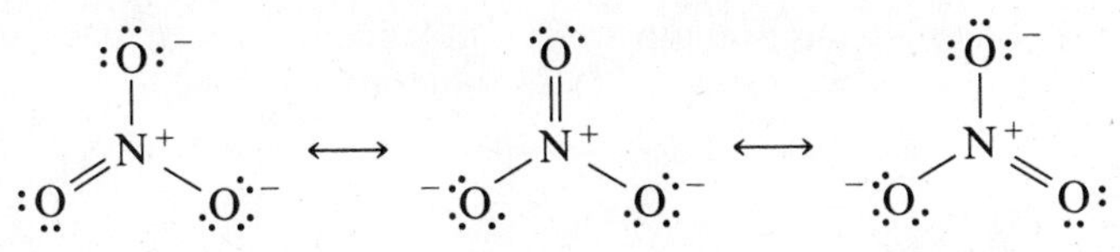

Any one of these three formulas by itself provides us with an inaccurate representation of the nitrate ion. We need all three to describe the electronic structure of the ion correctly. The structure of the real nitrate ion is thus a composite of all three formulas, more similar to the following formula:

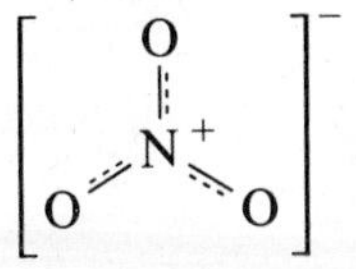

When a structure such as the nitrate ion can be represented by more than one line-bond formula in which the arrangement of the atoms is identical and only the positions of the electrons are different, the actual or real structure is a composite of all of the formulas. The real structure is called a **resonance hybrid**. The different formulas used to represent the hybrid are called **resonance formulas** or **resonance structures**.

If resonance structures are necessary to describe the electronic structure of a molecule accurately, we say the electrons are delocalized. The reason that a structure delocalizes its electrons instead of holding them in discrete bonds is that stability is gained by the spreading out of the negative electronic charge.

A compound bonded by only single bonds cannot have resonance formulas because the electrons are held tightly by the nuclei and cannot be delocalized. Only certain structures with double bonds or triple bonds can have resonance stabilization.

Resonance formulas are separated by a double-headed arrow (↔) to differentiate them from structures in equilibrium (two arrows: ⇄ or ⇌). Resonance formulas are not in equilibrium with each other because they are merely different formulas for the same compound.

When we draw resonance formulas for a structure we usually use small curved arrows to show how we have shifted the electrons to get from one formula to another. In these formulas the electrons are not truly shifting from one position to the other; we use these arrows only for keeping track of the electrons. Pay particular attention to how we use these curved arrows. Arrows can show electrons "shifting" only to adjacent bond positions or atoms. No atom in a resonance structure can carry more than its maximum number of valence electrons (eight for the period 2 elements N, O, C, etc.).

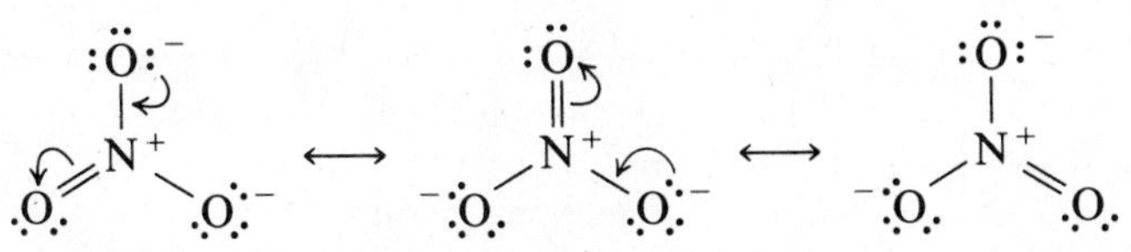

The carbonate ion is another ion that we represent by resonance formulas. As in the nitrate ion, all the bonds to O are identical. The real structure of the carbonate ion is a composite of the resonance structures.

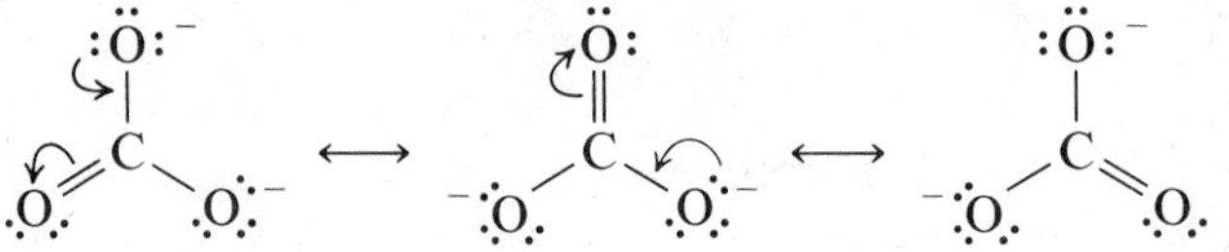

or

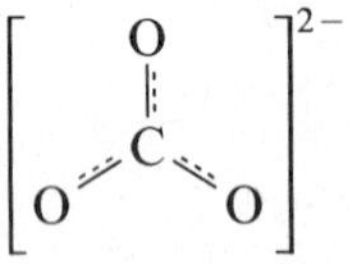

EXAMPLE

The following ion contains two identical carbon–oxygen bonds. Write resonance formulas for the ion.

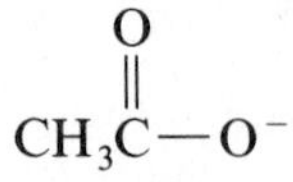

Acetate ion

Solution:

1. Rewrite the formula to show the valence electrons in the $-\overset{O}{\underset{\parallel}{C}}-O^-$ group.
2. Next, draw curved arrows to show how the negative charge could be moved to the second oxygen.

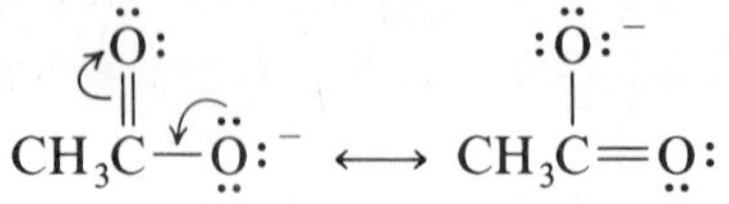

PROBLEM 1.6.   Write resonance formulas for the following structures:

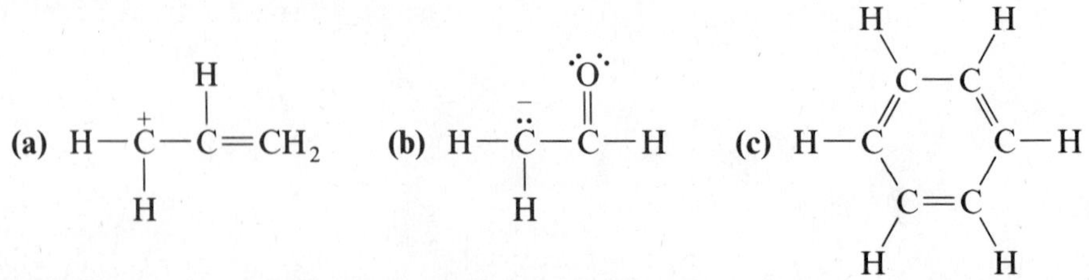

C. Three-Dimensional Formulas

We use three-dimensional formulas when we want to emphasize the three-dimensionality of a small molecule. A solid wedge represents a bond jutting out of the page toward the reader, a dashed wedge represents a bond angled into the page away from the reader, and a line bond represents a bond in the plane of the page.

A Three-Dimensional Formula:

Using three-dimensional formulas, we can represent other structures containing tetrahedral carbons.

Ethane   Ethanol   Propane

1.5 FUNCTIONAL GROUPS

Carbon–carbon single bonds and carbon–hydrogen bonds in organic compounds are generally nonreactive because they are nonpolar. Polar groups form the sites of reactivity in an organic molecule and are referred to as **functional groups**. For example, alcohols are one class of compounds that contain the **hydroxyl group** ($-OH$) bonded to a carbon. All alcohols undergo similar chemical reactions because they contain this functional group.

Examples of Some Alcohols:

$$CH_3OH \qquad CH_3CH_2OH \qquad \overset{\displaystyle OH}{\underset{\textstyle}{CH_3\overset{|}{C}HCH_3}}$$

Methanol   Ethanol   2-Propanol

Double bonds and triple bonds joining carbon atoms are also considered functional groups because the second and third bonds are more reactive than carbon–carbon single bonds.

Table 1.5 lists a few important functional groups and the names of the classes of compounds that contain these groups. A particular organic compound may contain

TABLE 1.5 SOME COMMON COMPOUND CLASSES AND THEIR FUNCTIONAL GROUPS

Compound class	Formula[a]	Functional group	Formula
alkene	$\begin{array}{ccc} R & & R \\ \diagdown & & \diagup \\ & C{=}C & \\ \diagup & & \diagdown \\ R & & R \end{array}$	carbon–carbon double bond	$\diagdown \diagup \\ C{=}C \\ \diagup \diagdown$
alkyne	$R{-}C{\equiv}C{-}R$	carbon–carbon triple bond	$-C{\equiv}C-$
alcohol	$R{-}OH$	hydroxyl group	$-OH$
ether	$R{-}OR$	alkoxyl group	$-OR$
aldehyde	$\overset{\textstyle O}{\overset{\|}{R{-}CH}}$	aldehyde group	$\overset{\textstyle O}{\overset{\|}{-CH}}$
ketone	$\overset{\textstyle O}{\overset{\|}{R{-}C{-}R}}$	keto group	$\overset{\textstyle O}{\overset{\|}{-C-}}$
carboxylic acid	$\overset{\textstyle O}{\overset{\|}{R{-}COH}}$	carboxyl group	$\overset{\textstyle O}{\overset{\|}{-COH}}, \quad -COOH,$ $or -CO_2H$
ester	$\overset{\textstyle O}{\overset{\|}{R{-}COR}}$	ester group	$\overset{\textstyle O}{\overset{\|}{-COR}}, \quad -COOR,$ $or -CO_2R$
amide	$\overset{\textstyle O}{\overset{\|}{R{-}CNH_2}}, \overset{\textstyle O}{\overset{\|}{RCNHR}},$ $or \overset{\textstyle O}{\overset{\|}{RCNR_2}}$	amide group	$\overset{\textstyle O}{\overset{\|}{-CNH_2}}, \quad \overset{\textstyle O}{\overset{\|}{-CNHR}},$ $or \overset{\textstyle O}{\overset{\|}{-CNR_2}}$
amine	$R{-}NH_2, R{-}NH{-}R,$ $or \; R{-}\overset{\textstyle }{\underset{\textstyle R}{N}}{-}R$	amino group	$-NH_2, -NHR,$ $or -NR_2$

[a] The letter R is used here to denote any carbon–hydrogen group, such as CH_3- or CH_3CH_2-.

no functional group, one functional group, or more than one functional group. The following structure contains many functional groups.

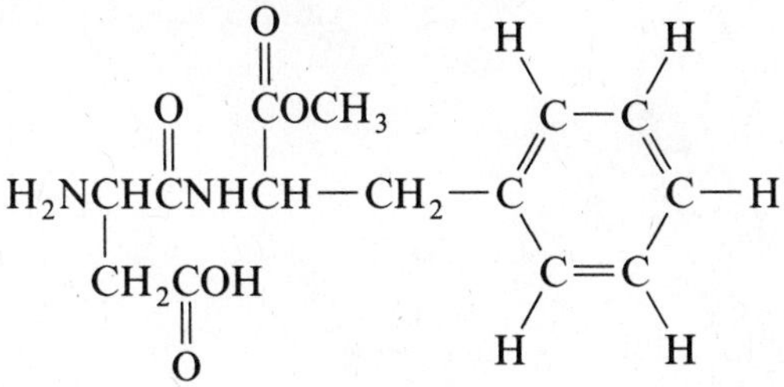

Aspartame
the synthetic sweetener Nutrasweet®

1.6 STRUCTURAL ISOMERISM

Compounds that have the same molecular formula but different structures (atoms bonded in a different order) are called **structural isomers** of one another. Ethanol and dimethyl ether are examples of structural isomers. These compounds have the same molecular formula (C_2H_6O), but they have different structures. Because they have different structures, they are different compounds with different physical, chemical, and physiological properties. Ethanol, a liquid at room temperature, is the drug in alcoholic beverages. Dimethyl ether, a gas at room temperature, has none of the intoxicating properties typical of ethanol.

The Two Structural Isomers for C_2H_6O:

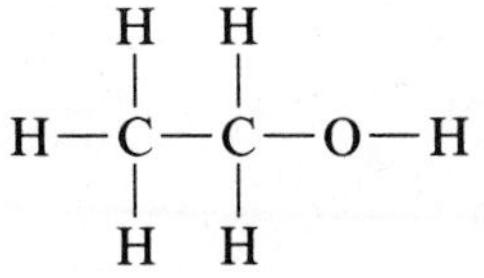 or CH_3CH_2OH

Ethanol

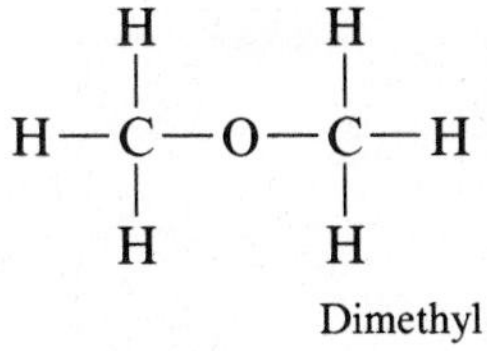 or CH_3OCH_3

Dimethyl ether

The molecular formulas CH_4, C_2H_6, and C_3H_8 represent only one compound each. If the rules of valence are followed, there is only one arrangement of the atoms that shows the correct number of covalent bonds.

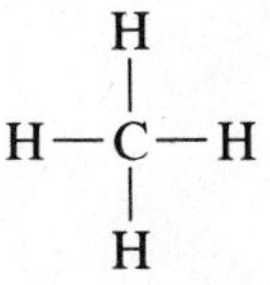

CH_4

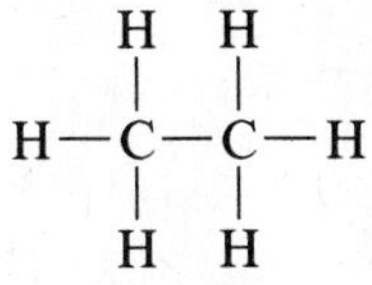

C_2H_6

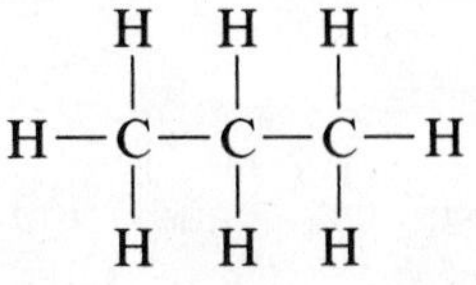

C_3H_8

However, molecules with an increasing number of atoms have an increasing number of possible isomers. The molecular formula C_4H_{10} represents two structural isomers, and the formula C_5H_{12} represents three structural isomers. The formula $C_{10}H_{22}$ represents 75 different structural isomers!

The Two Structural Isomers for C_4H_{10}:

$$CH_3CH_2CH_2CH_3 \qquad CH_3\overset{\overset{\displaystyle CH_3}{|}}{C}HCH_3$$

Butane   Methylpropane

a continuous-chain molecule   *a branched-chain molecule*

Structures containing atoms other than C and H have an even greater number of isomers. Although C_3H_8 represents only one compound, $CH_3CH_2CH_3$, the formula C_3H_8O represents three structural isomers.

EXAMPLE

Write condensed structural formulas for all structural isomers with the molecular formula C_3H_8O:

Solution: There are three possible structural isomers:

1. $CH_3CH_2OCH_3$ the same as $CH_3OCH_2CH_3$
2. $CH_3CH_2CH_2OH$ the same as $HOCH_2CH_2CH_3$
3. $CH_3\overset{\overset{\displaystyle OH}{|}}{C}HCH_3$ the same as $CH_3\underset{\underset{\displaystyle OH}{|}}{C}HCH_3$ the same as $\begin{matrix} H_3C \\ \\ H_3C \end{matrix}\!\!\!\diagdown\!\!\!\diagup CHOH$

Note that formulas in 3 can be written in a number of ways. These formulas do not represent structural isomers because the atoms are joined in the same sequence in each.

PROBLEM 1.7. Write formulas for all structural isomers with the molecular formula C_5H_{12}.

1.7 HYDROGEN BONDING

A. Formation of Hydrogen Bonds

Organic compounds containing $-OH$ or $\diagup NH$ groups can undergo **hydrogen bonding**, a relatively strong attraction between the hydrogen of the $-OH$ or $\diagup NH$

group and the unshared pair of electrons of another oxygen or nitrogen. Hydrogen bonding is possible because of the polarity of the O—H and N—H bonds and the small size of the hydrogen atom. These properties allow the partially positive hydrogen to be drawn very close to the partially negative oxygen or nitrogen.

A hydrogen bond is not a covalent bond; it is merely an electrostatic attraction between a positive hydrogen and an unshared pair of electrons. Only 2–7 kcal/mol is needed to break a hydrogen bond, while 75–100 kcal/mol is required to break a typical covalent bond.

Polarity of the —OH and ＞NH Groups:

electronegative atoms

$$\overset{\delta-}{\underset{\cdot\cdot}{\ddot{O}}}\!\!\diagup\overset{\delta+}{H}\qquad\overset{\delta-}{\underset{|}{\ddot{N}}}\!\!\diagdown\overset{\delta+}{H}$$

Hydrogen Bonds:

$$-\ddot{O}-H\text{----}:\ddot{O}-\quad\text{or}\quad-\ddot{O}-H\text{----}:\underset{|}{\overset{|}{N}}-$$

$$-\underset{|}{\ddot{N}}-H\text{----}:\ddot{O}-\quad\text{or}\quad-\underset{|}{\ddot{N}}-H\text{----}:\underset{|}{\overset{|}{N}}-$$

Two Examples of Hydrogen-Bonded Compounds Follow:

$$\cdots\text{H}-\underset{|}{\overset{\overset{\displaystyle H}{|}}{\ddot{O}}}:\cdots\text{H}-\underset{\underset{\displaystyle H}{|}}{\ddot{O}}:\cdots\qquad\cdots\text{H}-\underset{|}{\overset{\overset{\displaystyle CH_3}{|}}{\ddot{O}}}:\cdots\text{H}-\underset{\underset{\displaystyle CH_3}{|}}{\ddot{O}}:\cdots$$

         Water   Methanol

                                                    an alcohol

The hydrogen of a C—H group cannot form a hydrogen bond. Carbon is intermediate in electronegativity and, consequently, the C—H bond is not polar. Therefore, the hydrogen of the C—H group is not partially positive and is not attracted to unshared electrons.

$$CH_3\underset{\cdot\cdot}{\ddot{O}}:\quad\underset{\uparrow}{}\quad H-\underset{\underset{\displaystyle H}{|}}{\overset{\overset{\displaystyle H}{|}}{C}}-H$$

No hydrogen bond

PROBLEM 1.8. Show hydrogen bonds, if any, for the following pairs of compounds.

(a) $CH_3CH_2OH + H_2O$ (b) $CH_3OCH_3 + H_2O$

(c) $CH_3OCH_3 + CH_3\overset{\overset{\displaystyle O}{\|}}{C}CH_3$ (d) $CH_3OH + NH_3$

B. Properties of Hydrogen-Bonded Compounds

Boiling Point   If molecules did not attract one another, two different liquids with the same formula weight would have the same boiling point. However, if molecules of a liquid are strongly attracted to one another, the liquid will boil at a substantially higher temperature than a liquid with weaker molecular attractions. For example, ethanol is a liquid (bp, 78.5°C), while its isomer, dimethyl ether, is a gas (bp, −23.6°C) even though both compounds have the same formula weight, 46.08 atomic mass units (amu). The reason for this difference in boiling point is that extra energy must be supplied to break the hydrogen bonds in ethanol before the molecules can be driven off individually.

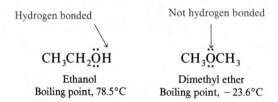

Hydrogen bonded	Not hydrogen bonded
$CH_3CH_2\overset{\cdot\cdot}{\underset{\cdot\cdot}{O}}H$	$CH_3\overset{\cdot\cdot}{\underset{\cdot\cdot}{O}}CH_3$
Ethanol	Dimethyl ether
Boiling point, 78.5°C	Boiling point, −23.6°C

Water Solubility   Because the —OH group can form hydrogen bonds with water, we say that the —OH group is **hydrophilic** ("water loving"). A C—H group cannot participate in hydrogen bonding; therefore, it is said to be **hydrophobic** ("water hating"). The water solubility of a compound is determined by its ratio of hydrophilic groups to hydrophobic groups. For example, alcohols containing one to three carbon atoms are miscible with (completely soluble in) water. In these small molecules, the one —OH group can overcome the hydrophobic character of the C—H groups.

Methanol (CH_3OH) is miscible with water

Alcohols containing more than three carbons are only partially soluble or insoluble in water because they have a sufficient number of hydrophobic C—H groups to overcome the hydrophilic effect of the —OH group.

$$CH_3CH_2CH_2CH_2OH \qquad CH_3CH_2CH_2CH_2CH_2OH$$

Solubility, 8.3 g/100 mL H_2O Solubility, 2.7 g/100 mL H_2O

Molecular Shape Many biological molecules, such as proteins, are held in their unique shapes by hydrogen bonding. When these compounds are heated the hydrogen bonds are broken, the structure loses its unique shape, and the biological activity is lost. Enzymes, for example, may lose their activity when heated. In part, the reason for cooking is to disrupt hydrogen bonding and make food easier to digest.

1.8 INTRODUCTION TO ACIDS AND BASES

Acid-base reactions play a major role throughout organic chemistry. In this introductory chapter we briefly review a few of the important concepts that we need to discuss organic reactions. In Chapters 8 and 11 we will discuss acids and bases in more detail.

A. Brønsted-Lowry Acids and Bases

In 1923 J. N. Brønsted, a Danish chemist, and T. M. Lowry, an English chemist, independently proposed the following useful concepts of acidity and basicity:

An **acid** is a proton donor.

A **base** is a proton acceptor.

The terms *proton* and *hydrogen ion* (H^+) are synonyms; both refer to a hydrogen atom that has lost its electron. A base is a molecule or an ion with an unshared pair of valence electrons that can form a covalent bond with a proton.

In an acid-base reaction, a proton is transferred from the acid to the base.

The Base Accepts the Proton from the Acid:

$$HO:^- \quad + \quad H—Cl: \longrightarrow HO—H \quad + \quad :Cl:^-$$

A base   An acid

In any acid-base reaction, the stronger acid donates a proton to the stronger base to yield a weaker acid and a weaker base. Let us rewrite the preceding equation to illustrate this statement.

$$HCl \quad + \quad ^-OH \longrightarrow H_2O \quad + \quad Cl^-$$

Stronger acid   Stronger base   Weaker acid   Weaker base
than H_2O than Cl^- than HCl   than $^-$OH

A list of acids in order of their relative acidities is shown inside the front cover of this book. We will discuss some of the reasons for variations in acid strengths in Chapter 11.

We can use a list of relative acidities to predict the products of an acid-base reaction. Consider our example, the reaction of hydrogen chloride and hydroxide ion. Hydrogen chloride ionizes into H^+ and Cl^-. In this reaction, the un-ionized acid (HCl) is referred to as the **conjugate acid** and the anion (Cl^-) the **conjugate base**.

$$H\text{—}Cl \rightleftharpoons H^+ + Cl^-$$

The conjugate acid   Its conjugate base

Because hydrogen chloride is more acidic than water, HCl appears above water in the relative acidity list. Therefore, HCl will donate a proton to the conjugate base of water (^-OH) to form the weaker acid (H_2O).

Water appears twice in the list of relative acidities—once as an acid and again as a conjugate base.

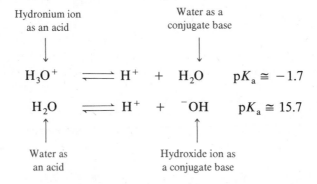

Will HCl donate a proton to water and form a hydronium ion (H_3O^+)? Because HCl is more acidic than H_3O^+ (above H_3O^+ in the list), HCl will donate a proton to H_2O to form H_3O^+.

$$HCl + H_2O \longrightarrow H_3O^+ + Cl^-$$

Stronger   Weaker
acid   acid

EXAMPLE

Will hydrogen cyanide (HCN) donate a proton to water?

Solution: Hydrogen cyanide (below water in the relative acidity list) is less acidic than water. Therefore, HCN will not donate a proton to H_2O to any extent because a stronger acid (H_3O^+) would be formed.

$$H_3O^+ \rightleftharpoons H^+ + H_2O \qquad pK_a \cong -1.7$$

$$HCN \rightleftharpoons H^+ + CN^- \qquad pK_a \cong 9.3$$

$$HCN + H_2O \longrightarrow \text{no appreciable acid-base reaction}$$

Acid-base reactions are reversible reactions that rapidly attain equilibrium. The position of the equilibrium (relative concentrations of the reactants and products) is determined by the relative acidity and basicity of the reactants. For some acid-base reactions the equilibrium lies so far toward the product side of the equation that we consider the reaction to be irreversible. The reaction of hydrogen chloride and hydroxide ions is an example of a reaction of this type. With other acid-base reactions the equilibrium lies so far on the reactant side of the equation that we consider no reaction to occur. The reaction of hydrogen cyanide and water is an example of a reaction of this type.

In order to describe common acids and bases a number of qualitative terms have been coined. **Strong acids** are acids that are almost completely ionized in water. ("Strong" in this context refers to the extent of ionization and not to the concentration. A solution of a strong acid can be dilute or concentrated.)

A Strong Acid is Almost Completely Ionized in Water:

$$H-Cl \ + \ H_2O \ \rightleftharpoons \ \underbrace{H_3O^+ \ + \ Cl^-}$$

The acid-base equilibrium
lies to the product
side of the equation

Weak acids, such as acetic acid (CH_3CO_2H), are ionized to only a small extent in water. These acids are weaker acids than is the hydronium ion; they lie below H_3O^+ in the list of relative acidities.

A Weak Acid is Almost All Un-Ionized in Water:

$$\underbrace{\overset{\displaystyle O}{\overset{\displaystyle \|}{CH_3COH}} \ + \ H_2O} \ \rightleftharpoons \ H_3O^+ \ + \ \overset{\displaystyle O}{\overset{\displaystyle \|}{CH_3CO^-}}$$

The acid-base equilibrium
lies to the reactant
side of the equation

The most common organic acids are **carboxylic acids**. Acetic acid is a typical carboxylic acid. All carboxylic acids contain the carboxyl group ($-CO_2H$). Most carboxylic acids are weak acids, ionized only about 1% in water.

Even though only a small percentage of carboxylic acid molecules are ionized in water, carboxylic acids can undergo complete reaction when treated with a base stronger than water.

$$\overset{\displaystyle O}{\overset{\displaystyle \|}{CH_3COH}} \ + \ {}^-OH \ \longrightarrow \ H_2O \ + \ \overset{\displaystyle O}{\overset{\displaystyle \|}{CH_3CO^-}}$$

Bases, too, can be strong or weak. Strong bases, such as sodium hydroxide ($Na^+ \ {}^-OH$) or potassium hydroxide ($K^+ \ {}^-OH$), are completely ionized in water.

Some bases, such as ammonia, can react with water in a hydrolysis reaction to yield an alkaline solution. Ammonia is considered a weak base because the equilibrium lies on the NH_3 side of the equation (the side of the weaker acid and weaker base).

Hydrolysis of a Base:

$$NH_3 \; + \; H_2O \; \rightleftharpoons \; NH_4{}^+ \; + \; OH^-$$

Stronger acid than H_2O	Stronger base than NH_3

The most common organic bases are **amines**, compounds whose structures are closely related to that of ammonia.

$H-\ddot{N}-H$	$H_3C-\ddot{N}-H$	$H_3C-\ddot{N}-H$	$H_3C-\ddot{N}-CH_3$
$\mid$	$\mid$	$\mid$	$\mid$
H	H	CH_3	CH_3
Ammonia	Methylamine	Dimethylamine	Trimethylamine

Amines are like ammonia in their basicity. In water, the amine is the base (proton acceptor) and water is the acid (proton donor). The resulting solution is alkaline because ^-OH ions are formed.

Reaction of an Amine with Water:

$$CH_3NH_2 \; + \; H_2O \rightleftharpoons CH_3\overset{+}{N}H_3 \; + \; {}^-OH$$

When treated with an acid stronger than water, amines can undergo complete reaction to yield salts, the ionic product of an acid-base reaction. Aqueous solutions of salts are neutral (or nearly neutral) because the concentrations of H^+ or ^-OH are equal.

Reaction of a Strong Acid and a Weak Base:

$$HCl \quad + CH_3NH_2 \rightleftharpoons \underbrace{CH_3\overset{+}{N}H_3 \, Cl^-}_{\text{A salt}}$$

Stronger acid than $CH_3\overset{+}{N}H_3$

PROBLEM 1.9. Which would act as an acid and which as a base when mixed with water? Explain.

(a)
```
    CH2CH2
   /      \    H
 CH2      N:—
   \      /
    CH2CH2
```

(b) $^-:\ddot{O}CH_3$

(c) $H\ddot{O}CH_2\overset{\displaystyle \ddot{O}}{\overset{\|}{C}}OH$

(d) $^+NH_4$

PROBLEM 1.10. Complete and balance the following equations for acid-base reactions:

(a)
```
        CH
      /    \\
   HC       CH
   ||        |       + HCl ⟶
   HC       CH
      \    /
        N
```

(b) $\underset{\parallel}{HOC} - \underset{\parallel}{COH} + 2 \ ^-OH \longrightarrow$

where O double bonds are above each C:

$$\underset{\substack{\parallel\\ O}}{HOC} - \underset{\substack{\parallel\\ O}}{COH} + 2\ ^-OH \longrightarrow$$

(c) $^-OCH_3 + H_2O \longrightarrow$

(d) $HOCH_2\underset{\substack{\parallel\\ O}}{C}OH + H_2O \longrightarrow$

B. Lewis Acids and Bases

About the same time that Brønsted and Lowry were publishing their concepts of acidity and basicity, G. N. Lewis first proposed more general definitions of acidity and basicity. He formalized his definitions in 1938.

A **Lewis acid** is an electron acceptor.

A **Lewis base** is an electron donor.

According to these definitions, a Lewis acid is an electron-deficient species; examples are H^+ and certain anhydrous metal salts, such as $FeBr_3$, $AlCl_3$, and $ZnCl_2$. A Lewis base is an ion or a molecule that has an unshared pair of electrons; ^-OH and NH_3 are examples.

A classical example of a Lewis acid-base reaction is the reaction of ammonia with boron trichloride.

$$:NH_3 \quad + \quad BCl_3 \quad \longrightarrow \quad H_3\overset{+}{N} - \overset{-}{B}Cl_3$$

A Lewis base   A Lewis acid

donating electrons   accepting electrons

In this text, almost all acid-base reactions you will encounter involve the transfer of a proton from an acid to a base (Brønsted-Lowry definition). Therefore, when we use the term *acid* we mean a proton donor. We will restrict our use of the term *Lewis acid* to refer to the anhydrous metal salts, which are important catalysts for many organic reactions.

PROBLEM 1.11. Although we do not emphasize Lewis acids and bases in this text, many organic reactions can be considered as Lewis acid-base reactions. Later in this book we will discuss the following reactions or reaction steps. Identify each reactant as a Lewis acid or a Lewis base.

(a) $FeBr_3 \ + \ Br_2 \rightleftharpoons \ ^-FeBr_4 \ + \ Br^+$

(b) $CH_3CH_2OH \ + \ H^+ \rightleftharpoons CH_3CH_2\overset{+}{O}H_2$

(c) $(CH_3)_3C^+ \ + \ Cl^- \longrightarrow (CH_3)_3C - Cl$

TABLE 1.6 SUMMARY OF ACIDS AND BASES

Brønsted-Lowry acid (proton donor):

Strong: H_2SO_4, HCl, HNO_3
Moderate: H_3PO_4

$$\overset{\displaystyle O}{\overset{\displaystyle \|}{\text{Weak: RCOH}}}$$

Very weak: H_2O, $HC\equiv N$

Brønsted-Lowry base (proton acceptor):

Strong: $^-:\overset{..}{\underset{..}{O}}H$, $^-:\overset{..}{\underset{..}{O}}R$

Weak: $:NH_3$, $R\overset{..}{N}H_2$, $R_2\overset{..}{N}H$, $R_3N:^a$

Lewis acid (electron acceptor): H^+, $AlCl_3$, $FeBr_3$, etc.
Lewis base (electron donor): same bases as Brønsted-Lowry bases

---

^{a} The symbol R, as in RNH_2, means a carbon–hydrogen group.

Table 1.6 summarizes the theories of acids and bases and provides some examples of each class.

SUMMARY

Organic chemistry is the chemistry of compounds that contain carbon.

The electrons of an atom are found in **atomic orbitals**, which are specified regions in space in the electron shells that surround the nucleus.

First shell: $1s$ orbital

Second shell: $2s$, $2p_x$, $2p_y$, $2p_z$ orbitals

Each orbital can contain a maximum of one pair of electrons.

Electronegativity is a measure of an atom's attraction for its valence (outer-shell) electrons. In general, electronegativity increases with decreasing atomic radius in a series of elements.

An **ionic bond**, which arises from electron transfer, is the electrostatic attraction between a cation (positive ion) and an anion (negative ion). A **covalent bond** is the *sharing* of one pair of electrons. The **bond dissociation energy** is the amount of energy per mole required to break a specified covalent bond.

Atoms share electrons to obtain a filled outer shell, usually eight electrons. The following elements generally form the given number of covalent bonds in stable neutral compounds.

1 covalent bond: H, F, Cl, Br, I

2 covalent bonds: O

3 covalent bonds: N

4 covalent bonds: C

Atoms that can form more than one covalent bond sometimes form multiple bonds, as in $CH_2{=}CH_2$ or $HC{\equiv}CH$.

Depending on electronegativity differences, covalent bonds can be **polar** or **nonpolar**.

Polar

$$H-\underset{\underset{H}{|}}{\overset{\overset{H}{|}}{C}}-\underset{\underset{H}{|}}{\overset{\overset{H}{|}}{C}}-H \qquad H{-}\overset{\overset{H}{|}}{\underset{\underset{H}{|}}{C}}{-}\overset{\delta-}{O}{-}H^{\delta+}$$

Relatively nonpolar

Formal charge is the ionic charge assigned to an atom in a molecule to allow the bonding to fit the rules of valence—the number of bonds the atom forms.

Covalent bonds arise from the merging of atomic orbitals to yield **molecular orbitals**. Carbon forms four tetrahedral single bonds with sp^3-hybridized orbitals.

Organic molecules can be represented by a variety of formulas: **empirical** (ratio of atoms), **molecular** (number of atoms), **structural** (showing the order of bonding of atoms), and **condensed** (*implying* the order of bonding). **Lewis formulas**, **line-bond formulas**, and **three-dimensional formulas** are structural formulas.

Structural Formulas for Methanol:

Three-dimensional	Lewis	Line-bond	Condensed

CH_3OH

Structural isomers are compounds with the same molecular formulas but different structures.

Resonance formulas are used when more than one line-bond formula can be drawn for a structure.

Hydrogen bonds are strong attractions between the H of OH or NH and the unshared valence electrons of another O or N. Hydrogen bonding increases boiling point and water solubility.

An **acid** is a proton donor. Strong acids, such as H_2SO_4 or HCl, are almost completely ionized in water, yielding H_3O^+ and an anion (Cl^-, HSO_4^-). Only a small percentage of weak acid molecules ionize in water. **Carboxylic acids**, compounds containing the $-CO_2H$ group, are usually weak acids.

A **base** is a proton acceptor. Strong bases, such as NaOH and KOH, are completely ionized in water. Weak bases, such as ammonia and **amines**, react to a limited extent with water to yield a cation and hydroxide ions.

$$
\begin{array}{llll}
& \quad\quad\quad\quad \overset{\displaystyle O}{\overset{\displaystyle \|}{}} & & \overset{\displaystyle O}{\overset{\displaystyle \|}{}} \\
\textit{A weak acid:} & CH_3COH & + \; H_2O \rightleftharpoons CH_3CO^- & + \; H_3O^+ \\
\textit{A weak base:} & CH_3NH_2 & + \; H_2O \rightleftharpoons CH_3\overset{+}{N}H_3 & + \; {}^-OH
\end{array}
$$

A **Lewis acid** is an electron acceptor and a **Lewis base** is an electron donor.

STUDY PROBLEMS

1.12. Sodium (Na) has the atomic number 11. Fill in the following diagram to show how many electrons are in each orbital of a sodium atom.

$$1s \quad 2s \quad 2p_x \; 2p_y \; 2p_z \quad 3s \quad 3p_x \; 3p_y \; 3p_z$$

1.13. Using a diagram similar to the one in the preceding problem, show the number of electrons in each orbital of **(a)** Na^+; **(b)** C, atomic number 6; **(c)** O, atomic number 8; and **(d)** O^{2-}.

1.14. Referring to the periodic table of the elements inside the cover of this book, predict which element has the shorter atomic radius. Explain the reason for your answers.
(a) N or P   **(b)** Br or I
(c) C or O   **(d)** F or S

1.15. Referring to the periodic table, predict which element is more electronegative.
(a) Li or B   **(b)** Si or S
(c) N or P   **(d)** Cl or I

1.16. Draw Lewis formulas for the following elements: **(a)** Mg, **(b)** B, **(c)** N, and **(d)** O.

1.17. Using Lewis formulas, write equations that show the formation of the follow ng ions from the elements: **(a)** Ca^{2+}, **(b)** O^{2-}, and **(c)** F^-.

1.18. Write Lewis formulas, showing all valence electrons, for the following covalent compounds.
(a) CH_3CH_3 **(b)** CH_3OH **(c)** $CH_2{=}CH_2$

(d) H_2O **(e)** CH_3Br **(f)** HCN

(g) CH_3OCH_3 **(h)** $H_2C{=}O$ **(i)** $CH_3C{\equiv}CH$

1.19. Write Lewis formulas for the following ionic compounds. Be sure to assign formal charges.
(a) $CH_3O^- \, Na^+$ **(b)** $^+NH_4 \, Cl^-$ **(c)** $Na^+ \; {}^-NH_2$

(d) $CH_3\overset{\displaystyle O}{\overset{\displaystyle \|}{C}}O^- \, K^+$ **(e)** $CH{\equiv}C^- \, Na^+$ **(f)** $(CH_3)_2N^- \, Na^+$

1.20. Rewrite the following formulas, adding any unshared pairs of valence electrons (as dots).

(a) $(CH_3)_2NH$ **(b)** CH_3OH **(c)** $\overset{\overset{\displaystyle O}{\|}}{HCOH}$

(d) $(CH_3)_3CO^- \; K^+$ **(e)** $H_2NCH_2\overset{\overset{\displaystyle O}{\|}}{C}O^- \; Na^+$

1.21. Circle the most electronegative atom in each formula.

(a) $CH_3CH_2CH_2OH$ **(b)** CH_3NHCH_3 **(c)** FCH_2CH_2OH

1.22. Rewrite the following formulas as line-bond formulas, showing unshared valence electrons and using the symbols $\delta+$ and $\delta-$ to show the polarity of any polar bonds.

(a) CH_3CH_2OH **(b)** CH_3NH_2 **(c)** $CH_3\overset{\overset{\displaystyle O}{\|}}{C}OH$

1.23. Calculate the formal charge, if any, on each atom except carbon and hydrogen in the following structures:

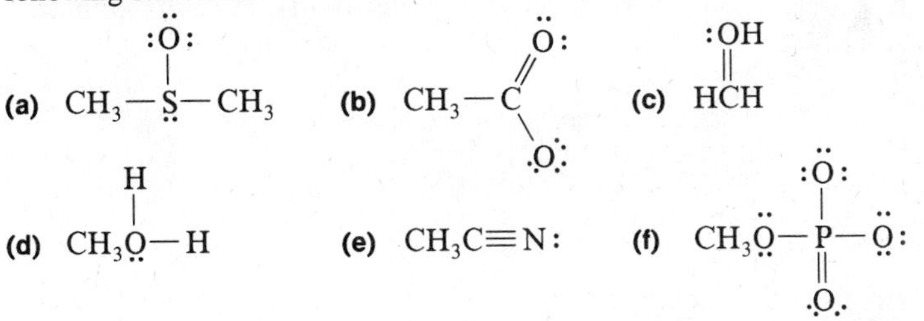

(a) $CH_3 - \overset{\overset{\displaystyle :\ddot{O}:}{|}}{\underset{\displaystyle \ddot{\ }}{S}} - CH_3$ **(b)** $CH_3 - \overset{\displaystyle \ddot{O}:}{\underset{\displaystyle \diagdown \;\dot{\ddot{O}}:}{C}}$ **(c)** $\overset{\overset{\displaystyle :OH}{\|}}{HCH}$

(d) $CH_3\overset{\overset{\displaystyle H}{|}}{\ddot{O}} - H$ **(e)** $CH_3C \equiv N:$ **(f)** $CH_3\ddot{O} - \overset{\overset{\displaystyle :\ddot{O}:}{|}}{\underset{\displaystyle \underset{\displaystyle \dot{\ddot{O}}\cdot}{\|}}{P}} - \ddot{O}:$

1.24. Calculate the formal charge, if any, on the central carbon in the following unstable species:

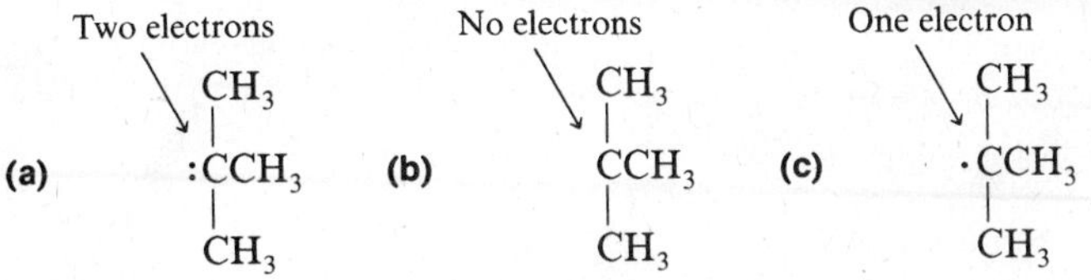

Two electrons

(a) $\overset{\diagdown \; CH_3}{\underset{\displaystyle \underset{\displaystyle CH_3}{|}}{\overset{\displaystyle |}{:CCH_3}}}$

No electrons

(b) $\overset{\diagdown \; CH_3}{\underset{\displaystyle \underset{\displaystyle CH_3}{|}}{\overset{\displaystyle |}{CCH_3}}}$

One electron

(c) $\overset{\diagdown \; CH_3}{\underset{\displaystyle \underset{\displaystyle CH_3}{|}}{\overset{\displaystyle |}{\cdot CCH_3}}}$

1.25. Circle each tetrahedral carbon in the following formulas:

(a) $CH_3CH_2CH_3$ **(b)** $CH_3\overset{\overset{\displaystyle O}{\|}}{C}OH$ **(c)** $CH_3CH = CHCH_3$

(d) 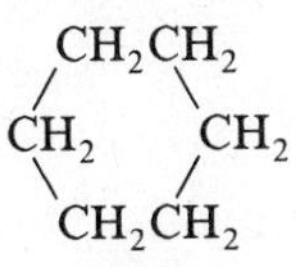 $\underset{\displaystyle \underset{\displaystyle CH_2CH_2}{\diagdown \quad \diagup}}{\overset{\displaystyle CH_2CH_2}{\diagup \quad \diagdown}} \; \begin{matrix} CH_2 & & CH_2 \end{matrix}$ **(e)** 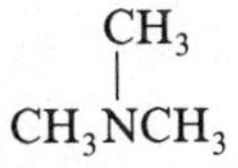 $CH_3\overset{\overset{\displaystyle CH_3}{|}}{N}CH_3$ **(f)** $CH_3CH_2\overset{\overset{\displaystyle O}{\|}}{C}H$

1.26. Tell which types of hybrid orbitals or atomic orbitals are used to form each bond in the following formula. For example, the bond in H_2 ($H-H$) is *s-s*.

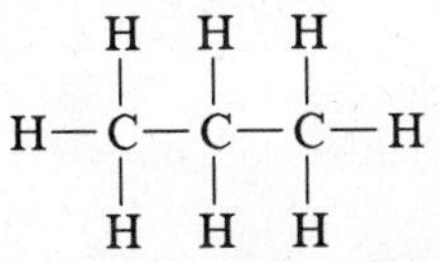

$$H - \overset{\overset{\displaystyle H}{|}}{\underset{\displaystyle \underset{\displaystyle H}{|}}{C}} - \overset{\overset{\displaystyle H}{|}}{\underset{\displaystyle \underset{\displaystyle H}{|}}{C}} - \overset{\overset{\displaystyle H}{|}}{\underset{\displaystyle \underset{\displaystyle H}{|}}{C}} - H$$

1.27. Show all principal resonance formulas for the following structures:

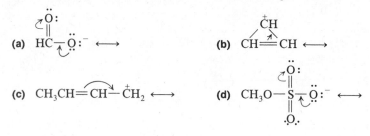

(a) $HC{-}\overset{..}{\underset{..}{O}}{:}^{-} \longleftrightarrow$

(b) $\overset{+}{C}H{-}CH{=}CH \longleftrightarrow$

(c) $CH_3CH{=}CH{-}\overset{+}{C}H_2 \longleftrightarrow$

(d) $CH_3O{-}S{-}\overset{..}{\underset{..}{O}}{:}^{-} \longleftrightarrow$

1.28. Write equations showing resonance formulas of the following compounds:

(a)
$$\begin{array}{c} CH \\ HC \diagup \quad \diagdown CH \\ \| \qquad \quad | \\ HC \diagdown \quad \diagup CH \\ N \end{array}$$

(b)
$$\begin{array}{c} CH \\ H_2C \diagup \quad \diagdown CH \\ | \qquad \quad | \\ H_2C \diagdown \quad \diagup CH \\ CH_2 \end{array}$$

1.29. Circle and name the functional groups in the following formula:

$$\begin{array}{c} O \\ \diagdown \\ C{-}CHCH_2CH{=}CHCH_2CH_2CH_2\overset{O}{\overset{\|}{C}}OH \\ H_2C \diagdown \quad \diagup CHCH{=}CHCHCH_2CH_2CH_2CH_2CH_3 \\ CH \qquad\qquad\qquad | \\ | \qquad\qquad\qquad OH \\ OH \end{array}$$

A prostaglandin
a hormone moderator in the body

1.30. Which of the following pairs represent the same compound and which represent structural isomers?

(a) CH_3NHCH_3 and $CH_3CH_2NH_2$

(b) $CH_3CH_2\overset{OH}{\overset{|}{C}}HCH_2CH_3$ and $CH_3CH_2\overset{CH_3CH_2}{\overset{|}{C}}HOH$

(c) $CH_3\overset{O}{\overset{\|}{C}}CH_2CH_2CH_3$ and $CH_3CH_2\overset{O}{\overset{\|}{C}}CH_2CH_3$

(d)
$$\begin{array}{c} OH \\ H_2C{-}C \diagup \\ \diagup \qquad \diagdown H \\ H_2C \qquad\quad OH \\ \diagdown \qquad C \diagdown \\ CH_2 \quad H \end{array}$$ and
$$\begin{array}{c} H_2C{-}CH_2 \\ H \diagup \quad H \quad \diagdown CH_2 \\ \diagdown C \qquad | \qquad \diagup \\ \quad \diagdown C \diagup \\ HO \quad | \\ OH \end{array}$$

1.31. Write condensed structural formulas for all the structural isomers of the following molecular formulas:

(a) C_4H_9Cl (four isomers)   (b) $C_4H_{10}O$ (seven isomers)

1.32. Which compound would have the higher boiling point? Explain your answers.

    (a) $CH_3CH_2CH_2OH$ or $CH_3CH_2CH_2CH_3$

    (b) $CH_3CH_2CH_2OH$ or $CH_3CH_2OCH_3$

    (c) $CH_3\overset{\displaystyle CH_3}{\underset{|}{N}}CH_3$ or $CH_3CH_2CH_2NH_2$

    (d) $HOCH_2CH_2OH$ or $CH_3CH_2CH_2OH$

1.33. Which of the following compounds can form hydrogen bonds in their pure state?

    (a) $CH_3\overset{\displaystyle O}{\overset{\|}{C}}H$ **(b)** $CH_3CH_2OCH_2CH_3$

    (c)
$$
\begin{array}{c}
H_2C-CH_2 \\
\diagup \qquad \diagdown \\
H_2C \qquad\qquad CH_2 \\
\diagdown \qquad \diagup \\
N \\
| \\
H
\end{array}
$$

1.34. Write formulas to show all types of hydrogen bonds in aqueous solutions of the following compounds:

    (a) CH_3CH_2OH **(b)** CH_3NHCH_3

    (c) $CH_3\overset{\displaystyle O}{\overset{\|}{C}}H$

1.35. Complete the following equations for acid-base reactions:

    (a) $CH_3CH_2CH_2\overset{\displaystyle O}{\overset{\|}{C}}OH + H_2O \rightleftharpoons$

    (b) $HO\overset{\displaystyle O}{\overset{\|}{C}}CH_2\overset{\displaystyle O}{\overset{\|}{C}}OH + 2Na^+\ {}^-OH \longrightarrow$

    (c) $CH_3\overset{\displaystyle CH_3}{\underset{|}{N}}CH_3 + HCl \longrightarrow$

    (d) $CH_3\overset{\displaystyle CH_3}{\underset{|}{N}}CH_3 + H_2O \rightleftharpoons$

    (e) $CH_3NH_2 + CH_3CH_2\overset{\displaystyle O}{\overset{\|}{C}}OH \longrightarrow$

    (f) $CH_3\overset{+}{N}H_3\ Cl^- + {}^-OH \longrightarrow$

    (g) $CH_3\overset{\displaystyle O}{\overset{\|}{C}}O^- + HCl \longrightarrow$

1.36. Identify the Lewis acid and the Lewis base in each of the following equations:

    (a) $CH_3NH_2 + BF_3 \longrightarrow CH_3\overset{+}{N}H_2-\overset{-}{B}F_3$

    (b) $CH_2{=}CH\overset{+}{C}H_2 + Br^- \longrightarrow CH_2{=}CHCH_2-Br$

$$\text{(c)} \quad CH_3\overset{\overset{\displaystyle O}{\|}}{C}CH_3 + {}^-CN \longrightarrow CH_3\overset{\overset{\displaystyle O^-}{|}}{\underset{\underset{\displaystyle CN}{|}}{C}}CH_3$$

1.37. Referring to Table 1.5, write all possible formulas with the following specifications:

  (a) an alcohol, $C_4H_{10}O$ **(e)** a ketone, $C_5H_{10}O$

  (b) an alkene, C_4H_8 **(f)** an aldehyde, $C_5H_{10}O$

  (c) a carboxylic acid, $C_4H_8O_2$ **(g)** an amine, $C_4H_{11}N$

  (d) an ether, $C_4H_{10}O$

1.38. Draw three-dimensional formulas (with solid and broken wedges) for the following compounds:

  (a) CH_2Cl_2 **(b)** CH_3CH_2Br

1.39. The nitrogen in $CH_3\overset{+}{N}H_3\,Cl^-$ is sp^3 hybridized and tetrahedral. Draw a three-dimensional formula depicting this formula.

1.40. The nitrogen in CH_3NH_2 is bonded to only three other atoms, yet it is sp^3 hybridized and tetrahedral. Using a three-dimensional formula showing the orbital containing the unshared valence electrons, explain the geometry.

POINT OF INTEREST 1

Alfred Nobel and the Nobel Prizes

Alfred Bernhard Nobel was a man who changed his will after reading his own obituary! Nobel, a Swedish chemist and industrialist, was born in Stockholm in 1833. His father was an inventor and munitions manufacturer whose operations were located in Russia. Young Nobel's health was poor and he received little formal academic training. His mother taught him to read and write. Later, his father provided Alfred and his three brothers with tutors. Alfred showed an early proficiency in language and, in later years, could speak and write six languages fluently.

Chemistry attracted the young Nobel, who assisted his father in the manufacture of gunpowder and munitions. At the age of 22, Nobel continued his chemical studies in the laboratories of Professor Zinin at St. Petersburg University. It was here that Nobel first became interested in the highly explosive and unpredictable nitroglycerin, which had been discovered in 1847 by the Italian chemist Ascanio Sobrero. Sobrero considered the explosive oil too dangerous for commercial use.

At the end of the Crimean war, the Russian need for munitions diminished. The Nobel family returned to Sweden, where Alfred conducted experiments on controlling the explosive force of nitroglycerin.

In 1863 Nobel developed his first major invention, the blasting cap, or detonator. This device consisted of a small capsule of easy-to-detonate mercury fulminate, $Hg(CNO)_2$, which could be inserted into a container of nitroglycerin. On combustion, the explosion of mercury fulminate caused the nitroglycerin to detonate. Although the blasting cap allowed nitroglycerin to be detonated at will, nitroglycerin remained extremely dangerous to transport and handle.

In Stockholm, the Nobel family rented a shed to manufacture nitroglycerin and blasting caps. In 1864 tragedy struck. A blast leveled the plant, killing Alfred

Nobel's youngest brother and four other people. Alfred's father suffered a stroke shortly thereafter and never recovered his health.

The deaths haunted Alfred. He strove to find a way to control the capricious nature of nitroglycerin and, in 1866, he succeeded. He discovered that nitroglycerin is much more stable when absorbed in diatomaceous earth. Yet the mixture could still be detonated with a blasting cap. He called the invention "dynamite."

Nobel built a vast international industrial empire that supplied explosive products to warring nations and to construction industries. Throughout his life, however, Nobel remained an inventor. He obtained 355 patents in different countries. One of his most significant inventions was smokeless powder.

In 1888, at age 55, Nobel received a shock. Another of his brothers had died, and a newspaper printed Alfred's obituary by mistake. Nobel read what the world would think of him after his death—"the dynamite king" and a "merchant of death." Nobel spent his remaining few years trying to change this image. To accomplish this goal he modified his will, leaving his immense fortune to establish the Nobel Prizes.

The Nobel Prizes   Nobel's will provided funds for the establishment and perpetuation of prizes in physics, chemistry, physiology or medicine, literature, and peace. In 1968 the Central Bank of Sweden created an additional prize, the Nobel Memorial Prize in Economic Sciences. Nobel's will stated that the prizes for physics and chemistry were to be awarded by the Swedish Academy of Science, the prize for physiology or medicine by the Caroline Institute in Stockholm, and the Literature Prize by the Swedish Academy. The Peace Prize became the responsibility of the Norwegian Parliament.

The first Nobel Prize in Chemistry was awarded in 1901 to J. H. van't Hoff, a Dutch chemist, for his work on osmotic pressure. The following is a partial list of chemistry Nobel laureates, many of whose discoveries are discussed in this text.

Some Nobel Laureates in Chemistry

1902   Emil Fischer (German): sugar and purine synthesis

1905   Adolf von Baeyer (German): dyes and hydroaromatic compounds

1910   Otto Wallach (German): alicyclic (aliphatic cyclic) compounds

1912   Victor Grignard (French): Grignard reagents
         Paul Sabatier (French): hydrogenation of organic compounds

1928   Adolf Windaus (German): steroids and vitamins

1937   Walter Haworth (British): carbohydrates and vitamin C
         Paul Karrer (Swiss): carotenoids, flavins, vitamins A and B_2

1938   Richard Kuhn (German): carotenoids and vitamins

1939   Adolf Butenandt (German): sex hormones
         Leopold Ruzicka (Swiss): polymethylenes and terpenes

1947   Sir Robert Robinson (British): plant products, alkaloids

1950   Otto Diels and Kurt Alder (German): Diels-Alder reaction

1954   Linus Pauling (American): chemical bonding

1958   Frederick Sanger (British): proteins, insulin

1961   Melvin Calvin (American): photosynthesis

1963   Karl Ziegler (German) and Giulio Natta (Italian): polymerization catalysts

1965   R. B. Woodward (American): organic synthesis, chlorophyll

1969   D. H. R. Barton (British) and Odd Hassel (Norwegian): stereochemistry

1975   J. W. Cornforth (Australian) and Vladimir Prelog (Czech-Swiss): stereochemistry

1979   H. C. Brown (American) and Georg Wittig (West German): temporary chemical links for complex molecules

1980   Paul Berg and Walter Gilbert (American) and F. Sanger (British): deoxyribonucleic acids

1983   H. Taube (American): work in the mechanism of electron transfer reactions, especially in metal complexes

1984   R. B. Merrifield (American): a simple and ingenious automated laboratory technique for synthesizing peptide chains

1987   D. J. Cram (American), C. J. Peterson (American), and Jean-Marie Lehn (French): wide-ranging research, including synthesis

Chapter 2

Alkanes and Cycloalkanes

An organic compound that contains only carbon and hydrogen is called a **hydrocarbon**. Today's society depends on hydrocarbons for fuels and for an inexpensive source of organic raw materials. Hydrocarbons are found in nature primarily as *natural gas* and *crude oil* (petroleum). The majority of our fuels, such as gasoline, kerosene, and fuel oil, are obtained from the refining of crude oil, which is a complex mixture consisting mostly of hydrocarbons. Natural gas, which contains 60–90% methane (CH_4), is a common fuel for heating homes and other buildings. Crude oil and natural gas also provide the raw materials for the **petrochemical industry**, the industry that manufactures over 90% of the organic chemicals produced in the United States, such as drugs, fertilizers, and plastics. (We will discuss petroleum in more detail in Point of Interest 2.)

2.1 CLASSIFICATION OF HYDROCARBONS

There is only one hydrocarbon with one carbon atom, methane (CH_4). All other hydrocarbons contain two or more carbon atoms bonded together by single, double, or triple bonds. Hydrocarbons can be classified according to the types of carbon–carbon bonds they contain. Hydrocarbons with only carbon–carbon single bonds are called **saturated hydrocarbons**. Hydrocarbons with one or more carbon–carbon double bonds or triple bonds are called **unsaturated hydrocarbons**. Although methane has no carbon–carbon bonds, for convenience and on the basis of its chemical reactivity, it is classified as a saturated hydrocarbon.

Saturated Hydrocarbons:

$$CH_4 \qquad CH_3CH_3 \qquad
\begin{array}{c}
H_2C-CH_2 \\
H_2C \qquad CH_2 \\
CH_2
\end{array}$$

Unsaturated Hydrocarbons:

$$CH_2{=}CH_2 \qquad CH{\equiv}CH$$

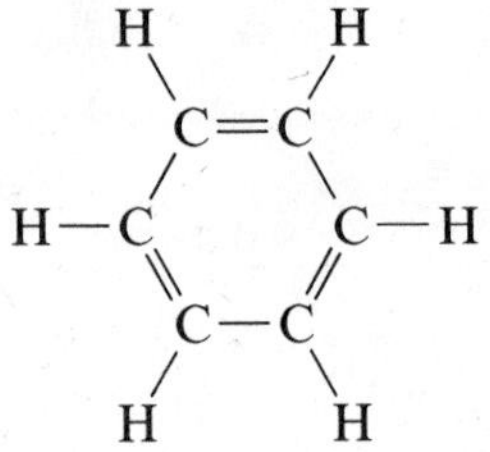

Hydrocarbons can have their carbons joined as chains or as rings. The saturated hydrocarbons with carbon atoms joined in a continuous chain or a branched chain are classified as **alkanes**. A continuous chain means that each carbon atom of the alkane is bonded to no more than two other carbons atoms. A branched-chain alkane contains at least one carbon atom bonded to three or more other carbon atoms.

Continuous-Chain Alkanes:

$$CH_3CH_2CH_2CH_3$$

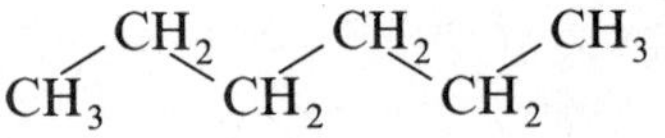

Branched-Chain Alkanes:

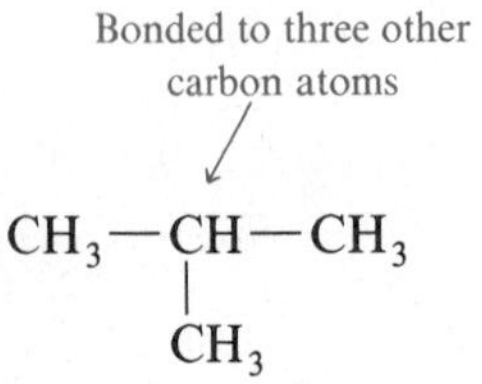

Bonded to three other carbon atoms

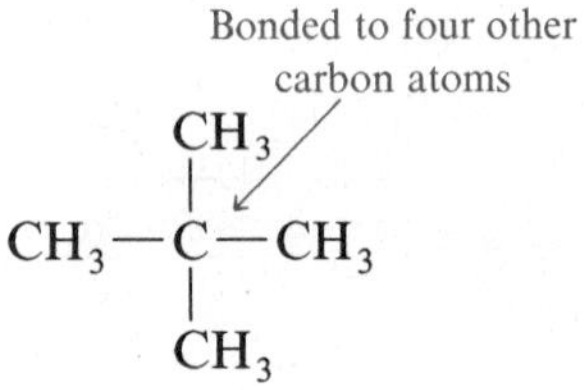

$$CH_3{-}\underset{\underset{CH_3}{|}}{CH}{-}CH_3$$

Bonded to four other carbon atoms

$$CH_3{-}\underset{\underset{CH_3}{|}}{\overset{\overset{CH_3}{|}}{C}}{-}CH_3$$

Saturated hydrocarbons with carbon atoms forming one or more rings are called **cycloalkanes**. Cycloalkanes are often considered to be a subclass of the alkanes.

Cycloalkanes:

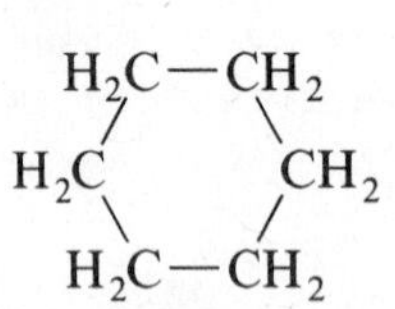

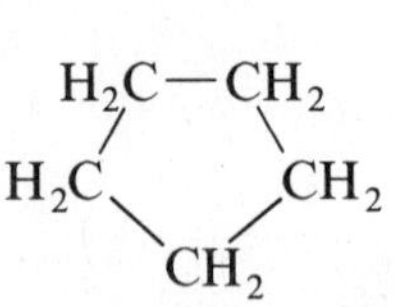

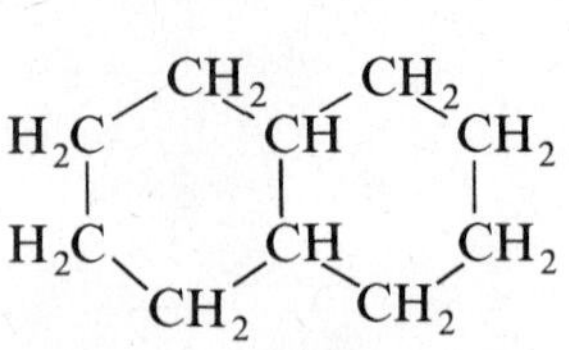

All open-chain (noncyclic) alkanes have the general formula C_nH_{2n+2}, where n is the number of carbon atoms. For example, methane contains one carbon ($n = 1$)

and four hydrogens ($2n + 2 = 4$). Two other examples follow:

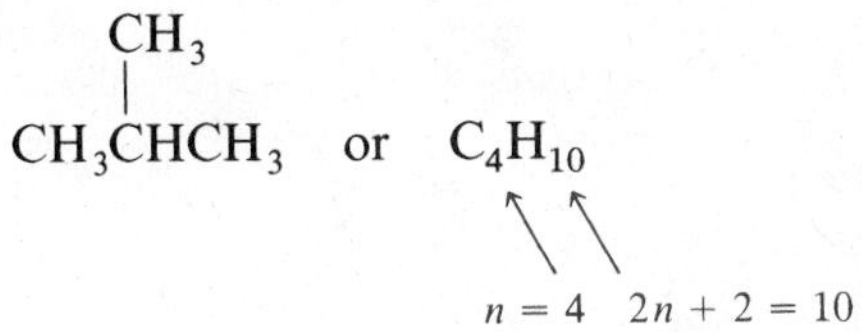

$$\overset{\displaystyle CH_3}{\underset{\displaystyle |}{CH_3CHCH_3}} \quad \text{or} \quad C_4H_{10}$$

$$n = 4 \quad 2n + 2 = 10$$

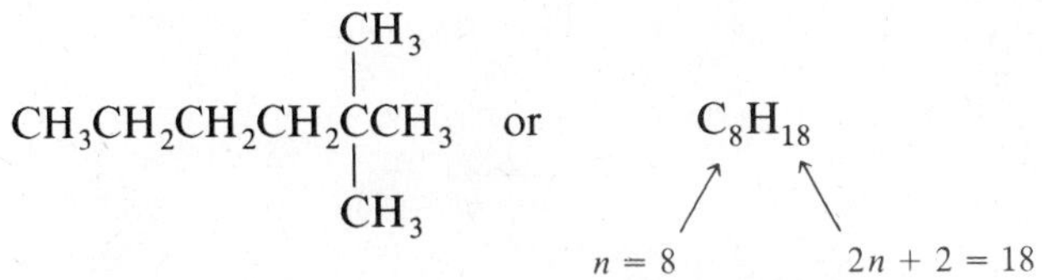

$$\overset{\displaystyle CH_3}{\underset{\displaystyle |}{CH_3CH_2CH_2CH_2CCH_3}} \quad \text{or} \quad C_8H_{18}$$
$$\underset{\displaystyle CH_3}{|}$$

$$n = 8 \qquad 2n + 2 = 18$$

A cycloalkane has two fewer hydrogen atoms than its open-chain counterpart. For example, a four-carbon open-chain alkane has a molecular formula of C_4H_{10} but the four-carbon cycloalkane has a molecular formula of C_4H_8. Therefore, a cycloalkane containing one ring has the general formula C_nH_{2n}.

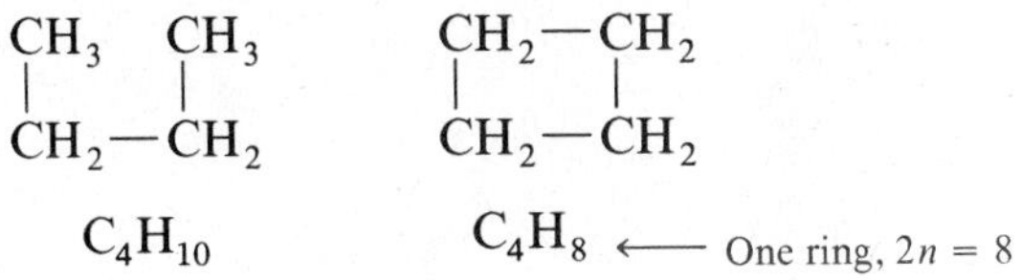

$$\begin{matrix} CH_3 & CH_3 \\ | & | \\ CH_2 & - & CH_2 \end{matrix} \qquad \begin{matrix} CH_2 & - & CH_2 \\ | & & | \\ CH_2 & - & CH_2 \end{matrix}$$

$$C_4H_{10} \qquad\qquad C_4H_8 \longleftarrow \text{One ring, } 2n = 8$$

PROBLEM 2.1. What are the general formulas for the following structures?

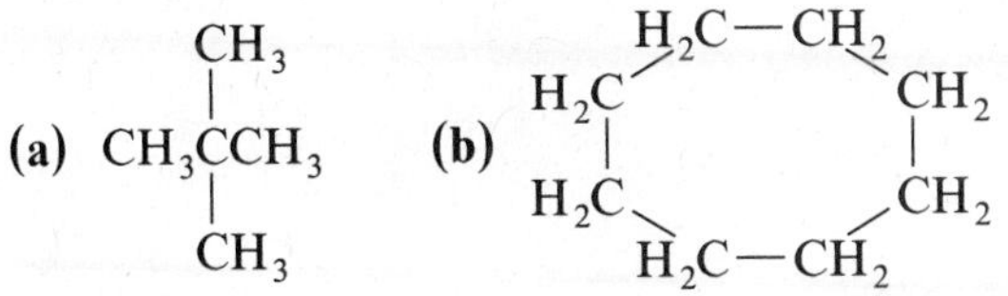

$$\textbf{(a)} \quad \overset{\displaystyle CH_3}{\underset{\displaystyle CH_3}{\overset{\displaystyle |}{\underset{\displaystyle |}{CH_3CCH_3}}}}$$

$$\textbf{(b)} \quad \begin{matrix} & H_2C-CH_2 & \\ H_2C & & CH_2 \\ | & & | \\ H_2C & & CH_2 \\ & H_2C-CH_2 & \end{matrix}$$

2.2 PHYSICAL PROPERTIES OF ALKANES

Because molecules of alkanes contain only nonpolar C—C and C—H bonds, alkane molecules have only small attractive forces among themselves. For this reason, alkanes are lower boiling than polar compounds of similar formula weight. For example, propane (FW 44.11), $CH_3CH_2CH_3$, boils at $-42.1°C$, slightly more than $19°$ below the boiling point of dimethyl ether (FW 46.07, bp $-23°C$), CH_3OCH_3. Table 2.1 shows the boiling points, densities, and physical states at room temperature of a few alkanes.

Compounds containing branched-chain molecules boil at slightly lower temperatures than their continuous-chain isomers. The reason for the lower boiling points is that branched molecules cannot align themselves as closely as can continuous-chain molecules and thus cannot develop as strong intermolecular attractive forces. Figure

TABLE 2.1 PHYSICAL PROPERTIES OF SOME ALKANES

Alkane	Bp (°C)	Density at 20°C (g/mL)	Physical state at room temperature
CH_4	−162	—	Gas
CH_3CH_3	−88.5	—	Gas
$CH_3CH_2CH_3$	−42	—	Gas
$CH_3(CH_2)_2CH_3$	0	—	Gas
$CH_3(CH_2)_3CH_3$	36	0.63	Liquid
$CH_3(CH_2)_4CH_3$	69	0.66	Liquid
$CH_3(CH_2)_5CH_3$	98	0.68	Liquid
$CH_3(CH_2)_6CH_3$	126	0.70	Liquid
$CH_3(CH_2)_7CH_3$	151	0.72	Liquid
$CH_3(CH_2)_8CH_3$	174	0.73	Liquid

2.1 illustrates how branching can hold molecules farther apart.

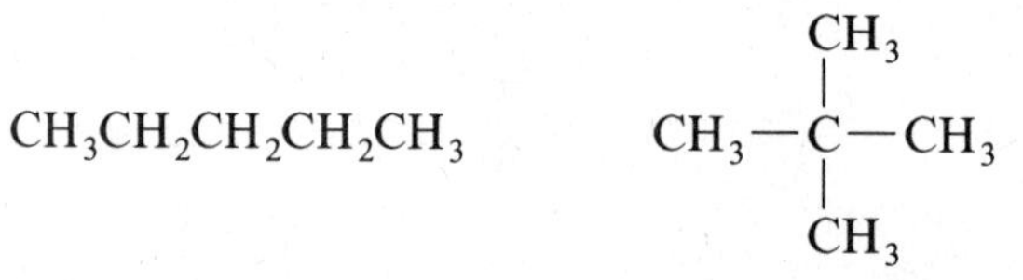

$$CH_3CH_2CH_2CH_2CH_3$$

$$CH_3-\overset{\overset{\displaystyle CH_3}{|}}{\underset{\underset{\displaystyle CH_3}{|}}{C}}-CH_3$$

Pentane, bp 36°C   2,2-Dimethylpropane, bp 10°C

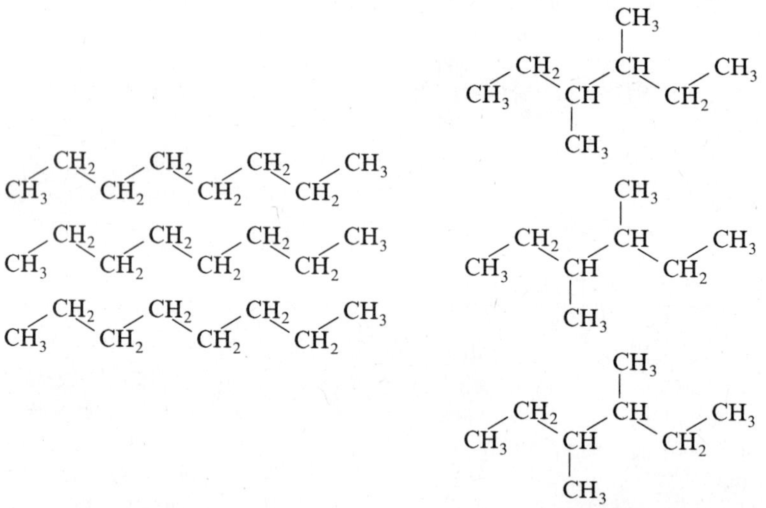

Continuous chains—
closer together;
stronger attractions

Branched chains—
farther apart;
weaker attractions

Figure 2.1 Continuous-chain compounds have greater intermolecular attractions than their branched-chain isomers.

Again, because of their lack of polarity, alkane molecules are not attracted to water molecules; therefore, alkanes are insoluble in water. Liquid alkanes are less dense than water (density of water, 1.0 g/mL) and thus float on water. An oil fire or a grease fire cannot be extinguished with water because the oil or grease (largely alkane in structure) will float on top of the water. Water actually helps spread such a fire!

2.3 NOMENCLATURE OF ALKANES

In the 1800s, when organic chemistry was an emerging science, the structures of almost all newly discovered compounds were unknown. These compounds had to be named for identification purposes. The chemists who named them usually chose a name to emphasize the properties, to identify the source, or simply to satisfy the whim of the discoverer. For example, the name for the flammable compound ethane (CH_3CH_3) is derived from the Greek word *aithein*, "to kindle or blaze." The name for the compound formic acid (HCO_2H) is derived from the Latin word *formica*, "ants," because at one time this compound was obtained by the distillation of red ants. Barbituric acid is named after the woman's name Barbara.

These types of names are referred to as **trivial names**, or **common names**. One problem with trivial names is that the name itself cannot usually be used to deduce the formula of the compound, nor can the formula be used to generate a unique name. Consequently, each structure and each name must be memorized. Millions of organic compounds are known; memorizing their individual names would be a formidable task! Another problem with trivial nomenclature is that two different compounds might inadvertently be given the same name.

To circumvent these problems, organic chemists in the late nineteenth century drew up systematic rules for organic nomenclature. These chemists devised a system that related the structure of a compound and its name. With this system it is possible to name any compound unambiguously, whether the compound is known or yet to be discovered, by inspection of its formula. Furthermore, it is possible to draw the structure of a compound from its name. These names are called **systematic names**, or **formal names**. The system that has been developed is now called the **IUPAC system of nomenclature**, after the International Union of Pure and Applied Chemistry, the group that revises and updates the system periodically.

Some of the original trivial names, such as ethane, were incorporated in the IUPAC system of nomenclature and are now considered to be systematic. In addition, many trivial names are still used by organic chemists, partly by habit but principally for convenience. Trivial nomenclature allows a chemist to give a short descriptive name to a compound whose systematic name is long and complex. Trivial nomenclature is also used today for naming a compound before its structure is known.

In this chapter we present the pertinent rules of systematic nomenclature. In subsequent chapters we will present the systematic nomenclature for each class of compounds when we discuss that class. When appropriate, we will also present trivial names that are in common use.

TABLE 2.2   THE FIRST 10 CONTINUOUS-CHAIN ALKANES

Number of carbons	Structure	Name
1	CH_4	Methane
2	CH_3CH_3	Ethane
3	$CH_3CH_2CH_3$	Propane
4	$CH_3(CH_2)_2CH_3$	Butane
5	$CH_3(CH_2)_3CH_3$	Pentane
6	$CH_3(CH_2)_4CH_3$	Hexane
7	$CH_3(CH_2)_5CH_3$	Heptane
8	$CH_3(CH_2)_6CH_3$	Octane
9	$CH_3(CH_2)_7CH_3$	Nonane
10	$CH_3(CH_2)_8CH_3$	Decane

A. Continuous-Chain Alkanes

The names and structures of the first 10 continuous-chain alkanes are listed in Table 2.2. The names of the continuous-chain alkanes provide the foundation of the IUPAC system of nomenclature. The names we present here will be modified by prefixes and suffixes to name other organic compounds.

In the table, each compound differs from the one preceding it by a *methylene group*, CH_2. Any series of organic compounds, such as the alkanes, that are listed in such a way that each member of the series differs from its neighbors by a CH_2 group is called a **homologous series**. The alkanes in Table 2.2 form a homologous series. The following alcohols also form a homologous series:

A Homologous Series:

$$CH_3CH_2OH \quad CH_3CH_2CH_2OH \quad CH_3CH_2CH_2CH_2OH \longrightarrow$$
Addition of one CH_2 group

The systematic names of the first four continuous-chain alkanes are based on the old trivial names for these compounds. The names of the fifth and following alkanes are based on Greek or Latin numbers; for example, the name of pentane, with five carbons, is from the Greek *penta*, "five." The ending *-ane* is used for all alkanes and cycloalkanes to denote a saturated hydrocarbon.

B. Branched Alkanes

When naming a branched-chain hydrocarbon, we consider the longest continuous chain in the structure as the *parent*, or *root*, for the name.

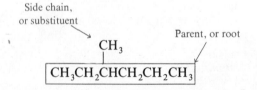

TABLE 2.3   THE FIRST SIX CONTINUOUS-CHAIN ALKYL GROUPS

Structure	Name
CH_3-	methyl
CH_3CH_2-	ethyl
$CH_3CH_2CH_2-$	propyl
$CH_3(CH_2)_2CH_2-$	butyl
$CH_3(CH_2)_3CH_2-$	pentyl
$CH_3(CH_2)_4CH_2-$	hexyl

The longest continuous chain may or may not be drawn as a straight line.

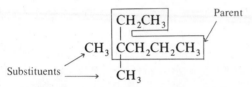

The name of a branched-chain alkane is formed by prefixing the names of side chains to the name of the parent. Side chains, such as CH_3- or CH_3CH_2-, are called **alkyl groups** from the name *alkane* with the ending changed to *-yl*. The names of the individual alkyl groups are derived from the names of the corresponding continuous-chain alkanes with the ending *-ane* changed to *-yl*. For example, meth*ane* is changed to meth*yl*. The names of the first six continuous-chain alkyl groups are listed in Table 2.3; the higher alkyl groups are named in an analogous fashion.

The letter R is used as a general symbol for an alkyl group. Thus, R in a formula can mean a methyl group, an ethyl group, or any other alkyl group.

$$CH_4 \qquad CH_3- \qquad RH \qquad R-$$

Methane   Methyl group   An alkane   An alkyl group

$R-$ can be CH_3-, CH_3CH_2-, $CH_3CH_2CH_2-$, etc.

The rules for naming a branched alkane follow.

Rules for Naming a Simple Branched Alkane

1. Determine the longest continuous chain of carbon atoms in the formula. This portion of the molecule is the parent chain, and its name is the principal part of the compound's name.
2. Number the carbons in the parent chain, beginning at the end nearer the substituent or branch.
3. Name the substituent.
4. Assemble the name by giving the position number (the number of the carbon with the branch), a hyphen, and the name of the substituent and parent combined into one word.

EXAMPLE

What is the name of the following compound?

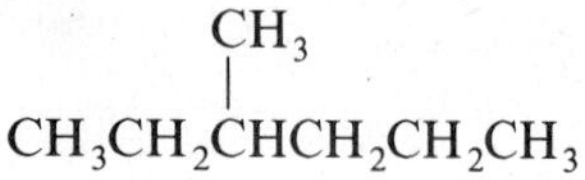

$$CH_3$$
$$|$$
$$CH_3CH_2CHCH_2CH_2CH_3$$

Solution:

1. The longest continuous chain is six carbons; the parent is *hexane*.
2. Number the chain, beginning at the end closer to the branch:

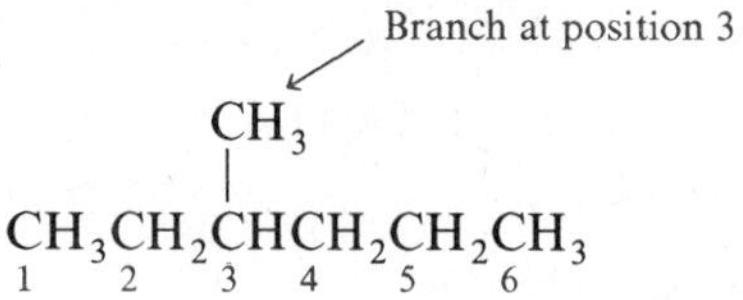

Branch at position 3

$$CH_3$$
$$|$$
$$CH_3CH_2CHCH_2CH_2CH_3$$
$$1\ \ \ 2\ \ \ 3\ \ \ 4\ \ \ 5\ \ \ 6$$

If the longest continuous chain had been numbered beginning at the other end of the chain, the branch would have been located at carbon 4. This numbering is incorrect because the chain must be numbered from the end that gives the position number the smallest value possible.

3. The name of the branch is *methyl*.
4. The systematic name of the compound is 3-methylhexane. Note that the word *methyl* and the word *hexane* are combined to form a single word, methylhexane.

PROBLEM 2.2.   Name the following branched alkanes:

$$CH_3$$
$$|$$
(a) 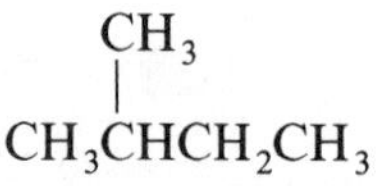 $CH_3CHCH_2CH_3$

$$CH_2CH_3$$
$$|$$
(b) 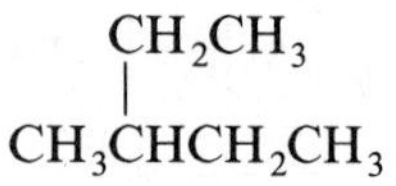 $CH_3CHCH_2CH_3$

$$CH_3$$
$$|$$
(c) 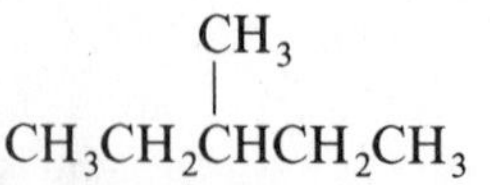 $CH_3CH_2CHCH_2CH_3$

$$CH_2CH_3$$
$$|$$
(d) 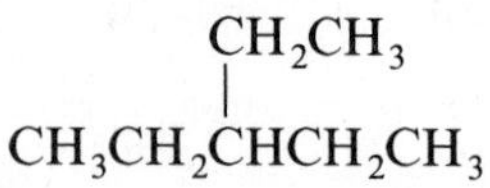 $CH_3CH_2CHCH_2CH_3$

Some functional groups are also named as prefixes in the IUPAC system of organic nomenclature. Table 2.4 lists the functional groups commonly named by prefixes along with their prefix designations. The following examples illustrate how

TABLE 2.4   SOME FUNCTIONAL GROUPS NAMED BY PREFIXES

Group	Prefix name
—F	fluoro-
—Cl	chloro-
—Br	bromo-
—I	iodo-
—NO_2	nitro-

these prefixes are used in naming compounds.

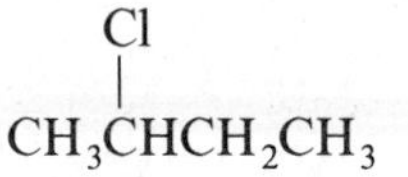

$$\underset{\text{2-Chlorobutane}}{CH_3\overset{\overset{\displaystyle Cl}{|}}{C}HCH_2CH_3}\qquad \underset{\text{Nitromethane}}{CH_3NO_2}$$

C. Branched Alkyl Groups

A side chain attached to a parent may be either a continuous chain or a branched chain. Consider a three-carbon alkyl group, the propyl group. A propyl group can be bonded to a parent chain either by an end carbon or by the central carbon. When the three-carbon group is bonded to a parent by an end carbon, the propyl group is a continuous chain and is called the **propyl group**. To emphasize that the three carbons form a continuous chain, we can add the prefix *n-* (meaning "normal"). The propyl group could thus also be called the ***n*-propyl group**. The prefix *n-* is optional.

A Continuous-Chain Three-Carbon Alkyl Group:

$$CH_3CH_2CH_2—$$

Propyl, or *n*-propyl, group

When the three carbons of a propyl are attached to a parent chain by the center carbon, the group is branched. The branched propyl group is called the **isopropyl group** (meaning the group *isomeric* with the *n*-propyl group). The prefix *iso-* is not optional.

A Branched Three-Carbon Alkyl Group:

$$CH_3\overset{\overset{\displaystyle CH_3}{|}}{C}H—$$

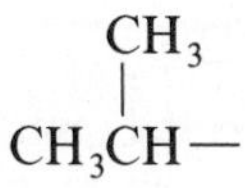

Isopropyl group

A four-carbon alkyl group can be bonded to a parent chain in four different ways.

$$CH_3CH_2CH_2CH_2-$$

Butyl, or *n*-butyl, group

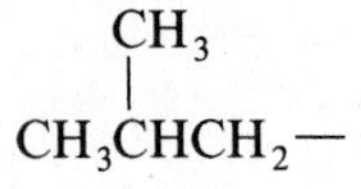

$$CH_3CHCH_2- \atop \overset{\displaystyle CH_3}{|}$$

Isobutyl group

$$CH_3CH_2CH- \atop \overset{\displaystyle CH_3}{|}$$

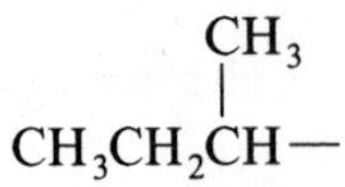

secondary-Butyl group
(abbreviated *sec*-butyl)

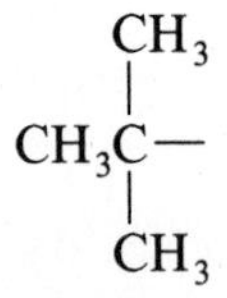

$$CH_3C- \atop \overset{\displaystyle CH_3}{|} \atop \underset{\displaystyle CH_3}{|}$$

tertiary-Butyl group
(abbreviated *tert*-butyl or
t-butyl)

The terms *secondary* and *tertiary* refer to the number of carbons bonded to the "head" carbon—the carbon that is bonded to the parent chain.

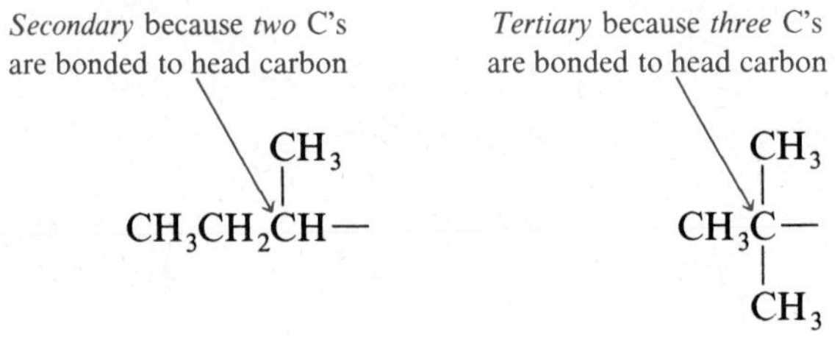

Secondary because *two* C's
are bonded to head carbon

$$CH_3CH_2CH- \atop \overset{\displaystyle CH_3}{|}$$

Tertiary because *three* C's
are bonded to head carbon

$$CH_3C- \atop \overset{\displaystyle CH_3}{|} \atop \underset{\displaystyle CH_3}{|}$$

The names of the branched alkyl groups are included in the names of alkanes just as are the names of other alkyl groups.

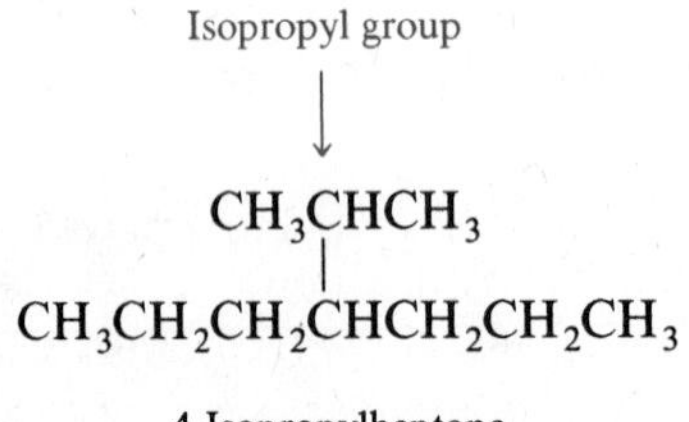

Isopropyl group

$$CH_3CHCH_3 \atop |$$
$$CH_3CH_2CH_2CHCH_2CH_2CH_3$$

4-Isopropylheptane

t-Butyl group

$$CH_3CCH_3 \atop \overset{\displaystyle CH_3}{|}$$
$$CH_3CH_2CH_2CH_2CHCH_2CH_2CH_2CH_3$$

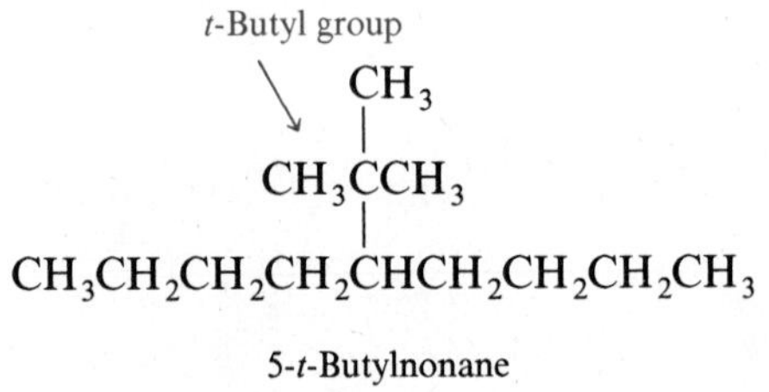

5-*t*-Butylnonane

D. Multiple Substituents

When a parent chain has more than one substituent, the substituent names, each with its position number, are listed alphabetically at the start of the name. In these names, each prefix is separated from its location number by a hyphen, and the different prefixes are also separated by hyphens as shown in the following examples.

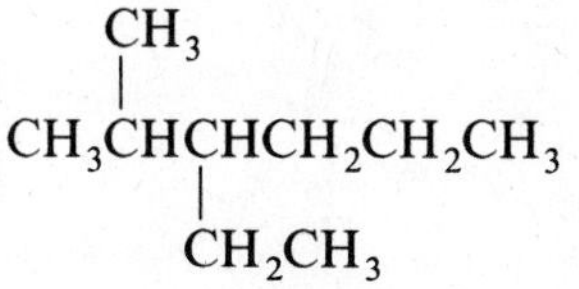

$$\underset{\text{3-Ethyl-2-methylhexane}}{\overset{\displaystyle \overset{CH_3}{|}}{CH_3CHCHCH_2CH_2CH_3}}$$
$$\underset{CH_2CH_3}{|}$$

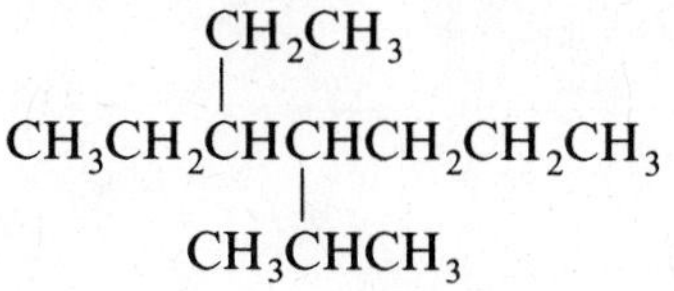

$$\underset{\text{3-Ethyl-4-isopropylheptane}}{\overset{\displaystyle \overset{CH_2CH_3}{|}}{CH_3CH_2CHCHCH_2CH_2CH_3}}$$
$$\underset{CH_3CHCH_3}{|}$$

Alphabetic listing

When a parent chain contains two or more identical substituents, such as two ethyl groups or three *t*-butyl groups, the prefix name is used only once. The position numbers of the identical substituents are grouped together and another prefix is used to denote the number of times the substituent is found in the structure. Table 2.5 lists the prefixes that denote number. Using these prefixes, let us present a few examples of this type of nomenclature.

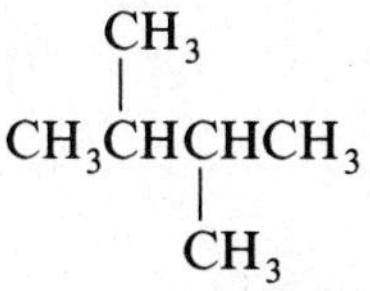

$$\overset{CH_3}{|}$$
$$CH_3CHCHCH_3$$
$$\overset{|}{CH_3}$$

2,3-Dimethylbutane

Two methyl groups

Two numbers, one for each methyl group, separated by comma

Hyphen between numbers and name

$$\overset{CH_3}{|} \quad \overset{CH_3}{|}$$
$$CH_3CHCHCHCH_3$$
$$\overset{|}{CH_3}$$

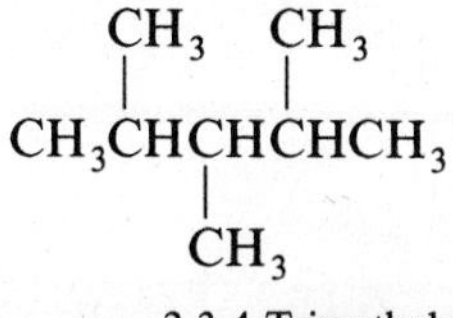

2,3,4-Trimethylpentane

Three numbers   Three methyl groups

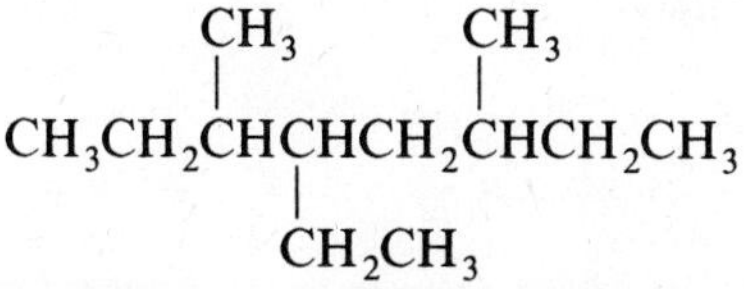

$$\overset{CH_3}{|} \qquad \overset{CH_3}{|}$$
$$CH_3CH_2CHCHCH_2CHCH_2CH_3$$
$$\overset{|}{CH_2CH_3}$$

4-Ethyl-3,6-dimethyloctane

Alphabetized as "methyl"

TABLE 2.5 PREFIXES DENOTING NUMBER

Number	Prefix
2	di-
3	tri-
4	tetra-
5	penta-
6	hexa-

PROBLEM 2.3. Name the following alkanes:

$$\text{(a)} \quad \underset{\displaystyle \overset{\displaystyle CH_3}{|}}{CH_3CH_2CHCH_2}\underset{\displaystyle \overset{|}{CH_2CH_3}}{\overset{\displaystyle \overset{CH_3}{|}}{CH}}$$

$$\text{(b)} \quad CH_3CH_2\underset{\displaystyle \overset{|}{CH_2CH_3}}{\overset{\displaystyle \overset{CH_2CH_3}{|}}{C}}CH_2CH_3$$

$$\text{(c)} \quad CH_3\underset{\displaystyle \overset{|}{CH_3}}{\overset{\displaystyle \overset{CH_3}{|}}{CH}}CHCH_2\underset{\displaystyle \overset{|}{CH_3}}{\overset{}{CH}}\underset{\displaystyle \overset{|}{CH_3}}{\overset{\displaystyle \overset{CH_3}{|}}{CH}}CHCH_3$$

$$\text{(d)} \quad CH_3CH_2CH_2\underset{\displaystyle \overset{|}{CH_2CH_2CH_3}}{\overset{\displaystyle \overset{CH(CH_3)_2}{|}}{CH}}CHCH(CH_3)_2$$

To write a formula for a structure from its systematic name, start with the parent and

1. Draw its continuous chain and number this chain.

2. Add the prefix groups to the parent.

3. Fill in the missing hydrogen atoms.

EXAMPLE

Draw the formula for 4-ethyl-3,4-dimethylheptane.

Solution:

1. Identify and draw the parent structure: heptane. Number the chain, starting at either end.

$$\underset{1}{C}-\underset{2}{C}-\underset{3}{C}-\underset{4}{C}-\underset{5}{C}-\underset{6}{C}-\underset{7}{C}$$

2. Add the prefix groups: 4-ethyl and 3,4-dimethyl.

$$\underset{1}{C}-\underset{2}{C}-\underset{\displaystyle \overset{3}{\underset{|}{C}}}{\overset{}{C}}-\underset{\displaystyle \overset{4}{\underset{|}{C}}}{\overset{\displaystyle \overset{C-C}{}}{C}}-\underset{5}{C}-\underset{6}{C}-\underset{7}{C}$$
$$\qquad\quad \overset{|}{C}\quad \overset{|}{C}$$

3. Fill in the missing hydrogens. Remember that each carbon can have no more than four bonds.

$$\begin{array}{c} CH_2CH_3 \\ | \\ CH_3CH_2CHCCH_2CH_2CH_3 \\ | \quad | \\ H_3C \quad CH_3 \end{array}$$

PROBLEM 2.4.   Write formulas for the following names:

(a) 2,4-dimethylhexane

(b) 5,5-di-*t*-butyldecane

(c) 2,2-dimethylpropane

(d) 3,3-diethyl-4,5,5-trimethyloctane

2.4 BONDING AND CONFORMATION OF ALKANES

Recall that a bond angle is the angle between any two bonds extending from one atom. In alkanes, each carbon atom has its bonds pointing toward the corners of a tetrahedron with bond angles of approximately 109.5°.

Bond angles
approx. 109.5°

A group attached by a single bond can rotate about that bond. The different shapes of ethane, butane, and other compounds that arise by simple rotation about the single bonds in a molecule are called **conformations** of those compounds. The study of shapes of molecules in three dimensions and how the shapes affect a compound's properties is called **stereochemistry**. The conformations of molecules form just one aspect of stereochemistry; you will discover later in this course that stereochemistry is important in both organic reactions and biochemical reactions.

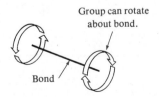

Group can rotate
about bond.

Bond

Rotation around a single bond does not affect the bond angles of the rotating carbon atoms. Only the distances between nonbonded atoms are affected. Let us use ethane to illustrate the rotation about a carbon–carbon single bond and to show the effect of rotation on the distances between nonbonded atoms on adjacent carbons. (We suggest that you construct a molecular model.)

To emphasize the staggered and eclipsed nature of the atoms, we use a formula called a **Newman projection**, which is an end-on view of a portion of a molecule. In a Newman projection we consider only two carbons in the molecule. One carbon is behind the other, and the bond joining these two carbons is hidden.

Newman Projection for Ethane:

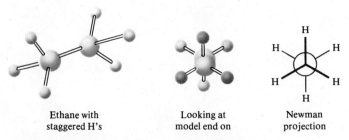

Front carbon	Rear carbon	Rear carbon behind front carbon

Relationships of Formulas for Ethane, CH$_3$CH$_3$:

Sawhorse	Ball-and-stick	Newman projection

Figure 2.2 shows how Newman projections are derived from molecular models. We strongly recommend that you use molecular models so that you can see the three-dimensionality of the conformations.

Ethane with staggered H's	Looking at model end on	Newman projection

Figure 2.2 Newman projections from molecular models.

Using a Newman projection, let us rotate the rear carbon about the carbon–carbon bond while holding the front carbon fixed. The Newman projection shows how staggered hydrogens become eclipsed, or aligned. (We show them not quite eclipsed so that the rear hydrogens will still be visible.)

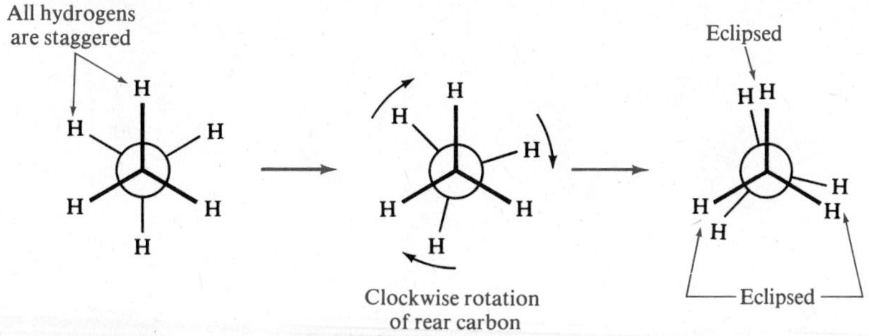

A pair of hydrogens in a molecule repel each other when they are near each other. An ethane molecule with six eclipsed hydrogens is about 3 kcal/mol higher in energy (less stable) than an ethane molecule with all its hydrogens staggered.* At room temperature, about 20 kcal/mol of energy is available to molecules; therefore, a molecule of ethane at room temperature can readily rotate from the staggered form to the eclipsed form and back to the staggered form. Because the staggered form is more stable, however, we would expect to find a greater percentage of staggered ethane molecules in a sample at any one time.

Rotation About Carbon–Carbon Bond of Ethane:

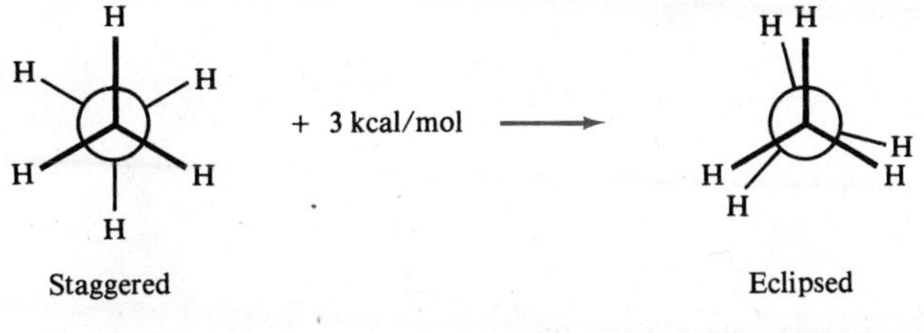

When groups bonded to carbon atoms are larger than hydrogen, there are greater repulsions between the groups in the eclipsed form than there are in the eclipsed form of ethane. Consider, for example, the rotation of butane around its carbon-2–carbon-3 bond. The most stable form of butane is that in which carbons 1 and 4 are as far apart from each other as possible. This conformation is called the **anti form**. To bring the methyl groups as close together as possible (eclipsed methyls) requires about 4.5 kcal/mol. Therefore, at any given time, most butane molecules are in the anti form. Figure 2.3 shows molecular models illustrating the anti and eclipsed-methyl conformations of butane.

---

*The unit joules (J) can be used instead of calories: 1 cal = 4.18 J and, therefore, 1 kcal = 4.18 kJ.

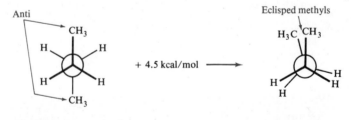

Anti methyls   Eclipsed methyls

Figure 2.3 Ball-and-stick models of butane.

Formulas for butane, CH_3CH_2—CH_2CH_3, using the indicated carbon-2–carbon-3 bond for the Newman projection:

Sawhorse   Ball - and - stick   Newman projection

Rotation About the Carbon-2–Carbon-3 Bond of Butane:

Anti

CH_3

$+ 4.5$ kcal/mol ⟶

Eclisped methyls

H_3C CH_3

The staggered form of butane
with opposite methyl groups
is called the anti form

EXAMPLE

Butane can exist in two eclipsed conformations and in two staggered conformations other than the ones shown. Beginning with the anti methyl conformation, draw Newman projections for all eclipsed and staggered conformations by rotating the rear carbon clockwise.

Solution:

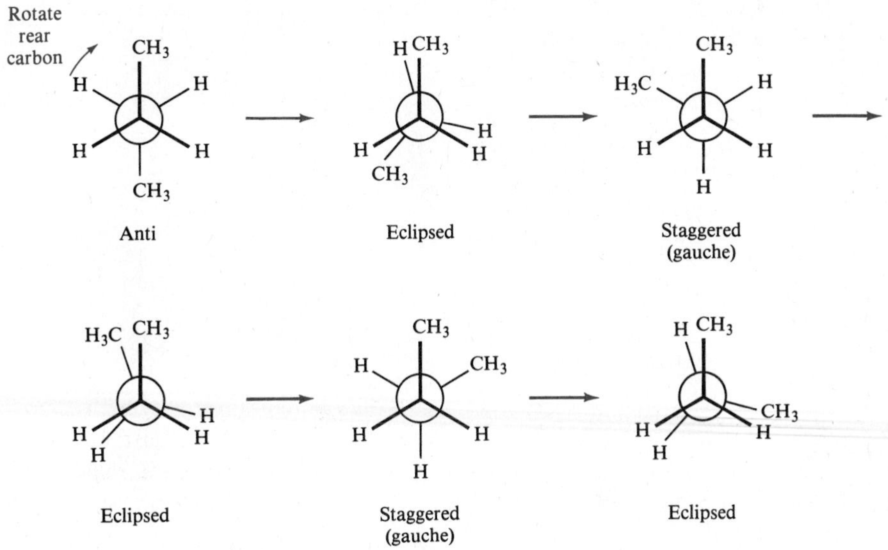

Anti — Eclipsed — Staggered (gauche) —

Eclipsed — Staggered (gauche) — Eclipsed

PROBLEM 2.5. Starting with a staggered conformation, draw Newman projections for all eclipsed and staggered conformations of 1-bromopropane, as we have just done for butane. Use carbons 1 and 2 for the Newman projection carbons.

2.5 REACTIONS OF ALKANES

Compared with most other organic compounds, alkanes are relatively inert. For example, alkanes do not react with acids, bases, or reducing agents. In fact, alkanes are sometimes called **paraffins** (Latin *parum affinis*, "slight affinity") to emphasize their lack of reactivity. Paraffin wax itself, used in candles and to seal jars of homemade jams, is a mixture of high-formula-weight alkanes. Petroleum jelly (Vaseline), a mixture of alkanes, is used to protect skin because it is nonreactive and will not dissolve in water. Alkanes are nonreactive because they are nonpolar compounds and have no functional group to attract an attacking reagent.

Two principal reactions that alkanes do undergo are *halogenation* and *oxidation*. We will discuss a third type of reaction, called *cracking*, in Point of Interest 2.

A. Halogenation

The reaction of an alkane with chlorine (Cl_2) or bromine (Br_2) is called a **halogenation reaction**. The more specific terms *chlorination* and *bromination* are also used to describe a particular halogenation reaction. Of the other halogens, fluorine reacts explosively with organic compounds, while iodine is not sufficiently reactive to react with alkanes.

Chlorination:

$$
\begin{array}{c}
\text{H lost} \\
\\
\text{H}-\overset{\displaystyle \text{H}}{\underset{\displaystyle \text{H}}{\text{C}}}-\text{H} \;+\; \text{Cl}-\text{Cl} \;\xrightarrow[\text{or heat}]{\text{light}}\; \text{H}-\overset{\displaystyle \text{H}}{\underset{\displaystyle \text{H}}{\text{C}}}-\text{Cl} \;+\; \text{H}-\text{Cl}
\end{array}
$$

Methane   Chlorine   Chloromethane   Hydrogen chloride

Halogenation of alkanes occurs only in light or at very high temperatures. When an alkane and a halogen (Cl_2 or Br_2) are mixed in the dark and kept cold, they do not react. To explain why this is so, we must discuss the **reaction mechanism**, the detailed description of how bonds are broken and remade as reactants are changed to products.

The first step in an alkane halogenation is cleavage of the halogen molecule into two neutral particles called **free radicals** or **radicals**. A radical is an atom or a group of atoms that contains one or more unpaired electrons. Chlorine radicals are simply neutral chlorine atoms—chlorine atoms that have neither a positive charge nor a negative charge.

Unpaired electron

$$
:\!\ddot{\text{C}}\text{l}\!:\!\ddot{\text{C}}\text{l}\!: \;\xrightarrow[\text{or heat}]{\text{light}}\; :\!\ddot{\text{C}}\text{l}\!\cdot \;+\; \cdot\!\ddot{\text{C}}\text{l}\!:
$$

Chlorine molecule   Chlorine radicals

In the equation, note the use of curved arrows to show electron movement. We use a half-headed arrow ($\rightharpoonup$) to show the direction in which one electron moves. Recall that we use full-headed arrows ($\rightarrow$) to represent the movement of electron pairs.

Cleavage of a Cl_2 or Br_2 molecule into radicals requires energy: 58 kcal/mol for Cl_2 and 46 kcal/mol for Br_2. This energy, provided by the light or heat, is absorbed by the halogen and causes the initial reaction step, called the **initiation step**.

Initiation:

$$Cl_2 + 58 \text{ kcal/mol} \longrightarrow 2 :\!\ddot{\text{C}}\text{l}\!\cdot$$

To explain the remaining steps in a free-radical halogenation, let us use a specific reaction, the chlorination of methane, for an example. A radical is a high-energy reactive particle. When a chlorine radical collides with a methane molecule, the radical abstracts (removes) a hydrogen atom ($H\cdot$) to yield $H-Cl$ and a new radical, the methyl radical ($\cdot CH_3$). Because a radical is produced as well as consumed, this step is called a **propagation step**.

Step 1 of a Propagation Cycle:

$$H-\overset{\overset{\displaystyle H}{|}}{\underset{\underset{\displaystyle H}{|}}{C}}-H \ + \ \cdot\ddot{\underset{..}{C}l}\colon \ + \ 1 \text{ kcal/mol} \longrightarrow H-\overset{\overset{\displaystyle H}{|}}{\underset{\underset{\displaystyle H}{|}}{C}}\cdot \ + \ H\colon\ddot{\underset{..}{C}l}\colon$$

| Methane | Chlorine radical | Methyl radical | Hydrogen chloride |

The methyl free radical, in turn, can collide with a chlorine molecule (Cl_2) to abstract a chlorine atom in another propagation step.

Step 2 of a Propagation Cycle:

$$H-\overset{\overset{\displaystyle H}{|}}{\underset{\underset{\displaystyle H}{|}}{C}}\cdot \ + \ \colon\ddot{\underset{..}{C}l}\colon\ddot{\underset{..}{C}l}\colon \longrightarrow H-\overset{\overset{\displaystyle H}{|}}{\underset{\underset{\displaystyle H}{|}}{C}}-\ddot{\underset{..}{C}l}\colon \ + \ \cdot\ddot{\underset{..}{C}l}\colon \ + \ 25.5 \text{ kcal/mol}$$

Methyl radical   Chlorine   Chloromethane (also called methyl chloride)   Chlorine radical

a gas used as a refrigerant

The observed products of a halogenation reaction are formed in the two steps of the propagation cycle. In our example, HCl is formed in the first step and CH_3Cl is formed in the second. Once the initial radicals are formed, the reaction is self-perpetuating. The new chlorine radical formed in the second propagation step can attack another methane molecule to keep the propagation cycle going. Because of its recurrent nature, this type of reaction is called a **free-radical chain reaction**.

A Self-Perpetuating Free-Radical Chain Reaction:

$$\colon\ddot{\underset{..}{C}l}\cdot \ + \ CH_4 \longrightarrow \ \cdot CH_3 \ + \ HCl$$

$$\cdot CH_3 \ + \ Cl_2 \longrightarrow CH_3Cl \ + \ \colon\ddot{\underset{..}{C}l}\cdot$$

The regeneration of Cl· results in a self-perpetuating chain reaction

A free-radical chain reaction continues until the reactants are consumed or until the radicals are destroyed. The reaction step in which radicals are destroyed is called a **termination step**. A termination step breaks the chain by removing a radical. Once the chain is broken, the propagation cycles cease and reaction products are no longer formed.

One way in which radicals can be destroyed is by the combination of two radicals to form a stable nonradical product in a **coupling reaction**. A coupling reaction can occur if two radicals collide.

A Coupling Termination Reaction:

$$
\begin{array}{ccc}
\text{H} & \text{H} & \text{H} \quad \text{H} \\
| & | & | \quad | \\
\text{H}-\text{C}\cdot \; + \; \cdot\text{C}-\text{H} & \longrightarrow & \text{H}-\text{C}-\text{C}-\text{H} \\
| & | & | \quad | \\
\text{H} & \text{H} & \text{H} \quad \text{H}
\end{array}
$$

Two methyl radicals   Ethane

Other radicals can also combine to terminate the reaction sequence. For example, $\cdot\text{CH}_3$ could combine with $:\!\ddot{\text{Cl}}\!\cdot$ to yield CH_3Cl.

One problem with free-radical reactions is that mixtures of products are often formed. For example, as the chlorination of methane proceeds, the concentration of methane decreases and the concentration of chloromethane increases. Thus, there is an increasingly greater chance that a chlorine radical will collide with a chloromethane molecule than with a methane molecule. When this happens, a new two-step propagation cycle leading to dichloromethane is begun.

The Propagation Cycle Leading to Dichloromethane:

$$
\begin{array}{ccc}
\text{H} & & \text{H} \\
| & & | \\
\text{Cl}-\text{C}\!:\!\text{H} \quad + \quad \cdot\ddot{\text{Cl}}\!: & \longrightarrow & \text{Cl}-\text{C}\cdot \quad + \quad \text{H}\!:\!\ddot{\text{Cl}}\!: \\
| & & | \\
\text{H} & & \text{H}
\end{array}
$$

Chloromethane   Chloromethyl
                                radical

$$
\begin{array}{ccc}
\text{H} & & \text{H} \\
| & & | \\
\text{Cl}-\text{C}\cdot \; + \; :\!\ddot{\text{Cl}}\!:\!\ddot{\text{Cl}}\!: & \longrightarrow & \text{Cl}-\text{C}-\ddot{\text{Cl}}\!: \quad + \quad :\!\ddot{\text{Cl}}\!\cdot \\
| & & | \\
\text{H} & & \text{H}
\end{array}
$$

Dichloromethane
(methylene chloride)
*used as a solvent and
degreasing agent*

Similarly, propagation cycles leading to trichloromethane (chloroform, CHCl_3, a toxic compound that once was used as an anesthetic) and to tetrachloromethane (carbon tetrachloride, CCl_4, a toxic solvent and reagent) are started as the concentrations of the required reactants increase.

The Free-Radical Chlorination of Methane Leads to a Mixture of Products:

$$CH_4 + Cl_2 \xrightarrow[\text{or light}]{\text{heat}} CH_3Cl + CH_2Cl_2 + CHCl_3 + CCl_4 + \text{coupling products}$$

PROBLEM 2.6. Write equations for the two-step propagation reactions for the formation of **(a)** $CHCl_3$ from CH_2Cl_2 and **(b)** CCl_4 from $CHCl_3$.

Alkanes larger than methane yield an even greater number of chlorination products. For example, the chlorination of ethane could yield varying percentages of nine products.

PROBLEM 2.7. Predict all possible monochlorination products (one Cl substituted for H) from the following compounds.

$$\begin{array}{ccc}
 & \overset{\displaystyle CH_3}{\underset{|}{}} & \\
\textbf{(a)}\ CH_3CH_2CH_3 & \textbf{(b)}\ CH_3CHCH_3 & \textbf{(c)}\
\end{array}$$

(c)
$$\begin{array}{c}
H_2C{-}CH_2 \\
H_2C\diagdown\quad\diagup CH_2 \\
CH_2
\end{array}$$

B. Oxidation

Alkanes are resistant to mild or moderately strong oxidizing agents, such as potassium permanganate, but they are readily oxidized by atmospheric oxygen when ignited. Rapid oxidation with oxygen, which gives off heat and light, is called burning, or **combustion**. Combustion provides most of the heat and energy used in our society for warmth and power.

Combustion:

$$CH_4 \quad + \quad 2O_2 \xrightarrow{\text{spark}} CO_2 \quad + \quad 2H_2O \quad + \quad 211\ \text{kcal/mol}$$

Methane   Oxygen   Carbon   Water
in natural gas   dioxide

$$CH_3CH_2CH_3 \quad + \quad 5O_2 \xrightarrow{\text{spark}} 3CO_2 \quad + \quad 4H_2O \quad + \quad 526\ \text{kcal/mol}$$

Propane
another fuel

A spark, flame, or red-hot object is needed to start a combustion reaction but once the reaction has started it is self-sustaining.

Incomplete combustion, combustion in a limited amount of oxygen, leads to carbon (the black component of soot) and the toxic gas carbon monoxide. For

TABLE 2.6 HEATS OF COMBUSTION FOR SOME ALKANES

$$RH + O_2 \longrightarrow CO_2 + H_2O + (-\Delta H)$$

Structure	Heat of combustion ($-\Delta H$, kcal/mol at 20°C)
CH_4	211
CH_3CH_3	368
$CH_3CH_2CH_3$	526
$CH_3(CH_2)_2CH_3$	688
$CH_3(CH_2)_3CH_3$	833
$CH_3(CH_2)_4CH_3$	990
$CH_3(CH_2)_5CH_3$	1150
$CH_3(CH_2)_6CH_3$	1303

example, when natural gas is burned in an insufficient oxygen supply, the following reactions might occur:

Incomplete Combustion:

$$CH_4 + O_2 \xrightarrow{\text{spark}} \underset{\text{Carbon}}{C} + 2H_2O$$

$$2CH_4 + 3O_2 \xrightarrow{\text{spark}} \underset{\substack{\text{Carbon} \\ \text{monoxide}}}{CO} + 4H_2O$$

Table 2.6 lists the **heat of combustion** (heat of reaction, or net heat liberated, for the complete combustion) of a few alkanes.

EXAMPLE

Calculate the number of kilocalories liberated per *gram* of **(a)** CH_4 and **(b)** $CH_3CH_2CH_2CH_2CH_3$.

Solution:

(a) FW of CH_4: $12.01 + 4(1.008) = 16.04$

$$\text{kcal/g} = 211 \text{ kcal/mol} \times 1 \text{ mol}/16.04 \text{ g}$$

$$= 13.2$$

(b) FW of $CH_3CH_2CH_2CH_2CH_3 = 72.15$

$$\text{kcal/g} = 833 \text{ kcal/mol} \times 1 \text{ mol}/72.15 \text{ g}$$

$$= 11.6$$

C. Bond Dissociation Energies and Heats of Reaction

The **dissociation energy** of a bond is the energy (in kilocalories per mole) required to break a molecule into two free radicals at that bond. For example, to cleave CH_4 into $\cdot CH_3$ and $\cdot H$ requires 104 kcal/mol. This value is the bond dissociation energy for a $C-H$ bond in methane.

Bond dissociation energy
for H_3C-H

$$H_3C \overgroup{\frown} H + 104 \text{ kcal/mol} \longrightarrow H_3C\cdot + \cdot H$$

Table 2.7 lists a few common bonds and their bond dissociation energies.

Using bond dissociation energies, we can calculate the **heat of reaction**, ΔH, the amount of energy in the form of heat that is either liberated or absorbed during the course of a reaction. To illustrate the procedure, let us calculate the heat of reaction for the monochlorination of methane using the bond dissociation energies listed in Table 2.7.

To carry out this calculation, we first write equations showing the cleavage of the reactants into radicals by breaking the bonds involved in the reaction. Next, we write equations that show the formation of the products from the radicals. The ΔH for each reaction step is placed to the right of its equation. The ΔH is written as a positive number if energy is required (absorbed) by the reaction step or a negative

TABLE 2.7 SELECTED BOND DISSOCIATION ENERGIES

Bond	Energy (kcal/mol)	Bond	Energy (kcal/mol)
Miscellaneous bonds		$C-Cl$ bonds	
$H-H$	104	CH_3-Cl	83.5
$N\equiv N$	226	CH_3CH_2-Cl	81.5
$F-F$	37	$(CH_3)_2CH-Cl$	81
$Cl-Cl$	58	$(CH_3)_3C-Cl$	78.5
$Br-Br$	46	$CH_2=CH-Cl$	84
$I-I$	36		
$H-F$	135	$C-C$ bonds	
$H-Cl$	103	CH_3-CH_3	88
$H-Br$	87	$CH_2=CH_2$	163
$H-I$	71	$CH\equiv CH$	230
$HO-OH$	35		
$C-H$ bonds		$C-Br$ bonds	
CH_3-H	104	CH_3-Br	70
CH_3CH_2-H	98	CH_3CH_2-Br	68
$(CH_3)_2CH-H$	94.5		
$(CH_3)_3C-H$	91		
$CH_2=CH-H$	108		

number if energy is released. The final ΔH for the overall reaction is calculated by summing the ΔH values for each step.

Calculation of ΔH for a Chlorination Reaction:

$$:\ddot{\text{Cl}}-\ddot{\text{Cl}}: \longrightarrow 2\,:\ddot{\text{Cl}}\cdot \qquad\qquad \Delta H = +58\ \text{kcal/mol}$$

$$\text{H}_3\text{C}-\text{H} \longrightarrow \text{H}_3\text{C}\cdot + \text{H}\cdot \qquad\qquad \Delta H = +104\ \text{kcal/mol}$$

$$\text{H}_3\text{C}\cdot + \cdot\ddot{\text{Cl}}: \longrightarrow \text{H}_3\text{C}-\ddot{\text{Cl}}: \qquad \Delta H = -83.5\ \text{kcal/mol}$$

$$\text{H}\cdot + \cdot\ddot{\text{Cl}}: \longrightarrow \text{H}-\ddot{\text{Cl}}: \qquad\qquad \underline{\Delta H = -103\ \text{kcal/mol}}$$

$$\Delta H \text{ for reaction } = -24.5\ \text{kcal/mol}$$

A reaction, such as the monochlorination of methane, that releases energy is said to be **exothermic** (heat liberating). A reaction that absorbs energy is said to be **endothermic** (heat absorbing). To determine whether a reaction is exothermic or endothermic, we inspect the final ΔH for the reaction. If the value is negative, the reaction is exothermic; if the value is positive, the reaction is endothermic.

Inspecting the final ΔH for the monochlorination of methane reaction, we can see that the reaction is exothermic and releases 24.5 kcal/mol. If we carry out the same type of calculation for the bromination of methane, we will find that the reaction is also exothermic but releases only 7 kcal/mol. Chlorination of an alkane is a more violent reaction, releasing far more energy per mole, than bromination.

PROBLEM 2.8.   Using the values in Table 2.7, calculate the ΔH for the reaction of CH_3CH_3 with Br_2 to yield $CH_3CH_2Br + HBr$.

D. Energy Requirements of Reactions

Figure 2.4 shows graphs depicting the energy requirements of simple one-step exothermic and endothermic reactions. (Halogenation is a multistep reaction; its energy curve would be more complex.) On either graph, we start at the left with the average energy of a mixture of the reactants. Moving to the right on either curve, we reach a peak. Energy (usually as heat, sometimes as light) must be supplied to the reactant molecules before a chemical reaction can occur. This energy is called the **activation energy**, or E_{act}, and is represented on the graphs by the difference between the average energy of the reactant molecules and the peak of the energy curve.

At the energy peak, the reactants are undergoing transition to products. This energy peak represents the **transition state** of the reaction—the point at which the structure could go back to reactants or on to products. When it goes on to products, energy is liberated.

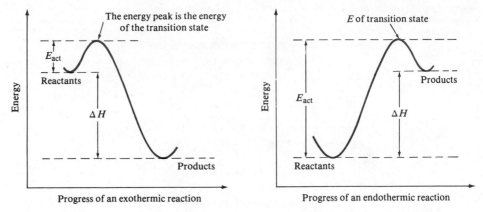

Figure 2.4 Potential energy diagrams for typical exothermic and endothermic reactions.

We can continue to follow each curve to the right until we reach the average energy of the products. The *net* heat liberated or absorbed is the difference in the average potential energy between the reactants and the products. This net heat is the heat of reaction, ΔH. A reaction that produces a lower-energy, more stable product relative to starting material is exothermic. A reaction with a higher-energy, less stable product is endothermic.

In any reaction, a small E_{act} means that less energy is needed to start the reaction. A larger E_{act} means that *more* energy must be supplied. If a reactant could undergo either of *two* irreversible reactions, the reaction whose E_{act} is the smaller will be favored and its products will predominate in the reaction mixture (see Figure 2.5).

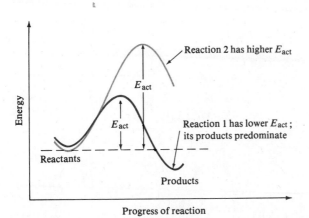

Figure 2.5 If a reactant can undergo two irreversible reactions, the reaction with the lower E_{act} predominates.

2.6 CYCLOALKANES

Cycloalkanes are saturated hydrocarbons with carbon atoms bonded together forming one or more rings. In nature, five- and six-membered rings are the most common, but other-sized rings are also encountered.

Some Cycloalkanes:

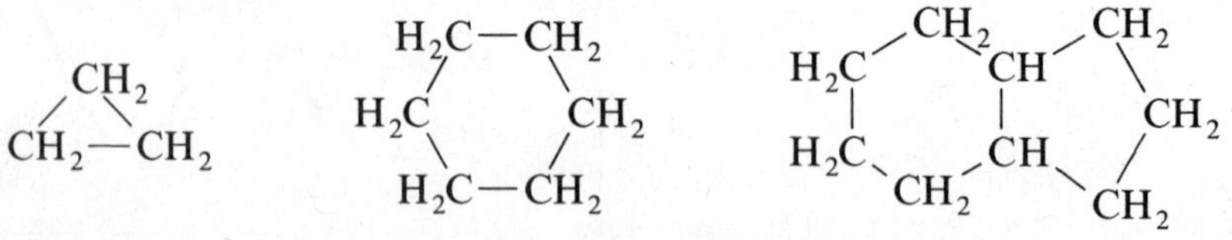

$$CH_2 \quad CH_2-CH_2 \quad CH_2$$

Usually, we use polygon formulas to represent cycloalkanes. The structures of the preceding cycloalkanes are more conveniently represented using the following formulas:

A carbon–carbon
single bond

A carbon with one H

A carbon with two H's

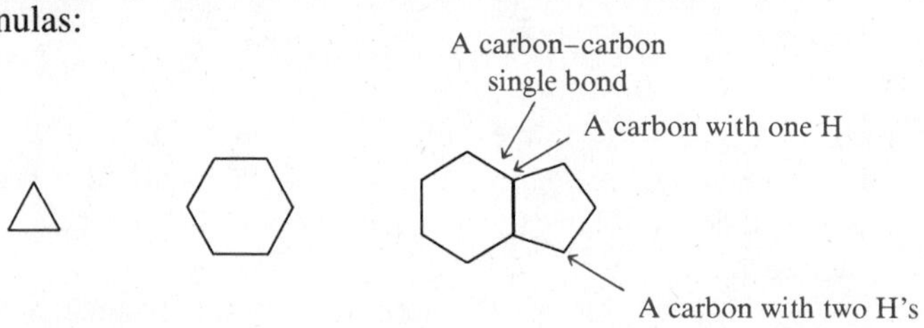

Functional groups and other substituents are incorporated into polygon formulas.

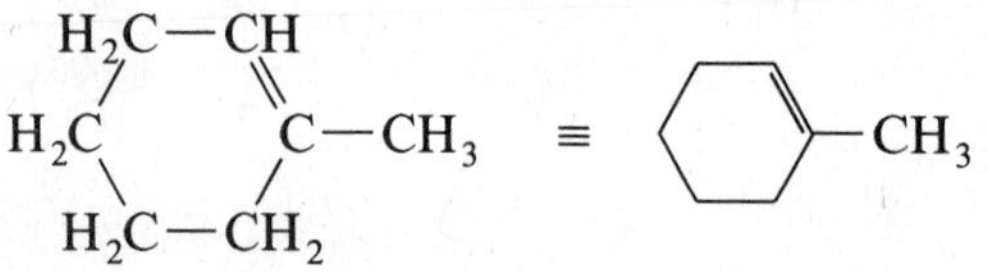

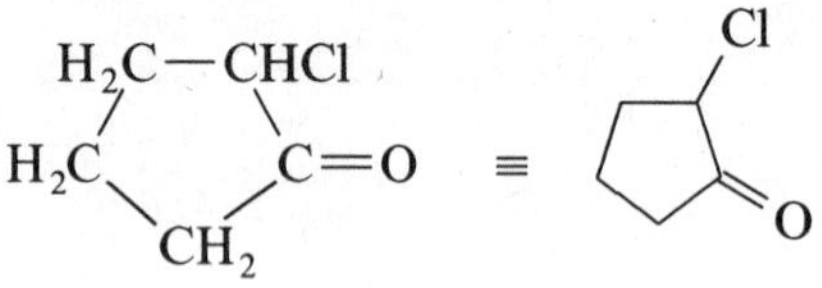

$$\begin{array}{c} H_2C-CH \\ H_2C \quad C-CH_3 \\ H_2C-CH_2 \end{array} \equiv \bigcirc\!\!-CH_3$$

$$\begin{array}{c} H_2C-CHCl \\ H_2C \quad C{=}O \\ CH_2 \end{array} \equiv \quad \begin{array}{c} Cl \\ \end{array}$$

PROBLEM 2.9.   Draw a polygon formula for each of the following formulas:

(a) 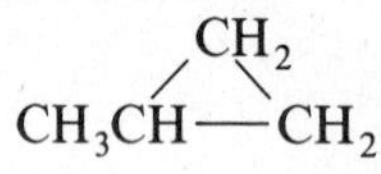 $CH_3CH{-}CH_2$ with CH_2

(b) 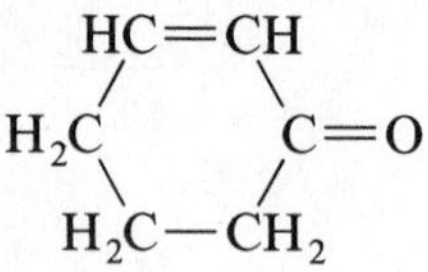 H_2C with $HC{=}CH$ and $C{=}O$ and H_2C-CH_2

(c) H_2C with H_2C-CH_2 and CH_2 and C with $\overset{O}{\overset{\|}{C}OH}$ and Br

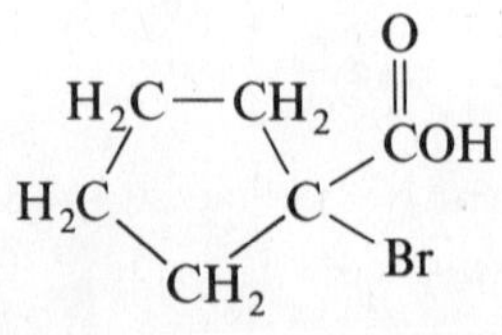

PROBLEM 2.10. Draw formulas showing the carbons and hydrogens for the following polygon formulas. (Remember that carbon forms four bonds.)

(a) [structure with Cl and CH₃ on cyclohexene ring]

(b) [structure with O atoms]

(c) [naphthalene structure]

A. Nomenclature

The names of cycloalkanes are taken directly from the names of the continuous-chain alkanes with the same number of carbons. The prefix *cyclo-* before the alkane name indicates a ring. For example, cyclopropane, an inhalation anesthetic, is a saturated hydrocarbon with three carbons in a ring; cyclobutane is a four-carbon cycloalkane; and cyclopentane is a five-carbon cycloalkane.

[triangle]   [square]   [pentagon]

Cyclopropane   Cyclobutane   Cyclopentane

Cycloalkanes with alkyl substituents are named similarly to branched alkanes. In most cases, the cycloalkane is considered the parent and the substituent names precede the parent name.

[cyclopentane with CH₂CH₃]   [cyclohexane with —C(CH₃)₃]

Ethylcyclopentane   *t*-Butylcyclohexane

If more than one substituent is present, the ring is numbered to give the substituents the lowest numbers.

[three cyclopentane structures with methyl groups numbered]

All represent 1,3-dimethylcyclopentane

If a cycloalkane ring is bonded to an alkane chain or other ring containing more carbon atoms than itself, the ring is named as a *cycloalkyl group*.

[cyclopropyl structure] —CHCH₂CH₂CH₃
                          |
                          CH₃

[cyclohexyl structure]
CH₃CH₂CHCHCH₂CH₂CH₃
          |
          CH₃

2-Cyclopropylpentane   3-Cyclohexyl-4-methylheptane

PROBLEM 2.11. Name the following substituted cycloalkanes:

(a) 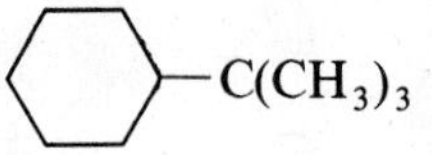
C(CH3)3

(b) 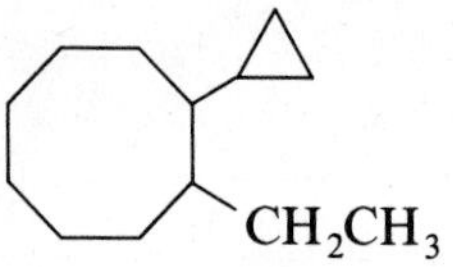
CH2CH3

(c) 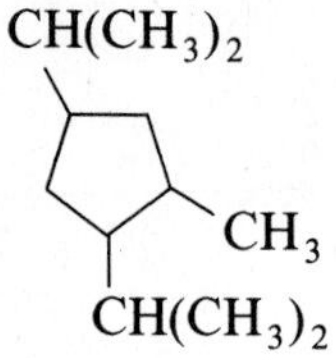
CH(CH3)2
CH3
CH(CH3)2

PROBLEM 2.12. Draw polygon formulas for the following compounds:

(a) *sec*-butylcyclopentane; **(b)** 1,2-diethyl-5-methylcyclooctane;
(c) 1-cyclobutyl-4-isopropylcyclohexane.

B. Cis-Trans Isomerism

The carbon atoms of a cycloalkane roughly form a plane, called the **plane of the ring**. A disubstituted cycloalkane can have its two substituents on the same side of the plane of the ring or on opposite sides. Two substituents on the same side are said to be **cis substituents** (Latin *cis*, "on this side").

Three Cis-Disubstituted Cycloalkanes:

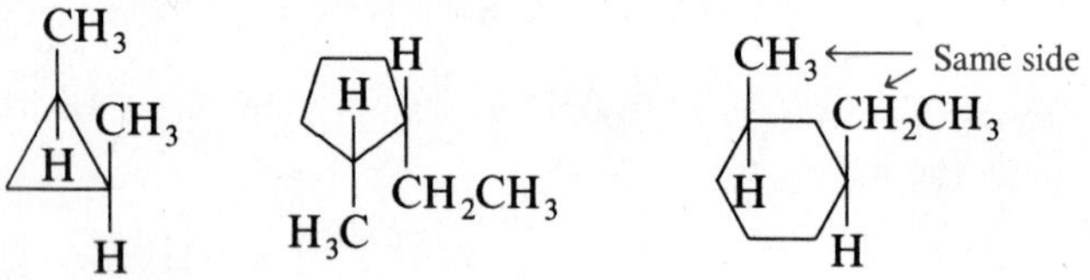
CH3
CH3
H
H
H
H
CH2CH3
H3C
CH3
Same side
CH2CH3
H
H

Two substituents on opposite sides of a ring are said to be **trans substituents** (Latin *trans*, "across").

Three Trans-Disubstituted Cycloalkanes:

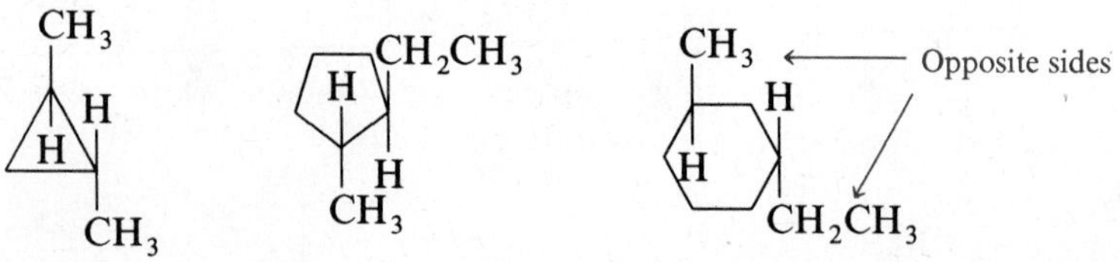
CH3
H
H
CH3
CH2CH3
H
H
CH3
CH3
Opposite sides
H
H
CH2CH3

In order to show two substituents as cis or trans, we draw the formula to emphasize the plane of the ring. In this type of formula, the bonds from ring carbons to the substituents are drawn as straight lines perpendicular to the plane. A substituent is located above the plane of the ring if it is positioned on the top of this line and below the plane of the ring if it is positioned on the bottom of the line.

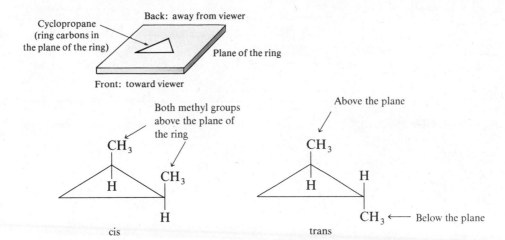

Another way we can represent the cis or trans relationships is by the use of wedges. A solid wedge represents a group above the plane of the ring, and a broken wedge represents a group below the plane.

Groups below the plane of the ring and cis

Groups above the plane of the ring and cis

Groups on opposite sides of the ring and trans

The cis and trans disubstituted compounds cannot be interconverted without breaking bonds. Therefore, these compounds are isomers, which we call **cis-trans isomers**. Note how the cis and trans designations are incorporated into the names.

A Pair of Cis-Trans Isomers:

is a cis-trans isomer of

cis-1,2-Dimethylcyclohexane

trans-1,2-Dimethylcyclohexane

Cis-trans isomers are not structural isomers because the atoms are not bonded in different orders. Instead, they belong to a class of isomers called **stereo-isomers**—isomers that differ only by the arrangement of their atoms in space. You will encounter other types of stereoisomerism in Chapters 3 and 6.

PROBLEM 2.13. Name the following compounds using cis and trans prefixes in the names:

(a)
CH_2CH_3
H
H
CH_2CH_3

(b)
CH_3CH_2 CH_3
H H

(c)
$CH(CH_3)_2$
H H
$CH_2CH(CH_3)_2$

PROBLEM 2.14. Draw polygon formulas for the following compounds:

(a) *cis*-1,4-di-*t*-butylcyclohexane

(b) *trans*-1-isobutyl-3-isopropylcyclopentane

C. Ring Size and Conformation

A carbon bonded by four single bonds is generally tetrahedral with bond angles of approximately 109.5°. In cyclopropane the geometry of the three-membered ring dictates that the C—C bond angles be 60°. The overlap of the tetrahedral sp^3-hybrid orbitals in cyclopropane is not maximal (see Figure 2.6). For this reason, cyclopropane is more reactive than the other cycloalkanes and undergoes many

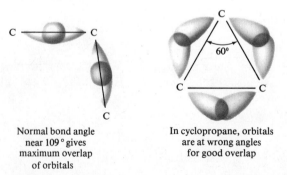

Normal bond angle near 109° gives maximum overlap of orbitals

In cyclopropane, orbitals are at wrong angles for good overlap

Figure 2.6 Maximum overlap cannot be achieved between the ring carbon atoms in cyclopropane.

reactions in which the ring opens to form products with more favorable bond angles.

$$\underset{CH_2-CH_2}{\overset{CH_2}{\diagup\diagdown}} + H_2 \xrightarrow{\text{catalyst}} CH_3CH_2CH_3$$

Other cycloalkanes $+ H_2 \xrightarrow{\text{catalyst}}$ no reaction

A cyclohexane ring would have bond angles of 120° if it were flat, but cyclohexane is not flat. It forms *puckered* rings with bond angles near the optimal value of 109°. The most stable shape a puckered cyclohexane can assume has the appearance of a lawn chair and, consequently, is called the **chair form** of cyclohexane.

All bond angles
approx. 109°

Chair form of cyclohexane

Besides the optimization of bond angles, another reason why cyclohexane exists as puckered rings is that the hydrogens are *staggered*. If the ring were flat, all the hydrogens would be eclipsed. As in the case of ethane and butane, the staggering of hydrogens adds stability to the molecule.

*Puckered ring
more stable*

*Flat ring,
less stable*

D. Equatorial and Axial Substituents

The chair form of cyclohexane contains two types of hydrogens. One type is roughly in the same plane as the carbons of the ring. These hydrogens are called **equatorial hydrogens**. Figure 2.7 shows the equatorial hydrogens. The second type of hydrogen atom is bonded to the ring by a bond that is perpendicular to the plane of the ring. These bonds are parallel to an axis drawn through the center of the ring; therefore, the hydrogens attached to these bonds are called **axial hydrogens**. The axial hydrogens are also shown in Figure 2.7.

Each carbon atom in cyclohexane has one equatorial hydrogen atom and one axial hydrogen. When a substituent replaces a hydrogen of a cyclohexane, the

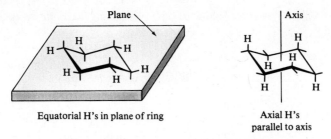

Figure 2.7 The equatorial and axial hydrogens of cyclohexane.

substituent will be either equatorial or axial.

An axial
chlorine atom

Cl

H

HO

H

An equatorial
hydroxyl group

PROBLEM 2.15. Indicate whether the following methyl groups are equatorial or axial.

(a)

CH₃

H

(b)

H

CH₃

(c)

H

CH₃

PROBLEM 2.16. Using the following carbon skeleton with the indicated position numbers, draw the formula for 1,2,4-trimethylcyclohexane with the methyl groups in equatorial positions.

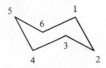

E. Conformations of Cyclohexane

The chair form of cyclohexane, in which all ring hydrogens are staggered, is the most stable conformation. However, at room temperature the cyclohexane ring is in a continuous state of flexing and we find that the ring can exist in two different chair forms.

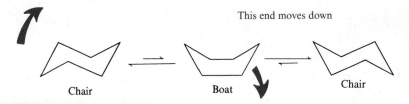

To flex from one chair form to another, a six-membered ring goes through a **boat form**, so named because it is shaped somewhat like a boat. The boat form of cyclohexane is less stable than either chair form because some of the hydrogens are eclipsed.

Eclipsed

H   H

H   H

Eclipsed

Boat form of cyclohexane

F. Substituted Cyclohexanes

When a cyclohexane ring is substituted with an alkyl group or other substituent, that group can be axial or equatorial. The conformation with the larger group *equatorial* is the more stable conformation. The reason for this stability difference is that there are fewer repulsions between atoms when the group is equatorial.

Atoms near each other—
repulsion   H
|
H) (H—C—H

H   H

H   H

Axial methyl group

Atoms farther apart—
less repulsion

H   H
|
C—H
|
H   H H

Equatorial methyl group

Thus, at any given time, a sample of methylcyclohexane or other monosubstituted cyclohexane would have most of its molecules in the conformation with the substituent equatorial. Because the molecules are in a continuous state of flexing, not all molecules would have an equatorial substituent, but each molecule would spend most of it time with the substituent equatorial.

PROBLEM 2.17. Predict the predominant conformation of the following compound and explain your choice.

(a) (b)

trans-1-*t*-Butyl-3-methylcyclohexane

SUMMARY

A **hydrocarbon** is a compound that contains only carbon and hydrogen. A **saturated** hydrocarbon, an **alkane** or a **cycloalkane**, contains no double or triple bonds. An unsaturated hydrocarbon, such as an alkene, contains one or more multiple bonds. Alkanes have the general formula C_nH_{2n+2}. Two hydrogens are removed for each ring in a cycloalkane.

In the **IUPAC nomenclature system** a name is based on the name of the parent (usually the longest chain). Substituents are prefixed to the parent name, with position numbers when necessary. Refer to Section 2.3 for the rules for systematic nomenclature.

Groups rotate about single bonds; the rotation leads to **conformations** in which the groups can be **eclipsed** or **staggered**. **Newman projections** (Section 2.4) are used to emphasize these conformations.

Because of their weak intermolecular attractions and lack of polarity, alkanes are low boiling and water insoluble. They tend to be nonreactive but do undergo **combustion** and **halogenation**. (Also, see Point of Interest 2.)

Combustion: $2CH_3CH_3 + 7O_2 \xrightarrow{\text{spark}} 4CO_2 + 6H_2O$

Halogenation: $CH_3CH_3 + Cl_2 \xrightarrow{\text{light}} CH_3CH_2Cl + HCl$

and other products

Halogenation is a **free-radical reaction**, characterized by an **initiation step** (formation of radicals); **propagation steps**, which yield products and new radicals; and **termination steps**, which end the reaction sequence by destruction of radicals. One termination step is a **coupling reaction**.

The **heat of reaction** (ΔH) can be calculated from **bond dissociation energies**. A reaction is either **exothermic** (heat releasing) or **endothermic** (heat absorbing). Energy may be required to initiate even an exothermic reaction; the minimum necessary energy is called the **activation energy** (E_{act}).

Cycloalkanes are usually represented by **polygon formulas**. Disubstituted cycloalkanes have cis-trans isomers, a type of **stereoisomer**.

Cycloalkanes exist in puckered conformations, such as the **chair form** of cyclohexane. Substituents on a cyclohexane ring can be **equatorial** or **axial**, the equatorial conformation being more stable.

STUDY PROBLEMS

2.18. Write condensed structural formulas for the following compounds:
  (a) 3-ethylheptane
  (b) 2-nitrobutane
  (c) 4-isopropylnonane
  (d) 2-chloro-3-methylbutane
  (e) cyclobutylcyclohexane
  (f) isobutylcyclobutane
  (g) *trans*-1,2-di-*n*-propylcyclobutane
  (h) 1,2,3-tribromoheptane
  (i) *cis*-1,3-dimethylcyclopentane
  (j) 1,1-dichloro-3-*sec*-butylcyclohexane

2.19. Name the following compounds.

(a) $CH_3\overset{\displaystyle Cl}{\underset{\displaystyle Cl}{C}}CH_3$

(b) $CH_3(CH_2)_5\overset{\displaystyle CH_3}{\underset{\displaystyle CH_3}{C}}CH_3$

(c) Cl_3CCH_2Cl

(d) with CH_2CH_3 / $CH(CH_2)_6CH_3$

(e) $\overset{\displaystyle CH_3}{CHCH_3}$

(f) $(CH_3)_3CH$

(g) $(CH_3)_2CH\overset{\displaystyle CH_2CH_3}{CHCH_3}$

(h) $C(CH_3)_3$

(i) $CH_3\overset{\displaystyle I}{CHC}(CH_3)_3$

2.20. Each of the following names is *incorrect*. Explain why each is incorrect, and write a correct name.
  (a) *t*-butylmethane
  (b) 6-methylheptane
  (c) 2,2-diethylhexane
  (d) 5-butyloctane
  (e) 1,4-dimethylcyclopentane

2.21. Draw condensed structural formulas and write the names for all the isomeric alkanes with the molecular formula C_7H_{16}.

2.22. Draw condensed structural (polygon) formulas and write the names for all the isomeric cycloalkanes containing five carbons.

2.23. Convert each of the following Newman projections to a three-dimensional formula with the same conformation.

(a)   (b)   (c)

2.24. Convert each of the following formulas to a Newman projection, using the indicated carbon as the front carbon.

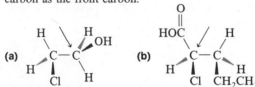

2.25. (a) Draw Newman projections for 1,2-dichloroethane showing the effects of rotation about the carbon–carbon bond. **(b)** Label each anti and eclipsed projection.

2.26. Which compound of each of the following pairs would have the higher boiling point? Explain your answers.

(a) $CH_3CH_2CH_2CH_2CH_3$ or $CH_3CHCH_2CH_3$ (with CH_3 substituent)

(b) $(CH_3)_2CHOCH(CH_3)_2$ or $CH_3CH_2CH_2OCH_2CH_2CH_3$

(c) $CH_3CH_2CH_2CH_3$ or $CH_3CH_2CH_2OH$

(d) $CH_3CH_2CH_2CH_3$ or $CH_3CH_2CH_2CH_2CH_3$

2.27. Write balanced equations for the complete combustion of **(a)** butane, **(b)** pentane, and **(c)** cyclopentane.

2.28. Complete the following equations showing all possible products. Name each product.

(a) $CH_3CH_2CH_2CH_3 + 2Cl_2 \xrightarrow{\text{light}}$

(b) $CH_3CHCH_3 + 1Br_2 \xrightarrow{\text{light}}$ (with CH_3 substituent)

(c) $CH_3CCH_3 + 2Br_2 \xrightarrow{\text{light}}$ (with CH_3 substituents)

2.29. Write equations for each step in the monobromination (one Br introduced) of cyclopentane.

2.30. Using the bond dissociation energies in Table 2.7, calculate the heats of reaction for the monochlorinations of the following alkanes at the position indicated:

(a) CH_3CH_3 **(b)** $(CH_3)_3CH$

2.31. In the preceding problem, which reaction is more exothermic?

2.32. Label each part of the following energy diagram for a one-step reaction.

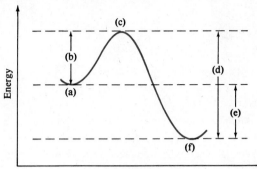

Progress of reaction

2.33. A gasoline that has the burning characteristics of a mixture of 11% heptane and 89% isooctane has an octane number of _____. (Refer to Point of Interest 2.)

2.34. Draw polygon formulas for the following cyclic compounds.

(a)
$$
\begin{array}{c}
\text{CH}_2 \\
\text{H}_2\text{C} \quad\quad \text{CHCl} \\
\text{H}_2\text{C} \quad\quad \text{C} \\
\text{CH}_2 \quad \text{O}
\end{array}
$$

(b)
$$
\begin{array}{c}
\text{HC—CH} \quad \text{O} \\
\text{HC} \quad\quad \text{C—CH} \\
\text{H}_2\text{C—CH}_2
\end{array}
$$

(c)
$$
\begin{array}{c}
\text{CH}_2 \quad \text{CH} \\
\text{H}_2\text{C} \quad \text{CH} \quad\quad \text{CH} \\
\text{H}_2\text{C} \quad \text{CH} \quad \text{CH}_2 \\
\text{CH} \\
\text{OH}
\end{array}
$$

2.35. Draw line-bond formulas (showing C's and H's) for the following polygon formulas.

(a) ⬡=CH₂   **(b)** furan–C(=O)CH   **(c)** pyrimidine ring with NH, N–H, O, O

2.36. Name each of the following cyclic compounds. Include a cis or trans designation in each name.

(a)
$$
\begin{array}{c}
\text{Br} \\
\text{H} \\
\text{H} \\
\text{Br}
\end{array}
$$

(b)
$$
\begin{array}{c}
\text{CH}_3 \\
\text{Br} \\
\text{H} \\
\text{H}
\end{array}
$$

(c)
$$
\begin{array}{c}
\text{CH}_2\text{CH}_3 \\
\text{H} \\
\text{H} \\
\text{CH}_3
\end{array}
$$

2.37. Draw polygon formulas for each of the following cyclic compounds:
(a) *trans*-1-chloro-2-ethylcyclopentane
(b) *cis*-1,4-diisopropylcyclooctane
(c) *cis*-1,3-dipropylcycloheptane

2.38. **(a)** Draw polygon formulas for all the structural isomers and cis-trans isomers that could be formed when chlorocyclohexane is monochlorinated. **(b)** Name each isomer.

2.39. Label each of the following indicated positions as equatorial or axial:

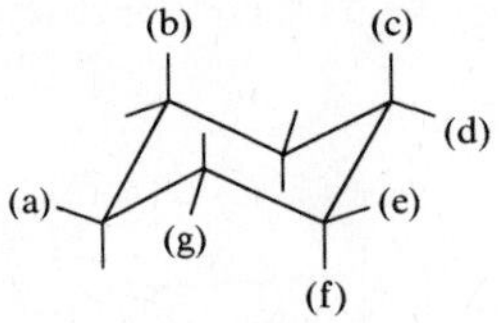

2.40. (a) Draw chair-form formulas of the most stable conformations of *cis*-1,2-dimethyl-cyclohexane and *trans*-1,2-dimethylcyclohexane. **(b)** Which isomer is more stable? Explain your answer.

2.41. (a) Draw chair-form formulas for *cis*- and *trans*-1,3-dimethylcyclohexane. **(b)** Which isomer is more stable?

2.42. The following chemical reaction is a one-step substitution reaction with an exothermic energy curve similar to the one in Figure 2.4. Draw a curve and label each part as it is related to this reaction specifically.

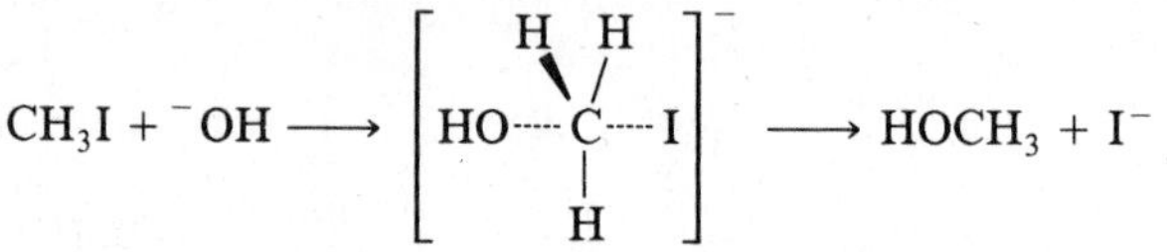

$$CH_3I + {}^-OH \longrightarrow \left[HO\text{----}\overset{\overset{\displaystyle H}{|}}{\underset{\underset{\displaystyle H}{|}}{C}}\text{----}I \right]^- \longrightarrow HOCH_3 + I^-$$

Transition state

2.43. An alkane with the molecular formula C_5H_{12} is subjected to chlorination. Four isomeric products with the molecular formula $C_5H_{11}Cl$ are obtained. From this information **(a)** deduce the structure of the original alkane and **(b)** write an equation for the chlorination.

POINT OF INTEREST 2

Petroleum and Its Uses

Petroleum, or **crude oil**, is a black evil-smelling liquid obtained from oil wells. It is composed primarily of alkanes, along with aromatic compounds, sulfur compounds, and nitrogen compounds. **Natural gas**, 60–90% methane, is also obtained from wells in the petroleum-producing regions of the world or is produced as a by-product of the petroleum industry.

The origin of petroleum and natural gas is still debated. One theory is that they were formed by the anaerobic decay (decomposition in the absence of oxygen) of plant material. Part of the evidence to support this theory is that methane is formed by the anaerobic decay of plant material in swamps and marshes (where it is called "marsh gas") and is also formed by the anaerobic decomposition of sewage and manure. Another theory is that the petroleum material was deposited by precipitation from a hydrocarbon atmosphere eons ago. Still another theory is that the hydrocarbons were deposited during formation of the planet.

Refining Crude oil, as it comes from the ground, is a complex mixture of alkanes and other compounds. In its crude state, it is not very useful. The conversion of

TABLE 2.8   TYPICAL FRACTIONS FROM THE STRAIGHT-RUN DISTILLATION OF PETROLEUM[a]

Boiling range (°C)	Number of carbons	Name	Use
under 30	1–4	Gas fraction	Heating fuel
30–180	5–10	Gasoline	Automobile fuel
180–230	11–12	Kerosene	Jet fuel, heating fuel
230–305	13–17	Light gas oil	Diesel fuel, heating fuel
305–405	18–25	Heavy gas oil	Heating fuel

[a]Residue: (1) Volatile oils: lubricating oils, paraffin wax, and petroleum jelly. (2) Nonvolatile material: asphalt and petroleum coke.

petroleum to fuels and other useful compounds is carried out in a series of processes collectively called **refining**. The first step in refining is *distillation*—the separation of the crude oil into fractions of different boiling ranges, such as shown in Table 2.8. The initial distillation of crude oil is often referred to as *straight-run distillation*.

Insufficient amounts of gasoline are obtained in the straight-run distillation to supply the world demand. Also, straight-run gasoline is a very low-quality fuel for modern automobile engines. To increase the quantity and quality of the gasoline and to synthesize other useful compounds, high-formula-weight compounds from higher-boiling fractions are cleaved into lower-formula-weight, lower-boiling compounds by processes called **cracking**.

In *catalytic cracking* the alkane mixture is heated under pressure in the presence of an aluminosilicate catalyst. Catalytic cracking cleaves large molecules into small ones and causes rearrangements in the carbon skeletons of the compounds. The rearrangements result in alkanes with more highly branched structures, which burn better in automobile engines than do continuous-chain alkanes.

In *steam cracking* a hydrocarbon fuel is mixed with steam and heated at a high temperature (about 800°C) for a short period of time. This process yields alkenes and aromatic compounds.

Other refining processes are used to combine low-formula-weight alkanes into more useful midweight alkanes, to convert open-chain compounds into cyclic ones, and to convert saturated compounds into unsaturated compounds. Some examples of the reactions that take place in refining follow:

Catalytic Cracking Yields Smaller Molecules:

$$CH_3(CH_2)_{15}CH_3 \xrightarrow[\text{catalyst}]{\text{heat}} CH_3(CH_2)_8CH_3 \ + \ CH_2{=}CH(CH_2)_4CH_3$$

Steam Cracking Yields Alkenes:

$$CH_3(CH_2)_{14}CH_3 \xrightarrow[\text{heat}]{H_2O}$$

$$H_2 + CH_4 + CH_2{=}CH_2 + CH_3CH{=}CH_2 + \text{higher-boiling hydrocarbons}$$

Alkylation (Combination) Yields Larger Molecules:

$$\underset{\overset{\displaystyle |}{CH_3}}{\overset{\overset{\displaystyle CH_3}{|}}{CH_3C-H}} + \underset{\overset{\displaystyle |}{CH_3}}{\overset{\overset{\displaystyle CH_3}{|}}{CH_2=C}} \xrightarrow{H_2SO_4} \underset{\overset{\displaystyle |}{CH_3}}{\overset{\overset{\displaystyle CH_3 \quad CH_3}{| \qquad |}}{CH_3C-CH_2CHCH_3}}$$

Catalytic Reforming Yields Aromatic Compounds:

$$CH_3(CH_2)_4CH_3 \xrightarrow[\text{catalyst}]{\text{heat}} \;\; \begin{array}{c} HC-CH \\ // \qquad \backslash\backslash \\ HC \qquad\quad CH \\ \backslash \qquad\quad / \\ HC=CH \end{array} \;\; + \;\; 4H_2$$

Benzene
an aromatic hydrocarbon

Octane Number The octane number of a gasoline is based on the difference in burning characteristics between a continuous-chain alkane (heptane) and a branched-chain alkane (2,2,4-trimethylpentane, "isooctane"). Heptane, which has poor burning characteristics in an automobile engine, is arbitrarily assigned the octane number 0, while the good-burning isooctane is assigned the octane number 100. A gasoline is ranked against a mixture of these two compounds; the gasoline's octane number is the percent isooctane in the test mixture that has the same burning characteristics as the gasoline. For example, an octane number of 85 indicates that a gasoline burns as well as a mixture of 85% isooctane and 15% heptane.

$$CH_3(CH_2)_5CH_3 \qquad\qquad \underset{\overset{\displaystyle |}{CH_3}}{\overset{\overset{\displaystyle CH_3 \quad CH_3}{| \qquad |}}{CH_3CCH_2CHCH_3}}$$

Heptane

2,2,4-Trimethylpentane
(trivial name: isooctane)

Octane numbers greater than 100 are common in gasoline additives, which are used to increase the octane number of a gasoline. For example, ethanol (CH_3CH_2OH, used in gasohol) and benzene both have octane numbers of 106, an indication that these compounds have better burning characteristics than isooctane.

Petrochemicals The starting materials, such as plastics and rubber, needed for the manufacture of many consumer products used in our society are obtained from **petrochemicals**, chemicals derived from petroleum or natural gas. Currently, about 175 different petrochemicals are manufactured in the United States.

Ethylene, $CH_2=CH_2$, formed in steam cracking, is a major starting material for the petrochemical industry. The amount of ethylene produced ranges from 20 billion to 30 billion pounds per year, depending on the world economy. This alkene is used to produce a wide range of intermediate products and consumer products, such as some plastics and synthetic rubbers. Figure 2.8 shows just a few of the interrelationships among ethylene and its derivatives in the petrochemical industry.

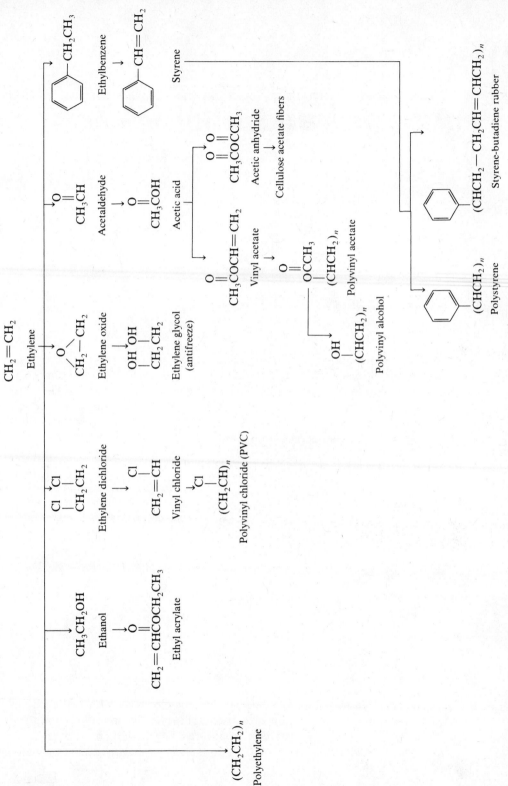

Figure 2.8 The petrochemical relationship between ethylene, $CH_2{=}CH_2$, and its products.

Chapter 3

Alkenes and Alkynes

Alkenes are hydrocarbons containing one double bond, and **alkynes** are hydrocarbons containing one triple bond. Hydrocarbons with more than one double bond will be discussed in Chapter 4. If an alkene is cyclic, it is called a **cycloalkene**. Cyclic alkynes are not stable unless the ring has at least eight carbons; therefore, we do not discuss them. Alkenes are more important than are alkynes; consequently, our emphasis in this chapter is on the alkenes.

Alkenes and Cycloalkenes:

$$CH_2\!=\!CH_2 \qquad CH_3CH_2CH\!=\!CH_2$$

| Ethene (ethylene) | 1-Butene | Cyclohexene | Cyclopentene |

Alkynes:

$$CH\!\equiv\!CH \qquad CH_3CH_2C\!\equiv\!CH$$

Ethyne
(acetylene)

1-Butyne

*used in
oxyacetylene torches*

Under the proper conditions, alkenes and alkynes react with hydrogen gas to yield alkanes, but alkanes do not react with hydrogen. This reaction allows hydrocarbons to be divided into two classes: saturated hydrocarbons and unsaturated hydrocarbons. Alkenes and alkynes are *unsaturated compounds* because they are not "saturated" with hydrogen.

Hydrogenation of Unsaturated Hydrocarbons:

$$\left.\begin{array}{c} H-H \\ + \\ CH_2{=}CH_2 \end{array}\right\} \xrightarrow{\text{Pt catalyst}} \overset{\displaystyle H}{\underset{\displaystyle |}{CH_2}}-\overset{\displaystyle H}{\underset{\displaystyle |}{CH_2}}$$

Ethane

$$\left.\begin{array}{c} H-H \\ + \\ CH{\equiv}CH \end{array}\right\} \xrightarrow{\text{Pt catalyst}} \overset{\displaystyle H}{\underset{\displaystyle |}{CH}}{=}\overset{\displaystyle H}{\underset{\displaystyle |}{CH}} \xrightarrow{H_2,\,Pt} \overset{\displaystyle H}{\underset{\displaystyle |}{CH}}-\overset{\displaystyle H}{\underset{\displaystyle |}{CH}}$$

An older name for alkenes is **olefins**, which means "oil-forming." The term arose because gaseous alkenes, such as ethylene, undergo reaction with chlorine (Cl_2) to yield oily liquids.

$$\left.\begin{array}{c} Cl-Cl \\ + \\ CH_2{=}CH_2 \end{array}\right\} \longrightarrow \overset{\displaystyle Cl}{\underset{\displaystyle |}{CH_2}}-\overset{\displaystyle Cl}{\underset{\displaystyle |}{CH_2}}$$

1,2-Dichloroethane
an oily liquid

Although alkanes and alkynes are rarely found in living systems, alkenes are fairly common. One example of a naturally occurring alkene is *muscalure*, the sex attractant for the common housefly (*Musca domestica*). More commonly, the carbon–carbon double bond is found in naturally occurring compounds containing other functional groups as well, such as oleic acid.

Muscalure
*the sex attractant for the
common housefly*

Oleic acid
from fats and oils

Industrially, alkenes are produced from alkanes in petroleum by cracking reactions; an example is shown in Point of Interest 2. A large percentage of these alkenes are used in the chemical manufacture of a variety of other organic compounds. One example follows.

$$CH_3CH{=}CH_2 + H_2O \xrightarrow{H^+} \overset{\displaystyle OH}{\underset{\displaystyle |}{CH_3}}CHCH_3$$

Propene
(propylene)

2-Propanol
(isopropyl alcohol)

in rubbing alcohol

Simple alkenes (one double bond and no rings) have the general formula C_nH_{2n}, the same formula as that for the monocyclic cycloalkanes.

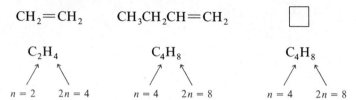

Each additional double bond or ring results in the subtraction of two more hydrogens from the general formula.

PROBLEM 3.1. Write the molecular formula and the general formula for $CH_2\!=\!CHC\!\equiv\!CH$.

PROBLEM 3.2. What structural information can you deduce about compounds with the following molecular formulas? (For example, you might say that a compound with the molecular formula C_8H_{14} contains two double bonds, or two rings, or one double bond and one ring, or one triple bond.)

(a) C_5H_{10} (b) C_5H_8 (c) C_6H_{10}

3.1 BONDING IN ALKENES

When a carbon atom forms a double bond, the carbon is in the sp^2-hybrid state—the sp^2-hybrid orbitals being formed by the hybridization of one $2s$ atomic orbital and two $2p$ atomic orbitals. One $2p$ atomic orbital of the carbon is not involved in the hybridization. Figure 3.1 shows the hybridization and energy relationships.

The three sp^2 orbitals form three sigma (σ) bonds, while the $2p$ orbital forms the second bond, called a **pi (π) bond**.

One bond of the double bond
is an sp^2-sp^2 sigma bond

$$\begin{array}{c} H \\ \diagdown \\ \\ C \\ \diagup \\ H \end{array} \!\!=\!\! \begin{array}{c} H \\ \diagup \\ C \\ \diagdown \\ H \end{array}$$

The other bond of the double
bond is a $2p$-$2p$ pi bond

Figure 3.1 When bonded to three other atoms, as in ethylene, $CH_2=CH_2$, carbon blends its atomic orbitals to form three sp^2-hybrid orbitals, which form sigma bonds to the three bonded atoms. One $2p$ orbital, containing one electron, remains.

A carbon atom with three sp^2-hybrid orbitals forms bonds to three other atoms and is called a **trigonal carbon**. The sigma bonds of a trigonal carbon lie in a plane, called the sigma-bond plane, with bond angles of approximately 120°, as shown in Figure 3.2. These bond angles allow the three groups bonded to the trigonal carbon to be at maximum distance from one another. The $2p$ orbital of a trigonal carbon is perpendicular to the sigma-bond plane—again, refer to Figure 3.2.

Each of the four second-level orbitals (three sp^2 orbitals and one $2p$ orbital) of a trigonal carbon contains one electron to contribute to bonding. Both carbon atoms in ethylene are in the sp^2-hybrid state. Two of the sp^2-hybrid orbitals on each carbon atom form sigma bonds by overlapping with the $1s$ orbital of hydrogen. The remaining two sp^2-hybrid orbitals, one on each carbon atom, form a carbon–carbon sigma bond by end-to-end overlap. Each sigma bond contains two electrons, one from each atom.

H   121° H
 \   /
   C—C)118°
 /   \
H   H

Sigma bonds of ethylene

In ethylene, each carbon atom also has one $2p$ orbital, which is perpendicular to the sigma-bond plane. Figure 3.3 shows how the two $2p$ orbitals undergo

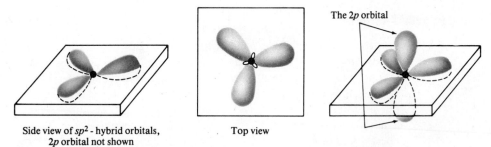

Side view of sp^2 - hybrid orbitals,   Top view   The $2p$ orbital
    $2p$ orbital not shown

Figure 3.2 A trigonal carbon showing the three sp^2-hybrid sigma orbitals in one plane and the $2p$ orbital perpendicular to this plane.

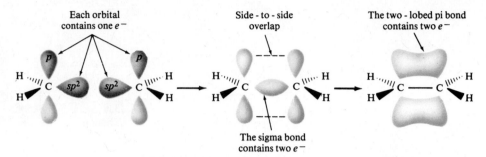

Figure 3.3 In ethylene, two trigonal carbon atoms are bonded by one sigma bond and one pi bond.

side-to-side overlap to form the second bond of the double bond. A bond formed by side-to-side overlap of *p* orbitals is a pi bond. Because the pi bond in ethylene is formed from two *p* orbitals, each with one electron, the pi bond, like a sigma bond, contains one pair of electrons.

Even though we write the formula for a compound with a double bond as if the two bonds were identical, keep in mind that any double bond contains one sigma bond and one pi bond—two very different bonds.

A sigma bond and a pi bond

$$CH_2{=}CH_2 \qquad CH_3CH{=}CHCH_3 \qquad CH_3\overset{\displaystyle O}{\overset{\displaystyle \|}{C}}H$$

A pi bond is weaker than a sigma bond. It is estimated that the carbon–carbon sigma-bond dissociation energy for ethylene is 95 kcal/mol, while the pi-bond dissociation energy is only 68 kcal/mol. Because it takes less energy to break a pi bond and because the pi-bond electrons are more exposed, the pi bond is much more reactive than a sigma bond. Because a carbon–carbon double bond has a pi bond, the double bond is considered to be a functional group, a site of chemical reactivity. Alkenes undergo many reactions that alkanes do not.

A. Cis-Trans Isomerism

A carbon–carbon double bond is rigid. The rigidity arises from the fact that the pi bonds are formed by the side-to-side overlap of *p* orbitals. Rotation can occur around sigma bonds because they are formed by the end-to-end overlap of orbitals. If the groups were forced to rotate around a pi bond, the bond would break because the *p* orbitals could no longer overlap (see Figure 3.4). Only about 20 kcal/mol is available to room-temperature molecules, and breaking a pi bond requires more energy than this.

Because of the rigidity of the carbon–carbon double bond, some disubstituted alkenes can exist as cis and trans isomers. An alkene with two groups on the same side of a double bond (and two hydrogens on the other side) is called the **cis isomer**.

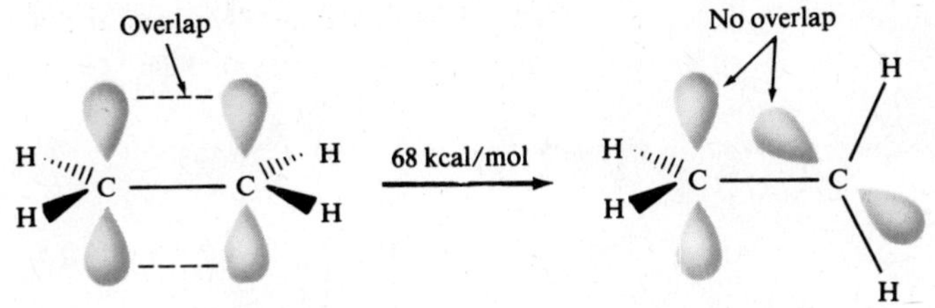

Figure 3.4 Rotation cannot occur about a carbon–carbon double bond unless the pi bond is broken.

An alkene with groups on opposite sides is called the **trans isomer**. A cis isomer cannot be converted to a trans isomer without rotation of the groups around the double bond.

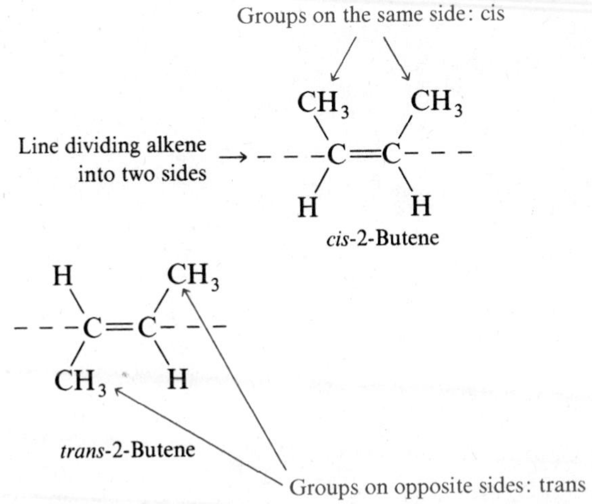

For cis-trans isomerism to be possible, an alkene must contain two different atoms or groups on each carbon of the double bond. The atoms or groups need not be the same on each carbon.

Cis-Trans Isomerism Possible:

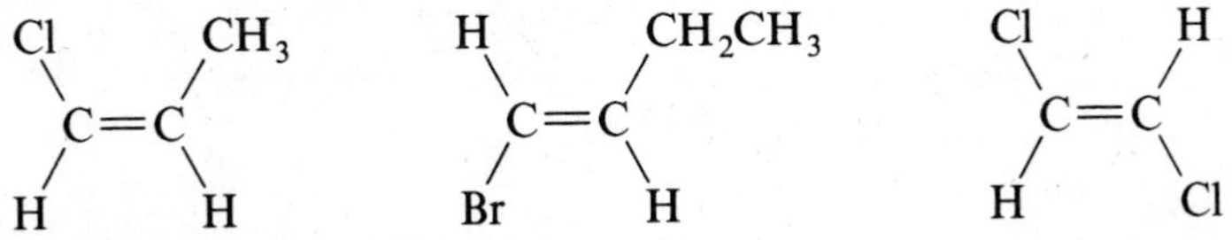

No Cis-Trans Isomerism Possible:

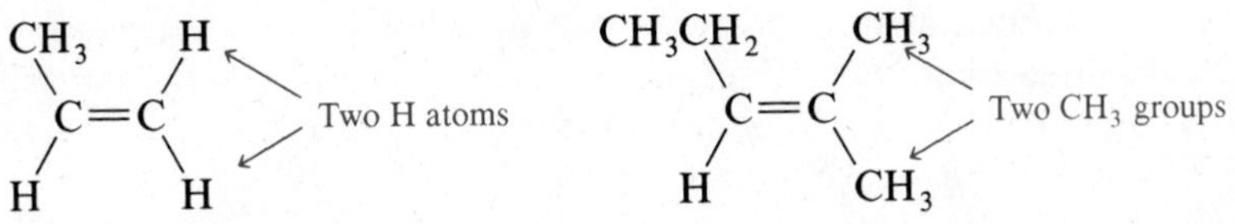

A molecule as a whole can move and rotate in space. Molecular rotation, however, does not cause cis isomers to change to trans isomers or vice versa.

The following formulas all represent cis-2-butene because the H's are on the same side of the double bond:

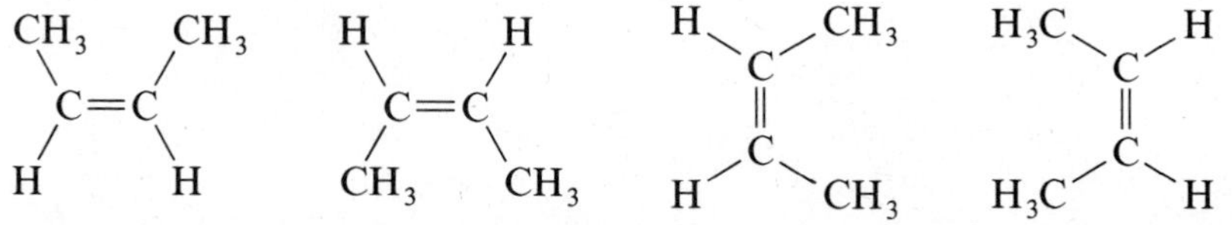

$$CH_3\diagdown C=C\diagup CH_3 \qquad H\diagdown C=C\diagup H \qquad H\diagdown C \underset{C}{\overset{\|}{}} CH_3 \qquad H_3C\diagdown C \underset{C}{\overset{\|}{}} H$$

The interconversion of cis and trans isomers cannot occur unless the pi bond is broken. For this reason, cis and trans isomers are different compounds, with different physical and chemical properties. Compare the boiling points of the cis and trans isomers of 2-butene and of 2-hexene.

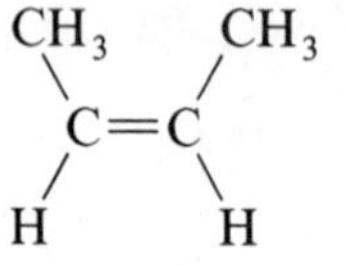

cis-2-Butene
(bp, 4°C)

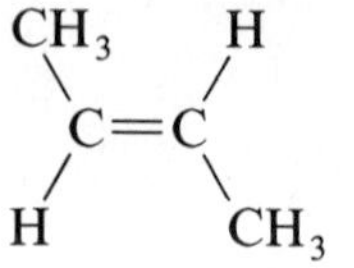

trans-2-Butene
(bp, 0.9°C)

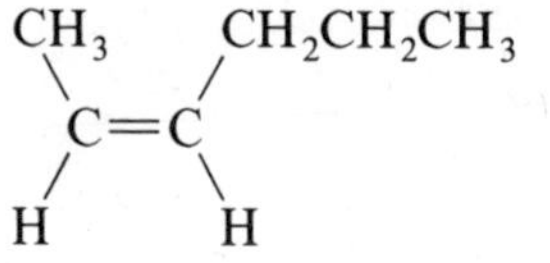

cis-2-Hexene
(bp, 69°C)

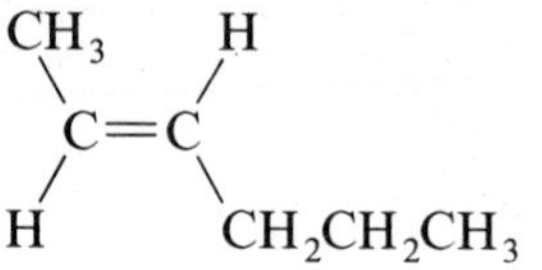

trans-2-Hexene
(bp, 68°C)

PROBLEM 3.3. Write formulas for the cis and trans isomers of the following compounds, if any are possible, and label each.

(a) $CH_3CH_2CH{=}CHCH_2CH_3$ (b) $ClCH{=}CHCH_3$

(c) $CH_2{=}CHCH{=}CHCH_3$ (d) $CH_2{=}CHCH_2CH_2Cl$

3.2 BONDING IN ALKYNES

If a carbon atom is bonded by a triple bond, it is in the *sp*-hybrid state. For this hybridization, the carbon uses its one $2s$ atomic orbital and one of its $2p$ atomic orbitals to form two *sp*-hybrid orbitals. Two unhybridized $2p$ orbitals remain, as shown in Figure 3.5.

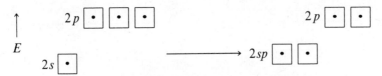

Figure 3.5 When bonded to two other atoms, as in acetylene, $CH \equiv CH$, carbon blends its atomic orbitals to form two *sp*-hybrid orbitals, which form a sigma bond between the two bonded atoms. Two $2p$ orbitals, each with one electron, remain.

In ethyne (acetylene, $CH \equiv CH$), each carbon atom uses its two *sp*-hybrid orbitals to form two sigma bonds, one with a hydrogen atom and the other with the other carbon atom. The two $2p$ orbitals of each carbon undergo side-to-side overlap to form two pi bonds, as shown in Figure 3.6.

A carbon in the *sp*-hybrid state forms *linear* bonds to maximize the distance between bonded groups.

180°

$$H—C \equiv C—H$$

180°

Multiple carbon–carbon bonds are shorter than single carbon–carbon bonds. Compare the following bond lengths:

$$CH_3—CH_3 \qquad CH_2 = CH_2 \qquad CH \equiv CH$$

1.54 Å   1.34 Å   1.20 Å

We will discuss the reason for this in Section 3.10E.

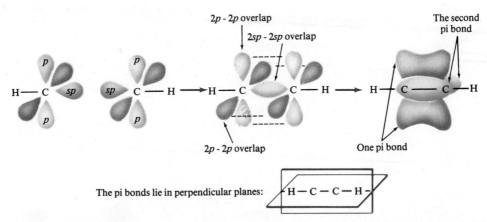

The pi bonds lie in perpendicular planes:

Figure 3.6 In acetylene, two carbon atoms are bonded by one sigma bond and two pi bonds.

TABLE 3.1   NAMES FOR SIMPLE ALKANES, ALKENES, AND ALKYNES

Formula	IUPAC name	Trivial name[a]
CH_3CH_3	Ethane	Ethane
$CH_2{=}CH_2$	Ethene	Ethylene
$CH{\equiv}CH$	Ethyne	Acetylene
$CH_3CH_2CH_3$	Propane	Propane
$CH_3CH{=}CH_2$	Propene	Propylene
$CH_3C{\equiv}CH$	Propyne	Methylacetylene

[a] These names are actually *semitrivial* instead of trivial because the structures can be deduced from the name.

3.3 NOMENCLATURE OF ALKENES AND ALKYNES

The name of an alkene is derived from the name of the parent alkane with the *-ane* ending changed to *-ene*. An alkyne is named similarly, but its ending is *-yne*. Table 3.1 shows the nomenclature relationships among two sets of simple alkanes, alkenes, and alkynes.

In general, a number is needed to indicate the position of the double bond or triple bond in the parent chain. The following rules illustrate the technique of naming a complex alkene or alkyne.

Rules for Naming an Alkene or Alkyne

1. Determine the longest continuous chain that *contains the double bond or triple bond*. This is the parent chain. (The parent chain of an alkene or alkyne is not necessarily the longest continuous chain in the structure.)
2. Number the carbon atoms in the parent chain beginning at the end *closer to the double bond or triple bond*, regardless of the positions of any branches.
3. Assemble the name by listing position numbers and names of side chains, the position number where the double bond or triple bond begins, the name of the parent with the ending *-ane* changed to *-ene* or *-yne*. Precede the entire name by *cis-* or *trans-* when applicable.

EXAMPLE

Name the following alkene

$$
\begin{array}{c}
CH_2 \\
\| \\
CH_3CH_2CCH_2CH_3
\end{array}
$$

Solution:

1. The longest continuous chain containing the double bond is four carbons. (The five-carbon chain is a longer continuous chain, but both carbons of the

double bond are not included in this chain.) The parent is butene (from butane with the ending changed to -ene.)

2. Number the parent chain, starting at the end closer to the double bond.

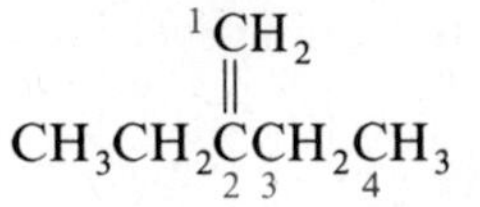

$$
\overset{1}{C}H_2
$$
$$
\parallel
$$
$$
CH_3CH_2\underset{2\ \ 3}{C}CH_2\underset{4}{CH_3}
$$

3. The name is 2-ethyl-1-butene (ethyl group at position 2; double bond beginning at position 1).

EXAMPLE

Name the following alkene:

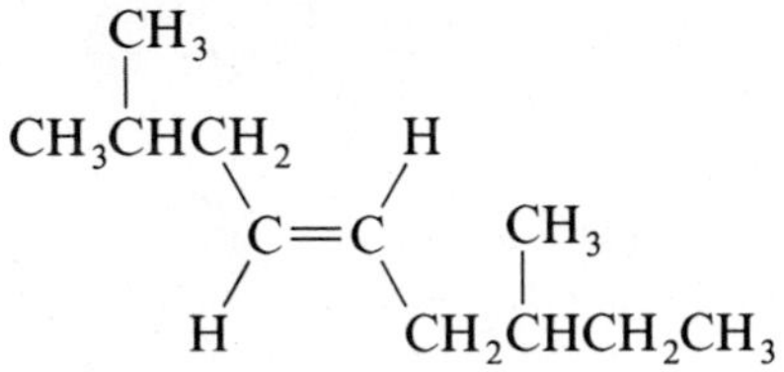

$$
\begin{array}{c}
CH_3 \\
| \\
CH_3CHCH_2 \quad H \\
\diagdown \qquad \diagup \\
C\!=\!C \qquad CH_3 \\
\diagup \qquad \diagdown \qquad | \\
H \qquad\quad CH_2CHCH_2CH_3
\end{array}
$$

Solution:

1. The longest continuous chain containing the double bond has nine carbons. The parent is nonene (derived from nonane).
2. Number the chain.

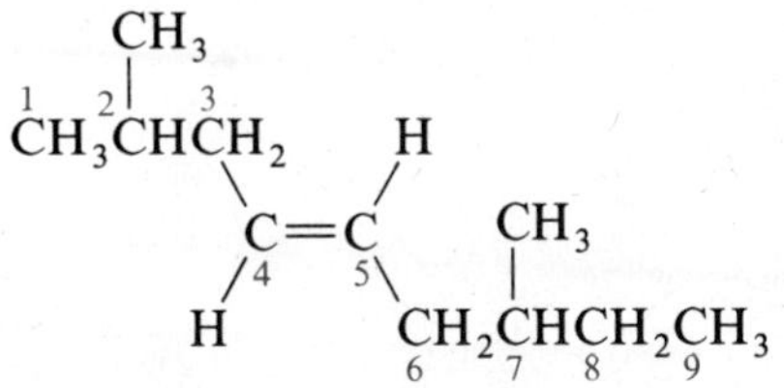

$$
\begin{array}{c}
CH_3 \\
| \\
\overset{1}{C}H_3\overset{2}{C}H\overset{3}{C}H_2 \quad H \\
\diagdown \qquad \diagup \\
\underset{4}{C}\!=\!\underset{5}{C} \qquad CH_3 \\
\diagup \qquad \diagdown \qquad | \\
H \qquad\quad \underset{6}{C}H_2\underset{7}{C}H\underset{8}{C}H_2\underset{9}{C}H_3
\end{array}
$$

3. The name is *trans*-2,7-dimethyl-4-nonene. (Note that *trans* is placed as the first prefix in the name.)

EXAMPLE

Name the following cycloalkene:

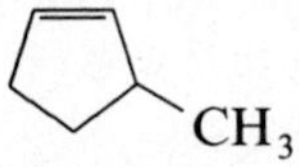

Solution:

1. The parent is cyclopentene (derived from cyclopentane).
2. Number the parent. With a simple cycloalkene, we assign the double bond positions 1 and 2. Then we number the ring toward a substituent in order to give it the lower number.

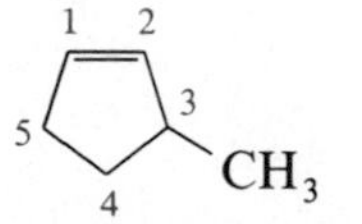

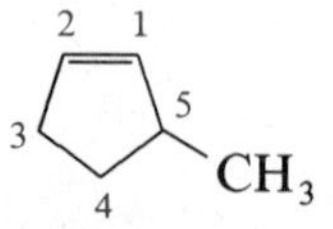

3. The name is 3-methyl-1-cyclopentene. The number 1 is commonly dropped in the name of a cycloalkene. Because the nomenclature rules require that the numbering start at the double-bond carbon, the name 3-methylcyclopentene is unambiguous. The number 1 can *not* be dropped in the name for most open-chain alkenes because numbering does not necessarily begin at the double bond.

PROBLEM 3.4.   Name the following compounds:

(a) 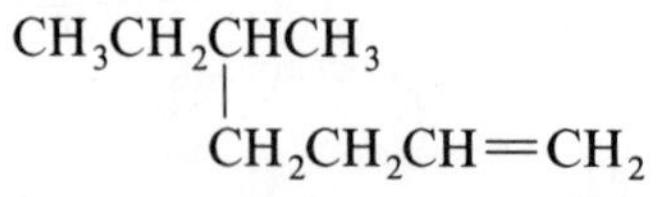 $CH_3CH_2CHCH_3$
         |
      $CH_2CH_2CH=CH_2$

(b)

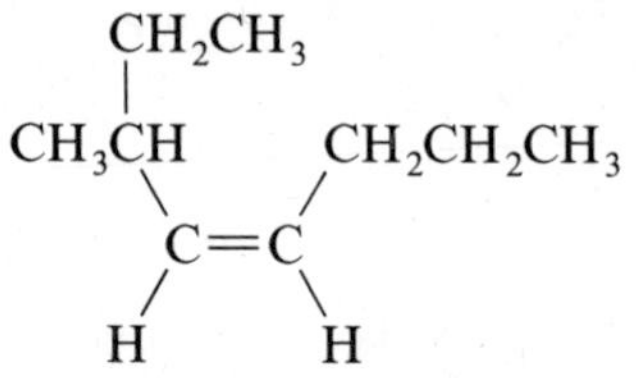

(c)

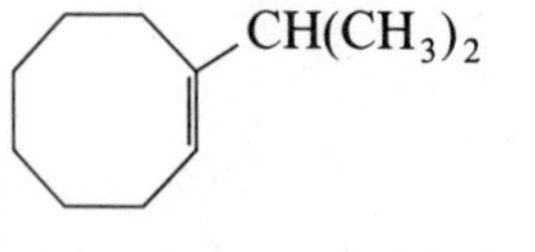

(d)

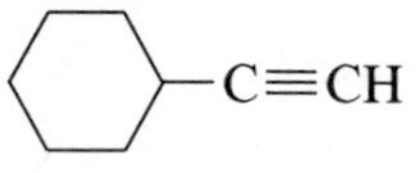

PROBLEM 3.5.   Write formulas for the following names:

(a) 2,4-dimethyl-1-pentene
(b) *cis*-5-methyl-2-hexene
(c) 2-ethyl-4-isopropylcyclohexene

3.4 PHYSICAL PROPERTIES OF ALKENES AND ALKYNES

Although alkenes and alkynes are chemically more reactive than are alkanes, they are not particularly polar. Consequently, their physical properties are similar to those of the alkanes. For example, the low-formula-weight alkenes and alkynes are gases at room temperature, just as the low-formula-weight alkanes are. Also like alkanes, alkenes and alkynes are water insoluble. The liquid members of all three

TABLE 3.2   SOME ALKENES AND THEIR BOILING POINTS

Formula	Name	Bp (°C)
$CH_2\!=\!CH_2$	Ethene (ethylene)	−104
$CH_3CH\!=\!CH_2$	Propene (propylene)	−47
$CH_3CH_2CH\!=\!CH_2$	1-Butene	−6
$\underset{H}{\overset{H_3C}{>}}C\!=\!C\underset{H}{\overset{CH_3}{<}}$	*cis*-2-Butene	4
$\underset{H}{\overset{H_3C}{>}}C\!=\!C\underset{CH_3}{\overset{H}{<}}$	*trans*-2-Butene	0.9
$CH_3\overset{\overset{\textstyle CH_3}{\vert}}{C}\!=\!CH_2$	Methylpropene	−7
$CH_3CH_2CH_2CH\!=\!CH_2$	1-Pentene	30

classes of compounds float on water. Table 3.2 lists the boiling points of some alkenes and alkynes.

3.5 PREPARATION OF ALKENES

In the laboratory, alkenes are usually prepared by **elimination reactions**—reactions in which part of a reactant molecule is lost (eliminated) as a small molecule such as H_2O or HBr.

A. Dehydration of Alcohols

When heated with a strong acid such as H_2SO_4 or H_3PO_4, an alcohol can eliminate water to yield an alkene. This reaction is called a **dehydration reaction**. (*Dehydration* means "loss of water.")

A Dehydration Reaction:

$$CH_3\!-\!\overset{\overset{\textstyle CH_3}{\vert}}{\underset{\underset{\textstyle CH_3}{\vert}}{C}}\!-\!OH \xrightarrow[\text{warm}]{\text{aqueous } H_2SO_4} CH_2\!=\!C\underset{CH_3}{\overset{CH_3}{<}} + H_2O$$

  t-Butyl alcohol   Methylpropene

Dehydration reactions will be discussed in more detail in Chapter 8.

TABLE 3.3   SUMMARY OF PREPARATIONS OF ALKENES

Reaction	Section reference

$$\underset{\text{An alcohol}}{\overset{\displaystyle H \quad OH}{R-\overset{\displaystyle |}{\underset{\displaystyle R}{C}}-\overset{\displaystyle |}{\underset{\displaystyle R}{C}}-R}} \xrightarrow[\text{heat}, -H_2O]{H^+} \overset{R}{\underset{R}{\diagdown}}C=C\overset{R}{\underset{R}{\diagup}}$$ 8.9

$$\underset{\text{A haloalkane}}{\overset{\displaystyle H \quad X}{R-\overset{\displaystyle |}{\underset{\displaystyle R}{C}}-\overset{\displaystyle |}{\underset{\displaystyle R}{C}}-R}} \xrightarrow[\text{heat}, -HX]{\overset{K^+ \;^-OH}{CH_3CH_2OH}} \overset{R}{\underset{R}{\diagdown}}C=C\overset{R}{\underset{R}{\diagup}}$$ 7.9

B. Dehydrohalogenation

A dehydrohalogenation reaction (loss of a hydrogen and a halogen from the reactant) of a haloalkane is another reaction that yields alkenes. In dehydrohalogenation, the halogen compound is heated with a strong base, such as potassium hydroxide (K^+ ^-OH) in ethanol (CH_3CH_2OH). The hydroxide ion abstracts a proton from the haloalkane to yield water. A halide ion is also lost from the starting haloalkane to yield the alkene.

A Dehydrohalogenation Reaction:

$$\underset{\text{2-Bromopropane}}{\overset{\displaystyle Br \quad H}{CH_3\overset{\displaystyle |}{C}H-\overset{\displaystyle |}{C}H_2}} + \,^-OH \xrightarrow[\text{ethanol}]{\text{heat}} \underset{\text{Propene}}{CH_3CH=CH_2} + \quad H-OH \quad + \quad Br^-$$

Dehydrohalogenation will be discussed in more detail in Chapter 7.
Table 3.3 summarizes the reactions used to prepare alkenes.

3.6 ELECTROPHILIC ADDITION REACTIONS OF ALKENES

The most characteristic reactions of alkenes are **addition reactions**, in which reagents add to the double bonds to yield saturated products.

Two Addition Reactions:

$$\left.\begin{array}{c} H-H \\ + \\ R_2C=CR_2 \end{array}\right\} \xrightarrow{\text{catalyst}} \overset{\displaystyle H \quad H}{R_2\overset{\displaystyle |}{C}-\overset{\displaystyle |}{C}R_2}$$

$$\left.\begin{array}{c} H-Br \\ + \\ R_2C=CR_2 \end{array}\right\} \longrightarrow \overset{\displaystyle H \quad Br}{R_2\overset{\displaystyle |}{C}-\overset{\displaystyle |}{C}R_2}$$

In addition reactions, the two carbons of the double bond are rehybridized from sp^2 to sp^3, and in each case a portion of the adding reagent becomes bonded to each carbon.

A. Addition of Hydrogen Halides

The reactions of alkenes with hydrogen halides (HCl, HBr, and HI) are quite general. Because all three reagents react in a similar manner, HX is often used as a general symbol to represent a hydrogen halide. The products of the addition reactions of hydrogen halides are haloalkanes (RX).

A general symbol
for a hydrogen halide

$$CH_3CH{=}CHCH_3 + HX \longrightarrow CH_3CH_2\overset{\overset{\displaystyle X}{|}}{C}HCH_3$$

cis- or *trans-*2-Butene   *A 2-halobutane*

An **electrophile** (literally, "electron lover") is a reagent that seeks negative electron-rich sites in molecules. Many positively charged ions, such as H^+, can act as electrophiles.

The addition reactions of alkenes with acidic reagents, such as HX, are initiated by the electrophilic attack of H^+ on the negatively charged pi electrons of the alkene. In the equation that follows, we use a curved arrow with a full arrowhead (⌒) to show the movement of the pair of pi electrons as they form a sigma bond with the electrophile H^+ and a second curved arrow to show the transfer of a pair of sigma-bond electrons to X^-.

Step 1, Electrophilic Attack of H^+:

The old pair of pi electrons now
forming a sigma bond

$$R_2C{=}CR_2 \ + \ H{-}\ddot{\underset{..}{X}}{:} \longrightarrow R_2\overset{\overset{\displaystyle H}{|}}{C}{-}\overset{+}{\ddot{C}}\,R_2 \ + \ {:}\ddot{\underset{..}{X}}{:}^-$$

A carbocation
(an unstable intermediate)

When the two pi electrons form a bond with H^+, the other carbon of the original carbon–carbon double bond becomes electron deficient. Because this carbon now has only six valence electrons, it is positively charged. The intermediate product of the electrophilic attack is a positive ion called a **carbocation**. (The term *carbonium ion* is also used to describe a carbocation.) Carbocations are usually intermediates, not true products, because they combine rapidly with halide ions to form neutral molecules. Carbocations themselves cannot be isolated from a reaction mixture.

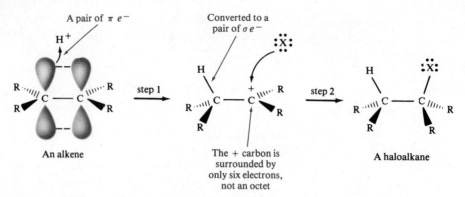

Figure 3.7 Electrophilic attack of H^+ on an alkene's pi electrons, followed by combination with a halide ion.

Step 2, Combination with the Halide Ion:

$$R_2CH\overset{+}{-}CR_2 \; + \quad :\overset{..}{\underset{..}{X}}:^- \quad \longrightarrow \quad R_2CH\overset{\displaystyle :\overset{..}{\underset{..}{X}}:}{-}CR_2$$

The carbocation   *A halide ion*   *A haloalkane*

Figure 3.7 provides a pictorial representation of the two steps of the reaction.

PROBLEM 3.6.   Complete the following equations, showing the two steps in each reaction:

(a) $CH_3CH{=}CHCH_3 + HBr \longrightarrow$

(b)   ⬡ + HI $\longrightarrow$

We can combine the two steps of an addition reaction into one *flow equation*, a chemical equation with more than one arrow.

A Flow Equation:

$$CH_3CH{=}CHCH_3 \xrightarrow{\; H^+ \;} CH_3\overset{+}{CH}{-}CH_2CH_3 \xrightarrow{\; Cl^- \;} CH_3\overset{\displaystyle Cl}{\overset{|}{CH}}{-}CH_2CH_3$$

If the mechanism of the reaction is not being stressed, the intermediate structures in a multistep reaction need not be shown:

$$CH_3CH{=}CHCH_3 + HCl \longrightarrow CH_3\overset{\displaystyle Cl}{\overset{|}{CH}}CH_2CH_3$$

B. Markovnikov's Rule

Alkenes can be classed as symmetrical or unsymmetrical.

Symmetrical (Double Bond Divides Molecule into Halves):

$$CH_2 = CH_2 \qquad CH_3CH = CHCH_3$$

Unsymmetrical (Double Bond Divides Molecule into Nonequivalent Pieces):

$$CH_3CH = CH_2 \qquad CH_3\overset{\overset{\displaystyle CH_3}{|}}{C} = CHCH_3 \qquad \text{⬡} - CH_3$$

When a hydrogen halide reacts with a symmetrical alkene, only one haloalkane can result, regardless of which carbon of the carbon–carbon bond forms the bond with the proton.

$$\overset{H^+}{} \qquad \qquad \overset{H^+}{}$$

$$CH_3CH = CHCH_3 \quad \text{or} \quad CH_3CH = CHCH_3 \longrightarrow$$

$$CH_3CH_2 - \overset{+}{C}HCH_3 \quad (\text{the same as } CH_3\overset{+}{C}H - CH_2CH_3) \overset{Cl^-}{\longrightarrow}$$

$$\overset{\overset{\displaystyle Cl}{|}}{CH_3CH_2CHCH_3}$$

Only possible haloalkane

When an unsymmetrical alkene undergoes this reaction, however, two halo-alkanes could result.

$$\overset{H^+}{} \qquad \qquad \overset{H^+}{}$$

$$CH_3CH = CH_2 \qquad \text{or} \quad CH_3CH = CH_2 \longrightarrow$$

$$CH_3\overset{+}{C}H - CH_3 \quad \text{or} \quad CH_3CH_2 - \overset{+}{C}H_2 \overset{Cl^-}{\longrightarrow}$$

Different carbocations

$$\overset{\overset{\displaystyle Cl}{|}}{CH_3CHCH_3} \quad \text{or} \quad CH_3CH_2CH_2Cl$$

Two possible haloalkanes

When the reaction is carried out in a laboratory, one product is formed in a larger amount than the other. In 1871, after having studied a number of alkene addition reactions, the Russian chemist Vladimir Markovnikov formulated the following empirical rule to predict the structure of the predominant product.

MARKOVNIKOV'S RULE.   When an unsymmetrical reagent adds to an unsymmetrical alkene, the positive part of the reagent (H^+ in HX) becomes bonded to the double-bond carbon that is already bonded to the greater number of hydrogens.

Markovnikov's rule allows us to predict which product will predominate in unsymmetrical addition reactions.

H^+ goes here (to the
C with more H's already)

$$CH_3CH{=\!}CH_2 + HCl \longrightarrow CH_3\overset{\overset{\displaystyle Cl}{|}}{C}HCH_3, \quad \text{not } CH_3CH_2CH_2Cl$$

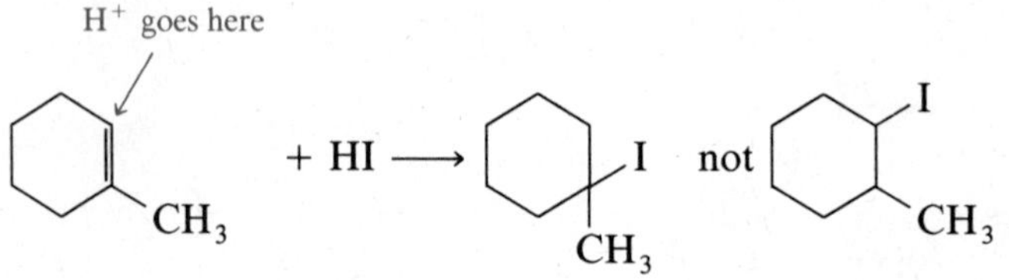

H^+ goes here

When dealing with cyclic structures, you may find it helpful to write out the atomic symbols to determine which carbon is bonded to the greater number of hydrogens.

One H; H^+ goes here

No H

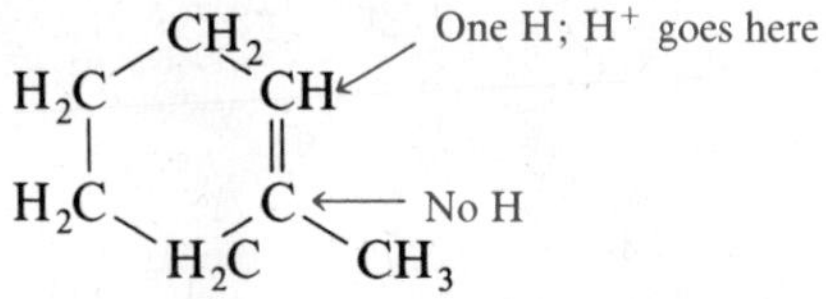

PROBLEM 3.7.   Predict the major products:

(a) $CH_3\overset{\overset{\displaystyle CH_3}{|}}{C}{=\!}CHCH_3 + HBr \longrightarrow$

(b) 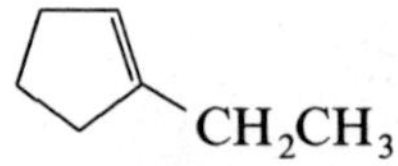 CH_2CH_3 + HCl $\longrightarrow$

C. The Reason for Markovnikov's Rule

The reason why Markovnikov's rule can be used to predict the predominant product in alkene addition reactions is related to the relative stabilities of the intermediate carbocations. In alkene addition reactions, three types of carbocations could be formed: **primary** (*one* R group bonded to the positively charged carbon), **secondary** (*two* R groups bonded to the positively charged carbon), or **tertiary** (*three* R groups bonded to the positively charged carbon).

The Three Principal Types of Carbocations:

one R group	two R groups	three R groups
$\downarrow$	$\downarrow$	$\downarrow$
$R\overset{+}{C}H_2$	$R_2\overset{+}{C}H$	$R_3\overset{+}{C}$
Primary (1°)	Secondary (2°)	Tertiary (3°)

In studies of many reactions, chemists have determined that tertiary carbocations are more stable than secondary carbocations, and secondary carbocations are much more stable than primary carbocations.

$$1° \ R\overset{+}{C}H_2 \ \lll \ 2° \ R_2\overset{+}{C}H \ < \ 3° \ R_3\overset{+}{C}$$

Increasing carbocation stability ⟶

The relative stabilities of the carbocations probably arise from the greater polarizability of alkyl groups than that of a hydrogen atom. Compared with a hydrogen atom, an alkyl group contains a greater number of electrons that can be drawn toward the positively charged atom; therefore, a greater number of atoms can help share the positive charge. The attraction of electron density toward a positive charge through sigma bonds in a molecule or ion is called the **inductive effect**.

A charge that is spread out over a number of atoms is said to be *delocalized*. An ion with a delocalized charge is more stable than an ion with a *localized* (less spread out) charge. Figure 3.8 illustrates the localization and delocalization of the positive charge of carbocations.

In addition reactions, the *more stable carbocation* is formed preferentially as an intermediate because less energy is required for its formation (see Figure 3.9). The

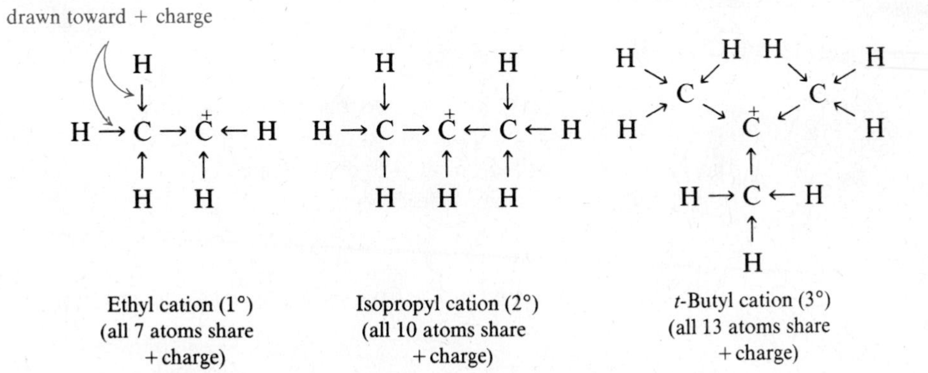

Arrows show σ electrons
drawn toward + charge

Ethyl cation (1°)
(all 7 atoms share
+ charge)

Isopropyl cation (2°)
(all 10 atoms share
+ charge)

t-Butyl cation (3°)
(all 13 atoms share
+ charge)

Figure 3.8 Carbocations are stabilized by the inductive effect.

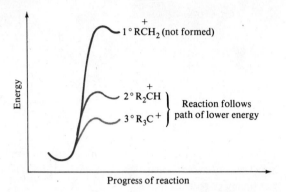

Figure 3.9 The E_{act} leading to a tertiary carbocation is less than that leading to a secondary carbocation. The E_{act} leading to a primary carbocation is prohibitive.

formation of the more stable carbocation leads to the observed Markovnikov products.

$$CH_3CH{=}CH_2 \xrightarrow{} CH_3\overset{+}{C}H{-}CH_3, \quad \text{not } CH_3CH_2{-}\overset{+}{C}H_2 \xrightarrow{Cl^-}$$

More stable   Less stable
2° carbocation   1° carbocation

$$\underset{\displaystyle CH_3CHCH_3}{\overset{\displaystyle Cl}{|}}, \quad \text{not } CH_3CH_2CH_2Cl$$

PROBLEM 3.8. Rewrite the equations in Problem 3.7 to show both possible intermediate carbocations. Indicate which carbocation is formed preferentially. Use the format that we used in the preceding equation.

D. Addition of Sulfuric Acid and Water

Other acidic compounds besides hydrogen halides can add to an alkene's double bond. Sulfuric acid (H_2SO_4) is an example. The product of the addition of sulfuric acid to an alkene is called an **alkyl hydrogen sulfate** ($ROSO_3H$). Some salts of alkyl hydrogen sulfates are used as detergents (see Point of Interest 12).

$$CH_3CH{=}CHCH_3 + H{-}OSO_3H \longrightarrow \underset{\displaystyle CH_3CH_2CHCH_3}{\overset{\displaystyle OSO_3H}{|}}$$

sec-Butyl hydrogen sulfate

The addition of sulfuric acid follows the same general pathway as does HX addition. The alkene is initially attacked by a proton to form the more stable carbocation. This intermediate carbocation is then attacked by the hydrogen sulfate ion ($HSO_4{}^-$) to yield the observed product. Because the more stable carbocation is an intermediate in the reaction, Markovnikov's rule is followed.

$$
CH_3CH{=}CH_2 \quad\xrightarrow{H^+}\quad \overset{2°,\ \text{more stable}}{CH_3\overset{+}{CH}{-}CH_3}, \quad \text{not}\ CH_3CH_2{-}\overset{+}{CH}_2
$$

Intermediate carbocation

$$
{}^-\!OSOH \qquad\qquad OSO_3H
$$

$$
\longrightarrow CH_3CHCH_3
$$

A mixture of sulfuric acid and water causes **hydration** of an alkene. *Hydration* means "addition of water." In a hydration reaction, sulfuric acid provides the proton for carbocation formation. In the second step, water instead of the hydrogen sulfate ion combines with the carbocation because of the greater concentration of water in the reaction mixture.

$$
CH_3CH{=}CH_2 \xrightarrow{H^+} CH_3\overset{+}{CH}CH_3 \xrightarrow{\ :OH\ (H)\ }
$$

A carbocation

$$
\underset{CH_3CHCH_3}{\overset{\overset{\textstyle H}{|}}{{}^+\!:\!OH}} \xrightarrow[-H^+]{\ \text{Means "loss of } H^+ \text{"}\ } \underset{CH_3CHCH_3}{\overset{\textstyle :\ddot{O}H}{|}}
$$

A protonated alcohol,　　　2-Propanol
an oxonium ion　　　*an alcohol*

In this reaction, addition of water to the carbocation results in a **protonated alcohol**, or an **oxonium ion**. (An *-onium ion* is a positively charged ion; therefore, an oxonium ion is a positively charged oxygen ion.) An oxonium ion loses a proton to water or other base in solution to yield an alcohol. Even though we know that H^+ is lost to a base, we often write $-H^+$ over the arrow of a deprotonation reaction to simplify the equation.

$$
\underset{\underset{H^+ \text{ is lost from the}}{\text{oxonium ion to the base } H_2O}}{R{-}\overset{\overset{\textstyle H}{|}}{\underset{\cdot\cdot}{O}}{-}H} \quad\overset{\overset{\textstyle H}{:\ddot{O}H}}{\phantom{R{-}O{-}H}}\quad \longrightarrow \quad \underset{\text{An alcohol}}{R\ddot{O}H} \quad \overset{\overset{\textstyle H}{HOH}}{\underset{+}{}}
$$

As in the addition of HX or H_2SO_4, the addition of H_2O to an alkene yields the Markovnikov product.

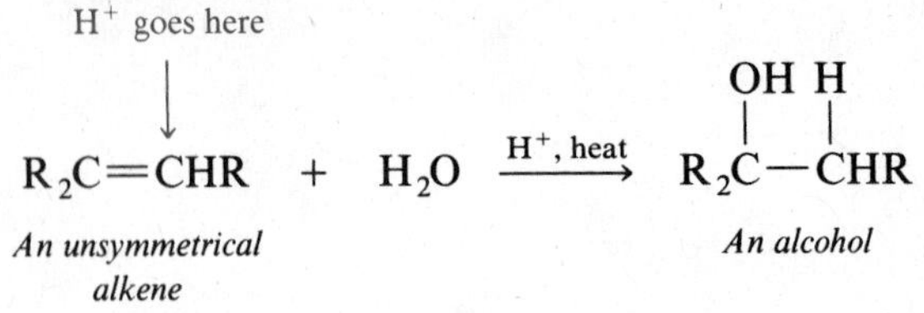

$$\overset{\underset{\displaystyle |}{\text{H}^+ \text{ goes here}}}{R_2C\!=\!CHR} \;+\; H_2O \xrightarrow{H^+,\ heat} R_2\overset{\displaystyle \text{OH}}{\underset{}{C}}\!-\!\overset{\displaystyle \text{H}}{\underset{}{C}}HR$$

An unsymmetrical alkene   *An alcohol*

Hydration reactions of alkenes are quite general. Solvent ethanol is synthesized industrially by this route, as is isopropyl alcohol. The equation showing the preparation of isopropyl alcohol was given in the introduction of this chapter.

$$CH_2\!=\!CH_2 + H_2O \xrightarrow{catalyst} CH_3CH_2OH$$
Ethanol

A hydration reaction is the reverse reaction of a dehydration reaction, in which an alcohol loses water to yield an alkene. In a hydration reaction we would treat an alkene with aqueous acid, providing an excess of water for the hydration. In a dehydration reaction we would heat the alkene with more-concentrated acid so that a minimum amount of water would be present in the reaction mixture.

Unfortunately, strong acids catalyze a variety of organic reactions and lead to mixtures of products. Therefore, other catalysts are sometimes used to convert alkenes to alcohols. In animal cells, hydration of double bonds is catalyzed by enzymes. For example, the following reaction occurs as one of many steps in the biological conversion of glucose to carbon dioxide, water, and energy.

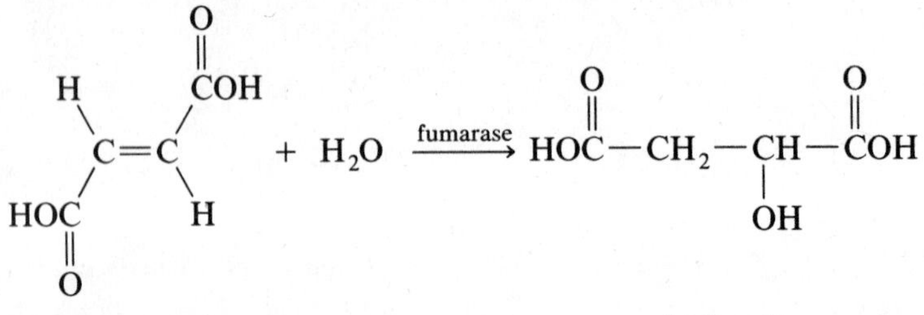

$$\underset{\text{Fumaric acid}}{\begin{array}{c}\text{H}\quad\quad\quad\overset{\displaystyle O}{\overset{\displaystyle \|}{\text{C}}}\text{OH}\\ \diagdown\quad\diagup\\ \text{C}\!=\!\text{C}\\ \diagup\quad\diagdown\\ \text{HOC}\quad\quad\text{H}\\ \underset{\displaystyle O}{\|}\end{array}} \;+\; H_2O \xrightarrow{fumarase} \underset{\text{Malic acid}}{\overset{\displaystyle O}{\overset{\displaystyle \|}{\text{HOC}}}\!-\!CH_2\!-\!\underset{\displaystyle \text{OH}}{\underset{|}{\text{CH}}}\!-\!\overset{\displaystyle O}{\overset{\displaystyle \|}{\text{COH}}}}$$

PROBLEM 3.9. Complete the following equations:

(a) $CH_3CH_2CH\!=\!CH_2 + H_2O \xrightarrow{H^+}$

(b) 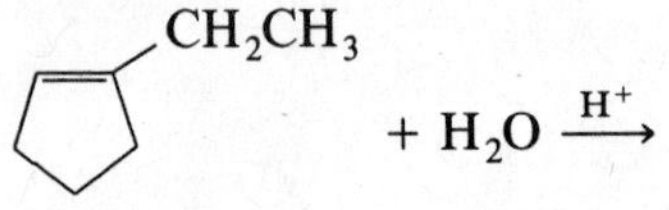 $\overset{\displaystyle CH_2CH_3}{}$ $+ H_2O \xrightarrow{H^+}$

PROBLEM 3.10. Write equations to show how you would prepare the following compounds:

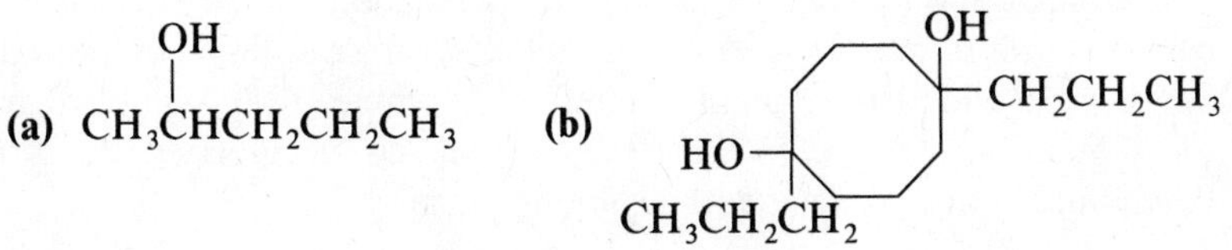

(a) $\overset{\displaystyle OH}{\underset{\displaystyle |}{CH_3CHCH_2CH_2CH_3}}$

(b)

E. Addition of Bromine

Bromine (Br_2) is not an acid, but it can add to a double bond because bromine molecules can be polarized. When the bromine molecule and the pi bond start to collide, the pi bond's electrons cause the bromine molecule to polarize so that one end of the bromine molecule is partially positive and the other end is partially negative. The positive end of the polarized bromine molecule can act as an electrophile and add to the double bond. Bromide ion (Br^-) is lost from the attacking bromine molecule. The result is that Br^+ is added to the double bond.

Step 1, Formation of Bromonium Ion:

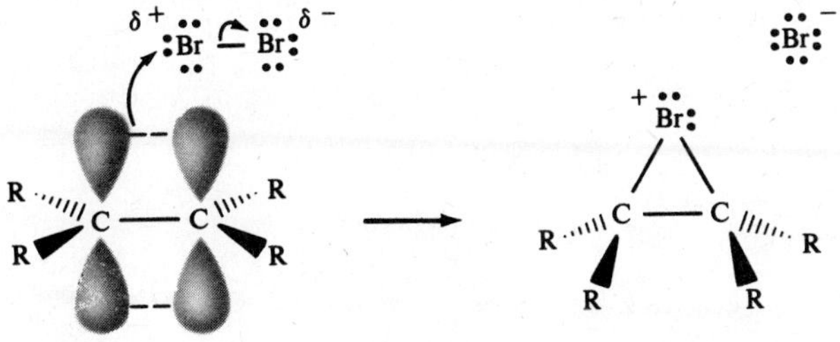

A bromonium ion

Step 2, Attack of Br^-:

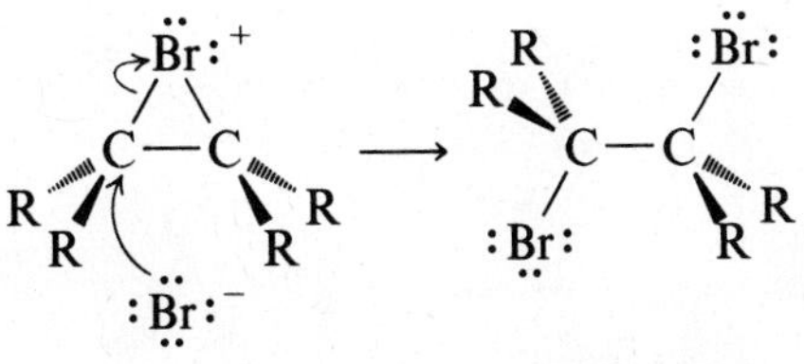

A dibromoalkane

When H^+ attacks a pi bond, a carbocation is formed. However, when Br^+ attacks a pi bond, a **bridged bromonium ion** is formed. (Evidence for this type of intermediate will be presented shortly.) The intermediate bromonium ion is like a carbocation and can be attacked by the bromide ion to yield the observed product, a dibromoalkane.

Bromination of alkenes is a general reaction and can be used as a test for carbon–carbon double bonds. In the test, a colorless compound of unknown structure is added to a small amount of bromine (red) dissolved in CCl_4. Dibromoalkanes are colorless; therefore, an alkene decolorizes the bromine. If the compound is not an alkene, the solution remains red.

$$CH_2{=}CH_2 + Br_2 \longrightarrow \overset{\displaystyle \overset{Br}{|}}{CH_2}{-}\overset{\displaystyle \overset{Br}{|}}{CH_2}$$

Colorless   *Red*   1,2-Dibromoethane

Colorless

1,2-Dibromoethane (ethylene dibromide, EDB) is produced industrially by the reaction of ethylene and bromine. This dibromo compound and 1,2-dichloroethane (ethylene dichloride) are used as coadditives in gasoline containing tetraethyllead, $(CH_3CH_2)_4Pb$. These dihaloethanes convert nonvolatile lead salts in the engine's cylinders to volatile lead halides, which can then be removed from the engine in the automobile exhaust.

Treatment of a cycloalkene with bromine yields a *trans*-1,2-dibromocycloalkane. The formation of the trans product is evidence for a bridged ion intermediate.

Mechanism by Way of a Bridged Bromonium Ion:

Br⁻ could attack either C but only from the side opposite the bridge (because the bridge blocks the other side)

Trans Br's

trans-1,2-Dibromocyclohexane

If a nonbridged carbocation intermediate had been formed, both the cis and trans products would have been observed. The fact that only the trans product is observed is evidence for a bridged intermediate.

Hypothetical Mechanism with a Carbocation Intermediate (Does Not Occur):

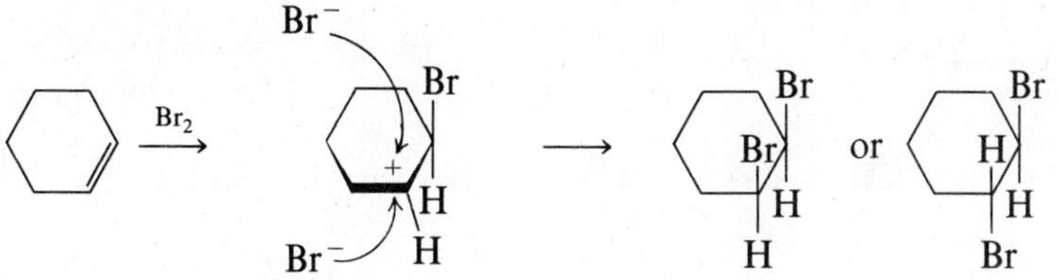

Br⁻ could attack   *Cis*   *Trans*
from above or below   *not observed*

Chlorine (Cl_2) also adds to a carbon–carbon double bond, but the mechanism is not so clear-cut.

PROBLEM 3.11. Write equations for the preparation of the following compounds from alkenes:

(a) $\underset{\underset{\displaystyle Br}{|}}{CH_3}\overset{\overset{\displaystyle Br}{|}}{C}H CHCH_2CH_3$ (b)

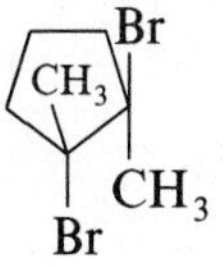

PROBLEM 3.12. The treatment of methylpropene with Br_2 in a solution containing Cl^- ions yields *two* dihaloalkane products. Using the mechanism we have discussed, show the structures of these products and how they are formed.

F. Addition of Halogens and Water

An alkene treated with an aqueous solution of Br_2 or Cl_2 undergoes *mixed addition* to yield a 1,2-halo alcohol, commonly called a **1,2-halohydrin**. In this context, the numbers 1 and 2 refer to any adjacent positions in a structure, not necessarily to positions 1 and 2 in the nomenclature numbering system.

$$CH_3CH{=}CH_2 + Br_2 + H_2O \longrightarrow \underset{\underset{\displaystyle \text{1-Bromo-2-propanol}}{}}{CH_3\overset{\overset{\displaystyle OH}{|}}{C}H{-}\overset{\overset{\displaystyle Br}{|}}{C}H_2} + HBr$$

1-Bromo-2-propanol

a 1,2-bromohydrin

The mechanism by which halohydrins are formed is directly analogous to the mechanism of the formation of dihaloalkanes.

Mechanism of the Formation of a Bromohydrin:

$$:\ddot{Br}-\ddot{Br}:$$

$$CH_3CH\!\!=\!\!CH_2 \xrightarrow{-Br^-} CH_3\underset{\underset{H}{\overset{\displaystyle |}{:\ddot{O}H}}}{CH}-CH_2 \longrightarrow CH_3\underset{\underset{H}{\overset{\displaystyle |}{\overset{+}{:}\ddot{O}H}}}{\overset{\overset{\displaystyle :\ddot{Br}:}{|}}{CH}}-CH_2 \xrightarrow{-H^+} CH_3\underset{\overset{\displaystyle |}{:\ddot{O}H}}{\overset{\overset{\displaystyle :\ddot{Br}:}{|}}{CH}}-CH_2$$

The first step in the mechanism is the formation of the bridged bromonium ion intermediate. In the second step, a water molecule, instead of the bromide ion Br^-, attacks the bridged intermediate. The product of this step is an oxonium ion, which loses a proton to the solvent (H_2O) to yield the bromohydrin.

In this reaction, water attacks the more positive carbon in the bridged ion. To determine which carbon is the more positive one, we use carbocation stability ($1°\ RCH_2^+ \ll 2°\ R_2CH^+ < 3°\ R_3C^+$).

$$\underset{2°}{CH_3\overset{+}{CH}-\overset{\overset{\displaystyle Br}{|}}{CH_2}} \quad not \quad \underset{1°}{CH_3\overset{\overset{\displaystyle Br}{|}}{CH}-\overset{+}{CH_2}}$$

In the bridged ion:
$$CH_3\underset{\delta+}{CH}\overset{Br^{\delta+}}{-\!\!\diagup\diagdown\!\!-}CH_2$$

More positive C,
H_2O attacks here

EXAMPLE

Predict the principal product of the reaction of 1-methylcyclopentene with Br_2 and H_2O.

Solution:

Br and CH_3 are
cis to each other.

The more positive carbon is the one with the greater number of C's bonded to it.

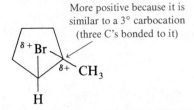

More positive because it is
similar to a 3° carbocation
(three C's bonded to it)

PROBLEM 3.13. Complete the following equations showing the bridged or protonated intermediates.

(a) $CH_3CH{=}\overset{\overset{\displaystyle CH_3}{|}}{C}CH_3 + Br_2 \xrightarrow{CCl_4}$

(b) $CH_3CH{=}\overset{\overset{\displaystyle CH_3}{|}}{C}CH_3 + Br_2 + H_2O \longrightarrow$

(c) $\langle \rangle + Br_2 \xrightarrow{CCl_4}$

(d) $\langle \rangle + Br_2 + H_2O \longrightarrow$

G. Addition of Borane

The addition of borane (BH_3) to an alkene occurs by the addition of hydrogen and boron across the double bond. Therefore, this reaction is referred to as **hydroboration**.

$$\left.\begin{array}{c} H{-}BH_2 \\ + \\ CH_2{=}CH_2 \end{array}\right\} \longrightarrow \overset{\overset{\displaystyle H}{|}}{C}H_2{-}\overset{\overset{\displaystyle BH_2}{|}}{C}H_2$$

Unlike hydrogens in the other reagents we have discussed, the hydrogens in BH_3 are the more electronegative element in the molecule. In hydroboration, the hydrogen does not act as an electrophile as it does in hydrogen halide addition.

The electrophile

$\overset{\delta+}{H}{-}\overset{\delta-}{Br}$ $\overset{\delta-}{H}{-}\overset{\delta+}{B}{-}H^{\delta-}$
                          $\underset{H^{\delta-}}{|}$

Because of its partial negative charge, the hydrogen from BH_3 becomes bonded to the double-bond carbon with *fewer* hydrogens (rather than to the carbon with more hydrogens, as in addition of HX). The $—BH_2$ group becomes bonded to the double-bond carbon with more hydrogens.

$$CH_3CH{=}CH_2 \longrightarrow CH_3CH{-}CH_2, \quad not \quad CH_3CH{-}CH_2$$

The initial product is a monoalkylorganoborane RBH_2, which still contains two partially negative hydrogens. Each of these hydrogens can react with another alkene molecule. Let us illustrate these reactions using the *n*-propylborane from the preceding equation.

$$CH_3CH_2CH_2BH_2 \xrightarrow{CH_3CH{=}CH_2} (CH_3CH_2CH_2)_2BH \xrightarrow{CH_3CH{=}CH_2}$$

n-Propylborane   di-*n*-Propylborane

$$(CH_3CH_2CH_2)_3B$$

tri-*n*-Propylborane

Trialkylorganoboranes (R_3B) can be converted to a wide variety of other compounds. For example, they can be converted to alcohols when they are oxidized with an alkaline solution of hydrogen peroxide (H_2O_2).

$$R_3B + H_2O_2 + 3\,^-OH \longrightarrow 3ROH + BO_3{}^{3-}$$

An alcohol   Borate ion

The utility of this reaction is that an unsymmetrical alkene yields the "anti-Markovnikov" alcohol, a product that cannot be obtained from acid-catalyzed hydration reactions.

$$CH_3CH{=}CH_2$$

$\xrightarrow{H—OH, H^+}$

$$CH_3CH{-}CH_3$$
$$\overset{OH}{|}$$

2-Propanol

Markovnikov product

$\xrightarrow[\text{(2) } H_2O_2,\ ^-OH]{\text{(1) } BH_3}$

$$CH_3CH_2CH_2$$
$$\overset{OH}{|}$$

1-Propanol

anti-Markovnikov product

Organoborane reactions are sufficiently useful in organic chemistry that the 1979 Nobel Prize in Chemistry was awarded to their developer, Professor H. C. Brown, at Purdue University.

PROBLEM 3.14. Show by flow equations how you would convert 1-butene to the following alcohols.

        OH
        |
(a) $CH_3CHCH_2CH_3$ (b) $CH_3CH_2CH_2CH_2OH$

3.7 OXIDATION AND REDUCTION OF ORGANIC COMPOUNDS

Our next topics are the oxidation and reduction of alkenes. Before continuing with our discussion of alkene reactions, let us pause and discuss briefly the concepts of oxidation and reduction as they apply to organic compounds.

Oxidation is defined as the loss of electrons, and reduction is defined as the gain of electrons. In inorganic equations, it is relatively easy to identify which elements or ions have been oxidized and which have been reduced; we merely inspect the valences.

$$S^{2-} + 2Fe^{3+} \longrightarrow S^0 + 2Fe^{2+}$$

(oxidation: $S^{2-} \to S^0$; reduction: $2Fe^{3+} \to 2Fe^{2+}$)

These definitions of oxidation and reduction cannot be applied directly to organic reactions because of the difficulty in assigning electrons to individual atoms in covalent compounds. To avoid this problem, we say that an organic oxidation is the net loss of H atoms or the net gain of O atoms. An organic reduction is the net gain of H atoms or the net loss of O atoms.

Oxidation, the Loss of H or Gain of O:

$$CH_3CH_2OH \longrightarrow CH_3\overset{\displaystyle O}{\overset{\|}{C}}H \qquad \text{Loss of two H}$$

$$CH_3\overset{\displaystyle O}{\overset{\|}{C}}H \longrightarrow CH_3\overset{\displaystyle O}{\overset{\|}{C}}OH \qquad \text{Gain of one O}$$

Reduction, the Gain of H or Loss of O:

$$CH_2{=}CH_2 \longrightarrow CH_3CH_3 \qquad \text{Gain of two H}$$

$$CH_3\overset{\displaystyle O}{\overset{\|}{C}}OH \longrightarrow CH_3\overset{\displaystyle O}{\overset{\|}{C}}H \qquad \text{Loss of one O}$$

PROBLEM 3.15. Identify each of the following conversions as an oxidation of an organic compound, a reduction of the compound, or neither. Explain your answers.

(a) $\longrightarrow$ $\underset{\displaystyle \|}{\overset{\displaystyle O}{HOC}}(CH_2)_4\underset{\displaystyle \|}{\overset{\displaystyle O}{C}}OH$

(b) $\longrightarrow$

(c) $CH_2{=}CH_2 \longrightarrow CH_3CH_2OH$

(d) $CH_3CH_2C{\equiv}CH \longrightarrow CH_3CH_2CH{=}CH_2$

3.8 ADDITION OF HYDROGEN TO ALKENES

The addition of hydrogen gas (H_2) to an alkene double bond is called **hydrogenation** of the alkene. The hydrogenation of an alkene is a reduction because the reactant gains hydrogen.

$$CH_2{=}CH_2 + H_2 \xrightarrow{\text{Pt}} CH_3CH_3$$

Hydrogenation is a general reaction of compounds containing carbon–carbon double bonds and is a reaction with commercial applications. For example, hydrogenation is used to "harden" vegetable oils to a buttery consistency for margarine or peanut butter.

$CH_2O\overset{\displaystyle O}{\overset{\displaystyle \|}{C}}(CH_2)_{14}CH_3$

$CHO\overset{\displaystyle O}{\overset{\displaystyle \|}{C}}(CH_2)_7CH{=}CH(CH_2)_7CH_3 \qquad \xrightarrow[\text{Pt}]{2H_2}$

$CH_2O\overset{\displaystyle O}{\overset{\displaystyle \|}{C}}(CH_2)_7CH{=}CHCH_2CH{=}CH(CH_2)_4CH_3$

A typical vegetable oil

$CH_2O\overset{\displaystyle O}{\overset{\displaystyle \|}{C}}(CH_2)_{14}CH_3$

$CHO\overset{\displaystyle O}{\overset{\displaystyle \|}{C}}(CH_2)_7CH{=}CH(CH_2)_7CH_3$

$CH_2O\overset{\displaystyle O}{\overset{\displaystyle \|}{C}}(CH_2)_{16}CH_3$

A typical fat

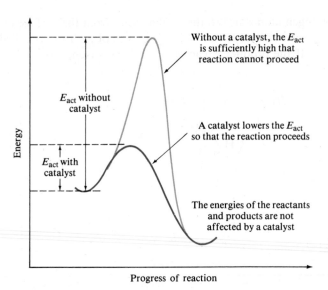

Figure 3.10 A catalyst lowers the activation energy, or energy barrier, for hydrogenation of an alkene.

A. Mechanism of Hydrogenation

Hydrogenation occurs by a type of mechanism different from those of the reactions we have discussed previously. Although a hydrogenation reaction is exothermic, it has a high activation energy and requires an appropriate catalyst, such as finely divided metallic palladium, platinum, or nickel. Any catalyst, such as a hydrogenation catalyst, lowers the activation energy for a reaction but does not affect the potential energy of either the reactant or the product. Figure 3.10 is a pair of superimposed energy diagrams showing the effect the hydrogenation catalyst has on the activation energy.

Experimental evidence suggests that H_2 molecules are adsorbed onto the surface of the metal and that the H—H bonds are weakened by their interaction with the metal. When the pi electrons of an alkene molecule encounter the hydrogen atoms, reaction occurs. Figure 3.11 shows a diagram for this mechanism.

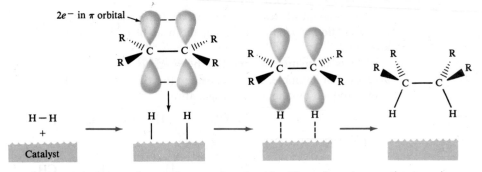

Figure 3.11 A hydrogenation catalyst acts by adsorbing H_2 and causing reaction to occur on its surface.

TABLE 3.5   PERCENTAGES OF s AND p CHARACTER IN HYBRID ORBITALS

Hybrid orbital	Atomic orbitals used to form the hybrid orbital	s (%)	p (%)
sp	s, p	50	50
sp^2	s, p, p	33	67
sp^3	s, p, p, p	25	75

PROBLEM 3.22. Complete the following equations, showing the structure of the intermediate enol as well as the structure of the organic product.

(a) $CH_3C\equiv CCH_3 + H_2O \xrightarrow[\text{HgSO}_4]{\text{H}_2\text{SO}_4}$

(b) 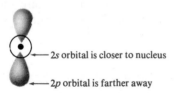 $-C\equiv CH + H_2O \xrightarrow[\text{HgSO}_4]{\text{H}_2\text{SO}_4}$

E. Reaction with Strong Base

Hybrid orbitals can be ranked according to their percentage of s character and percentage of p character, as shown in Table 3.5.

The s orbital is closer to the nucleus than is a $2p$ orbital. Therefore, the greater percentage of s character in a hybrid orbital, the closer are the hybrid orbital's electrons to the nucleus.

— $2s$ orbital is closer to nucleus

— $2p$ orbital is farther away

$$\begin{array}{ccccc} & s & sp & sp^2 & sp^3 \\ \text{% } s \text{ character:} & 100\% & 50\% & 33\% & 25\% \end{array}$$
Increasing distance from nucleus

Because the electrons in an sp orbital are closer to the nucleus than those in sp^2 or sp^3 orbitals, an sp-hybridized carbon is more electronegative than sp^2- or sp^3-hybridized carbons. This increased electronegativity affects the reactivity of alkynes. For example, the $C\equiv CH$ group can lose a proton in an acid-base reaction to an extremely strong base, such as the amide ion ($^-NH_2$). Amide ions are stronger bases than hydroxide ions, which cannot react with the $C\equiv C-H$ proton.

$$CH_3C\equiv C-H \;+\; ^-:\ddot{N}H_2 \quad \xrightarrow{\text{liq. } NH_3} \quad CH_3C\equiv C:^- \;+\; :NH_3$$

Amide ion   *An acetylide ion*   Ammonia
(from $Na^+\ ^-NH_2$)

The acidity of the $C-H$ bond of the terminal alkyne (an alkyne with a triple bond at the end of a carbon chain) is an exception to the general rule that $C-H$ bonds are inert in acid-base reactions. While an amide ion can react with the terminal alkyne proton, it does not abstract other $C-H$ protons.

Only the terminal alkyne proton is sufficiently acidic to react with amide ions

The increasing percentage of s character in the series sp^3, sp^2, sp also explains why carbon–carbon single bonds (sp^3) are the longest and carbon–carbon triple bonds (sp) are the shortest in the series alkanes, alkenes, alkynes.

PROBLEM 3.23.   Predict the products, if any:

(a) $CH_3CH_2C\equiv CH + NaNH_2 \xrightarrow{\text{liq. } NH_3}$

(b) $CH_3CH_2C\equiv CCH_3 + NaNH_2 \xrightarrow{\text{liq. } NH_3}$

(c) $CH_3CH_2CH=CH_2 + NaNH_2 \xrightarrow{\text{liq. } NH_3}$

(d) $CH_3CH_2C\equiv CH + NaOH \longrightarrow$

SUMMARY

Simple noncyclic alkenes are unsaturated hydrocarbons containing one double bond and having the general formula C_nH_{2n}. Each additional pi bond or ring removes two hydrogens. A simple alkyne, which contains a carbon–carbon triple bond, has the general formula C_nH_{2n-2}.

A carbon–carbon double bond is formed from an sp^2–sp^2 sigma bond and a $2p$–$2p$ pi bond, while a carbon–carbon triple bond is formed from an sp–sp sigma bond and two $2p$–$2p$ pi bonds.

Two π

Because groups cannot rotate about a carbon–carbon double bond at room temperature, cis-trans isomerism is possible in disubstituted alkenes.

$$
\underset{\text{cis}}{\overset{\displaystyle R \diagdown \qquad \diagup R'}{\underset{\displaystyle H \diagup \qquad \diagdown H}{C=C}}}
\qquad
\underset{\text{trans}}{\overset{\displaystyle R \diagdown \qquad \diagup H}{\underset{\displaystyle H \diagup \qquad \diagdown R'}{C=C}}}
$$

Alkenes are named after the parent alkane with the *-ane* ending changed to *-ene*. Alkyne names are given the ending *-yne*.

The physical properties of alkenes and alkynes are similar to those of alkanes.

Alkenes can be prepared by the **dehydration** of an alcohol or by the **dehydrohalogenation** of a haloalkane, as shown in Table 3.3.

Alkenes undergo **electrophilic addition reactions** with a variety of reagents. These reactions proceed by way of the more stable **carbocations**, where carbocation stability is $3° > 2° \gg 1°$.

$$
R_2C{=}CHR \xrightarrow{H^+} R_2\overset{+}{C}{-}CH_2R \xrightarrow{X^-} R_2\overset{\displaystyle X}{\overset{|}{C}}{-}CH_2R
$$

where X = Cl, Br, or I

Because of this mechanism, addition reactions of unsymmetrical alkenes follow **Markovnikov's rule**: The electrophile (H^+) attacks the carbon that is already bonded to the greater number of hydrogens. This and some other general examples of alkene addition reactions are listed in Table 3.4.

Alkenes can be reduced to alkanes by **catalytic hydrogenation**.

Hydrogenation: $R_2C{=}CR_2 + H_2 \xrightarrow{Pt} R_2CH{-}CHR_2$

Hydrogenation data show that more highly substituted alkenes are more stable than less substituted alkenes and that trans alkenes are more stable than cis alkenes.

Alkenes can be oxidized to diols, ketones, aldehydes, or carboxylic acids, depending on their structure and on the oxidizing conditions.

Oxidation: $R_2C{=}CHR$

cold MnO_4^-, ^-OH → $R_2\overset{\displaystyle OH\ OH}{\overset{|\quad\ |}{C}}{-}CHR$

(1) hot MnO_4^-, (2) H^+
or (1) O_3, (2) H_2O_2 → $R_2C{=}O + HO\overset{\displaystyle O}{\overset{\|}{C}}R$

(1) O_3, (2) Zn, H_2O, H^+ → $R_2C{=}O + H\overset{\displaystyle O}{\overset{\|}{C}}R$

Alkynes undergo the same types of reactions as alkenes; however, the addition of water yields an aldehyde or a ketone.

$$RC\equiv CR + 2X_2 \longrightarrow RCX_2{-}CX_2R$$

$$+ 2HX \longrightarrow RCX_2{-}CH_2R$$

$$+ H{-}OH \xrightarrow{Hg^{2+}} \underset{\text{An enol}}{RCH{=}\overset{\overset{\displaystyle OH}{|}}{C}R} \longrightarrow RCH_2{-}\overset{\overset{\displaystyle O}{\|}}{C}R$$

Alkynes with a $-C\equiv CH$ group react with very strong bases to yield acetylides.

$$RC\equiv CH + {}^-\!:\!\ddot{N}H_2 \xrightarrow{\text{liq. }NH_3} RC\equiv C\!:^- + :NH_3$$

STUDY PROBLEMS

3.24. State whether each of the following alkenes is cis, trans, or incapable of cis-trans isomerism.

(a)

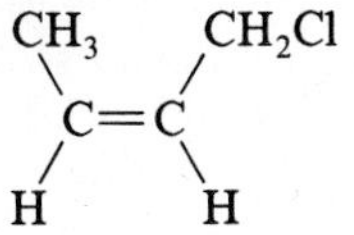

(b)

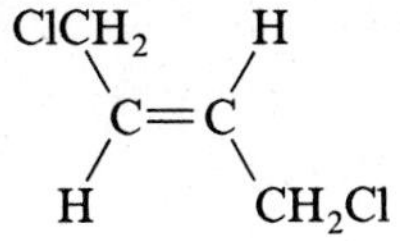

(c)

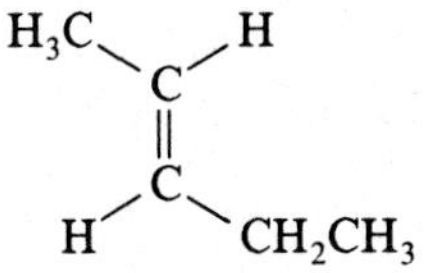

(d)

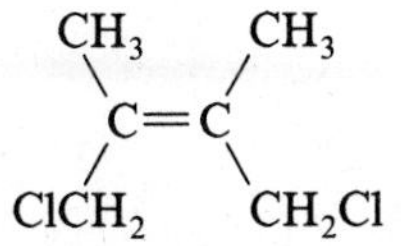

(e)

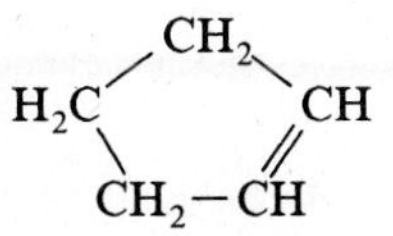

(f)

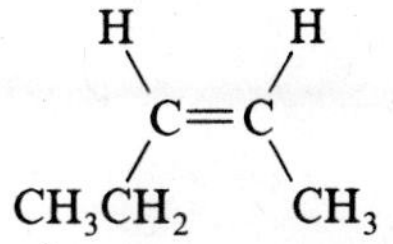

3.25. Name each alkene in the preceding problem, including the cis or trans designation where appropriate.

3.26. Name the following alkenes or alkynes:

(a) $CH_2{=}CHCH_2CH_2CH_3$

(b) 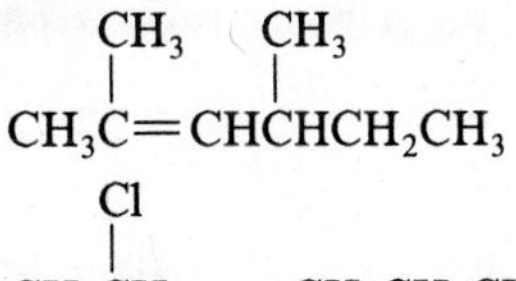 $CH_3\overset{\overset{\displaystyle CH_3}{|}}{C}{=}CH\overset{\overset{\displaystyle CH_3}{|}}{C}HCH_2CH_3$

(c) $CH_3C\equiv CCH_3$

(d)

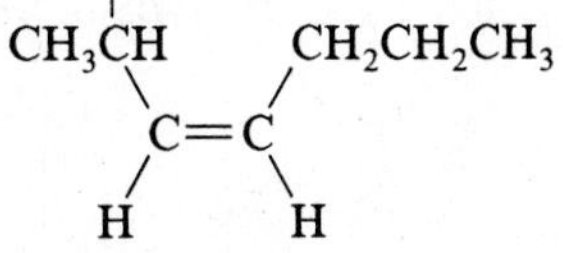

(e)

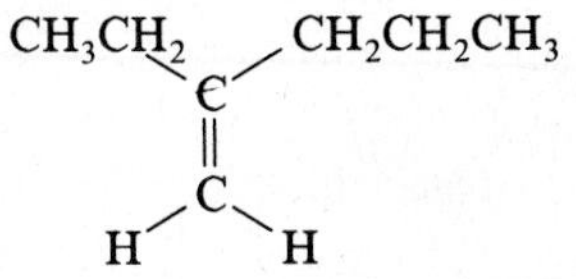

(f)

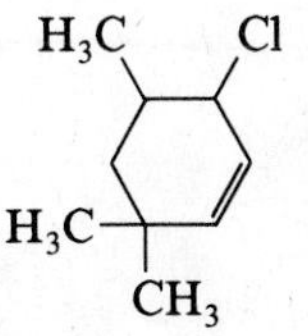

3.27. Write condensed structural formulas or polygon formulas for the following compounds:

(a) 2-methyl-2-heptene

(b) *trans*-3-nonene

(c) 1-chloropropene

(d) 3-chloropropene

(e) 4-methyl-2-hexyne

(f) *cis*-1,4-dichloro-2-pentene

(g) 3-ethyl-1-pentyne

(h) 3-isopropylcyclopentene

(i) 1,2-dimethylcycloheptene

(j) *trans*-3,4-dimethylcyclohexene

3.28. Write structural formulas and names for all the isomeric alkenes with the molecular formula C_5H_{10}.

3.29. Draw polygon formulas for all possible cycloalkenes containing five- or six-membered rings and having the molecular formula C_6H_{10}.

3.30. Complete the following equations showing all possible alkene products.

(a) $\underset{\overset{|}{OH}}{CH_3CHCH_3} \xrightarrow[\text{heat}]{\text{conc. } H_2SO_4}$

(b) $\xrightarrow[\text{heat}]{\text{conc. } H_2SO_4}$

(c) $\underset{\overset{|}{I}}{CH_3CHCH_2CH_2CH_3} + {}^-OH \xrightarrow{\text{heat}}$

(d) $+ {}^-OH \xrightarrow{\text{heat}}$

3.31. Write the formula for the most likely carbocation intermediate when each of the following alkenes is treated with strong acid.

(a) $CH_3CH_2CH_2CH{=}CH_2$

(b) $CH_3CH{=}C(CH_3)_2$

(c) $={}CH_2$

3.32. Complete the following equations. If more than one organic product is possible, show only the major one.

(a) $CH_3CH_2CH{=}CH_2 + Br_2 \longrightarrow$

(b) $CH_3CH{=}CHCH_3 + HBr \longrightarrow$

(c) $CH_3CH{=}CH_2 + \text{conc. } H_2SO_4 \longrightarrow$

(d) $(CH_3)_2C{=}C(CH_3)_2 + H_2O \xrightarrow{H_2SO_4}$

(e) $-CH_3 + HCl \longrightarrow$

(f) $+ H_2O \xrightarrow{H_2SO_4}$

(g) $(CH_3)_2C{=}CH_2 + HCl \longrightarrow$

(h) $-CH_2CH_3 + Br_2 \longrightarrow$

(i) 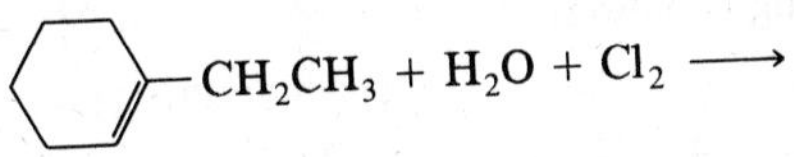 —$CH_2CH_3 + H_2O + Cl_2 \longrightarrow$

(j) 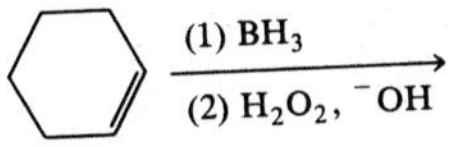 $\xrightarrow[\text{(2) } H_2O_2, \ ^-OH]{\text{(1) } BH_3}$

(k) $3CH_3CH{=}CHCH_3 + BH_3 \longrightarrow$

(l) $1CH_3CH{=}CHCH_3 + BH_3 \longrightarrow$

(m) 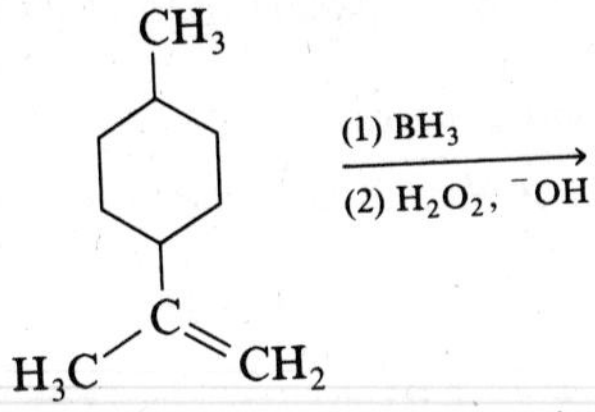 $\xrightarrow[\text{(2) } H_2O_2, \ ^-OH]{\text{(1) } BH_3}$

3.33. Write flow equations to show how you would prepare the following alcohols from alkenes.

(a) 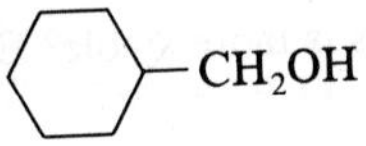 —CH_2OH (b) 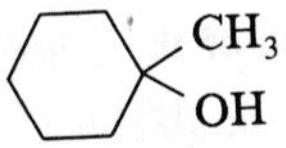 (two routes)

3.34. Identify each of the following biochemical conversions as oxidation or reduction.

(a) 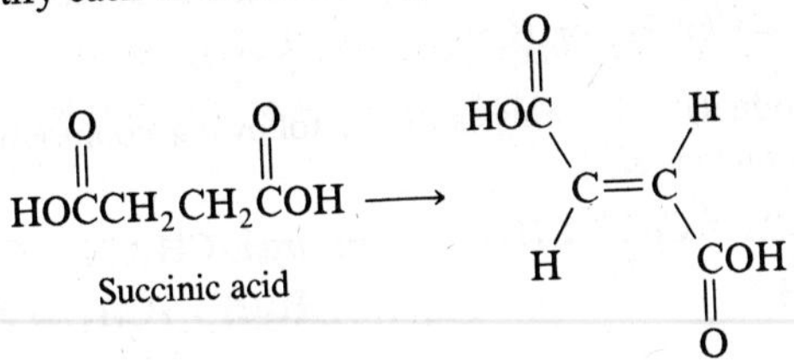 $HOCCH_2CH_2COH \longrightarrow$

Succinic acid

Fumaric acid

(b) 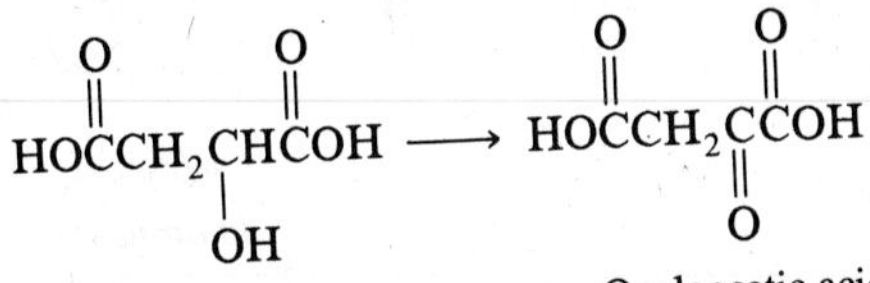 $HOCCH_2CHCOH \longrightarrow HOCCH_2CCOH$

Malic acid

Oxaloacetic acid

(c) 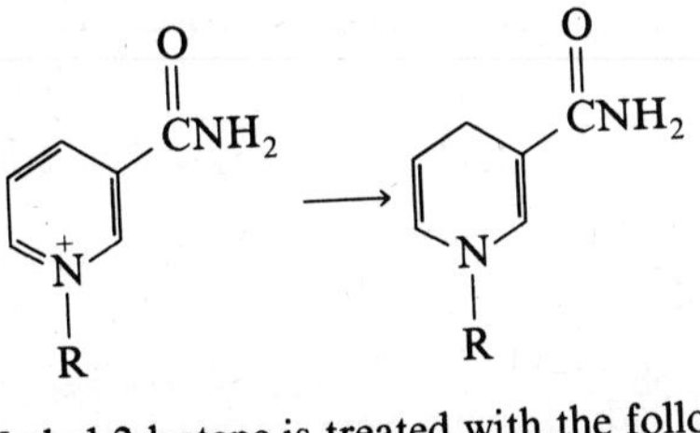 $\longrightarrow$

3.35. 2-Methyl-2-butene is treated with the following reagents. Predict the organic product in each case.

(a) H_2, Pt

(b) (1) O_3, (2) Zn, H_2O, H^+

(c) cold aqueous K^+ MnO_4^-

(d) (1) hot aqueous K^+ MnO_4^-, (2) H^+

ized, allylic cations are of lower energy and are more stable than ordinary alkyl carbocations. We can show the two carbon atoms sharing the positive charge by conventional line-bond formulas as well, but we must use two formulas—resonance formulas—just as we did for the nitrate ion and the carbonate ion in Section 1.4B. We will discuss allylic cations in more detail in Section 4.3.

$$CH_2{=}CH{-}\overset{+}{C}H{-}CH_2Br \longleftrightarrow \overset{+}{C}H_2{-}CH{=}CH{-}CH_2Br$$

Resonance formulas for the carbocation

The 1,2-addition reaction is completed by attack of Br^- on the positively charged carbon in the first resonance formula.

$$CH_2{=}CH{-}\overset{+}{C}H{-}CH_2Br \longrightarrow CH_2{=}CH{-}CHCH_2Br$$

$$:\!\overset{..}{Br}\!:^- \qquad \qquad \qquad :\!\overset{..}{Br}\!:$$

The 1,2-addition product

The 1,4-addition reaction is completed by attack of Br^- on the positively charged carbon in the second resonance formula.

$$:\!\overset{..}{Br}\!:^- \quad \overset{+}{C}H_2{-}CH{=}CHCH_2Br \longrightarrow :\!\overset{..}{Br}CH_2CH{=}CHCH_2Br$$

The 1,4-addition product

An equation for the overall reaction of addition of 1 mol of Br_2 to a conjugated diene, showing both 1,2- and 1,4-addition products, follows. These products are not necessarily formed in equal amounts; the proportions of the two products depend, in part, on the experimental conditions.

Addition of 1 mol of Br_2:

$$CH_2{=}CHCH{=}CH_2 \xrightarrow{Br_2} \overset{\overset{\displaystyle Br}{|}}{CH_2}{-}\overset{\overset{\displaystyle Br}{|}}{CH}CH{=}CH_2 + BrCH_2CH{=}CHCH_2Br$$

3,4-Dibromo-1-butene	1,4-Dibromo-2-butene
from 1,2-addition	*from 1,4-addition*

Addition of a second mole of Br_2 to either dibromoalkene in the previous equation yields the same product, a tetrabromoalkane.

Addition of Second Mole of Br_2:

$$\overset{\overset{\displaystyle Br \quad\; Br}{|\qquad |}}{CH_2{-}CHCH}{=}CH_2 \quad or \quad BrCH_2CH{=}CHCH_2Br \xrightarrow{Br_2}$$

$$\overset{\overset{\displaystyle Br \quad\; Br \quad\; Br \quad\; Br}{|\qquad |\qquad |\qquad |}}{CH_2{-}CH{-}CH{-}CH_2}$$

1,2,3,4-Tetrabromobutane

Dienes with isolated double bonds cannot undergo 1,4-addition reactions because there is no interaction between the pi bonds, and therefore these dienes cannot form allylic carbocations.

PROBLEM 4.3. Predict the 1,2- and 1,4-addition products:

(a) $CH_3CH{=}CHCH{=}CH_2 + Br_2 \longrightarrow$

(b) $CH_3CH{=}CHCH{=}CH_2 + HBr \longrightarrow$

(c) $CH_2{=}CHCH_2CH{=}CH_2 + Br_2 \longrightarrow$

(d)  ⬡ $+ Br_2 \longrightarrow$

(e) 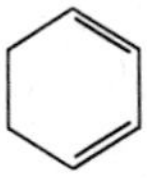 ⬡ $+ HBr \longrightarrow$

4.3 ALLYLIC CARBOCATIONS

Any carbocation containing a positive charge on the carbon adjacent to a double bond is an **allylic carbocation**, or **allylic cation**. The term allylic arises from the trivial name for the 2-propenyl group, $(CH_2{=}CHCH_2{-})$, commonly called the **allyl group**.

Some Allylic Carbocations:

$$CH_2{=}CH\overset{+}{C}H_2 \qquad \qquad \qquad$$

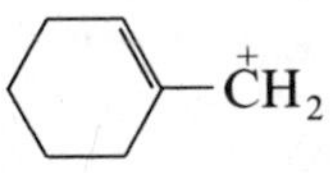

Because of the *p*-orbital interactions shown in Figure 4.2, any allylic cation carries the positive charge on two carbons. Because this positive charge is delocalized, an allylic cation is stabilized. Therefore, although a primary carbocation is very unstable, an allylic carbocation is a relatively stable intermediate, even if it is primary.

$$1°\ R\overset{+}{C}H_2 \quad 2°\ R_2\overset{+}{C}H \quad 3°\ R_3\overset{+}{C} \quad \text{allylic}$$

Increasing stability of carbocation intermediates

The delocalization of pi electrons (which, in turn, delocalizes the positive charge in the allylic cation) is why a single line-bond formula cannot describe certain ions and molecules adequately. The additional resonance formulas for a structure show how the pi electrons are delocalized. Recall that the actual structure of a resonance hybrid is a composite of all the resonance structures.

The Allyl Cation:

Resonance formulas   Resonance hybrid

$$[CH_2\!=\!CH\!-\!\overset{+}{C}H_2 \longleftrightarrow \overset{+}{C}H_2\!-\!CH\!=\!CH_2] \equiv \overset{\delta+}{CH_2}\!\cdots\!CH\!\cdots\!\overset{\delta+}{CH_2}$$

*A resonance hybrid is a real structure that is described
with two or more resonance formulas.*

Resonance Formulas For Two Other Allylic Cations:

$$CH_3CH\!=\!CH\!-\!\overset{+}{C}HCH_3 \longleftrightarrow CH_3\overset{+}{C}H\!-\!CH\!=\!CHCH_3$$

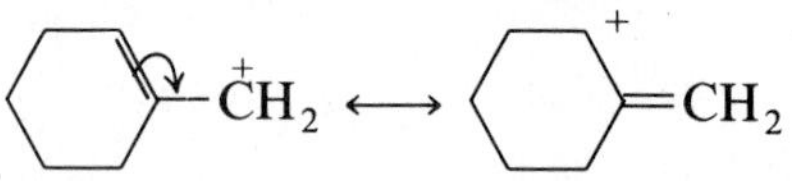

EXAMPLE

Write resonance formulas, if any, for the following carbocations:

(a) $CH_3CH_2\overset{+}{C}HCH\!=\!CH_2$ (d)

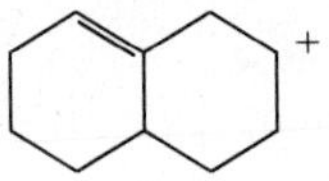

(b) $CH_3\overset{+}{C}HCH_2CH\!=\!CH_2$ (e)

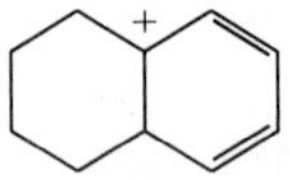

(c)

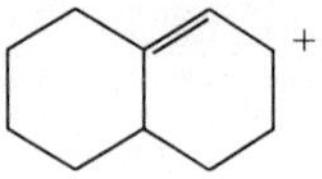

Solution:

(a) Expand the formula to show appropriate single bonds.

Shift pi electrons to + carbon

$$CH_3CH_2\!-\!\overset{+}{C}H\!-\!CH\!=\!CH_2 \longleftrightarrow CH_3CH_2\!-\!CH\!=\!CH\!-\!\overset{+}{C}H_2$$

Expanded formula

(b) A resonance structure is not possible because it would have a five-bonded carbon.

Would be forced to form a five-
bonded carbon atom

$$CH_3\!-\!\overset{+}{C}H\!-\!CH_2\!-\!CH\!=\!CH_2$$

This structure has no resonance formulas because the pi bond is not on the carbon adjacent to the positive carbon. This carbocation is not an allylic cation.

(c)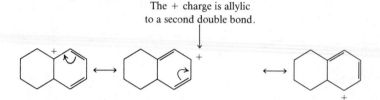

(d) , no resonance structures; see **(b)**

(e) Three carbons help share the positive charge.

The + charge is allylic
to a second double bond.

PROBLEM 4.4. Write resonance structures for the following compounds or ions:

(a) $CH_3CH_2CH=CH\overset{+}{C}HCH_3$ **(b)** $CH_2=CHCH=CH\overset{+}{C}HCH_3$

(c)

(d)

4.4 LINE FORMULAS

Cyclic compounds are conveniently represented by polygon formulas. Similarly, open-chain compounds can be represented by *line formulas*. A few examples follow:

means $\begin{array}{c}CH_2\\ \| \\ CH_2\end{array}$

means $H_3C\overset{CH}{\diagup}\overset{\displaystyle =}{}\overset{}{\diagdown}CH_2$

means $\begin{array}{c}HC\overset{\displaystyle =CH_2}{\diagup}\\ | \\ HC\diagdown CH_2\end{array}$

Methyl group

Isopropyl group

means

$$CH_3$$
$$|$$
$$CH$$
$$H_2C \qquad CH_2$$
$$H_2C \qquad CH_2$$
$$CH$$
$$|$$
$$CH$$
$$H_3C \qquad CH_3$$

We often use these line formulas for the sake of consistency when describing reactions in which open-chain compounds are converted to cyclic compounds.

PROBLEM 4.5.   Convert the following line formulas to condensed structural formulas:

(a)   (b)   (c)

4.5 THE DIELS – ALDER REACTION

A **cycloaddition reaction** is a reaction in which alkenes, dienes, or polyenes undergo an addition reaction to yield a cyclic compound.

A Cycloaddition Reaction:

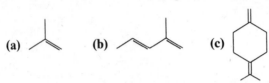

$$CH_2 \qquad CH_2$$
$$\| \quad + \quad \| \quad \xrightarrow{\text{light}} \quad \square$$
$$CH_2 \qquad CH_2$$

Ethylene   Cyclobutane

or, using line formulas,

$$\| \quad + \quad \| \quad \xrightarrow{\text{light}} \quad \square$$

One of the most useful cycloaddition reactions is the reaction of a conjugated diene with an alkene or an alkyne, called the **dienophile** ("diene-lover"), to yield a cyclohexene ring.

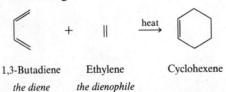

1,3-Butadiene   Ethylene   Cyclohexene

the diene   *the dienophile*

This type of reaction is called the **Diels–Alder reaction** after Otto Diels and Kurt Alder, the German chemists who developed it. These chemists jointly received the 1950 Nobel Prize in Chemistry for their work in this area. A Diels–Alder reaction is a type of 1,4-addition reaction in which the double bonds of the diene are converted to one double bond at the central position of the four-carbon diene system.

The Diels–Alder reaction is versatile. A large number of different types of six-membered rings can be formed.

Three Diels–Alder Reactions:

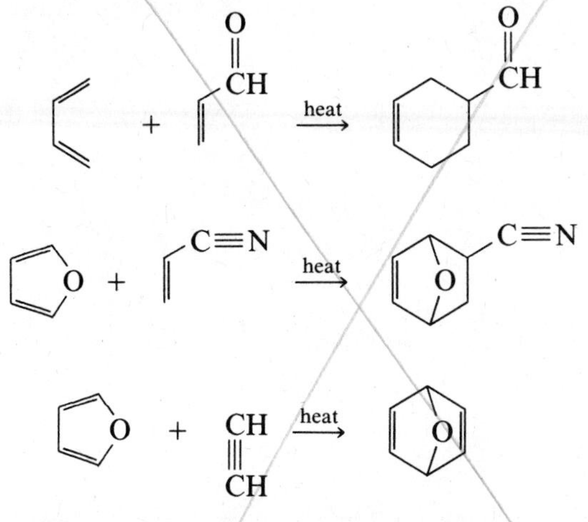

In analyzing any Diels–Alder reaction, we first look for the conjugated diene structure, which may be part of a more complex structure, as in the second and third examples given above. The diene structure adds to one pi bond of an alkene or alkyne structure, which also may be part of a more complex structure. A cyclohexene ring is formed.

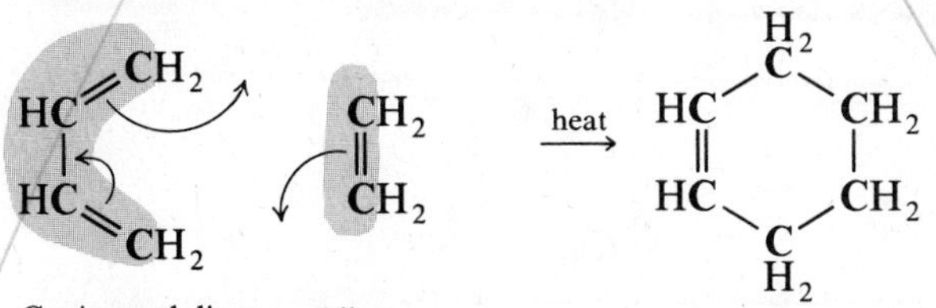

Conjugated diene   Alkene

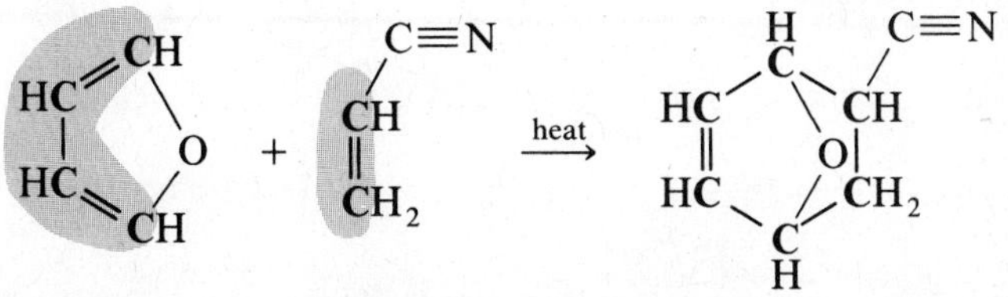

PROBLEM 4.6. Predict the products. (*Hint:* Begin by rewriting the formulas in the proper position for cyclization.)

(a) 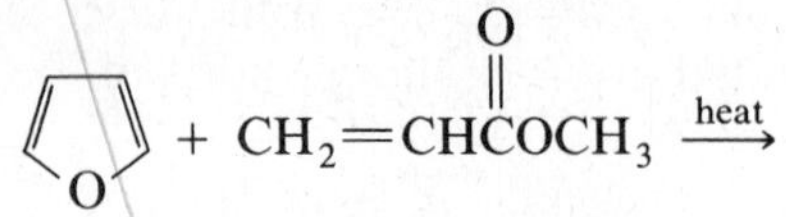 [furan] + CH$_2$=CHCOCH$_3$ $\xrightarrow{\text{heat}}$

(b) 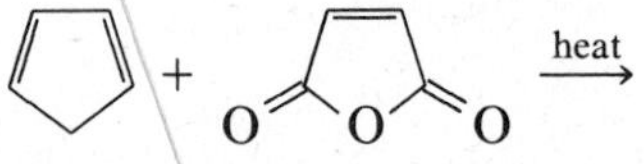 [cyclopentadiene] + [maleic anhydride] $\xrightarrow{\text{heat}}$

To determine the reactants needed to synthesize a cyclohexene by a Diels–Alder reaction, first identify the cyclohexene ring and then circle the part of the structure that came from a diene.

(1) 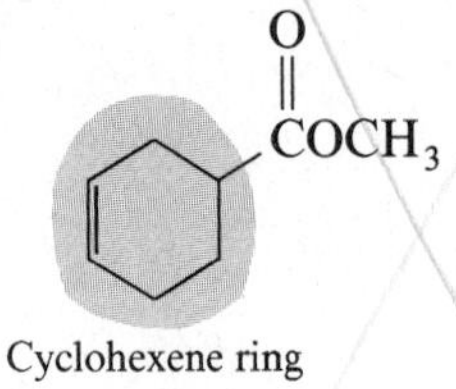 [structure with COCH$_3$]

Cyclohexene ring

(2) 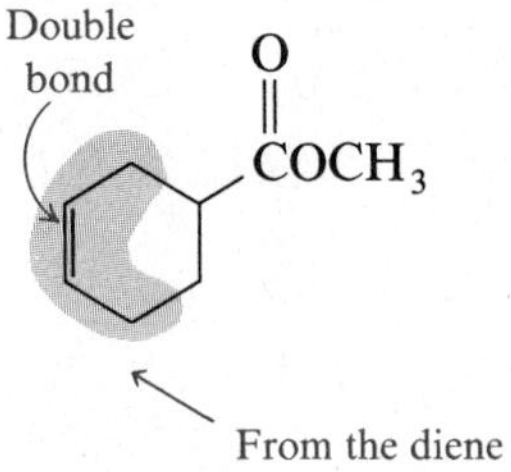 Double bond [structure with COCH$_3$]

From the diene

(3) [structure with COCH$_3$]

The bonds that are formed

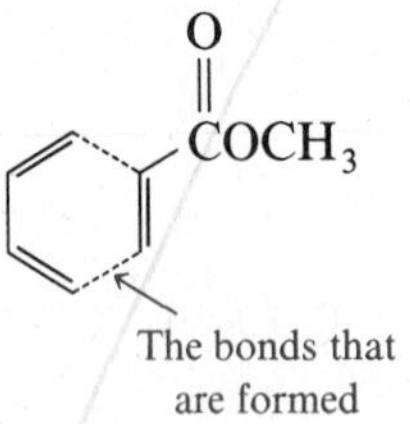

(4) The reaction: [diene] + [dienophile with COCH$_3$] $\xrightarrow{\text{heat}}$

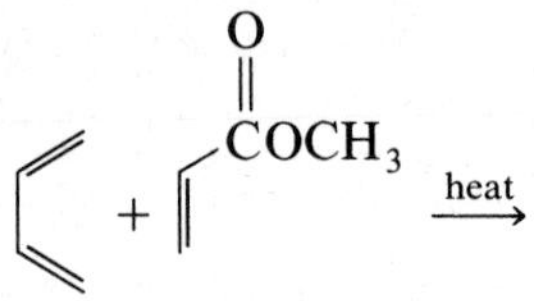

PROBLEM 4.7. Write equations for the Diels–Alder syntheses of the following compounds.

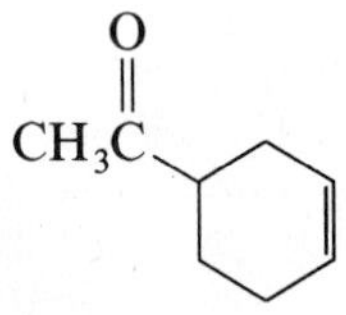

(a) 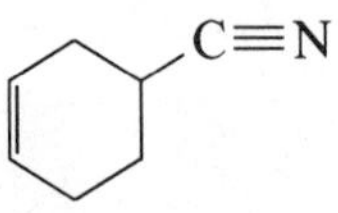 [structure with C≡N]

(b) CH$_3$C [structure]

(c) [bicyclic structure]

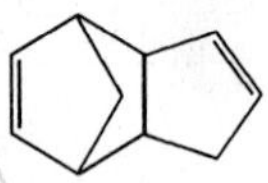

4.6 POLYMERIZATION OF DIENES

Conjugated dienes can be polymerized by 1,4-addition reactions.

$$n\text{CH}_2=\text{CH}-\text{CH}=\text{CH}_2 \xrightarrow{\text{catalyst}} (\text{CH}_2\text{CH}=\text{CHCH}_2)_n$$

1,3-Butadiene   Polybutadiene

Natural rubber, the all-cis form of polyisoprene, is discussed in Point of Interest 4. The first practical synthetic rubber, polychloroprene (Neoprene), was developed in 1932. Neoprene is an oil-resistant, chemically inert polymer used to manufacture washers, flexible tubing, hoses, and containers for petroleum products. It is made by the 1,4-polymerization of 2-chloro-1,3-butadiene.

$$n\,CH_2\!\!=\!\!CHC\!\!=\!\!CH_2 \xrightarrow{\text{catalyst}} \left(\!\!\!-CH_2CH\!\!=\!\!CCH_2\!-\!\!\!\right)_n$$
$$\overset{\displaystyle Cl}{|} \qquad\qquad\qquad\qquad \overset{\displaystyle Cl}{|}$$

2-Chloro-1,3-butadiene   Polychloroprene
(Chloroprene)   (Neoprene)

SUMMARY

Hydrocarbons with more than one double bond are called **dienes**, **trienes**, etc., up to **polyenes**. They are named with the ending -*adiene*, -*atriene*, and so forth.

Two or more double bonds can be **isolated** or **conjugated**. Pi-electron interactions occur between conjugated double bonds but not between isolated double bonds.

Conjugated double bonds are more stable than isolated double bonds.

Conjugated dienes can undergo **1,4-addition reactions**, as well as **1,2-addition reactions**, as shown in Section 4.2. The reason for 1,4-addition is that either addition reaction proceeds by way of an **allylic carbocation**, a cation containing $C\!\!=\!\!C\!-\!\overset{+}{C}$. Allylic cations are more stable than alkyl carbocations because of resonance stabilization.

Open-chain compounds can be represented by **line formulas**, which are similar to polygon formulas for cyclic compounds.

A **cycloaddition reaction** is an addition of unsaturated compounds that yields a cyclic product. A **Diels–Alder reaction** is a cycloaddition reaction of a conjugated diene and an unsaturated compound called a **dienophile**. The product is a cyclohexene.

Conjugated dienes can undergo 1,4-addition **polymerization reactions**.

Table 4.1 summarizes the reactions of conjugated dienes.

TABLE 4.1   SUMMARY OF REACTIONS OF CONJUGATED DIENES

Reactants		Products	Section reference
$CH_2=CHCH=CH_2 + X_2$	$\xrightarrow{\text{1,2-addition}}$	$\overset{\displaystyle X \quad X}{\underset{\displaystyle \mid \quad \mid}{CH_2CHCH=CH_2}}$	4.2
$CH_2=CHCH=CH_2 + X_2$	$\xrightarrow{\text{1,4-addition}}$	$\overset{\displaystyle X \qquad X}{\underset{\displaystyle \mid \qquad \mid}{CH_2CH=CHCH_2}}$	4.2
$CH_2=CHCH=CH_2 + 2X_2$	$\xrightarrow{\hspace{1cm}}$	$\overset{\displaystyle X\ X\quad X\ X}{\underset{\displaystyle \mid\ \mid\quad \mid\ \mid}{CH_2CH-CHCH_2}}$	4.2
⧵⧸ + ‖	$\xrightarrow[\text{Diels–Alder}^a]{\text{heat}}$	⬡	4.5
$n\,CH_2=\overset{\displaystyle R}{\underset{\displaystyle \mid}{C}}-CH=CH_2$	$\xrightarrow{\text{polymerization}^b}$	$\left(\!\!-CH_2\overset{\displaystyle R}{\underset{\displaystyle \mid}{C}}=CHCH_2-\!\!\right)_{\!n}$	4.6

^{a} The diene and the dienophile are usually substituted.

^{b} The diene may contain other substituents.

STUDY PROBLEMS

4.8. Draw formulas for the stereoisomers (cis-trans isomers) of 2,4-hexadiene.

4.9. Name the following compounds:

(a) $\overset{\displaystyle CH_3 \qquad CH=CH_2}{\underset{\displaystyle H \qquad\quad H}{C=C}}$

(b) $CH_3\overset{\displaystyle Cl}{\underset{\displaystyle \mid}{C}}=CHCH=CH_2$

(c) $\begin{array}{c} H_3C \diagdown \quad \diagup CH_2 \\ C \\ \mid \\ C \\ H_3C \diagup \quad \diagdown CH_2 \end{array}$

(d) H_3C⬡

(e) $CH_2=CH\overset{\displaystyle CH_3}{\underset{\displaystyle \mid}{C}}=CHCH_2\overset{\displaystyle CH_3}{\underset{\displaystyle \mid}{C}}=CH_2$

(f) ⬡

4.10. Draw condensed structural formulas for the following compounds:
   (a) 1,3-butadiene
   (b) isoprene
   (c) 1-chloro-1,4-cyclohexadiene
   (d) 2,3,4,5-tetrachloro-1,3,5-hexatriene

4.11. Which of the following compounds contain conjugated double bonds and which contain isolated double bonds?

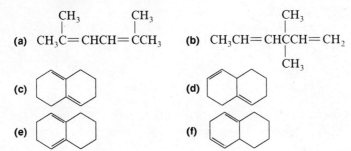

(a) $CH_3\overset{\underset{\displaystyle |}{CH_3}}{C}=CHCH=\overset{\underset{\displaystyle |}{CH_3}}{C}CH_3$

(b) $CH_3CH=CH\overset{\underset{\displaystyle |}{CH_3}}{\underset{\underset{\displaystyle |}{CH_3}}{C}}CH=CH_2$

(c)

(d)

(e)

(f)

4.12. Draw the formula for an isomer of each of the following dienes in which the two double bonds are conjugated:

(a) $CH_2=CH\overset{\underset{\displaystyle |}{CH_3}}{C}HCH=CH_2$

(b) $CH_2=CH\overset{\underset{\displaystyle |}{CH_2CH=CH_2}}{C}HCH_2CH_3$

(c)

(d)

4.13. Complete the following equations showing both the 1,2- and 1,4-addition products:

(a) $CH_2=CH-$⬡$\ + 1Br_2 \longrightarrow$

(b) ⬡⬡ $+ 1Br_2 \longrightarrow$

(c) ⬠ $+ 1Br_2 \longrightarrow$

(d) ⬡⬡⬡ $+ 1Br_2 \longrightarrow$

(e) ⬡⬡⬡ $+ 2Br_2 \longrightarrow$

(f) CH_3O-⬡$-CH=CHCH=CH-$⬡$-OCH_3 + 1HBr \longrightarrow$

4.14. Write resonance structures for the following carbocations:

(a) ⬡⬡$^+$

(b) ⬠$^+$

(c) $\overset{+}{C}H_2CH=CHCH=CH_2$

(d) ⬠$_+$

4.15. Write flow equations showing each step in the mechanisms of the following reactions. (Be sure to show the formation of both 1,2- and 1,4-addition products.)

(a) $CH_2{=}\overset{\displaystyle CH_3}{\underset{\displaystyle |}{C}}{-}CH{=}CH_2$ + 1.0 mol Br_2 $\longrightarrow$

(b)  + 1.0 mol HBr $\longrightarrow$

4.16. Draw a condensed structural formula for each of the following line formulas. Be sure to retain the proper stereochemistry in your answer.

(a) 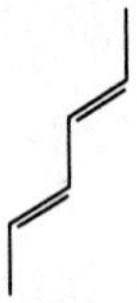   (b)

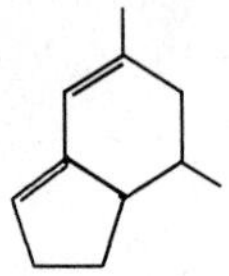

CH_2OH

(c) 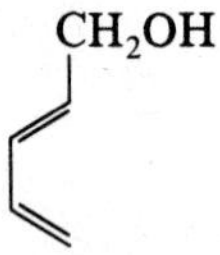   (d)

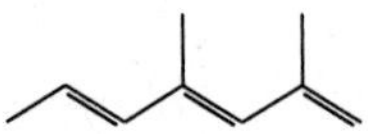

4.17. Convert each of the following formulas to a line formula:

(a) 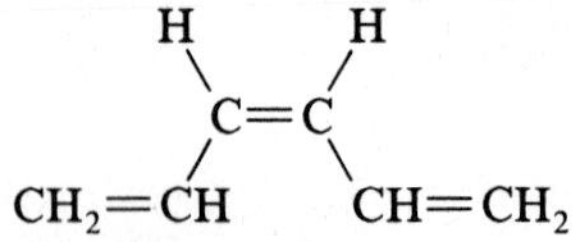
$$\overset{\displaystyle H}{\underset{\displaystyle CH_2{=}CH}{\diagdown}}C{=}C\overset{\displaystyle H}{\underset{\displaystyle CH{=}CH_2}{\diagup}}$$

(b) 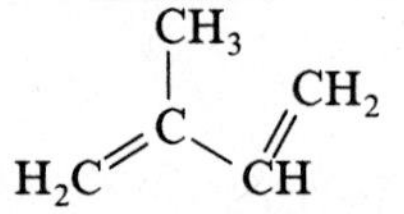
$$\underset{\displaystyle H_2C}{\overset{\displaystyle CH_3}{\diagdown}}C\overset{\displaystyle CH_2}{\underset{\displaystyle CH}{\diagup}}$$

(c) $CH_3\overset{\displaystyle O}{\overset{\displaystyle \|}{C}}CH_3$

(d) $CH_2{=}CH\overset{\displaystyle O}{\overset{\displaystyle \|}{C}}OCH_2CH_3$

4.18. Complete the following equations showing the Diels–Alder products.

(a) 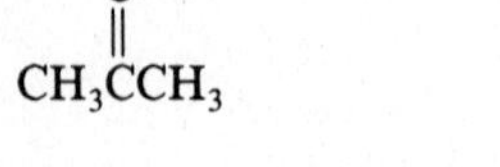 + $\xrightarrow{\text{heat}}$

(b) 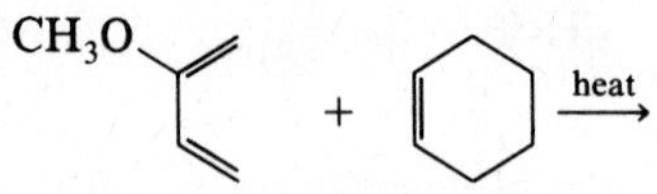 + $\xrightarrow{\text{heat}}$

(c)  + $CH_3O\overset{\displaystyle O}{\overset{\displaystyle \|}{C}}C{\equiv}C\overset{\displaystyle O}{\overset{\displaystyle \|}{C}}OCH_3$ $\xrightarrow{\text{heat}}$

(d) 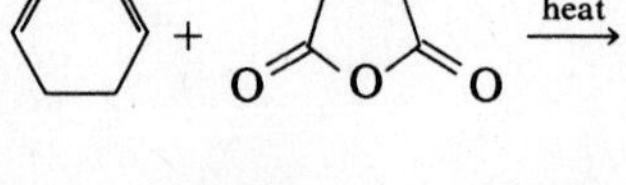 + $\xrightarrow{\text{heat}}$

4.19. What diene and dienophile would be needed to prepare each of the following substituted cyclohexenes by a Diels–Alder reaction?

(a)

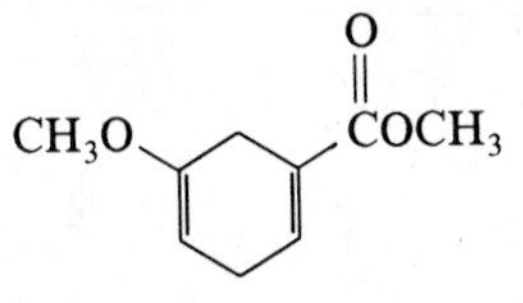

(b)

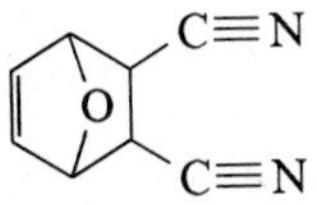

(c)

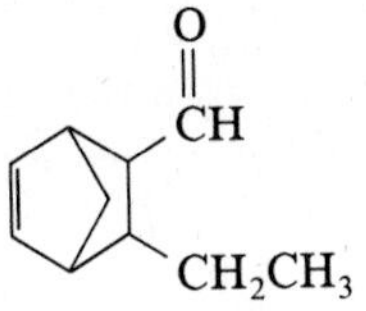

(d)

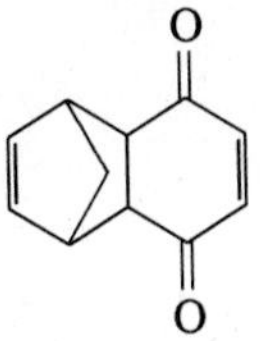

4.20. Circle the isoprene units in the following terpenes and substituted terpenes. (Terpenes are discussed in Point of Interest 4.)

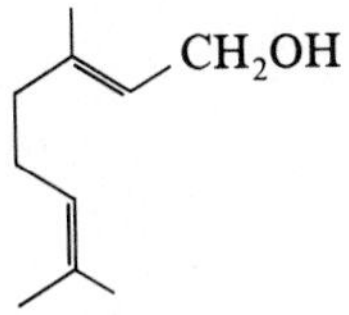

Geraniol

in roses

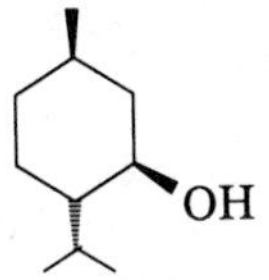

Menthol

in mint

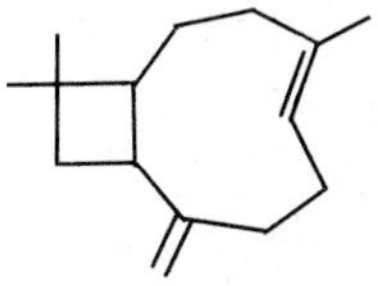

Caryophyllene

in oil of cloves

POINT OF INTEREST 4

Terpenes and Natural Rubber

Terpenes are hydrocarbons composed of repeating units of isoprene's carbon skeleton.

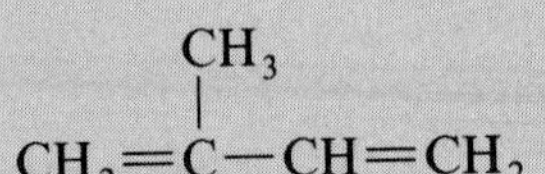

2-Methyl-1,3-butadiene

(isoprene, C_5H_8)

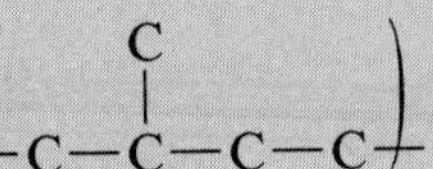

The carbon skeleton

of an isoprene unit

Terpenes are found in both plant and animal tissues but are more abundant in plants. The term terpene is derived from the Greek word *terebinthos*, meaning "turpentine tree." Terpenes are the compounds that confer the characteristic odors to pine, eucalyptus, camphor, and citrus trees and plants such as mint. *Essential oils*, which are the odorous steam-volatile components from plants and flowers and are used in perfume and colognes, are mixtures of hydrocarbons, alcohols, ethers, aldehydes, and esters, most of which are terpenes.

Carbon 1 of isoprene is called the head carbon; carbon 4 is called the tail carbon. In terpenes, the isoprene units are joined together from the head carbon of

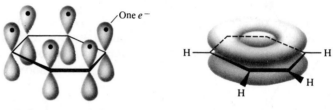

A flat benzene ring showing sigma bonds only

Each C has a *p* orbital perpendicular to the ring (H's not shown)

The *p* orbitals overlap equally with their neighbors

Figure 5.1 The bonding in benzene.

result of this overlap is a molecular orbital shaped like two doughnuts, one on top of the other. Because the electronic charge is evenly distributed, the molecular orbital is referred to as an **aromatic pi cloud**.

Because we cannot represent benzene's structure adequately with conventional line-bond formulas, you might encounter any of the following formulas for the benzene ring. These formulas are equivalent, and all refer to the same structure. In this text we use two types of benzene formulas—those with circles in the ring emphasizing the aromatic pi cloud and those with lines for bonds (the Kekulé symbols) when we want to keep track of electrons.

Formulas for Benzene:

The circle represents the aromatic pi cloud.

C. Stability of the Benzene Ring

Because delocalization of electronic charge stabilizes a structure, the benzene ring is more stable (lower energy) than a hypothetical cyclic triene. The energy of stabilization for an aromatic compound, called the **resonance energy**, is 36 kcal/mol for benzene.

The value for the resonance energy of an aromatic compound can be determined experimentally. Cyclohexene liberates 28.6 kcal/mol when hydrogenated. The hypothetical cyclohexatriene would therefore theoretically liberate 3×28.6, or 85.8, kcal/mol. However, benzene liberates only 49.8 kcal/mol. The difference, 36.0 kcal/mol, is the resonance energy.

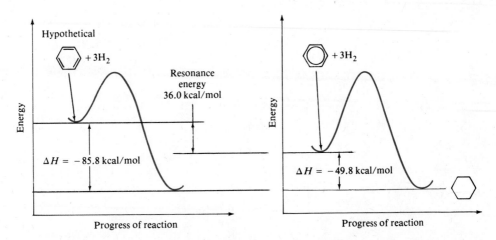

Figure 5.2 shows these energy changes graphically. The resonance stabilization of the benzene ring is the principal reason that benzene does not undergo alkene addition reactions. If a reagent is added to the ring, the aromaticity would be lost

Figure 5.2 Benzene is lower in energy (more stable) than the hypothetical cyclohexatriene. The hydrogenation of benzene thus liberates less energy than would the hydrogenation of cyclohexatriene.

and a less stable product would result.

 does not occur

---

PROBLEM 5.2. Anthracene liberates 116.2 kcal/mol when hydrogenated. What is the resonance energy of anthracene?

Anthracene

D. The Hückel 4n + 2 Rule

To be aromatic, a compound must contain sp^2-hybridized atoms (usually carbon or nitrogen) in a ring system. This hybridization gives rise to the alternate single and double bonds in the line-bond formula. However, not all cyclic compounds containing sp^2-hybridized atoms in a ring system are aromatic. For example, the following two compounds are *not* aromatic.

Cyclobutadiene   Cyclooctatetraene

In 1931 the German chemist Erich Hückel put forth the following rule to summarize which compounds are aromatic and which are not.

HÜCKEL RULE. To be aromatic, a monocyclic compound with alternate single and double bonds must contain $4n + 2$ pi electrons in the cyclic conjugated system, where n is an integer (0, 1, 2, 3, etc.).

According to the Hückel rule, a compound containing 2, 6, 10, or 14 pi electrons can be aromatic, but a compound with 4, 8, or 12 pi electrons cannot.

Aromatic   Not aromatic

6 pi electrons   8 pi electrons
$4n + 2$, where $n = 1$ $4n$, where $n = 2$

The Hückel rule works very well for most common aromatic compounds. Why the Hückel rule works can be explained by modern molecular orbital theories, which we do not discuss in this text.

5.2 NOMENCLATURE OF SUBSTITUTED BENZENES

The original trivial names of many common aromatic compounds have been accepted by the IUPAC as systematic names. A few of these are listed in Table 5.1.

Aside from the trivial names, we usually name monosubstituted benzenes with -*benzene* as the parent. The name of the substituent is prefixed to the parent name.

Ethylbenzene   Nitrobenzene   Chlorobenzene

A disubstituted benzene can be named with prefix numbers or by the **ortho, meta, para system**, which shows the positional relationships of the two groups to each other on the ring.

ortho (o)   meta (m)   para (p)
  (1,2)   (1,3)   (1,4)

TABLE 5.1 TRIVIAL NAMES OF SOME COMMON SUBSTITUTED BENZENES

Formula	Name	Formula	Name
$\langle\bigcirc\rangle$—CH_3	Toluene	$\langle\bigcirc\rangle$—OH	Phenol
H_3C—$\langle\bigcirc\rangle$—CH_3	*para*-Xylene	$\langle\bigcirc\rangle$—NH_2	Aniline
$\langle\bigcirc\rangle$—$CH{=}CH_2$	Styrene	$\langle\bigcirc\rangle$—$\overset{\overset{O}{\|\|}}{C}OH$	Benzoic acid

The following examples show how these prefixes are used.

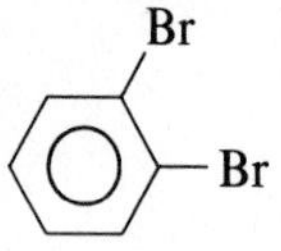

| *o*-Dibromobenzene | *m*-Chlorophenol | *p*-Methylaniline |
| (1,2-dibromobenzene) | (3-chlorophenol) | (4-methylaniline) |

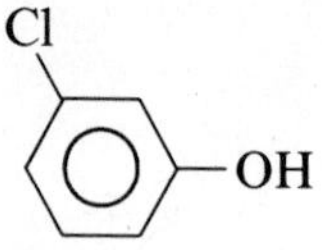

In the second and third examples, a substituted benzene is used as the parent. In these compounds, the principal substituent of the parent ($-OH$ in phenol and $-NH_2$ in aniline) is considered to be at position 1.

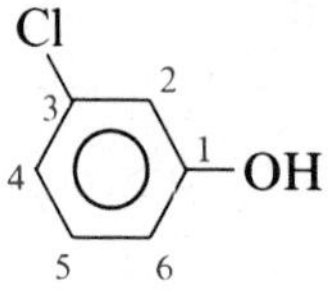

We need not use "1" in the name 3-chlorophenol.

The o, m, p system is used only for disubstituted benzenes and never for substituted cyclohexanes or other compounds.

PROBLEM 5.3. Referring to Table 5.1, name the following compounds by the o, m, p system.

(a)   (b) 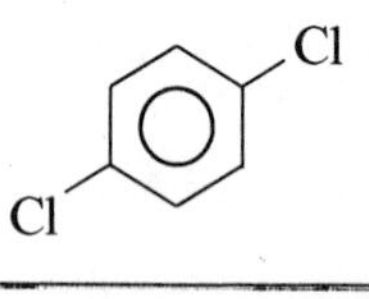   (c)

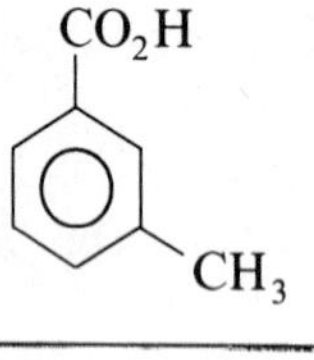

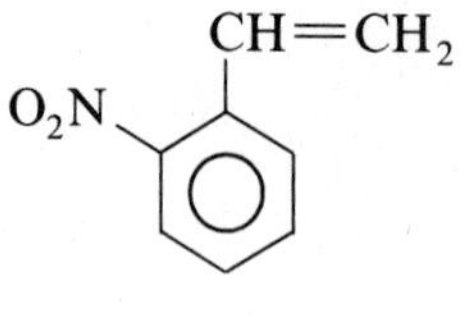

PROBLEM 5.4. Referring to Table 5.1, draw structural (polygon) formulas for the following compounds:

(a) *p*-diethylbenzene

(b) *o*-xylene

(c) *p*-aminobenzoic acid (the amino group is $-NH_2$)

A benzene with more than two substituents is more conveniently named using prefix numbers to specify the location of the groups on the ring. Again, if phenol, aniline, or other substituted benzene is used as the parent, the principal substituent is considered to be at position 1.

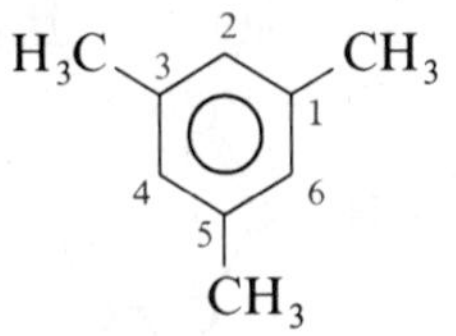

H₃C — CH₃ (positions 2, 3, 1, 4, 5, 6) CH₃

H₃C — OH (positions 6, 5, 1, 4, 3, 2) CH₂CH₃

1,3,5-Trimethylbenzene

(*not a dimethyltoluene,
which would be an inconsistent
grouping of the methyls*)

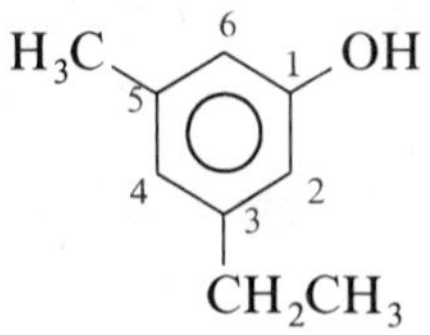

3-Ethyl-5-methylphenol

When a benzene ring is bonded to a chain of more than six carbons or to a more important parent, the ring is referred to as a **phenyl group**.

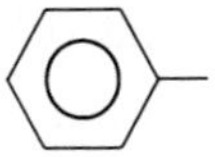

$$CH_3$$
$$\big|$$
$$CHCH_2CH_2CH_2CH_2CH_3$$

The phenyl group

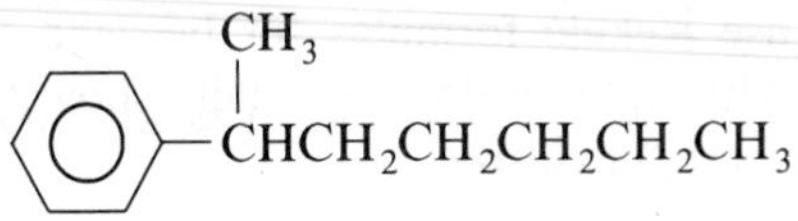

2-Phenylheptane

PROBLEM 5.5. Name the following compounds:

(a) 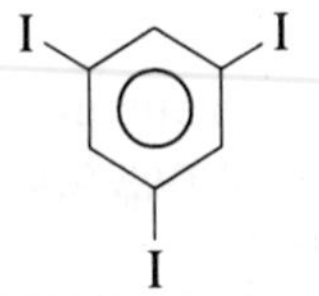 I — ring — I / I

(b) 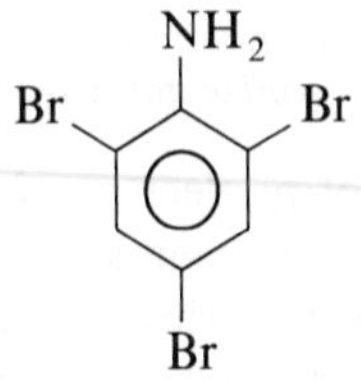 Br — ring(NH₂) — Br / Br

(c) 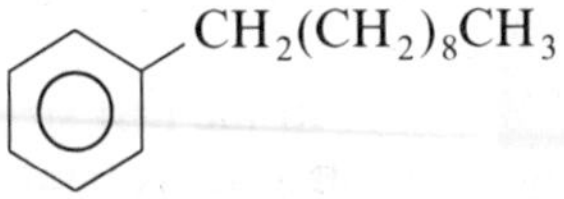 ring — CH₂(CH₂)₈CH₃

PROBLEM 5.6. Draw structures for the following names:

(a) 2,6-dibromophenol   (b) phenylacetylene

(c) *o*-chloroaniline   (d) 2,4,6-trinitrophenol

5.3 ELECTROPHILIC AROMATIC SUBSTITUTION REACTIONS

Benzene does not undergo addition reactions as alkenes do, but substitution reactions of benzene are common. In the equation that follows, a Br has been substituted for a ring H, hence the name *substitution reaction*. Because the substitution has occurred on an aromatic ring, the reaction is referred to as an **aromatic substitution reaction**.

Energy

arranged product—again, isopropylbenzene.

$$\langle\bigcirc\rangle + CH_3CH_2CH_2Cl \xrightarrow{AlCl_3}$$

1-Chloropropane

mainly $\langle\bigcirc\rangle—CH(CH_3)_2$ and very little $\langle\bigcirc\rangle—CH_2CH_2CH_3$

PROBLEM 5.8. Treatment of benzene with a mixture of propylene ($CH_3CH{=}CH_2$) and HCl yields an alkylbenzene. Write steps for the mechanism of this reaction. (*Hint:* The first step is the generation of an electrophile; see Section 3.6.)

E. Acylation

Friedel and Crafts developed a reaction similar to the alkylation reaction just presented. This type of reaction is called a **Friedel–Crafts acylation** because an *acyl group*, not an alkyl group, is substituted on the benzene ring.

Acylation:

$$\langle\bigcirc\rangle + \underset{\text{An acid chloride}}{R\overset{\overset{\displaystyle O}{\|}}{C}Cl} \xrightarrow[\text{warm}]{AlCl_3} \underset{\text{An acylbenzene}}{\langle\bigcirc\rangle-\overset{\overset{\displaystyle O}{\|}}{C}R} + HCl$$

An acyl group

Acylation reactions do not proceed by way of a carbocation and no rearrangement occurs. By combining acylation reactions with a reaction that reduces the carbonyl group ($C{=}O$) to CH_2, a chemist can synthesize an *n*-alkylbenzene without worry of rearrangements. For example, *n*-propylbenzene can be prepared by the following sequence of reactions.

$$\langle\bigcirc\rangle + CH_3CH_2\overset{\overset{\displaystyle O}{\|}}{C}Cl \xrightarrow[\text{warm}]{AlCl_3} \langle\bigcirc\rangle-\overset{\overset{\displaystyle O}{\|}}{C}CH_2CH_3 + HCl$$

$$\Bigg\downarrow \begin{array}{l}\text{Zn/Hg, HCl}\\ \text{(a reducing agent)}\\ \text{heat}\end{array}$$

$$\langle\bigcirc\rangle-CH_2CH_2CH_3$$

n-Propylbenzene

TABLE 5.2 SUMMARY OF ELECTROPHILIC
AROMATIC SUBSTITUTION REACTIONS OF BENZENE

Reaction	Reactants	Products
Halogenation	$\bigcirc$ + X_2 $\xrightarrow{FeX_3}$	$\bigcirc$—X + HX
Nitration	$\bigcirc$ + HNO_3 $\xrightarrow{H_2SO_4}$	$\bigcirc$—NO_2 + H_2O
Sulfonation	$\bigcirc$ + H_2SO_4 $\underset{}{\overset{SO_3}{\rightleftarrows}}$	$\bigcirc$—SO_3H + H_2O
Alkylation	$\bigcirc$ + RCl $\xrightarrow{AlCl_3}$	$\bigcirc$—R + HCl
Acylation	$\bigcirc$ + $\overset{\overset{O}{\|\|}}{R}CCl$ $\xrightarrow{AlCl_3}$	$\bigcirc$—$\overset{\overset{O}{\|\|}}{C}R$ + HCl

Table 5.2 summarizes the electrophilic aromatic substitution reactions of benzene.

5.4 SECOND SUBSTITUTION REACTIONS

A. Position of Substitution

A substituted benzene can undergo a second substitution to yield a disubstituted benzene. *The structure of the first substituent determines the position on the ring of the second substitution.* For example, a methyl group on the ring directs a second incoming substituent mainly to the ortho and para positions, while a nitro group on the ring directs a second incoming substituent mainly to the meta position.

o,p-Director

$\bigcirc$—CH_3 $\xrightarrow[H_2SO_4]{HNO_3}$ $\overset{NO_2}{\bigcirc}$—CH_3 + O_2N—$\bigcirc$—CH_3

Toluene o-Nitrotoluene (60%) p-Nitrotoluene (40%)

m-Director

$\bigcirc$—NO_2 $\xrightarrow[H_2SO_4]{HNO_3}$ $\overset{O_2N}{\bigcirc}$—NO_2 + 7% o isomer

Nitrobenzene m-Dinitrobenzene (93%)

TABLE 5.3 DIRECTIVE INFLUENCE OF COMMON SUBSTITUENTS[a]

Ortho,para-directors (activating)	Meta-directors (deactivating)
$-\overset{..}{N}H_2, -\overset{..}{N}HR, -\overset{..}{N}R_2$ $-\overset{..}{O}H, -\overset{..}{O}R$ $\overset{O}{\overset{\|}{-\overset{..}{N}HCR}}$ benzene ring $-R$ $-\overset{..}{\underset{..}{X}} :$ (deactivating)	$\overset{O}{\overset{\|}{-CH,}} \overset{O}{\overset{\|}{-CR,}} \overset{O}{\overset{\|}{-COH,}} \overset{O}{\overset{\|}{-COR}}$ $-C\equiv N$ $\overset{O}{\overset{\|}{-\underset{\underset{\|}{O}}{S}OH}}$ $-\overset{+}{N}\overset{O^-}{\underset{O}{\diagdown}}$ $-\overset{+}{N}R_3$

[a] The activating and deactivating influence of the substituents will be discussed in Sections 5.4D and 5.4E.

A substituent on a benzene ring can be classified as either an **ortho,para-director** or a **meta-director**. As the names imply, an ortho,para-director directs the next substituent onto the ring at a position either ortho or para to the director. A methyl group is an ortho,para-director. Meta-directors cause the next substituent on the ring to be positioned meta to the director.

Table 5.3 lists the common substituents and their directive effects. Note that all the o,p-directors except alkyl and aryl groups contain at least one pair of unshared valence electrons on the atom bonded directly to the ring.

Ortho,para-directors Circled:

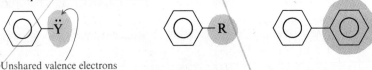

Unshared valence electrons

All the meta-directors have an atom with a partial positive charge or an atom with a full positive ionic charge bonded to the ring.

Meta-directors Circled:

Positive or partial positive charge

PROBLEM 5.9. Without referring to Table 5.3, predict the products of the second substitutions.

(a) ⟨O⟩—OCH_3 + Br_2 $\xrightarrow{FeBr_3}$

(b) ⟨O⟩—$\overset{\displaystyle O}{\overset{\|}{C}}OCH_3$ + Cl_2 $\xrightarrow{FeCl_3}$

(c) ⟨O⟩—$O\overset{\displaystyle O}{\overset{\|}{C}}CH_3$ + HNO_3 $\xrightarrow{H_2SO_4}$

B. Mechanism for Ortho,Para Substitution

All the o, p-directors are electron donors, either by resonance or by the inductive effect. For example, phenol contains the o, p-directing —OH group. Phenol can be brominated at the ortho position and the para position; very little meta substitution occurs. By examining the resonance structures of the intermediates, we can see why substitution occurs at o, p-positions and not at the m-position. Figure 5.4 shows these resonance structures.

In ortho or para substitution, the oxygen of the —OH group helps share the positive charge and thus stabilizes the intermediate by helping delocalize the charge. If substitution occurs in the meta position, the oxygen cannot help share the charge. The meta intermediate and the transition state leading to it are higher in energy than are those leading to o, p-substitution. Since the activation energy is higher, reaction by the meta path is less likely. The energy profiles for these reaction paths are shown in Figure 5.5.

In substitution reactions of alkylbenzenes, electron density is donated to the ring by the inductive effect to help stabilize the ortho and para intermediates.

⟨O⟩—Ÿ   ⟨O⟩—$\overset{\displaystyle H}{\underset{\displaystyle H}{C}}$← H

Electron donation   *Electron donation*
by resonance   *by inductive effect*

Ortho attack:

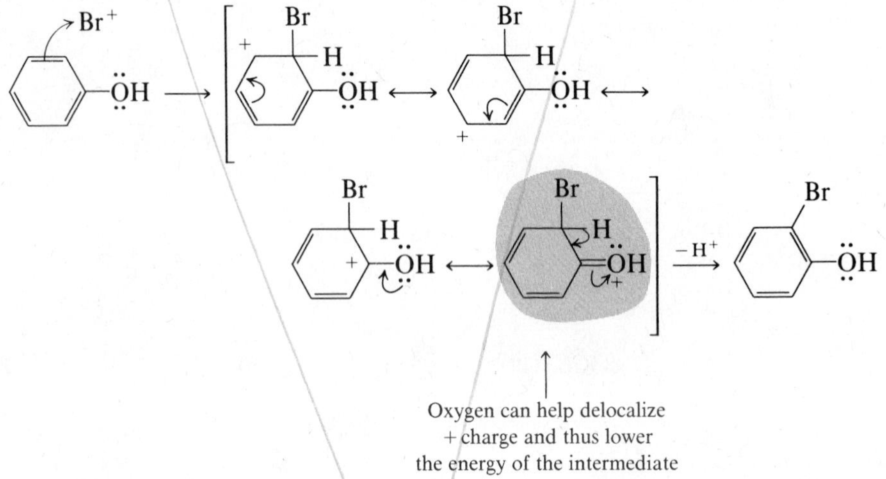

Oxygen can help delocalize
+ charge and thus lower
the energy of the intermediate

Para attack:

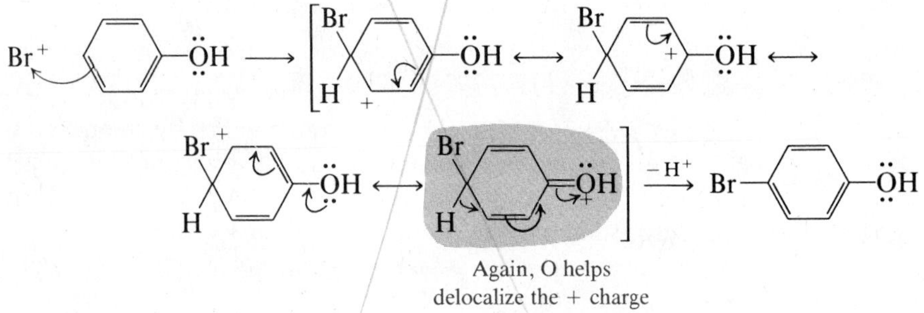

Again, O helps
delocalize the + charge

Meta attack:

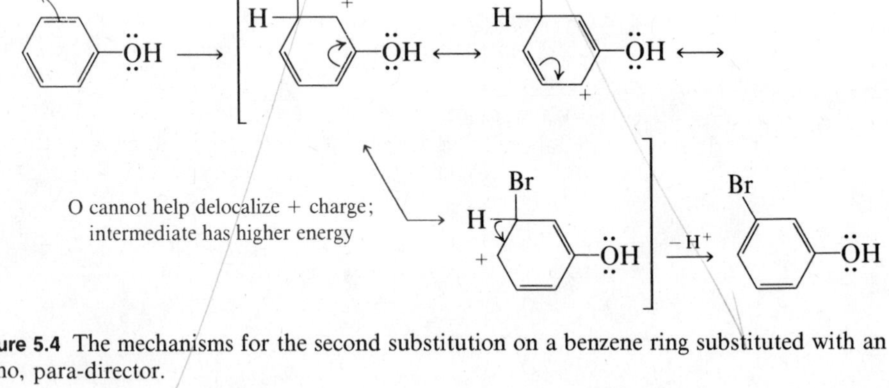

O cannot help delocalize + charge;
intermediate has higher energy

Figure 5.4 The mechanisms for the second substitution on a benzene ring substituted with an ortho, para-director.

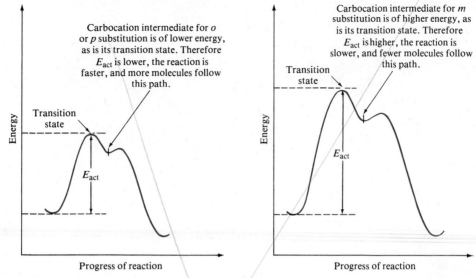

Figure 5.5 Energy diagrams showing why benzene substituted with an o, p-director undergoes very little meta substitution.

PROBLEM 5.10.   Write an equation, including resonance structures of the intermediate, to show the role of the phenyl group in the para bromination of biphenyl.

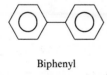

Biphenyl

C. Mechanism for Meta Substitution

A meta-director has a positive or partially positive atom bonded to the benzene ring. Figure 5.6 shows resonance structures for the intermediates in the second substitution of nitrobenzene, which contains the meta-directing nitro group.

In substitution reactions of nitrobenzene, the nitro group adds no stabilization to any intermediate. In fact, the intermediates for ortho or para substitution and transition states leading to them are *less* stable (higher energy) because one resonance structure contains positive charges on adjacent atoms. Attack, therefore, occurs preferentially at the meta position because its transition state and intermediate contain no adjacent positive charges. With this pathway, the positive charge is less localized and the intermediate is of lower energy than those for attack at the other positions.

Meta attack:

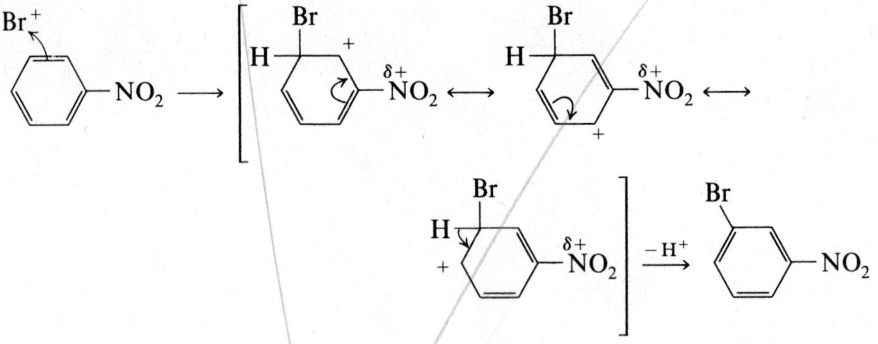

Ortho attack:

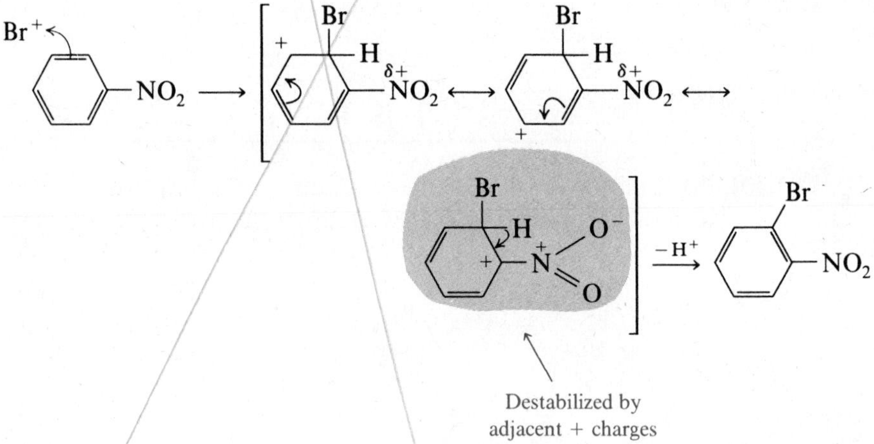

Destabilized by
adjacent + charges

Para attack:

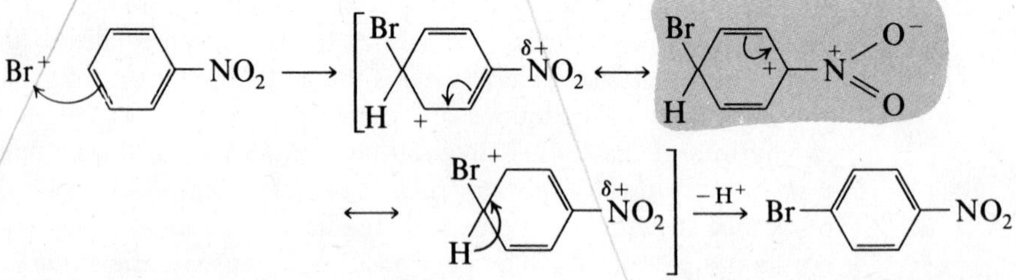

Figure 5.6 The mechanisms for the second substitution on a benzene ring substituted with a meta-director.

D. Rate of the Second Substitution

Besides influencing the position of a second substitution, the first substituent influences the rate, or speed, at which the second substitution reaction takes place. All ortho,para-directors except the halogens increase the rate of electrophilic substitution compared to the rate for benzene itself. The increased rate allows the reaction to proceed under milder conditions (lower temperature or no catalyst).

o,p-Directors Activate the Ring:

$$\text{C}_6\text{H}_5\text{—NH}_2 + 3\text{Br}_2 \xrightarrow[-3\text{HBr}]{\text{H}_2\text{O}} \text{tribromoaniline}$$

No catalyst needed

All *o* and *p* positions readily brominated

By contrast, all meta-directors deactivate the ring—the second substitution is slower than that of benzene itself.

m-Directors Deactivate the Ring:

$$\text{C}_6\text{H}_5\text{—NO}_2 + \text{HNO}_3 \xrightarrow[\text{heat}]{\text{H}_2\text{SO}_4} \text{dinitrobenzene}$$

Much slower to nitrate than is benzene

A meta-directing substituent may deactivate the ring sufficiently that a second substitution does not occur. For example, nitrobenzene does not undergo Friedel–Crafts alkylations.

$$\text{C}_6\text{H}_5\text{—NO}_2 + \text{CH}_3\text{CH}_2\text{Cl} \xrightarrow{\text{AlCl}_3} \text{no reaction}$$

Although halogen substituents are *o, p*-directors, they deactivate the benzene ring toward electrophilic substitution.

Halogens Deactivate the Ring:

$$\text{C}_6\text{H}_5\text{—X} + \text{HNO}_3 \xrightarrow[\text{heat}]{\text{H}_2\text{SO}_4} \text{o-X-nitrobenzene} + \text{O}_2\text{N—C}_6\text{H}_4\text{—X}$$

Slower to nitrate than benzene

TABLE 5.4 REACTIVITY OF SUBSTITUTED BENZENES

Substituent	Position of second substitution	Reactivity
o, p-Director	*o* and *p*	Activated
m-Director	*m*	Deactivated
Halogen (X)	*o* and *p*	Deactivated

Table 5.4 summarizes the reactivity of substituted benzenes.

PROBLEM 5.11. When aniline is treated with a strong acid, such as a mixture of HCl and $AlCl_3$, and then is brominated, the meta-brominated product is formed instead of 2,4,6-tribromoaniline. Explain why this occurs.

E. Reasons for the Differences in Reactivity of the Substituted Benzenes

Let us examine the reasons for the relative reactivities of various substituted benzenes in electrophilic aromatic substitution reactions. In the mechanism of the reaction, the first step (attack of the electrophile) is the slow step; the second step (loss of H^+) is fast. Because a multistep reaction sequence cannot proceed faster than its slowest step, the rate of the overall reaction depends on the rate of the first step only. The slowest step in any reaction sequence is called the **rate-limiting step**, or **rate-determining step**.

The overall rate of reaction depends on the rate of this step.

A benzene ring substituted with an ortho,para-director other than a halogen is more reactive than benzene because the intermediate carbocation and the transition state leading to the carbocation are stabilized relative to an unsubstituted intermediate. The lower E_{act} for the first step means an increased rate of reaction. (Refer to Figure 5.4 to see how the unshared valence electrons of the *o, p*-directing group add additional resonance stabilization.)

A benzene ring substituted with a meta-director is less reactive than benzene is because of the positive or partially positive atom adjacent to the positively charged ring of the intermediate carbocation. The nearness of these positive charges raises the energy of the intermediate and also the activation energy of step 1 in the reaction. The higher E_{act} means a decreased rate.

A halogen substituent, like other ortho,para-directors, directs a second substituent to the ortho or para position because of donation of an unshared pair of valence electrons by resonance.

The other common ortho,para-directors ($-OH$, $-OR$, $-NR_2$, etc.) contain an oxygen or a nitrogen atom bonded to a carbon of a benzene ring. These elements, like carbon, contain their valence electrons in the second shell. When an oxygen or a nitrogen forms a pi bond with carbon, the same-sized orbitals can overlap readily.

Chlorine, bromine, and iodine contain their valence electrons in the third, fourth, and fifth shells, respectively. These orbitals are farther from the nucleus than carbon's second-shell orbitals. The overlap of p orbitals between carbon and chlorine, bromine, or iodine is not as good as the overlap between the p orbitals of carbon and nitrogen or oxygen.

With poorer overlap, the halobenzene intermediates have less resonance stabilization. Therefore, resonance stabilization is relatively less important and the electronegativity of the halogen substituent assumes relatively greater importance. Withdrawal of electron density from the ring by the inductive effect destabilizes the intermediate and thus deactivates the ring toward electrophilic attack. The E_{act} is thus higher and the rate is lower.

5.5 ALKYLBENZENES

A carbon adjacent to an aromatic ring is called a **benzylic carbon**. Although alkyl groups are not generally reactive, atoms and groups on a benzylic carbon exhibit unusual reactivity.

The reason for the reactivity at this position is that a benzylic carbocation or a benzylic free radical is resonance stabilized.

Because carbocations and free radicals are common intermediates in organic reactions, the activation energies for a number of reactions are lower for benzylic positions. For example, alkylbenzenes containing at least one benzylic hydrogen are readily oxidized at this position to yield benzoic acid. The reaction is called **side-chain oxidation**.

Side-Chain Oxidation:

B. Aromatic He

In bi
are c
adjec
elem

Free-radical bromination also occurs at the benzylic position. (This type of bromination is a free-radical reaction and does not occur under the same reaction conditions as electrophilic bromination of the ring.)

$$\text{C}_6\text{H}_5-\text{CH}_3 \xrightarrow[\text{light}]{\text{Br}_2} \text{C}_6\text{H}_5-\text{CH}_2\text{Br} + \text{HBr}$$

$$\text{C}_6\text{H}_5-\text{CH}_2\text{CH}_3 \xrightarrow[\text{light}]{\text{Br}_2} \text{C}_6\text{H}_5-\overset{\text{Br}}{\underset{|}{\text{C}}}\text{HCH}_3 + \text{HBr}$$

$$\text{C}_6\text{H}_5-\text{CH}_2\text{CH}_2\text{CH}_3 \xrightarrow[\text{light}]{\text{Br}_2} \text{C}_6\text{H}_5-\overset{\text{Br}}{\underset{|}{\text{C}}}\text{HCH}_2\text{CH}_3 + \text{HBr}$$

T
F
in be
withd
to be
pyridi
Anotl
conta

It is interesting that the allylic position is similarly activated toward free-radical halogenation.

$$\text{CH}_2{=}\text{CH}{-}\overset{\text{H}}{\underset{|}{\text{CH}_2}} \xrightarrow[\text{light}]{\text{Br}_2} \text{CH}_2{=}\text{CH}{-}\overset{\text{Br}}{\underset{|}{\text{CH}_2}} + \text{HBr}$$

Propene   3-Bromo-1-propene

Resonance Structures of the Allyl Free-Radical Intermediate:

$$\overset{\cdot}{\text{CH}_2}{=}\text{CH}{-}\overset{\cdot}{\text{CH}_2} \longleftrightarrow \overset{\cdot}{\text{CH}_2}{-}\text{CH}{=}\text{CH}_2$$

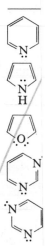

PROBLEM 5.12.   Predict the organic products of the following reactions:

(a) [naphthalene with CH(CH₃)₂ substituent] $\xrightarrow[\text{(2) H}^+]{\text{(1) KMnO}_4, \text{ heat}}$

(b) $\text{C}_6\text{H}_5-\text{CH(CH}_3)_2 \xrightarrow[\text{light}]{\text{Br}_2}$

(c) $\text{C}_6\text{H}_5-\text{CH(CH}_3)_2 \xrightarrow[\text{FeBr}_3]{\text{Br}_2}$

"Ring sy

5.6 OTHE

A. Polycyc

PROBLEM 6.2. Which of the following formulas represent enantiomeric pairs, which represent the same compound, and which represent structural isomers? (*Hint:* You must move these structures around in space in order to see if they are superimposable or are nonsuperimposable mirror images. Therefore, we recommend that you use molecular models.)

(a)

$$CH_3$$
$$H-C\text{\tiny IIIII}CH_2Cl$$
$$H$$

$$CH_3$$
$$H-C\text{\tiny IIIII}H$$
$$CH_2Cl$$

(b)

$$\overset{O}{\underset{\parallel}{C}}OH$$
$$HO-C\text{\tiny IIIII}H$$
$$CH_2OH$$

$$\overset{O}{\underset{\parallel}{C}}OH$$
$$H-C\text{\tiny IIIII}CH_2OH$$
$$OH$$

(c)

$$CH_3$$
$$HO-C\text{\tiny IIIII}H$$
$$CH_2OH$$

$$CH_2OH$$
$$HO-C\text{\tiny IIIII}H$$
$$CH_3$$

C. Representations of Chiral Molecules

In order to represent three-dimensional molecules on two-dimensional paper, we must adopt certain conventions. We have already been using one method for representing three-dimensionality of molecules (see Section 1.4C).

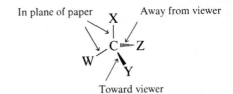

In plane of paper   Away from viewer

Toward viewer

In addition to this method, you may encounter formulas in which a horizontal bond means that the group is projected out of the plane *toward* the viewer. In these formulas, a vertical bond means that the group is projected *away* from the viewer.

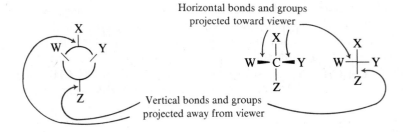

Horizontal bonds and groups projected toward viewer

Vertical bonds and groups projected away from viewer

Formulas in which chiral carbons are shown only as the center of a cross and not as the letter C, as in the last formula, are called **Fischer projections**, after the German chemist Emil Fischer (1852–1919), who developed this type of formula. Fischer has been called "the father of modern organic chemistry" because of his numerous contributions to the development of structural theory, especially concerning carbohydrates and proteins. Fischer was awarded the Nobel Prize in Chemistry in 1902.

Three-Dimensional Formulas for the Enantiomers of Glyceraldehyde:

$$
\begin{array}{cc}
\overset{\displaystyle O}{\underset{\displaystyle \|}{}} & \overset{\displaystyle O}{\underset{\displaystyle \|}{}} \\
CH & CH \\
H\text{—}C\text{—}OH & HO\text{—}C\text{—}H \\
CH_2OH & CH_2OH
\end{array}
$$

Fischer Projections for the Same Pair:

$$
\begin{array}{cc}
\overset{\displaystyle O}{\underset{\displaystyle \|}{}} & \overset{\displaystyle O}{\underset{\displaystyle \|}{}} \\
CH & CH \\
H\text{—}\!\!+\!\!\text{—}OH & HO\text{—}\!\!+\!\!\text{—}H \\
CH_2OH & CH_2OH
\end{array}
$$

6.2 PROPERTIES OF ENANTIOMERS

Enantiomers have structures so similar that they have almost identical physical and chemical properties. For example, they have the same melting point, the same boiling point, and the same solubility in ordinary solvents. In fact, all their properties are the same with only two exceptions: how they behave in a chiral environment and how they interact with plane-polarized light.

A. Behavior in Chiral Environments

A reaction in chiral environment means a reaction with a chiral reactant, reaction with the aid of a chiral catalyst, or reaction in a chiral solvent. Biological systems constitute chiral environments because they are composed almost entirely of chiral organic molecules. Therefore, we can expect enantiomers to undergo different reactions in living systems. Indeed, the following two enantiomeric amino acids behave entirely differently in the body: (*S*)-alanine is incorporated into protein molecules, while (*R*)-alanine is oxidized to a keto acid and metabolized. [The designations (*R*) and (*S*) are used to identify the configurations of the two enantiomers; their definitions will be discussed in Section 6.3A.]

(*S*)-Alanine   (*R*)-Alanine

PROBLEM 6.3.   The following formula represents carvone, a compound found in caraway seed and dill seed oils. Carvone has the odor of caraway seeds, while its enantiomer has the odor of spearmint. Draw a polygon formula for the enantiomer and star the chiral carbon.

Carvone

in oil of caraway

PROBLEM 6.4.   Are the odor receptors in the nose chiral or achiral? Explain.

B. Interaction with Plane-Polarized Light

Enantiomers interact equally but in an opposite manner with plane-polarized light. This property of enantiomers aided chemists in determining the structures of carbon compounds in the nineteenth century. Before we can discuss the interaction further, we must first discuss plane-polarized light.

Ordinary light consists of waves vibrating in all planes perpendicular to the direction in which the light beam is traveling. Plane-polarized light consists of waves vibrating in only one of these planes. Figure 6.5 depicts ordinary light and plane-polarized light.

Ordinary light is polarized by passing it through a filtering device that blocks out all but one plane of vibration. A *Nicol prism*, a type of calcium carbonate crystal, is an effective filter. Another filter is a Polaroid sheet, made from properly oriented crystals of an organic compound embedded in a plastic film. (Polaroid sunglasses are also made from Polaroid sheets.)

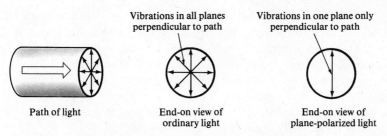

Figure 6.5 Ordinary light vibrates in all planes perpendicular to its path; plane-polarized light vibrates in only one plane, called the plane of polarization.

In 1815 the French chemist Jean-Baptiste Biot discovered that when plane-polarized light is passed through a solution of some substances, the plane of polarization changes. For example, when plane-polarized light passes through a solution of cane sugar, the plane is rotated about 66° to a new plane. By contrast, a solution of table salt has no effect on the plane of polarization. Compounds that can rotate the plane of polarization of plane-polarized light are called **optically active compounds**. Cane sugar (sucrose) is optically active; table salt (sodium chloride) is not.

The instrument used to measure the rotation of the plane of polarization, called a **polarimeter**, is diagrammed in Figure 6.6. The observed angle of rotation is called α (alpha). This angle of rotation may be to the right or to the left of the original value of $\alpha = 0°$. A compound that rotates the plane of polarization to the right is said to be **dextrorotatory** (Latin *dexter*, "right") and is given a positive ($+$) value for α. A compound that rotates the plane to the left is said to be **levorotatory** (Latin *laevus*, "left") and is given a negative ($-$) value for α.

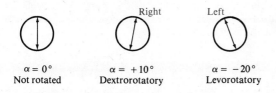

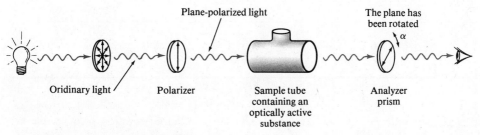

Figure 6.6 Diagram of a polarimeter.

The magnitude of the observed rotation is dependent on a number of factors—nature of the compound, concentration of the solution, cell length, wavelength of light, temperature, and solvent. To compare the values of measurements made under differing conditions, we convert the observed value for the rotation to a **specific rotation, $[\alpha]_D^T$**, where D refers to the wavelength of light usually used (the sodium D line, 589.3 nm) and T refers to the temperature. We also specify the solvent. We can calculate the specific rotation by the following equation. (Here, we assume the sodium D line and a temperature of 20°C.)

$$[\alpha]_D^{20} = \frac{\alpha}{l \times c} \quad \text{(solvent)}$$

where α = observed rotation
   l = length of sample cell in decimeters (1 dm = 0.1 m)
   c = concentration of solution in grams per milliliter

EXAMPLE

Given the following information, calculate the specific rotation of cocaine: 1.80 g of sample is dissolved in 10.0 mL of chloroform and placed in a sample cell 5.00 cm long. At 20°C (sodium D line), the observed rotation is −14.4°.

Solution:

1. Convert the data given to the proper units:

$\alpha = -14.4°$

$l = 5.00 \text{ cm} \times \dfrac{1 \text{ dm}}{10 \text{ cm}} = 0.500 \text{ dm}$

$c = 1.80 \text{ g}/10.0 \text{ mL} = 0.180 \text{ g/mL}$

2. Substitute the information into the equation and solve.

$$[\alpha]_D^{20} = \frac{-14.4°}{0.500 \times 0.180} \quad \text{(chloroform)}$$

$$= -160.0° \quad \text{(chloroform)}$$

PROBLEM 6.5.   What is the specific rotation of each of the following samples (sodium D line, 20°C)?

(a) 1.00 g of sample in 5.00 mL of ethanol solution in a sample tube 5.00 cm long exhibits an observed rotation of +23.0°.

(b) 0.500 g of sample in 6.00 mL of ethanol solution in a sample tube 5.00 cm long exhibits an observed rotation of −3.5°.

C. Racemic Mixtures

In 1848 Louis Pasteur observed that sodium ammonium tartrate forms two types of crystals—mirror images of each other (see Point of Interest 6). Pasteur separated the crystal types and measured the specific rotations of their solutions. He observed that one solution rotated the plane of polarization of plane-polarized light to the right and the other to the left by the same number of degrees. A solution of equal amounts of the two crystal types did not rotate the plane of polarization at all. Let us restate Pasteur's observations in modern terminology.

1. The dextrorotatory enantiomer of a pair rotates the plane of polarization of plane-polarized light to the right.
2. The levorotatory enantiomer rotates the plane of polarization to the left by the same number of degrees.
3. A **racemic mixture**, a 50 : 50 mixture of an enantiomeric pair, does not rotate the plane of polarization because the rotation of one enantiomer is canceled by the rotation of the other.

A dextrorotatory enantiomer is designated by the prefix ($+$) before the name; a levorotatory enantiomer's name is prefixed with ($-$). For example, the main odorous enantiomer in oil of caraway is ($+$)-carvone and the main odorous compound in oil of spearmint is ($-$)-carvone. The name of a racemic mixture is often prefixed with ($\pm$); therefore, a mixture of one part ($+$)-carvone and one part ($-$)-carvone would be ($\pm$)-carvone.

6.3 CONFIGURATION OF ENANTIOMERS

The prefix ($+$) or ($-$) tells us the direction in which an enantiomer rotates the plane of polarization of plane-polarized light when that physical constant is determined in a polarimeter. However, these prefixes do not tell us the configuration around a chiral carbon. Unfortunately, there is no simple relationship between direction of rotation and configuration.

Glyceraldehyde

One enantiomer is dextrorotatory and one is levorotatory. Which is which?

In 1951 J. M. Bijvoet at the University of Utrecht subjected ($+$)-glyceraldehyde to x-ray diffraction, which can reveal molecular configuration. As a result of his studies, he reported the following configuration for ($+$)-glyceraldehyde:

$$O$$
$$\parallel$$
$$CH$$
$$|$$
$$C \text{---} CH_2OH$$

$$H \diagup \diagdown OH$$

($+$)-Glyceraldehyde

Therefore, the
($-$) enantiomer is

$$O$$
$$\parallel$$
$$CH$$
$$|$$
$$HOCH_2 \text{---} C$$
$$\diagup \diagdown H$$
$$OH$$

($-$)-Glyceraldehyde

Over the years, other optically active compounds have been related chemically to ($+$)-glyceraldehyde or have been studied instrumentally to reveal their configurations. For example, when the —CHO group of ($+$)-glyceraldehyde is oxidized to a —CO$_2$H group, the configuration around the chiral carbon is not changed. Therefore, the configuration of the chiral carbon of the carboxylic acid must be the same as that of the starting aldehyde.

Known configuration →

$$O$$
$$\parallel$$
$$CH$$
$$|$$
$$C \text{---} CH_2OH$$
$$H \diagup \diagdown OH$$

($+$)-Glyceraldehyde

$$\xrightarrow{\text{mild oxidizing agent}}$$

Configuration same as that of reactant →

Oxidized ↙

$$O$$
$$\parallel$$
$$COH$$
$$|$$
$$C \text{---} CH_2OH$$
$$H \diagup \diagdown OH$$

($-$)-Glyceric acid

This particular reaction demonstrates the difficulty in attempting to relate the configuration of the chiral carbon to the sign of rotation of an enantiomer. The structures and configurations of ($+$)-glyceraldehyde and ($-$)-glyceric acid are very similar, yet these two compounds have opposite signs of rotation of plane-polarized light.

A. The (R) and (S) System of Nomenclature for Chiral Compounds

Because it is inconvenient to draw a formula each time we want to specify a configuration, we need a convention for designating the configuration of chiral carbons in the name of a compound. The system we use today is called the **(R) and (S) system** or the **Cahn-Ingold-Prelog system** after the chemists who developed it.

Ranking the Groups In the (R) and (S) system, the four different atoms or groups bonded to a chiral carbon are ranked one through four in order of priority. The rules for assigning the priority of the groups are summarized in Figure 6.7.

(1) Higher atomic number of atom bonded to the chiral carbon means higher priority.

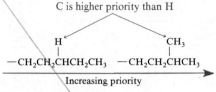

(2) If two atoms are the same, proceed to the second atom or even farther along the chain to the *first point of difference*.

C is higher priority than H

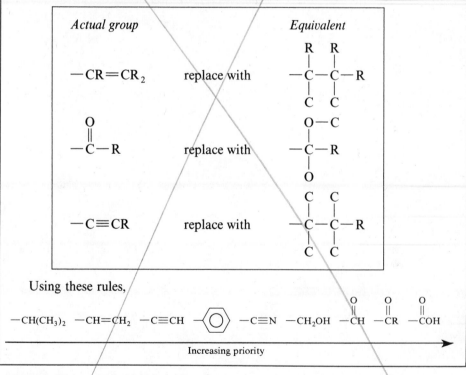

(3) Groups with pi bonds are given single-bond equivalents by duplication or triplication of multiply bonded atoms.

Figure 6.7 Sequence rules for the (*R*) and (*S*) system.

PROBLEM 6.6. Rank the following groups in order of priority according to the (*R*) and (*S*) system:

(a) —N(CH₃)₂   (b) —OCH₂CH₃   (c) —SH   (d) —SCH₃

PROBLEM 6.7. Rank the following groups in order of priority:

(a) $-\overset{\overset{\displaystyle O}{\|}}{O}CCH_3$ (b) $-\overset{\overset{\displaystyle O}{\|}}{C}OCH_2CH_2CH_3$

(c) $-\overset{\overset{\displaystyle O}{\|}}{C}CHI_2$ (d) $-\overset{\overset{\displaystyle O}{\|}}{C}OCH(CH_3)_2$

Assigning (R) or (S) Configuration to a Chiral Carbon   After the groups bonded to the chiral carbon are ranked one through four, the chiral carbon is assigned the (R) or (S) configuration by the following procedure.

RULES FOR DETERMINING (R) AND (S)

    1. Place the molecule so that *the lowest-priority group is in the rear*, projected away from the viewer. (You may find it helpful to use molecular models to

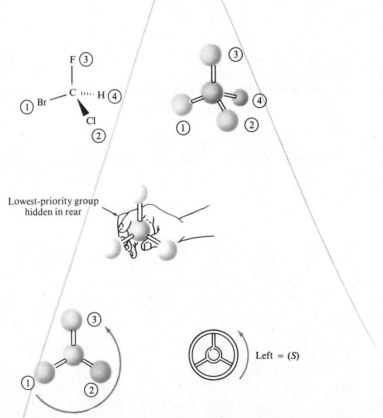

Figure 6.8 Using a molecular model to determine (R) or (S) configuration.

accomplish this task. See Figure 6.8.)

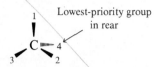

Lowest-priority group
in rear

2. Draw a curved arrow from the highest-priority group to the second-highest group. If the arrow curves in a clockwise direction, the configuration is (*R*) (Latin *rectus*, "right"). If the arrow curves in a counterclockwise direction, the configuration is (*S*) (*sinister*, "left"). An easy way to remember these directions is to equate them with the steering wheel of a car: Clockwise is the same as turning a steering wheel for a *right* turn. Counterclockwise is the same as turning a steering wheel for a *left* turn.

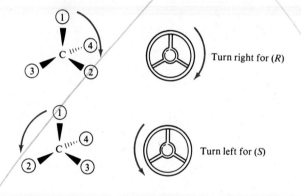

Turn right for (*R*)

Turn left for (*S*)

EXAMPLE

Is the configuration of the following compound (*R*) or (*S*)?

$$
\begin{array}{c}
\text{OH} \\
\mid \\
\text{H}_3\text{C} \diagdown \text{C} \text{\textbardbl}\text{\textemdash} \text{H} \\
\diagdown \\
\text{CH}_2\text{CH}_3
\end{array}
$$

Solution:

1. By atomic number we determine that, of the four atoms bonded to the chiral carbon, —H is lowest priority and the O of the —OH is highest. The two carbons are intermediate; the priority of these groups must be determined using another sequence rule.

$$
\xrightarrow{\quad \text{H} \quad \text{C} \quad \text{C} \quad \text{O} \quad}
$$
Increasing priority

2. To rank the two remaining groups in order of priority, we proceed to the next atoms bonded to these two carbons.

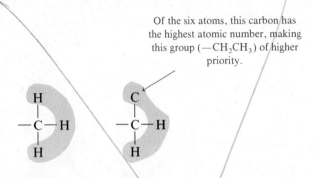

Of the six atoms, this carbon has the highest atomic number, making this group ($-CH_2CH_3$) of higher priority.

The priority ranking of the four groups is:

$$\underrightarrow{\quad -H \quad -CH_3 \quad -CH_2CH_3 \quad -OH \quad}$$
Increasing priority

3. We place the group of lowest priority ($-H$) in the rear, projected away from the viewer.
4. We draw a curved arrow from the highest-priority group to the second-highest-priority group.
5. Because the arrow curves clockwise, the compound has the (R) configuration.

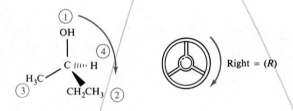

Right = (R)

EXAMPLE

Is the following compound, ($-$)-glyceraldehyde, (R) or (S)?

Solution:

1. By atomic number, we determine that —H is lowest priority and —OH is highest.

2. To differentiate the carbons, we proceed to the first point of difference.

$$\begin{array}{c} O \\ \parallel \\ CH \\ | \\ HOCH_2 \diagdown \overset{C \cdots H}{} \\ OH \end{array}$$

3. Because one group contains a double bond, we use the single bond equivalence shown in Figure 6.7.

$$\begin{array}{ccc} O-H & O & O-C \\ | & \parallel & | \\ -C-H & -CH \equiv -C-H \\ | & & | \\ H & & O \longleftarrow \text{ Higher priority} \end{array}$$

Both groups of three have an O and an H. They differ in the third atom, O in the right group and H in the left group. The O has a higher atomic number than H, making the aldehyde group of higher priority.

The overall ranking is:

$$\xrightarrow[\text{Increasing priority}]{\quad -H \quad -CH_2OH \quad -\overset{\displaystyle O}{\overset{\parallel}{CH}} \quad -OH \quad}$$

4. To complete the assignment of configuration, we draw the compound with the lowest-priority group projected to the rear and draw a curved arrow from the highest-priority group to the second-highest-priority group.

$$\begin{array}{c} O \\ \parallel \\ CH\,② \\ | \\ HOCH_2 \diagdown \overset{C \cdots H}{}④ \\ ③ \quad OH \,① \end{array} \qquad \text{Left} = (S)$$

The compound is (S)-$(-)$-glyceraldehyde.

PROBLEM 6.8. Designate each of the following compounds as (*R*) or (*S*):

(a)
$$
\begin{array}{c}
CH_2CH_3 \\
| \\
HO{-}C{\cdots\cdots}CH_3 \\
| \\
CH_2CH_2CH_3
\end{array}
$$

(b)
$$
\begin{array}{c}
C{\equiv}N \\
| \\
H{\cdots\cdots}C \\
H_3C{\diagup}\quad{\diagdown}COCH_3 \\
\quad\qquad \| \\
\quad\qquad O
\end{array}
$$

(c)
$$
\begin{array}{c}
CH{=}CH_2 \\
| \\
H_3C{\diagup}C{\cdots\cdots}CH_2CH_3 \\
\quad{\diagdown}Cl
\end{array}
$$

(d)
$$
\begin{array}{c}
CH_2Cl \\
| \\
H_3C{\cdots\cdots}C \\
\quad{\diagup}\quad{\diagdown}CH_2CHCl_2 \\
H
\end{array}
$$

B. The (*E*) and (*Z*) System of Nomenclature for Alkenes

The sequence rules are useful for other situations besides chirality. Some alkenes capable of stereoisomerism cannot be designated as cis or trans. For example, the following substituted alkene cannot be called either cis or trans because groups cannot be identified as "same side" or "opposite sides." Yet this alkene can exist as a pair of stereoisomers.

$$
\begin{array}{ccc}
H_3C & & CH_2Cl \\
\diagdown & & \diagup \\
& C{=}C & \\
\diagup & & \diagdown \\
CH_3CH_2 & & CH(CH_3)_2
\end{array}
$$

We cannot call this alkene cis or trans.

A system of nomenclature using the priority sequence rules has been developed to name stereoisomers of this type. The system can be used to name any cis-trans isomeric pair of alkenes, but it must be used when the cis-trans designations are unsuitable.

In the system called the (*E*) and (*Z*) system of nomenclature the two groups on each double-bond carbon are ranked as shown in Figure 6.7. If the higher-priority groups are on the same side of the double bond, the prefix letter (*Z*) (from the German *zusammen*, "together") is prefixed to the compound's name. If the higher-priority groups are on opposite sides of the double bond, the prefix letter (*E*) (*entgegen*, "across") is used. (See Figure 6.9.)

Same side = (*Z*)

$$
\begin{array}{ccc}
{}^1H_3C & & CH_2CH_3{}^1 \\
\diagdown & & \diagup \\
& C{=}C & \\
\diagup & & \diagdown \\
{}^2H & & H^2
\end{array}
$$

(*Z*)-2-Pentene

Opposite sides = (*E*)

$$
\begin{array}{ccc}
{}^2H & & CH_2CH_3{}^1 \\
\diagdown & & \diagup \\
& C{=}C & \\
\diagup & & \diagdown \\
{}^1H_3C & & H^2
\end{array}
$$

(*E*)-2-Pentene

(1) To determine whether an alkene or other compound containing a carbon–carbon double bond is (E) or (Z), rank the substituents on each doubly bonded carbon.

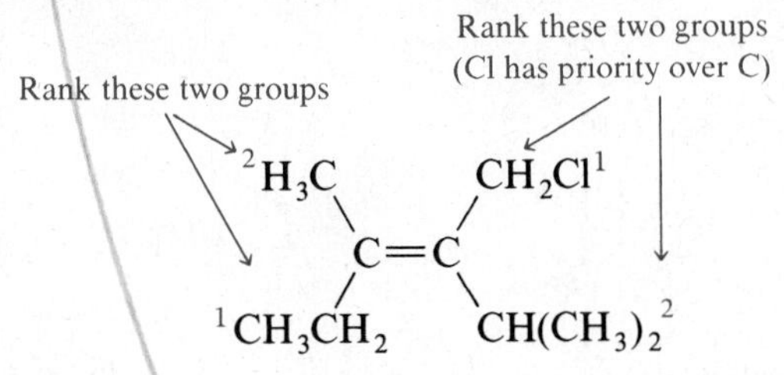

Rank these two groups
(Cl has priority over C)

Rank these two groups

$$^2H_3C \qquad CH_2Cl^1$$
$$C=C$$
$$^1CH_3CH_2 \qquad CH(CH_3)_2^{\,2}$$

(2) Determine whether the two higher-priority groups are on the same side (Z) or opposite sides (E) of the double bond.

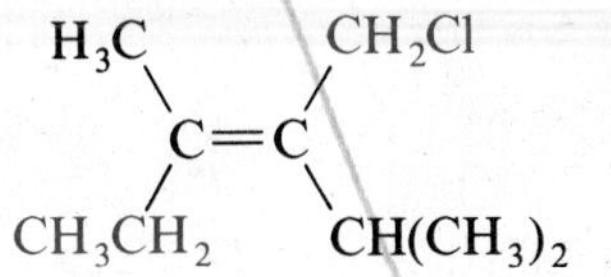

$$H_3C \qquad CH_2Cl$$
$$C=C \qquad\qquad \text{Opposite sides} = (E)$$
$$CH_3CH_2 \qquad CH(CH_3)_2$$

The other isomer would be the (Z) isomer.

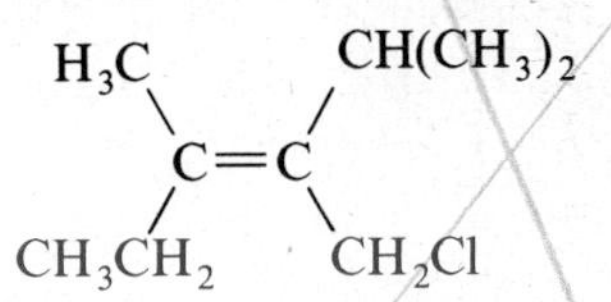

$$H_3C \qquad CH(CH_3)_2$$
$$C=C \qquad\qquad \text{Same side} = (Z)$$
$$CH_3CH_2 \qquad CH_2Cl$$

Figure 6.9 Determination of (E) and (Z) geometry of alkenes.

PROBLEM 6.9. Name the following alkenes, using the (E) and (Z) system of nomenclature.

(a) 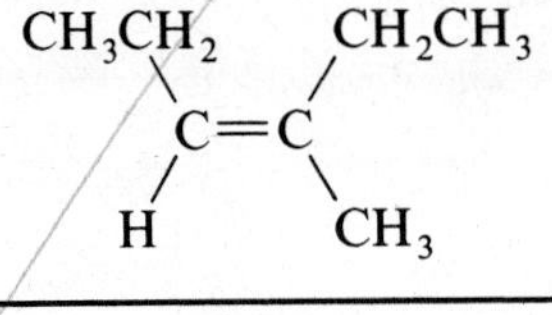
$$CH_3CH_2 \qquad CH_2CH_3$$
$$C=C$$
$$H \qquad CH_3$$

(b) 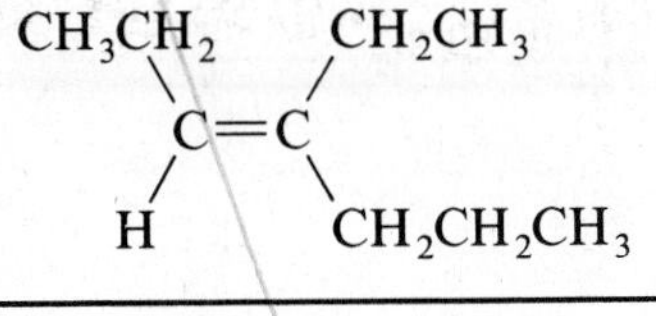
$$CH_3CH_2 \qquad CH_2CH_3$$
$$C=C$$
$$H \qquad CH_2CH_2CH_3$$

6.4 MORE THAN ONE CHIRAL CARBON

Organic compounds can have more than one chiral carbon. For example, the following hydroxy aldehyde has two chiral carbons. To compare the configurations of compounds with more than one chiral carbon, we draw the structures as stretched-out chains instead of true three-dimensional formulas.

A Compound with Two Chiral Carbons:

Group with lowest nomenclature number at top

Horizontal bonds projected toward viewer

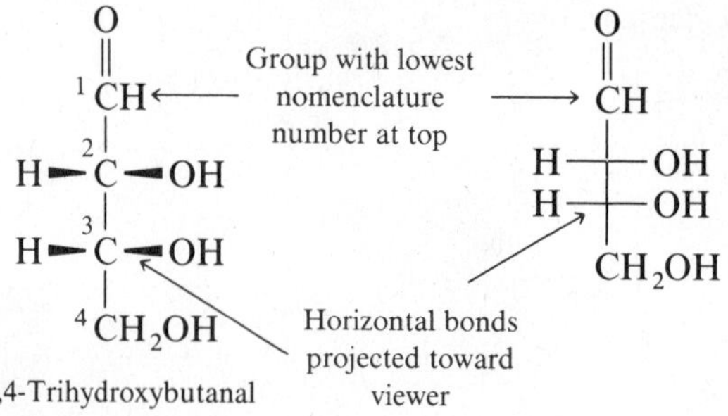

2,3,4-Trihydroxybutanal

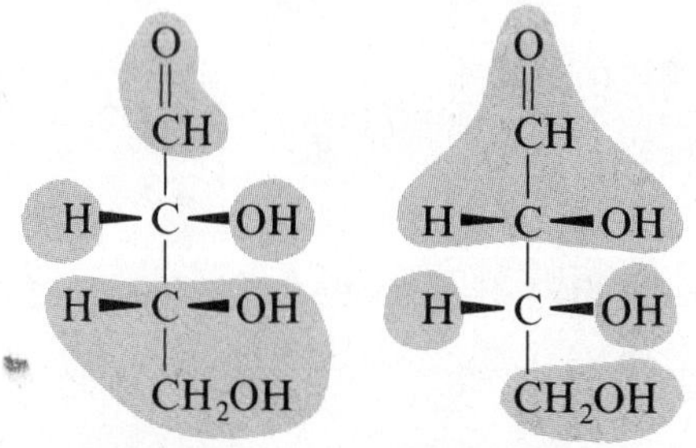

The four groups on carbon 2

The four groups on carbon 3

This structure has two chiral carbons, carbon 2 and carbon 3. Carbon 2 can be either (R) or (S). Carbon 3 can also be (R) or (S). Therefore, four different combinations of configuration are possible.

The Four Possible Combinations of Configuration for a Compound with Two Chiral Carbons:

Individual configuration		Combined configuration
Carbon 2	Carbon 3	
(R)	(R)	$(2R,3R)$
(R)	(S)	$(2R,3S)$
(S)	(R)	$(2S,3R)$
(S)	(S)	$(2S,3S)$

Formulas showing the configurations of these four stereoisomers follow:

$$
\begin{array}{cccc}
\underset{\text{CH}}{\overset{\displaystyle \text{O} \atop \|}{}} & \underset{\text{CH}}{\overset{\displaystyle \text{O} \atop \|}{}} & \underset{\text{CH}}{\overset{\displaystyle \text{O} \atop \|}{}} & \underset{\text{CH}}{\overset{\displaystyle \text{O} \atop \|}{}}
\end{array}
$$

H—C—OH	HO—C—H	H—C—OH	HO—C—H
H—C—OH	HO—C—H	HO—C—H	H—C—OH
CH$_2$OH	CH$_2$OH	CH$_2$OH	CH$_2$OH
(2R,3R)	(2S,3S)	(2R,3S)	(2S,3R)

Enantiomers	*Enantiomers*

To calculate the maximum number of stereoisomers we use the formula 2^n, where n is the number of chiral carbons. For example, a compound with two chiral carbons can form up to 2^2, or four, stereoisomers. A compound with three chiral carbons can exist as up to 2^3, or eight, stereoisomers.

A. Diastereomers

Because enantiomers are mirror images, they must exist as pairs. Looking at the preceding formulas, you can see that the (2R,3R) compound and the (2S,3S) compound are enantiomers, as are the (2R,3S) and (2S,3R) compounds. However, the (2R,3R) compound is not an enantiomer of the (2R,3S) compound, nor is it an enantiomer of the (2S,3R) compound. Stereoisomers that are not enantiomers are called **diastereomers**. The (2R,3S) and (2S,3R) compounds are diastereomers of both the (2R,3R) and (2S,3S) compounds.

Diastereomers for 2,3,4-Trihydroxybutanal:

(2R,3S) and (2R,3R)   (2R,3S) and (2S,3S)
(2S,3R) and (2R,3R)   (2S,3R) and (2S,3S)

Diastereomers are isomers that have different physical and chemical properties, whereas enantiomers are isomers that have the same physical and chemical properties (except for the interactions with other chiral substances and with plane-polarized light).

PROBLEM 6.10.   Identify each of the following pairs as enantiomers or as diastereomers:

(a)

$$
\begin{array}{c}
\overset{\text{O}}{\overset{\|}{\text{C}}}\text{OH} \\
\text{H}\!-\!\text{C}\!-\!\text{OH} \\
\text{H}\!-\!\text{C}\!-\!\text{OH} \\
\overset{\text{C}}{\underset{\|}{\text{O}}}\text{OH}
\end{array}
\qquad
\begin{array}{c}
\overset{\text{O}}{\overset{\|}{\text{C}}}\text{OH} \\
\text{HO}\!-\!\text{C}\!-\!\text{H} \\
\text{H}\!-\!\text{C}\!-\!\text{OH} \\
\overset{\text{C}}{\underset{\|}{\text{O}}}\text{OH}
\end{array}
$$

(b)

$$
\begin{array}{c}
\text{CH}_2\text{Cl} \\
\text{H}\!-\!\text{C}\!-\!\text{Cl} \\
\text{H}\!-\!\text{C}\!-\!\text{Cl} \\
\text{CH}_3
\end{array}
\qquad
\begin{array}{c}
\text{CH}_2\text{Cl} \\
\text{Cl}\!-\!\text{C}\!-\!\text{H} \\
\text{Cl}\!-\!\text{C}\!-\!\text{H} \\
\text{CH}_3
\end{array}
$$

(c)

$$
\begin{array}{c}
\text{CH}_2\text{OH} \\
\text{HO}\!-\!\text{C}\!-\!\text{H} \\
\text{H}\!-\!\text{C}\!-\!\text{Cl} \\
\text{CH}_2\text{OH}
\end{array}
\qquad
\begin{array}{c}
\text{CH}_2\text{OH} \\
\text{H}\!-\!\text{C}\!-\!\text{OH} \\
\text{Cl}\!-\!\text{C}\!-\!\text{H} \\
\text{CH}_2\text{OH}
\end{array}
$$

(d)

$$
\begin{array}{c}
\overset{\text{O}}{\overset{\|}{\text{C}}}\text{H} \\
\text{HO}\!-\!\text{C}\!-\!\text{H} \\
\text{H}\!-\!\text{C}\!-\!\text{OH} \\
\text{H}\!-\!\text{C}\!-\!\text{OH} \\
\text{CH}_2\text{OH}
\end{array}
\qquad
\begin{array}{c}
\overset{\text{O}}{\overset{\|}{\text{C}}}\text{H} \\
\text{H}\!-\!\text{C}\!-\!\text{OH} \\
\text{HO}\!-\!\text{C}\!-\!\text{H} \\
\text{HO}\!-\!\text{C}\!-\!\text{H} \\
\text{CH}_2\text{OH}
\end{array}
$$

B. Assigning (*R*) or (*S*)

To assign the (*R*) or (*S*) configuration to each chiral carbon in a compound with more than one chiral carbon, we must treat each chiral carbon separately.

EXAMPLE

Assign (R) or (S) to each chiral carbon in the following molecule:

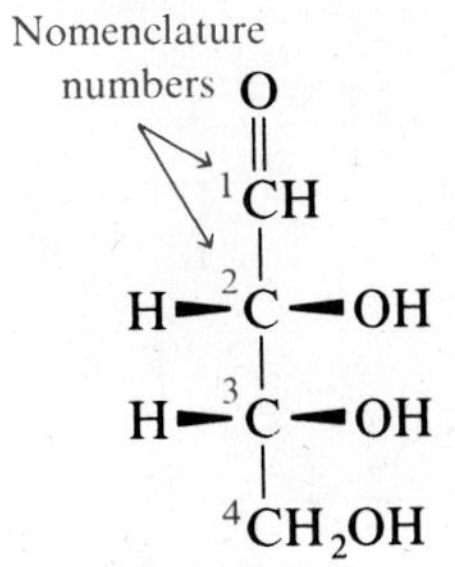

Nomenclature
numbers

$$
\begin{array}{c}
\overset{O}{\underset{\|}{}} \\
^{1}CH \\
H \!-\! \overset{2}{C} \!-\! OH \\
H \!-\! \overset{3}{C} \!-\! OH \\
^{4}CH_2OH
\end{array}
$$

Carbons 2 and 3 are chiral.

Solution: Looking at carbon 2:

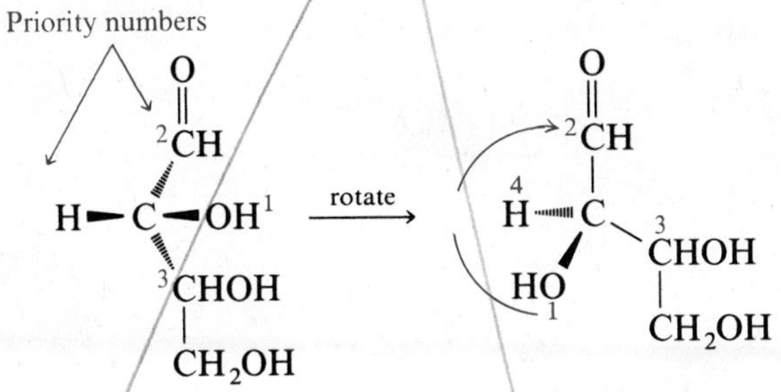

Priority numbers

$$
\begin{array}{c}
\overset{O}{\underset{\|}{}} \\
^{2}CH \\
H \!-\! C \!-\! OH^{1} \\
^{3}CHOH \\
CH_2OH
\end{array}
$$
⟶ rotate ⟶
$$
\begin{array}{c}
\overset{O}{\underset{\|}{}} \\
^{2}CH \\
^{4}H \cdots C \; \overset{3}{CHOH} \\
HO_{1} \\
CH_2OH
\end{array}
$$

(2R)

Looking at carbon 3:

$$
\begin{array}{c}
\overset{O}{\underset{\|}{}} \\
CH \\
^{2}CHOH \\
^{4}H \!-\! C \!-\! OH^{1} \\
\underset{3}{CH_2OH}
\end{array}
$$
⟶ rotate ⟶
$$
\begin{array}{c}
\overset{O}{\underset{\|}{}} \\
CH \\
^{2}CHOH \\
^{4}H \cdots C \; \overset{3}{CH_2OH} \\
HO_{1}
\end{array}
$$

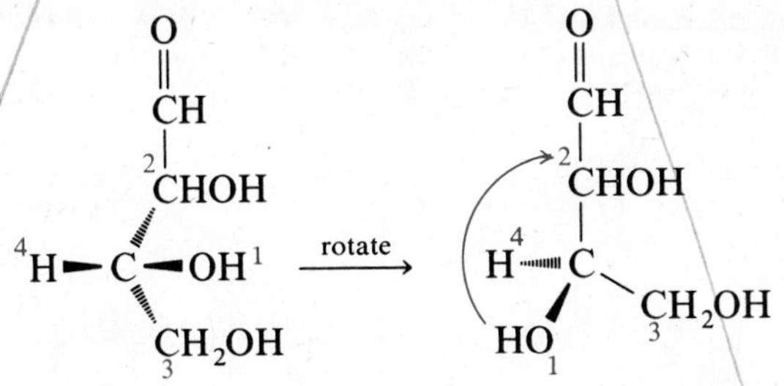

(3R)

This compound is therefore the (2R,3R) stereoisomer.

PROBLEM 6.11. Determine the (R) or (S) configuration of each chiral carbon. (We suggest that you use molecular models.)

(a)

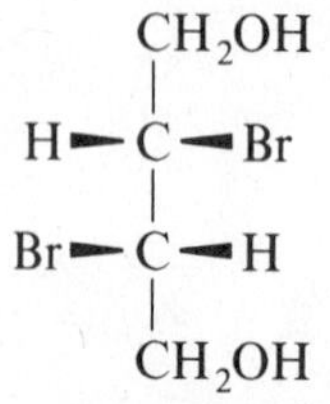

$$CH_2OH$$
$$H-C-Br$$
$$Br-C-H$$
$$CH_2OH$$

(b)

$$\overset{O}{\overset{\|}{CH}}$$
$$HO-C-H$$
$$CH_2$$
$$H-C-OH$$
$$CH_2OH$$

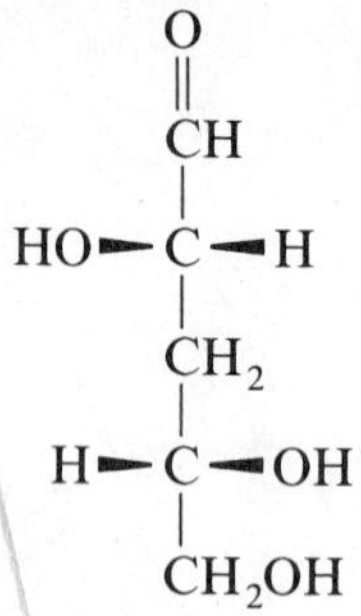

C. Meso Compounds

We have stated that a compound with more than one chiral carbon can have a maximum of 2^n stereoisomers. A few compounds, however, have fewer than 2^n stereoisomers. Tartaric acid is a compound with two chiral carbons and only three stereoisomers instead of the predicted maximum of four.

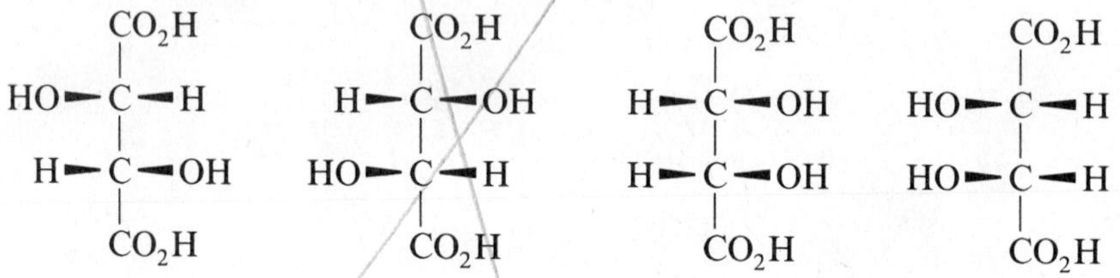

$$CO_2H$$
$$HO-C-H$$
$$H-C-OH$$
$$CO_2H$$
($2S,3S$)

$$CO_2H$$
$$H-C-OH$$
$$HO-C-H$$
$$CO_2H$$
($2R,3R$)

$$CO_2H$$
$$H-C-OH$$
$$H-C-OH$$
$$CO_2H$$
($2R,3S$)

$$CO_2H$$
$$HO-C-H$$
$$HO-C-H$$
$$CO_2H$$
($2S,3R$)

A pair of enantiomers

Not a pair of enantiomers,
but the same compound

Although the ($2S,3S$) and ($2R,3R$) stereoisomers of tartaric acid are enantiomers, the other two apparent enantiomers are superimposable on each other. They are not isomers of each other but are two different drawings of the same compound. This tartaric acid is called *meso*-tartaric acid.

Slide 180° in the
plane of the paper

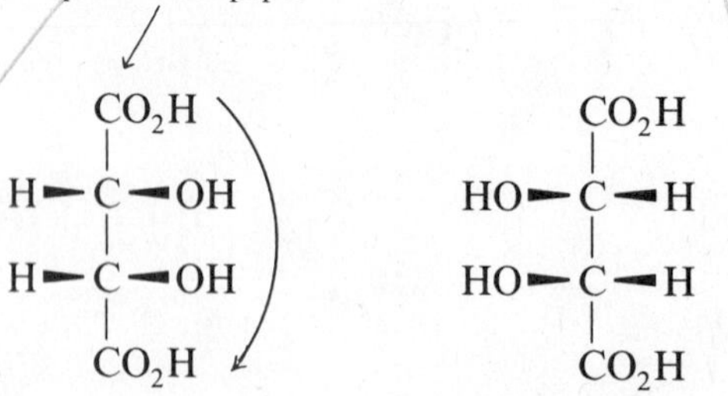

$$CO_2H$$
$$H-C-OH$$
$$H-C-OH$$
$$CO_2H$$

$$CO_2H$$
$$HO-C-H$$
$$HO-C-H$$
$$CO_2H$$

Same compound as
its mirror image

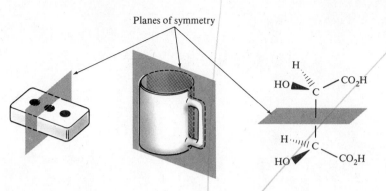

Planes of symmetry

Figure 6.10 A plane of symmetry is a hypothetical plane that divides an object into a pair of mirror-image halves. A symmetrical object is achiral (not chiral).

meso-Tartaric acid has an internal plane of symmetry, a plane that divides the structure into two mirror-reflective halves (see Figure 6.10). Any molecule with an internal plane of symmetry can be superimposed on its mirror image. Therefore, the molecule as a whole is not chiral even though it does contain chiral carbons. The *meso* form of tartaric acid does not rotate the plane of polarization of plane-polarized light—it is optically inactive.

Any structure that has an internal plane of symmetry in at least one conformation is optically inactive. If the structure also has more than one chiral carbon, it is called a **meso form**.

Two Meso Compounds:

$$
\begin{array}{ccc}
& CO_2H && CO_2H \\
& | && | \\
H\!\!\blacktriangleright\!\!&C\!\!\blacktriangleleft\!\!OH & H\!\!\blacktriangleright\!\!&C\!\!\blacktriangleleft\!\!OH \\
& | && | \\
-\;-\;-\;H\!\!\blacktriangleright\!\!&C\!\!\blacktriangleleft\!\!OH\;-\;-\;-\;HO\!\!\blacktriangleright\!\!&C\!\!\blacktriangleleft\!\!H\;-\;-\;-\;- & \text{Plane of symmetry} \\
& | && | \\
H\!\!\blacktriangleright\!\!&C\!\!\blacktriangleleft\!\!OH & H\!\!\blacktriangleright\!\!&C\!\!\blacktriangleleft\!\!OH \\
& | && | \\
& CO_2H && CO_2H \\
\end{array}
$$

6.5 CYCLIC COMPOUNDS

The cis and trans isomers of some cyclic compounds are chiral. To determine whether a cyclic compound is chiral, you might construct models of the compound and its mirror image and determine whether the two models are superimposable. Generally, it is easier to determine whether a plane of symmetry can be drawn through the polygon formula. If the formula has a plane of symmetry, the compound is not optically active.

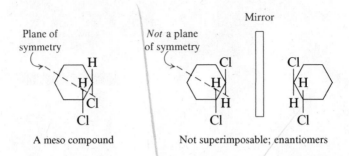

We can redraw these formulas to show the plane of symmetry more clearly.

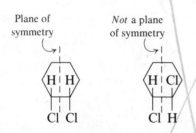

PROBLEM 6.12.   Determine whether each of the following cyclic compounds is chiral or achiral. If a plane of symmetry can be drawn through a formula, show it.

(a)   (b)   (c)

PROBLEM 6.13.   Which of the following compounds are meso compounds and which contain no chiral carbons?

(a)   (b)   (c)

6.6 RESOLUTION OF A RACEMIC MIXTURE

Most laboratory reactions lead to achiral or racemic products. The following addition to a double bond is an example.

$$CH_3CH_2CH=CH_2 \xrightarrow{H^+} CH_3CH_2-\overset{+}{C}\text{-----}H \longrightarrow$$

achiral

Cl⁻ can attack the planar carbocation from the top or the bottom as shown (equal probability).

(S)-2-Chlorobutane

(R)-2-Chlorobutane

Equal amounts of (R) and (S): racemic mixture

Pure enantiomers of certain compounds are necessary in biochemical studies and are also invaluable tools for the study of organic chemistry (see Chapter 7). One procedure for obtaining a pure enantiomer of a compound is to start with a pure enantiomer of a reactant and choose reactants that will not cause racemization. The problems with this method are finding the correct reactant and finding workable reactions.

A second way to obtain an optically active product is by using a chiral catalyst. Some alkenes can be hydrogenated with a chiral organic-metal complex catalyst to yield a pure enantiomeric product instead of a racemic mixture.

In biological systems, enzymes (the biological catalysts) cause reactions to yield pure stereoisomers by providing chiral reaction sites. We will discuss enzymes in more detail in Chapter 18. The use of enzymes and other chiral catalysts in the laboratory is an area of current research.

A third way to obtain a pure enantiomer is to separate a pair of enantiomers. This separation of a racemic mixture into pure (+) and (−) enantiomers is called **resolution** of the racemic mixture. Pasteur's separation of the enantiomers of sodium ammonium tartrate is an example of a resolution (see Point of Interest 6).

Because the physical properties, except for interaction with plane-polarized light, of a pair of enantiomers are the same, most enantiomeric pairs cannot be separated by a direct physical process, such as crystallization or distillation. Instead, an indirect method must be used. One such indirect method is the treatment of the racemic mixture with a chiral reagent to convert the enantiomers to diastereomers. Because diastereomers differ in their physical and chemical properties, they can be separated by a physical process. Once the diastereomers are separated from each other, each can be converted back to one of the enantiomers, now free from the other.

Let us illustrate the process by outlining the resolution of a racemic carboxylic acid. In our outline, we will use the general formula ($\pm$)-RCO_2H to represent any racemic carboxylic acid, such as the following specific example.

Chiral carbon

$$CH_3CHCOH$$
(with O double-bonded above COH, Br below the chiral carbon)

A carboxylic acid can lose a proton to a base, such as an amine (R_3N:), to yield a salt.

$$RCO-H \ + \ :NR_3 \longrightarrow RCO:^- \ \ H-\overset{+}{N}R_3$$

A salt

If the racemic carboxylic acid is treated with a pure ($+$) or ($-$) amine, a mixture of two salts is formed.

$$
\begin{array}{c}
(+)\text{-RCOH} \\
+ \\
(-)\text{-RCOH}
\end{array}
\xrightarrow{(+)\text{-}R_3N:}
\left\{
\begin{array}{c}
(+)(+)\text{-salt} \\
+ \\
(-)(+)\text{-salt}
\end{array}
\right.
$$

These two salts are diastereomers, not enantiomers. The enantiomer of the ($+$)($+$)-salt would be the ($-$)($-$)-salt, an impossible product because we started with the ($+$)-amine. A pair of diastereomers can be separated by physical means such as fractional crystallization (repeated crystallization, each procedure resulting in purer product). Once the salts are separated, they can be treated with an acid, such as HCl, to liberate the now-separated enantiomeric carboxylic acids.

In One Flask:

$$(+)(+)\text{-salt} \xrightarrow{H^+} (+)\text{-RCOH} \ + \ (+)\text{-}H\overset{+}{N}R_3$$

Resolved products

In a Second Flask:

$$(-)(+)\text{-salt} \xrightarrow{H^+} (-)\text{-RCOH} \ + \ (+)\text{-}H\overset{+}{N}R_3$$

PROBLEM 6.14. Amphetamine, a synthetic amine, can be resolved by treatment with (+)-tartaric acid followed by treatment with 10% H_2SO_4.

$$\langle \bigcirc \rangle - CH_2\overset{\overset{\displaystyle CH_3}{|}}{C}HNH_2$$

(*R*)- and (*S*)-Amphetamine

Write flow equations using the general formulas RNH_2 and (+)-RCO_2H to diagram this resolution.

SUMMARY

Stereochemistry is the study of molecules in three dimensions. A **chiral molecule** is one that cannot be superimposed on its mirror image. Nonsuperimposable mirror-image molecules are called **enantiomers**.

Chirality usually arises from the presence of one or more **chiral carbons**— carbons bonded to four different substituents. The arrangement of the substituents about the chiral carbon is called the **configuration** of that carbon. Enantiomers have opposite configurations.

Mirror

Opposite configurations

Enantiomers have the same chemical and physical properties with two exceptions: behavior in a chiral environment and interaction with plane-polarized light. **Optically active compounds** are those that rotate the plane of polarization of plane-polarized light. A **dextrorotatory** compound rotates the plane to the right. Its **levorotatory** enantiomer rotates the plane to the left the same number of degrees. The **specific rotation**, $[\alpha]_D^T$, is calculated from the rotation observed in the polarimeter. A **racemic mixture**, a 50:50 mixture of a pair of enantiomers, is optically inactive.

Configuration around a chiral carbon is designated by (*R*) or (*S*). The rules of priority and how these are used to specify the configuration are discussed in Section 6.3A. Assigning (*R*) and (*S*) to compounds with more than one chiral carbon is discussed in Section 6.4B.

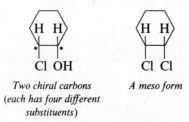

How the rules of priority are used to specify the stereochemistry of alkenes is discussed in Section 6.3B.

A compound with more than one chiral carbon can have up to 2^n stereoisomers, where n is the number of chiral carbons. Stereoisomers that are not enantiomers are called **diastereomers**. When a molecule contains more than one chiral carbon, a stereoisomer may contain an internal plane of symmetry. A molecule that has such a plane, called a **meso form**, is superimposable on its mirror image and is optically inactive.

	CO_2H			CO_2H				CO_2H		
HO—	C	—H	H—	C	—OH		H—	C	—OH	
H—	C	—OH	HO—	C	—H	- - - - - -	H—	C	—OH	— Plane of symmetry
	CO_2H			CO_2H				CO_2H		

Enantiomers   *A meso form*

Cyclic compounds may contain chiral carbons.

⟨H H⟩   ⟨H H⟩
Cl OH   Cl Cl

Two chiral carbons   *A meso form*
(each has four different
substituents)

A pair of enantiomers can be treated with a chiral reagent and thus converted to a pair of diastereomers, which can be separated. Each diastereomer can then be converted back to a single enantiomer. Separation of a racemic mixture into pure enantiomers is called **resolution** of the mixture.

STUDY PROBLEMS

6.15. Identify with asterisks the chiral carbons in the following structures:

(a)

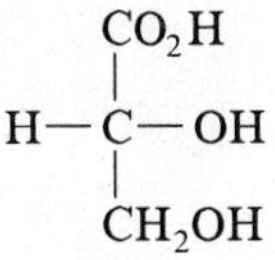

$$
\begin{array}{c}
CO_2H \\
| \\
H-C-OH \\
| \\
CH_2OH
\end{array}
$$

(b) $\langle\bigcirc\rangle\!-\!CH_2CHCH_3$ with NH_2 above

(c) $CH_3CHCH_2CH_3$ with Br above

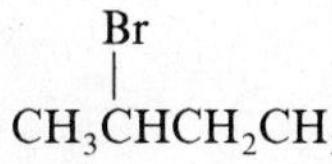

(d) 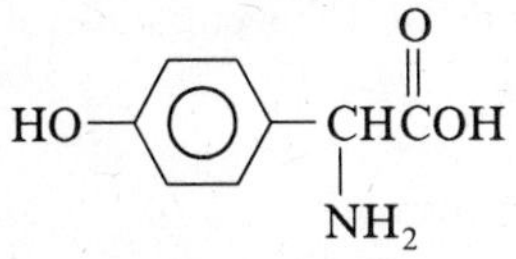 $HO\!-\!\langle\bigcirc\rangle\!-\!CHCOH$ with $\overset{O}{\overset{||}{}}$ and NH_2 below

(e)

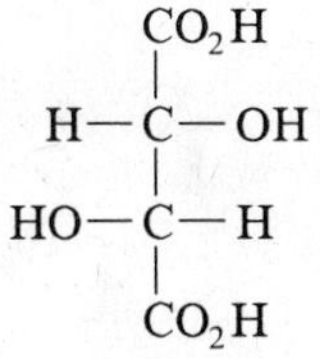

$$
\begin{array}{c}
CO_2H \\
| \\
H-C-OH \\
| \\
HO-C-H \\
| \\
CO_2H
\end{array}
$$

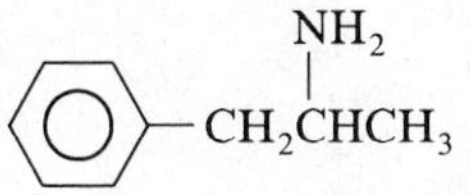

(f)

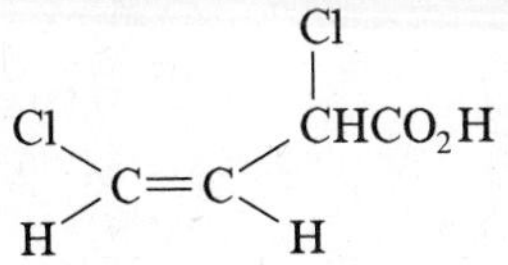

$$
\begin{array}{c}
Cl \\
| \\
CHCO_2H \\
Cl \quad \\
\backslash C=C \diagup \\
H \qquad \quad H
\end{array}
$$

(g)

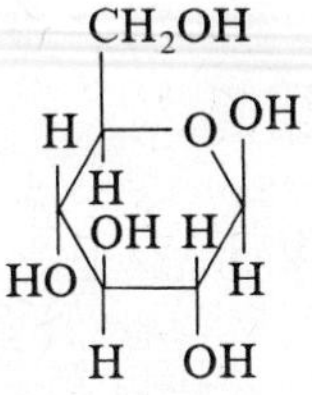

$$
\begin{array}{c}
CH_2OH \\
H \quad\quad O \quad OH \\
\quad H \\
OH \quad H \\
HO \quad\quad\quad H \\
H \quad OH
\end{array}
$$

Glucose

(h)

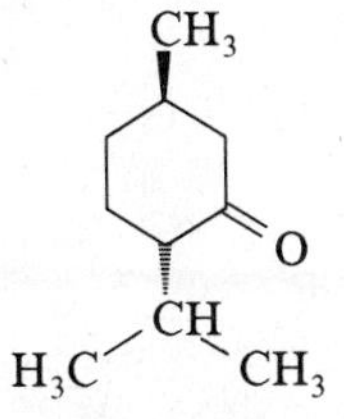

$$
\begin{array}{c}
CH_3 \\
\\
\\
CH \\
H_3C \quad CH_3
\end{array}
$$

Menthone

(i)

$$
CH_3OPSCH \begin{array}{c} \overset{O}{\overset{||}{COCH_2CH_3}} \\ \\ \overset{}{\underset{||}{COCH_2CH_3}} \\ O \end{array}
$$

with $\overset{S}{\overset{||}{}}$ and OCH_3

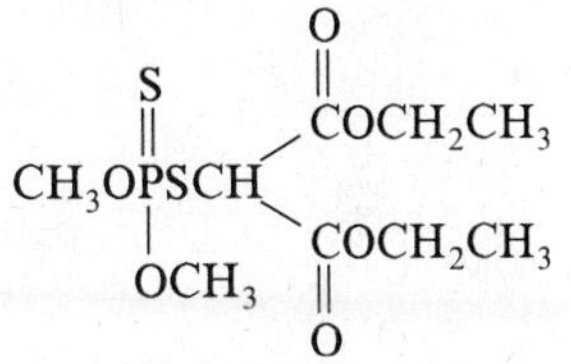

Malathion

6.16. Draw formulas for both enantiomers for each of the following chiral compounds. (It may be necessary to redraw the structures to emphasize the chiral carbons.)

(a) 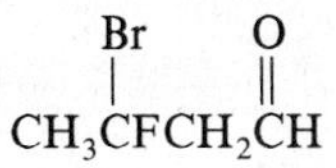 CH_3CFCH_2CH with Br above first C and $\overset{O}{\overset{||}{}}$ above CH

(b)

$$
\begin{array}{c}
CH_2OCH_3 \\
| \\
CH_2 \\
| \\
CHOCH_3 \\
| \\
CH_3
\end{array}
$$

(c)

$$
\begin{array}{c}
O \\
|| \\
CH \\
| \\
H-C-OH \\
| \\
CH_2OH
\end{array}
$$

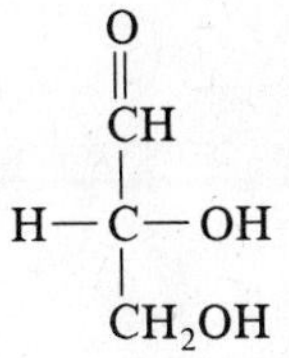

(d)

$$
\begin{array}{c}
CH_2OH \\
| \\
H\!-\!C\!\!\blacktriangleleft\!OH \\
| \\
Cl\!\blacktriangleright\!C\!\!\blacktriangleleft\!H \\
| \\
CH_3
\end{array}
$$

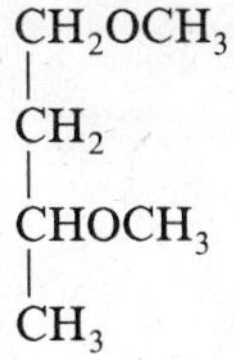

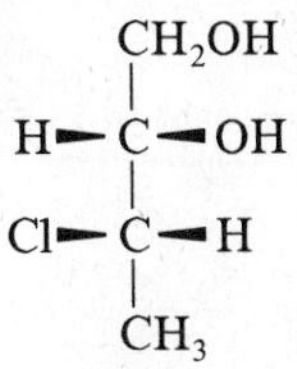

(e) 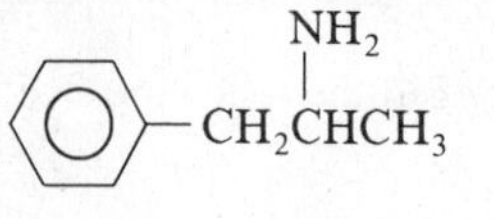 [benzene ring]—CH$_2$CHCH$_3$ with NH$_2$ on the middle carbon

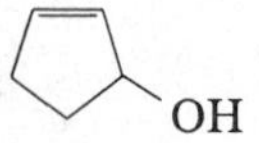

(f) 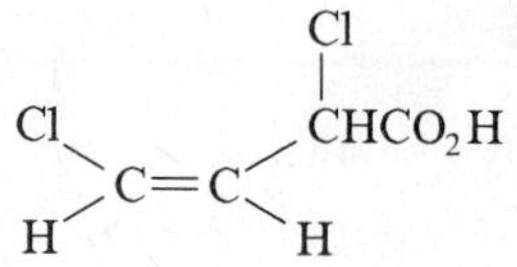 Cl and H on left carbon of C=C, right carbon has CHCO$_2$H with Cl above, and H below

(g) [cyclopentene ring]—OH

(h) [cyclohexanone ring] with =O and Cl substituent

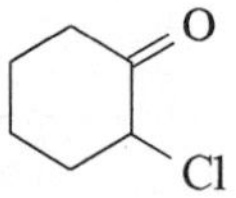

(i) 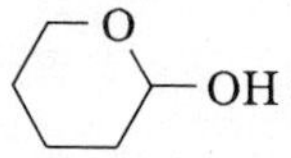 [tetrahydropyran ring]—OH with O in ring

6.17. Draw formulas for all compounds with the following molecular formulas that have at least one chiral carbon:

(a) $C_4H_{10}O$ (b) $C_4H_8Br_2$

6.18. Calculate the specific rotation for each sample:

(a) *Sample 1:* 2.00 g of the compound was dissolved in 15.00 mL of ethanol and placed in a sample cell 15.0 cm long at 20°C. The observed rotation (sodium D line) is +28.7°.

(b) *Sample 2:* 6.96 g of the compound was dissolved in 25.00 mL of ethanol and placed in a sample cell 5 cm long at 20°C. The observed rotation (sodium D line) is −20°.

6.19. Pure (*R*)-2-butanol has a specific rotation of −13.5° at 25°C (sodium D line). (a) What is the specific rotation of pure (*S*)-2-butanol?   (b) What is the rotation of a 50 : 50 mixture of (*R*)-2-butanol and (*S*)-2-butanol?

6.20. Identify the higher-priority group (Cahn–Ingold–Prelog system) of each of the following pairs:

(a) $-CH_2\overset{\displaystyle O}{\overset{\|}{C}}OH, \ -\overset{\displaystyle O}{\overset{\|}{C}}OH$ (b) $-C\equiv CH, \ -NH_2$

(c) $-Br, \ -OH$ (d) 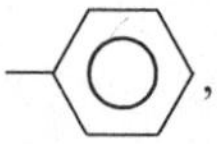 $-[benzene\ ring], \ -\overset{\displaystyle CH_3}{\overset{|}{C}}=CH_2$

6.21. Rank the following substituents in terms of priority by the Cahn–Ingold–Prelog system (*lowest* priority first).

(a) $-Cl, \ -Br, \ -F, \ -I$

(b) $-CH_3, \ -NH_2, \ -H, \ -OCH_3$

(c) $-CH_3, \ -CH_2CH_3, \ -CH(CH_3)_2, \ -OCH_3$

(d) $-CH_3, \ -\overset{\displaystyle O}{\overset{\|}{C}}H, \ -CH=CH_2, \ -OCH_3$

6.22. Assign the chiral carbons in the following structures as (*R*) or (*S*).

(a) 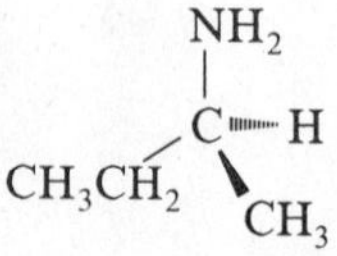 CH_3CH_2—C with NH$_2$ up, H wedge, CH$_3$ down

(b) 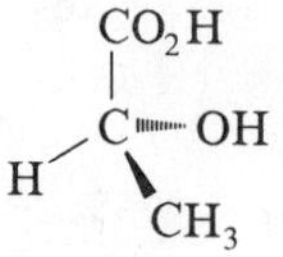 C with CO$_2$H up, H on left, OH wedge, CH$_3$ down

(c)

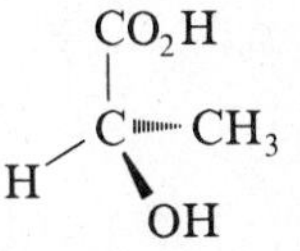

CO_2H
C—CH$_3$
H OH

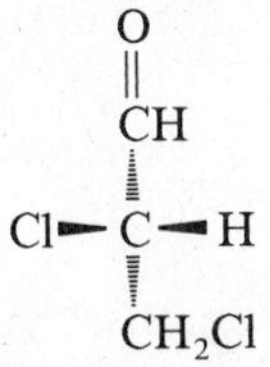

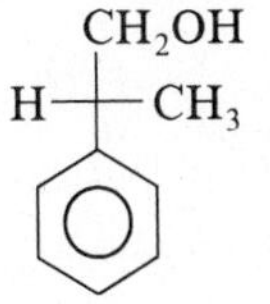

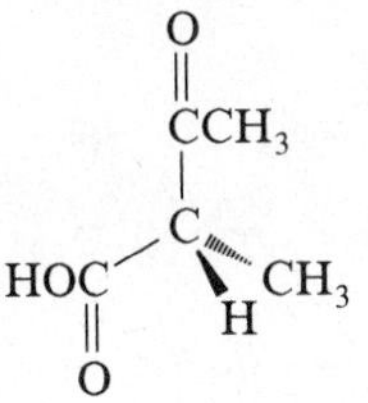

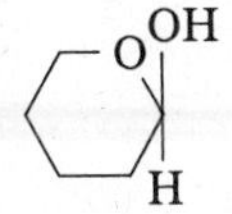

(d)

O
||
CH
Cl—C—H
CH$_2$Cl

(e)

CH$_2$OH
H—CH$_3$
(benzene ring)

(f)

O
||
CCH$_3$
C
HOC CH$_3$
|| H
O

(g)

O OH
H

(h)

O O
H
Cl

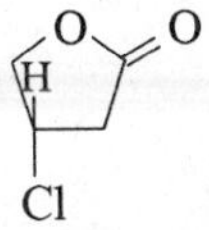

6.23. Draw three-dimensional formulas for the following compounds:
  (a) (R)-2-iodobutane   **(c)** (R)-1-chloro-1-phenylethane
  (b) (S)-2-iodobutane   **(d)** (S)-1-chloro-1-phenylethane

6.24. For each of the following structures, draw formulas for two stereoisomers and label each as (E) or (Z).

  (a) 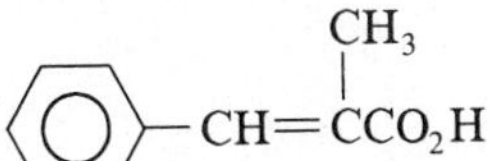

  (phenyl)—CH=CCO$_2$H
              |
              CH$_3$

  (b) FClC=ClBr

  (c) CH$_3$CCl=CHCH
                    ||
                    O

  (d) HOCH$_2$CH=CCH$_2$\overset{+}{N}(CH$_3$)$_3$ Cl$^-$
                      |
                      CH$_3$

6.25. Draw formulas for the following compounds:
  (a) (Z)-3,4-diethyl-2-methyl-3-heptene
  (b) (1E,3Z)-1,5-dichloro-1,3-pentadiene
  (c) (2E,6E)-octadiene
  (d) (2Z,4E)-hexadiene

6.26. Assign each chiral carbon in the following compounds as (R) or (S).

(a)

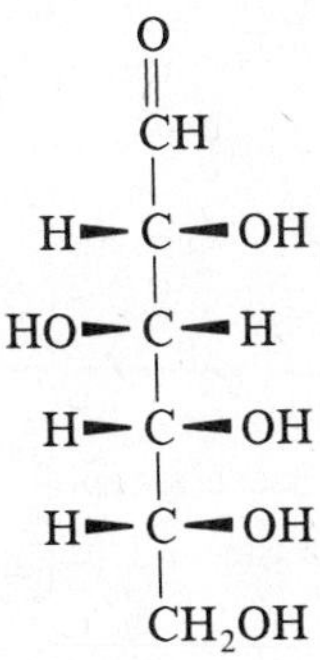

O
||
CH
H—C—OH
HO—C—H
H—C—OH
H—C—OH
CH$_2$OH

(b)

O
||
COH
H—C—Cl
Cl—C—H
CH$_3$

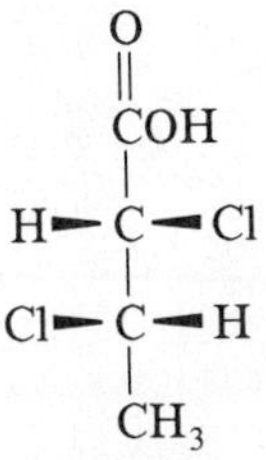

(c)

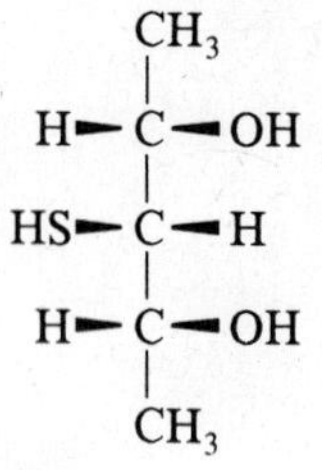

$$CH_3$$
H—C—OH
HS—C—H
H—C—OH
$$CH_3$$

(d)

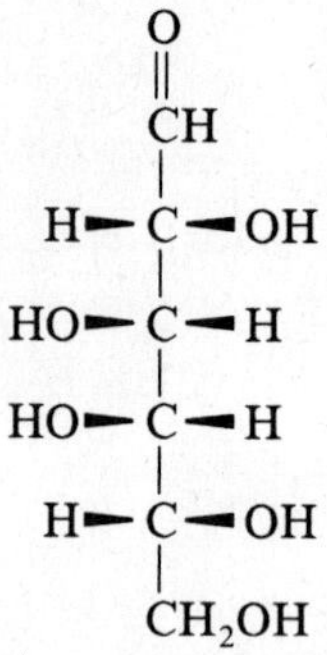

O
‖
CH
H—C—OH
HO—C—H
HO—C—H
H—C—OH
$$CH_2OH$$

6.27. Which of the following formulas represents a meso compound?

(a)

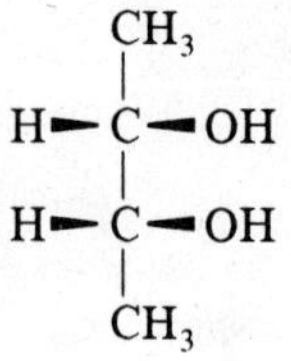

$$CH_3$$
H—C—OH
H—C—OH
$$CH_3$$

(b)

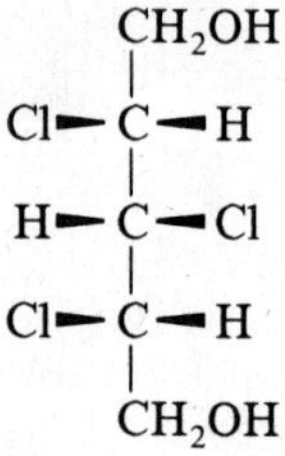

$$CH_2OH$$
Cl—C—H
H—C—Cl
Cl—C—H
$$CH_2OH$$

(c)

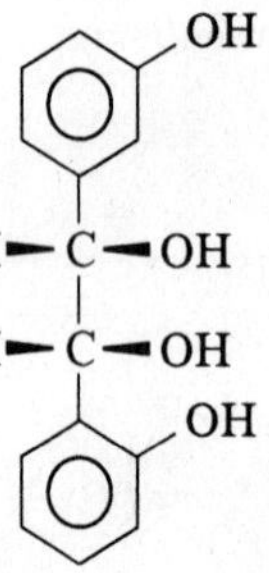

OH (benzene ring)
H—C—OH
H—C—OH
OH (benzene ring)

(d)

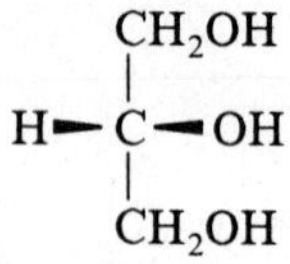

$$CH_2OH$$
H—C—OH
$$CH_2OH$$

(e)

$$CH_2OH$$
H—C—OH
HO—C—H
H—C—OH
$$CH_2OH$$

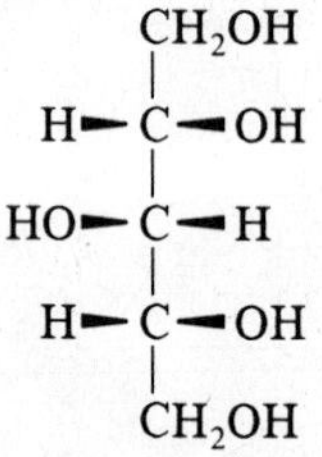

Xylitol
*a sweetener used
for the prevention
of dental cavities*

(f)

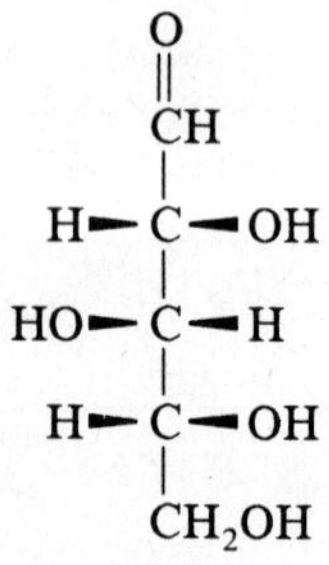

O
‖
CH
H—C—OH
HO—C—H
H—C—OH
$$CH_2OH$$

Xylose
a sugar

6.28. Draw formulas for the enantiomeric pair and meso form of 2,3-dibromobutane.

6.29. Draw formulas for three meso diols with the molecular formula $C_6H_{14}O_2$.

6.30. Draw formulas for all meso compounds for the following structures, and indicate the plane of symmetry through each structure.

(a)

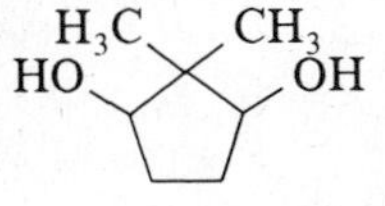

(b) 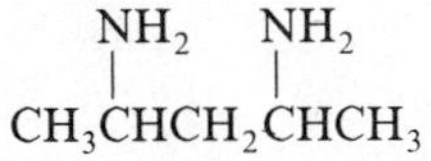 $CH_3CHCH_2CHCH_3$ (with NH_2 groups on carbons 2 and 4)

(c)

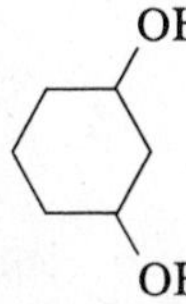

(d)

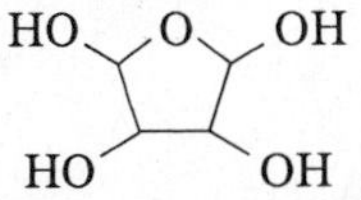

6.31. Star the chiral carbon (if any) in the following structures. Label each as (R) or (S).

(a)

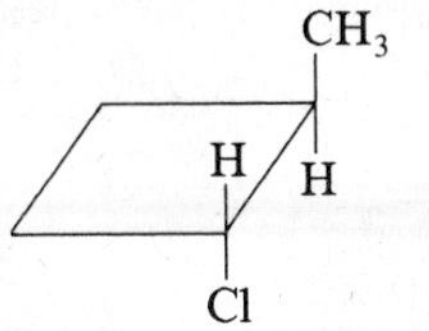

(b)

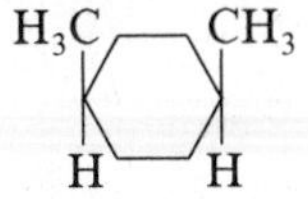

(c)

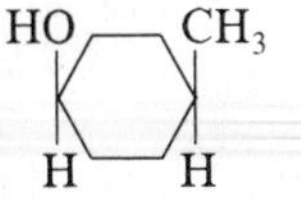

6.32. In structure determinations of sugars, Emil Fischer oxidized the following sugar, arabinose, to a dicarboxylic acid that was optically active, not meso. What are the possible configurations of the hydroxy groups at carbons 2 and 3 in arabinose? (Use wedge formulas.)

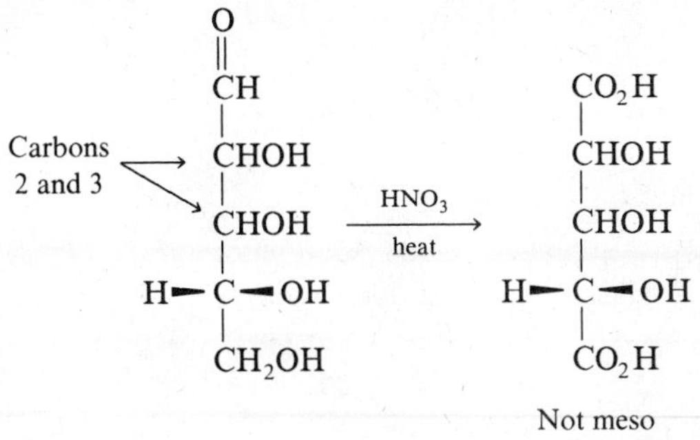

Not meso

POINT OF INTEREST 6

Louis Pasteur: The Resolution of a Racemic Salt

Louis Pasteur was born in 1822. He attended college at Besançon in France, where his performance was creditable but not brilliant. At age 24, Pasteur completed his doctoral work at the École Normale in Paris. His first postdoctoral position was that of laboratory assistant at his graduate school, a position that allowed him freedom and opportunity for individual research.

To further his laboratory skills, Pasteur decided to repeat some previous studies on the crystals of tartaric acid—a project that led to his first major scientific discovery.

At this time, only the dextrorotatory (+) isomer (called tartaric acid) and the optically inactive racemic mixture (called "paratartaric acid" because it was thought to be a single compound) were known. The levorotatory (−) and meso isomers of tartaric acid, as well as the concept of a racemic mixture, were unknown. The year was 1848, many years before even the acceptance of a tetrahedral carbon.

Previous workers suggested that (+)-tartaric acid and paratartaric acid had the same structure, but Pasteur was convinced that there must be a correlation between molecular structure and the ability of a compound to rotate the plane of polarization of plane-polarized light.

He carefully crystallized the sodium ammonium salts of the two known tartaric acids and, to his surprise, detected two types of crystals. Paratartaric acid had crystallized into mirror-image crystals, which Pasteur called right-handed and left-handed. However, (+)-tartaric acid formed only right-handed crystals.

A Pair of Mirror-Image Crystals:

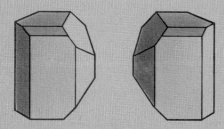

The fact that the paratartaric acid is actually a mixture of equal parts of (+)-tartaric acid and the unknown (−) isomer must have come to Pasteur in a flash of intuition because all his subsequent acts were aimed at obtaining physical evidence for this idea. Using a hand lens and tweezers, Pasteur separated the right-handed crystals from the left-handed crystals. Then he determined that a solution of the right-handed crystals rotated the plane of polarization of plane-polarized light to the right just as did the salt of (+)-tartaric acid. The solution of the left-handed crystals rotated the plane to the left by the same number of degrees. Pasteur completed the experiment by dissolving equal weights of the two crystal types and observing that this solution did not rotate the plane of polarization.

At this point, Pasteur is said to have jumped up from the polarimeter and dashed into the hall shouting, "I have it! I have made a great discovery." Indeed, he had. At the age of 26, Pasteur had just completed one of the classical experiments of chemistry—the first resolution of a racemic mixture.

The scientific world of Paris was soon buzzing with talk about Pasteur's experiment. The news reached Jean Baptiste Biot, who at 74 was a "grand old man" of French science. Biot himself had discovered the optical activity of some organic compounds, including (+)-tartaric acid.

Biot summoned Pasteur and, by voice and gesture, made it known that he was skeptical of Pasteur's claim that he could separate paratartaric acid into optically active substances. To personally verify the claim, Biot provided the chemicals Pasteur would need to repeat the experiment, including paratartaric acid previously ·

determined to be optically inactive. Pasteur mixed the ingredients to make a solution of sodium ammonium tartrate in an evaporating dish. Biot then took the dish from Pasteur and dismissed him. Two days later, crystals had formed and Biot sent for Pasteur.

With Biot looking on, Pasteur separated the crystals.

> "So you affirm," said Biot, "that your right-handed crystals will deviate to the right the plane of polarization, and your left-hand ones will deviate it to the left?"
> "Yes," said Pasteur.
> "Well, let me do the rest."

Biot again dismissed Pasteur and prepared solutions containing weighed amounts of the crystals for the polarimeter. When all was ready, he again summoned Pasteur.

Biot selected the solution that Pasteur said would rotate the plane of polarization to the left and placed it in the polarimeter. This solution, according to Pasteur, contained the heretofore unknown levorotatory isomer of sodium ammonium tartrate. In Pasteur's presence, Biot peered into the polarimeter and instantly saw that the solution was strongly levorotatory. He turned, grabbed Pasteur's hand, and said:

> "My dear child, I have loved science so much throughout my life that this makes my heart throb."

Luck had smiled on Pasteur that spring of 1848. From all possible salts, he chose to study sodium ammonium tartrate. It was Pasteur's good fortune that he did. The spontaneous crystallization of enantiomers from a racemic mixture to yield mirror-image crystals is an extremely rare occurrence. There are only nine compounds known that will form enantiomeric crystals large enough to be separated with tweezers.

Luck smiled more than once on Pasteur. Sodium ammonium tartrate forms enantiomeric crystals only when crystallized at temperatures below 26°C (79°F). Had Biot's laboratory been warmer, the demonstration would have failed and Pasteur's reputation would have been tarnished. But the demonstration did not fail and Biot became Pasteur's sponsor, publicly lauding him and launching him on his brillant scientific career.

Chapter 7

Organohalogen Compounds

An **organohalogen compound** is any compound containing a carbon–halogen bond. Organohalogen compounds are commonly found as natural products in marine life, such as algae, but they are rarely found in terrestrial plants and animals. One important exception is *thyroxine*, a component of the thyroid hormone *thyroglobulin*.

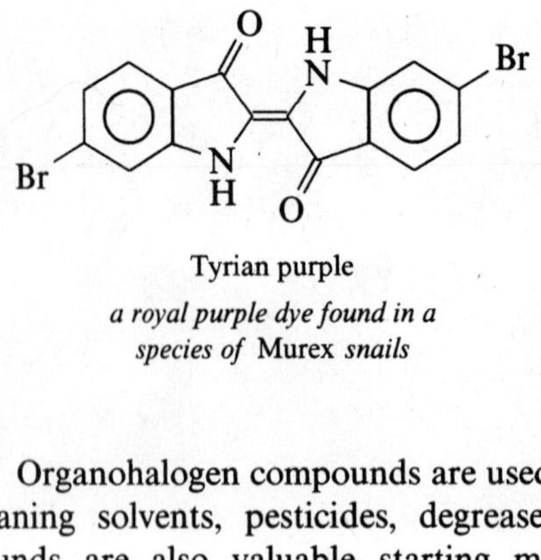

Tyrian purple

a royal purple dye found in a species of Murex *snails*

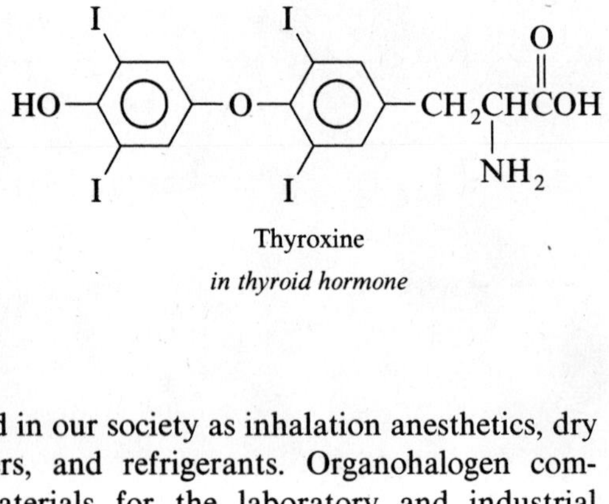

Thyroxine

in thyroid hormone

Organohalogen compounds are used in our society as inhalation anesthetics, dry cleaning solvents, pesticides, degreasers, and refrigerants. Organohalogen compounds are also valuable starting materials for the laboratory and industrial synthesis of other organic compounds.

7.1 BONDING AND STRUCTURE OF ORGANOHALOGEN COMPOUNDS

Halogen atoms, frequently generalized as X, form single covalent bonds with carbon atoms in a variety of compounds. An alkane substituted with a halogen (RX) is called a **haloalkane** or an **alkyl halide**. In these compounds, the halogen is bonded to an sp^3-hybridized tetrahedral carbon.

Haloalkanes (Alkyl Halides):

$$CH_3CH_2Cl$$

Chloroethane
(ethyl chloride)

a topical anesthetic
(bp 12°C) that numbs the skin
by evaporation and chilling

$$\overset{\displaystyle Cl}{\underset{|}{CF_3CHBr}}$$

1-Bromo-1-chloro-
2,2,2-trifluoroethane
(halothane)

an inhalation anesthetic

Halogen atoms can also be bonded to sp^2-hybridized double-bond carbons. These compounds are called **vinylic halides**, from "vinyl," the trivial name for the CH_2=CH— group.

Vinylic Halides:

$$CH_2{=}CH{-}Cl$$

Chloroethene
(vinyl chloride)

monomer for polyvinyl
chloride (PVC), a plastic
used for phonograph records,
trash bags, etc.

$$CF_2{=}CF_2$$

Tetrafluoroethene

monomer for Teflon,
used in gaskets and cookware

An **aryl halide** is an organohalogen compound in which a halogen is bonded to an sp^2-hybridized aromatic-ring carbon.

Aryl Halides:

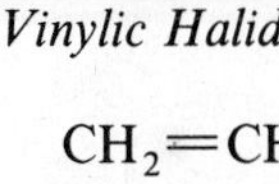

Bromobenzene

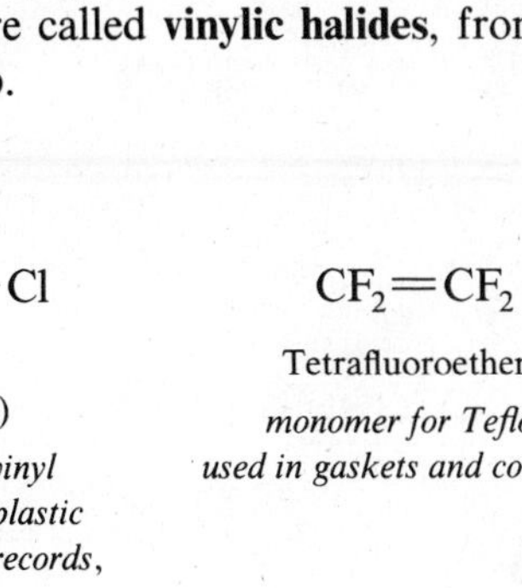

A polychlorobiphenyl (PCB)

one member of the class of toxic
compounds formerly used as heat
transfer fluids in large
electric transformers

Organohalogen compounds in which the halogen is bonded to an sp-hybridized carbon are unstable; therefore, we will not discuss them in this text.

Allylic and benzylic halides are compounds in which the halogen is bonded to an sp^3-hybridized carbon that is adjacent to an sp^2-hybridized carbon. The names for these classes are taken from the names of the *allyl group* (CH_2=$CHCH_2$—) and the *benzyl group* ($C_6H_5CH_2$—).

Allylic and Benzylic Halides:

Adjacent to sp^2-hybrid carbon

CH_2=$CHCH_2Cl$ ⬡—CH_2Cl

3-Chloropropene   Benzyl chloride
(allyl chloride)

Table 7.1 summarizes the different types of organohalogen compounds.

TABLE 7.1   TYPES OF ORGANOHALOGEN COMPOUNDS

Type of compound	General formula[a]
Haloalkane (alkyl halide)	R_3C—X H or alkyl group
Vinylic halide	R, X on C=C, R, R
Aryl halide	⬡—X or other aryl group
Allylic halide	R, CH_2X on C=C, R, R
Benzylic halide	⬡—CH_2X or other aryl group

[a] In these formulas, X is used to represent a halogen: F, Cl, Br, or I.

PROBLEM 7.1. Classify each of the following compounds as an alkyl halide, a vinylic halide, an allylic halide, or a benzylic halide:

(a)

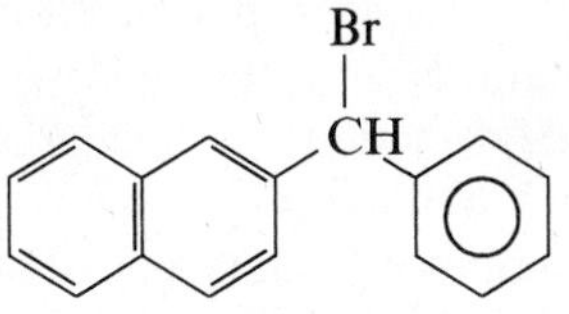

(b) 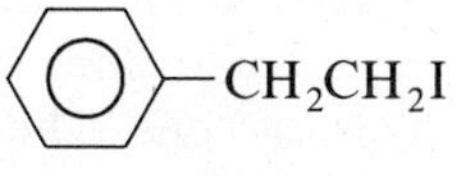 CH_2CH_2I

(c) $CH_2=CHCH_2CH_2I$

(d) $CH_3CH=CHCl$

(e)

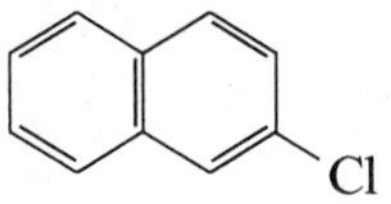

(f)

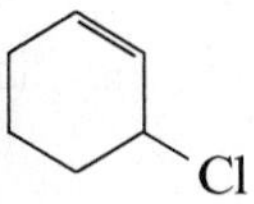

(g)

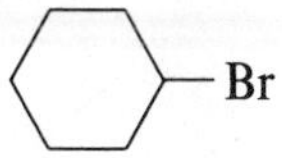

7.2 PHYSICAL PROPERTIES OF ORGANOHALOGEN COMPOUNDS

Except for fluorine, halogen atoms are much heavier than the other atoms commonly found in organic compounds. Because of their mass, halogen atoms add to the density of a compound (see Table 7.2). A bottle of carbon tetrachloride is surprisingly heavy. The common halogenated solvents (CH_2Cl_2, $CHCl_3$, CCl_4,

TABLE 7.2   PHYSICAL PROPERTIES OF SOME HALOGENATED METHANES

Formula	Name[a]	Bp (°C)	Density (g/mL) at 20°C
CH_3Cl	Chloromethane (methyl chloride)	−24	Gas
CH_2Cl_2	Dichloromethane (methylene chloride)	40	1.34
$CHCl_3$	Trichloromethane (chloroform)	61	1.49
CCl_4	Tetrachloromethane (carbon tetrachloride)	77	1.60
CH_3I	Iodomethane (methyl iodide)	43	2.28

[a] Trivial names are in parentheses.

$Cl_2C=CHCl$) are all denser than water and sink to the bottom of a container when mixed with water. This behavior is in direct contrast to that of alkanes, which float on water.

Large halogen atoms are easily polarized. This polarization causes a charge separation to develop in the organohalogen molecule and thus causes an increase in intermolecular attractions. The increased attractions among organohalogen molecules result in higher melting points and boiling points than are observed with comparable hydrocarbons. The boiling points of some haloalkanes are listed in Table 7.2. Note the increasing boiling points in the series CH_3Cl, CH_2Cl_2, $CHCl_3$, CCl_4.

Polarizability and electronegativity
cause a charge separation—a partially
positive C and a partially negative X.
The charge separation increases
attractions among molecules.

$$\overset{\delta+ \quad \delta-}{R_3C-X}$$

Organohalogen compounds are not sufficiently polar to dissolve in water. They are, however, soluble in less polar solvents such as ethanol (CH_3CH_2OH) and benzene.

7.3 NOMENCLATURE OF ORGANOHALOGEN COMPOUNDS

In the IUPAC system, a halogen is a substituent named by a prefix, just as an alkyl group is. The prefixes are straightforward: **fluoro-** ($-F$), **chloro-** ($-Cl$), **bromo-** ($-Br$), and **iodo-** ($-I$).

In naming an organohalogen compound, we number the longest continuous chain beginning at the end nearer the substituents. The name is assembled with the substituent position number, the substituent name, and the parent name.

$$\overset{\displaystyle Cl}{\underset{\displaystyle |}{CH_3CHCH_2CH_3}} \qquad \overset{\displaystyle Cl}{\underset{\displaystyle |}{CH_3CCH_2CH_2CH_3}}$$
$$\underset{\displaystyle CH_3}{|}$$

2-Chlorobutane   2-Chloro-2-methylpentane

alphabetic listing

If more than one halogen of the same type is present, the prefixes di-, tri-, and so on are used to group the halogen in the name. This procedure is analogous to the

TABLE 7.3   NAMES OF SOME ORGANOHALOGEN COMPOUNDS

Formula	Systematic name	Trivial name
CH_3I	Iodomethane	Methyl iodide
CH_3CH_2Br	Bromoethane	Ethyl bromide
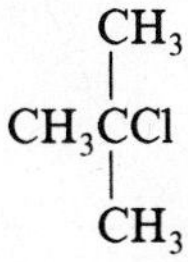 $\begin{matrix} CH_3 \\ \vert \\ CH_3CCl \\ \vert \\ CH_3 \end{matrix}$	2-Chloro-2-methylpropane	*t*-Butyl chloride
$CH_2{=}CHCl$	Chloroethene	Vinyl chloride
$CH_2{=}CHCH_2Cl$	3-Chloropropene	Allyl chloride
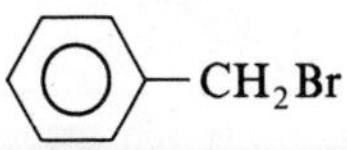 ⬡—CH_2Br	—	Benzyl bromide

procedure we use to group alkyl groups (e.g., dimethyl, triethyl) in the name.

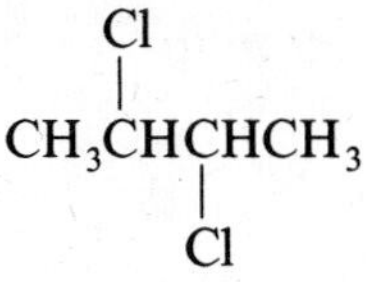

$$\begin{matrix} & Cl \\ & \vert \\ CH_3 & CHCHCH_3 \\ & \vert \\ & Cl \end{matrix}$$

2,3-Dichlorobutane

Nonsystematic names are encountered frequently. A few of the more common trivial names are listed in Table 7.3. Note that these trivial names are generally composed of the name of the organic group followed by the name of the halide. We will use many of these trivial names in this text because of convenience and because they are in such common use.

PROBLEM 7.2.   Name the following halogen compounds by the IUPAC system.

(a) 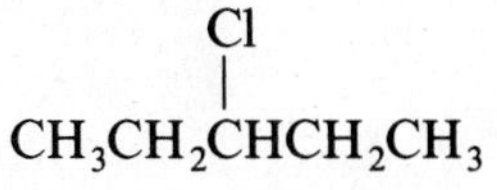 $\begin{matrix} & Cl \\ & \vert \\ CH_3CH_2 & CHCH_2CH_3 \end{matrix}$ (b) 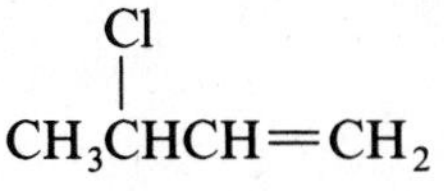 $\begin{matrix} & Cl \\ & \vert \\ CH_3 & CHCH{=}CH_2 \end{matrix}$

(c) 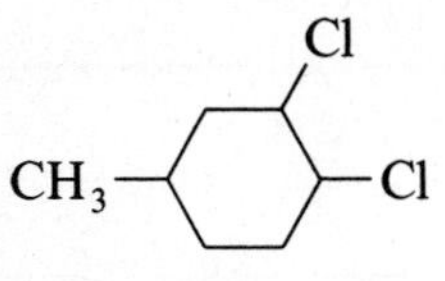 CH_3-⬡$-Cl$ (with Cl at top)

PROBLEM 7.3. Write formulas for the following compounds:

(a) 1,1-dichloro-6-methylheptane

(b) 1,2,3-trichlorobenzene

(c) 3-bromo-1-ethylcyclopentene

(d) 3-methylcyclohexyl iodide

(e) isopropyl bromide

7.4 CLASSIFICATION OF ALKYL HALIDES

Alkyl halides (RX) can be classified by the number of carbon atoms bonded to the halogen's carbon. We will use these classifications when we discuss the reactions of alkyl halides. A **methyl halide** (CH_3X) contains no other carbons bonded to the C—X carbon. A **primary (1°) alkyl halide** contains *one* carbon bonded to the C—X carbon. A **secondary (2°) alkyl halide** contains *two* carbons bonded to the C—X carbon. A **tertiary (3°) alkyl halide** contains *three* carbons bonded to the C—X carbon. This classification is very similar to that for carbocations.

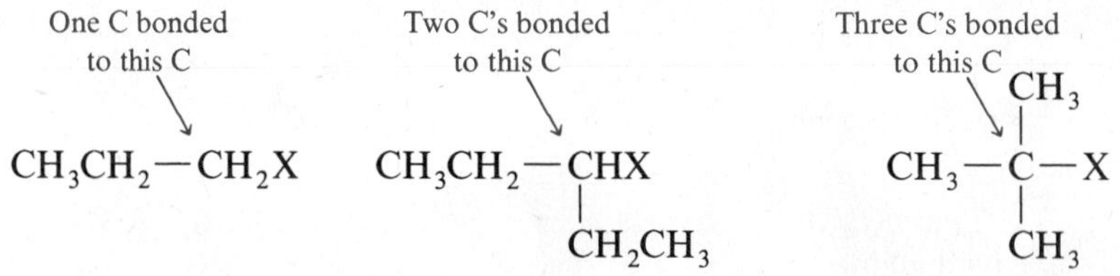

One C bonded to this C

Two C's bonded to this C

Three C's bonded to this C

$CH_3CH_2—CH_2X$

$CH_3CH_2—CHX$
$\quad\quad\quad\quad CH_2CH_3$

$\quad\quad\quad CH_3$
$CH_3—C—X$
$\quad\quad\quad CH_3$

Primary (1°)

Secondary (2°)

Tertiary (3°)

PROBLEM 7.4. Classify each of the following halogen compounds as methyl, primary, secondary, or tertiary:

(a) 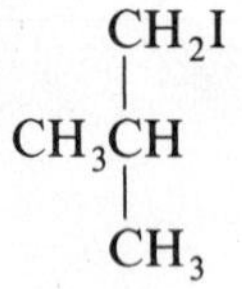
$\quad CH_2I$
$\quad |$
CH_3CH
$\quad |$
$\quad CH_3$

(b) 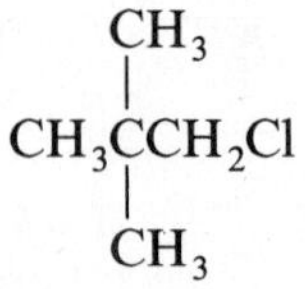
$\quad CH_3$
$\quad |$
CH_3CCH_2Cl
$\quad |$
$\quad CH_3$

(c) 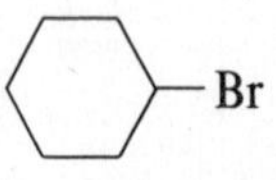 ⬡—Br

(d) 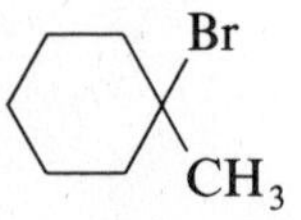 ⬡ with Br and CH_3

7.5 PREPARATION OF ORGANOHALOGEN COMPOUNDS

A. Direct Halogenation

We have already discussed the common halogenation reactions. *Free-radical halogenation* of alkanes (see Section 2.5A) is more commonly carried out as an industrial process than as a laboratory reaction.

A Free-Radical Chlorination:

$$CH_3CH_3 \; + \; Cl_2 \; \xrightarrow{\text{light}} \; CH_3CH_2Cl \; + \; HCl$$

Ethane   Chloroethane

(*and other products*)

The allylic and benzylic positions undergo ready halogenation because of resonance stabilization of the radical intermediate. Bromine is more selective than chlorine in these reactions and thus gives a better yield of allylic or benzylic halide. In allylic halogenation, we must avoid an excess of bromine to prevent the reaction of the halogen with the double bond.

Allylic and Benzylic Free-Radical Halogenation:

Allylic position

$$CH_3CH{=}CH_2 \; + \; Br_2 \; \xrightarrow{\text{light}} \; \overset{\overset{\displaystyle Br}{|}}{CH_2}CH{=}CH_2 \; + \; HBr$$

Propene   3-Bromopropene
                                     (allyl bromide)

Benzylic position

$$\langle\!\bigcirc\!\rangle{-}CH_2CH_3 \; + \; Br_2 \; \xrightarrow{\text{light}} \; \langle\!\bigcirc\!\rangle{-}\overset{\overset{\displaystyle Br}{|}}{C}HCH_3 \; + \; HBr$$

Ethylbenzene   (1-Bromoethyl)benzene

Aromatic halogenation reactions are used to obtain the aryl halides. These halogenations are not free-radical reactions but electrophilic aromatic substitution reactions (see Section 5.3A).

An Aromatic Bromination Reaction:

$$\langle\!\bigcirc\!\rangle \; + \; Br_2 \; \xrightarrow{\text{FeBr}_3} \; \langle\!\bigcirc\!\rangle{-}Br \; + \; HBr$$

Benzene   Bromobenzene

B. Addition of Hydrogen Halides and Halogens to Alkenes

Addition of hydrogen halides (HX) to alkenes generally follows Markovnikov's rule (see Section 3.6B).

$$\underset{\text{Propene}}{CH_3CH=CH_2} \;+\; HCl \;\longrightarrow\; \underset{\text{2-Chloropropane}}{CH_3\overset{\displaystyle Cl}{\overset{|}{C}}HCH_3}$$

H⁺ goes here

Addition of halogens (X_2) yields 1,2-dihaloalkanes. This reaction can be carried out with Br_2 or Cl_2 but not with F_2 or I_2 (see Section 3.6E).

$$CH_3CH=CH_2 + Br_2 \longrightarrow \underset{\text{1,2-Dibromopropane}}{CH_3\overset{\displaystyle Br}{\overset{|}{C}}H-\overset{\displaystyle Br}{\overset{|}{C}}H_2}$$

C. Substitution Reactions with Alcohols

In Section 8.8 we will present a detailed discussion of the substitution reactions of alcohols (ROH). In general, we find that alcohols react with HI or HBr readily to yield alkyl iodides or alkyl bromides. A zinc chloride ($ZnCl_2$) catalyst is sometimes required for reaction with HCl.

$$\underset{\text{\textit{t}-Butyl alcohol}}{(CH_3)_3C-OH} \;+\; HBr \;\longrightarrow\; \underset{\text{\textit{t}-Butyl bromide}}{(CH_3)_3C-Br} \;+\; H_2O$$

The use of other halogenating agents, such as phosphorus tribromide (PBr_3) or thionyl chloride ($SOCl_2$), to prepare alkyl halides from alcohols will be discussed in Section 8.8B.

The different methods by which organohalogen compounds can be synthesized are summarized in Table 7.4.

PROBLEM 7.5. Write an equation for each of the following syntheses:

(a) 2-bromopropane from an alkene

(b) 2-iodobutane from an alcohol

(c) 2,3-dichlorobutane from an alkene

(d) [naphthalene structure with CH₂Br substituent] from a hydrocarbon

TABLE 7.4 SUMMARY OF PREPARATIONS OF ORGANOHALOGEN COMPOUNDS

Reaction	Section reference

Direct halogenation

$$RH + X_2 \xrightarrow{\text{light}} RX + HX$$ 2.5A

$-CH_2R + Br_2 \xrightarrow{\text{light}}$ $-\overset{\overset{\displaystyle Br}{|}}{C}HR + HBr$ 5.5

$$R_2C{=}\overset{\overset{\displaystyle }{|}}{\underset{\underset{\displaystyle R}{|}}{C}}CH_2R + Br_2 \xrightarrow{\text{light}} R_2C{=}\overset{\overset{\displaystyle Br}{|}}{\underset{\underset{\displaystyle R}{|}}{C}}CHR + HBr$$ 5.5

$+ X_2 \xrightarrow{\text{FeX}_3}$ $-X + HX$ 5.3A

Addition to alkenes

$$R_2C{=}CR_2 + HX \longrightarrow R_2CH{-}\overset{\overset{\displaystyle X}{|}}{C}R_2$$ 3.6B

$$R_2C{=}CR_2 + X_2 \longrightarrow R_2\overset{\overset{\displaystyle X}{|}}{C}{-}\overset{\overset{\displaystyle X}{|}}{C}R_2$$ 3.6E

Substitution reactions of alcohols

$$ROH \xrightarrow[\text{or HCl}(+\text{ZnCl}_2)]{\text{HI, HBr,}} RX + H_2O$$ 8.8A

$$ROH \xrightarrow{\text{PBr}_3} RBr$$ 8.8B

$$ROH \xrightarrow{\text{SOCl}_2} RCl$$ 8.8B

7.6 SUBSTITUTION REACTIONS — THE S$_N$2 MECHANISM

A. The Nucleophile

A **nucleophile** ("nucleus lover") is an electron-rich ion or molecule that reacts at a positively charged site in a compound.

Nu:$^-$ represents a negatively charged nucleophile, such as $^-:\!\overset{..}{\underset{..}{O}}H$ or $^-:\!C{\equiv}N:$

Nu: represents an uncharged nucleophile, such as $H_2\overset{..}{O}:$ or $R\overset{..}{\underset{..}{O}}H$

TABLE 7.5 SOME NUCLEOPHILES THAT REACT WITH ALKYL HALIDES

Formula	Name
Strong or moderately strong	
$^-:\ddot{O}H$	Hydroxide ion
$^-:\ddot{O}R$	Alkoxide ion
$^-:\ddot{S}H$	Hydrogen sulfide ion
$^-:\ddot{S}R$	Alkyl sulfide ion
$^-:C\equiv N:$	Cyanide ion
$^-:C\equiv CR$	Acetylide ion
$:\ddot{I}:^-$	Iodide ion
$R_3N:$	Amine
$CH_3\overset{\overset{\displaystyle O}{\|}}{C}\ddot{O}:^-$	Acetate ion
Weak	
$H_2\ddot{O}:$	Water
$R\ddot{O}H$	Alcohol

The C—X carbon of an alkyl halide is a partially positive site that can be attacked by a variety of nucleophiles to yield substitution products. Table 7.5 lists some nucleophiles that can react with alkyl halides.

General Equation for a Substitution Reaction:

$$Nu:^- \curvearrowright \overset{\delta+}{C}H_2 \curvearrowleft \overset{\delta-}{\ddot{X}:} \longrightarrow Nu-\underset{\underset{R}{|}}{C}H_2 + :\ddot{X}:^-$$

$$\underset{R}{|}$$

Specific Example of a Substitution Reaction:

$$H\ddot{O}:^- + \curvearrowright \overset{}{C}H_2 \curvearrowleft \ddot{B}r: \longrightarrow H\ddot{O}-\underset{\underset{CH_3}{|}}{C}H_2 + :\ddot{B}r:^-$$

$$\underset{CH_3}{|}$$

A 1° alkyl halide

Because attack is by a nucleophile, these reactions are called **nucleophilic substitution reactions,** a term that is contrasted to *electrophilic* substitution (attack by E$^+$).

PROBLEM 7.6.   Predict the products:

(a) $(CH_3)_2CHI + K^+ \ ^-:C\equiv N: \longrightarrow$

In **(a)** the C of ^-CN is more nucleophilic than the N.

(b) $CH_3I + CH_3CH_2\ddot{S}:^- Na^+ \longrightarrow$

(c) $(CH_3)_2CHCH_2CH_2Cl + CH_3CH_2\ddot{O}:^- Na^+ \longrightarrow$

(d)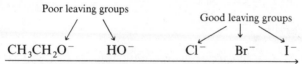

B. The Leaving Group

In a nucleophilic substitution reaction of an alkyl halide, the halide ion that is displaced is called the **leaving group.** Halide ions are excellent leaving groups because they are such weak bases and thus are quite stable in solution. A more basic ion is more difficult to displace.

Poor leaving groups

Good leaving groups

$$CH_3CH_2O^- \qquad HO^- \qquad Cl^- \qquad Br^- \qquad I^-$$

Decreasing basicity; increasing ability to be displaced
in a nucleophilic substitution reaction

The iodide ion is a better leaving group than the bromide ion or the chloride ion because it is the least basic of the halide ions. It is the least basic because it is more stable—its larger size allows the negative charge to be more delocalized.

C. The S$_N$2 Mechanism

Two principal types of substitution paths for alkyl halides are known—the S$_N$2 mechanism and the S$_N$1 mechanism. We discuss the S$_N$2 mechanism here and defer the discussion of the S$_N$1 mechanism until Section 7.7.

The more useful of these reaction types is the **bimolecular nucleophilic substitution reaction,** or **S$_N$2 reaction.** The symbol S$_N$2 stands for "substitution, nucleophilic, bimolecular." The term *biomolecular* will be defined when we discuss the mechanism of this reaction.

The reaction of bromoethane with hydroxide ion is an example of a reaction that proceeds by an S$_N$2 path. The hydroxide ion approaches the carbon bonded to the halogen from the rear—the opposite side of the carbon from the halogen. As the carbon–oxygen bond is forming, the carbon–bromine bond is breaking. As these bonds are forming and breaking, the other three atoms or groups bonded to the carbon flatten out, forming a plane, and then move to the opposite side of the carbon, away from the incoming nucleophile. We say that the molecule undergoes an **inversion of configuration**, called a **Walden inversion** after the chemist who first reported it in 1895.

S_N2 Mechanism, Backside Attack with Inversion:

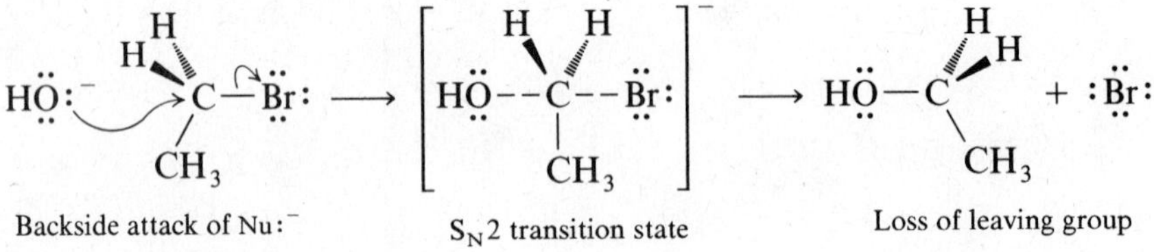

Backside attack of Nu:⁻ S_N2 transition state   Loss of leaving group

The S_N2 mechanism is a one-step backside displacement of a leaving group by a nucleophile. It is called a **bimolecular** reaction because *two* particles (molecules or ions) are involved in the transition state of the slowest (rate-limiting) reaction step, the only step in this particular reaction.

General S_N2 Mechanism for Alkyl Halides:

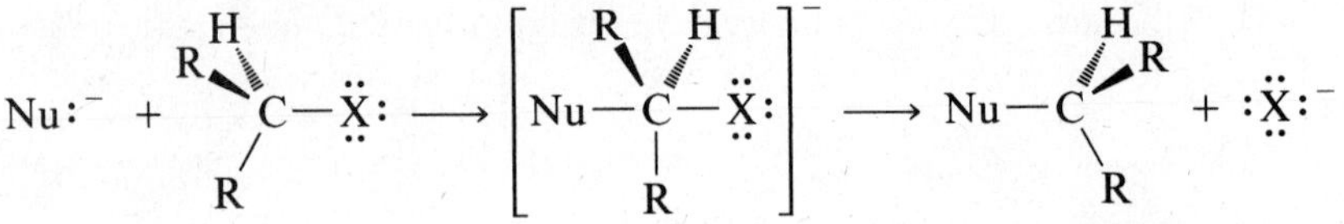

In the reaction of bromoethane with hydroxide ion, we cannot detect inversion of configuration because both the starting material and the product are achiral. If a pure enantiomer of a chiral alkyl halide is used as the starting material, however, the inversion can be detected because of the change in configuration.

Inversion of Configuration of a Chiral Starting Alkyl Halide:

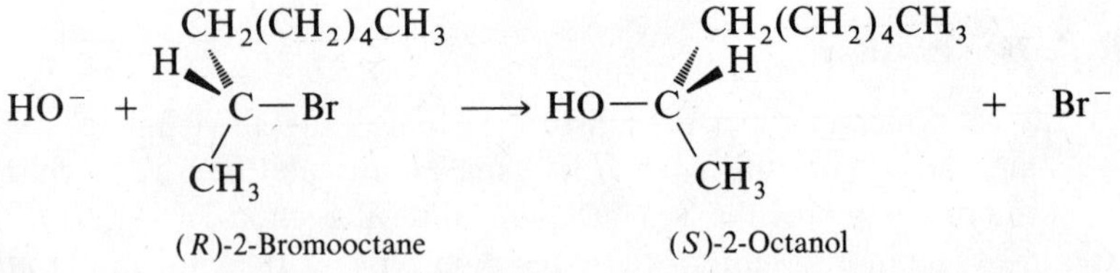

(R)-2-Bromooctane   (S)-2-Octanol

PROBLEM 7.7. Write an equation for the reaction of 1-chloropropane with Na⁺ ⁻:ÖCH₃ showing the structure of the S_N2 transition state.

D. Transition State and Energy in an S$_N$2 Reaction

Recall that a transition state is the highest-energy arrangement of atoms and molecules as they pass from being reactants to being products. Also recall that a transition state is not an intermediate as is a carbocation. An S_N2 reaction is a one-step reaction with a single transition state. The S_N2 transition state has the reacting carbon bonded equally to the attacking nucleophile and to the leaving group.

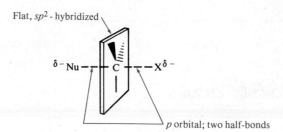

This transition state represents the highest-energy state in the course of the S_N2 reaction. Figure 7.1 is an energy diagram for an S_N2 reaction.

E. Rate of an S$_N$2 Reaction

One way we know that *two* particles (the nucleophile and the alkyl halide) are involved in the transition state of an S_N2 reaction is that, when other variables (such as temperature and solvent) are held constant, the rate of reaction is directly proportional to the concentrations of *two* reactants.

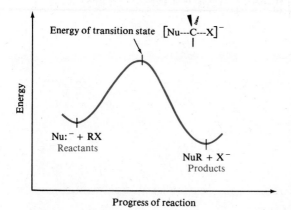

Figure 7.1 Energy diagram for an S_N2 reaction, showing the one step and a single transition state.

$$S_N2 \text{ reaction rate} = k[\text{Nu:}^-][\text{RX}]$$

where k = a proportionality constant, called the **rate constant**
$[\text{Nu:}^-]$ = molarity of Nu:^-
$[\text{RX}]$ = molarity of RX

If the concentration of either reactant (the nucleophile or the alkyl halide) is doubled, the reaction rate doubles. If *both* concentrations are doubled, the reaction rate quadruples (four times the original).

PROBLEM 7.8.   What would happen to the rate of the reaction of bromoethane with hydroxide ion if the concentration of hydroxide ion, $[^-\text{OH}]$, were changed from 0.010 M to 0.0010 M?

F. Reactivity in an S_N2 Reaction

Only methyl halides, primary alkyl halides, and secondary alkyl halides undergo reaction by the S_N2 mechanism. Tertiary alkyl halides (R_3CX) do not undergo reaction by this path because of crowding and repulsion of atoms, called **steric hindrance**, in the transition state.

Figure 7.2 shows the increasing amount of steric hindrance in the series $CH_3X < 1°\ RX < 2°\ RX < 3°\ RX$. This increasing amount of steric hindrance

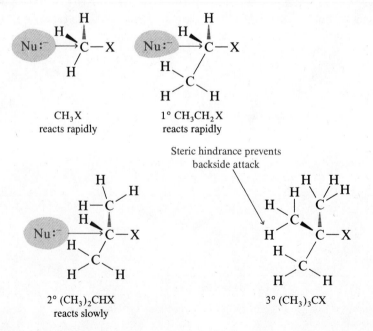

Figure 7.2 Steric hindrance increases in the series CH_3X, 1° RX, 2° RX, and 3° RX.

explains the laboratory observation that the rate of S$_N$2 reaction increases in the series 2° RX, 1° RX, and CH$_3$X and that the rate is negligible for 3° RX.

Relative Reactivities of Alkyl Halides in an S$_N$2 Reaction:

	$(CH_3)_3CX$	$(CH_3)_2CHX$	CH_3CH_2X	CH_3X
Average relative S$_N$2 rates:	0	1	40	1200

$$\underrightarrow{\text{(3° RX)   2° RX   1° RX   CH}_3\text{X}}$$
Decreasing steric hindrance; increasing S$_N$2 rate

Aryl and vinylic halides do not undergo S$_N$2 reactions, partly because the sp^2-hybrid orbitals of carbon form slightly stronger bonds than do sp^3-hybrid orbitals and partly because the geometry of the molecule prevents backside attack by the nucleophile.

Aryl and Vinylic Halides Do Not Undergo S$_N$2 Reactions:

$\langle\hspace{-4pt}\bigcirc\hspace{-4pt}\rangle$—Br + ⁻OH $\longrightarrow$ no reaction

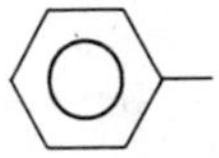

$CH_2{=}CHCl$ + ⁻OH $\longrightarrow$ no reaction

EXAMPLE

Rank the following halogen compounds in order of expected reactivity under S$_N$2 conditions:

$$\text{CH}_3$$
$$|$$
(a) $CH_3CHCH_2CH_2Cl$

$$\text{Cl}$$
$$|$$
(b) $CH_3CH_2C{=}CHCH_3$

$$\text{Cl}$$
$$|$$
(c) $CH_3CH_2CH_2CHCH_3$

Solution: **(a)** is primary (fast reaction), **(b)** is vinylic (no reaction), and **(c)** is secondary (slow reaction). Therefore, the ranking is as follows:

$$\underrightarrow{\textbf{(b)   (c)   (a)}}$$
Increasing reactivity

PROBLEM 7.9. The following compounds are warmed with 0.010 M KOH in ethanol as the solvent. Rank these compounds in order of S_N2 reactivity, beginning with any compounds that do not undergo reaction by an S_N2 path.

G. Syntheses with S$_N$2 Reactions

The S_N2 reactions of alkyl halides are useful reactions for synthesizing other types of organic compounds. Table 7.6 lists some of these compounds. As we have pointed out, not all alkyl halides react with all nucleophiles. For a synthesis to succeed, the starting alkyl halide should be methyl or primary. Secondary alkyl halides can also be used, but yields of product will be lower because of side reactions (discussed in Section 7.9).

TABLE 7.6 SOME TYPES OF COMPOUNDS THAT CAN BE SYNTHESIZED FROM METHYL, 1°, AND 2° ALKYL HALIDES

Reactants		Products	
$RX + {}^-:\ddot{O}H$	$\longrightarrow$	$R\ddot{O}H$	alcohols
$RX + {}^-:\ddot{O}R'$	$\longrightarrow$	$R\ddot{O}R'^{a}$	ethers
$RX + {}^-:\ddot{S}H$	$\longrightarrow$	$R\ddot{S}H$	thiols
$RX + {}^-:\ddot{S}R'$	$\longrightarrow$	$R\ddot{S}R'^{a}$	sulfides
$RX + {}^-:C\equiv N:$	$\longrightarrow$	$RC\equiv N:$	nitriles
$RX + {}^-:C\equiv CR'$	$\longrightarrow$	$RC\equiv CR'^{a}$	alkynes
$RX + {}^-:\ddot{I}:$	$\longrightarrow$	$R\ddot{I}:$	alkyl iodides
$RX + :NR'_3$	$\longrightarrow$	$R\overset{+}{N}R'^{b}_3$	amine salts

[a] R' may be alkyl, aryl, vinyl, etc. In the preparation of ethers, R' cannot be vinyl. ("Vinyl alkoxides," $CH_2\text{=}CHO^-$, called *enolates*, react differently; see Section 13.1.)

[b] The substitution reactions of ammonia and amines are discussed in Section 14.5A.

One of the best general synthetic reactions for preparing ethers (ROR' or ROAr, where Ar is an aryl group) is the reaction of an alkoxide, such as Na$^+$ $^-$OCH$_3$, or a phenoxide with an alkyl halide. This type of reaction is called the **Williamson ether synthesis,** named after the British chemist Alexander Williamson, who discovered the reaction in 1850.

General Equation for Williamson Ether Synthesis:

$$R-X \; + \; ^-OR' \; \text{or} \; ^-OAr \; \longrightarrow \; \underline{R-OR' \; \text{or} \; R-OAr} \; + \; X^-$$

As the Na$^+$ or K$^+$ salt   *An ether*

The preparation of alkoxides ($^-$OR') and phenoxides ($^-$OAr), used as starting materials in the ether synthesis, is discussed in Section 8.7.

EXAMPLE

Write equations for the preparation of the following ethers by S_N2 reactions of organohalogen compounds:

(a) $CH_3CH_2OCH_2CH_2CH_3$

(b)

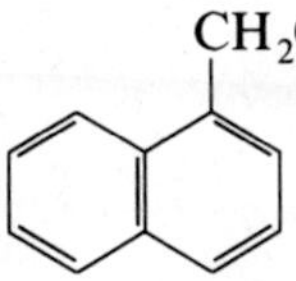

CH_2OCH_3

(c)

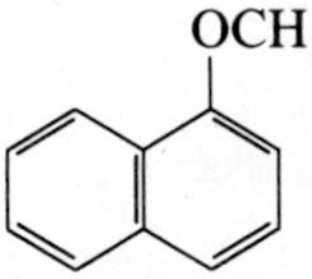

OCH_3

Solution:

(a) Either of the following reactions could be used. RI or RCl could be used instead of RBr.

(1) $CH_3CH_2Br + {}^-OCH_2CH_2CH_3$ ⟶

                         $CH_3CH_2OCH_2CH_2CH_3 + Br^-$

(2) $CH_3CH_2O^- + BrCH_2CH_2CH_3$ ⟶

(b)

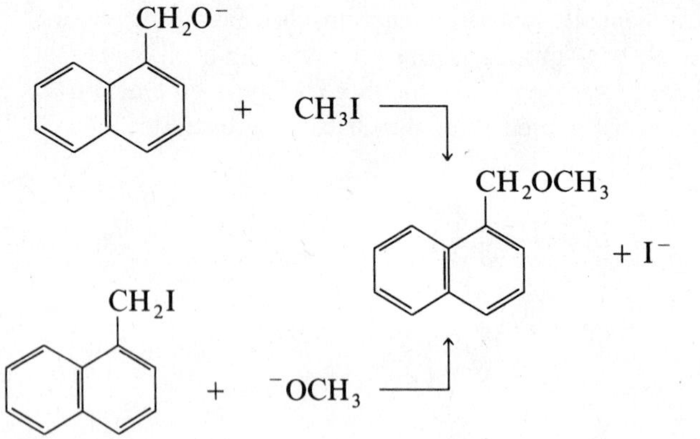

(c)

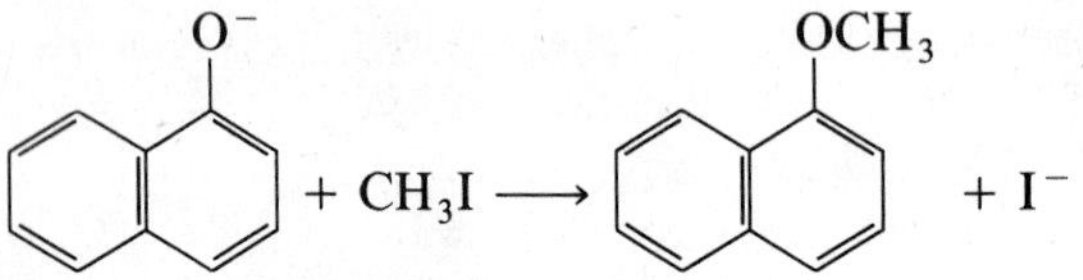

In **(c)**, the methyl halide must be used because aryl halides do not undergo reaction by an S_N2 path.

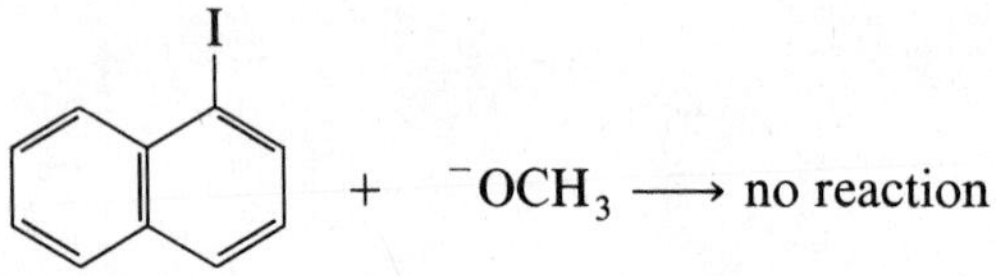

PROBLEM 7.10. Complete the following equations for substitution reactions. (Note that we are not including electron dots in these equations. You may find it helpful to insert them.)

(a) $CH_3CH_2CH_2Br + Na^+\ {}^-C{\equiv}CH \longrightarrow$

(b) $CH_3CH_2CH_2Cl + K^+\ {}^-C{\equiv}N \longrightarrow$

(c) $CH_3\overset{\overset{\displaystyle I}{|}}{C}HCH_3 + Na^+\ {}^-OCH_3 \longrightarrow$

(d) $CH_3I + CH_3\overset{\overset{\displaystyle O^-\ K^+}{|}}{C}HCH_3 \longrightarrow$

PROBLEM 7.11. Write an equation representing the synthesis of each of the following compounds from an organohalogen compound:

(a) $CH_3CH_2SCH_2CH_3$

(b) ⟨○⟩—CH_2OH

(c) $CH_2{=}CHCH_2C{\equiv}N$

(d) ⟨⟩—CH_2OCH_3

7.7 SUBSTITUTION REACTIONS — THE S$_N$1 MECHANISM

Tertiary alkyl halides do not undergo substitution by an S$_N$2 path because of steric hindrance. However, when treated with a very weak nucleophile, such as water or an alcohol, tertiary alkyl halides (but not aryl or vinylic halides) can undergo substitution by a different path. The **S$_N$1 mechanism** (substitution, nucleophilic, unimolecular) is the ionization of an organic compound into a carbocation and the leaving group, followed by the combination of the carbocation with the weak nucleophile.

General Equation for an S$_N$1 Reaction:

$$R_3C\ddot{X}{:} \xrightarrow[\text{step 1}]{-\,{:}\ddot{X}{:}} R_3C^+ \xrightarrow[\text{step 2}]{{:}Nu^-} R_3CNu$$

A carbocation

Specific Examples of S$_N$1 Reactions:

$$(CH_3)_3C{-}\ddot{C}l{:} \xrightarrow[\text{step 1}]{-\,{:}\ddot{C}l{:}^-} (CH_3)_3\overset{+}{C} \xrightarrow[\text{step 2}]{H_2\ddot{O}{:}}$$

A 3° alkyl halide

$$(CH_3)_3C{-}\overset{+}{\underset{\underset{H}{|}}{\ddot{O}H}} \overset{-H^+}{\rightleftharpoons} (CH_3)_3C{-}\ddot{O}H$$

A 3° alochol

$$(CH_3)_3C{-}\ddot{C}l{:} \xrightarrow[\text{step 1}]{-\,{:}\ddot{C}l{:}^-} (CH_3)_3\overset{+}{C} \xrightarrow[\text{step 2}]{CH_3\ddot{O}H}$$

$$(CH_3)_3C{-}\overset{+}{\underset{\underset{H}{|}}{\ddot{O}CH_3}} \overset{-H^+}{\rightleftharpoons} (CH_3)_3C{-}\ddot{O}CH_3$$

An ether

Because these S_N1 reactions of tertiary alkyl halides are reactions with the solvent (such as water or an alcohol), they are called **solvolysis reactions** (from "solvent" and the Greek *lysis*, "loosening" or "breaking").

Reaction of an alkyl halide by an S_N1 path proceeds only with a weak nucleophile because the basicity of stronger nucleophiles causes elimination reactions, which yield alkenes. Even the weak nucleophiles cause some elimination. We discuss elimination reactions later in this chapter. Because they yield mixtures of products, S_N1 reactions of alkyl halides are not particularly useful as a general synthetic technique.

A. The S$_N$1 Mechanism

Let us inspect the S_N1 mechanism for the reaction of *t*-butyl chloride and water.

The S_N1 Mechanism:

Step 1, Ionization (the Slow Step in the Sequence):

$$(CH_3)_3C\!-\!\ddot{C}l\!: \xrightarrow{slow} [(CH_3)_3C\!-\!-\!\ddot{C}l\!:] \longrightarrow (CH_3)_3C^+ \;+\; :\!\ddot{C}l\!:^-$$

A 3° alkyl halide   *Transition state*   *A carbocation intermediate*

Step 2, Combination:

$$(CH_3)_3C^+ \;+\; :\!\overset{\displaystyle H}{\underset{\displaystyle\cdot\cdot}{O}}\!H \xrightarrow{fast} (CH_3)_3C\!-\!\overset{\displaystyle H}{\overset{+}{\underset{\displaystyle\cdot\cdot}{O}}}\!H$$

A protonated alcohol

Step 3, Loss of H^+ to Solvent—An Acid-Base Reaction:

$$(CH_3)_3C\!-\!\overset{\displaystyle \overset{H}{|}}{\overset{+}{\underset{\displaystyle\cdot\cdot}{O}}}\!H \;\underset{}{\overset{fast}{\rightleftharpoons}}\; (CH_3)_3C\!-\!\ddot{O}H \;+\; H^+$$

A 3° alcohol

The final step in the solvolysis of an alkyl halide is the loss of a proton by a protonated alcohol or ether. This reaction is an acid-base reaction and is not actually part of the S_N1 mechanism. The S_N1 path is a two-step sequence: (1) ionization of the alkyl halide to yield the intermediate carbocation and (2) combination of the carbocation with the nucleophile.

When an S_N2 reaction occurs at a chiral carbon, the configuration of that carbon becomes inverted in the product. In an S_N1 reaction at the chiral carbon of an optically active alkyl halide, **racemization** takes place. Racemization is the conversion of a single enantiomer to a racemic mixture. Examination of the first step of the reaction reveals why this occurs.

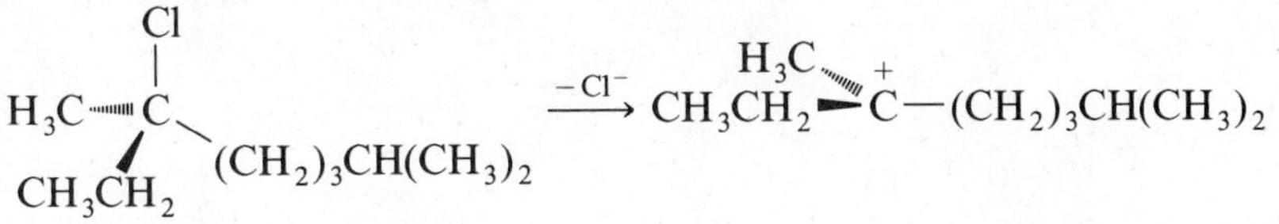

(R)-3-Chloro-3,7-dimethyloctane   *Planar, achiral carbocation*

The ionization in step 1 leads to a planar achiral carbocation. Attack by a nucleophile can occur at either side of the positively charged carbon to yield the product. Some of the carbocations react to form the (R) product while others react to form the (S) product. The result is the mixture of (R) and (S), a racemic product.

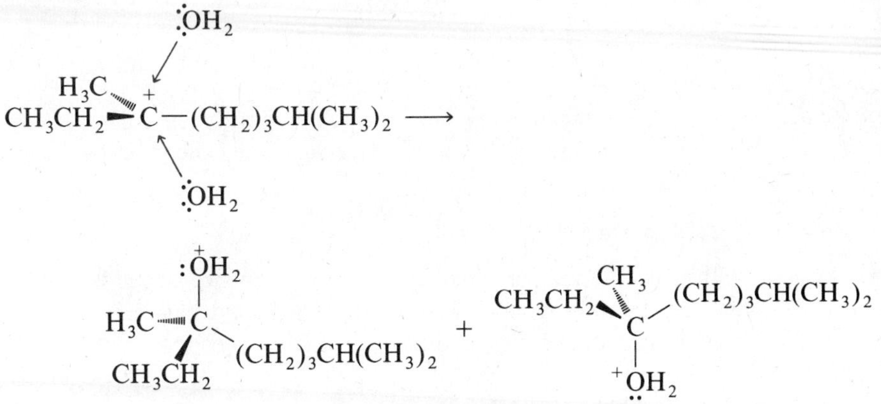

(R)   (S)

PROBLEM 7.12. Write equations for the steps in the following S$_N$1 reactions. (Use electron dots and curved arrows to show the movement of electron pairs.)

(a) 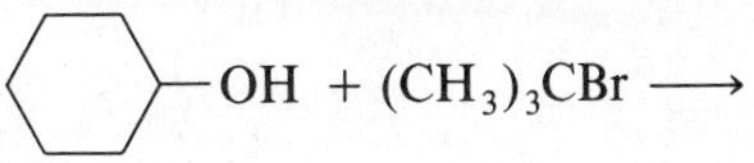 —OH + (CH$_3$)$_3$CBr ⟶

(b) 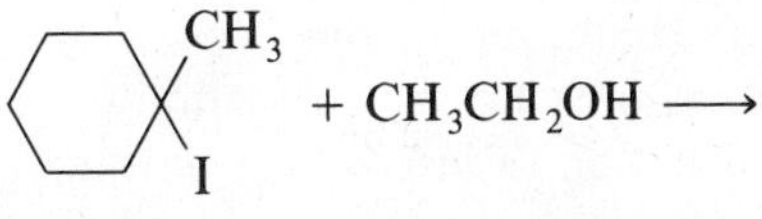 + CH$_3$CH$_2$OH ⟶

B. Energy in an S$_N$1 Reaction

Figure 7.3 is an energy profile for the two steps of an S$_N$1 mechanism—ionization and the subsequent combination of the intermediate carbocation with a nucleophile. Because the ionization is the slow step in the reaction, the energy of its transition

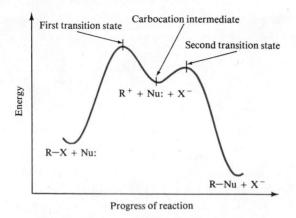

Figure 7.3 Energy diagram for an S_N1 reaction, showing two steps: ionization and combination.

state is the high-energy point on the diagram. The carbocation intermediate—high in energy and reactive—shows as a dip on the energy curve.

C. Rate of an S_N1 Reaction

Because the rate of an entire reaction sequence depends on the rate of its slowest step, the rate of an S_N1 reaction of an alkyl halide is determined by the rate of ionization of the alkyl halide.

$$\textit{Step 1:} \qquad R_3CX \xrightarrow{\text{slow}} \left[\overset{\delta+}{R_3C} \text{--} \overset{\delta-}{X} \right] \longrightarrow R_3C^+ \quad + \quad X^-$$

Transition state
for ionization

The only reactant in this step is the alkyl halide itself; therefore, the reaction rate is proportional to the concentration of alkyl halide alone (assuming sufficient nucleophile is present for reaction to proceed).

$$S_N1 \text{ reaction rate} = k[RX]$$

Because only one particle is involved in the transition state of the rate-limiting step, this reaction is *unimolecular* (Latin *unus*, "one").

Other factors, such as temperature and the structure of the solvent, affect the rate of reaction. The type of solvent has a large effect on the S_N1 reaction rate. A polar solvent, such as water or a solution of an organic solvent in water, increases the S_N1 rate by stabilizing the intermediate ions.

D. Reaction of Alkyl Halides in an S_N1 Reaction

The solvolysis reactions of methyl halides and primary and secondary alkyl halides are extremely slow when compared to the solvolysis of a tertiary alkyl halide.

Relative Rates of Solvolysis in Water:

CH_3Br	CH_3CH_2Br	$(CH_3)_2CHBr$	$(CH_3)_3CBr$
1	1	12	1,200,000

The differences in these relative rates can be attributed to carbocation stability (see Section 3.6C). Methyl halides and primary halides do not undergo step 1, the ionization step, of the S_N1 path, and secondary halides ionize very slowly.

Increasing stability
————————————————————→

$(\overset{+}{C}H_3)$ $(CH_3\overset{+}{C}H_2)$ $(CH_3)_2\overset{+}{C}H$ $(CH_3)_3\overset{+}{C}$

Not formed   *Formed very slowly*   *Formed fairly rapidly*

Based on carbocation stabilities, the relative reactivities of alkyl halides in an S_N1 reaction are:

$(CH_3X$ and 1° RX)   2° RX   3° RX
————————————————————————————————→
Increasing S_N1 rate

Interestingly, allylic and benzylic halides, even if they are primary, undergo solvolysis reactions readily—almost as rapidly as tertiary halides and, in some cases, more rapidly. Again, carbocation stability is the reason. Allylic and benzylic halides can undergo ionization because their carbocations are resonance stabilized.

Solvolysis of Allyl Chloride:

$CH_2\text{=}CHCH_2Cl \xrightarrow{-Cl^-} [CH_2\text{=}CH\text{-}\overset{+}{C}H_2 \longleftrightarrow \overset{+}{C}H_2\text{-}CH\text{=}CH_2] \xrightarrow{H_2O}$

Resonance-stabilized allyl cation

$$\left\{ \begin{array}{c} \overset{+}{:}OH_2 \\ | \\ CH_2\text{=}CH\text{-}CH_2 \\ \text{same as} \\ \overset{+}{:}OH_2 \\ | \\ CH_2\text{-}CH\text{=}CH_2 \end{array} \right\} \xrightarrow{-H^+} \begin{array}{c} OH \\ | \\ CH_2\text{=}CHCH_2 \\ \text{Allyl alcohol} \end{array}$$

PROBLEM 7.13. Arrange the following list in order of increasing reactivity toward water. (Begin with the least reactive, or nonreactive, halide.)

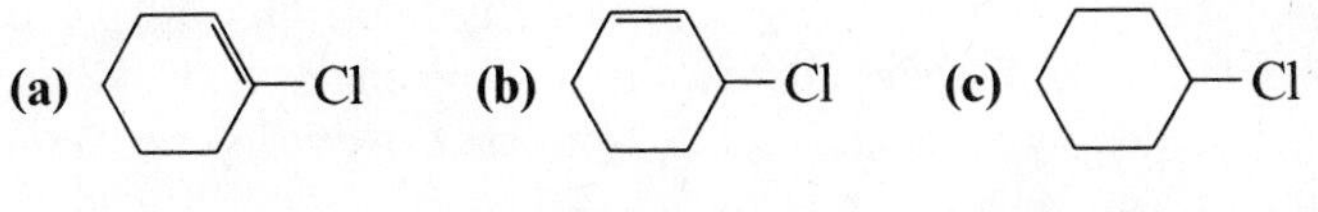

(a) [cyclohexenyl]—Cl   (b) [cyclohexenyl]—Cl   (c) [cyclohexyl]—Cl

7.8 ELIMINATION REACTIONS — THE E1 MECHANISM

S_N1 substitution reactions of alkyl halides are accompanied by elimination reactions that yield alkenes. These reactions are called **dehydrohalogenation reactions**. In these elimination reactions, the nucleophile is acting as a base—a proton acceptor.

$$
\underset{\substack{\text{t-Butyl bromide}}}{\underset{\displaystyle CH_3}{\overset{\displaystyle CH_3}{CH_3CBr}}} + H_2O \xrightarrow{25°C} \underset{\substack{\text{t-Butyl alcohol}\\(70\%)}}{\underset{\displaystyle CH_3}{\overset{\displaystyle CH_3}{CH_3COH}}} + \underset{\substack{\text{Methylpropene}\\(30\%)}}{\underset{\displaystyle CH_3}{\overset{\displaystyle CH_2}{CH_3C}}}
$$

This elimination reaction proceeds by an **E1 mechanism** (elimination, unimolecular). The first step in an E1 reaction is the ionization of the alkyl halide, the same first step as in reaction by the S_N1 path. The second step in an E1 reaction is the loss of a proton (H^+) to the solvent. The electrons in the C—H bond are used to form a pi bond. The product is a stable uncharged alkene. (In the following equations, we do not show the transition states.)

E1 Mechanism:

Step 1, Ionization:

$$(CH_3)_3C \overset{\frown}{-} \ddot{B}r\colon \xrightarrow{\text{slow}} (CH_3)_3C^+ + \colon\!\ddot{B}r\colon^-$$

Step 2, Loss of H $^+$:

$$
\underset{\substack{\displaystyle\text{Solvent}\curvearrowright}}{\underset{\displaystyle H}{CH_3\underset{+}{C}\overset{CH_3}{\underset{\curvearrowleft}{-}}CH_2}} \xrightarrow{\text{fast}} \underset{\displaystyle CH_3}{\overset{CH_3}{CH_3C}}\!=\!CH_2 + H\!-\!\text{solvent}^+
$$

7.9 ELIMINATION REACTIONS — THE E2 MECHANISM

Some nucleophiles, such as ^-OH and ^-OR, are also strong bases. A tertiary alkyl halide cannot undergo an S_N2 backside displacement because of steric hindrance; however, when heated with a strong base, usually K^+ ^-OH dissolved in ethanol, a tertiary alkyl halide undergoes an elimination reaction to yield an alkene. This elimination proceeds by a different path from that of the E1 mechanism, called an

E2 mechanism (elimination, bimolecular).

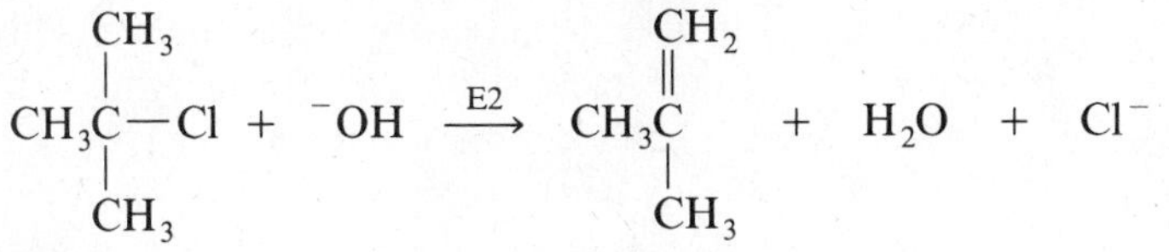

$$CH_3C{-}Cl \;+\; {}^-OH \;\xrightarrow{E2}\; CH_3C{=}CH_2 \;+\; H_2O \;+\; Cl^-$$

(with CH_3 substituents)

t-Butyl chloride   Methylpropene

A. The E2 Mechanism

An E2 reaction is a one-step reaction, just as is an S_N2 reaction. The strong base abstracts a proton from the alkyl halide, the electron pair forms a pi bond, and the halide ion leaves, all in one step.

E2 Mechanism:

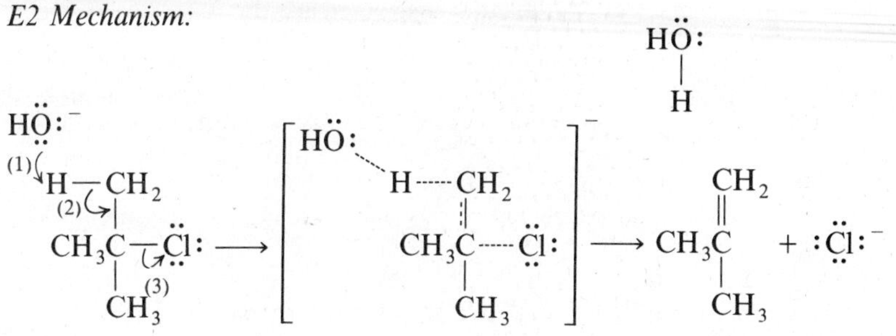

E2 transition state

(1) The ^-OH abstracts H^+.
(2) The two electrons form a pi bond.
(3) The Cl^- leaves with the pair of sigma electrons.

A reaction proceeding by an E2 mechanism is a **bimolecular elimination** because two particles (^-OH and RX) are involved in the transition state of the only step (thus the rate-limiting step) of the reaction.

$$\text{E2 reaction rate} = k[\text{base}][\text{RX}]$$

PROBLEM 7.14. Write an equation for the following reaction. In your equation, use curved arrows to show electron movement and show the structure of the transition state.

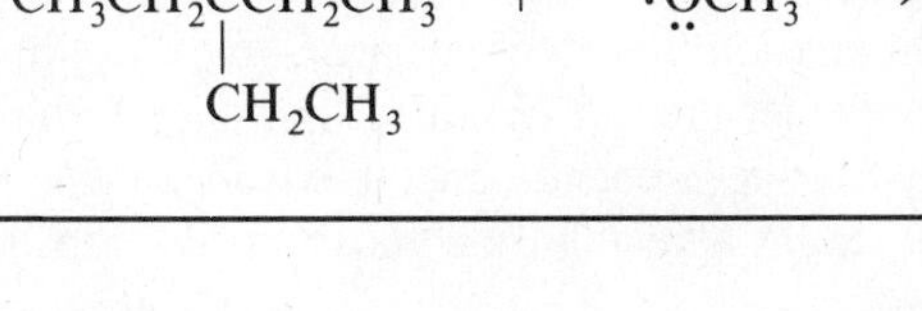

$$CH_3CH_2\overset{\displaystyle :\!\ddot{B}r:}{\underset{\displaystyle CH_2CH_3}{C}}CH_2CH_3 \;+\; {}^-:\ddot{O}CH_3 \;\xrightarrow{E2}$$

B. Reactivity in an E2 Reaction

Tertiary alkyl halides undergo elimination by an E2 path readily, secondary alkyl halides undergo E2 elimination more slowly, and primary alkyl halides undergo the slowest E2 reaction. Methyl halides cannot undergo this type of elimination.

$$\underset{\text{Increasing E2 rate}}{\underline{\text{1° RX}\qquad\text{2° RX}\qquad\text{3° RX}}}\longrightarrow$$

Primary alkyl halides are not usually subjected to E2 reactions because the substitution reaction is more likely to be formed. (Remember, S_N2 reactivity is 3° RX ≪ 2° RX < 1° RX.)

C. Alkene Mixtures

In an E2 reaction the proton abstracted by the base is a β (beta) hydrogen:

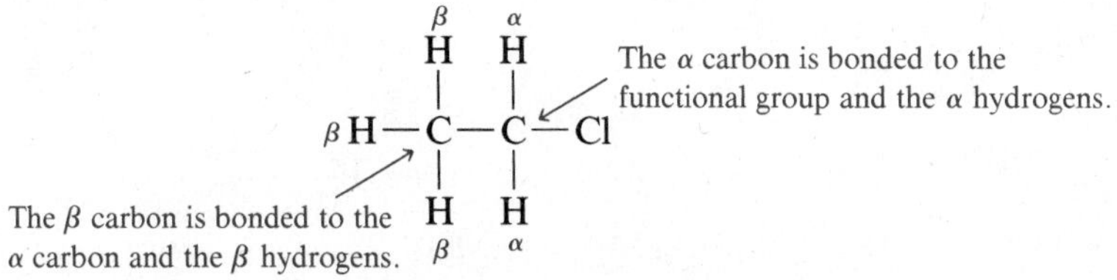

The α carbon is bonded to the functional group and the α hydrogens.

The β carbon is bonded to the α carbon and the β hydrogens.

In an E2 Reaction, a β Hydrogen is Abstracted:

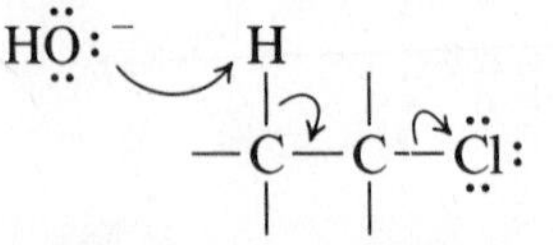

In the elimination reaction of a *t*-butyl halide, only one alkene can be formed because all the β hydrogens are equivalent. However, other alkyl halides may contain nonequivalent β hydrogens; these alkyl halides yield mixtures of alkenes when subjected to an elimination reaction.

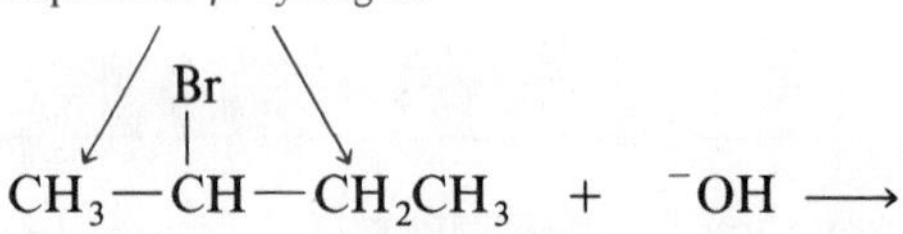

Nonequivalent β hydrogens

$$CH_3-CH-CH_2CH_3 \;+\; {}^-OH \longrightarrow$$

$$\underset{\text{1-Butene}}{CH_2{=}CHCH_2CH_3} \;+\; \underset{\textit{cis-} \text{ and } \textit{trans-}\text{2-Butene}}{CH_3CH{=}CHCH_3} \;+\; H_2O + Br^-$$

Three possible alkenes

When a typical elimination reaction is carried out in the laboratory, equal amounts of all possible alkenes are *not* obtained. In general, the most stable alkene is formed; however, the product depends on the nature of the halide and base. In simple cases, the most stable alkene is the most highly substituted trans alkene

(Section 3.8B). For example, in the elimination of 2-bromobutane, the principal product is *trans*-2-butene.

$$CH_3CHCHCH_3 \ + \ ^-OH \longrightarrow \quad C=C \quad + \ H_2O + Br^-$$

More highly substituted and trans

PROBLEM 7.15.   Rewrite the following formulas, and circle *all* hydrogens that are β to the halogen.

(a) $CH_3CH_2CCH_2CH_2CH_3$ (b)

PROBLEM 7.16.   If the alkyl halides in the preceding problem were heated with K^+ ^-OH in ethanol, what possible alkenes could be formed? (Show structures for all of them.)

PROBLEM 7.17.   In your answer to the preceding problem, circle the alkene that would probably predominate in the product mixture.

7.10 COMPETING REACTIONS

We have presented a number of reactions of alkyl halides with nucleophiles and bases. Because more than one reaction could be occurring at once, the reactions are called **competing reactions**, and mixtures of products are common.

Table 7.7 summarizes the most likely reaction paths for the different types of alkyl halides. In general, primary, allylic, and benzylic halides yield substitution products. These reactions are good preparative reactions in the laboratory.

$$1° \ RX \ + \ Nu:^- \longrightarrow RNu + X^-$$

(or allylic
or benzylic)

Tertiary alkyl halides yield alkenes when treated with strong base. These reactions are also useful. Secondary alkyl halides are also used for substitution and elimination reactions but are likely to give mixtures of products and, therefore, reduced yields of the desired products.

TABLE 7.7 SUMMARY OF REACTION MECHANISMS FOR ALKYL HALIDE REACTIONS

Halide	S_N2^a	S_N1 (E1)^b	E2^c
Methyl or 1°	Rapid with good Nu:	—	With very strong base
2°	With strong Nu:⁻	Slow with weak Nu:	With strong base
3°	—	With weak Nu:	With strong base
Allylic and benzylic	Rapid with good Nu:⁻	With weak Nu:	—

^{a}Best with high concentration of good nucleophile, such as ⁻C≡N or ⁻OH.

^{b}Best with polar solvent and weak nucleophile, such as H_2O or ROH.

^{c}Best with high concentration of strong base, such as ⁻OH or ⁻OR, and heating.

EXAMPLE

Although mixtures of products can be obtained in substitution and elimination reactions, we can often predict the most likely major product. The following examples show how we do this.

(a)

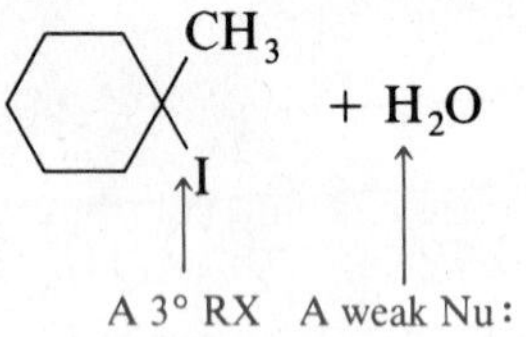

A 3° RX   A weak Nu:

We would predict reaction by an S_N1 path.

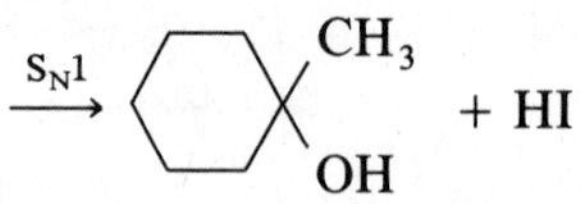

(b)

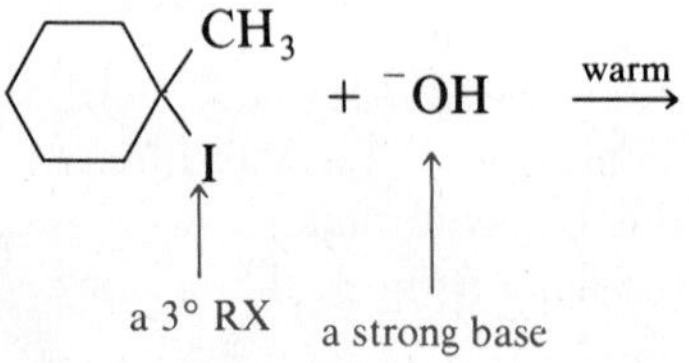

a 3° RX   a strong base

We would predict reaction by an E2 path.

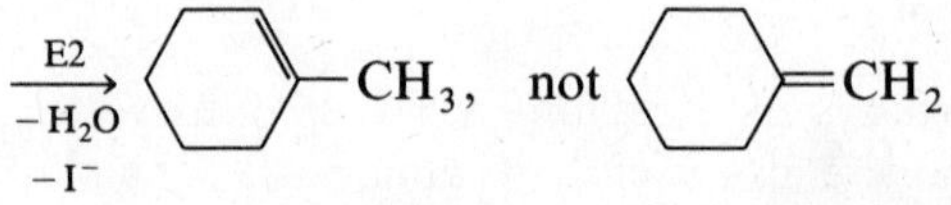

PROBLEM 7.18. (1) Predict the most likely products, (2) state the mechanistic path, and (3) explain your reasoning.

(a) $CH_3CH_2CH_2Br + K^+ \ ^-OH \xrightarrow[\text{warm}]{CH_3CH_2OH}$

(b) $(CH_3CH_2CH_2)_3CBr + K^+ \ ^-OH \xrightarrow[\text{warm}]{CH_3CH_2OH}$

(c) $(CH_3CH_2CH_2)_3CBr + H_2O \xrightarrow{25°C}$

(d) $(CH_3CH_2CH_2)_2CHBr + K^+ \ ^-OH \xrightarrow[\text{warm}]{CH_3CH_2OH}$

(e) $CH_2{=}CHCH_2I + K^+ \ ^-OH \xrightarrow{CH_3CH_2OH}$

(f) $CH_2{=}CHCH_2I + H_2O \xrightarrow{25°C}$

7.11 ORGANOMETALLIC COMPOUNDS

Carbon can form covalent bonds with metallic elements, as in tetraethyllead. A compound in which carbon is bonded directly to a metal is called an **organometallic compound**. A compound in which carbon is bonded to magnesium is called a **Grignard reagent**, after the French chemist Victor Grignard (pronounced Grin′-yard), who received a Nobel Prize in 1912 for his discovery and the chemical development of these compounds.

Two Organometallic Compounds:

$$CH_3CH_2{-}\overset{\displaystyle CH_2CH_3}{\underset{\displaystyle CH_2CH_3}{\overset{|}{\underset{|}{Pb}}}}{-}CH_2CH_3 \qquad CH_3CH_2{-}MgBr$$

Tetraethyllead   Ethylmagnesium bromide

a gasoline additive   *a Grignard reagent*

Grignard reagents are formed by the reaction of organohalogen compounds with magnesium metal in anhydrous (dry) diethyl ether ($CH_3CH_2OCH_2CH_3$) or other ethers. The halogen compound can be of almost any type—alkyl, aryl, and so forth.

Formation of Grignard Reagents (RMgX):

$$CH_3CH_2CH_2Cl + \quad Mg: \quad \xrightarrow{\text{diethyl ether}} \quad CH_3CH_2CH_2MgCl$$

Magnesium   *n*-Propylmagnesium

metal   chloride

$$\langle O \rangle - I + \quad Mg: \quad \xrightarrow{\text{diethyl ether}} \quad \langle O \rangle - MgI$$

Phenylmagnesium

iodide

In most organic compounds, such as ROH and RX, a carbon is bonded to hydrogen or to a more electronegative atom (O, X, or N). In a Grignard reagent, a carbon is bonded to an electropositive metal; therefore, the carbon bonded to the magnesium in a Grignard reagent is electronegative with respect to magnesium. This carbon carries a partial negative charge.

C is $\delta+$ C is $\delta-$

$\overset{\delta+}{R_3C} - \overset{\delta-}{X}$ $\overset{\delta-}{R_3C} - \overset{\delta+}{MgX}$

Grignard reagents are extremely strong bases, much stronger than the hydroxide ion. They are readily destroyed by reaction with water, alcohols, and many other compounds.

$$\left. \begin{array}{l} R_3C{-}MgX \\ \quad\quad\quad \\ H{-}\ddot{O}R \end{array} \right\} \longrightarrow \quad \begin{array}{l} R_3C{-}H \quad {}^+MgX \\ \quad\quad \\ \quad {}^-{:}\ddot{O}R \end{array}$$

Some Compounds That Can Lose Protons to RMgX:

HO—H	H$_2$N—H	RC≡C—H
RO—H	RNH—H	RCO$_2$—H
ArO—H	R$_2$N—H	RS—H

Because of this extreme basicity, all equipment, solvents, and reagents used in Grignard reactions are specially dried. Reagents containing hydroxyl groups (HOH, ROH, ArOH) or other groups in the list cannot be used in Grignard reactions. An exception would be when a hydrocarbon is the desired product. For example, deuterated products can be prepared by the reaction of a Grignard reagent with deuterium oxide (D$_2$O, or ^{2}H$_2$O).

$$CH_2{=}CHCH_2Br \xrightarrow[\text{ether}]{Mg} CH_2{=}CHCH_2MgBr \xrightarrow{D_2O} CH_2{=}CHCH_2D$$

Grignard reagents are extremely useful in synthesis. We will discuss the reactions of these reagents in future chapters.

SUMMARY

Organohalogen compounds can be classed as **haloalkanes** or **alkyl halides**, **vinylic halides**, **aryl halides**, **allylic halides**, or **benzylic halides** (see Table 7.1). These compounds are denser and higher boiling than comparable hydrocarbons.

In IUPAC names, halogens are named with a *halo-* prefix. In the trivial names, the hydrocarbon group is named (methyl, ethyl, etc.), followed by the second word *halide*. Alkyl halides can be classified as **primary** (1°), **secondary** (2°), or **tertiary** (3°).

Alkyl halides can be prepared by free-radical halogenation of an alkane, by free-radical halogenation at an allylic or benzylic position, by aromatic electrophilic halogenation, by addition of HX or X_2 to an alkene, or by a substitution reaction between an alcohol and HX (see Table 7.4).

The halide ion is a good **leaving group** because it is a very weak base. Methyl, primary, secondary, allylic, and benzylic alkyl halides undergo **substitution reactions** with **nucleophiles**, such as ^-OH or ^-OR. Reaction is by an S_N2 **path** (backside displacement, one step). Tertiary alkyl halides cannot undergo substitution by an S_N2 path because of **steric hindrance**. Aryl and vinylic halides do not undergo this reaction because their bonding is not appropriate.

$$RCH_2X \quad (\text{or } R_2CHX) + :Nu^- \xrightarrow{S_N2} RCH_2-Nu \quad (\text{or } R_2CH-Nu) + X^-$$

Williamson Ether Synthesis:

$$R-X \ + \ ^-OR' \ (\text{or } ^-OAr) \ \longrightarrow \ R-OR' \ (\text{or } R-OAr) \ + \ X^-$$

Not 3°, aryl,
or vinylic

Secondary and tertiary alkyl halides with a β hydrogen undergo **elimination of HX**, or **dehydrohalogenation**, when heated with a strong base such as a hydroxide. This reaction proceeds by an **E2 path**.

2° or 3°

The most highly
substituted
trans alkene

Secondary, tertiary, allylic, and benzylic alkyl halides also undergo S_N1 reactions (solvolysis) and E1 reactions with a weakly nucleophilic solvent, such as H_2O or an alcohol.

S_N1 *Reaction:*

$$R_3C-X \xrightarrow[\text{step 1}]{-X^-} R_3C^+ \xrightarrow[\text{step 2}]{H_2O} R_3C-\overset{+}{O}H_2 \xrightarrow{-H^+} R_3COH$$

A carbocation
intermediate

E1 Reaction:

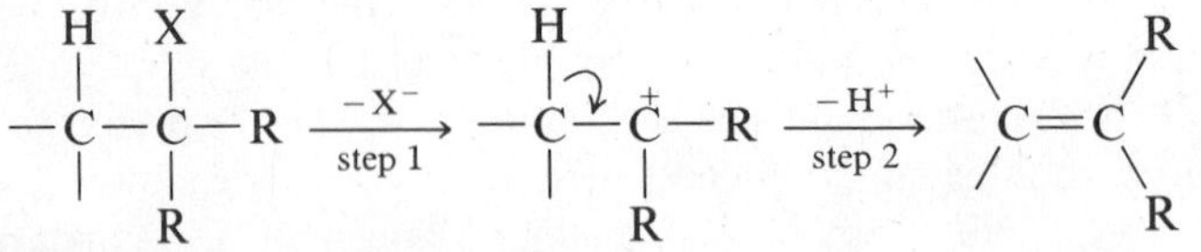

Primary alkyl halides (unless allylic or benzylic) cannot undergo S_N1 or E1 reactions because they do not form carbocations.

Alkyl halides of all varieties, aryl halides, and vinylic halides react with magnesium in a dry ether solvent to yield organomagnesium compounds (RMgX), called **Grignard reagents**.

STUDY PROBLEMS

7.19. Classify each halogen in the following structures as 1°, 2°, 3°, vinylic, aryl, benzylic, or allylic. If more than one category applies, name all of them.

(a) 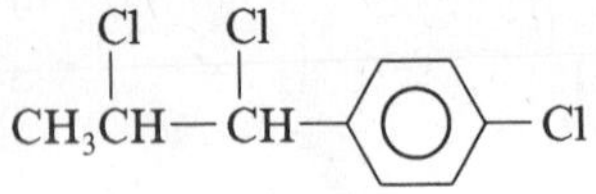
(b) 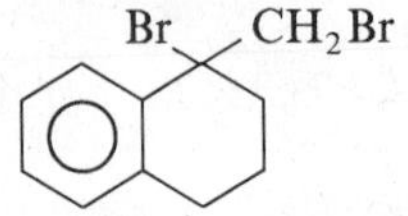
(c) 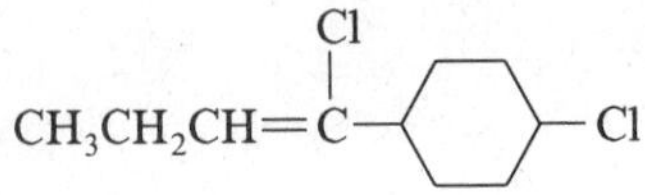
(d) 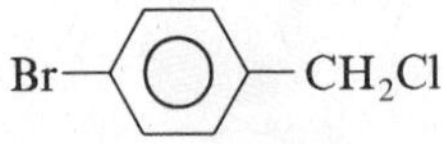
(e) 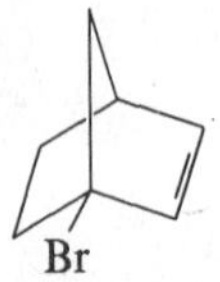
(f)

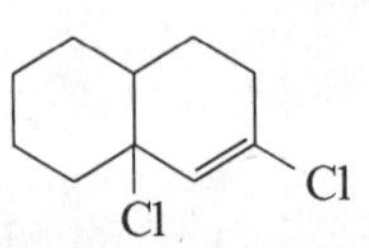

7.20. Name the following organohalogen compounds by the IUPAC system:

(a) $(CH_3)_2CHCH_2CHCH_3$

(b) $CH_3CH_2CH-CHCH_3$

(c) $BrCH_2CH_2CH=CH_2$

(d) $CH_3(CH_2)_6CBr_3$

(e)

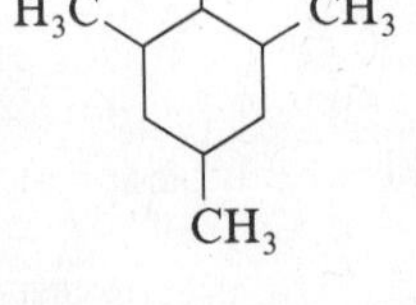

(f)

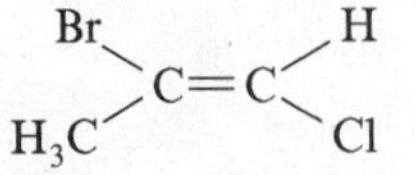

(g) Cl—⟨benzene ring with Cl⟩—OH

(h) [cyclopentane ring with H, H, Cl, Cl substituents]

(i) Cl—⟨benzene ring with Cl, Cl⟩—NH$_2$

7.21. Write formulas for the following compounds:
- **(a)** *trans*-1-*t*-butyl-4-chlorocyclohexane
- **(b)** *sec*-butyl chloride
- **(c)** 1,1,1-trichloroethane
- **(d)** benzyl iodide

7.22. Write equations showing how each of the following compounds could be prepared by the synthetic method indicated:

(a) ⟨benzene⟩—CH(Br)—⟨benzene⟩   by direct halogenation

(b) ⟨benzene⟩—CHCH$_3$ (Br)   by the addition of HX to an alkene

(c) [cyclopentane ring with Br]   by a substitution reaction of an alcohol

(d) [naphthalene ring with CHBrCH$_3$]   by direct halogenation

(e) [cyclopentane ring with Br]   by direct halogenation

(f) Br—⟨benzene⟩—CH$_3$ by direct halogenation

(g) ICH$_2$CH$_2$CH$_2$CH$_2$CH$_2$I   by a substitution reaction of an alcohol

7.23. Which of the following ions and molecules are nucleophiles? Explain.
- **(a)** I$^-$
- **(b)** $^-$CN
- **(c)** CH$_3$CH$_3$
- **(d)** H$_2$O
- **(e)** NH$_3$
- **(f)** H$^+$

7.24. Complete the following equations:

(a) ⟨benzene⟩—CH$_2$Br + CH$_3$S$^-$ ⟶

(b) CH$_3$CH$_2$CH$_2$CH$_2$Br + $^-$OCCH$_3$ (=O) ⟶

(c) CH$_3$C≡C$^-$ + CH$_3$I ⟶

(d) $(CH_3)_3C-I + NH_3 \longrightarrow$

(e) 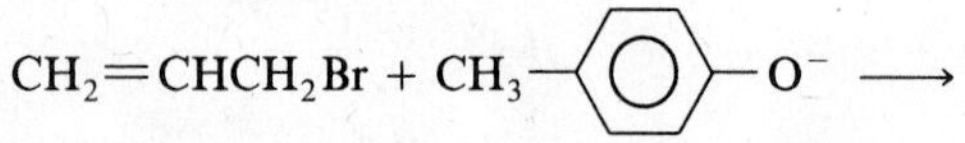 $CH_2=CHCH_2Br + CH_3-\langle\bigcirc\rangle-O^- \longrightarrow$

7.25. Which of the following groups would be good leaving groups in substitution reactions? Explain.

(a) CH_3CH_2-OH

(b) 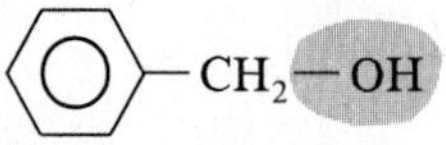 $\langle\bigcirc\rangle-CH_2-OH$

(c) $CH_3-\overset{\overset{\displaystyle O}{\|}}{O}CCH_3$

(d) $CH_3CH_2CH_2-I$

(e) $CH_3CH_2-OCH_2CH_3$

(f) CH_3CH_2-Br

7.26. Draw the structure of the transition state, showing inversion at the reacting center, for the following two S_N2 reactions. (*Hint:* First redraw the formulas.)

(a) (S)-$CH_3CH_2\overset{\overset{\displaystyle D}{|}}{\underset{\underset{\displaystyle Br}{|}}{C}}CH_3 + I^- \longrightarrow$

(b) 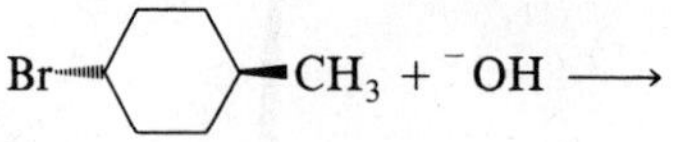 $Br\cdots\langle\bigcirc\rangle\blacktriangleleft CH_3 + {}^-OH \longrightarrow$

7.27. Write equations showing how each of the following ethers could be prepared by an S_N2 reaction:

(a) 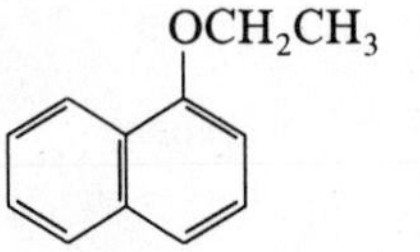 [naphthalene with OCH_2CH_3 substituent]

(b) 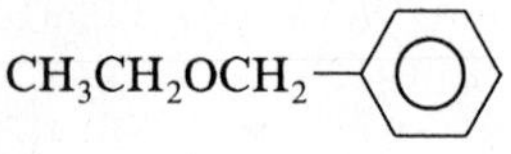 $CH_3CH_2OCH_2-\langle\bigcirc\rangle$ (two methods)

7.28. Rank the following compounds in order of probable S_N2 reactivity toward K^+ ^-OH (least reactive first):

(a) $CH_3CH_2CH_2Br$

(b) $(CH_3CH_2)_3CBr$

(c) $(CH_3CH_2)_2CHBr$

(d) CH_3Br

(e) 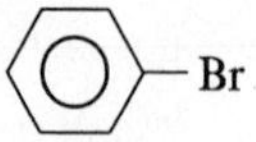 $\langle\bigcirc\rangle-Br$

7.29. Write an equation for each of the following S_N1 reactions showing the intermediate carbocation and the final product:

(a) $CH_3CH_2\overset{\overset{\displaystyle CH_3}{|}}{\underset{\underset{\displaystyle CH_3}{|}}{C}}Br + H_2O \longrightarrow$

(b) 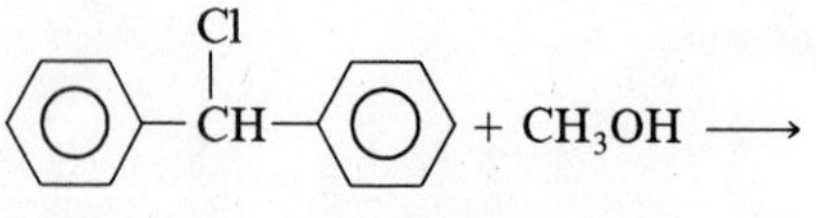 $\langle\bigcirc\rangle-\overset{\overset{\displaystyle Cl}{|}}{C}H-\langle\bigcirc\rangle + CH_3OH \longrightarrow$

(c) $\left(\langle\bigcirc\rangle\right)_3CCl + CH_3\overset{\overset{\displaystyle O}{\|}}{C}OH \longrightarrow$

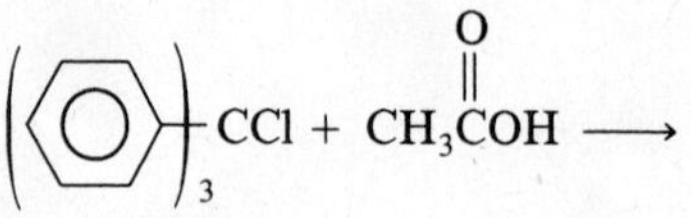

(d) $(CH_3)_3CCl +$ $-OH \longrightarrow$

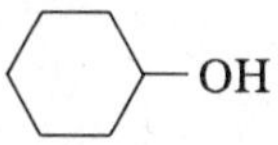

7.30. For each of the following pairs, state which member would react faster by an S_N1 pathway. Briefly explain your choices.

(a) $CH_3CH_2CH_2Br$ or $CH_3\overset{\overset{\displaystyle Br}{|}}{C}HCH_3$

(b) 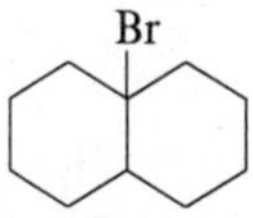 or

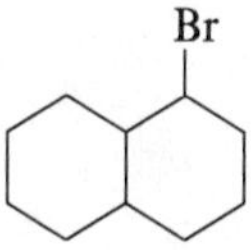

(c) 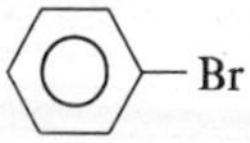 $-Br$ or $-Br$

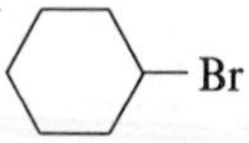

(d) $CH_2{=}CHCH_2Cl$ or $CH_2{=}CHCH_2CH_2Cl$

7.31. Write an equation for the preparation of each of the following alkenes by an E2 reaction:

(a) 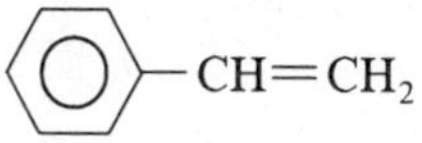 $-CH{=}CH_2$ **(b)** 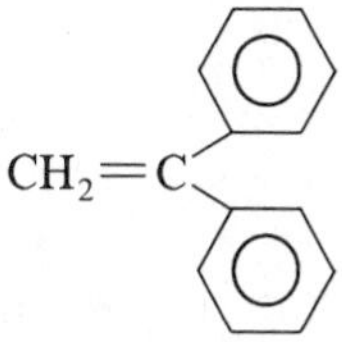 $CH_2{=}C\Big\langle$

(c) $={C}{\overset{\displaystyle \diagup CH_3}{\diagdown CH_3}}$

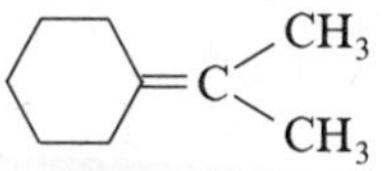

7.32. Predict the principal product of each of the following elimination reactions:

(a) $CH_3CH_2\overset{\overset{\displaystyle I}{|}}{C}(CH_3)_2 + {}^-OH \xrightarrow{\text{heat}}$

(b) 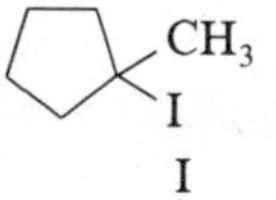 $\overset{\displaystyle CH_3}{\underset{\displaystyle I}{\diagup}} + {}^-OH \xrightarrow{\text{heat}}$

(c) $CH_3\overset{\overset{\displaystyle I}{|}}{C}H\underset{\underset{\displaystyle CH_3}{|}}{C}HCH_3 + {}^-OH \xrightarrow{\text{heat}}$

7.33. For each of the following alkyl halides, write an equation showing the S_N1 and the E1 products, if any.

(a) 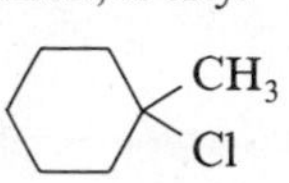 $\overset{\displaystyle CH_3}{\underset{\displaystyle Cl}{\diagup}} + H_2O \longrightarrow$

(b) 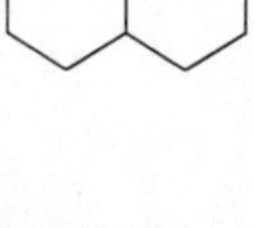 $+ CH_3OH \longrightarrow$

(c) $CH_3\overset{\displaystyle CH_2CH_3}{\underset{\displaystyle CH_2CH_2CH_3}{\overset{|}{\underset{|}{C}}Cl}}$ $+ H_2O \longrightarrow$

(d) ⬡—$\overset{\displaystyle Cl}{\overset{|}{C}H}CH_2$—⬡ $+ H_2O \longrightarrow$

7.34. Write equations showing how you would prepare the following Grignard reagents:

(a) CH_3MgI (b) ⬡—$MgBr$ (c) [naphthalene]—MgI

7.35. Write equations showing how you would make the following conversions:

(a) ⬡—$OCH(CH_3)_2$ from 2-bromopropane

(b) 2-butene from 2-bromopropane

(c) $(CH_3CH_2CH_2)_2S$ from 1-bromopropane

(d) $CH_3CH_2CH_2CN$ from 1-bromopropane

7.36. Predict which reaction in each pair would give the larger percent of substitution as opposed to elimination. Give reasons for your answers.

(a) Treatment of $CH_3CH_2CH_2Br$ or $CH_3CH_2\overset{\displaystyle CH_3}{\underset{\displaystyle CH_3}{\overset{|}{\underset{|}{C}}Br}}$ with ^-OH

(b) Treatment of ⬡$\overset{\displaystyle Cl}{\underset{\displaystyle CH_3}{<}}$ or ⬡—CH_2Cl with ^-OH

7.37. The sodium salt of 2,4-dichlorophenoxyacetic acid (2,4-D, in Point of Interest 7) is an example of a selective herbicide that overstimulates the growth of broad-leafed plants. Suggest starting materials that would be needed to prepare 2,4-D by an S_N2 reaction.

7.38. If (S)-2-iodohexane were treated with sodium iodide in acetone solvent, racemization would occur. Why?

7.39. Write equations for the following conversions. (More than one step may be required.)

(a) $CH_3CH_2CH{=}CH_2 \longrightarrow CH_3CH_2\overset{\displaystyle CN}{\overset{|}{C}H}CH_3$

(b) $CH_3CH_2CH{=}CH_2 \longrightarrow CH_3CH_2CH_2CH_2OCH(CH_3)_2$

(c) ⬡—CH$_3$ ⟶ BrMg—⬡—CH$_3$

(d) ⬡—CH$_3$ ⟶ ⬡—CH$_2$OH

7.40. Explain why the following compound does not undergo reaction with nucleophiles under either S$_N$1 or S$_N$2 conditions.

Br

POINT OF INTEREST 7

Polyhalogen Compounds

Polyhalogenated alkanes and alkenes, which are aliphatic hydrocarbons substituted with more than one halogen atom, have been widely used in our society for a variety of purposes. Part of the reason for their usefulness is the relative inertness of the carbon–halogen bond, especially the carbon–chlorine and carbon–fluorine bonds. Compounds containing chlorine and especially fluorine are less reactive than compounds containing bromine and iodine.

Chlorinated Methanes Except for chloromethane, which boils at −24°C, the chlorinated methanes are useful solvents. *Carbon tetrachloride* (tetrachloromethane, CCl_4), bp 77°C, is a colorless, water-insoluble, nonflammable liquid. It has been used as a solvent for nonpolar compounds and, as recently as the 1950s, in fire extinguishers. Its use as a solvent has declined because long-term inhalation of its vapors causes liver and kidney damage. This compound is no longer used in fire extinguishers because of its toxicity and also because it can form the deadly gas phosgene ($Cl_2C{=}O$) when used to extinguish electrical fires.

Chloroform (trichloromethane, $CHCl_3$), bp 61°C, is a sweet-smelling, water-insoluble, nonflammable liquid. Chloroform is somewhat more polar than carbon tetrachloride and is an excellent solvent for a variety of types of compounds. Along with nitrous oxide (N_2O) and diethyl ether ($CH_3CH_2OCH_2CH_3$), it was one of the first inhalation anesthetics. It is no longer used as an anesthetic and even its solvent use has decreased because inhalation of large amounts of its vapor can cause heart and lung damage, and long-term inhalation can cause liver damage. Recently, chloroform has been listed as carcinogenic (cancer-causing) by the Environmental Protection Agency (EPA). Halothane ($CF_3CHClBr$), a much safer anesthetic, is used today.

Methylene chloride (dichloromethane, CH_2Cl_2), bp 40°C, also a colorless nonflammable liquid, is not as toxic as the other liquid chlorinated hydrocarbons and is widely used as a solvent in such products as paint removers.

Chlorofluorocarbons   The chlorofluorocarbons (CFCs), often called **Freons** (a trademark registered by the Du Pont Company), are gases at room temperature and are colorless, practically odorless, water insoluble, nontoxic, nonflammable, and noncorrosive. Freon-12 (dichlorodifluoromethane, Cl_2CF_2) boils at $-30°C$ and is stable to $550°C$. CFCs are widely used as refrigerants in air conditioners and in deep-freeze units. Because of their lack of toxicity and lack of flammability, CFCs were used for years as aerosol propellants for pesticides, hair sprays, underarm deodrants, paints, and virtually every other sprayable product.

In the 1970s environmental chemists expressed concern about the effects of CFCs on the stratospheric ozone layer. Under the influence of strong ultraviolet radiation, CFCs can decompose to yield chlorine, which can catalyze the destruction of ozone—reactions that might be a factor in ozone loss from the stratosphere.

The ozone layer protects the earth's inhabitants by absorbing the strongest and most harmful rays of the sun's ultraviolet radiation. Depletion of this layer by CFCs or other agents could lead to an increase in the incidence of skin cancer, to changes in the climate, and to losses of ultraviolet-susceptible plant species. The use of CFCs in aerosol cans was banned in the United States in 1979 because of these environmental concerns.

Other Aliphatic Polyhalogen Compounds   At one time, carbon tetrachloride was used as a dry-cleaning solvent. Because of its toxicity, other less toxic halogenated compounds, such as trichloroethylene ($Cl_2C{=}CHCl$) and tetrachloroethylene ($Cl_2C{=}CCl_2$), are used today. Like carbon tetrachloride, these compounds can dissolve oil and grease and are water insoluble.

Teflon $+CF_2CF_2+_n$, is a polymer made from tetrafluoroethylene ($CF_2{=}CF_2$). This polymer is resistant to high temperatures and almost all chemicals. Teflon is used to make gaskets and stopcocks and is also used to coat cookware, such as nonstick frying pans. Thin layers of Teflon are used as stain protectors for upholstery fabric.

Aromatic Polyhalogen Compounds   Aromatic halogen compounds have been used in mothballs (*p*-dichlorobenzene), insecticides, and herbicides. The broad-leaf herbicide 2,4-D is commonly found in weed killers. The similar compound 2,4,5-T has been banned because it is often contaminated with the toxic chemical *dioxin*, a by-product of its manufacture.

"Dichlorodiphenyl-
trichloroethane" (DDT)

*an insecticide no longer
marketed in the United States*

Sodium salt of (2,4-dichloro-
phenoxy)acetic acid (2,4-D)

a broad-leaf herbicide

(2,4,5-Trichlorophenoxy)-
acetic acid (2,4,5-T)

a broad-leaf herbicide

2,3,7,8-Tetrachlorodibenzodioxin
(2,3,7,8-TCDD, or "dioxin")

*a toxic by-product of
the manufacture of 2,4,5-T*

Polychlorinated biphenyls (PCBs, see Section 7.1) and polybrominated biphenyls (PBBs) have been used for a variety of purposes. Because of their toxicity and carcinogenic properties, the use of PCBs has declined in recent years. The use of PBBs has been banned as a result of an accidental contamination of livestock feed in Michigan in 1973 and subsequent loss of thousands of farm animals.

Chapter 8

Alcohols, Thiols, and Phenols

Alcohols and phenols both contain **hydroxyl groups** (—OH). In an alcohol the hydroxyl group is bonded to an sp^3-hybridized tetrahedral carbon, and in a phenol it is bonded to the sp^2-hybridized carbon of an aromatic ring.

Alcohols (ROH):

Phenols (ArOH):

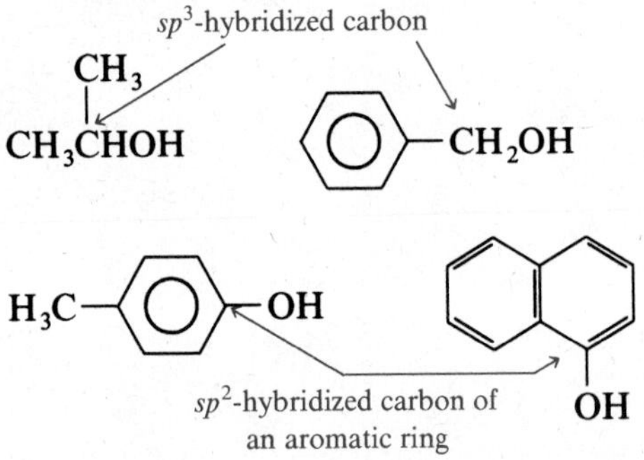

Alcohols and phenols are widely used in our industrial society and are commonly found in nature. Ethanol is used as a solvent and in beverages. Isopropyl alcohol is used as rubbing alcohol to chill and disinfect the skin. Cholesterol (Section 17.4), implicated in hardening (clogging) of the arteries, is an alcohol found in animals. Menthol, used in cigarettes and throat lozenges, is an alcohol found in mint plants. Urushiols are dihydroxybenzenes that are the principal irritants in poison ivy and poison oak.

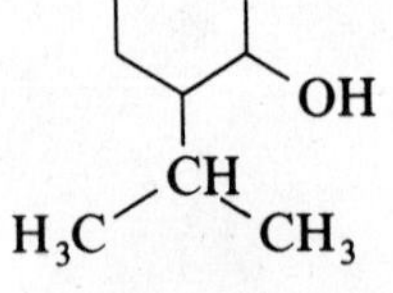

Menthol

in mint

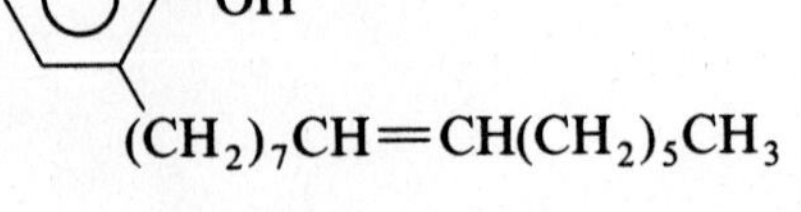

A urushiol

(*an irritant in poison ivy*)

8.1 BONDING AND STRUCTURE OF ALCOHOLS AND PHENOLS

In water, the alcohols, and the phenols, the oxygen is sp^3-hybridized and has two pair of unshared valence electrons.

$$\overset{\cdot\cdot}{\underset{\cdot\cdot}{O}} \quad \overset{\cdot\cdot}{\underset{\cdot\cdot}{O}} \quad \overset{\cdot\cdot}{\underset{\cdot\cdot}{O}}$$

H $\quad$ H $\qquad$ R $\quad$ H $\qquad$ Ar $\quad$ H

Water $\qquad\qquad$ *An alcohol* $\qquad\qquad$ *A phenol*

Because oxygen is electronegative with respect to carbon or hydrogen, alcohols and phenols, like water, are polar molecules.

$$\overset{\delta-}{O} \qquad \overset{\delta-}{O} \qquad \overset{\delta-}{O}$$

$\overset{\delta+}{H} \quad H^{\delta+} \qquad R \quad H^{\delta+} \qquad Ar \quad H^{\delta+}$

EXAMPLE

Water can undergo ionization to yield trace amounts of H_3O^+ and ^-OH. Alcohols can undergo a similar ionization. Write the equations for the ionization of water and methanol (CH_3OH).

Solution:

$$H-\overset{\cdot\cdot}{\underset{\cdot\cdot}{O}}-H \rightleftharpoons H-\overset{\cdot\cdot}{\underset{\cdot\cdot}{O}}:^- + H^+$$

$$CH_3-\overset{\cdot\cdot}{\underset{\cdot\cdot}{O}}-H \rightleftharpoons CH_3-\overset{\cdot\cdot}{\underset{\cdot\cdot}{O}}:^- + H^+$$

The more polar bond to
oxygen breaks, and the
electrons go to the oxygen.

When water undergoes ionization to H^+ and ^-OH, the proton is not actually released as a free proton. Instead, it is transferred to a second molecule of water, which is the proton acceptor. When methanol undergoes ionization, its proton is transferred similarly to another molecule of methanol. However, to simplify our equations, we usually do not show the proton acceptors.

Ionization of Water and Methanol:

$$HO-H \; + \; HO-H \rightleftharpoons HO^- \; + \; H-\overset{+}{\underset{H}{O}}-H$$

$$CH_3O-H \; + \; CH_3O-H \rightleftharpoons CH_3O^- \; + \; CH_3\overset{+}{\underset{H}{O}}H$$

8.2 PHYSICAL PROPERTIES OF ALCOHOLS AND PHENOLS

Like water, alcohols and phenols are capable of forming hydrogen bonds. Figure 8.1 shows how the —OH groups of methanol and phenol are hydrogen bonded. Because of hydrogen bonding, alcohols and phenols are higher boiling than other compounds with similar formula weights. Even the lowest-formula-weight alcohol, methanol (CH_3OH), is a liquid at room temperature.

Alkyl groups, alkenyl groups, and benzene rings are relatively nonpolar and contribute to the water insolubility of a compound. These groups are hydrophobic (water hating).

Because hydroxyl groups can form hydrogen bonds with water, they are said to be hydrophilic (water loving). The low-formula-weight alcohols are miscible with (infinitely soluble in) water. With these compounds, the influence of the hydrophilic —OH groups outweighs the influence of the hydrophobic character of the hydrocarbon portion of the molecule. 1-Butanol, $CH_3CH_2CH_2CH_2OH$, is the first member of the homologous series that is not miscible with water because of the hydrophobic nature of the *n*-butyl group. Higher-formula-weight alcohols are even less soluble.

CH_3OH	CH_3CH_2OH	$CH_3CH_2CH_2OH$	$CH_3CH_2CH_2CH_2OH$
Miscible	*Miscible*	*Miscible*	8.3 g/100 mL H_2O

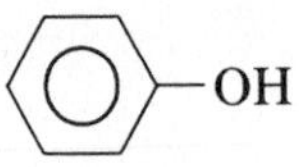

6 g/100 mL 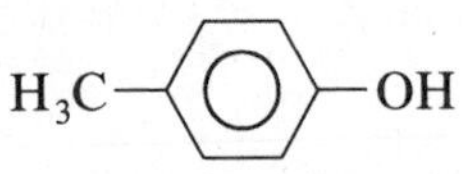   2.5 g/100 mL

Branching of a hydrocarbon chain increases water solubility by interfering with intermolecular attractions among the carbon–hydrogen groups. *t*-Butyl alcohol, $(CH_3)_3COH$, is miscible with water even though 1-butanol is not.

Table 8.1 lists some alcohols and phenols and their physical properties.

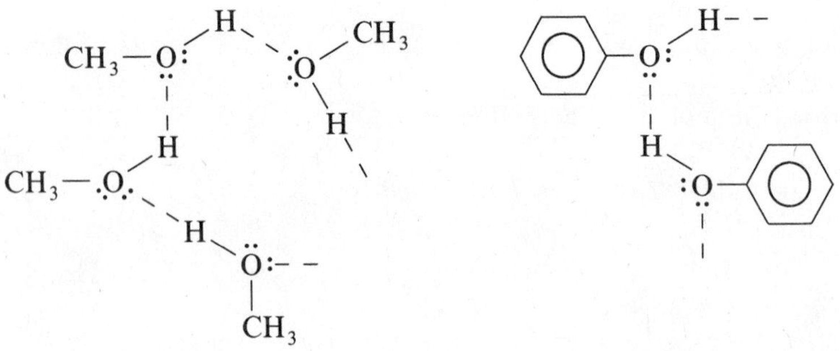

Methanol   Phenol

Figure 8.1 The hydrogen bonding in methanol and phenol.

TABLE 8.1   PHYSICAL PROPERTIES OF SOME ALCOHOLS AND PHENOLS

Formula	Name	Bp (°C)	Solubility in water at 20°C (g / 100 mL)
Alcohols			
CH_3OH	Methanol	65	∞^a
CH_3CH_2OH	Ethanol	78.5	∞
$CH_3CH_2CH_2OH$	1-Propanol	97	∞
$CH_3(CH_2)_2CH_2OH$	1-Butanol	117	8.3
$CH_3(CH_2)_3CH_2OH$	1-Pentanol	137.5	2.7 (22°C)
Phenols			
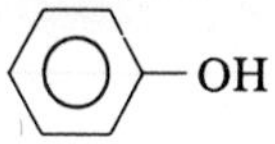 —OH	Phenol	182^b	6.7
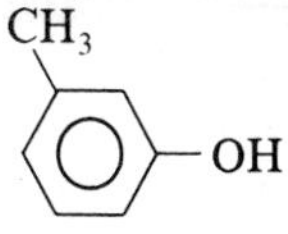 —OH	*m*-Cresol	202	2.5

^{a}Infinitely soluble, or miscible.

^{b}Melting point, 43°C.

PROBLEM 8.1.   Rank the following compounds in order of increasing water solubility (least soluble first). Explain your answer.

(a) $CH_3(CH_2)_5CH_2OH$ (b) $HOCH_2(CH_2)_5CH_2OH$

(c) $CH_3CH_2\underset{\underset{\displaystyle OH}{|}}{\overset{\overset{\displaystyle CH_2CH_3}{|}}{C}}CH_2CH_3$ (d) $HOCH_2\underset{\underset{\displaystyle OH}{|}}{CH}(CH_2)_3\underset{\underset{\displaystyle OH}{|}}{CH}CH_2OH$

8.3 NOMENCLATURE AND CLASSIFICATION OF ALCOHOLS

A. Systematic Nomenclature

The IUPAC names of alcohols are taken from the name of the parent alkane with the -*ane* ending changed to -*anol*. A prefix number is usually needed to specify the position of the hydroxyl group.

TABLE 8.2 NAMES OF SOME ALCOHOLS

Formula	IUPAC name	Trivial name
CH_3OH	Methanol	Methyl alcohol
CH_3CH_2OH	Ethanol	Ethyl alcohol
$CH_3CH_2CH_2OH$	1-Propanol	n-Propyl alcohol
$CH_3\overset{\overset{\textstyle OH}{\textstyle \|}}{C}HCH_3$	2-Propanol	Isopropyl alcohol
$CH_3\overset{\overset{\textstyle CH_3}{\textstyle \|}}{\underset{\underset{\textstyle CH_3}{\textstyle \|}}{C}}OH$	2-Methyl-2-propanol	t-Butyl alcohol
$CH_2{=}CHCH_2OH$	2-Propenol	Allyl alcohol
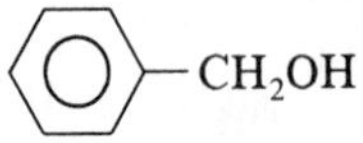 ⬡—CH_2OH	—	Benzyl alcohol
$HOCH_2CH_2OH$	1,2-Ethanediol	Ethylene glycol
$HOCH_2\overset{}{\underset{\underset{\textstyle OH}{\textstyle \|}}{C}}HCH_2OH$	1,2,3-Propanetriol	Glycerol

8.4 NOMENCLATURE OF PHENOLS

Phenol itself is the simplest member of the class of phenols. Simple phenols are usually named using *phenol* as the parent. A phenol with one other ring substituent can be named by the *o, m, p* system. If the ring contains more than two substituents, numbers are used to designate the positions of the groups. The ring is numbered starting with the hydroxyl carbon as position 1.

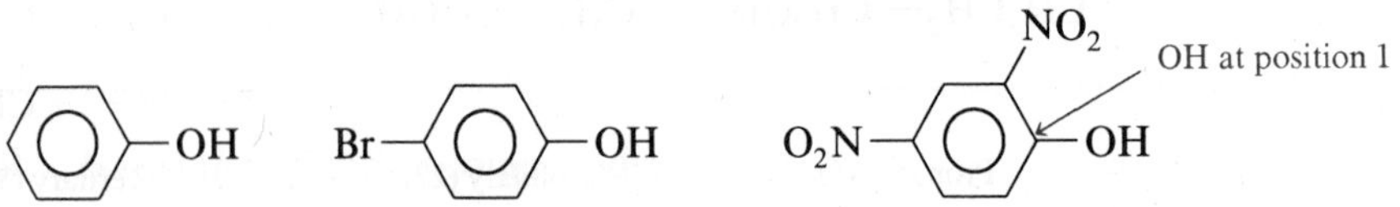

Phenol

p-Bromophenol
(4-bromophenol)

2,4-Dinitrophenol
used as a wood preservative, an insecticide, and an acid-base indicator

OH at position 1

Many phenols have trivial names. Some of these will be mentioned in Section 8.13. There is no classification system for the phenols like that for the alcohols.

PROBLEM 8.5.   Name the following phenols:

(a) H_3C—⬡—OH   (b) Cl—⬡—OH (with Cl substituent)

(c) H_3C—⬡—OH (with NO_2 substituent)

PROBLEM 8.6.   Draw polygon formulas for the following names:

(a) the acid-base indicator *p*-nitrophenol
(b) the insecticide and wood preservative pentachlorophenol ("Penta")
(c) the food preservative 2,6-di-*t*-butyl-4-methylphenol ("butylated hydroxy-toluene," or BHT)

8.5 PREPARATION OF ALCOHOLS

We have already discussed a number of reactions leading to alcohols. These reactions and others discussed later are summarized here.

A. Addition Reactions of Alkenes

In the presence of an acidic catalyst, water can add to the double bond of an alkene. The reaction follows Markovnikov's rule: the hydrogen of the adding reagent (H_2O) goes to the double-bond carbon with the greater number of hydrogens.

Hydration of Alkenes (Section 3.6D):

$$\left. \begin{array}{c} HO-H \\ + \\ CH_3CH{=}CH_2 \end{array} \right\} \xrightarrow{H^+} \underset{\text{2-Propanol}}{CH_3\overset{\displaystyle OH}{\underset{\displaystyle |}{C}}H-\overset{\displaystyle H}{\underset{\displaystyle |}{C}}H_2}$$

The addition of borane (BH_3), followed by oxidation of the intermediate trialkylborane with alkaline hydrogen peroxide (H_2O_2), yields an alcohol in which the hydrogen goes to the double-bond carbon with the lesser number of hydrogens.

Hydroboration-Oxidation of Alkenes (Section 3.6G):

$$3CH_3CH{=}CH_2 + BH_3 \longrightarrow (CH_3CH_2CH_2)_3B$$

Tripropylborane
a trialkylborane

$$\xrightarrow[^-OH]{H_2O_2} 3CH_3CH_2CH_2OH + BO_3{}^{3-}$$

1-Propanol

B. Substitution Reactions of Alkyl Halides

Primary alkyl halides, allylic halides, and benzylic halides undergo substitution reactions with hydroxides to yield alcohols. Secondary alkyl halides can be used, but the yield of alcohol is lower (unless the halide is also allylic or benzylic). Tertiary halides cannot be used in this reaction because they undergo elimination to yield alkenes rather than undergoing substitution.

$$1°: \quad CH_3CH_2CH_2I + {}^-OH \longrightarrow CH_3CH_2CH_2OH + Cl^-$$

2°, and allylic:

2-Cyclohexenol

PROBLEM 8.7.   Write flow equations showing how you could make the following conversions.

(a) $CH_3CH_2CH_2CH_2Cl \longrightarrow CH_3CH_2CH_2CH_2OH$

(b) $CH_3CH_2CH{=}CH_2 \longrightarrow CH_3CH_2CH_2CH_2OH$

(c) $CH_3CH_2CH{=}CH_2 \longrightarrow CH_3CH_2\underset{\underset{\textstyle CH_3}{|}}{C}HOH$

C. Grignard Reactions

Grignard reagents ($R{-}MgX$) can be used to synthesize alcohols. The syntheses involve reactions of Grignard reagents with carbonyl compounds (compounds containing a $C{=}O$ group) and are discussed in Chapters 10 and 12, where carbonyl chemistry is covered. In Chapter 9 we discuss the reaction of Grignard reagents with oxiranes. For completeness, we have summarized the Grignard reactions that lead to alcohols in Table 8.3, along with the other methods of preparation.

TABLE 8.3   SUMMARY OF PREPARATIONS OF ALCOHOLS

Reaction	Section reference

Addition to alkenes

$$R_2C{=}CHR + H_2O \xrightarrow{H^+} R_2\overset{\displaystyle OH}{\underset{\displaystyle |}{C}}{-}CH_2R \qquad\qquad\qquad 3.6D$$

$$R_2C{=}CHR \xrightarrow[\text{(2) } H_2O_2]{\text{(1) } BH_3} R_2CH{-}\overset{\displaystyle OH}{\underset{\displaystyle |}{C}}HR \qquad\qquad\qquad 3.6G$$

Substitution with 1°, 2°, allylic, and benzylic halides

$$R{-}X + {}^-OH \longrightarrow ROH \qquad\qquad\qquad 7.6,\ 7.7$$

Grignard reactions

$$H_2C{=}O \xrightarrow[\text{(2) } H_2O,\ H^+]{\text{(1) } RMgX} RCH_2OH \qquad 1°\ ROH \qquad\qquad 10.6$$

$$RCH{=}O \xrightarrow[\text{(2) } H_2O,\ H^+]{\text{(1) } R'MgX} \underset{\displaystyle \underset{\displaystyle R'}{|}}{RCHOH} \qquad 2°\ ROH \qquad\qquad 10.6$$

$$R\overset{\displaystyle O}{\overset{\displaystyle \|}{C}}R \xrightarrow[\text{(2) } H_2O,\ H^+]{\text{(1) } R'MgX} \underset{\displaystyle \underset{\displaystyle R'}{|}}{R\overset{\displaystyle OH}{\underset{\displaystyle |}{C}}R} \qquad 3°\ ROH \qquad\qquad 10.6$$

$$R\overset{\displaystyle O}{\overset{\displaystyle \|}{C}}OR'' \xrightarrow[\text{(2) } H_2O,\ H^+]{\text{(1) } 2R'MgX} \underset{\displaystyle \underset{\displaystyle R'}{|}}{R\overset{\displaystyle OH}{\underset{\displaystyle |}{C}}R'} \qquad 3°\ ROH \qquad\qquad 12.3C$$

$$\overset{\displaystyle O}{\overset{\diagup\ \diagdown}{CH_2{-}CH_2}} \xrightarrow[\text{(2) } H_2O,\ H^+]{\text{(1) } RMgX} RCH_2CH_2OH \qquad 1°\ ROH \qquad\qquad 9.5B$$

Reduction of carbonyl compounds

$$RCH{=}O \xrightarrow[\text{(2) } H_2O,\ H^+]{\text{(1) } NaBH_4} RCH_2OH \qquad 1°\ ROH \qquad\qquad 10.8$$

$$R_2C{=}O \xrightarrow[\text{(2) } H_2O,\ H^+]{\text{(1) } NaBH_4} R_2CHOH \qquad 2°\ ROH \qquad\qquad 10.8$$

A Typical Grignard Reaction (Section 10.6):

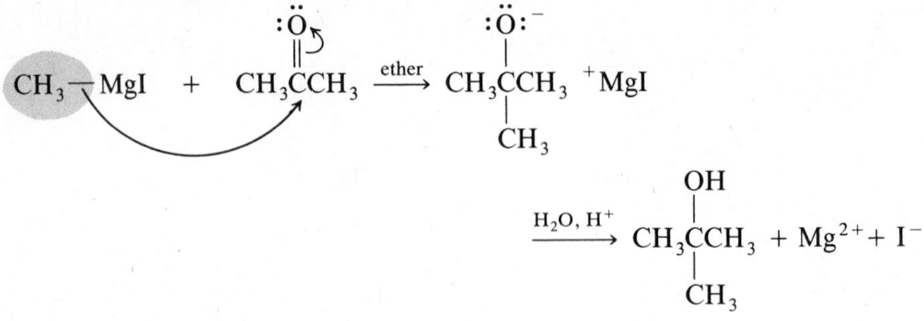

$$CH_3 \!-\! MgI \; + \; CH_3\overset{\displaystyle :\!\ddot{O}:}{\overset{\|}{C}}CH_3 \xrightarrow{\text{ether}} CH_3\underset{\underset{\displaystyle CH_3}{|}}{\overset{\overset{\displaystyle :\ddot{O}:^-}{|}}{C}}CH_3 \; {}^+MgI$$

$$\xrightarrow{\text{H}_2\text{O, H}^+} CH_3\underset{\underset{\displaystyle CH_3}{|}}{\overset{\overset{\displaystyle OH}{|}}{C}}CH_3 \; + \; Mg^{2+} + \; I^-$$

D. Reduction of Carbonyl Compounds

Hydrogen (H_2) can add to a carbon–oxygen double bond, just as it can to a carbon–carbon double bond, to yield an alcohol. The reaction is a reduction because hydrogen is gained. Other reducing agents, such as sodium borohydride ($Na^+ \; {}^-BH_4$), also convert carbonyl compounds to alcohols. We discuss these reactions in Chapter 10.

Reduction of Carbonyl Compounds (Section 10.8):

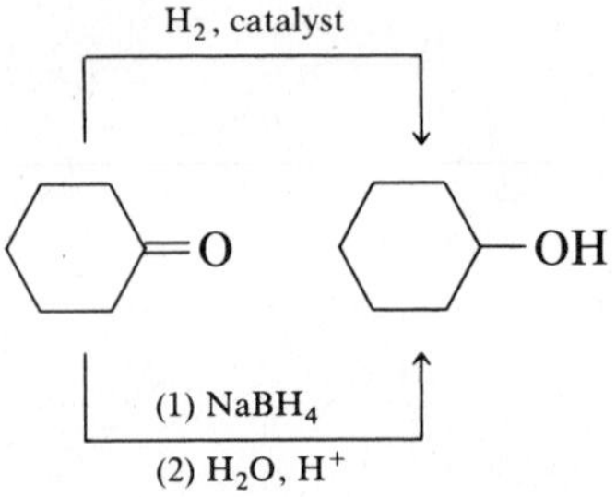

8.6 SOME IMPORTANT ALCOHOLS

Methanol (methyl alcohol, CH_3OH) is a colorless, water-soluble liquid. Methanol is toxic; ingestion of even small amounts or chronic inhalation of methanol can cause blindness. Death from ingestion of less than 30 mL has been reported.

Historically, methanol was called *wood alcohol*. Until 1923 this alcohol was obtained from the destructive distillation (distillation in the absence of air) of hardwoods. Today, methanol is obtained from the reduction of carbon monoxide.

$$CO \; + \; 2H_2 \; \xrightarrow[\text{260°C, 100–150 atm}]{\text{Cu catalyst}} \; CH_3OH$$

Most methanol produced today is used to synthesize formaldehyde ($H_2C{=}O$) and other chemicals. Some methanol is used for fuel, some for gas-line antifreeze, and some for solvent.

Ethanol (ethyl alcohol, "alcohol," CH_3CH_2OH), a colorless, water-soluble liquid, is sometimes called *grain alcohol* because it can be obtained by the fermentation of grain. Actually, the fermentation of any carbohydrate-containing material, such as grapes, molasses, and potatoes, as well as grain, yields ethanol.

$$\text{Carbohydrates} \xrightarrow{\text{enzymes}} \underset{\text{Glucose}}{C_6H_{12}O_6} \xrightarrow{\text{enzymes}} 2CH_3CH_2OH + 2CO_2$$

in fruits, vegetables, grain, or molasses

Ethanol for use in beverages and gasohol is still made by fermentation. Ethanol for use as a solvent is synthesized by the hydration of ethylene, a petrochemical obtained from petroleum cracking reactions.

$$CH_2{=}CH_2 + H_2O \xrightarrow[\text{heat}]{H^+} CH_3CH_2OH$$

Fermentation reactions cease when the alcohol concentration reaches 12–14%. (Alcohol in greater concentrations inhibits the enzymes that cause fermentation.) Stronger beverages are made by distillation of the fermentation mixture. When an aqueous solution of ethanol is distilled, it forms a **low-boiling azeotrope**, a mixture that distills without changing its composition. The azeotrope of ethanol and water, consisting of 95% ethanol and 5% water, boils at 78.15°C, slightly lower than either pure ethanol (bp 78.5°C) or pure water (bp 100°C). Because of these boiling points, pure ethanol cannot be obtained by distillation from water (unless the ethanol contains less than 5% water). For this reason, ethanol used as a solvent or a fuel is usually 95% ethanol.

The term **proof** is used to describe the concentration of beverage alcohol. The proof of an alcoholic beverage is twice its percent by volume of ethanol. Thus, 100-proof spirits contain 50% ethanol. The word "proof" supposedly comes to us from rum runners in the days of sailing ships. A purchaser tested the alcoholic strength by pouring the beverage on a pile of gunpowder and igniting the mixture. After the ethanol burned off, ignition of the gunpowder was "proof" that the beverage had not been watered down, leaving the gunpowder too wet to ignite.

Anhydrous ethanol, also called *absolute ethanol*, can be prepared by removing the water from 95% ethanol by a chemical reaction. For example, calcium oxide (CaO) can be used as a drying agent for ethanol because it undergoes reaction with the water to form ethanol-insoluble calcium hydroxide, $Ca(OH)_2$. Industrially, distillation of 95% ethanol with benzene removes water because a benzene-water-ethanol azeotrope distills first.

Denatured alcohol is ethanol to which a toxic substance such as benzene has been added to make the alcohol unfit for consumption. In most countries, ethanol used in industry and commerce is denatured because ethanol that is *not* denatured is subject to liquor excise taxes.

Isopropyl alcohol (2-propanol), $(CH_3)_2CHOH$, like methanol and ethanol, is a colorless, water-soluble liquid. Isopropyl alcohol, like ethanol, is made by the hydration of an alkene obtained from petroleum cracking reactions.

$$CH_3CH{=}CH_2 \ + \ H_2O \ \xrightarrow[\text{heat}]{\text{catalyst}} \ \underset{\underset{\text{Isopropyl alcohol}}{}}{CH_3\overset{\overset{\displaystyle OH}{|}}{C}HCH_3}$$

Propene

A 70% aqueous solution of isopropyl alcohol is used as rubbing alcohol. Like methanol, isopropyl alcohol is toxic if ingested.

Ethylene glycol (1,2-ethanediol, $HOCH_2CH_2OH$) is a colorless, water-soluble liquid manufactured by the air oxidation of ethylene followed by hydration of the intermediate ethylene oxide.

$$CH_2{=}CH_2 \ \xrightarrow{O_2} \ \overset{O}{\overset{\diagup \ \diagdown}{CH_2{-}CH_2}} \ \xrightarrow[H^+]{H_2O} \ \underset{\underset{\text{Ethylene glycol}}{}}{\overset{\overset{\displaystyle OH \quad OH}{| \qquad |}}{CH_2{-}CH_2}}$$

Ethylene glycol is used as an industrial solvent, as a starting material for such products as Dacron (Section 12.3C), and as an automobile radiator antifreeze. It is sweet-tasting but quite toxic; one teaspoon can kill a cat and less than half a cup can kill an adult human. Because of its toxicity, antifreeze containing ethylene glycol should never be left in an open container, and spills should be cleaned up.

Glycerol (glycerin; 1,2,3-propanetriol) is a colorless, syrupy, sweet-tasting, non-toxic liquid. This triol is obtained from the alkaline hydrolysis of fats as a by-product of soapmaking (see Point of Interest 12) or from the petrochemical propene.

Glycerol is used as a solvent, as an emollient (an agent that softens or smooths as does a lotion), as a sweetener in confectioneries and liqueurs, as an antifreeze, as a chemical reactant, and for many other purposes. One source lists 1583 different uses for glycerol!

8.7 ACIDITY AND BASICITY OF ALCOHOLS

A. Acidity

Water acts as an acid by losing a proton to a strong base. Alcohols undergo a similar reaction. The product anion from the loss of a proton from an alcohol (^-OR) is called an **alkoxide ion**. The salts are called *metal alkoxides*, for example, sodium methoxide ($Na^+ \ ^-OCH_3$).

Water as an Acid:

$$B\overset{..}{:}{}^- \ + \ H{-}\overset{..}{\underset{..}{O}}H \ \rightleftharpoons \ B{-}H \ + \ {}^-\overset{..}{\underset{..}{:O}}H$$

A strong base   Hydroxide ion

An Alcohol as an Acid:

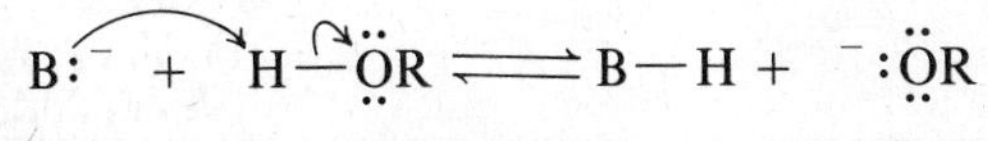

$$B:^- \; + \; H\!-\!\ddot{O}R \; \rightleftharpoons \; B\!-\!H \; + \; ^-\!:\!\ddot{O}R$$

A strong   An alkoxide
base   ion

Most alkoxides are slightly stronger bases than hydroxides. Therefore, the reaction of hydroxide ion with an alcohol does not generally yield a significant amount of the alkoxide. A base stronger than hydroxide must be used in the preparation of an alkoxide.

The easiest way to prepare an alkoxide is to treat the alcohol with an alkali metal, such as sodium or potassium metal, or with sodium hydride, a stronger base than alkoxides.

$$CH_3CH_2OH \; + \; Na \; \longrightarrow \; CH_3CH_2O^-\,Na^+ \; + \; \tfrac{1}{2}H_2$$

Ethanol   Sodium ethoxide

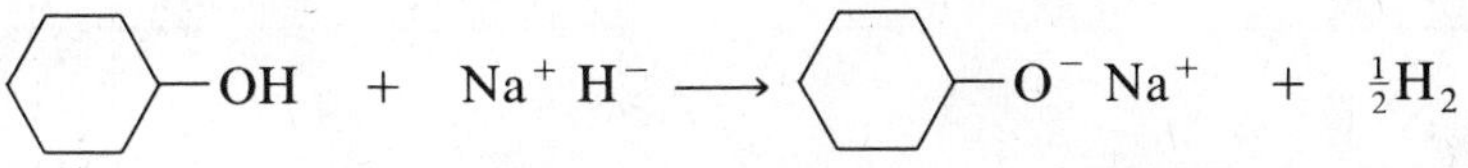

Cyclohexanol   Sodium cyclohexoxide

Alkoxides are useful for the preparation of ethers (Williamson ether synthesis; see Section 7.6).

B. Basicity

Like water, an alcohol can act as a base by accepting a proton from a strong acid.

Water as a Base:

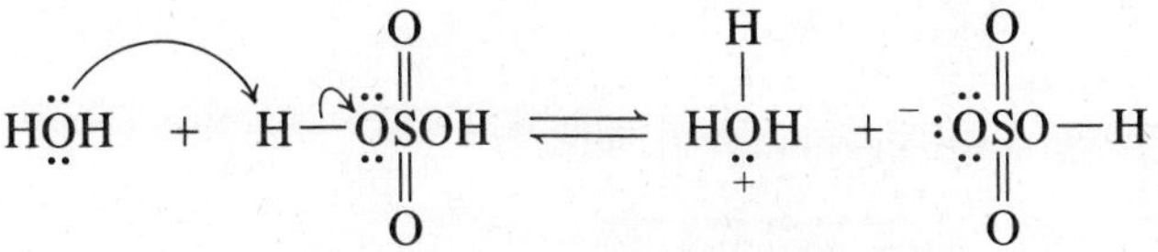

Hydronium
ion

An Alcohol as a Base:

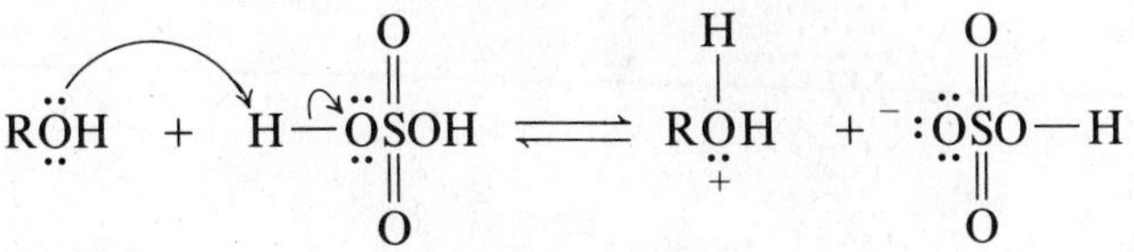

*A protonated
alcohol, or
oxonium ion*

In the next section, we will see that the protonation of alcohols is an important facet of their reactivity, activating the hydroxyl group so that other reactions can occur.

PROBLEM 8.8. Complete and balance the following equations. In **(b)** and **(d)**, show only the acid-base reactions.

(a) $CH_3CH_2OH + Na^+ \ ^-NH_2$ (a stronger base than alkoxides) $\longrightarrow$

(b) $CH_3CH_2OH + $ conc. $HCl \longrightarrow$

(c) ⬡$-OH + K \longrightarrow$

(d) $CH_3CH_2\overset{\overset{\displaystyle OH}{|}}{C}HCH_3 + HI \longrightarrow$

8.8 SUBSTITUTION REACTIONS OF ALCOHOLS

Unlike an alkyl halide, an alcohol cannot undergo nucleophilic substitution in neutral or alkaline solution. If an alcohol did undergo substitution, the leaving group of the alcohol would be a hydroxide ion (^-OH), an ion that is a strong base and, therefore, a very poor leaving group.

Very weak base
good leaving group

$$CH_3CH_2 \overset{\frown}{-} \overset{..}{\underset{..}{Br}}: \ + \ ^- :\overset{..}{\underset{..}{O}}H \longrightarrow CH_3CH_2\overset{..}{\underset{..}{O}}H + :\overset{..}{\underset{..}{Br}}:^-$$

Strong base,
poor leaving group

$$CH_3CH_2 - \overset{..}{\underset{..}{O}}H \ + \ Nu:^- \longrightarrow \text{no reaction}$$

In strongly acidic solution, however, alcohols undergo substitution readily because a protonated alcohol contains a very weakly basic leaving group—water.

Very weak base,
good leaving group

$$CH_3CH_2 - \overset{..}{\underset{..}{O}}H \xrightarrow{H^+} CH_3CH_2 - \overset{\overset{\displaystyle H}{|}}{\underset{\underset{\displaystyle +}{..}}{O}}H \xrightarrow{:\overset{..}{\underset{..}{Br}}:^-} CH_3CH_2 - \overset{..}{\underset{..}{Br}}: + :\overset{\overset{\displaystyle H}{|}}{\underset{..}{O}}H$$

A. Reaction with Hydrogen Halides

All alcohols undergo reaction readily with HI or HBr to yield alkyl halides.

$$1°, 2°, \text{ or } 3° \text{ ROH} + \text{HI (HBr)} \longrightarrow \text{RI (RBr)} + H_2O$$

PROBLEM 8.9. Complete the following equations for alcohol substitution reactions:

(a) $CH_2{=}CHCH_2OH + HBr \longrightarrow$

(b) ⬡—OH + HI ⟶

(c) ⬡—CH_2OH + HBr ⟶

(d) $(CH_3)_2CHOH + HI \longrightarrow$

Tertiary alcohols, allylic alcohols, and benzylic alcohols react readily with HCl; primary and secondary alcohols are less reactive and require a catalyst, such as zinc chloride ($ZnCl_2$).

$$3°: \qquad (CH_3)_3C{-}OH + HCl \xrightarrow{25°C} (CH_3)_3C{-}Cl + H_2O$$

$$2°: \qquad (CH_3)_2CH{-}OH + HCl \xrightarrow[25°C]{ZnCl_2} (CH_3)_2CH{-}Cl + H_2O$$

$$1°: \qquad CH_3CH_2{-}OH + HCl \xrightarrow[\text{heat}]{ZnCl_2} CH_3CH_2{-}Cl + H_2O$$

We can summarize the reactivity of the different types of alcohols:

$$CH_3OH \quad 1° \text{ ROH} \quad 2° \text{ ROH} \quad 3° \text{ ROH} \quad \text{allylic and benzylic ROH}$$
$$\xrightarrow{\text{Increasing reactivity toward HX}}$$

The mechanism of the reaction of alcohols with hydrogen halides depends on the structure of the alcohol. Allylic, benzylic, secondary, and tertiary alcohols react by an S_N1 path—by way of a carbocation. If the starting alcohol is optically active, racemization occurs, just as it does during the solvolysis of a tertiary alkyl halide (Section 7.7).

Mechanism of the Reaction of HX with 2°, 3°, Allylic, and Benzylic Alcohols:

Step 1, Protonation and Loss of Water:

$$R{-}\overset{..}{\underset{..}{O}}H \underset{\text{fast}}{\overset{H^+}{\rightleftharpoons}} R{-}\overset{+}{\underset{..}{O}}H_2 \xrightarrow[\text{slow}]{-H_2\overset{..}{\underset{..}{O}}} R^+$$

Protonated   *A carbocation*
(planar and achiral)

Step 2, Combination with Nucleophile:

$$R^{+} \ + \ :\ddot{X}:^{-} \xrightarrow{\text{fast}} \ R-\ddot{X}:$$

An alkyl halide

Because primary carbocations do not form under usual laboratory conditions, primary alcohols do not react by an S_N1 path. Instead, their reaction is by an S_N2 path—backside displacement.

Mechanism of the Reaction of HX with 1° Alcohols:

Step 1, Protonation:

$$CH_3CH_2-\ddot{O}H \ + \ H^{+} \rightleftharpoons CH_3CH_2\overset{+}{\ddot{O}}H_2$$

A protonated alcohol

Step 2, Attack of Nucleophile:

$$:\ddot{Br}:^{-} \ + \ CH_3CH_2-\overset{+}{\ddot{O}}H_2 \xrightarrow{S_N2} CH_3CH_2-\ddot{Br}: \ + \ H_2\ddot{O}:$$

An alkyl halide

PROBLEM 8.10. Complete the following equations, and rank the reactions in order of rate of reaction. (List the slowest reaction first.)

(a) 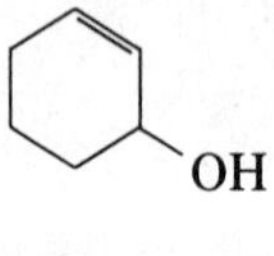 [cyclohexene ring with OH] $+ \ HI \longrightarrow$

(b) $CH_3CH_2CH_2OH + \overline{HCl} \longrightarrow$

(c) $CH_3CH_2CH_2OH + HBr \longrightarrow$

(d) $(CH_3)_3COH + HBr \longrightarrow$

B. Reaction with Other Halogenating Agents

With primary and secondary alcohols, a higher yield of an alkyl halide can be obtained if a halogenating agent such as thionyl chloride ($SOCl_2$) or phosphorus tribromide (PBr_3) is used in place of a hydrogen halide.

For 1° and 2° alcohols:

$$CH_3CH_2OH \quad + \quad \underset{\text{Thionyl chloride}}{Cl-\overset{\overset{\displaystyle O}{\|}}{S}-Cl} \quad \longrightarrow \quad CH_3CH_2-Cl \quad + \quad \underset{\substack{\text{Sulfur} \\ \text{dioxide}}}{O=\overset{\overset{\displaystyle O}{\|}}{S}} \quad + \quad HCl$$

$$3(CH_3)_2CHOH \quad + \quad \underset{\substack{\text{Phosphorus} \\ \text{tribromide}}}{PBr_3} \quad \longrightarrow \quad 3(CH_3)_2CH-Br \quad + \quad \underset{\substack{\text{Phosphorous} \\ \text{acid}}}{H_3PO_3}$$

PROBLEM 8.11.   Predict the organohalogen product:

(a)   ⬡—OH + $SOCl_2$ ⟶

(b)   ⬡—CH_2OH + PBr_3 ⟶

8.9 ELIMINATION REACTIONS OF ALCOHOLS

When heated with a strong acid, an alcohol with a β hydrogen can undergo **dehydration** (loss of water) to yield an alkene. This reaction is the reverse of the hydration of an alkene. Sulfuric acid (H_2SO_4) and phosphoric acid (H_3PO_4) are commonly used as acids for dehydration reactions.

Dehydration of Alcohols:

β hydrogen

$$3°: \quad \underset{\substack{| \\ CH_3}}{\overset{\overset{\displaystyle H \quad OH}{| \quad |}}{CH_2-CCH_3}} \quad \xrightarrow[\text{60°C}]{H_2SO_4} \quad CH_2=C\overset{\diagup CH_3}{\underset{\diagdown CH_3}{}} \quad + H_2O$$

$$2°: \quad \overset{\overset{\displaystyle H \quad OH}{| \quad |}}{CH_2-CHCH_3} \quad \xrightarrow[\text{100°C}]{H_2SO_4} \quad CH_2=CHCH_3 + H_2O$$

$$1°: \quad \overset{\overset{\displaystyle H \quad OH}{| \quad |}}{CH_2-CH_2} \quad \xrightarrow[\text{170°C}]{\text{conc. } H_2SO_4} \quad CH_2=CH_2 + H_2O$$

We can summarize the reactivity of primary, secondary, and tertiary alcohols toward dehydration when treated with strong acid.

$$\text{1° ROH} \qquad \text{2° ROH} \qquad \text{3° ROH}$$

Increasing ease of dehydration

Dehydration reactions are elimination reactions similar to elimination reactions of alkyl halides in strong base. Secondary and tertiary alcohols undergo dehydration by an E1 path, by way of carbocations. (Primary alcohols probably undergo reaction by an E2 path.) Tertiary alcohols are easier to dehydrate because they form lower-energy carbocations.

Mechanism of a Dehydration Reaction:

Step 1, Protonation and Loss of Water:

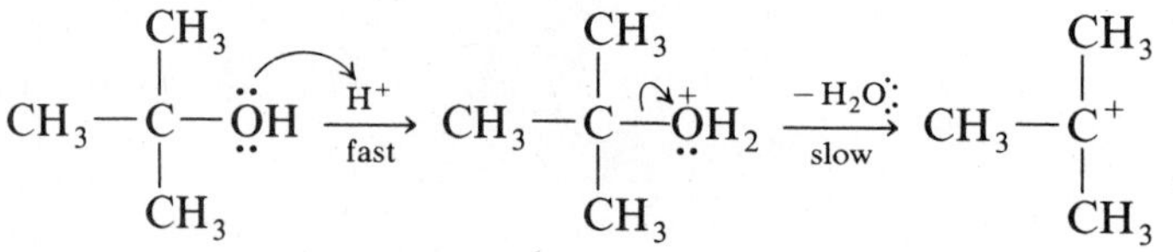

Step 2, Loss of β Hydrogen:

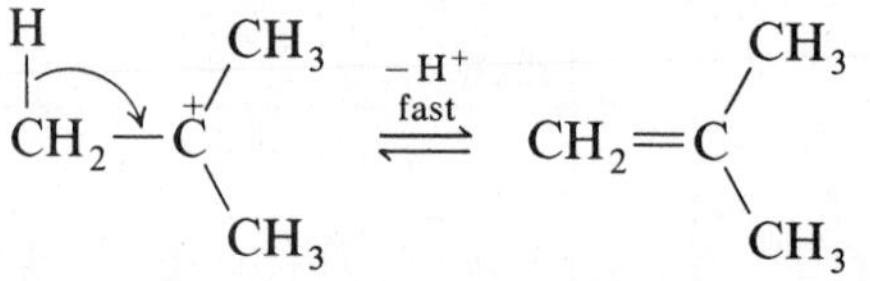

Like the elimination reactions of alkyl halides, the dehydration reactions of alcohols with more than one type of β hydrogen lead predominantly to the more stable alkene—generally the more substituted trans alkene.

Two β hydrogens

$$CH_2{=}CH{-}CH_2CH_2CH_3 \ +$$

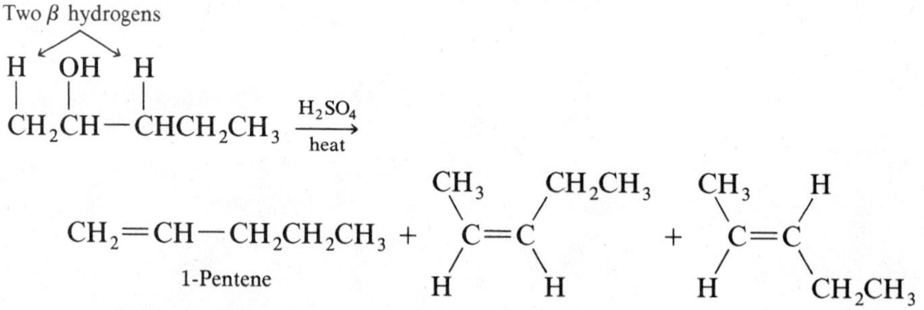

1-Pentene

cis-2-Pentene

trans-2-Pentene

*more substituted
trans alkene;
major product*

PROBLEM 8.12. Redraw the following formulas, and circle all hydrogens β to the hydroxyl group, if any.

(a) $CH_3\overset{\displaystyle OH}{\underset{\displaystyle |}{C}}HCH_2CH_3$ (b) $(CH_3CH_2)_3COH$ (c)

PROBLEM 8.13. Write equations for the elimination reactions of the alcohols in the preceding problem with an acidic catalyst. In your answer show all possible alkene products, and circle the most probable predominant product.

8.10 ESTERS OF ALCOHOLS

Under the proper reaction conditions, alcohols and acids react, losing water, to yield esters. Reaction of an alcohol with a carboxylic acid yields a **carboxylic ester**, also called simply an **ester**. These compounds and the reactions leading to them will be discussed in Sections 11.7 and 12.3.

An Esterification Reaction:

$$\underset{\text{A carboxylic acid}}{\overset{\displaystyle O}{\overset{\displaystyle \|}{RC}}-OH} + \underset{\text{An alcohol}}{H-OR'} \overset{H^+, \text{ heat}}{\rightleftharpoons} \underset{\text{An ester}}{\overset{\displaystyle O}{\overset{\displaystyle \|}{RC}}-OR'} + HOH$$

Reaction of an alcohol with an inorganic acid or its chloride, such as thionyl chloride ($SOCl_2$), can yield an **inorganic ester of an alcohol**, a compound in which the $HO-$ of the inorganic acid is replaced by $RO-$ of the alcohol.

$$\underset{\text{Nitric acid}}{HO-NO_2} \qquad \underset{\text{An alkyl nitrate}}{RO-NO_2}$$

$$\underset{\text{Phosphoric acid}}{\overset{\displaystyle O}{\overset{\displaystyle \|}{HO-P}}\underset{\displaystyle |}{\underset{\displaystyle OH}{OH}}} \qquad \underset{\text{An alkyl phosphate}}{\overset{\displaystyle O}{\overset{\displaystyle \|}{RO-P}}\underset{\displaystyle |}{\underset{\displaystyle OH}{OH}}}$$

Most inorganic ester groups, such as $-OPO_3H_2$, are good leaving groups. Phosphate groups are common leaving groups in biochemical reactions; phosphate esters are biological synthetic intermediates and energy storehouses.

The specific reaction of an alcohol with an inorganic acid depends on the relative amounts of reactants and also on the temperature of the reaction mixture. For example, when heated with sulfuric acid, an alcohol undergoes dehydration. At 0°C, however, primary alcohols yield alkyl hydrogen sulfates or dialkyl sulfates. At moderate temperatures, ethers can be formed.

$$\xrightarrow{0°C} CH_3CH_2O\overset{\overset{\displaystyle O}{\|}}{\underset{\underset{\displaystyle O}{\|}}{S}}OH + CH_3CH_2O\overset{\overset{\displaystyle O}{\|}}{\underset{\underset{\displaystyle O}{\|}}{S}}OCH_2CH_3 + H_2O$$

Alkyl sulfates (inorganic esters)

$$CH_3CH_2OH + H_2SO_4 \xrightarrow{140°C} CH_3CH_2OCH_2CH_3 + H_2O$$

An ether

$$\xrightarrow{170°C} CH_2{=}CH_2 + H_2O$$

An alkene

8.11 OXIDATION OF ALCOHOLS

Depending on the conditions of the reaction, primary alcohols can be oxidized to aldehydes or to carboxylic acids. Secondary alcohols, however, can be oxidized only to ketones. Tertiary alcohols resist oxidation in alkaline solutions; in acidic solution, tertiary alcohols undergo dehydration to yield alkenes, which then are oxidized (Section 3.9).

$$1°: \quad RCH_2OH \xrightarrow[\text{loss of 2 H}]{[O]} R\overset{\overset{\displaystyle O}{\|}}{C}H \xrightarrow[\text{gain of O}]{[O]} R\overset{\overset{\displaystyle O}{\|}}{C}OH$$

$$\qquad\qquad\qquad\qquad\qquad \textit{An aldehyde} \qquad\quad \textit{A carboxylic acid}$$

$$2°: \quad R\overset{\overset{\displaystyle OH}{|}}{C}HR \xrightarrow[\text{loss of 2 H}]{[O]} R\overset{\overset{\displaystyle O}{\|}}{C}R$$

$$\qquad\qquad\qquad\qquad\qquad \textit{A ketone}$$

$$3°: \quad R_3COH \xrightarrow{[O]} \text{no reaction}$$

A. Oxidation to Aldehydes

The initial oxidation product of a primary alcohol is an aldehyde ($RCH{=}O$). Aldehydes, however, are readily oxidized to carboxylic acids. Therefore, treatment of a primary alcohol with a strong oxidizing agent results in a carboxylic acid, not the intermediate aldehyde. Special reagents must be used when the intermediate aldehyde is the desired product. In the laboratory, pyridinium chlorochromate (PCC) is one reagent that oxidizes alcohols to aldehydes but does not oxidize aldehydes to carboxylic acids.

$$ClCrO_3^-$$

Pyridinium chlorochromate (PCC)

Oxidation of Primary Alcohols to Aldehydes:

General Reaction:

$$\underset{\text{An alcohol}}{\overset{\displaystyle \overset{OH}{|}}{RCH_2}} \xrightarrow[\text{CH}_2\text{Cl}_2]{\text{PCC}} \underset{\text{An aldehyde}}{\overset{\displaystyle \overset{O}{\|}}{RCH}}$$

Specific Example:

$$CH_3(CH_2)_5CH_2OH \xrightarrow[\text{CH}_2\text{Cl}_2]{\text{PCC}} \underset{\text{Heptanal}}{CH_3(CH_2)_5\overset{\displaystyle \overset{O}{\|}}{C}H}$$

1-Heptanol

B. Oxidation to Carboxylic Acids

The common strong oxidizing agents oxidize primary alcohols to carboxylic acids.

Common Oxidizing Agents:

Hot aqueous $KMnO_4 + {}^-OH$, followed by acidification

Hot aqueous $CrO_3 + H_2SO_4$ (Jones' reagent)

Oxidation of Primary Alcohols to Carboxylic Acids:

General Reaction:

$$RCH_2OH \xrightarrow{\text{[O]}} \underset{\text{A carboxylic acid}}{R\overset{\displaystyle \overset{O}{\|}}{C}OH}$$

Specific Example:

$$CH_3(CH_2)_8CH_2OH \xrightarrow[\text{heat}]{CrO_3,\ H_2SO_4} \underset{\text{Decanoic acid}}{CH_3(CH_2)_8\overset{\displaystyle \overset{O}{\|}}{C}OH}$$

1-Decanol

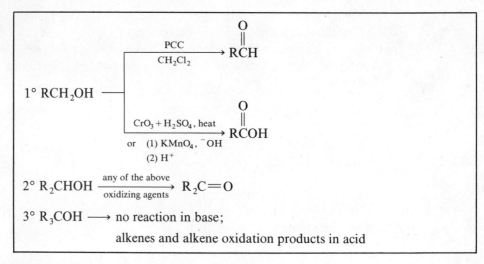

Figure 8.2 Summary of oxidation reactions of alcohols.

C. Oxidation to Ketones

Secondary alcohols are oxidized by any relatively strong oxidizing agent to ketones.

Oxidation of Secondary Alcohols to Ketones:

General Reaction:

$$\underset{\text{RCHR}}{\overset{\text{OH}}{|}} \xrightarrow{[O]} \underset{\text{RCR}}{\overset{O}{\|}}$$

A ketone

Specific Example:

$$\underset{\text{2-Octanol}}{CH_3(CH_2)_5\overset{\overset{\displaystyle OH}{|}}{C}HCH_3} \xrightarrow{CrO_3,\ H_2SO_4} \underset{\text{2-Octanone}}{CH_3(CH_2)_5\overset{\overset{\displaystyle O}{\|}}{C}CH_3}$$

Figure 8.2 summarizes the oxidation reactions of alcohols.

PROBLEM 8.14. Predict the organic products:

(a) $CH_3\overset{\overset{\displaystyle CH_3}{|}}{C}HCH_2OH \xrightarrow[\text{heat}]{CrO_3,\ H_2SO_4}$

(b) ⟨O⟩—$CH_2OH \xrightarrow[\text{heat}]{CrO_3,\ H_2SO_4}$

(c) —CH$_2$OH $\xrightarrow[\text{CH}_2\text{Cl}_2]{\text{PCC}}$

(d) $\xrightarrow[\text{heat}]{\text{KMnO}_4,\ ^-\text{OH}}$

Menthol

PROBLEM 8.15. Write equations for the preparation of the following compounds from alcohols or diols:

(a)

(b) $(CH_3)_2\text{CHCH}$ with a C=O above (O double bonded)

$$\text{(b)}\quad (CH_3)_2CH\overset{\displaystyle O}{\overset{\|}{C}}H$$

(c) —COH (with C=O)

$$\text{(d)}\quad CH_3CH_2\overset{\displaystyle O}{\overset{\|}{C}}CH_2\overset{\displaystyle O}{\overset{\|}{C}}OH$$

D. Biological Oxidation of Ethanol

Enzymatic oxidation of ethanol in human liver cells yields acetaldehyde. This aldehyde is further oxidized to acetate ion, the anion of acetic acid. The acetate ion then undergoes esterification with **coenzyme A**, a complex structure abbreviated CoA or HSCoA (to emphasize the —SH group). The product of the esterification is a **thioester**, an ester formed from a carboxylic acid or its anion and a thiol (RSH). In this form, the acetyl group can be converted to other compounds needed within the biological system or it can be converted to CO_2, H_2O, and biological energy. These reactions are outlined in Figure 8.3. (Some additional reactions of acetylcoenzyme A are discussed in Point of Interest 13.)

8.12 THIOLS

A **thiol** (RSH) is the sulfur analog of an alcohol. The term "thiol" is a combination of the terms *thio-*, which refers to sulfur, and *-ol*, which refers to an alcohol.

CH$_3$OH   CH$_3$SH

Methanol   Methanethiol

Thiols are also called the older name **mercaptans** because they react with, or capture,

$$CH_3CH_2OH \xrightarrow[\text{dehydrogenase}]{\text{alcohol}} \underset{\substack{\text{Ethanal} \\ \text{(acetaldehyde)}}}{CH_3\overset{\overset{\displaystyle O}{\|}}{C}H} \xrightarrow{[O]} \underset{\text{Acetate ion}}{CH_3\overset{\overset{\displaystyle O}{\|}}{C}O^-}$$

$$\xrightarrow{\text{HSCoA}} \underset{\substack{\text{Acetylcoenzyme A} \\ \textit{a thio ester}}}{CH_3\overset{\overset{\displaystyle O}{\|}}{C}-SCoA} \longrightarrow \begin{array}{l}\text{fats and other compounds} \\ \text{or } CO_2 + H_2O + \text{energy}\end{array}$$

Figure 8.3 Flow equation for the biological oxidation of ethanol, a sequence that occurs primarily in the liver.

mercury(II) ions to yield insoluble precipitates.

$$2RSH + Hg^{2+} \longrightarrow (RS)_2Hg\downarrow + H^+$$

Water insoluble

The functional group of a thiol ($-SH$) is called a **sulfhydryl group**, **thiol group**, or **mercapto group**.

The outstanding property of low-formula-weight thiols is their unpleasant odor. For example, the odor of a skunk's spray arises from thiols.

Two Odorous Thiols in Skunk Spray:

$$\underset{\text{3-Methyl-1-butanethiol}}{CH_3\overset{\overset{\displaystyle CH_3}{|}}{C}HCH_2CH_2SH} \qquad \underset{\text{2-Butene-1-thiol}}{CH_3CH=CHCH_2SH}$$

Sulfur is just below oxygen in the periodic table; therefore, sulfur compounds exhibit chemical behavior similar to that of analogous oxygen compounds. Sulfur atoms are larger and less electronegative than oxygen atoms. For this reason, thiols form far weaker hydrogen bonds than do alcohols. Because of the lack of hydrogen bonding, thiols are less water soluble and lower boiling than the corresponding alcohols. For example, methanol boils at 65°C, but methanethiol boils at 6°C.

Just as H_2S is more acidic than H_2O, thiols are more acidic than alcohols. The reason for the greater acidity is that the larger sulfur atom can better disperse the negative charge of the anion.

Thiols can be prepared in the laboratory by nucleophilic substitution reactions (S_N2 path) of alkyl halides and sodium hydrogen sulfide ($Na^+ \, ^-SH$).

Preparation of Thiols:

excellent nucleophile

$$CH_3CH_2-\ddot{B}r: + \quad ^-:\ddot{S}H \xrightarrow{S_N2} CH_3CH_2-\ddot{S}H + :\ddot{B}r:^-$$

Bromoethane   Ethanethiol

One important reaction of thiols is their conversion to **disulfides** ($RS-SR$) when treated with a mild oxidizing agent such as O_2 or H_2O_2.

$$CH_3CH_2S-H + H-SCH_2CH_3 \xrightarrow{\frac{1}{2}O_2} CH_3CH_2S-SCH_2CH_3 + H_2O$$

Diethyl disulfide

Treatment of the disulfide with a reducing agent yields the corresponding thiol.

$$CH_3CH_2S-SCH_2CH_3 + Zn + CH_3CO_2H \longrightarrow$$

$$2CH_3CH_2SH + Zn^{2+} + CH_3CO_2^-$$

The thiol $-SH$ groups in proteins can be oxidized to disulfide groups. The disulfide group can bond two protein molecules and hold them in their necessary shapes. Hair is rich in sulfur-containing protein; the positions of the disulfide links determine whether hair is straight or curly. In permanent waving of hair, these disulfide links are broken and then reformed in different positions.

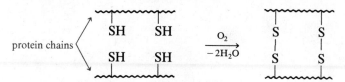

The precipitation of enzymes and other soluble proteins containing $-SH$ groups by heavy metal ions such as Hg^{2+} is a principal reason that compounds containing these ions are poisonous.

8.13 PHENOLS

A phenol is a compound in which the $-OH$ group is bonded to a carbon of an aromatic ring. Table 8.4 lists some common phenols.

The simplest member of the class, phenol itself, is a low-melting solid slightly soluble in water. Phenol was once called *carbolic acid* because it is slightly acidic, a property that is discussed in this section.

In the 1800s, the British surgeon Joseph Lister introduced phenol as a hospital antiseptic. Although phenol is a good antiseptic, it is also caustic and toxic; a lethal dose can be absorbed through the skin! Chronic exposure of smaller amounts of phenol leads to liver and kidney damage. Less toxic phenols, such as *n*-hexyl-resorcinol, are used as antiseptics today (see Point of Interest 8).

TABLE 8.4 SOME IMPORTANT PHENOLS

Structure	Name	Occurrence or use
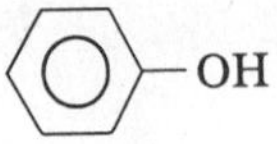 ⬡—OH	Phenol	Disinfectant and in the synthesis of resins, dyes, and other types of compounds
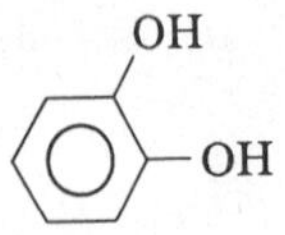 OH ⬡—OH	Catechol	Topical antiseptic
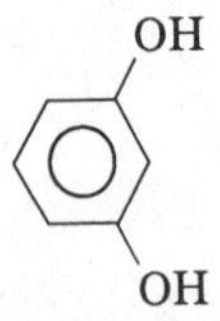 OH ⬡ OH	Resorcinol	In tanning; as an antiseptic; and in the synthesis of dyes, explosives, and cosmetics
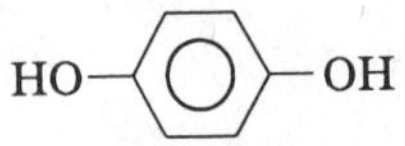 HO—⬡—OH	Hydroquinone	In photography and as an antioxidant
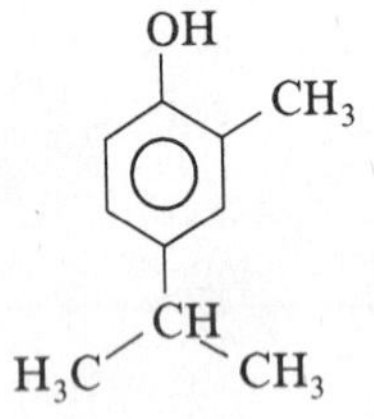 OH ⬡ CH₃ H₃C—CH—CH₃	Thymol	In thyme plants; used as a fungicide and in embalming fluid

A. Acidity

Although phenols are less acidic than carboxylic acids, they are more acidic than alcohols or water because phenoxide ions are resonance stabilized.

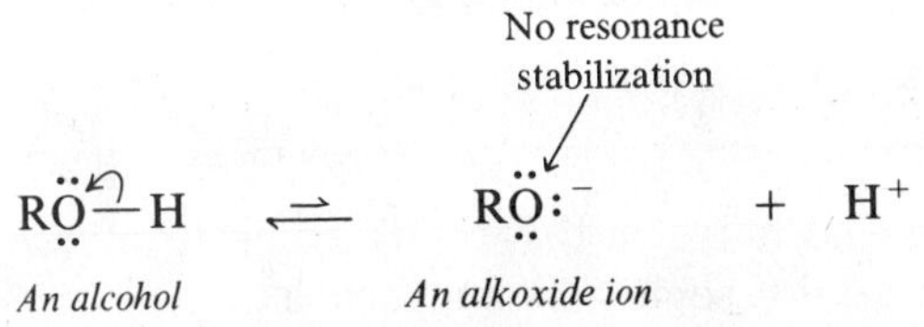

No resonance
stabilization

$$R\ddot{O}-H \;\rightleftharpoons\; R\ddot{O}:^- \;+\; H^+$$

An alcohol   *An alkoxide ion*

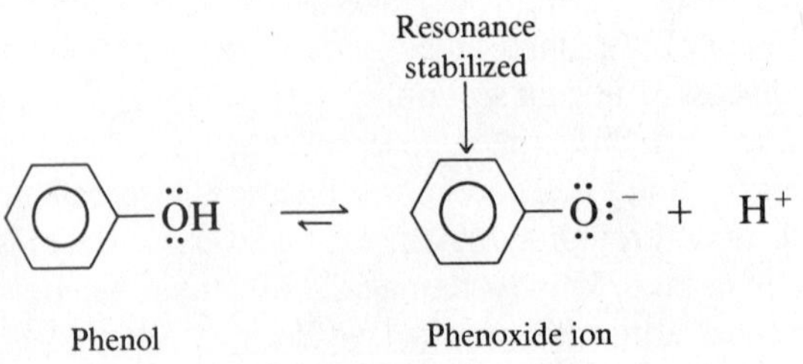

Resonance
stabilized

$$⬡-\ddot{O}H \;\rightleftharpoons\; ⬡-\ddot{O}:^- \;+\; H^+$$

Phenol   Phenoxide ion

Resonance Formulas for the Phenoxide Ion:

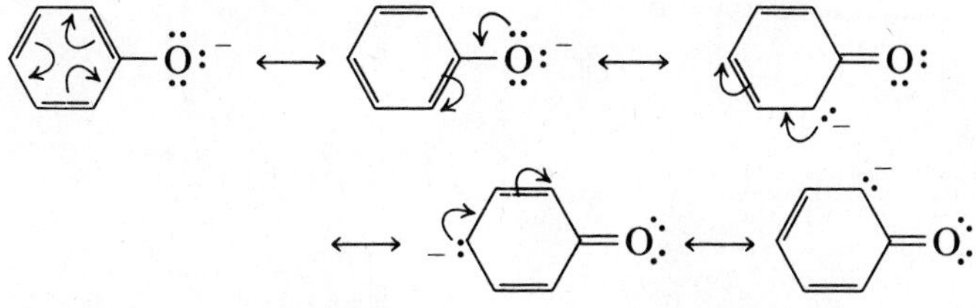

Unlike an alcohol, a phenol can be converted to a phenoxide ion by treatment with aqueous NaOH. However, most phenols do not react with weaker bases such as sodium bicarbonate ($NaHCO_3$), a base capable of forming salts with carboxylic acids.

Relative Acidity of Alcohols, Phenols, and Carboxylic Acids
($ROH < ArOH < RCO_2H$):

$$ROH \quad + \quad Na^+\ {}^-OH \longrightarrow \text{no appreciable reaction}$$

$$ArOH \quad + \quad Na^+\ {}^-OH \longrightarrow ArO^-\ Na^+ \quad + \quad H_2O$$

$$ArOH \quad + \quad Na^+\ HCO_3^- \longrightarrow \text{no appreciable reaction}$$

$$RCO_2H \quad + \quad Na^+\ HCO_3^- \longrightarrow RCO_2^-\ Na^+ \quad + \quad CO_2 \quad + \quad H_2O$$

Phenoxide ions formed by treatment of phenols with NaOH can be used in the Williamson ether synthesis (Sections 7.6 and 9.3).

$$ArO^- \quad + \quad RBr \longrightarrow ArOR \quad + \quad Br^-$$

The acidity of phenols allows them to be separated from less acidic compounds, such as alcohols. If an organic solution containing a mixture of a water-insoluble phenol and other water-insoluble compounds is shaken in a separatory funnel with a dilute aqueous solution of sodium hydroxide, the phenol is converted to a sodium phenoxide ($ArO^-\ Na^+$). The phenoxide is ionic and, therefore, dissolves in the aqueous sodium hydroxide layer. The aqueous layer is then separated from the organic layer, which retains the other compounds. Acidification of the aqueous layer allows isolation of the purified phenol.

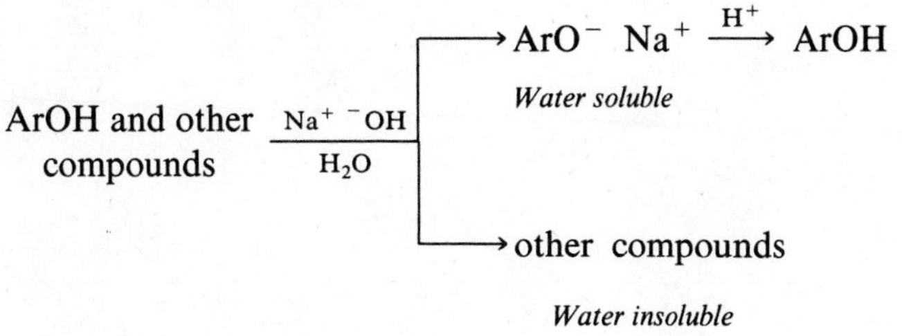

PROBLEM 8.16. Write equations that illustrate the separation of *m*-cresol, a toxic disinfectant, from toluene.

m-Cresol   Toluene

B. Oxidation

The oxidation of simple phenols results in complex mixtures of products. However, *catechol* (*o*-dihydroxybenzene) and *hydroquinone* (*p*-dihydroxybenzene) are readily oxidized by mild oxidizing agents such as Ag^+ or Fe^{3+} to dicarbonyl products called *quinones*. These oxidations are reversible; the quinone products are easily reduced back to the dihydroxy compounds.

Substituted quinones and hydroquinones form an important part of the electron transport system in biological organisms. These compounds are involved in the cellular interconversions of Fe^{3+} to Fe^{2+}, reactions that are necessary for the utilization of oxygen gas. Hydroquinone is also used in photographic development to reduce silver ions to silver metal.

Hydroquinone   1,4-Benzoquinone
(quinone)

PROBLEM 8.17. Predict the organic products formed when the following compounds are treated with an acidic aqueous solution of CrO_3.

C. Antioxidants

Many organic compounds, including alkenes in gasoline, rubber in tires, and nutrients in food, can be attacked by atmospheric oxygen. Oxygen (O_2) is a diradical, a radical with two unpaired electrons. Oxygen reacts with many organic compounds to yield hydroperoxides (Section 9.4A) that decompose to lower-formula-weight substances. This decomposition causes gasoline to become gummy, rubber to stiffen and crack, and fat to turn rancid.

$$R_3C-H + O_2 \xrightarrow{\text{light}} R_3C-OOH \longrightarrow R_3C\ddot{O}\cdot + \cdot\ddot{O}H$$

A hydroperoxide   *Radicals*

$$\downarrow$$

A mixture
of products

Substituted phenols called **antioxidants** are commonly used to block these free-radical reactions. In the food industry, the nontoxic phenols used as radical inhibitors are called **preservatives**.

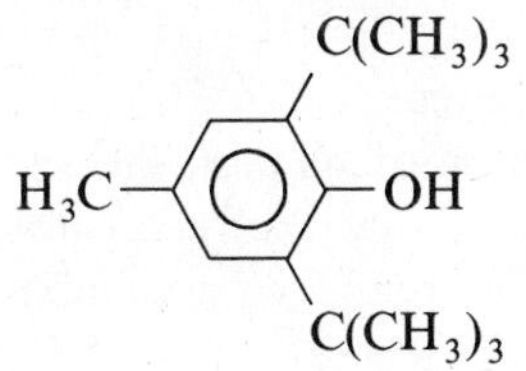

"Butylated hydroxytoluene" (BHT)

a food preservative

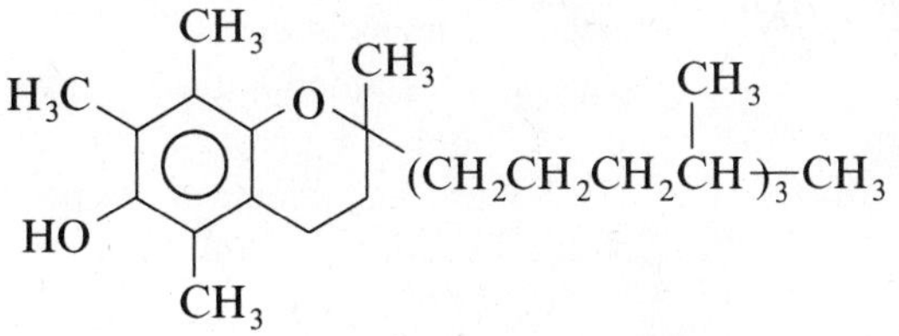

α-Tocopherol (vitamin E)

a natural antioxidant

Phenols are effective antioxidants because they react with radical intermediates to yield stable, nonreactive phenolic radicals. The formation of these nonreactive radicals terminates the undesirable radical oxidation process.

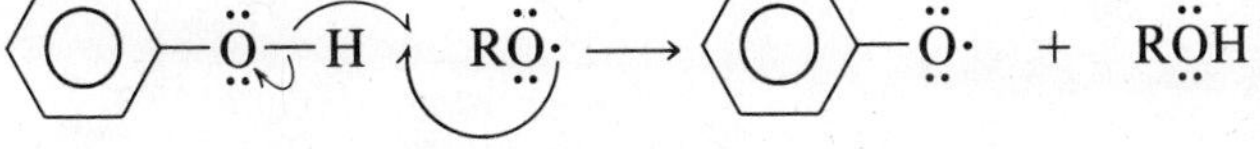

Resonance
stabilized

Resonance Formulas for the Phenol Radical:

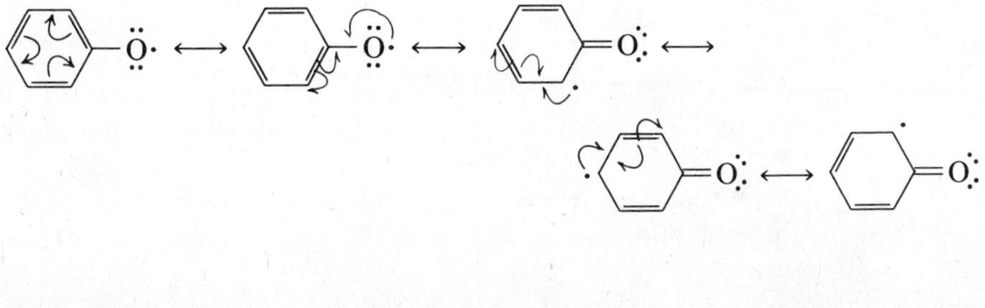

> **PROBLEM 8.18.**   BHA, "butylated hydroxyanisole," is a food preservative similar in structure to BHT. Instead of the ring methyl group, BHA contains a methoxyl group (CH_3O-). Write an equation for the reaction of BHA with $R\ddot{O}\cdot$, and then draw resonance formulas for the BHA radical.

SUMMARY

An **alcohol** (ROH) contains a hydroxyl group bonded to an sp^3-hybridized carbon. A **phenol** (ArOH) contains a hydroxyl group bonded to the sp^2-hybridized carbon of an aromatic ring.

Compounds with hydroxyl groups can undergo hydrogen bonding; therefore, these compounds are higher boiling and more water soluble than hydrocarbons or organohalogen compounds.

In the IUPAC system, alcohols are named with an *-ol* suffix. In the trivial names, the word *alcohol* is preceded by the name of the alkyl group bonded to the oxygen. For example, CH_3OH can be called methanol or methyl alcohol. Simple phenols are named with *phenol* as the parent.

Alcohols are prepared by the hydration of alkenes; the hydroboration-oxidation of alkenes; the reaction of primary, secondary, allylic, or benzylic halides with hydroxide ion; Grignard reactions of carbonyl compounds; or the reduction of carbonyl compounds.

An alcohol can lose a proton to a strong base or an alkali metal. An alcohol can gain a proton from a strong acid.

$$R\ddot{O}H \begin{cases} \xrightarrow{\text{K}} R\ddot{\underset{\cdot\cdot}{O}}\colon^- K^+ & \textit{An alkoxide} \\[2em] \xrightarrow{H_2SO_4} R\overset{+}{\underset{\cdot\cdot}{O}}H_2 & \textit{A protonated alcohol} \end{cases}$$

Protonated alcohols can undergo substitution reactions with nucleophiles, such as halide ions. Primary alcohols undergo reaction by the S_N2 path; secondary and tertiary alcohols, by the S_N1 path.

$$ROH + HX \longrightarrow RX + H_2O$$

Tertiary, allylic, and benzylic halides are the most reactive in this reaction. Other halogenating agents include thionyl chloride ($SOCl_2$) and the phosphorus halides (PX_3, PX_5).

When heated with a strong acid, alcohols undergo loss of water, or **dehydration**, to yield alkenes. Tertiary alcohols are the most reactive in these elimination reactions.

$$R-\overset{\overset{\displaystyle OH}{|}}{\underset{\underset{\displaystyle R}{|}}{C}}-\overset{\overset{\displaystyle H}{|}}{\underset{\underset{\displaystyle R}{|}}{C}}-R \xrightarrow[\text{heat}]{H_2SO_4} \overset{\displaystyle R}{\underset{\displaystyle R}{\diagdown}}C=C\overset{\displaystyle R}{\underset{\displaystyle R}{\diagup}} + H_2O$$

The most highly
substituted
trans alkene

Alcohols undergo **esterification** to yield *carboxylic esters* or *inorganic esters*.

$$R'\overset{\overset{\displaystyle O}{\|}}{C}-OH + H-OR \xrightarrow{H^+} R'\overset{\overset{\displaystyle O}{\|}}{C}-OR + H-OH$$

$$RO-H + HO-SO_3H \longrightarrow ROSO_3H + H-OH$$

Alcohols can be oxidized to aldehydes, ketones, or carboxylic acids. These reactions are summarized in Figure 8.2.

Thiols (RSH), the sulfur analogs of alcohols, can be oxidized to **disulfides** (RSSR).

Phenols are more acidic than alcohols because the phenoxide ion is resonance stabilized. Some phenols are used as food preservatives because they are **free-radical traps**.

Dicarbonyl compounds called **quinones** can be obtained from the oxidation of *o*- and *p*-dihydroxybenzenes and other similar aromatic compounds.

STUDY PROBLEMS

8.19. Write an equation to show the ionization of 2-propanol in water.

8.20. Write formulas showing all types of hydrogen bonds in an aqueous solution of 2-propanol.

8.21. Which compound in each of the following pairs would be the more water soluble? Explain.

(a) $\overset{\overset{\displaystyle OH}{|}}{CH_2}\overset{\overset{\displaystyle OH}{|}}{CH}-\overset{\overset{\displaystyle OH}{|}}{CH_2}$ or $CH_3CH_2CH_2OH$

(b) $CH_3(CH_2)_4CH_2OH$ or $CH_3(CH_2)_6CH_2OH$

(c) $CH_3CH_2CH_2CH_2OH$ or $(CH_3)_3COH$

(d) $CH_3CH_2CH_2CH_2OH$ or $CH_3CH_2CH_2Br$

8.22. Name the following compounds by the IUPAC system:

(a) $CH_3\overset{\overset{\displaystyle CH_3}{|}}{CH}CH_2CH_2OH$

(b) $CH_3\overset{\overset{\displaystyle Br}{|}}{CH}\overset{\underset{\underset{\displaystyle Br}{|}}{}}{CH}CH_2OH$

(c) [structure: benzene ring with Br and OH]

(d) [structure: cyclohexane with OH and ---CH$_2$CH$_3$]

(e) [structure: benzene–CH(OH)–benzene]

(f) Br—[benzene ring with Cl]—OH

8.23. Write trivial names for the following alcohols:

(a) $(CH_3)_2CHOH$ **(b)** CH_3CH_2OH **(c)** $CH_3\overset{\underset{|}{CH_3}}{C}HCH_2OH$

(d) $CH_3CH_2\overset{\underset{|}{OH}}{C}HCH_3$ **(e)** $(CH_3)_3COH$ **(f)** [cyclopentane with OH]

8.24. Draw formulas for the following compounds:
  (a) 3,5-dichloro-2-heptanol
  (b) 3-hydroxy-1-phenyl-1-propanone
  (c) 2,4-dimethylphenol
  (d) 3-hexen-1-ol
  (e) (*R*)-1,2-butanediol
  (f) *trans*-4-methylcyclohexanol
  (g) 2,4,6-trimethylphenol
  (h) *trans*-3-bromocyclopentanol

8.25. Write formulas for and name all the isomeric alcohols with the molecular formula $C_5H_{12}O$. (Include stereoisomers in your answer.)

8.26. Classify the following alcohols as primary, secondary, tertiary, allylic, or benzylic. (Choose all classifications that apply.)

(a) $(CH_3)_2\overset{\underset{|}{OH}}{C}CH_2CH_3$ **(b)** $CH_3\overset{\underset{|}{OH}}{C}HC(CH_3)_3$

(c) $(CH_3)_3CCH_2OH$ **(d)** [structure: decalin with OH at ring junction]

(e) [structure: tetrahydronaphthalenol with OH]   **(f)** [structure: cyclohexene with CH$_2$OH]

8.27. Write flow equations to show how you would make the following conversions:

(a) $(CH_3)_2C{=}CHCH_3 \longrightarrow (CH_3)_2\overset{\underset{|}{OH}}{C}CH_2CH_3$

(b) $(CH_3)_2C{=}CHCH_3 \longrightarrow (CH_3)_2CH\overset{\underset{|}{OH}}{C}HCH_3$

(c) $\overset{\overset{\displaystyle Cl}{|}}{CH_3CH=CHCHCH_3} \longrightarrow \overset{\overset{\displaystyle OH}{|}}{CH_3CH=CHCHCH_3}$

(d) 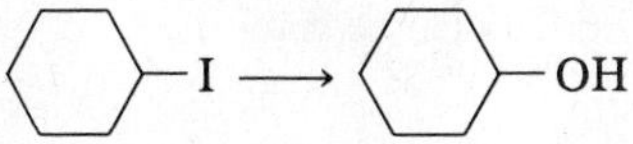 ⬡—I ⟶ ⬡—OH

(e) 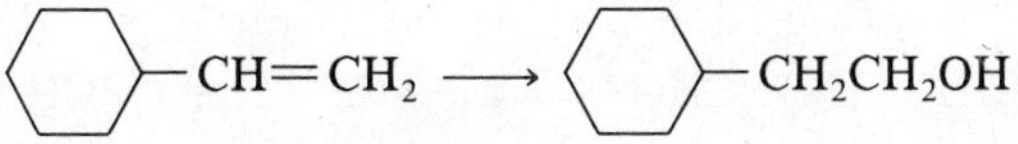 ⬡—CH=CH₂ ⟶ ⬡—CH₂CH₂OH

8.28. Complete the following equations:

(a) $(CH_3)_3COH + K \longrightarrow$

(b) $CH_3CH_2OH + conc.\ H_2SO_4 \underset{cold}{\rightleftharpoons}$

(c) $CH_3CH_2OH + conc.\ H_2SO_4 \xrightarrow{170°C}$

(d) $CH_3CH_2CH_2OH + Na \longrightarrow$

(e) $(CH_3)_3COH + Na^+\ H^- \longrightarrow$

(f) $CH_3CH_2CH_2OH + HI \longrightarrow$

(g) $CH_3CH_2CH_2CH_2OH + HCl \longrightarrow$

(h) $(CH_3CH_2)_3COH + HCl \longrightarrow$

(i) $(CH_2=CH)_2CHOH + HCl \longrightarrow$

(j) 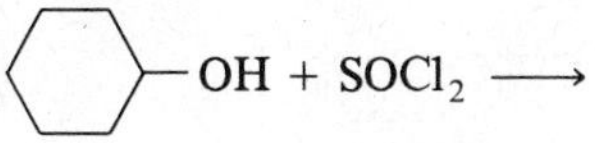 ⬡—OH + $SOCl_2 \longrightarrow$

(k) 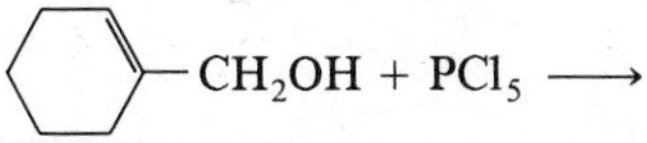 ⬡—CH₂OH + $PCl_5 \longrightarrow$

8.29. Predict the relative reactivities toward HBr (least reactive first).

(a) $CH_3CH_2\overset{\overset{\displaystyle OH}{|}}{C}HCH_2CH_3$ (b) $CH_3CH_2\overset{\overset{\displaystyle OH}{|}}{C}H_2$ (c) $(CH_3CH_2)_3COH$

8.30. Write flow equations for the mechanisms of the following reactions.

(a) 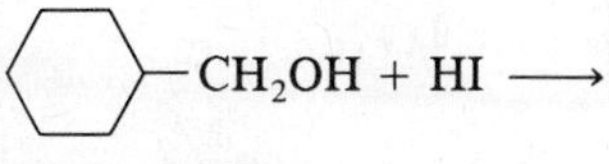 ⬡—CH₂OH + HI ⟶

(b) $CH_3CH_2\overset{\overset{\displaystyle OH}{|}}{C}HCH_3 + HI \longrightarrow$

(c) 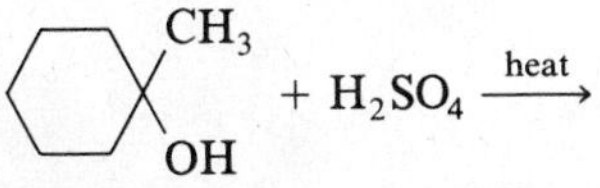 (cyclohexane ring with CH₃ and OH) + $H_2SO_4 \xrightarrow{heat}$

8.31. Predict the product of the treatment of 1-heptanol with the following reagents:

(a) CrO_3, H_2SO_4, heat   (b) pyridinium chlorochromate (PCC)

(c) HBr   (d) PBr_3

(e) excess $CH_3\overset{\overset{\displaystyle O}{\|}}{C}OH$, H^+, heat   (f) $KMnO_4$, ^-OH, heat

8.32. How would you make the following conversions?

(a) $CH_3CH_2OH \longrightarrow CH_3CH_2OCH_2CH_3$

(b) 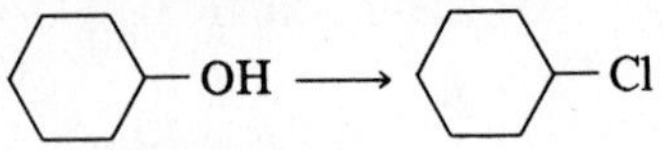 ⬡—OH $\longrightarrow$ ⬡—Cl   (three routes)

(c) $\underset{\underset{OH}{|}}{CH_3CHCH_2CH_2CH_2CH_3} \longrightarrow$ *trans*-$CH_3CH=CHCH_2CH_2CH_3$

(d) $CH_3CH_2CH_2OH \longrightarrow CH_3CH_2\overset{\overset{O}{\|}}{C}H$

(e) $CH_3OH \longrightarrow CH_3CH_2CH_2\overset{\overset{O}{\|}}{C}OCH_3$

(f) 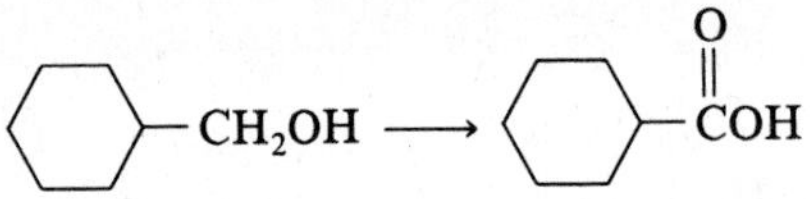 ⬡—$CH_2OH \longrightarrow$ ⬡—$\overset{\overset{O}{\|}}{C}OH$

(g) 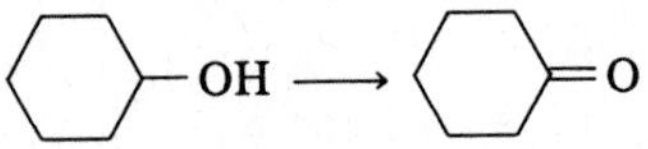 ⬡—OH $\longrightarrow$ ⬡$=O$

8.33. Which of the following compounds would be the stronger acid? Use equations to explain your answer.

(a) $CH_3CH_2CH_2S-H$ (b) $CH_3CH_2CH_2O-H$

8.34. Complete the following equations:

(a) $(CH_3)_2CHSH + Hg^{2+} \longrightarrow$

(b) $(CH_3)_2CHSH + H_2O_2 \longrightarrow$

(c) $(CH_3)_2CHSH + Na^+ \ ^-OH \longrightarrow$

(d) $(CH_3)_2CHSSCH(CH_3)_2 + Zn \xrightarrow{CH_3CO_2H}$

8.35. Predict the organic products, if any, when 2-naphthol is treated with the following reagents:

(a) $KHCO_3$ (b) KOH (c) KOH followed by CH_3I

8.36. Write resonance formulas for the following structures:

(a) 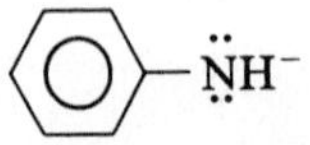 ⬡—$\ddot{N}H^-$ (b) 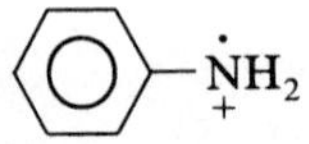 ⬡—$\ddot{N}H_2^+$ (c) 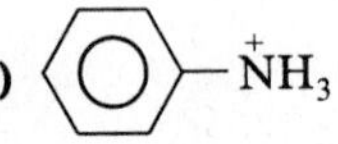 ⬡—$\overset{+}{N}H_3$

8.37. Write flow equations for the conversion of 1-butanol to the following compounds:

(a) $CH_3CH_2CH_2\overset{\overset{O}{\|}}{C}H$

(b) $CH_3CH_2CH_2\overset{\overset{O}{\|}}{C}OH$

(c) $CH_3CH_2CH_2\overset{\overset{O}{\|}}{C}OCH_3$

(d) $CH_3CH_2CH_2CH_2I$

(e) $CH_3CH_2CH=CH_2$

(f) $CH_3CH_2CH_2CH_2CN$

(g) $\underset{\underset{OH}{|}}{CH_3CH_2CHCH_3}$

(h) $CH_3CH_2CH_2CH_2SH$

(i) $CH_3CH_2CH_2CH_2O$—

8.38. *p*-Nitrophenol (pK_a about 7) is far more acidic than phenol ($pK_a = 10$). Use resonance theory to explain why this is so.

8.39. Allyl disulfide ($CH_2\!\!=\!\!CHCH_2SSCH_2CH\!\!=\!\!CH_2$) is a contributor to the odor of garlic. How would you prepare this compound from allyl chloride?

POINT OF INTEREST 8

Antiseptics and Disinfectants

Antiseptics are substances that, when applied to living tissue, kill or prevent the growth of microorganisms. **Disinfectants** are chemicals that destroy pathogenic microorganisms, usually on inanimate objects.

Antiseptics and disinfectants were being used at the beginning of recorded history. Ancient Persian law, for example, required that drinking water be stored in bright copper vessels, and as far back as Hippocrates, vinegar and wine were recommended for use with wound dressings. These early users of antiseptics had no idea why they were effective.

Even in the mid-1800s, antiseptics were used effectively, but for the wrong reasons. In 1844 the Viennese physician Ignaz Semmelweis believed that odor was the cause of the high incidence of childbed fever. He required medical students to wash their hands in chlorine water, which destroys odors, when passing from the autopsy room to the obstetrics ward in a Vienna hospital and was rewarded with a dramatic reduction in the disease.

Louis Pasteur was the first to suggest a relationship between bacteria and disease. The British surgeon Joseph Lister recognized the significance of Pasteur's suggestion. In 1867 he announced his method of sterilization with carbolic acid, now called phenol.

Since then, numerous antiseptics have been introduced. Table 8.5 lists a few antiseptics in current use.

The compounds listed in Table 8.5 illustrate the major classes of antiseptics and disinfectants. Hydrogen peroxide, chlorine, and sodium hypochlorite are oxidizing

TABLE 8.5   SOME COMMON ANTISEPTICS AND DISINFECTANTS

Benzalkonium chloride	Mercurochrome
Ethanol	Merthiolate
Hexylresorcinol	Pine oil
Hexachlorophene	Phenol
Hydrogen peroxide	Soap
Iodine	Sodium hypochlorite (chlorine bleach)
Isopropyl alcohol	Chlorine (added to water)

agents. Mercurochrome and Merthiolate are both organomercury compounds. Ethanol and isopropyl alcohol are alcohols; pine oil also contains alcohols. Benzalkonium chloride and soap are surface-active agents, with detergent activity. The most important antiseptics, however, are phenols, phenol itself, hexylresorcinol, and hexachlorophene. (Hexachlorophene can be absorbed through the skin and is no longer sold for home use.)

Mercurochrome   Merthiolate   α-Terpineol

                                                        found in
                                                        pine oil

Probably more is known about phenols than about the other classes of antiseptics. Phenol kills all types of cells by binding to and precipitating their proteins. Phenol does not form a strong protein complex; consequently, it is easily released from the precipitate. Then the phenol can bond with other proteins within the cell or can pass to another cell.

When applied to the skin, phenol itself acts as an antiseptic, but it can also cause burns. Phenol is too corrosive for use as a general antiseptic; however, it finds current use in combination with other ingredients in over-the-counter topical salves.

The principal test for antiseptic activity is the **phenol coefficient test**, which provides a ranking of phenol and other antiseptics or disinfectants. The phenol coefficient is obtained by dividing the antiseptic's greatest effective dilution by the greatest dilution of phenol that shows the same test results. The larger the coefficient, the greater the effectiveness of the antiseptic.

4-*n*-Hexylresorcinol		Hexachlorophene
Phenol coefficient:	about 50	about 125

Chapter 9

Ethers and Epoxides

In an ether, two alkyl groups, two aryl groups, or one of each are bonded to an oxygen. The functional groups in ethers are named as **alkoxy groups** ($-OR$) or **phenoxy groups** ($-OAr$).

Ethers (ROR', ROAr, or ArOAr'):

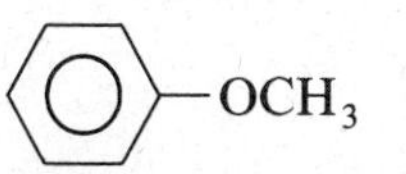

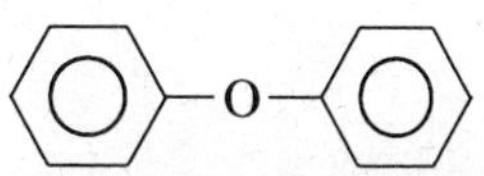

Many naturally occurring compounds contain ether groups in addition to other functional groups. Tetrahydrocannabinol, the principal active ingredient in marijuana, contains a cyclic ether group, a phenol hydroxyl group, and a carbon–carbon double bond. The painkiller codeine contains two ether oxygens, one of which is part of a ring.

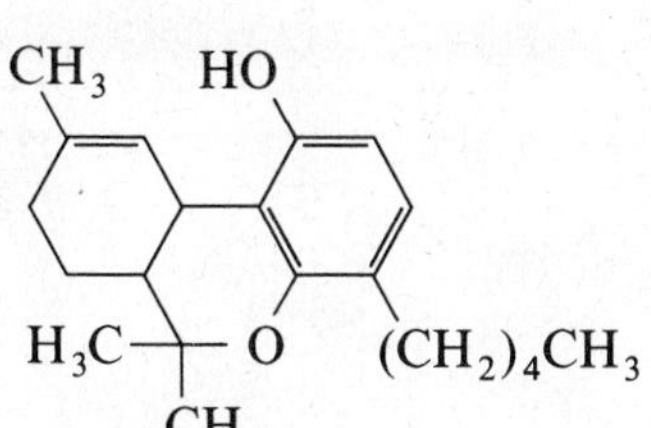

Tetrahydrocannabinol (THC) 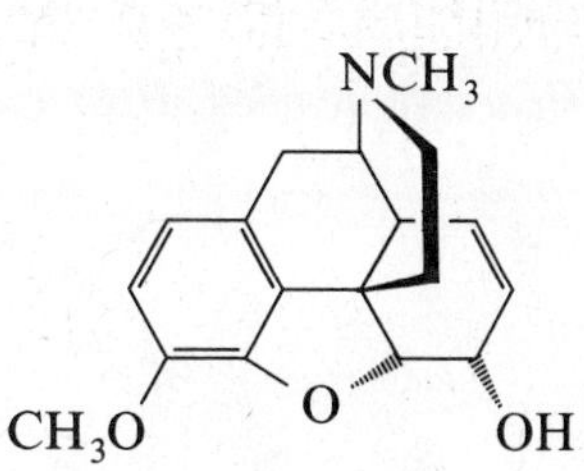   Codeine

An **epoxide** is a three-membered-ring ether. Because of their three-membered rings, epoxides are far more reactive than other cyclic or open-chain ethers. Because of this reactivity, only a few natural products contain an epoxide ring. An interesting example of such a naturally occurring epoxide is the sex attractant of the female gypsy moth, a compound called *disparlure*.

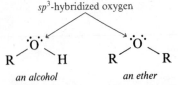

Disparlure

gypsy moth sex attractant

A chemical such as disparlure that is given off by one member of a species to attract or communicate with another member of the same species is called a **pheromone**. Insects are especially sensitive to pheromones. A typical female moth may carry only 10^{-10} gram of sex attractant but may be able to lure male moths from miles away!

9.1 BONDING AND PHYSICAL PROPERTIES OF ETHERS

Water, alcohols, phenols, and ethers all contain an sp^3-hybridized oxygen atom. The difference among these classes of compounds is that an ether has *two* organic groups bonded to the oxygen.

sp^3-hybridized oxygen

$$R \overset{\cdot\cdot}{\underset{\cdot\cdot}{O}} H \qquad R \overset{\cdot\cdot}{\underset{\cdot\cdot}{O}} R$$

an alcohol   *an ether*

Ethers do not have H atoms bonded to their oxygens; therefore, in their pure states, ethers do *not* undergo hydrogen bonding. Because of the lack of hydrogen bonding, ethers have boiling points more comparable to those of alkanes than to those of alcohols.

Hydrogen Bonded, Higher Boiling:

$CH_3CH_2CH_2CH_2OH$ H_3C—⟨O⟩—OH

1-Butanol (bp, 117°C)   *p*-Methylphenol (bp, 202°C)
                                 (*p*-cresol)

Not Hydrogen Bonded, Lower Boiling:

$CH_3CH_2OCH_2CH_3$ ⟨O⟩—OCH_3

Diethyl ether (bp, 34.5°C)   Methoxybenzene (bp, 155°C)
                               (anisole)

Although ethers cannot undergo hydrogen bonding in the pure state, they can form hydrogen bonds with other compounds containing —OH groups, such as water, alcohols, or phenols, or with compounds containing the ⟩NH group, such as the amines.

Hydrogen bond between
an ether and water

$$R-\ddot{O}\text{:}\cdots\cdots H-\underset{\underset{R}{|}}{\overset{\overset{H}{|}}{\ddot{O}\text{:}}}$$

Because ethers can form hydrogen bonds with water, their water solubility is about the same as that of their isomeric alcohols.

$CH_3CH_2CH_2CH_2OH$ $CH_3CH_2OCH_2CH_3$

1-Butanol   Diethyl ether
8.3 g/100 mL H_2O 8 g/100 mL H_2O

Cyclic ethers are more water soluble than the corresponding open-chain ethers because the hydrophobic CH_2 groups are held away from the oxygen, allowing less sterically hindered hydrogen bonds to form. Tetrahydrofuran (THF), a convenient solvent for many organic compounds, is miscible with water.

$$\begin{array}{c} H_2C-CH_2 \\ H_2C \qquad CH_2 \\ \ddot{O} \end{array} \quad \text{or} \quad \text{(structure)}$$

Tetrahydrofuran (THF)

Table 9.1 lists some ethers and their physical properties.

TABLE 9.1   PHYSICAL PROPERTIES OF SOME ETHERS

Formula	Name	Bp (°C)	Solubility in water at 20°C (g / 100 mL)
CH_3OCH_3	Dimethyl ether	−24	7
$CH_3CH_2OCH_2CH_3$	Diethyl ether	35	8
CH_2-CH_2 (epoxide with O)	Ethylene oxide	11	∞
(cyclic O structure)	Tetrahydrofuran	66	∞
(benzene ring)—OCH_3	Anisole	155.5	Insoluble

9.2 NOMENCLATURE OF ETHERS

Simple ethers are usually referred to by their trivial names. In these names, we name the alkyl or aryl groups bonded to the ether oxygen and add the word *ether*.

$$CH_3CH_2{-}O{-}CH_2CH_3 \qquad CH_3O{-}\langle\bigcirc\rangle$$

Diethyl ether

(*also called ethyl ether or simply "ether"*)

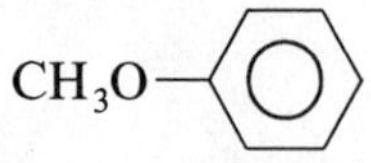

Methyl phenyl ether

(*also called anisole*)

In more complex structures, the IUPAC system is followed. In this system, the —OR group is named *alkoxy-*, with the ending of the parent alkane names changed from *-ane* to *-oxy*.

Alkoxyl Groups:

$$-OCH_3 \qquad -OCH_2CH_3 \qquad \overset{\displaystyle CH_3}{\underset{\displaystyle -OCHCH_3}{|}}$$

Methoxy-   Ethoxy-   Isopropoxy-

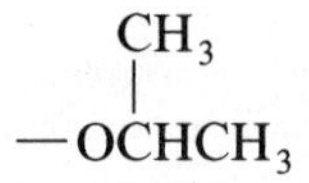

An alkoxyl group is named as a prefix substituent, just as a methyl group or a halogen is named.

$$CH_3OCH_2CH_2OCH_3$$

1,2-Dimethoxyethane
(DME, or "glyme")

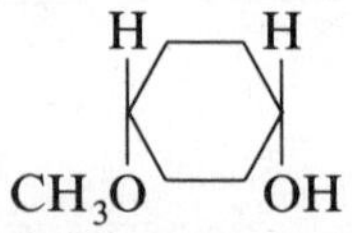

cis-4-Methoxycyclohexanol

Cyclic ethers have individual names, which can be a parent name.

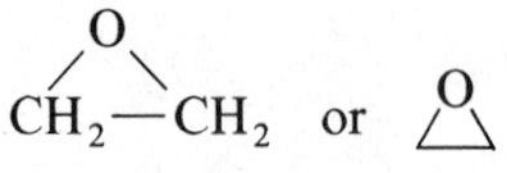

$$\overset{\displaystyle O}{CH_2{-}CH_2} \quad \text{or} \quad \triangle$$

Oxirane or
ethylene oxide

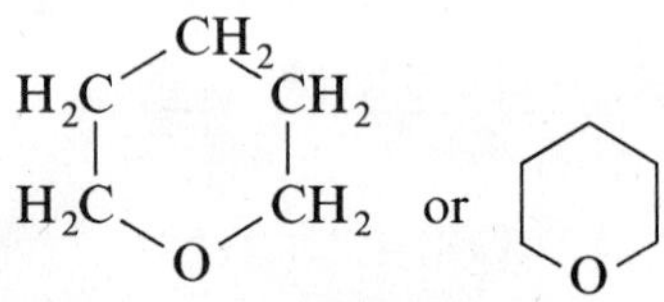

Tetrahydropyran

PROBLEM 9.1. Write formulas for the following ethers: **(a)** diisopropyl ether, **(b)** methyl *n*-propyl ether, **(c)** *t*-butyl phenyl ether.

PROBLEM 9.2. Name the following ethers by the alkoxy system:

(a)  $CH_3O-\bigcirc-OCH_3$ (b) $CH_3OCH_2CH_2CH_3$

(c) $(CH_3)_2CHOCHCH_2CH_3$ with CH_3 substituent   (d) $CH_3OCH_2CH_2CH_2OH$

9.3 PREPARATION OF ETHERS

Diethyl ether is widely used as a solvent and has been used as an anesthetic. Industrially, this ether is prepared from ethanol (see Section 8.10).

$$2CH_3CH_2OH \xrightarrow[140°C]{H_2SO_4} CH_3CH_2-O-CH_2CH_3 + H_2O$$

Diethyl ether

In the laboratory, a more useful synthetic route to ethers is the *Williamson ether synthesis*—treatment of an alkyl halide with an alkoxide (RO^-) or a phenoxide (ArO^-). The reaction is an S_N2 substitution reaction. Consequently, the halides that can be used in this reaction are limited to 1°, 2°, allylic, and benzylic.

Williamson Ether Synthesis:

1°, 2°, allylic, or benzylic
(but not 3°, vinylic, or aryl)

$$R-X + R'O^- \longrightarrow R-OR' + X^-$$

1°, 2°, 3°, allylic,
benzylic, or aryl

PROBLEM 9.3. Write equations for two Williamson syntheses for each of the following ethers.

(a) $CH_3OCH_2CH_3$ (b) 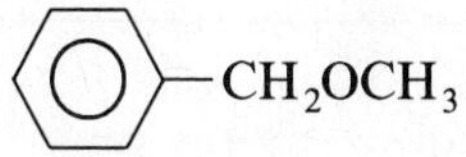 $\bigcirc-CH_2OCH_3$

PROBLEM 9.4.   There is only one Williamson synthesis leading to each of the following ethers. Write an equation for each, and explain why this reaction is the only possible route to the ether.

(a) ⟨O⟩—OCH_2CH_3 (b) $CH_3OC(CH_3)_3$

9.4 REACTIONS OF ETHERS

Ordinary ethers do not react with most reagents used in the organic laboratory. This lack of reactivity is one reason that diethyl ether is a widely used solvent. Lack of reactivity is also a reason that ethers can be used as inhalation anesthetics; nonreactive compounds are generally less toxic than reactive ones.

Ethers are not totally devoid of reactivity. The two principal reactions that ethers do undergo are *oxidation* (combustion and autoxidation) and *substitution* (acidic cleavage).

A. Oxidation

Combustion   Almost all organic compounds can undergo combustion. Ethers are no exception; they burn readily. The low-formula-weight ethers, commonly used as solvents and anesthetics, are a special fire hazard because of their volatility.

$$CH_3CH_2OCH_2CH_3 + 6O_2 \xrightarrow{\text{spark}} 4CO_2 + 5H_2O$$

Autoxidation   Ethers also undergo a slow reaction with oxygen, a reaction called **auto-oxidation** or **autoxidation**, to yield **hydroperoxides**, compounds containing —OOH groups. Autoxidation reactions are free-radical reactions, initiated by oxygen ($\cdot O—O\cdot$), a diradical.

An Autoxidation Reaction:

Attack on either
C adjacent to O

$$CH_3CH_2OCH_2CH_3 + O_2 \longrightarrow CH_3CH_2OCHCH_3$$
$$\overset{\displaystyle OOH}{\underset{\displaystyle |}{}}$$

a hydroperoxide

Hydroperoxides of ethers can explode when heated. Therefore, any hydroperoxides present should be destroyed before an ether is distilled.

B. Substitution

Although ethers are less reactive than alcohols, they can undergo substitution reactions when heated with a strong acid such as HI or HBr. Since the reaction results in the cleavage of the ether molecule, it is also referred to as **acid cleavage**.

The mechanism of the acid cleavage of an ether is very similar to the mechanism of substitution of an alcohol with HI or HBr—protonation followed by substitution by an S_N1 or S_N2 path.

S_N2 Substitution of an Ether:

$$R-OR' + 2HI \longrightarrow RI + R'I + H_2O$$

Mechanism:

Step 1, Protonation:

$$CH_3CH_2\overset{..}{\underset{..}{O}}CH_2CH_3 + H^+ \xrightarrow{fast} CH_3CH_2\overset{\overset{\displaystyle H}{|}}{\underset{+}{\overset{..}{O}}}CH_2CH_3$$

Step 2, Substitution:

$$CH_3CH_2-\overset{\overset{\displaystyle H}{|}}{\underset{+}{O}}CH_2CH_3 \xrightarrow{slow} CH_3CH_2 + :\overset{\overset{\displaystyle H}{|}}{\underset{..}{O}}CH_2CH_3$$

$$\underset{:\overset{..}{\underset{..}{I}}:^-}{} \qquad \underset{:\overset{..}{\underset{..}{I}}:}{} \quad \textit{Intermediate alcohol}$$

Step 3, Reaction of the Intermediate Alcohol with HX:

$$CH_3CH_2OH + HI \xrightarrow{S_N2} CH_3CH_2I + H_2O$$

The products of the substitution reaction (step 2) of diethyl ether with HI are iodoethane and ethanol. The ethanol is an intermediate in this acid cleavage and undergoes further reaction with the HI present in the reaction mixture to yield another molecule of iodoethane (step 3).

EXAMPLE

Predict the products when anisole (methyl phenyl ether) is heated with an excess of HBr.

Solution: Write a flow equation for the steps in the reaction.

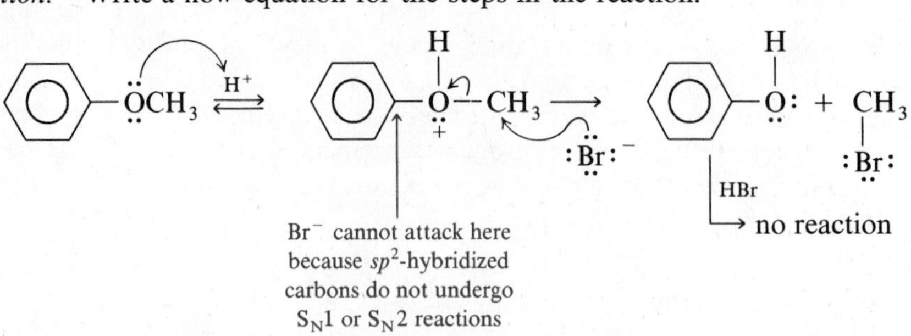

Br⁻ cannot attack here
because sp^2-hybridized
carbons do not undergo
S_N1 or S_N2 reactions

HBr
→ no reaction

The products are phenol and bromomethane. Phenol does not react further with HBr because phenols do not undergo alcohol substitution reactions.

PROBLEM 9.5.   Predict the products:

(a) $CH_3CH_2OCH_2CH_2CH_3$ + excess HI $\xrightarrow{\text{heat}}$

(b) $CH_3CH_2OCH(CH_3)_2$ + excess HBr $\xrightarrow{\text{heat}}$

(c) $(CH_3)_2CHOCH(CH_3)_2$ + O_2 $\xrightarrow{\text{cold}}$

9.5 EPOXIDES

Oxiranes, also called **epoxides**, are three-membered-ring ethers in which the ring contains two carbons and one oxygen. If the two carbons of the oxirane ring contain different substituents, cis and trans isomers are possible.

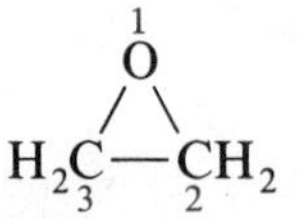

Oxirane
(ethylene oxide)

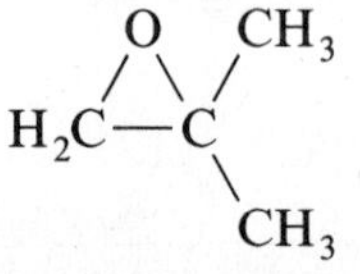

2,2-Dimethyloxirane

*No cis-trans
isomers are possible*

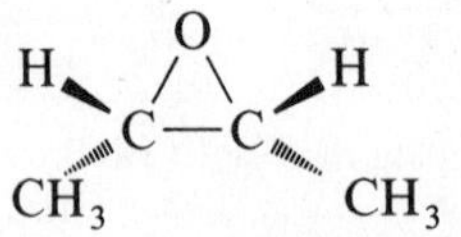

cis-2,3-Dimethyloxirane

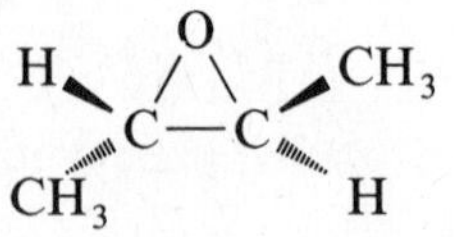

trans-2,3-Dimethyloxirane

A. Reactions with Acid or Base

Epoxides are more reactive than other ethers because the C—C and C—O orbitals cannot undergo maximum overlap.

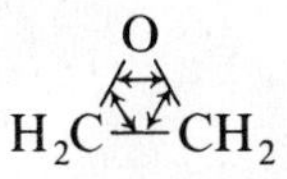

$$\text{H}_2\text{C} \overset{\text{O}}{\triangle} \text{CH}_2$$

Bond angles are 60° instead of the more stable 109.5° typical of sp^3-hybridized atoms.

Epoxides undergo substitution reactions readily in acidic or alkaline solution. These substitution reactions result in opening of the strained three-membered ring; therefore, the products are more stable than the original epoxide.

In alkaline solution, ring opening occurs by nucleophilic attack at the epoxide ring carbon displacing the oxygen. An acid-base reaction generates a β-substituted alcohol.

General Reaction in Base:

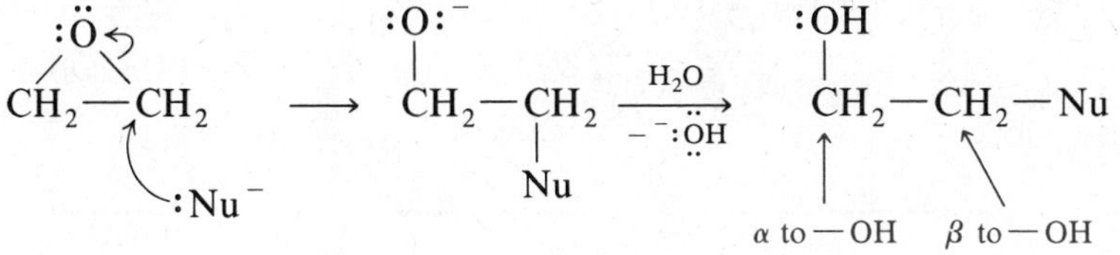

Specific Examples:

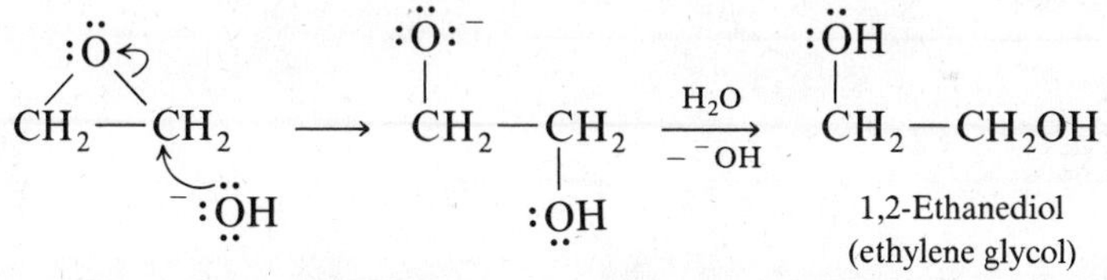

1,2-Ethanediol
(ethylene glycol)

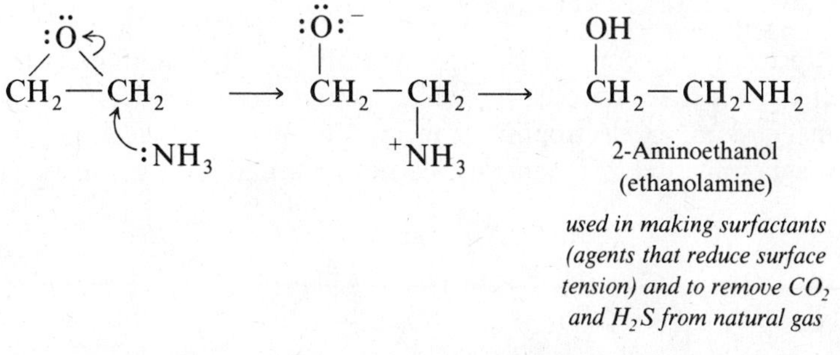

2-Aminoethanol
(ethanolamine)

used in making surfactants (agents that reduce surface tension) and to remove CO_2 and H_2S from natural gas

In acidic solution, the oxygen is first protonated; then reaction proceeds in a similar fashion as in base.

General Reaction in Acid:

$$\underset{\text{CH}_2-\text{CH}_2}{\overset{:\ddot{\text{O}}:}{\triangle}} \;\overset{\text{H}^+}{\rightleftharpoons}\; \underset{\text{CH}_2-\text{CH}_2}{\overset{:\overset{+}{\text{O}}\text{H}}{\triangle}} \longrightarrow \underset{\underset{\text{Nu}}{|}}{\overset{:\ddot{\text{O}}\text{H}}{\overset{|}{\text{CH}_2-\text{CH}_2}}}$$

$$:\text{Nu}^-$$

Specific Examples:

$$\underset{\text{CH}_2-\text{CH}_2}{\overset{\text{O}}{\triangle}} + \text{H}_2\text{O} \xrightarrow{\text{H}^+} \underset{}{\overset{\text{OH}}{\overset{|}{\text{CH}_2-\text{CH}_2\text{OH}}}}$$

$$\underset{\text{CH}_2-\text{CH}_2}{\overset{\text{O}}{\triangle}} + \text{CH}_3\text{OH} \xrightarrow{\text{H}^+} \underset{}{\overset{\text{OH}}{\overset{|}{\text{CH}_2-\text{CH}_2\text{OCH}_3}}}$$

2-Methoxyethanol
a toxic industrial solvent

$$\underset{\text{CH}_2-\text{CH}_2}{\overset{\text{O}}{\triangle}} + \text{HOCH}_2\text{CH}_2\text{OH} \xrightarrow{\text{H}^+} \underset{}{\overset{\text{OH}}{\overset{|}{\text{CH}_2\text{CH}_2-\text{OCH}_2\text{CH}_2\text{OH}}}}$$

"Diethylene glycol"
an antifreeze and solvent

PROBLEM 9.6.   Predict the products:

(a) $\underset{\text{CH}_2-\text{CH}_2}{\overset{\text{O}}{\triangle}} + \text{CH}_3\text{OH} \xrightarrow{\text{CH}_3\text{O}^-}$

(b) $\langle\!\!\!\bigcirc\!\!\text{O} + \text{H}_2\text{O} \xrightarrow{\text{H}^+}$

B. Reaction with Grignard Reagents

Most organic compounds, such as ROH and RX, contain a carbon bonded to an electronegative atom (O, X, or N). A Grignard reagent contains a carbon bonded to magnesium, an electropositive metal. This carbon is electronegative with respect to magnesium and, consequently, carries a partial negative charge (see Section 7.11).

$$\underset{\overset{\delta+}{\text{R}_3\text{C}}-\overset{\delta-}{\text{X}}}{\text{C is }\delta+} \qquad \underset{\overset{\delta-}{\text{R}_3\text{C}}-\overset{\delta+}{\text{MgX}}}{\text{C is }\delta-}$$

The negative carbon in a Grignard reagent is nucleophilic and can attack an epoxide-ring carbon just as other nucleophiles can. The product of the nucleophilic reaction is a magnesium salt of an alcohol (an alkoxide). Like other alkoxides, magnesium alkoxides yield alcohols when treated with water or with dilute aqueous acid. This reaction is shown as the second step in the following sequence.

Step 1, Reaction with the Grignard Reagent:

$$\overset{\overset{\ddot{\text{O}}}{\diagup\diagdown}}{\text{CH}_2\text{—CH}_2} \ + \ \overset{\delta-\quad\delta+}{\text{R—MgX}} \ \longrightarrow \ \overset{\overset{:\ddot{\text{O}}:^-}{|}}{\underset{\underset{\text{R}}{|}}{\text{CH}_2\text{—CH}_2}} + \ {}^+\text{MgX}$$

A new carbon–
carbon bond

Step 2, Hydrolysis:

$$\overset{\overset{:\ddot{\text{O}}:^-}{|}}{\underset{\underset{\text{R}}{|}}{\text{CH}_2\text{—CH}_2}} + \ {}^+\text{MgX} \ + \ \text{H}^+ \ \xrightarrow{\text{H}_2\text{O}} \ \overset{\overset{\text{OH}}{|}}{\text{CH}_2\text{CH}_2\text{R}} \ + \ \text{Mg}^{2+} \ + \ \text{X}^-$$

An alcohol

When we write equations for the steps in a Grignard reaction, we usually combine the two steps over and under the reaction arrow.

$$\text{R—MgX} \ \xrightarrow[\text{(2) H}_2\text{O, H}^+]{\text{(1) }\overset{\text{O}}{\overset{\diagup\diagdown}{\text{CH}_2\text{—CH}_2}}} \ \overset{\overset{\text{OH}}{|}}{\text{RCH}_2\text{CH}_2}$$

Specific Example:

$$\text{C}_6\text{H}_5\text{—MgBr} \ \xrightarrow[\text{(2) H}_2\text{O, H}^+]{\text{(1) }\overset{\text{O}}{\overset{\diagup\diagdown}{\text{CH}_2\text{—CH}_2}}} \ \text{C}_6\text{H}_5\text{—CH}_2\text{CH}_2\text{OH}$$

The reaction of a Grignard reagent with ethylene oxide is a synthetic reaction used to add two carbons to a chain or ring.

$$\text{R—X} \ \xrightarrow[\text{ether}]{\text{Mg}} \ \text{R—MgX} \ \xrightarrow[\text{(2) H}_2\text{O, H}^+]{\text{(1) }\overset{\text{O}}{\overset{\diagup\diagdown}{\text{CH}_2\text{—CH}_2}}} \ \text{R—}\underline{\text{CH}_2\text{CH}_2}\text{OH}$$

Extended by two carbons

EXAMPLE

Write a flow equation for the following conversion:

$$(\text{CH}_3)_2\text{CHBr} \ \longrightarrow \ (\text{CH}_3)_2\text{CHCH}_2\text{CH}_2\text{OH}$$

Solution: Your first step should be to determine the changes in structure. In this conversion, we wish to add two carbons to the reactant and to obtain an alcohol product. The Grignard reaction with ethylene oxide is the method of choice.

$$(CH_3)_2CHBr \xrightarrow[\text{ether}]{Mg} (CH_3)_2CHMgBr \xrightarrow[\text{(2) }H_2O,\ H^+]{\text{(1) }CH_2\!-\!CH_2\ (O)} (CH_3)_2CHCH_2CH_2OH$$

PROBLEM 9.7. Complete the following equations:

(a) $CH_2{=}CHI \xrightarrow[\text{ether}]{Mg}$ _____ $\xrightarrow[\text{(2) }H_2O,\ H^+]{\text{(1) }CH_2\!-\!CH_2\ (O)}$ _____

(b) 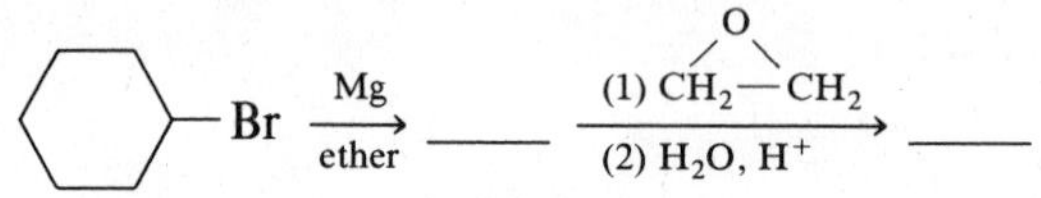 ⬡—Br $\xrightarrow[\text{ether}]{Mg}$ _____ $\xrightarrow[\text{(2) }H_2O,\ H^+]{\text{(1) }CH_2\!-\!CH_2\ (O)}$ _____

(c) 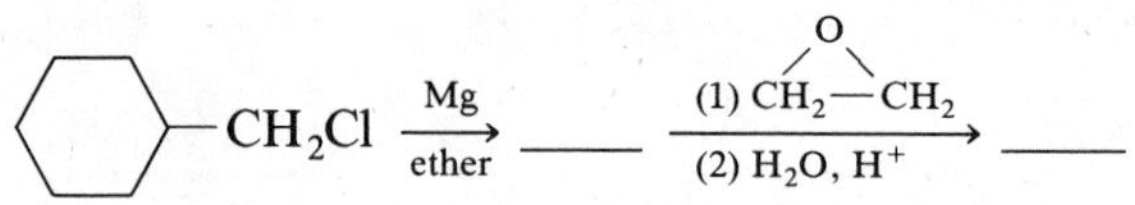 ⬡—$CH_2Cl \xrightarrow[\text{ether}]{Mg}$ _____ $\xrightarrow[\text{(2) }H_2O,\ H^+]{\text{(1) }CH_2\!-\!CH_2\ (O)}$ _____

PROBLEM 9.8. Write equations for the syntheses of the following compounds by Grignard reactions with ethylene oxide.

(a) $CH_3CH_2CH_2CH_2OH$ (b) 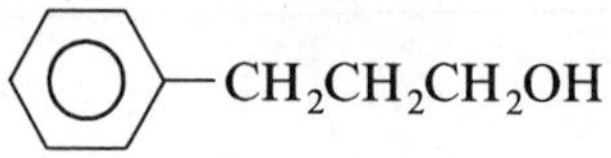 ⬡—$CH_2CH_2CH_2OH$

(c) 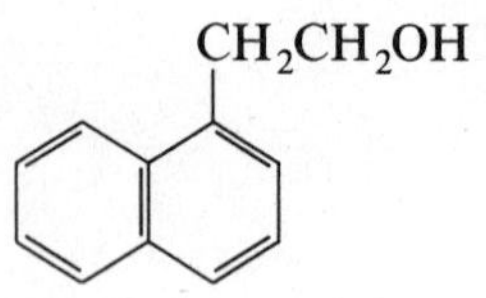 [naphthalene]—CH_2CH_2OH

9.6 CROWN ETHERS

Crown ethers are cyclic ethers containing repeating units of $-OCH_2CH_2-$ in their rings.

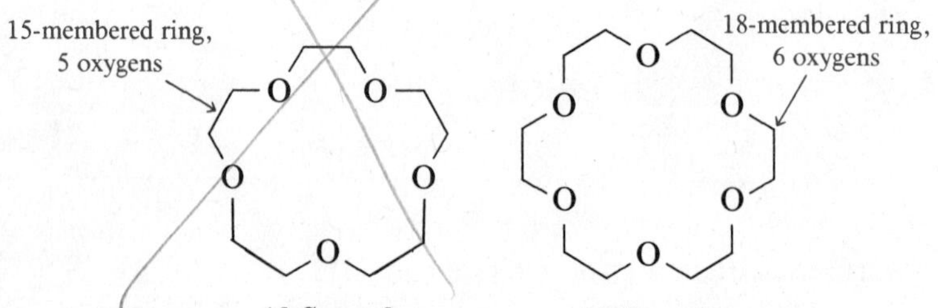

15-membered ring, 5 oxygens   18-membered ring, 6 oxygens

15-Crown-5   18-Crown-6

These compounds are named as x-crown-y, where x is the number of ring atoms and y is the number of oxygens in ring.

Crown ethers are used as *chelating agents* (from Greek *chele*, "crab's claw"), agents that hold other molecules or ions in a cagelike structure. Crown ethers chelate metal ions. A unique property of crown ethers is that different-sized rings chelate different metal ions, depending on the size of the cation and the size of the cavity in the ring.

Larger cavity, larger ion

Smaller cavity, smaller ion

MnO_4^-

I^-

18-Crown-6

chelates K^+

15-Crown-5

chelates Na^+

Ionic compounds are generally insoluble in inorganic solvents, and most organic compounds are insoluble in water. Reactions between organic compounds and inorganic reagents can be slow if reaction occurs only at the interface of an organic solution and the aqueous solution. Crown ethers can help solve this problem by acting as **phase-transfer agents**. A chelated metal ion can be dissolved in an organic solvent, the nonsolvated anion being carried along to maintain the required ionic balance. Once in the organic solution, the anion can cause substitution, oxidation, or other reactions.

"Purple benzene" is an example of a reagent containing a crown ether as a phase-transfer agent. This reagent is a solution of the purple potassium permanganate (K^+ MnO_4^-) in benzene. The MnO_4^- in the benzene solution is very reactive because it is not solvated with water and thus can be used to oxidize water-insoluble organic compounds.

SUMMARY

Ethers can be alkyl, aryl, or cyclic.

ROR'   ROAr   ArOAr'

Dialkyl   Alkyl aryl   Diaryl

Open-chain ethers

A cyclic ether

An epoxide

Ethers cannot form hydrogen bonds in the pure state and thus are lower boiling than comparable alcohols. Because ethers can form hydrogen bonds with compounds containing hydroxyl groups, their water solubility is similar to that of comparable alcohols.

Open-chain ethers are named by listing the alkyl or aryl groups and adding the word *ether*. Alternatively, ether groups may be named as *alkoxy-* groups.

The **Williamson ether synthesis** is the most common laboratory reaction leading to ethers.

$$R—X + R'O^- \xrightarrow{\text{S}_\text{N}2} R—OR' + X^-$$

Ethers are relatively unreactive. They undergo combustion, autoxidation, and substitution with cleavage when heated with a strong acid.

$$R—O—CHR'_2 \begin{cases} \xrightarrow[\text{spark}]{O_2} CO_2 + H_2O \\ \\ \xrightarrow{O_2} R—\overset{\overset{\displaystyle OOH}{|}}{O}CR'_2 \\ \\ \xrightarrow[\text{heat}]{2HI} RI + ICHR'_2 + H_2O \end{cases}$$

Unlike most other ethers, **epoxides** (oxiranes) are relatively reactive. These compounds undergo substitution in either acidic or alkaline solutions.

$$R_2\overset{\overset{\displaystyle O}{\diagup \diagdown}}{C}—CHR \begin{cases} \xrightarrow{Nu:^-} R_2\overset{\overset{\displaystyle OH}{|}}{C}—\underset{\underset{\displaystyle Nu}{|}}{C}HR \quad \text{where } Nu:^- = {}^-OH, NH_3, \text{ etc.} \\ \\ \xrightarrow[H^+]{Nu:H} R_2\overset{\overset{\displaystyle OH}{|}}{C}—\underset{\underset{\displaystyle Nu}{|}}{C}HR \quad \text{where } Nu:H = H_2O, R'OH, \text{ etc.} \end{cases}$$

Ethylene oxide reacts with Grignard reagents (RMgX) to yield alcohol products with two more carbons than the Grignard reagent contains (RCH_2CH_2OH).

Crown ethers are cyclic ethers containing repeating units of $—OCH_2CH_2—$. These compounds are unique chelating agents that are used as *phase-transfer agents*.

STUDY PROBLEMS

9.9. In each pair, which would have the higher boiling point? Why?
    (a) $CH_3OCH_2CH_2OCH_3$ or $CH_3CH_2OCH_2CH_3$
    (b) $CH_3OCH_2CH_3$ or $CH_3CH_2CH_2OCH_2CH_3$

(c) $CH_3OCH_2CH_2OCH_3$ or $CH_3OCH_2CH_2OH$

(d) $CH_3OCH_2CH_2CH_2CH_3$ or $CH_3OC(CH_3)_3$

9.10. In each pair, which would you expect to be the more water soluble? Why?

(a) $CH_3OCH_2CH_2CH_3$ or $CH_3OCH_2CH_2OCH_3$

(b) $CH_3OCH_2CH_2OCH_3$ or $CH_3OCH_2CH_2CH_2OH$

(c)   or $CH_3CH_2OCH_2CH_3$

9.11. Explain why tetrahydrofuran (bp 66°C) has a boiling point only slightly higher than that of cyclopentane (bp 49°C) yet, unlike cyclopentane, is miscible with water. Use formulas in your answer.

9.12. Name the following compounds:

(a) $(CH_3)_2CHOCH_2CH_3$

(b)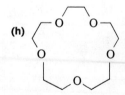

(c) $(CH_3)_2CHO-$ ⬡   (two names)

(d) $CH_3OCH_2\overset{\overset{\displaystyle OCH_3}{|}}{C}HCH_2OH$

(e) $\overset{\displaystyle O}{\overset{\diagup\diagdown}{CH_2-CH_2}}$

(f) $\underset{H}{\overset{H_3C}{>}}C-C\underset{CH_2CH_2CH_3}{\overset{H}{<}}$ (with epoxide O bridging the two central carbons)

(g) (crown ether with four O)

(h) (crown ether with five O)

9.13. Write formulas for the following compounds:

(a) di-*n*-propyl ether

(b) cyclohexyl methyl ether

(c) methoxycyclohexane

(d) *cis*-3-chloromethoxycyclohexane

(e) 2,2-diethyloxirane

(f) *trans*-2-methyl-3-isopropyloxirane

(g) cyclopropyl methyl ether

(h) propylene oxide

(i) (*S*)-2-methoxyhexane

(j) 21-crown-7

9.14. Draw formulas for and name all isomeric ethers with the molecular formula $C_5H_{12}O$.

9.15. Write equations or flow equations showing how you would prepare the following ethers by Williamson ether syntheses. (Start with organic bromo compounds, and show all possible routes.)

(a) $(CH_3)_2CHCH_2OCH_2CH_3$

(b) $(CH_3CH_2CH_2)_2O$

(c) $\underset{H}{\overset{H_3C}{>}}C=C\underset{H}{\overset{CH_2OC(CH_3)_3}{<}}$

(d) (tetrahydrofuran ring with O)

(e) (cyclohexane with OCH$_3$ and CH$_3$ substituents)

(f) ⬡$-OCH_2CH_2OH$

9.16. Predict the products when diisopropyl ether is treated with **(a)** atmospheric oxygen and sunlight; **(b)** atmospheric oxygen and a flame; **(c)** excess concentrated hydriodic acid and heat; **(d)** hot aqueous sodium hydroxide.

9.17. Predict the products when propylene oxide is treated with the following reagents:
   (a) water containing sulfuric acid
   (b) Na^+ $^-OCH_3$ in methanol
   (c) CH_3MgI, followed by aqueous acid
   (d) ammonia
   (e) aqueous concentrated HCl

9.18. Write equations or flow equations showing how you would make the following conversions:

   (a) $CH_3CH_2Br \longrightarrow CH_3CH_2CH_2CH_2OH$

   (b) $CH_3CH_2CH_2CH_2Br \longrightarrow CH_3CH_2CH_2CH_2OCH_3$

   (c) $\overset{\overset{\displaystyle OH}{|}}{CH_3CH_2CHCH_3} \longrightarrow \overset{\overset{\displaystyle CH_3}{|}}{CH_3CH_2CHCH_2CH_2OH}$

   (d) $\longrightarrow ICH_2CH_2CH_2CH_2I$

   (e)

POINT OF INTEREST 9

Inhalation Anesthetics

In the past century, general anesthetic agents have contributed greatly to human welfare by allowing painless surgery. The ancient Assyrians asphyxiated children by strangulation before circumcision surgery. In other cultures, patients were struck on the head with heavy blows. Narcotics, such as alcohol, opium, and hashish, have also been used to insensitize surgical patients.

In 1772 Joseph Priestley, the chemist who discovered oxygen, first made nitrous oxide, now sometimes called "laughing gas." He did not report or did not observe the effects of nitrous oxide on the central nervous system. In 1800 Sir Humphry Davy was studying the physiological effects of inhaling all varieties of gases. Davy's experiments nearly cost him his life, but he did discover the curious effects of nitrous oxide—initial excitation followed by sedation. Davy suggested to the medical profession that nitrous oxide be used during surgery; however, the suggestion was ignored. Instead, nitrous oxide became popular as a recreational drug.

In 1844 the American dentist Horace Wells introduced nitrous oxide as a dental anesthetic. Unfortunately, during a demonstration, the patient woke up early, screaming in pain. Wells was hissed out of the room. In 1846, at Massachusetts General Hospital, dentist William Morton and surgeon John Warren performed the first public demonstration of surgery, removal of a jaw tumor, under ether anesthesia. This demonstration was extraordinarily successful, and inhalation anesthesia for surgery quickly became popular.

Diethyl ether is still used as an anesthetic. This compound is quite safe physiologically, but it has the disadvantages of causing nausea and tissue irritation. In addition, its vapors are explosive. Other anesthetics have become popular in recent years. These include divinyl ether, a rapid-acting, short-term anesthetic; cyclopropane; and "halothane."

Some Inhalation Anesthetic Agents:

$$CH_2{=}CH{-}O{-}CH{=}CH_2 \qquad \begin{array}{c} CH_2 \\ \diagup \; \diagdown \\ CH_2{-}CH_2 \end{array} \qquad \begin{array}{c} Br \\ | \\ CF_3CHCl \end{array}$$

| Divinyl ether | Cyclopropane | Halothane |

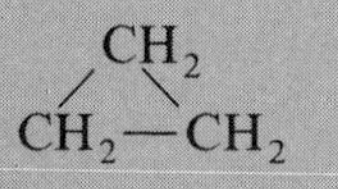

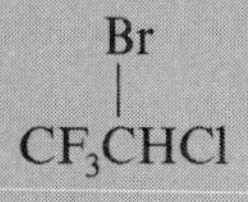

The volatile anesthetics act by initial excitation, followed by depression, of the central nervous system. How these compounds do this is not completely understood. Currently, it is believed that anesthetics such as halothane act by specific interaction with a protein in brain nerve membranes or perhaps by bonding to the membranes themselves.

A combination of inhalation anesthetics and preoperative sedatives, such as the barbiturate Pentothal and scopolamine, are often used today to initiate anesthesia and control its depth and time span.

Preoperative Sedatives:

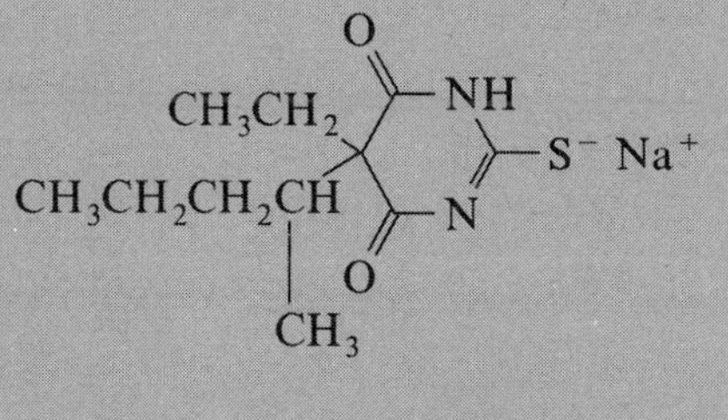

Sodium thiopental
(Pentothal sodium)

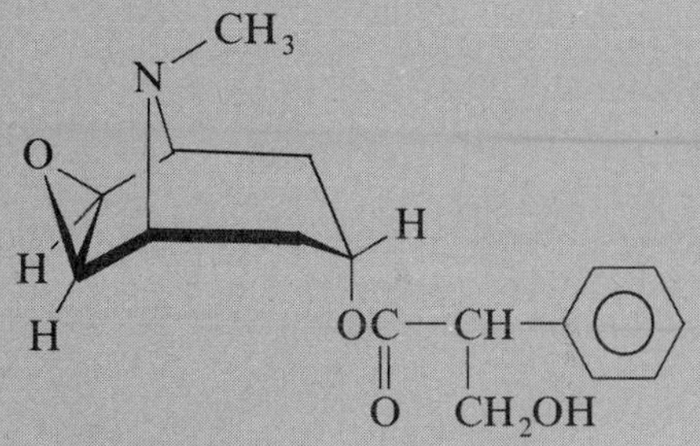

Scopolamine

Chapter 10

Aldehydes and Ketones

An **aldehyde** is an organic compound containing a carbonyl group bonded to one or two hydrogens.

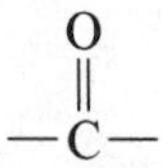

$$\begin{array}{c} O \\ \parallel \\ -C- \end{array}$$

The carbonyl group
found in both aldehydes and ketones

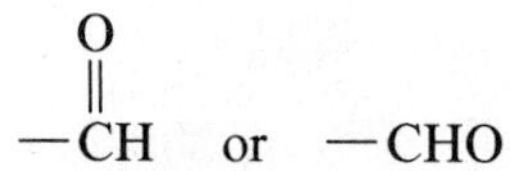

$$\begin{array}{c} O \\ \parallel \\ -CH \end{array} \quad \text{or} \quad -CHO$$

The aldehyde group

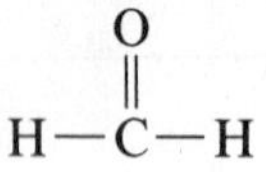

$$\begin{array}{c} O \\ \parallel \\ H-C-H \end{array}$$

Formaldehyde

$$\begin{array}{c} O \\ \parallel \\ R-C-H \end{array} \quad \text{or} \quad \begin{array}{c} O \\ \parallel \\ Ar-C-H \end{array}$$

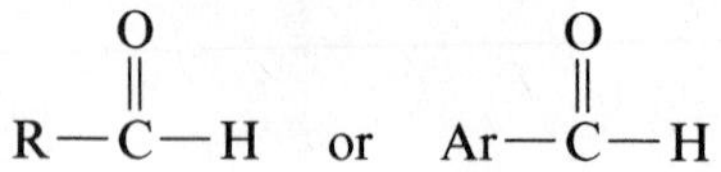

Any other aldehyde

A **ketone** is an organic compound that has a carbonyl group bonded to two alkyl groups, two aryl groups, or one of each. A ketone does not contain a hydrogen bonded to the carbonyl group.

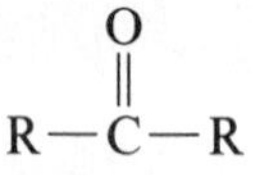

$$\begin{array}{c} O \\ \parallel \\ R-C-R \end{array} \qquad \begin{array}{c} O \\ \parallel \\ Ar-C-Ar \end{array} \qquad \begin{array}{c} O \\ \parallel \\ Ar-C-R \end{array}$$

A dialkyl ketone 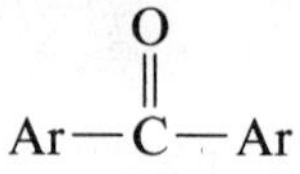   *A diaryl ketone* 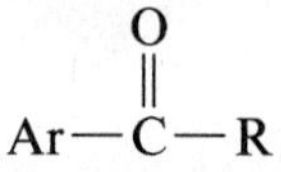   *An aryl alkyl ketone*

Aldehydes and ketones are common naturally occurring compounds. Figure 10.1 shows the structures of a few of these compounds.

10.1 STRUCTURE AND BONDING IN ALDEHYDES AND KETONES

The carbon of a carbonyl group is sp^2 hybridized. As in alkenes, the sp^2-hybridized carbon forms three sigma bonds that lie in a plane. The bond angle between any two of the sigma bonds is approximately 120°. In the carbonyl group, one of the

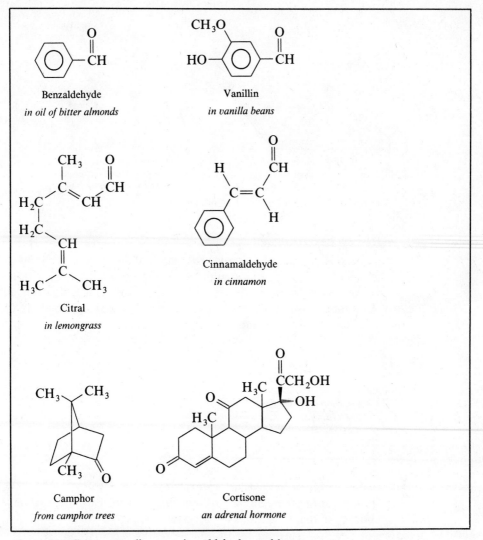

Figure 10.1 Some naturally occurring aldehydes and ketones.

atoms bonded to the carbon by a sigma bond is an oxygen atom. The carbon and the oxygen are also joined by a pi bond.

Figure 10.2 illustrates the bonding in the carbonyl group in more detail.

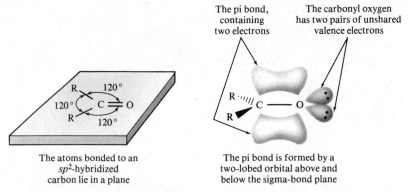

Figure 10.2 The bonding in a carbonyl group.

Because oxygen is more electronegative than carbon, a carbon–oxygen bond is polar. The carbon–oxygen double bond in a carbonyl group, with its sigma and pi electrons, is more polar than the carbon–oxygen single bond in an alcohol. The reason for this is that the pi electrons of the carbonyl's pi bond are farther from the nuclei, less tightly held, and more easily polarized.

Pi electrons are more easily drawn toward the oxygen than are sigma electrons.

The carbonyl oxygen has an unshared pair of valence electrons that can react with a proton to increase the positive charge of the carbonyl carbon. Protonation is important in many reactions of carbonyl groups.

e^- drawn toward positive oxygen

10.2 PHYSICAL PROPERTIES OF ALDEHYDES AND KETONES

Because aldehydes and ketones do not contain a hydrogen bonded to an oxygen, they cannot undergo hydrogen bonding as do alcohols. On the other hand, aldehydes and ketones are polar and can form relatively strong electrostatic attractions between molecules, the positive portion of one molecule being attracted to the

TABLE 10.1   NAMES AND BOILING POINTS OF SOME ALDEHYDES

Formula	IUPAC name	Trivial name	Bp (°C)
HCHO	Methanal	Formaldehyde	−21
CH_3CHO	Ethanal	Acetaldehyde	20
CH_3CH_2CHO	Propanal	Propionaldehyde	49
$CH_3CH_2CH_2CHO$	Butanal	Butyraldehyde	76
⬡—CHO	Benzaldehyde	Benzaldehyde	178
OH ⬡—CHO	—	Salicylaldehyde	197

negative portion of another.

$$\overset{\delta+}{\underset{/}{\overset{\backslash}{C}}}=\overset{\delta-}{O}\cdots\cdots\overset{\delta+}{\underset{/}{\overset{\backslash}{C}}}=\overset{\delta-}{O}$$

↑

Electrostatic attraction

For this reason, aldehydes and ketones have boiling points intermediate between those of hydrogen-bonded compounds and nonpolar compounds.

$$CH_3CH_2CH_2{-}CH_3 \qquad CH_3CH_2CH{=}O \qquad CH_3CH_2CH_2{-}OH$$

bp, °C: −0.5   49   97

——————————————————————————————————→

Increasing boiling point

Although pure aldehydes and ketones cannot form hydrogen bonds, these compounds can form hydrogen bonds with the hydrogen of water or alcohols.

Hydrogen bond

$$\underset{R}{\overset{R}{\diagdown}}C{=}\ddot{O}\text{:}\cdots\text{H}\diagdown\underset{R}{\overset{..}{\underset{..}{O}}}$$

Because of this type of hydrogen bonding, the water solubility of aldehydes and ketones parallels that of alcohols.

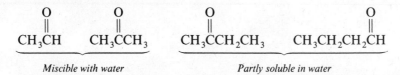

$$\underset{CH_3CH}{\overset{O}{\overset{\|}{}}} \qquad \underset{CH_3CCH_3}{\overset{O}{\overset{\|}{}}} \qquad \underset{CH_3CCH_2CH_3}{\overset{O}{\overset{\|}{}}} \qquad \underset{CH_3CH_2CH_2CH}{\overset{O}{\overset{\|}{}}}$$

Miscible with water   *Partly soluble in water*

TABLE 10.2   NAMES AND BOILING POINTS OF SOME KETONES

Formula	IUPAC name	Trivial name	Bp (°C)
CH_3CCH_3 (O)	Propanone	Acetone	56
$CH_3CCH_2CH_3$ (O)	Butanone	Methyl ethyl ketone	80
$CH_3CH_2CCH_2CH_3$ (O)	3-Pentanone	Diethyl ketone	101
⬡=O	Cyclohexanone	—	156
⬡—CCH₃ (O)	—	Acetophenone	202

Tables 10.1 and 10.2 list some aldehydes and ketones and their boiling points.

PROBLEM 10.1.   Draw formulas that show how acetone, $(CH_3)_2C{=}O$, can form hydrogen bonds with water.

10.3 NOMENCLATURE OF ALDEHYDES AND KETONES

A. Aldehydes

In the IUPAC system, simple aldehydes are named after the parent alkane with the *-ane* ending changed to *-anal* (*-an-* for a saturated chain and *-al* for aldehyde). The carbonyl carbon in a simple aldehyde must be carbon 1; therefore, this number is omitted in the name. Other substituents, such as alkyl branches, are named by prefixes in the usual manner.

The carbonyl carbon
is carbon 1.

$$\underset{4}{CH_3}\underset{3}{CH}\underset{2}{CH_2}\underset{1}{CH}$$

with CH_3 branch and O (carbonyl)

3-Methylbutanal

Common aldehydes are often called by their trivial names. Many of these names are derived from the trivial name of the corresponding carboxylic acid with the *-ic acid* (or *-oic acid*) ending changed to *-aldehyde*. Some of these names, such as benzaldehyde, are accepted by the IUPAC. (The trivial names of the carboxylic acids are discussed in Section 11.3.)

$$CH_3\overset{\displaystyle O}{\overset{\|}{C}}OH \qquad CH_3\overset{\displaystyle O}{\overset{\|}{C}}H$$

Acetic acid   Acetaldehyde

Benzoic acid   Benzaldehyde

With trivial names, Greek letters *alpha* (α), *beta* (β), *gamma* (γ), and so forth are used instead of numbers to denote the position of a substituent. The alpha carbon is the one bonded to the carbonyl group, and the beta carbon is the next in line. The Greek letters are not used in IUPAC names, only in trivial names.

$$CH_3CH_2\overset{\displaystyle O}{\overset{\|}{C}}H \qquad CH_3\underset{\underset{\textstyle Cl}{|}}{C}H\overset{\displaystyle O}{\overset{\|}{C}}H \qquad ClCH_2CH_2\overset{\displaystyle O}{\overset{\|}{C}}H$$

| *IUPAC:* | Propanal | 2-Chloropropanal | 3-Chloropropanal |
| *Trivial:* | Propionaldehyde | α-Chloropropionaldehyde | β-Chloropropionaldehyde |

EXAMPLE

Name the following aldehyde by the IUPAC system:

$$\underset{\underset{\textstyle CH_3CH_2}{|}\quad \underset{\textstyle CH_3}{|}}{CH_3CHCHCHCH_2}\overset{\overset{\textstyle CH_2CH_3}{|}\quad \overset{\textstyle O}{\|}}{CH}$$

Solution: Number the longest continuous chain containing the aldehyde group. Be sure to assign position 1 to the carbonyl carbon.

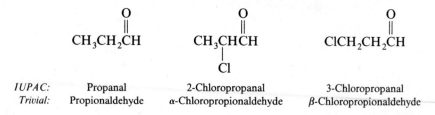

The seven-carbon chain is a *heptanal*. List the prefix substituents in the usual manner.

   4-Ethyl-3,5-dimethylheptanal

PROBLEM 10.2. Name the following aldehydes:

(a) 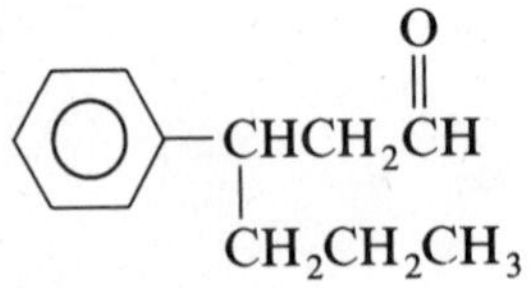 [benzene ring]—CHCH$_2$CH (with O double bonded to the terminal CH)
         |
         CH$_2$CH$_2$CH$_3$

(b) 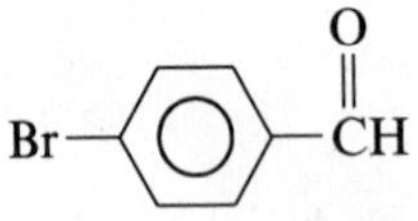 Br—[benzene ring]—CH (with O double bonded)

         CH$_3$ O
          | ‖
(c) 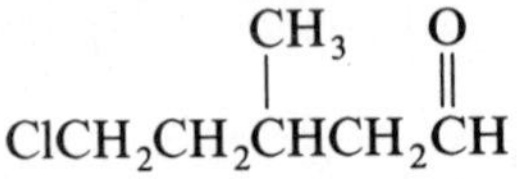 ClCH$_2$CH$_2$CHCH$_2$CH

PROBLEM 10.3. Draw formulas for the following aldehydes:

(a) 3-ethyl-3-hydroxypentanal
(b) 3-chlorohexanal
(c) 2-chloroethanal
(d) 2,4-dichlorobenzaldehyde

B. Ketones

In the IUPAC system, simple ketones are named with the ending -*anone*. Most ketone names require a position number to specify the location of the carbonyl group because it can be at different positions within the carbon chain.

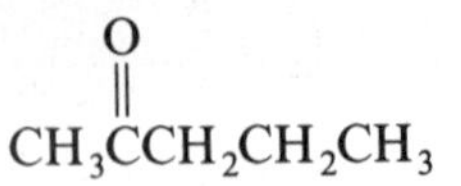

        O   O
        ‖   ‖
CH$_3$CCH$_2$CH$_2$CH$_3$ CH$_3$CH$_2$CCH$_2$CH$_3$ [cyclohexane ring]=O

   2-Pentanone 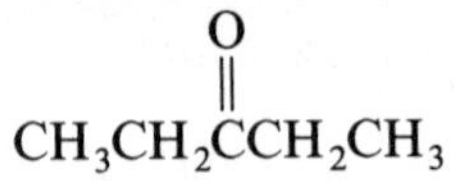   3-Pentanone 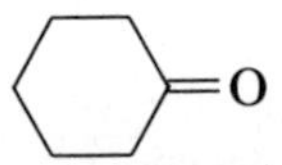   Cyclohexanone
                                                       (no number needed)

Several types of trivial nomenclature systems for ketones are used. The simplest system is naming the two groups bonded to the carbonyl carbon and adding the word *ketone*.

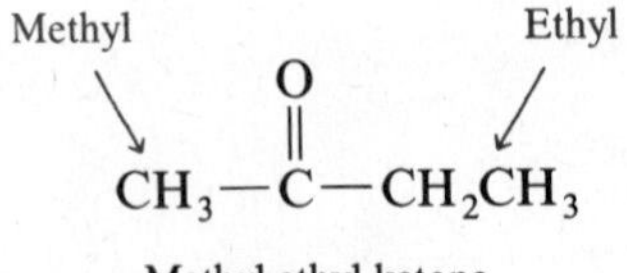

   Methyl   Ethyl
      \   O   /
       \   ‖   /
       CH$_3$—C—CH$_2$CH$_3$
        Methyl ethyl ketone

Other common names are acetone, acetophenone, and benzophenone.

Aceto group

Phenone group

$$CH_3\overset{O}{\overset{\|}{C}}CH_3 \qquad CH_3\overset{O}{\overset{\|}{C}}-\bigcirc \qquad \bigcirc-\overset{O}{\overset{\|}{C}}-\bigcirc$$

    Acetone   Acetophenone   Benzophenone

The Greek letters alpha (α) and alpha prime (α') are used to denote substituent positions in the trivial names of some ketones. One alpha carbon is designated as the α carbon and the other is designated the α' carbon in order to differentiate them.

$$Cl_2CH\overset{O}{\overset{\|}{C}}CH_3 \qquad\qquad ClCH_2\overset{O}{\overset{\|}{C}}CH_2Cl$$

IUPAC:	1,1-Dichloro-2-propanone	1,3-Dichloro-2-propanone
Trivial:	α,α-Dichloroacetone	α,α'-Dichloroacetone

PROBLEM 10.4.   Name the following ketones by the IUPAC system:

(a) $CH_3\underset{\underset{\displaystyle CH_2CH_3}{|}}{CH}\overset{O}{\overset{\|}{C}}CH_2CH_3$ (b) $(CH_3)_2CH\overset{O}{\overset{\|}{C}}CH(CH_3)_2$

(c) $\bigcirc-\overset{O}{\overset{\|}{C}}CH_2CH_3$

PROBLEM 10.5.   Draw formulas for the following ketones:

(a) 4-hydroxy-2-pentanone

(b) 2,2-dichlorocyclohexanone

(c) α,β'-dibromodiethyl ketone

10.4 SOME IMPORTANT ALDEHYDES AND KETONES

Formaldehyde ($H_2C{=}O$) is a flammable, colorless, toxic gas with a pungent suffocating odor. It is prepared commercially by the oxidation of methanol.

A water solution containing 37% formaldehyde (with methanol added as a stabilizer) is known as *formalin*. Formalin is used as a disinfectant; as an insecticide; as a fumigant; in embalming fluid; and in the manufacture of explosives, resins, plastics, fabrics, dyes, and other compounds.

When an aqueous solution of formaldehyde is concentrated, a white polymer called *paraformaldehyde* is formed. This polymer is used as a disinfectant and in manufacturing. When warmed, paraformaldehyde gives off formaldehyde gas.

$$ n \quad \overset{H}{\underset{H}{\diagup}}C{=}O \quad \underset{heat}{\overset{concentrate}{\rightleftharpoons}} \quad \left\{\!\!\left\{ \overset{H}{\underset{H}{-\overset{|}{\underset{|}{C}}}}-O-\overset{H}{\underset{H}{\overset{|}{\underset{|}{C}}}}-O-\overset{H}{\underset{H}{\overset{|}{\underset{|}{C}}}}-O-\!\!\right\}\!\!\right\} $$

Paraformaldehyde

Acetaldehyde ($CH_3CH{=}O$) is a flammable, colorless, water-soluble liquid with a pungent odor. This volatile compound is a narcotic and a toxic irritant. If large doses of its vapors are inhaled, death can occur by respiratory failure. Acetaldehyde is used in the manufacture of dyes, plastics, synthetic rubber, and other compounds.

Acetaldehyde can be polymerized to the cyclic compounds *paraldehyde* and *metaldehyde*.

$$ 3 \text{ or } 4 \text{ } CH_3\overset{O}{\overset{||}{CH}} \xrightarrow{H^+} $$

Paraldehyde
(bp 125°C)

a hypnotic

or

Metaldehyde
(mp 246°C)

used as a snail poison

Acetone, $(CH_3)_2C{=}O$, is a colorless, volatile, flammable liquid with a sweet pungent odor. It is prepared commercially by the oxidation of isopropyl alcohol (2-propanol) and as a by-product in the manufacture of phenol.

Isopropylbenzene
(cumene)

a petrochemical

A hydroperoxide

Phenol   Acetone

Acetone is relatively nontoxic, it is miscible with water and most other organic solvents, and it can dissolve most organic compounds. For these reasons, acetone is a widely used solvent.

The breath and urine of a person with uncontrolled diabetes sometimes smell of acetone. In the body's cells, acetone arises from acetoacetate ion (see Point of Interest 13).

Methyl ethyl ketone (MEK, 2-butanone) is less water soluble and higher boiling than acetone. Methyl ethyl ketone is used as a solvent and in the synthesis of other chemicals.

10.5 PREPARATION OF ALDEHYDES AND KETONES

A. Aldehydes

One laboratory reaction used to prepare an aldehyde is the oxidation of a primary alcohol. Most oxidizing agents cannot be used because they also oxidize aldehydes to carboxylic acids. Chromic oxide–pyridine complexes, such as pyridinium chlorochromate (PCC, Section 8.11A), are oxidizing agents that convert primary alcohols to aldehydes without oxidizing the aldehyde to the carboxylic acid.

Most Oxidizing Agents Convert 1° Alcohols to Carboxylic Acids:

$$RCH_2OH \xrightarrow{[O]} \overset{\displaystyle O}{\overset{\displaystyle \|}{RCH}} \xrightarrow{[O]} \overset{\displaystyle O}{\overset{\displaystyle \|}{RCOH}}$$

Chromic Oxide–Pyridine Complexes Convert 1° Alcohols to Aldehydes:

$$CH_3CH_2CH_2CH_2OH \xrightarrow[CH_2Cl_2]{PCC} \overset{\displaystyle O}{\overset{\displaystyle \|}{CH_3CH_2CH_2CH}}$$

Carboxylic acids can be reduced to primary alcohols but not to aldehydes. However, the chlorides of carboxylic acids can be reduced to aldehydes by partially deactivated aluminum hydride reagents, such as the following one.

The reducing portion

$$Li^+ \quad H-\underset{\underset{OC(CH_3)_3}{|}}{\overset{\overset{OC(CH_3)_3}{|}}{Al^-}}-OC(CH_3)_3$$

Lithium tri-*t*-butoxyaluminum hydride

$$\overset{\displaystyle O}{\overset{\displaystyle \|}{CH_3CH_2C}}-Cl \xrightarrow[\text{(2) } H_2O,\, H^+]{\text{(1) } Li^+\ ^-AlH[OC(CH_3)_3]_3} \overset{\displaystyle O}{\overset{\displaystyle \|}{CH_3CH_2CH}}$$

Propanoyl chloride

the acid chloride of
$CH_3CH_2CO_2H$

Propanal

an aldehyde

B. Ketones

The most common method for ketone preparation is the oxidation of a secondary alcohol. Almost any strong oxidizing agent can be used. Typical reagents are chromic oxide (CrO_3), pyridinium chlorochromate, sodium dichromate ($Na_2Cr_2O_7$), and potassium permanganate ($KMnO_4$).

$$
\underset{\text{A 2° alcohol}}{CH_3\overset{\underset{\displaystyle |}{OH}}{C}HCH_3} \xrightarrow{CrO_3 + H_2SO_4} \underset{\text{A ketone}}{CH_3\overset{\displaystyle O}{\overset{\displaystyle \|}{C}}CH_3}
$$

Diaryl ketones ($Ar\overset{\displaystyle O}{\overset{\displaystyle \|}{C}}Ar$) and aryl alkyl ketones ($Ar\overset{\displaystyle O}{\overset{\displaystyle \|}{C}}R$) are prepared by Friedel-Crafts acylation reactions—the reaction of benzene or other aromatic compound with an acid chloride (see Section 5.3E).

$$
\underset{\text{Benzene}}{\bigcirc} + \underset{\text{Acetyl chloride}}{CH_3\overset{\displaystyle O}{\overset{\displaystyle \|}{C}}-Cl} \xrightarrow[\text{heat}]{AlCl_3} \underset{\text{Acetophenone}}{\bigcirc-\overset{\displaystyle O}{\overset{\displaystyle \|}{C}}CH_3} + HCl
$$

Aldehyde and ketone preparations are summarized in Table 10.3.

TABLE 10.3   PREPARATION OF SIMPLE ALDEHYDES AND KETONES

Aldehydes

$$
1° \ RCH_2OH \xrightarrow[CH_2Cl_2]{PCC} RCHO
$$

$$
R\overset{\displaystyle O}{\overset{\displaystyle \|}{C}}Cl \xrightarrow[\text{(2) } H_2O, H^+]{\text{(1) } LiAlH[OC(CH_3)_3]_3} RCHO
$$

Ketones

$$
2° \ R_2CHOH \xrightarrow{CrO_3 + H_2SO_4} R_2C{=}O
$$

Aryl ketones

$$
\bigcirc + R\overset{\displaystyle O}{\overset{\displaystyle \|}{C}}Cl \xrightarrow{AlCl_3} \bigcirc-\overset{\displaystyle O}{\overset{\displaystyle \|}{C}}R
$$

PROBLEM 10.6. Write flow equations for the following conversions:

(a) 2-hexanol to 2-hexanone

(b) 1-hexanol to hexanal

(c) benzene to 2-methyl-1-phenyl-1-propanone

10.6 ADDITION REACTIONS OF ALDEHYDES AND KETONES

Reagents can add to a carbonyl double bond, just as they can to a carbon–carbon double bond. For example, carbonyl compounds can undergo hydrogenation. In this reaction, an aldehyde is reduced to a primary alcohol, while a ketone is reduced to a secondary alcohol.

General Reaction:

$$\underset{\substack{\text{An aldehyde} \\ \text{or ketone}}}{\overset{\displaystyle O}{\underset{R}{\overset{\|}{\underset{\diagdown}{C}}}\diagup R}} \;+\; \overset{H}{\underset{H}{\overset{|}{}}} \;\xrightarrow{\text{catalyst}}\; \underset{\substack{a \; 1° \; or \; 2° \; alcohol}}{R-\overset{OH}{\underset{R}{\overset{|}{\underset{|}{C}}}}-H}$$

Specific Example:

$$\underset{\substack{\text{Cyclohexanone} \\ \text{A ketone}}}{\bigcirc\!\!=\!\!O} \;+\; H_2 \;\xrightarrow[50°C, \, 65 \, atm]{Ni}\; \underset{\substack{\text{Cyclohexanol} \\ a \; 2° \; alcohol}}{\bigcirc\!\!-\!\!OH}$$

A carbonyl group is polar and, unlike a carbon–carbon double bond, can be attacked by either a nucleophile at the carbonyl carbon or an electrophile at the carbonyl oxygen.

Addition Reaction with Strong Nucleophile:

$$\underset{(strong)}{\overset{\displaystyle \ddot{O}\!:^{\delta-}}{\underset{\underset{:Nu^-}{\curvearrowleft}}{\overset{\|}{\underset{\delta+}{C}}}}} \;\longrightarrow\; \underset{Nu}{\overset{:\ddot{O}:^-}{\underset{|}{\overset{|}{-C-}}}} \;\xrightarrow{H^+}\; \underset{Nu}{\overset{:\ddot{O}H}{\underset{|}{\overset{|}{-C-}}}}$$

Addition Reaction with Electrophile (H$^+$) Followed by Reaction with a Weak Nucleophile:

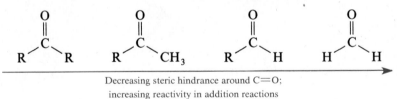

In either type of reaction, aldehydes react faster or more completely than ketones. One reason for this difference in reactivity is that ketones are more stable than aldehydes. A ketone's greater stability is due to inductive delocalization of the carbonyl carbon's positive charge.

No R groups to donate electrons by the inductive effect

One R group to donate electrons and help share + charge

Two R groups to donate electrons

Another reason for the lesser reactivity of ketones is steric hindrance in the addition product and the transition state leading to it. The aldehyde's carbonyl carbon is more exposed and its addition products are less hindered.

Decreasing steric hindrance around C=O; increasing reactivity in addition reactions

---

PROBLEM 10.7. Rank the following aldehydes and ketones in order of reactivity toward addition reactions (least reactive first):

(a) CH_3CCH_3 (b) $CH_3CH_2CCH_3$

(c) CH_3CH_2CH (d) $CH_3CH_2CCH_2CH_3$

A. Addition of Water

Water is a weak nucleophile; therefore, a trace of acid is necessary to catalyze its addition to a carbonyl group. The addition is a reversible reaction. Only the most reactive aldehydes, such as formaldehyde and chloral ($Cl_3CCH{=}O$), form stable hydrates.

The first step in the addition is the protonation of the carbonyl oxygen. The weakly nucleophilic water then attacks the carbonyl carbon. Finally, the protonated hydrate loses a proton to the solvent water.

$$
\begin{array}{c}
\ddot{O}: \\
\parallel \\
HCH
\end{array}
\xrightleftharpoons{H^+}
\begin{array}{c}
{}^+\!\ddot{O}H \\
\parallel \\
HCH
\end{array}
\xrightleftharpoons{H\ddot{O}H}
\begin{array}{c}
:\ddot{O}H \\
| \\
HCH \\
| \\
H{-}\underset{\overset{+}{}}{\ddot{O}}H
\end{array}
\xrightleftharpoons{-H^+}
\begin{array}{c}
:\ddot{O}H \\
| \\
HCH \\
| \\
:\ddot{O}H
\end{array}
$$

Formaldehyde

A stable hydrate in solution (formalin)

$$
\begin{array}{c}
O \\
\parallel \\
Cl_3CCH
\end{array}
+ H_2O
\xrightleftharpoons{H^+}
\begin{array}{c}
OH \\
| \\
Cl_3CCH \\
| \\
OH
\end{array}
$$

Chloral

Chloral hydrate
a stable hydrate
(a hypnotic used in
"Mickey Finns")

$$
\begin{array}{c}
O \\
\parallel \\
CH_3CCH_3
\end{array}
+ H_2O
\xrightleftharpoons{H^+}
\begin{array}{c}
OH \\
| \\
CH_3CCH_3 \\
| \\
OH
\end{array}
$$

The equilibrium lies on this side of the equation

Not stable

PROBLEM 10.8. Suggest a reason why chloral can form a stable hydrate.

PROBLEM 10.9. Write equations for hydrate formation by the following compounds. Show by arrow lengths where you would expect the equilibrium to lie.

$$
\textbf{(a)} \begin{array}{c} O \\ \parallel \\ CH_3CH \end{array}
\qquad
\textbf{(b)} \begin{array}{c} O \\ \parallel \\ CH_3CH_2CCH_3 \end{array}
\qquad
\textbf{(c)} \begin{array}{c} O \\ \parallel \\ Cl_3CCCCl_3 \end{array}
$$

B. Addition of Alcohols

The addition of alcohols to aldehydes and ketones is similar to the addition of water. Alcohols are weak nucleophiles, and a trace of acid is necessary to catalyze the reversible reaction. The product of the addition of 1 mole of alcohol to 1 mole of aldehyde is a **hemiacetal**, a compound with —OH and —OR on one carbon.

Addition of First Mole of R'OH:

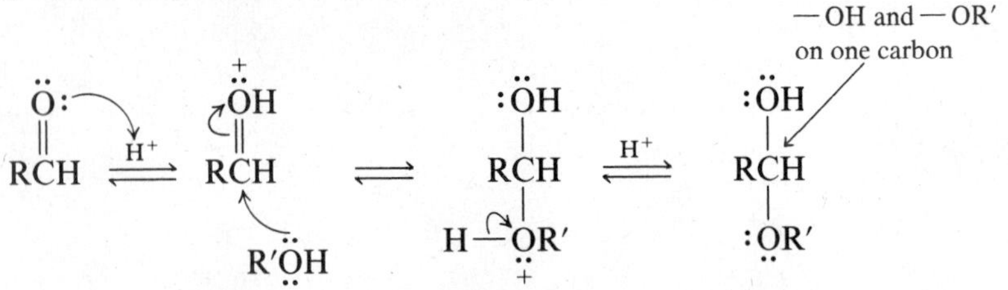

A hemiacetal can react further with an alcohol to yield an **acetal**, a compound in which two —OR groups are bonded to one carbon.

Substitution of Second Mole of R'OH:

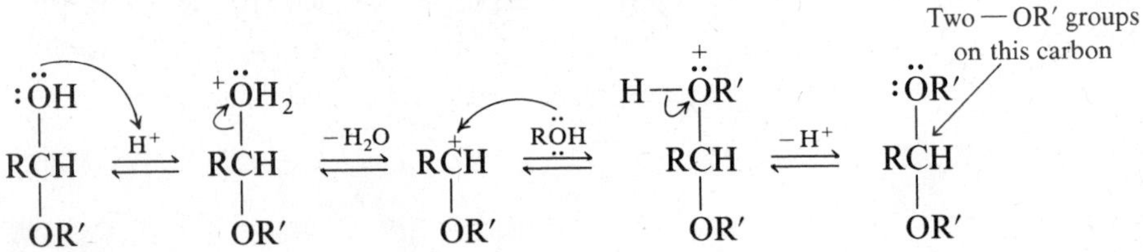

For convenience, we can combine the general equations for hemiacetal and acetal formation into one flow equation.

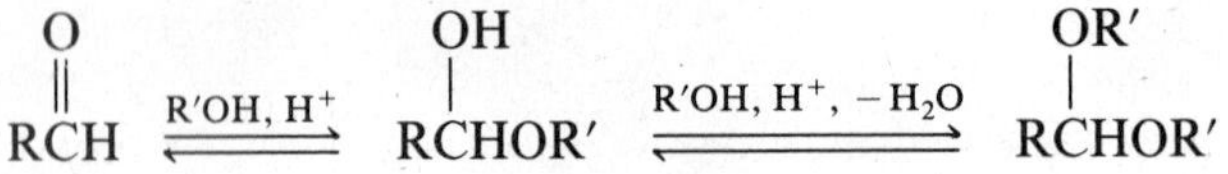

$$\underset{\text{A hemiacetal}}{\overset{O}{\underset{\|}{RCH}}} \xrightleftharpoons{R'OH, H^+} \underset{\text{A hemiacetal}}{\overset{OH}{\underset{|}{RCHOR'}}} \xrightleftharpoons{R'OH, H^+, -H_2O} \underset{\text{An acetal}}{\overset{OR'}{\underset{|}{RCHOR'}}}$$

Reactions leading to hemiacetals and acetals are reversible reactions, and the equilibrium generally lies on the aldehyde side of the equation. For the less reactive ketones, the equilibrium favors the carbonyl side of the equation even more.

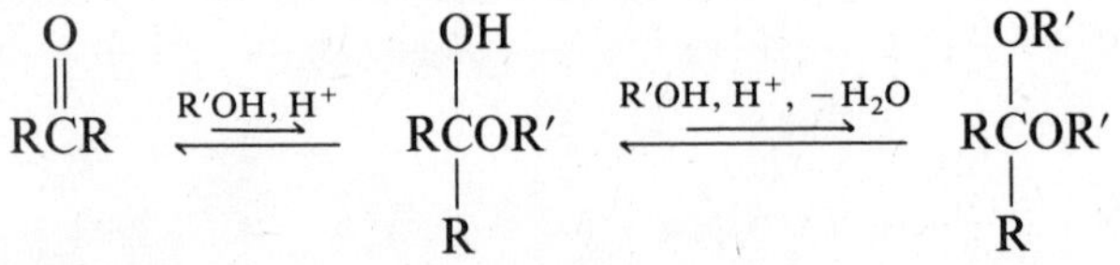

$$\underset{\text{A ketone}}{\overset{O}{\underset{\|}{RCR}}} \xrightleftharpoons{R'OH, H^+} \underset{\text{A hemiketal}}{\overset{OH}{\underset{\underset{R}{|}}{RCOR'}}} \xrightleftharpoons{R'OH, H^+, -H_2O} \underset{\text{A ketal}}{\overset{OR'}{\underset{\underset{R}{|}}{RCOR'}}}$$

The equilibria among aldehydes, hemiacetals, and acetals usually favor the aldehyde. However, if a five-membered-ring or six-membered-ring cyclic hemiacetal or acetal can be formed, the cyclic product is favored. One way we can obtain a cyclic product is by using 1,2-ethanediol (ethylene glycol) as the alcohol.

General Equation:

This — OH reacts with
the hemiacetal carbon.

$$
\underset{\text{RCH}}{\overset{\text{O}}{\|}} \quad \underset{\xrightarrow{\text{HOCH}_2\text{CH}_2\text{OH, H}^+}}{\rightleftharpoons} \quad \underset{\underset{\text{OCH}_2\text{CH}_2}{\text{RCH} \quad \text{OH}}}{\overset{\text{OH}}{|}} \quad \underset{\xrightarrow{\text{H}^+, -\text{H}_2\text{O}}}{\rightleftharpoons} \quad \text{RCH} \overset{\text{O}}{\underset{\text{O}}{<}} \overset{\text{CH}_2}{\underset{\text{CH}_2}{|}}
$$

A hemiacetal   *A cyclic acetal*

Specific Example:

$$
\underset{\text{CH}_3\text{CH}}{\overset{\text{O}}{\|}} \quad \underset{\xrightarrow{\text{HOCH}_2\text{CH}_2\text{OH, H}^+}}{\rightleftharpoons} \quad \underset{\underset{\text{OCH}_2\text{CH}_2}{\text{CH}_3\text{CH} \quad \text{OH}}}{\overset{\text{OH}}{|}} \quad \underset{\xrightarrow{-\text{H}_2\text{O}}}{\rightleftharpoons} \quad \text{CH}_3\text{CH} \overset{\text{O}}{\underset{\text{O}}{<}} \overset{\text{CH}_2}{\underset{\text{CH}_2}{|}}
$$

Another way we can form cyclic products is by using a starting aldehyde that contains a hydroxyl group in position 4 or 5.

—OH in position 4

$$
\underset{4\quad3\quad2\quad1}{\overset{\text{OH}}{\underset{\text{CH}_2\text{CH}_2\text{CH}_2\text{CH}}{|}}} \underset{\overset{\text{O}}{\|}}{} \quad \underset{\xrightarrow{\text{H}^+}}{\rightleftharpoons} \quad \underset{\text{H}_2\text{C}-\text{CH}_2}{\text{H}_2\text{C} \overset{\text{O}}{\diagup \diagdown} \text{CHOH}}
$$

A five-membered-ring
cyclic hemiacetal

In Chapter 16 you will see that the monosaccharide sugars, which are polyhydroxy aldehydes and ketones, are in equilibrium with their cyclic hemiacetals and hemiketals in solution.

Cyclization of Glucose:

—OH in position 5

Aldehyde

Hemiacetal group

PROBLEM 10.10.   Complete the following equations, showing the reaction steps where appropriate.

(a) $\underset{\substack{\|\\ \text{O}}}{\text{CH}_3\text{CH}}$ + 2CH$_3$CH$_2$OH $\overset{\text{H}^+}{\rightleftharpoons}$

(b)   —$\underset{\substack{\|\\ \text{O}}}{\text{CH}}$ + HOCH$_2$CH$_2$OH $\overset{\text{H}^+}{\rightleftharpoons}$

(c)   CH$_3$$\underset{\substack{\|\\ \text{O}}}{\text{C}}CH_2CH_2CH_2$OH $\overset{\text{H}^+}{\rightleftharpoons}$

PROBLEM 10.11.   What reagents would you need to prepare the following acetal?

CH$_3$CH$_2$—$\overset{\text{O}\rceil}{\underset{\text{O}\rfloor}{\big\langle}}$

C. Addition of Hydrogen Cyanide

Hydrogen cyanide (H—C≡N:, bp 26°C), a toxic liquid or gas at room temperature, is a very weak acid. In the presence of cyanide ion, hydrogen cyanide adds to the carbonyl group of an aldehyde or a ketone to yield a **cyanohydrin**, a compound with —OH and —CN bonded to the same carbon.

$$\underset{\substack{\text{An aldehyde}\\ \text{or a ketone}}}{\underset{\substack{\|\\ \text{O}}}{\text{RCR}}} \ + \ \text{H—CN} \ \underset{}{\overset{^-\text{CN}}{\rightleftharpoons}} \ \underset{\substack{\text{A cyanohydrin}}}{\underset{\substack{|\\ \text{CN}}}{\overset{\substack{\text{OH}\\ |}}{\text{RCR}}}} \quad \substack{\text{—OH and —CN}\\ \text{on the same carbon}}$$

Because of its toxicity and low boiling point, HCN is usually prepared in the reaction mixture by adding a mineral acid, such as H$_2$SO$_4$ or HCl, to NaCN or KCN.

Cyanohydrin formation is initiated by attack on the carbonyl group by the strongly nucleophilic cyanide ion; no protonation of the carbonyl oxygen is necessary (see step 2 in the following equations). The intermediate alkoxide is a sufficiently strong base that it undergoes an acid-base reaction with the undissociated HCN in solution to release ⁻CN ion (see step 3 below).

If ⁻CN is not present in the reaction mixture, the cyanohydrin is not formed. Hydrogen cyanide is so weak an acid that neither H⁺ nor ⁻CN is present in sufficient amounts to initiate the reaction. To ensure that some cyanide ions are present in the reaction mixture, less than a molar amount of acid is added to the cyanide salt when the HCN is prepared in the reaction vessel (see step 1 in the following equations).

Step 1, Formation of HCN:

Some free cyanide needed for reaction

$$HCl \ + \ excess \ Na^+ \ {}^-CN \longrightarrow HCN \ + \ Na^+ \ {}^-Cl \ + \ Na^+ \ {}^-CN$$

Step 2, Addition of Cyanide Ion:

$$R-\overset{\overset{\displaystyle :\ddot{O}}{\|}}{C}-R \quad \overset{}{\rightleftharpoons} \quad R-\overset{\overset{\displaystyle :\ddot{O}:^-}{|}}{\underset{\underset{\displaystyle C \equiv N:}{|}}{C}}-R$$

$$:C \equiv N:$$

Step 3, Cyanohydrin Formation:

$$R-\overset{\overset{\displaystyle :\ddot{O}:^-}{|}}{\underset{\underset{\displaystyle C \equiv N:}{|}}{C}}-R \quad \xrightarrow[\text{H}-\text{C} \equiv \text{N:}]{} \quad R-\overset{\overset{\displaystyle :\ddot{O}H}{|}}{\underset{\underset{\displaystyle C \equiv N:}{|}}{C}}-R \ + \ {}^-:C \equiv N:$$

Cyanohydrins are found in nature. For example, the millipede *Apheloria corrugata* (a many-legged arthropod similar to the centipede) carries the cyanohydrin mandelonitrile in defensive glands. When this millipede is attacked, the mandelonitrile is decomposed enzymatically to a mixture of benzaldehyde and hydrogen cyanide (HCN), which is squirted at the predator. A single millipede can produce sufficient hydrogen cyanide to kill a mouse!

$$\overset{\overset{\displaystyle OH}{|}}{\underset{}{\text{C}_6\text{H}_5-CH-CN}} \quad \xrightarrow{\text{enzyme}} \quad \overset{\overset{\displaystyle O}{\|}}{\underset{}{\text{C}_6\text{H}_5-CH}} \ + \ HCN$$

Mandelonitrile   Benzaldehyde

PROBLEM 10.12.   Write an equation for the preparation of the following cyanohydrin:

$$\overset{\overset{\displaystyle OH}{|}}{\underset{\underset{\displaystyle CH_3}{|}}{\text{(cyclopentyl)}-C-C \equiv N}}$$

D. Addition of Grignard Reagents

Grignard reagents (RMgX), discussed in Section 7.11, are powerful nucleophiles that add to aldehydes and ketones to yield alkoxides. Acidification of an alkoxide yields an alcohol as the product.

$$
\underset{}{R-\overset{\overset{\displaystyle :\ddot{O}:}{\|}}{C}-R} \;+\; \overset{\delta-}{R'}-\overset{\delta+}{MgX} \longrightarrow R-\underset{\underset{\displaystyle R'}{|}}{\overset{\overset{\displaystyle :\ddot{O}:^{-}}{|}}{C}}-R\ ^{+}MgX \xrightarrow{H_2O,\ H^+}
$$

An alkoxide

$$
R-\underset{\underset{\displaystyle R'}{|}}{\overset{\overset{\displaystyle :\ddot{O}H}{|}}{C}}-R \;+\; Mg^{2+} + X^{-}
$$

An alcohol

Grignard reagents react with formaldehyde to yield primary alcohols, with other aldehydes to yield secondary alcohols, and with ketones to yield tertiary alcohols.

From HCHO

$$
\underset{\text{Formaldehyde}}{\overset{\overset{\displaystyle O}{\|}}{HCH}} \xrightarrow[\text{(2) }H_2O,\ H^+]{\text{(1) }CH_3CH_2CH_2\,MgBr,\ ether} CH_3CH_2CH_2-\overbrace{CH_2OH}
$$

1-Butanol
a 1° alcohol

$$
\underset{\text{Acetaldehyde}}{\overset{\overset{\displaystyle O}{\|}}{CH_3CH}} \xrightarrow[\text{(2) }H_2O,\ H^+]{\text{(1) }\bigcirc-MgBr,\ ether} CH_3\underset{\underset{}{}}{\overset{\overset{\displaystyle OH}{|}}{CH}}-\bigcirc
$$

1-Phenylethanol
a 2° alcohol

$$
\underset{\text{Acetone}}{\overset{\overset{\displaystyle O}{\|}}{CH_3CCH_3}} \xrightarrow[\text{(2) }H_2O,\ H^+]{\text{(1) }\bigcirc-MgBr,\ ether} CH_3\underset{\underset{\displaystyle CH_3}{|}}{\overset{\overset{\displaystyle OH}{|}}{C}}-\bigcirc
$$

2-Cyclohexyl-2-propanol
a 3° alcohol

EXAMPLE

Write a flow equation to show how you would prepare cyclohexylmethanol by a Grignard reaction.

Solution: Write the formula, determine the alcohol type, and then determine the reagents.

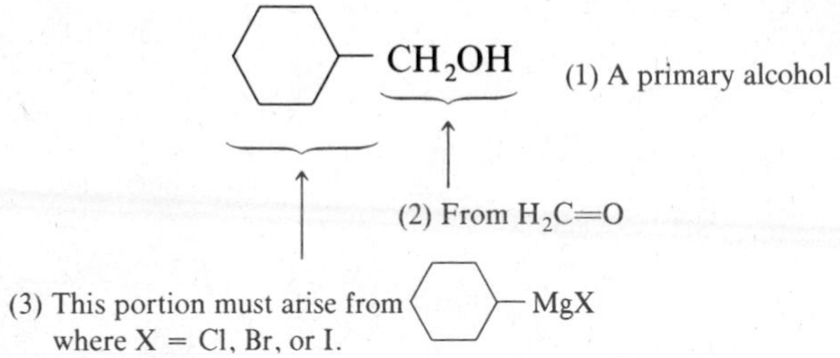

Write the flow equation:

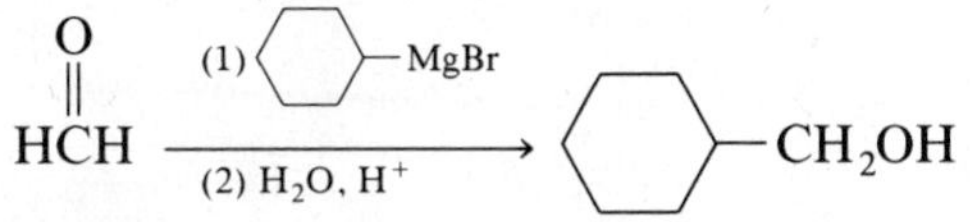

EXAMPLE

Write a flow equation for the preparation of 3-methyl-3-hexanol by a Grignard reaction.

Solution: Analyze the formula.

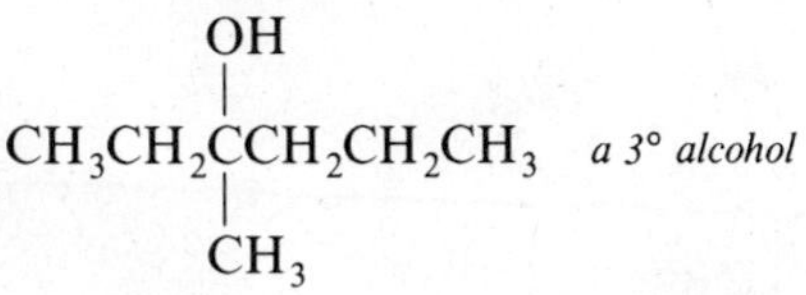

$$
\begin{array}{c}
\text{OH} \\
| \\
\text{CH}_3\text{CH}_2\text{CCH}_2\text{CH}_2\text{CH}_3 \quad \textit{a 3° alcohol} \\
| \\
\text{CH}_3
\end{array}
$$

To obtain a tertiary alcohol, we need a ketone and a Grignard reagent. Three routes to this compound are possible.

$$\underset{\underset{CH_3}{|}}{\overset{\overset{OH}{|}}{CH_3CH_2CCH_2CH_2CH_3}} \qquad \text{from } CH_3MgI + \underset{\overset{\parallel}{O}}{CH_3CH_2CCH_2CH_2CH_3}$$

$$\underset{\underset{CH_3}{|}}{\overset{\overset{OH}{|}}{CH_3CH_2CCH_2CH_2CH_3}} \qquad \text{from } CH_3CH_2MgBr + \underset{\overset{\parallel}{O}}{CH_3CCH_2CH_2CH_3}$$

$$\underset{\underset{CH_3}{|}}{\overset{\overset{OH}{|}}{CH_3CH_2CCH_2CH_2CH_3}} \qquad \text{from } CH_3CH_2CH_2MgCl + \underset{\overset{\parallel}{O}}{CH_3CH_2CCH_3}$$

Any of the following flow equations would be a suitable synthesis.

$$\underset{\overset{\parallel}{O}}{CH_3CH_2CCH_2CH_2CH_3} \xrightarrow[\text{(2) } H_2O, H^+]{\text{(1) } CH_3MgI}$$

$$\underset{\overset{\parallel}{O}}{CH_3CCH_2CH_2CH_3} \xrightarrow[\text{(2) } H_2O, H^+]{\text{(1) } CH_3CH_2MgBr} \qquad \underset{\underset{CH_3}{|}}{\overset{\overset{OH}{|}}{CH_3CH_2CCH_2CH_2CH_3}}$$

$$\underset{\overset{\parallel}{O}}{CH_3CCH_2CH_3} \xrightarrow[\text{(2) } H_2O, H^+]{\text{(1) } CH_3CH_2CH_2MgCl}$$

Lithium reagents (RLi) are reactants similar to Grignard reagents. Because lithium is more electropositive than magnesium, these reagents are more reactive than Grignard reagents and thus are useful synthetic reagents.

Preparation of Lithium Reagent:

$$RX + 2Li \longrightarrow RLi + Li^+ + X^-$$

Reaction of Lithium Reagent, Followed by Hydrolysis:

$$\underset{\overset{\parallel}{O}}{R'CR''} \xrightarrow{RLi} \underset{\underset{R}{|}}{\overset{\overset{O^- Li^+}{|}}{R'CR''}} \xrightarrow{H_2O, H^+} \underset{\underset{R}{|}}{\overset{\overset{OH}{|}}{R'CR''}}$$

Table 10.4 summarizes the addition reactions of aldehydes and ketones.

TABLE 10.4   SUMMARY OF THE ADDITION REACTIONS OF ALDEHYDES AND KETONES

Simple addition reactions

$$\underset{\substack{\nearrow \\ A\ reactive\ aldehyde}}{\overset{\overset{\displaystyle O}{\|}}{RCH}} + H_2O \underset{}{\overset{H^+}{\rightleftharpoons}} \overset{\overset{\displaystyle OH}{|}}{RCHOH} \quad \text{a hydrate}$$

$$\overset{\overset{\displaystyle O}{\|}}{RCH}\ (\overset{\overset{\displaystyle O}{\|}}{RCR}) + R'OH \overset{H^+}{\rightleftharpoons} R-\overset{\overset{\displaystyle OH}{|}}{\underset{\underset{\displaystyle OR'}{|}}{C}}-H \quad \text{a hemiacetal or a hemiketal}$$

Can react further
to yield an acetal
or a ketal

$$\overset{\overset{\displaystyle O}{\|}}{RCR}\ (\overset{\overset{\displaystyle O}{\|}}{RCH}) + HCN \overset{^-CN}{\rightleftharpoons} R-\overset{\overset{\displaystyle OH}{|}}{\underset{\underset{\displaystyle CN}{|}}{C}}-R \quad \text{a cyanohydrin}$$

Grignard reactions

$$H_2C{=}O \xrightarrow[\text{(2) } H_2O,\ H^+]{\text{(1) } R'MgX} R'CH_2OH \quad \text{a 1}° \text{ alcohol}$$

$$\overset{\overset{\displaystyle O}{\|}}{RCH} \xrightarrow[\text{(2) } H_2O,\ H^+]{\text{(1) } R'MgX} \overset{\overset{\displaystyle OH}{|}}{RCHR'} \quad \text{a 2}° \text{ alcohol}$$

$$\overset{\overset{\displaystyle O}{\|}}{RCR} \xrightarrow[\text{(2) } H_2O,\ H^+]{\text{(1) } R'MgX} \overset{\overset{\displaystyle OH}{|}}{\underset{\underset{\displaystyle R'}{|}}{RCR}} \quad \text{a 3}° \text{ alcohol}$$

PROBLEM 10.13.   Complete the following equations:

(a) $CH_3CH_2\overset{\overset{\displaystyle O}{\|}}{CH} + CH_3CH_2MgBr \xrightarrow{\text{diethyl ether}}$

(b) ⬡$=O + CH_3MgI \xrightarrow{\text{diethyl ether}}$

(c) ⬡$-MgBr + CH_3\overset{\overset{\displaystyle O}{\|}}{C}CH_3 \xrightarrow{\text{diethyl ether}}$

PROBLEM 10.14. Write flow equations to show how you would prepare the following compounds by Grignard reactions.

(a) 4-methyl-2-pentanol

(b) 2-methyl-2-butanol

(c) 1-heptanol

10.7 ADDITION-ELIMINATION REACTIONS OF ALDEHYDES AND KETONES

Compounds that contain $-NH_2$ groups such as primary amines (RNH_2 or $ArNH_2$) and hydrazines (NH_2NH_2, $ArNHNH_2$, etc.) can undergo addition reactions with aldehydes and ketones. However, the addition products are unstable and undergo elimination of water.

Addition:

$$R-\overset{\overset{\displaystyle ::O}{\|}}{C}-R + H_2\ddot{N}-R' \rightleftharpoons R-\overset{\overset{\displaystyle ::\ddot{O}:^-}{|}}{\underset{\underset{\displaystyle H}{\overset{\displaystyle |}{H-\overset{+}{N}-R'}}}{C}}-R \xrightarrow[\text{transfer}]{\text{proton}} R-\overset{\overset{\displaystyle :\ddot{O}H}{|}}{\underset{\underset{\displaystyle \text{Addition product}}{\displaystyle H\ddot{N}R'}}{C}}-R$$

H$^+$

Addition product (unstable)

Elimination:

$$R-\overset{\overset{\displaystyle :\ddot{O}H}{|}}{\underset{\underset{\displaystyle H\ddot{N}R'}{|}}{C}}-R \overset{H^+}{\rightleftharpoons} R-\overset{\overset{\displaystyle :\overset{+}{O}H_2}{|}}{\underset{\underset{\displaystyle H-\ddot{N}R'}{|}}{C}}-R \overset{-H^+}{\rightleftharpoons} R-\overset{\overset{\displaystyle ||}{}}{\underset{\underset{\displaystyle \ddot{N}R'}{}}{C}}-R$$

$H_2\ddot{O}:$

Elimination product

If ammonia is one of the reactants, the product is called an **imine**. If a primary amine is one of the reactants, the product is also called an imine but is frequently referred to as a **Schiff base**.

A variety of $-NH_2$ compounds can react with aldehydes and ketones. Table 10.5 lists a few of these compounds and the products of their reaction. We will show just two examples here.

The product of the reaction of *hydroxylamine* ($HO-NH_2$) with an aldehyde or a ketone is called an **oxime**.

TABLE 10.5 ADDITION-ELIMINATION PRODUCTS FROM CARBONYL COMPOUNDS AND NITROGEN COMPOUNDS

Reagent	Name	Product of reaction with $R_2C=O$	
$R'NH_2$	1° amine	$R_2C=NR'$	imine (Schiff base)
H_2NOH	Hydroxylamine	$R_2C=NOH$	oxime
H_2NNH_2	Hydrazine	$R_2C=NNH_2$	hydrazone
$H_2NNH-\!\!\bigcirc$	Phenylhydrazine	$R_2C=NNH-\!\!\bigcirc$	phenylhydrazone
$H_2NNH-\!\!\bigcirc\!-NO_2$, NO_2	2,4-Dinitrophenylhydrazine	$R_2C=NNH-\!\!\bigcirc\!-NO_2$, NO_2	
			2,4-dinitrophenylhydrazone

When a *hydrazine* ($R'—NHNH_2$) reacts with an aldehyde or a ketone, the product is a **hydrazone**.

$$CH_3\overset{O}{\overset{\|}{C}}CH_3 + H_2NNH-\hspace{-2pt}\bigcirc \underset{}{\overset{H^+}{\rightleftharpoons}} CH_3\overset{NNH-\bigcirc}{\overset{\|}{C}}CH_3 + H_2O$$

Phenylhydrazine

A phenylhydrazone
(mp 42°C)

The products of these addition-elimination reactions are often solids. In the laboratory, these reactions are used to prepare solid derivatives of liquid or low-melting aldehydes and ketones. Solids are easily characterized by melting point and are thus useful for identification of aldehydes and ketones of unknown structure.

PROBLEM 10.15. Predict the products of the reactions of cyclohexanone with the following reagents:

(a) $HONH_2$ (b) $\bigcirc\!\!-\!NH_2$ (c) $O_2N-\!\bigcirc\overset{NO_2}{}\!-NHNH_2$

PROBLEM 10.16. What reactants would you need to prepare the following products?

(a) $CH_3CH_2CH\!=\!NNH_2$ (b) $CH_3CH_2CH\!=\!NOH$

(c) $CH_3CH_2CH\!=\!NNH-\!\bigcirc$

10.8 OXIDATION OF ALDEHYDES

Aldehydes are easily oxidized. Only very mild oxidizing agents are needed to convert aldehydes to carboxylic acids (RCO_2H) or carboxylates (RCO_2^-).

$$R\overset{O}{\overset{\|}{C}}H \xrightarrow[\text{(addition of O)}]{[O]} R\overset{O}{\overset{\|}{C}}OH \quad \text{or} \quad R\overset{O}{\overset{\|}{C}}O^-$$

In acid   *In base*

Many tests for aldehydes are based on their ease of oxidation. For example, the **Tollens test** is based on the oxidation of an aldehyde with a solution of silver ions (Ag^+), a very weak oxidizing agent, in aqueous alkaline ammonia solution. If an aldehyde is present, the products are the carboxylate and silver metal. The silver metal forms a black precipitate; or, if the reaction vessel is scrupulously clean, the silver plates on the glass to form a silver mirror.

Tollens Test:

$$\underset{\text{An aldehyde}}{\overset{\displaystyle \overset{O}{\underset{\displaystyle \|}{}}}{RCH}} \;+\; \underbrace{2Ag(NH_3)_2{}^+ \;+\; 3{}^-OH}_{\text{Tollens reagent}} \longrightarrow \underset{\substack{\text{Silver}\\\text{mirror}}}{\overset{\displaystyle \overset{O}{\underset{\displaystyle \|}{}}}{RCO^-}} \;+\; 2Ag\downarrow \;+\; 4NH_3 \;+\; 2H_2O$$

$$\overset{\displaystyle \overset{O}{\underset{\displaystyle \|}{}}}{RCR} \;+\; \text{Tollens reagent} \longrightarrow \text{no reaction}$$

The **Fehling test** and the **Benedict test** depend on the oxidation of an aldehyde by copper(II) ions. In these tests, the deep blue color of Cu^{2+} changes to a red precipitate of Cu_2O.

Fehling Test or Benedict Test:

$$\underset{\text{Blue}}{\overset{\displaystyle \overset{O}{\underset{\displaystyle \|}{}}}{RCH}} \;+\; 2Cu^{2+} \;+\; 5{}^-OH \longrightarrow \underset{\text{Red}}{\overset{\displaystyle \overset{O}{\underset{\displaystyle \|}{}}}{RCO^-}} \;+\; Cu_2O\downarrow \;+\; 3H_2O$$

Aldehydes are also oxidized by atmospheric oxygen at room temperature. For this reason, opened containers of aldehyde are often contaminated with their corresponding carboxylic acid.

$$2\overset{\displaystyle \overset{O}{\underset{\displaystyle \|}{}}}{RCH} \;+\; O_2 \longrightarrow 2\overset{\displaystyle \overset{O}{\underset{\displaystyle \|}{}}}{RCOH}$$

PROBLEM 10.17. Predict the organic products, if any, of the following reactions.

(a) $CH_3CH_2\overset{\displaystyle \overset{O}{\underset{\displaystyle \|}{}}}{CH} \;\xrightarrow[{}^-OH]{Ag(NH_3)_2{}^+}$

(b) 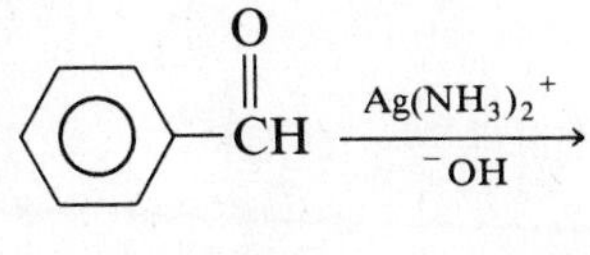 ⬡—$\overset{\displaystyle \overset{O}{\underset{\displaystyle \|}{}}}{CH} \;\xrightarrow[{}^-OH]{Ag(NH_3)_2{}^+}$

(c) 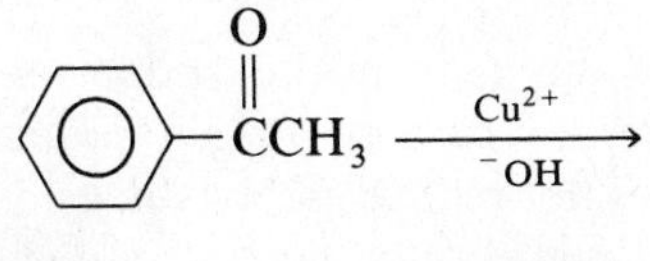 ⬡—$\overset{\displaystyle \overset{O}{\underset{\displaystyle \|}{}}}{C}CH_3 \;\xrightarrow[{}^-OH]{Cu^{2+}}$

10.9 REDUCTION OF ALDEHYDES AND KETONES

Aldehydes and ketones can be reduced by the catalytic addition of hydrogen gas, as shown at the start of Section 10.6. In the laboratory, a chemist is more likely to use a metal hydride reducing agent for convenience. The most commonly used of these is **sodium borohydride** (Na^+ $^-BH_4$).

General Equation:

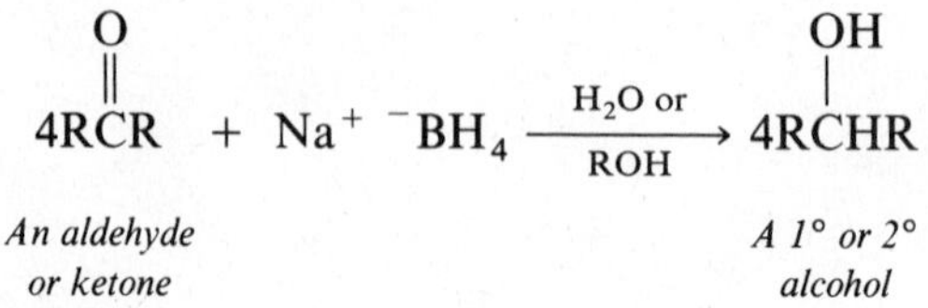

$$\underset{\substack{\text{An aldehyde}\\\text{or ketone}}}{4R\overset{\overset{\textstyle O}{\|}}{C}R} + Na^+\ ^-BH_4 \xrightarrow[\text{ROH}]{H_2O\text{ or}} \underset{\substack{A\ 1°\ or\ 2°\\\text{alcohol}}}{4R\overset{\overset{\textstyle OH}{|}}{C}HR}$$

Because a BH_4^- ion contains four hydrides, it can reduce four molecules of aldehyde or ketone. The reaction proceeds by successive transfers of hydride ions ($H:^-$) from the boron to four different carbonyl carbons.

Step 1, Reduction:

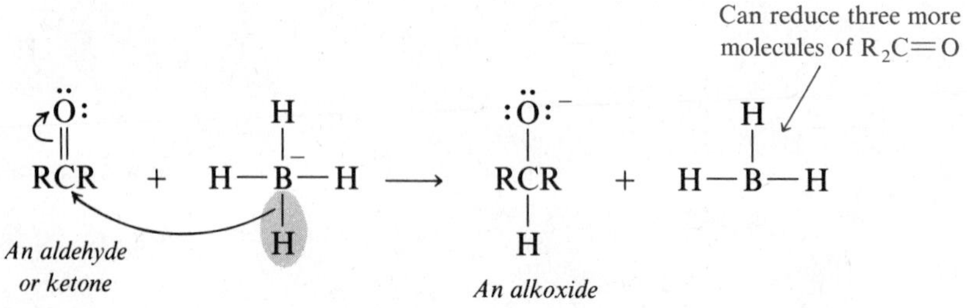

Water or an alcohol, used as the solvent, destroys the intermediate alkoxide ion and produces the product alcohol. If water or an alcohol is *not* used as the solvent, aqueous acid can be added after the reduction to convert the alkoxide to the product alcohol.

Step 2, Hydrolysis or Alcoholysis:

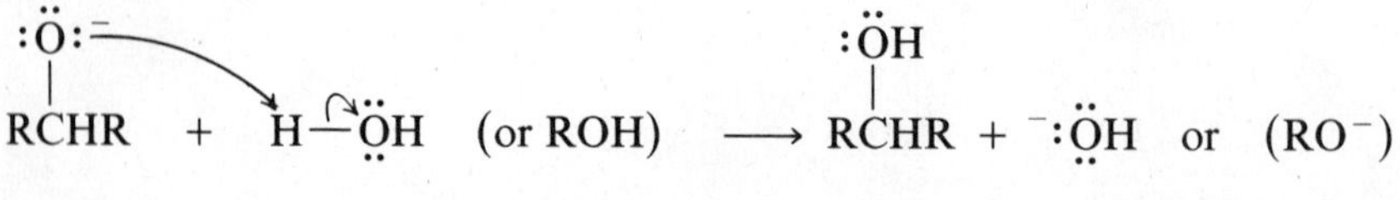

Although sodium borohydride readily reduces the carbon–oxygen double bond of an aldehyde or a ketone, it reduces the carbon–oxygen double bond of an ester very slowly. Sodium borohydride does not reduce carboxylic acids or alkenes. Thus,

In any equilibrium mixture, the relative quanti[...]
on the relative stabilities of the structures. Bec[...]
enol form, 2,4-pentanedione exists as an equili[...]
enol and only 20% dione.

EXAMPLE

Write equations for the tautomerism of the following carbon[...]
possible tautomers in your answer.

(a) CH_3CH_2CH with $=O$

(b) cyclohexanone with CH_3 and $=O$

(c) pyrrolidinone structure with N–H

Solution: First, circle all alpha hydrogens. In (b) and (c)[...]
redrawn to show the alpha hydrogens.

one type of α hydrogen:
one tautomer

two types:
two tautomers

(a) CH_3CH_2CH with $=O$

(b) cyclohexanone with H, CH_3, H, H, $=O$

(c) [...]

Second, write equations for the transfer of any types of[...]

(a) $CH_3\overset{H}{\underset{|}{C}H}-\overset{O}{\underset{||}{C}}H \rightleftharpoons CH_3CH=\overset{OH}{\underset{|}{C}}H$

(b) cyclohexanone structure $\rightleftharpoons$ enol structures $+$

(c) pyrrolidinone $\rightleftharpoons$ enol $+$ enol

In (c), note that an alpha hydrogen on the nitrogen [...]
tautomerization.

---

sodium borohydride allows a chemist to reduce portions of a molecule selectively.

Reduced readily   Ester not
reduced readily

$HCCH_2CH_2COCH_2CH_3 \xrightarrow[\text{(2) H}_2\text{O, H}^+]{\text{(1) NaBH}_4} H_2CCH_2CH_2COCH_2CH_3$

Not reduced

$CH_3CCH_2CH=CHCH_2CCH_3 \xrightarrow[\text{(2) H}_2\text{O, H}^+]{\text{(1) NaBH}_4} CH_3CHCH_2CH=CHCH_2CHCH_3$

When hydrogen gas is used as the reducing agent for an unsaturated aldehyde
or ketone, the carbon–carbon double bonds are also reduced.

$CH_2=CHCH_2CH \xrightarrow[\text{catalyst}]{2H_2} CH_3CH_2CH_2CH_2$

Another metal hydride that reduces carbonyl groups is **lithium aluminum
hydride** (Li^+ $^-AlH_4$). This reagent is a more powerful reducing reagent than
sodium borohydride and reduces the carbonyl groups in aldehydes, ketones, carboxylic acids, esters, and others. Like sodium borohydride, lithium aluminum
hydride does not reduce isolated carbon–carbon double bonds. Unlike sodium
borohydride, lithium aluminum hydride reacts violently with water or alcohols;
therefore, an ether or other inert solvent must be used in its reactions.

$CH_3CCH_3 \xrightarrow[\text{(2) H}_2\text{O, H}^+]{\text{(1) LiAlH}_4} CH_3CHCH_3$

PROBLEM 10.18. Write equations for the borohydride syntheses of the following
alcohols from aldehydes or ketones.

(a) cyclohexyl $-CH_2\overset{OH}{\underset{|}{C}}HCH_3$

(b) menthol structure with CH_3, OH, CH, H_3C, CH_3

Menthol

10.10 TAUTOMERISM OF ALDEHYDES A

An aldehyde or a ketone with at least o[ ]
enol form—an isomer that contains a [ ]
hydroxyl group (*-ol*).

α hydrogen

α carbon

$$CH_2-CH \rightleftharpoons -C=$$

Aldehyde   *Enol*
or keto form

These structures are a special type of is[ ]
same"). The enol tautomer differs fro[ ]
location of the double bond and a hydr[ ]
the same order, tautomers are *not* reso[ ]
equilibrium. (Recall that resonance str[ ]
atoms bonded in the same order but [ ]
tautomerism to occur, a structure must [ ]
atom other than carbon, usually oxygen[ ]

$$\underset{\text{Acetaldehyde}}{\overset{\text{H}\quad\text{O}}{CH_2-CH}} \rightleftharpoons \underset{\text{Enol form}}{\overset{\text{OH}}{CH_2=CH}}$$

Two types of
α hydrogens

$$\underset{\text{Butanone}}{CH_3-\overset{O}{\overset{\|}{C}}-CH_2CH_3} \rightleftharpoons CH_3C$$

Most simple aldehydes and ketones [ ]
form. For example, in a sample of acet[ ]
enol form for every million molecules in [ ]

In some 1,3-diones (β-diketones) [ ]
carbonyl group beta to the aldehyde or [ ]
intramolecular hydrogen bonding.

No hydrogen bond

$$CH_3C\underset{CH_2}{\diagdown}\overset{}{\diagup}CCH_3 \rightleftharpoons$$

2,4-Pentanedione

a β-diketone

10.33. Complete the following equations. Show both the organic products and the inorganic products.

(a) [naphthalene-CHO] $+ \text{Ag(NH}_3)_2^+ + {}^-\text{OH} \longrightarrow$

(b) [benzene with two CHO groups] $+ \text{excess Cu}^{2+} + {}^-\text{OH} \longrightarrow$

(c) $\underset{}{CH_2CH_2CH=CHCH} \overset{OH}{}\overset{O}{} + \text{excess Ag(NH}_3)_2^+ + {}^-\text{OH} \longrightarrow$

10.34. Write the formula for the lithium aluminum hydride reduction product of each of the following aldehydes or ketones:

(a) [phenyl]—CH with O   (b) $CH_3CH_2CCH_3$ (with O)

(c) $\begin{array}{c} CH_2OH \\ | \\ C=O \\ | \\ HO-C-H \\ | \\ H-C-OH \\ | \\ H-C-OH \\ | \\ CH_2OH \end{array}$ (d) [phenyl]—C(=O)—[phenyl]

10.35. Write equations for the tautomerization of the following compounds:

(a) [phenyl]—CCH$_3$ (with O)

(b) [five-membered ring structure with HO, OH, CH$_2$—CH, OH, OH, O, O]

(c) $CH_3C\overset{O\cdots H\cdots O}{}\diagup\diagdown C$ ... CH$_3$ with OH and CH$_3$

(d) $(CH_3)_2CHCH$ (with O)

10.36. Write equations for the reactions of butanal with the following reagents:

(a) excess $CH_3CH_2CH_2OH$, H^+

(b) (1) $CH_3CH_2CH_2MgBr$; (2) H_2O, H^+

(c) $C_6H_5NH_2$, H^+

(d) HCN, $^-$CN

(e) 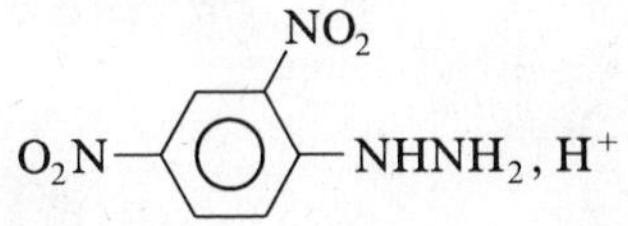 O_2N—⟨benzene ring with NO_2 substituent⟩—$NHNH_2$, H^+

(f) H_2, Ni, heat, pressure

(g) CrO_3, H_2SO_4

(h) $Ag(NH_3)_2^+$ ^-OH

(i) $NaBH_4$, CH_3OH

10.37. Write equations for the reaction of cyclopentanone with each of the sets of reactants in the preceding problem.

10.38. Write equations for the mechanisms of the following reactions:

(a) $CH_3\overset{O}{\overset{\|}{C}}CH_3$ + HCN $\xrightarrow{\ ^-CN\ }$

(b) $H\overset{O}{\overset{\|}{C}}H$ + $HOCH_2CH_2OH$ $\xrightarrow{\ H^+\ }$

(c) 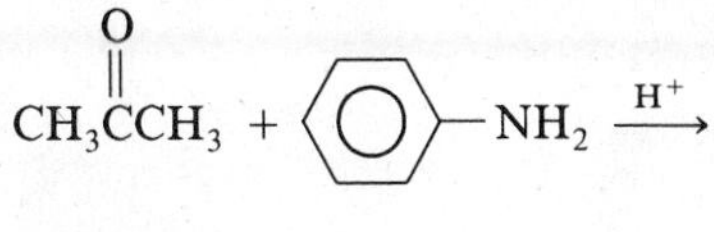 $CH_3\overset{O}{\overset{\|}{C}}CH_3$ + ⟨benzene ring⟩—NH_2 $\xrightarrow{\ H^+\ }$

10.39. Write equations or flow equations for the following conversions:

(a) 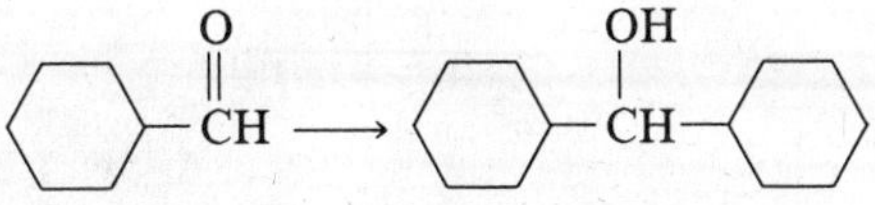 ⟨cyclohexyl⟩—$\overset{O}{\overset{\|}{C}}H$ $\longrightarrow$ ⟨cyclohexyl⟩—$\overset{OH}{\overset{|}{C}}H$—⟨cyclohexyl⟩

(b) $H\overset{O}{\overset{\|}{C}}CH_2\overset{O}{\overset{\|}{C}}H$ $\longrightarrow$ $NC\overset{OH}{\overset{|}{C}}HCH_2\overset{OH}{\overset{|}{C}}HCN$

(c) 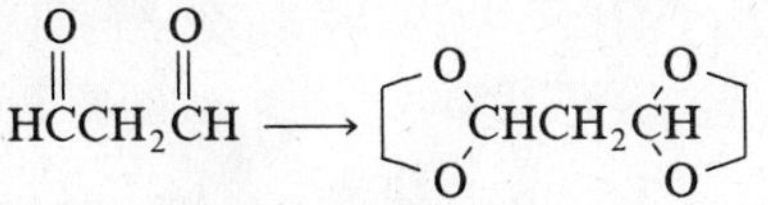 $H\overset{O}{\overset{\|}{C}}CH_2\overset{O}{\overset{\|}{C}}H$ $\longrightarrow$ $\left[\ \overset{O}{\underset{O}{\diagdown}}CHCH_2\overset{O}{\underset{O}{\diagup}}CH\ \right]$

(d) $CH_3CH_2\overset{O}{\overset{\|}{C}}CH_2CH_3$ $\longrightarrow$ $CH_3CH_2CH{=}CHCH_3$

(e) $CH_3CH_2CH_2I$ $\longrightarrow$ $CH_3CH_2CH_2CH_2OH$

(f)

(g)

10.40. Pyridoxine (commercial vitamin B_6) undergoes oxidation of the $-CH_2OH$ group at position 4 to form pyridoxal, an aldehyde. On reaction with an α-amino acid, pyridoxal forms a Schiff base, pyridoxamine, an important intermediate in the metabolism of amino acids. What is the structure of pyridoxal and the general structure of pyridoxamine?

Pyridoxine   An α-amino acid

10.41. Ninhydrin is an analytical reagent used for the detection of α-amino acids obtained from the hydrolysis of proteins. The initial steps in the reaction leading to a colored product are (1) loss of water from ninhydrin and (2) formation of a Schiff base. Write equations showing these reactions.

Ninhydrin

POINT OF INTEREST 10

An Imine in the Vision Process

Imines are biologically indispensable. For example, black-and-white vision or vision in dim light depends on a portion of the eye called *rods*. The rods contain a compound called *rhodopsin* (visual purple), an imine of retinal and the $-NH_2$ group of a protein molecule called *opsin*. Rhodopsin is responsible for the detection of light.

When a photon of light hits a rhodopsin molecule, one of the carbon–carbon double bonds is isomerized from cis to trans, as shown in Figure 10.3. The trans compound no longer fits in its protein pocket. The formerly protected imine link becomes exposed and undergoes hydrolysis. The all-trans-retinal then activates enzymes that change the ion permeability of the cell and thus its electrical properties. These changes, in turn, initiate a nerve impulse that travels to the brain. The exact mechanisms of these reactions are still under investigation.

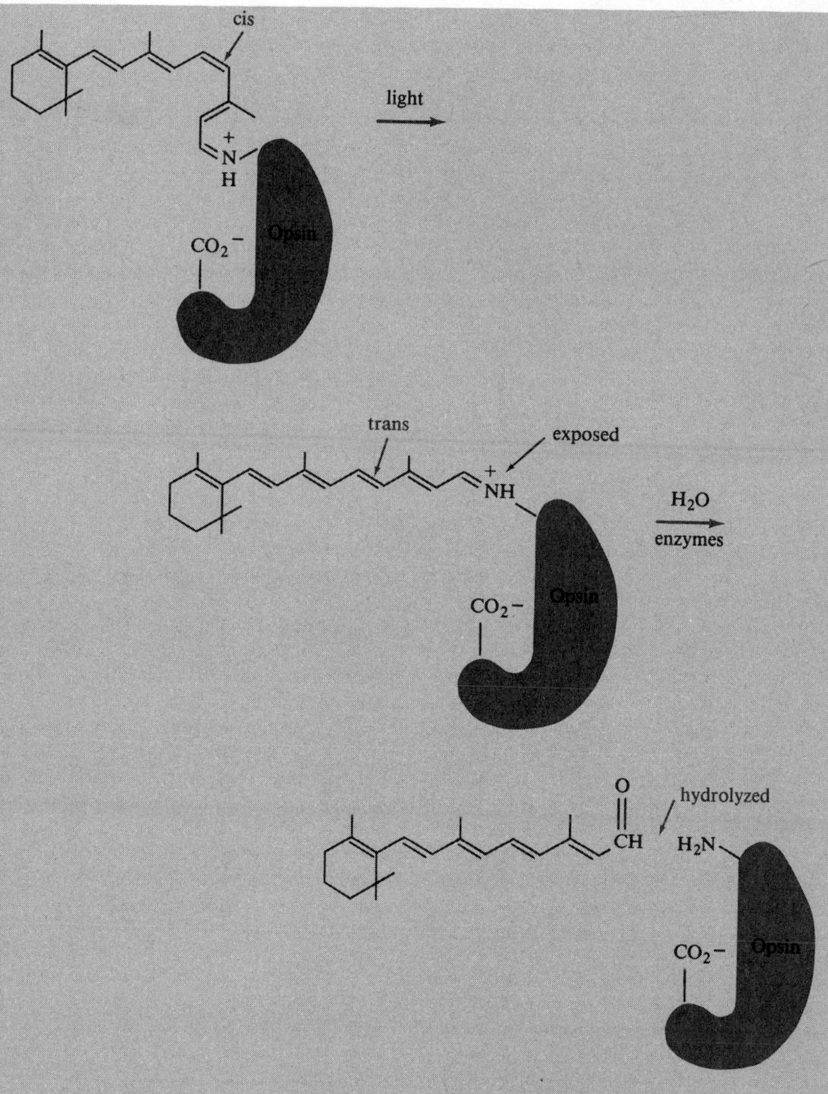

Figure 10.3 Changes in rhodopsin when it is hit by a photon of light.

Chapter 11

Carboxylic Acids

A **carboxylic acid** is a compound containing the **carboxyl group**, a term derived from *carb*onyl and hydr*oxyl*.

The Carboxyl Group:

$$-\overset{\overset{\textstyle O}{\|}}{C}-OH \qquad \text{also written} \qquad -\overset{\overset{\textstyle O}{\|}}{C}OH, \quad -CO_2H, \quad -COOH$$

The group bonded to the carboxyl group in a carboxylic acid can be almost any grouping, even another carboxyl group.

Some Carboxylic Acids:

$$CH_3\overset{\overset{\textstyle O}{\|}}{C}OH \qquad \langle\text{benzene}\rangle-\overset{\overset{\textstyle O}{\|}}{C}OH \qquad CH_2{=}CH\overset{\overset{\textstyle O}{\|}}{C}OH \qquad HO\overset{\overset{\textstyle O}{\|}}{C}-\overset{\overset{\textstyle O}{\|}}{C}OH$$

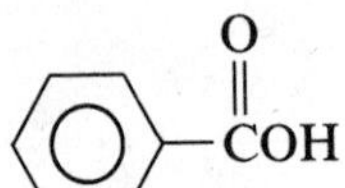

| Acetic acid | Benzoic acid | Acrylic acid | Oxalic acid |

Carboxylic acids are common in nature and in commerce. Figure 11.1 shows a few interesting and important examples.

11.1 STRUCTURE AND BONDING IN CARBOXYLIC ACIDS

The carbonyl carbon in a carboxyl group is sp^2 hybridized. Each oxygen atom carries two pairs of unshared valence electrons. These oxygen atoms are electronegative with respect to the carbonyl carbon and the hydroxyl hydrogen. Thus, the

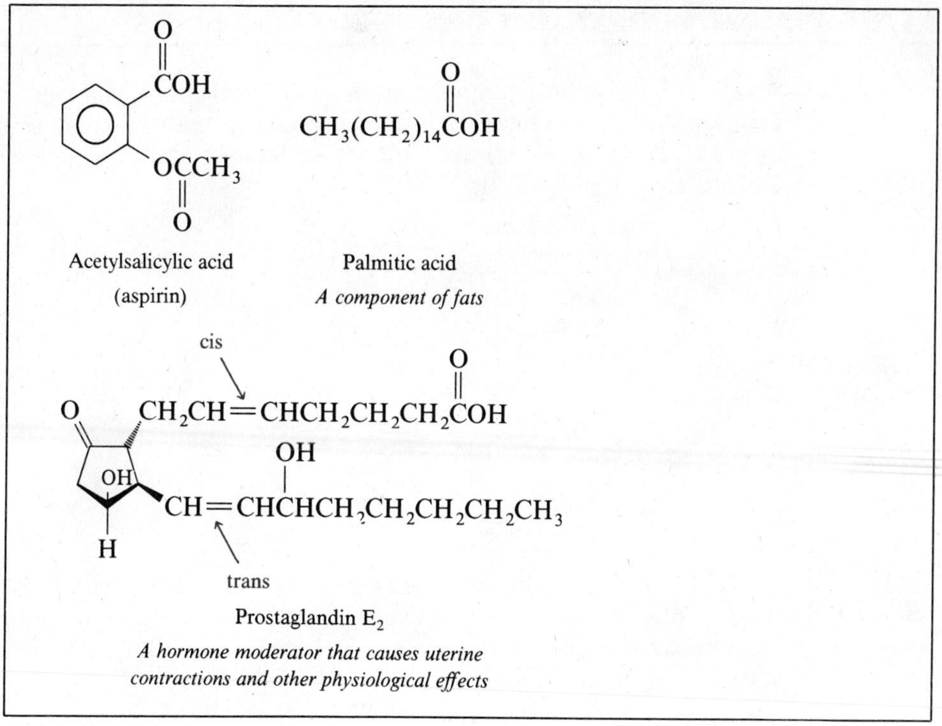

Acetylsalicylic acid
(aspirin)

Palmitic acid
A component of fats

$$CH_3(CH_2)_{14}COH$$

cis

Prostaglandin E_2

*A hormone moderator that causes uterine
contractions and other physiological effects*

Figure 11.1 Examples of carboxylic acids.

carboxyl group is polar.

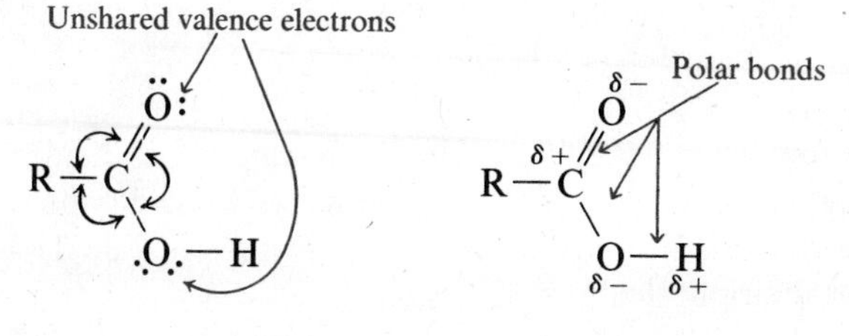

Because of the polarity of the O—H bond and because the carboxylate ion (RCO_2^-) is resonance stabilized, carboxylic acids can lose a proton to a strong or moderately strong base. The acidity of carboxylic acids will be discussed further in Section 11.6.

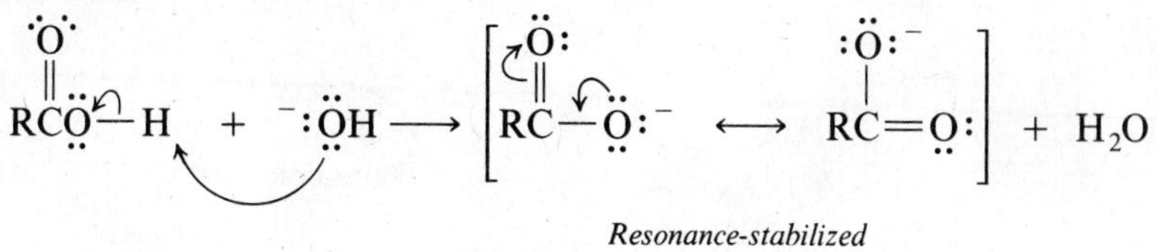

*Resonance-stabilized
carboxylate ion*

11.2 PHYSICAL PROPERTIES OF CARBOXYLIC ACIDS

A carboxylic acid molecule contains an —OH group and thus can form hydrogen bonds with water. Because of hydrogen bonding, the carboxylic acids containing one to four carbons are miscible with water. Many of those with a larger number of carbons are partially soluble.

Hydrogen bond from carbonyl oxygen to water

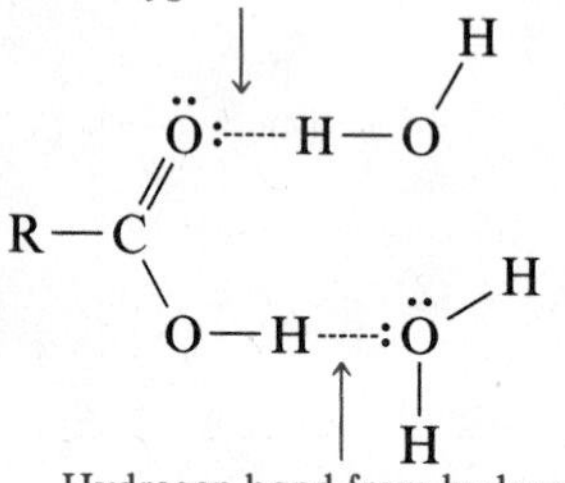

Hydrogen bond from hydroxyl hydrogen to water

Carboxylic acids also hydrogen bond with other carboxylic acid molecules by forming *two* hydrogen bonds between the two carboxyl groups. In non-hydrogen-bonding solvents, carboxylic acids exist as pairs of associated molecules, called *dimers* ("two parts").

A Carboxylic Acid Dimer:

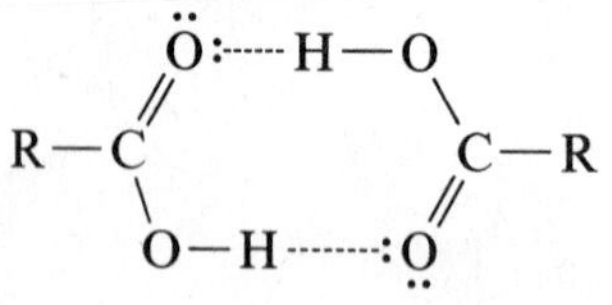

Because of the strong attractions between carboxylic acid molecules in the dimer form, the melting points and boiling points of these compounds are relatively high—higher even than those of comparable alcohols. Table 11.1 lists some carboxylic acids and their physical properties.

TABLE 11.1 PHYSICAL PROPERTIES OF SOME CARBOXYLIC ACIDS

Formula	IUPAC name	Mp (°C)	Bp (°C)
HCO_2H	Methanoic	8	101
CH_3CO_2H	Ethanoic	17	118
$CH_3CH_2CO_2H$	Propanoic	−22	141
$CH_3CH_2CH_2CO_2H$	Butanoic	−5	163
$CH_3(CH_2)_3CO_2H$	Pentanoic	−34	187
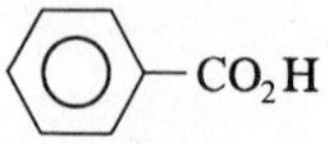 ⬡—CO_2H	Benzoic	122	250

PROBLEM 11.1.   Draw formulas to show all types of hydrogen bonds you might find in a solution of benzoic acid in ethanol.

11.3 NOMENCLATURE OF CARBOXYLIC ACIDS

The IUPAC name of a simple carboxylic acid is taken from the name of the parent alkane with the *-ane* ending changed to *-anoic acid*. The carboxyl carbon is carbon 1, just as the aldehyde carbonyl carbon is.

$$\underset{\text{HCOH}}{\overset{\overset{\displaystyle O}{\|}}{}} \qquad \underset{\underset{3}{C}H_3\underset{2}{C}H_2\underset{1}{C}OH}{\overset{\overset{\displaystyle O}{\|}}{}} \qquad \underset{\underset{\displaystyle CH_3}{|}}{\overset{\overset{\displaystyle O}{\|}}{\underset{3}{C}H_3\underset{2}{C}H\underset{1}{C}OH}}$$

IUPAC: Methanoic acid   Propanoic acid   2-Methylpropanoic acid

The low-formula-weight carboxylic acids are usually referred to by their trivial names, which often are derived from the acid's source or odor. Table 11.2 lists some acids, their names, and the derivations of these names.

The name *formic acid* is derived from the Latin word for ants—*formica*. At one time, this acid was prepared by the distillation of red ants. The name *acetic acid* is derived from the Latin *acetum*, "vinegar." Vinegar is a 5–6% aqueous solution of

TABLE 11.2   TRIVIAL NAMES AND DERIVATIONS OF SOME CARBOXYLIC ACIDS

Number of carbons	Formula	Name	Derivation
1	HCO_2H	Formic	Ants (L. *formica*)
2	CH_3CO_2H	Acetic	Vinegar (L. *acetum*)
3	$CH_3CH_2CO_2H$	Propionic	Milk, butter, and cheese (Gr. *protos*, "first"; *pion*, "fat")
4	$CH_3(CH_2)_2CO_2H$	Butyric	Butter (L. *butyrum*)
5	$CH_3(CH_2)_3CO_2H$	Valeric	Valerian root (L. *valere*, "to be strong")
6	$CH_3(CH_2)_4CO_2H$	Caproic	Goat (L. *caper*)
7	$CH_3(CH_2)_5CO_2H$	Enanthic	(Gk. *oenanthe*, "vine blossom")
8	$CH_3(CH_2)_6CO_2H$	Caprylic	Goat
9	$CH_3(CH_2)_7CO_2H$	Pelargonic	Its ester is found in the geranium *Pelargonium roseum.*
10	$CH_3(CH_2)_8CO_2H$	Capric	Goat

acetic acid. *Butyric acid* is the principal odorous ingredient of rancid butter. *Valeric acid* (Latin *valere*, "to be strong") is not a strong acid but is a very strong-smelling acid. Of particular interest are the trivial names of the 6-, 8-, and 10-carbon acids, which are foul-smelling compounds found in goat sweat. The names of all three acids are derived from *caper*, "goat."

As in aldehyde names, Greek prefixes (α, β, γ, etc.) can be used to indicate a substituent's position in reference to the carbonyl group. These Greek prefixes are used only in trivial names; in the IUPAC system, numbers are used.

$$\underset{\substack{| \\ \text{OH}}}{\text{CH}_3\text{CH}\overset{\overset{\text{O}}{\|}}{\text{C}}\text{OH}} \qquad \text{ClCH}_2\text{CH}_2\text{CH}_2\overset{\overset{\text{O}}{\|}}{\text{C}}\text{OH}$$

IUPAC:	2-Hydroxypropanoic acid	4-Chlorobutanoic acid
Trivial:	α-Hydroxypropionic acid	γ-Chlorobutyric acid
	also called lactic acid	
	(Latin *lac*, "milk")	

When a carboxyl group is bonded to a ring carbon, the cyclic portion of the molecule is named and the ending -*carboxylic acid* is added.

Cyclohexanecarboxylic acid   *trans*-3-Hydroxycyclopentanecarboxylic acid

Substituted benzoic acids are named in the same fashion as other substituted benzenes, with *benzoic acid* as the parent name. A second substituent can be *ortho*, *meta*, or *para* to the carboxyl group. If the ring contains more than two substituents, position numbers must be used. If numbers are used, the number 1 can be omitted from the name.

p-Aminobenzoic acid   3,5-Dibromobenzoic acid
or 4-aminobenzoic acid

used in sunscreens

A diacid is named as an -*anedioic acid* in the IUPAC system. (The alkane -*e*- is retained before a consonant.) The trivial names of these compounds are more frequently encountered than the IUPAC names. Some of these trivial names are

TABLE 11.3 NAMES OF SOME DIACIDS

Formula	IUPAC name	Trivial name[a]
HO_2C-CO_2H	Ethanedioic	Oxalic
$HO_2CCH_2CO_2H$	Propanedioic	Malonic
$HO_2C(CH_2)_2CO_2H$	Butanedioic	Succinic
$HO_2C(CH_2)_3CO_2H$	Pentanedioic	Glutaric
$HO_2C(CH_2)_4CO_2H$	Hexanedioic	Adipic
$HO_2C(CH_2)_5CO_2H$	Heptanedioic	Pimelic

[a] The classical mnemonic device for memorization of this series of diacids is that the first letters (o, m, s, g, a, p) fit the sentence: "Oh my, such good apple pie."

listed in Table 11.3.

$$\underset{\text{HOC}-\text{COH}}{\overset{\text{O}\quad\text{O}}{\|\quad\|}} \qquad \underset{\text{HOCCH}_2\text{COH}}{\overset{\text{O}\qquad\text{O}}{\|\qquad\|}}$$

IUPAC: Ethanedioic acid   Propanedioic acid

Trivial: Oxalic acid   Malonic acid

PROBLEM 11.2. Write formulas for the following carboxylic acids:

(a) 4-methylpentanoic acid

(b) *m*-nitrobenzoic acid

(c) chloroacetic acid

(d) *cis*-1,4-cyclohexanedicarboxylic acid

PROBLEM 11.3. Name the following carboxylic acids by the IUPAC system:

(a) $Cl_2CHCH_2\overset{\overset{\text{O}}{\|}}{\text{C}}OH$

(b) $CH_3CH_2\overset{\overset{\text{H}_3\text{C}}{|}}{\underset{\underset{CH_2CH_2CH_3}{|}}{\text{C}}}-\overset{\overset{\text{O}}{\|}}{\text{C}}OH$

(c) $HO\overset{\overset{\text{O}}{\|}}{\text{C}}\overset{}{\underset{\underset{CH_2CH_2CH_3}{|}}{\text{C}}}H\overset{\overset{\text{O}}{\|}}{\text{C}}OH$

(d) [cyclooctane ring with H H at top carbon and HOC COH (with C=O double bonds) substituents at the bottom]

(e) [structure: cyclohexane ring with O_2N substituent and COH (C=O) group]

(f) [structure: cyclopentane ring with COH (C=O) group, H, and Cl substituents]

11.4 SOME IMPORTANT CARBOXYLIC ACIDS

Acetic acid (CH_3CO_2H) is by far the most important carboxylic acid in commerce, industry, and the laboratory. Its pure form is called *glacial acetic acid* because this compound solidifies to an icy-looking solid when chilled. Glacial acetic acid is a colorless, flammable liquid (mp 17°C, bp 118°C) with a pungent biting odor. It is miscible with water and with most organic solvents. In liquid form or as vapor, glacial acetic acid is highly corrosive to skin and other tissues.

Commercial glacial acetic acid is made by reaction of methanol with carbon monoxide or by the oxidation of ethylene. The starting materials for these reactions are synthesized from natural gas, petroleum, or coal.

$$CH_3OH + CO \xrightarrow[\text{heat, pressure}]{Rh^{3+}, HI, H_2O} CH_3COH \ (C=O)$$

$$CH_2{=}CH_2 \xrightarrow[PdCl_4^-, Cu^{2+}]{O_2 \text{ (air)}} CH_3CH \ (C=O) \xrightarrow{O_2} CH_3COH \ (C=O)$$

Acetic acid is used in the synthesis of many industrial products, including acetate fibers and some plastics. It is also used as a solvent and as a reagent for synthesis in the laboratory.

Vinegar, a 3–6% aqueous solution of acetic acid, is prepared by the fermentation of apple juice, other fruit juice, or wine or by the dilution of synthetic acetic acid.

$$C_6H_{12}O_6 \xrightarrow[-CO_2]{\text{yeast enzymes}} 2CH_3CH_2OH \xrightarrow[O_2]{\substack{Acetobacter \\ \text{enzymes}}} 2CH_3COH \ (C=O)$$

Glucose
in fruit

Ethanol

Acetic acid

Pickled food products are foods preserved in vinegar. Examples are dill pickles, pickled beets, and pickled pigs' feet. The acetic acid in vinegar is sufficiently dilute that it is not considered corrosive; however, chronic exposure of teeth to pickled foods and similar acidic products can damage the tooth enamel.

Oxalic acid ($HO_2C—CO_2H$) is a corrosive, toxic solid (mp 102°C) that is a moderately strong acid. It is present in small amounts in many plants, such as oxalis and spinach, as the sodium or calcium salt. Oxalic acid is used as a rust remover, as a reagent for the manufacture of dyes, and for many other purposes.

(+)-Tartaric acid is a water-soluble solid (mp 170°C) that occurs in many fruits. Commercially, it is obtained as a by-product of winemaking.

$$
\begin{array}{c}
CO_2H \\
| \\
H\!\!-\!\!C\!\!-\!\!OH \\
| \\
HO\!\!-\!\!C\!\!-\!\!H \\
| \\
CO_2H
\end{array}
$$

(+)-Tartaric acid

Tartaric acid is used as an acidifying agent in soft drinks, candy, and other foods. It is also used in photography, ceramics, tanning, and manufacturing processes. Potassium hydrogen tartrate (cream of tartar) is used primarily as a baking powder.

11.5 PREPARATION OF CARBOXYLIC ACIDS

A variety of reactions can be used to synthesize carboxylic acids. Included in these reactions are hydrolysis of carboxylic acid derivatives, reactions of Grignard reagents with carbon dioxide, and oxidation reactions.

A. Hydrolysis of Carboxylic Acid Derivatives

Hydrolysis means cleavage of a molecule by water. (*Hydro-* is a combining form meaning "water," and *-lysis* means "loosing," or "breaking.") A compound that yields a carboxylic acid when it is hydrolyzed is called a **derivative of a carboxylic acid**.

Derivatives of Carboxylic Acids:

$$
\begin{array}{cc}
\overset{\displaystyle O}{\overset{\displaystyle \|}{R-C-Cl}} & \overset{\displaystyle O \qquad\quad O}{\overset{\displaystyle \| \qquad\quad \|}{R-C-O-C-R}}
\end{array}
$$

An acid chloride   *An acid anhydride*

$$
\begin{array}{ccc}
\overset{\displaystyle O}{\overset{\displaystyle \|}{R-C-OR'}} & \overset{\displaystyle O}{\overset{\displaystyle \|}{R-C-NH_2}} & R-C\equiv N
\end{array}
$$

An ester   *An amide*   *A nitrile*

The carbonyl group of a carboxylic acid derivative (except nitriles) has two bonds. One of these bonds is to a hydrogen, an alkyl group, or an aryl group. The other bond is to an electronegative atom: X, O, or N.

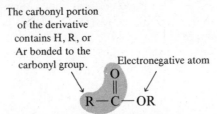

The carbonyl portion
of the derivative
contains H, R, or
Ar bonded to the
carbonyl group.

Electronegative atom

R—C—OR

When the derivative is heated in aqueous acid or base, the electronegative atom can be cleaved from the carbonyl carbon. If the cleavage is carried out in acidic solution, the carbonyl portion of the acid derivative is converted to a carboxylic acid. If the reaction is carried out in base, the carbonyl portion forms a carboxylate ion. Carboxylates can be converted to carboxylic acids by acidification.

General Reactions for Hydrolysis:

$$\text{In acid:}\quad \underset{\|}{\overset{O}{RC}}-Y + H-OH \xrightarrow{H^+} \underset{\|}{\overset{O}{RC}}-OH + H-Y$$

A carboxylic
acid

$$\text{In base:}\quad \underset{\|}{\overset{O}{RC}}-Y + {}^-OH \longrightarrow \underset{\|}{\overset{O}{RC}}-O^- + H-Y$$

A carboxylate

$$\xrightarrow{H^+}\ \underset{\|}{\overset{O}{RCOH}}$$

Specific Examples:

$$\underset{\|}{\overset{O}{CH_3C}}-OCH_2CH_3 + H_2O \underset{}{\overset{H^+}{\rightleftharpoons}} \underset{\|}{\overset{O}{CH_3C}}-OH + HOCH_2CH_3$$

Ethyl acetate   Acetic acid   Ethanol

$$\underset{\|}{\overset{O}{CH_3C}}-OCH_2CH_3 + Na^+\ {}^-OH \longrightarrow \underset{\|}{\overset{O}{CH_3CO}}{}^-\ Na^+ + HOCH_2CH_3$$

Sodium acetate

These hydrolysis reactions will be discussed in detail in Chapter 12.

Recall from Chapter 7 that many alkyl halides can be converted to nitriles when treated with cyanide ion ($^-:C\equiv N:$), which is a strong nucleophile. Primary, allylic, and benzylic halides give the best yields of nitrile; secondary alkyl halides can also be used. (Tertiary alkyl halides yield elimination products instead of nitriles.) Hydrolysis of a nitrile yields the carboxylic acid, just as hydrolysis of the other carboxylic acid derivatives does. This reaction sequence is a fairly standard technique for converting halides to carboxylic acids containing one additional carbon.

$$CH_3CH_2Br \xrightarrow[-Br^-]{^-C\equiv N} CH_3CH_2-C\equiv N \xrightarrow[heat]{H_2O,\ H^+} CH_3CH_2-\overset{\displaystyle O}{\overset{\displaystyle \|}{C}}OH$$

Bromoethane	Propanenitrile	Propanoic acid
(*two carbons*)	(*three carbons*)	(*three carbons*)

B. Oxidation of Primary Alcohols and Aldehydes

Strong oxidizing agents such as chromic oxide (CrO_3) or potassium permanganate ($KMnO_4$) convert primary alcohols to carboxylic acids in acidic solution or to carboxylates in alkaline solution. These reactions were discussed in Section 8.11.

$$CH_3(CH_2)_8CH_2OH \xrightarrow{CrO_3 + H_2SO_4} CH_3(CH_2)_8\overset{\displaystyle O}{\overset{\displaystyle \|}{C}}OH$$

1-Decanol   Decanoic acid

$K^+\ MnO_4^-$
$Na^+\ {}^-OH$

$$\longrightarrow CH_3(CH_2)_8\overset{\displaystyle O}{\overset{\displaystyle \|}{C}}O^-\ Na^+$$

$H_2O,\ H^+$

Sodium decanoate

Oxidation of an aldehyde with a strong or a weak oxidizing agent also yields a carboxylic acid or a carboxylate. Recall from Section 10.8 that the Tollens test for aldehydes is an oxidation using the silver-ammonia complex ion as the oxidizing agent. In synthetic reactions, a less expensive oxidizing agent, such as permanganate, would be chosen.

$$\underset{\text{Cyclohexanecarbaldehyde}}{\bigcirc\!-\!\overset{\displaystyle O}{\overset{\displaystyle \|}{C}}H} \xrightarrow{\text{aqueous } K^+ MnO_4^-} \bigcirc\!-\!\overset{\displaystyle O}{\overset{\displaystyle \|}{C}}O^- \xrightarrow{H^+} \underset{\substack{\text{Cyclohexanecarboxylic}\\ \text{acid}}}{\bigcirc\!-\!\overset{\displaystyle O}{\overset{\displaystyle \|}{C}}OH}$$

C. Oxidation of Alkenes

Oxidation of an alkene containing one hydrogen on one of the double-bond carbons ($=CH-R$) with a strong oxidizing agent yields a carboxylic acid. Alkene oxidations are discussed in Section 3.9.

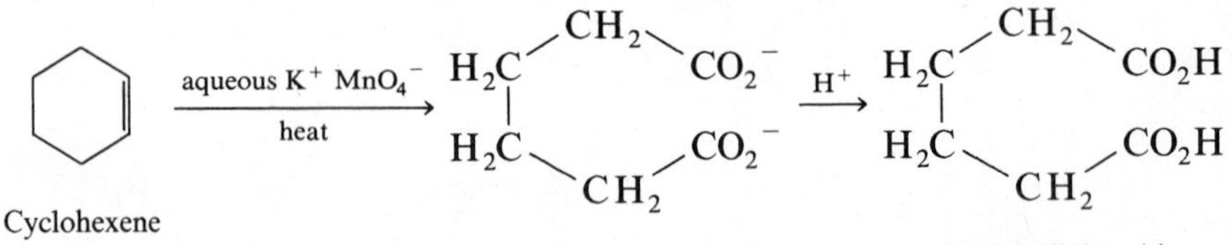

Cyclohexene

$$\xrightarrow[\text{heat}]{\text{aqueous K}^+ \text{MnO}_4{}^-}$$

$$\xrightarrow{\text{H}^+}$$

Hexanedioic acid

(adipic acid)

used in nylon synthesis

PROBLEM 11.4. Predict the organic products:

(a) $CH_3CH{=}CHCH_2CH_3 \xrightarrow[\text{heat}]{CrO_3 + H_2SO_4}$

(b) 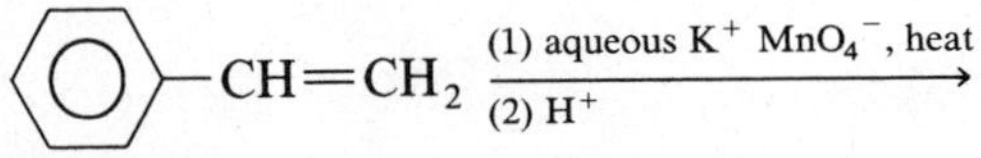 $-CH{=}CH_2 \xrightarrow[\text{(2) H}^+]{\text{(1) aqueous K}^+ \text{MnO}_4{}^-, \text{heat}}$

(c) $CH_3CH_2CH_2CH_2OH \xrightarrow[\text{heat}]{CrO_3 + H_2SO_4}$

(d) $CH_3CH_2CH_2\overset{\overset{\displaystyle O}{\|}}{CH} \xrightarrow[\text{heat}]{CrO_3 + H_2SO_4}$

D. Oxidation of Alkyl Aromatic Compounds

Aromatic carboxylic acids can be prepared by oxidation of alkylbenzenes. Aromatic rings, such as benzene or pyridine rings, without alkyl side chains are not as easily oxidized.

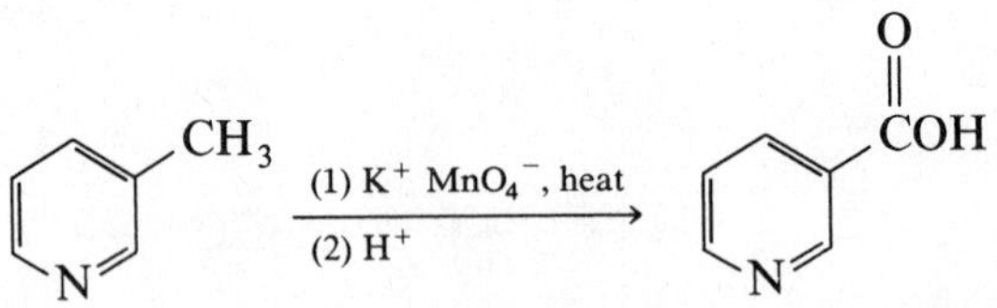

$$\xrightarrow[\text{(2) H}^+]{\text{(1) K}^+ \text{MnO}_4{}^-, \text{heat}}$$

Nicotinic acid (niacin)

a B vitamin

PROBLEM 11.5. Suggest syntheses for the following carboxylic acids from hydro-carbons.

(a)

$$\text{HOC} - \langle \bigcirc \rangle - \text{COH}$$

(b)

E. Grignard Reactions

Recall that the negative carbon of a Grignard reagent (RMgX) can attack the positive carbon of a carbonyl group (Section 10.6D).

$$\text{R'MgBr} + \text{R} - \overset{\overset{\displaystyle \ddot{O}}{\|}}{C} - \text{R} \longrightarrow \text{R} - \overset{\overset{\displaystyle :\ddot{O}:^- \ {}^+\text{MgBr}}{|}}{\underset{\underset{\displaystyle R'}{|}}{C}} - \text{R}$$

Carbon dioxide contains a carbonyl group and, like other carbonyl compounds, its carbon can be attacked by a Grignard reagent. The addition product is a magnesium carboxylate, which precipitates from the reaction mixture and does not react further. Acidification of the mixture yields a carboxylic acid.

Step 1, Nucleophilic Attack of RMgX:

$$\text{R} - \text{MgX} + \ :\text{O}{=}\text{C}{=}\ddot{\text{O}}: \ \xrightarrow{\text{diethyl ether}} \ \text{R} - \overset{\overset{\displaystyle \ddot{O}}{\|}}{C} - \ddot{\text{O}}:^- + {}^+\text{MgX}$$

A carboxylate

insoluble in diethyl ether

Step 2, Acidification:

$$\overset{\overset{\displaystyle O}{\|}}{\text{RCO}}\!:^- \ {}^+\text{MgX} + \text{H}^+ \ \xrightarrow{\text{H}_2\text{O}} \ \overset{\overset{\displaystyle O}{\|}}{\text{RCOH}} + \text{Mg}^{2+} + \text{X}^-$$

For convenience, dry ice (solid CO_2) is used instead of gaseous carbon dioxide as the reagent. The Grignard reagents can be of almost any type—alkyl, aryl, etc.

PROBLEM 11.6. Show by flow equations how you would synthesize the following carboxylic acids using Grignard reactions. (Start with organohalogen compounds.)

(a) $CH_3CH_2CH_2CO_2H$ (b) 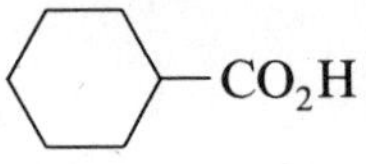 ⬡$-CO_2H$

Table 11.4 summarizes the syntheses of carboxylic acids.

TABLE 11.4 SUMMARY OF PREPARATIONS OF CARBOXYLIC ACIDS

Reaction	Section reference

Hydrolysis of derivatives

$$RC(=O)-Cl,\ RC(=O)-OCR(=O),\ RC(=O)-OR',\ RC(=O)-NR'_2, \quad \text{or} \quad RC{\equiv}N \xrightarrow{H_2O,\ H^+} RC(=O)OH \qquad \text{11.5A}$$

Oxidation of 1° alcohols or aldehydes

$$RCH_2OH \xrightarrow{\text{strong oxidizing agent}^a} RC(=O)OH \qquad \text{8.11}$$

$$RCH(=O) \xrightarrow{\text{strong or weak oxidizing agent}^b} RC(=O)OH \qquad \text{10.8}$$

Oxidation of alkenes

$$RCH{=}CHR \xrightarrow{\text{strong oxidizing agent}^a} RC(=O)OH + HOC(=O)R \qquad \text{3.9}$$

must have hydrogen
to form carboxylic acid

Oxidation of alkyl aromatic compounds

$$Ar-CHR'_2 \xrightarrow{\text{strong oxidizing agent}^a} Ar-C(=O)OH \qquad \text{5.5}$$

Grignard reactions

$$RMgX \quad \text{or} \quad ArMgX \xrightarrow[\text{(2) H}_2\text{O, H}^+]{\text{(1) CO}_2} RC(=O)OH \quad \text{or} \quad ArC(=O)OH \qquad \text{11.5E}$$

[a] Typical strong oxidizing agents are hot aqueous $KMnO_4$, $Na_2Cr_2O_7$, or $CrO_3 + H_2SO_4$.

[b] A typical weak oxidizing agent is Tollens reagent (see Section 10.8).

EXAMPLE

Show by flow equations how you would make the following conversions. If two routes are possible, show both of them.

(a) H$_3$C—⟨O⟩—Br ⟶ H$_3$C—⟨O⟩—C(=O)OH

(b) CH$_3$CHCH$_2$I (with CH$_3$ on second carbon) ⟶ CH$_3$CHCH$_2$COH (with CH$_3$ and O)

Solution:

(a) The only techniques presented here for the synthesis of an aromatic carboxylic acid are the reaction of an aryl Grignard reagent with CO_2 and the oxidation of an alkyl aromatic compound. Because the required starting material here is an aryl halide, a Grignard reaction is the correct choice.

H$_3$C—⟨O⟩—Br $\xrightarrow[\text{ether}]{\text{Mg}}$ H$_3$C—⟨O⟩—MgBr $\xrightarrow[\text{(2) H}_2\text{O, H}^+]{\text{(1) CO}_2}$

H$_3$C—⟨O⟩—C(=O)OH

(b) The product contains one more carbon than the starting material. The two possible syntheses are by a Grignard reaction and by a nitrile sequence. Because the starting halide is primary, either sequence could be used.

CH$_3$CHCH$_2$I $\xrightarrow{^-\text{CN}}$ CH$_3$CHCH$_2$C≡N $\xrightarrow[\text{heat}]{\text{H}_2\text{O, H}^+}$ CH$_3$CHCH$_2$COH

CH$_3$CH$_2$CH$_2$I $\xrightarrow[\text{ether}]{\text{Mg}}$ CH$_3$CH$_2$CH$_2$MgI $\xrightarrow[\text{(2) H}_2\text{O, H}^+]{\text{(1) CO}_2}$ CH$_3$CH$_2$CH$_2$COH

PROBLEM 11.7. Complete the following equations:

(a) CH$_3$CH=CHCH$_3$ $\xrightarrow[\text{(2) H}_2\text{O, H}^+]{\text{(1) aqueous K}^+ \text{MnO}_4^-\text{, heat}}$

(b) CH$_3$CH$_2$OH $\xrightarrow[\text{heat}]{\text{CrO}_3 + \text{H}_2\text{SO}_4}$

(c) CH_3I $\xrightarrow[\text{(3) H}_2\text{O, H}^+]{\begin{array}{l}\text{(1) Mg, ether}\\\text{(2) CO}_2\end{array}}$

(d) ⬡$-C{\equiv}N + H_2O + H^+ \longrightarrow$

PROBLEM 11.8. Suggest syntheses for the following carboxylic acids from organo-halogen compounds. (Write equations for as many syntheses as you can.)

(a) ⬡$-CH_2\overset{\displaystyle O}{\overset{\|}{C}}OH$

(b) $CH_3\underset{\underset{\displaystyle OH}{|}}{C}H\overset{\displaystyle O}{\overset{\|}{C}}OH$

(c) $\underset{CH_3CH_2}{\overset{CH_3CH_2}{\text{⬡}}}-\overset{\displaystyle O}{\overset{\|}{C}}OH$

11.6 ACIDITY OF CARBOXYLIC ACIDS

A. Measuring Acid Strength

In water, carboxylic acids are in equilibrium with carboxylate ions and hydronium ions.

$$RC\overset{\cdot\cdot}{\overset{\displaystyle \cdot\cdot O\cdot}{\overset{\|}{}}}\!\!-H \;+\; H\overset{\cdot\cdot}{\underset{\cdot\cdot}{O}}H \;\rightleftharpoons\; RC\overset{\cdot\cdot}{\overset{\displaystyle \cdot\cdot O\cdot}{\overset{\|}{}}}\!\!:^{-} \;+\; H\overset{H}{\underset{\overset{\cdot\cdot}{+}}{\overset{|}{O}}}H$$

A carboxylic acid   *A carboxylate*   *Hydronium ion*
   (weaker)   *ion*   *(stronger)*

One measure of the strength of an acid is the degree of which it ionizes in water. The greater the amount of ionization, the stronger is the acid. Carboxylic acids are, for the most part, *weaker* acids than H_3O^+; in aqueous solution, most of the carboxylic acid molecules are un-ionized.

The strength of an acid is expressed as the **acidity constant** K_a, the equilibrium constant for the ionization reaction in water. (The concentration of water is omitted from this equation. Its concentration changes by such a small amount that it is considered a constant and is included in K_a.)

$$K_a = \frac{[RCO_2^-]\,[H_3O^+]}{[RCO_2H]} \quad \text{or, simplified,} \quad \frac{[RCO_2^-]\,[H^+]}{[RCO_2H]}$$

where $[RCO_2H]$ = molarity of RCO_2H

$[RCO_2^-]$ = molarity of RCO_2^-

$[H_3O^+]$ or $[H^+]$ = molarity of H_3O^+

A larger value for K_a means that the acid is stronger because the concentrations of RCO_2^- and H^+ are larger.

EXAMPLE

The K_a for acetic acid is 1.8×10^{-5}. What is the concentration of H^+ in a 0.100 M solution of acetic acid?

Solution:

1. Write the chemical equation.

$$CH_3CO_2H \rightleftharpoons CH_3CO_2^- + H^+$$

2. Write the mathematical equation.

$$K_a = \frac{[CH_3CO_2^-][H^+]}{[CH_3CO_2H]} = 1.8 \times 10^{-5}$$

3. Let $[H^+] = x$. Then $[CH_3CO_2^-]$ must also equal x because the ionization produces equal numbers of anions and cations.
   Although $[CH_3CO_2H] = 0.100 - x$, we can round this value to 0.100 because only a very small percentage of CH_3CO_2H molecules undergo ionization.

4. Substitute and solve.

$$K_a = \frac{x \cdot x}{0.100} = 1.8 \times 10^{-5}$$

$$x^2 = (1.8 \times 10^{-5})(0.100)$$

$$x^2 = 1.8 \times 10^{-6}$$

$$x = \sqrt{1.8 \times 10^{-6}} = \sqrt{1.8} \times \sqrt{10^{-6}}$$

$$x = [H^+] = 1.34 \times 10^{-3} \quad \text{or} \quad 0.00134 \ M$$

Because exponential numbers are often inconvenient, K_a values are often converted to **pK_a values**, where $pK_a = -\log K_a$. In the following expressions, the pK_a value is the negative of the exponent in K_a.

This number
is the pK_a.

If $K_a = 1.0 \times 10^{-3}$, then $pK_a = 3$

If $K_a = 1.0 \times 10^{-7}$, then $pK_a = 7$

If $K_a = 1.0 \times 10^{2}$, then $pK_a = -2$

TABLE 11.5   ACIDITY CONSTANTS OF SOME CARBOXYLIC ACIDS

Formula	Name	K_a	pK_a
HCO_2H	Formic	1.8×10^{-4}	3.75
CH_3CO_2H	Acetic	1.8×10^{-5}	4.74
$CH_3CH_2CO_2H$	Propanoic	1.3×10^{-5}	4.87
$CH_3(CH_2)_2CO_2H$	Butanoic	1.5×10^{-5}	4.81
$ClCH_2CO_2H$	Chloroacetic	1.4×10^{-3}	2.85
Cl_2CHCO_2H	Dichloroacetic	3.3×10^{-2}	1.48
Cl_3CCO_2H	Trichloroacetic	2×10^{-1}	0.70
$CH_3CHClCO_2H$	2-Chloropropanoic	1.5×10^{-3}	2.83
$ClCH_2CH_2CO_2H$	3-Chloropropanoic	1.0×10^{-4}	3.98

As K_a increases, pK_a decreases; therefore, a smaller pK_a means a stronger acid.

K_a:	10^{-5}	10^{-3}	10^0	10^3	10^5
pK_a:	5	3	0	-3	-5

Increasing acid strength →

Table 11.5 lists the K_a values and pK_a values for several carboxylic acids. A more complete listing of comparative acidities is inside the front cover.

PROBLEM 11.9.   List the following acids in order of increasing acid strength (least acidic first).

(a) HCl   (b) H_3PO_4 (c) CH_3CO_2H
K_a: ~ 10^7 7.5×10^{-3} 1.8×10^{-5}

PROBLEM 11.10.   List the following compounds in order of increasing acid strength (least acidic first).

pK_a: 2.9   3.8   4.0

PROBLEM 11.11.   Given the following pK_a values, determine the K_a values.

(a) $pK_a = 4$ (b) $pK_a = -4$ (c) $pK_a = 9$

B. Resonance and Acid Strength

As we have mentioned, the principal reason that a carboxylic acid is acidic is the resonance stabilization of the carboxylate ion.

Resonance stabilized

The two resonance structures of a carboxylate ion are equivalent; the negative charge is shared equally by the two oxygens.

This delocalization of negative charge explains why carboxylic acids are more acidic than phenols. Although the phenoxide ion is resonance stabilized, the principal contributing resonance structures have the negative charge localized on one atom.

Principal resonance structures because aromaticity is retained and because the most electronegative atom carries the charge.

Alcohols are less acidic than either carboxylic acids or phenols because the negative charge in the anion (RO^-) is not delocalized at all.

C. Inductive Effect and Acid Strength

Other factors besides resonance stabilization of the carboxylate ion affect the acidity of a compound. Further delocalization of the carboxylate ion's negative charge stabilizes the anion relative to the acid. This added stabilization of the anion increases the acidity of the acid. For example, chlorine is electronegative. In chloroacetic acid, the chlorine withdraws electron density from the carboxyl group toward itself. This electron withdrawal causes a further delocalizing of the negative charge, thus stabilizing the anion and increasing the acid strength of the acid. Chloroacetic acid is a stronger acid than acetic acid.

Cl helps share the negative charge
by withdrawing electron density.

$$Cl \leftarrow CH_2 \leftarrow \overset{\overset{\displaystyle O}{\|}}{C} \leftarrow O^-$$

$$\overset{\overset{\displaystyle O}{\|}}{ClCH_2C}OH \rightleftharpoons \overset{\overset{\displaystyle O}{\|}}{ClCH_2C}O^- + H^+$$

Stabilized

The greater the electron withdrawal by the inductive effect, the stronger is the acid. Dichloroacetic acid contains *two* electron-withdrawing chlorine atoms and is a stronger acid than chloroacetic acid. Trichloroacetic acid contains *three* chlorine atoms and is an even stronger acid than dichloroacetic acid.

$$CH_3CO_2H \quad ClCH_2CO_2H \quad Cl_2CHCO_2H \quad Cl_3CCO_2H$$

Increasing acid strength

PROBLEM 11.12. The electron-withdrawing power by the inductive effect diminishes as the distance between the electron-withdrawing substituent and the carboxyl group increases. Rank the following carboxylic acids in order of increasing acid strength (least acidic first).

(a) $ClCH_2CH_2CH_2CO_2H$ (b) $CH_3CH_2CH_2CO_2H$

(c) $CH_3CH_2\underset{\underset{\displaystyle Cl}{|}}{C}HCO_2H$

D. Salts of Carboxylic Acids

Water is a very weak base—too weak to remove a large percentage of protons from most carboxylic acids. Stronger bases such as sodium hydroxide undergo complete reaction with carboxylic acids to yield salts called **carboxylates**. The reaction is a typical acid-base neutralization reaction.

Reaction with Water Lies on the Un-ionized Side:

$$\overset{\overset{\displaystyle O}{\|}}{CH_3C}OH + H_2O \rightleftharpoons \overset{\overset{\displaystyle O}{\|}}{CH_3C}O^- + H_3O^+$$

Reaction with NaOH Goes to Completion:

$$CH_3\overset{\overset{\cdot\cdot}{\overset{\displaystyle O}{\|}}}{\underset{}{C}}\overset{\cdot\cdot}{\underset{\cdot\cdot}{O}} \!- H \;\; + \;\; Na^+ \; \;^-\!\!:\!\overset{\cdot\cdot}{\underset{\cdot\cdot}{O}}H \;\longrightarrow\; CH_3\overset{\overset{\cdot\cdot}{\overset{\displaystyle O}{\|}}}{\underset{}{C}}\overset{\cdot\cdot}{\underset{\cdot\cdot}{O}}\!:^- \; Na^+ \; + \; H\overset{\cdot\cdot}{\underset{\cdot\cdot}{O}}H$$

　　　Acetic acid   Sodium acetate

Carboxylates are salts and behave much like inorganic salts; they are odorless, relatively high melting, and often water soluble. Being ionic, they are insoluble in organic solvents. The sodium salts of long-hydrocarbon-chain carboxylic acids (fatty acids) are called *soaps*. These compounds and the mode of their action will be discussed in Point of Interest 12.

　　　Carboxylates are named similarly to inorganic salts. The cation is named first, followed by the name of the anion as a separate word. The name of the carboxylate anion is taken from the name of the parent carboxylic acid with the *-oic acid* ending changed to *-oate*.

　　　Benzoic acid   Sodium benzoate

Carboxylic acids react with sodium bicarbonate (Na^+ HCO_3^-) to yield a sodium carboxylate and carbonic acid (H_2CO_3). Carbonic acid is unstable and forms carbon dioxide gas and water. (You can perform this reaction in your kitchen by mixing vinegar, which is dilute acetic acid, and baking soda, which is sodium bicarbonate.) Alcohols and most phenols do not form salts when treated with $NaHCO_3$ because they are less acidic than carbonic acid.

$$CH_3\overset{\overset{\displaystyle O}{\|}}{C}OH \;\; + \;\; Na^+ \, HCO_3^- \;\longrightarrow\; CH_3\overset{\overset{\displaystyle O}{\|}}{C}O^- \, Na^+ \; + \;\; H_2CO_3$$

　　Stronger acid　　　　　Sodium　　　　　　　　　　Carbonic acid
　　　　　　　　　　　　bicarbonate　　　　　　　　　　*weaker acid*
　　　　　　　　　　　　　　　　　　　　　　　　　　　　　↓
　　　　　　　　　　　　　　　　　　　　　　　　　　　$H_2O + CO_2$

$$ROH \text{ and } ArOH + Na^+ \, HCO_3^- \;\longrightarrow\; \text{No reaction}$$

Carboxylic acids also react with ammonia and amines to yield ammonium carboxylates. The reaction with amines is especially important in protein chemistry because protein molecules are rich in carboxyl groups and amino groups.

$$\underset{\text{Acetic acid}}{CH_3\overset{\overset{\displaystyle O}{\|}}{C}OH} + \underset{\text{Ammonia}}{:NH_3} \longrightarrow \underset{\text{Ammonium acetate}}{CH_3\overset{\overset{\displaystyle O}{\|}}{C}O^- \; {}^+NH_4}$$

$$\underset{\text{Methylamine}}{CH_3\overset{\overset{\displaystyle O}{\|}}{C}OH + H_2\ddot{N}CH_3} \longrightarrow \underset{\text{Methylammonium acetate}}{CH_3\overset{\overset{\displaystyle O}{\|}}{C}O^- \; H_3\overset{+}{N}CH_3}$$

Treatment of a carboxylate with a strong or moderately strong acid converts the salt back to the carboxylic acid.

$$R\overset{\overset{\displaystyle O}{\|}}{C}O^- \, Na^+ + HCl \longrightarrow R\overset{\overset{\displaystyle O}{\|}}{C}OH + Na^+ \, Cl^-$$

PROBLEM 11.13.   Complete the following equations. If no reaction would occur, write "no reaction."

(a) ⬡—$\overset{\overset{\displaystyle O}{\|}}{C}OH$ + $CH_3CH_2NHCH_2CH_3$ $\longrightarrow$

(b) ⎔—$\overset{\overset{\displaystyle O}{\|}}{C}OH$ + $Na^+\,HCO_3^-$ $\longrightarrow$

(c) ⎔—OH + $Na^+\,HCO_3^-$ $\longrightarrow$

(d) ⬡—OH + $Na^+\,HCO_3^-$ $\longrightarrow$

(e) ⬡—$\overset{\overset{\displaystyle O}{\|}}{C}O^-\,Na^+$ + HCl $\longrightarrow$

(f) ⎔—$O^-\,Na^+$ + $CH_3\overset{\overset{\displaystyle O}{\|}}{C}OH$ $\longrightarrow$

(g) ⎔—OH + $CH_3\overset{\overset{\displaystyle O}{\|}}{C}O^-\,Na^+$ $\longrightarrow$

PROBLEM 11.14. Alanine is an amino acid that can be obtained from proteins. Using an equation, explain why this organic compound is water soluble but insoluble in diethyl ether.

$$\begin{array}{c} \quad\quad\; O \\ \quad\quad\; \| \\ CH_3CHCOH \\ \quad | \\ \quad NH_2 \end{array}$$

Alanine

E. Buffer Solutions

A **buffer solution** is a solution that resists a change in pH when small amounts of acid or base are added to it. A mixture of a weak carboxylic acid and a carboxylate is one type of buffer solution.

Consider a solution containing equal parts of acetic acid and sodium acetate. If a small amount of strong acid, such as HCl, is added to this solution, the added hydrogen ions combine with the acetate ions to yield acetic acid. In this way, excess hydrogen ions are removed from the solution.

Add H^+ to a Solution of CH_3CO_2H and $CH_3CO_2^-$ Na^+:

$$\begin{array}{cc} O & O \\ \| & \| \\ CH_3CO^- + H^+ \longrightarrow & CH_3COH \end{array}$$

If a base is added to the solution containing acetic acid and sodium acetate, the acetic acid reacts with the base and is converted to acetate ions. Thus, excess base can also be removed from the solution.

Add ^-OH to a Solution of CH_3CO_2H and $CH_3CO_2^-$ Na^+:

$$\begin{array}{cc} O & O \\ \| & \| \\ CH_3COH + {}^-OH \longrightarrow & CH_3CO^- + H_2O \end{array}$$

Control of pH is imperative for biological systems. Buffer mixtures in the body fluids include $HCO_3^- + H_2CO_3$ and $HPO_4^{2-} + H_2PO_4^-$.

EXAMPLE

Write equations showing how the blood buffers **(a)** the bicarbonate buffer and **(b)** the biphosphate buffer react with (1) acid and (2) hydroxide ion.

Solution: For the bicarbonate system,

$$HCO_3^- + H^+ \longrightarrow H_2CO_3 \longrightarrow H_2O + CO_2$$

$$H_2CO_3 + {}^-OH \longrightarrow HCO_3^- + H_2O$$

For the biphosphate system,

$$HPO_4^{2-} + H^+ \longrightarrow H_2PO_4^-$$

$$H_2PO_4^- + {}^-OH \longrightarrow HPO_4^{2-} + H_2O$$

11.7 OTHER REACTIONS OF CARBOXYLIC ACIDS

Two other important reactions of carboxylic acids are their reduction to primary alcohols and their conversion to esters.

A. Reduction

Carboxylic acids are not reduced by catalytic hydrogenation. Other unsaturated groups in a molecule can be reduced without concurrent reduction of the carboxyl group.

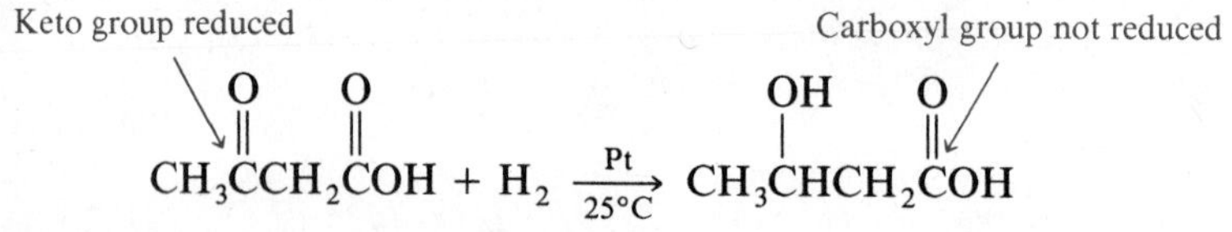

Keto group reduced   Carboxyl group not reduced

$$CH_3CCH_2COH + H_2 \xrightarrow[25°C]{Pt} CH_3CHCH_2COH$$

Carboxylic acids are readily reduced to primary alcohols, however, with the very reactive reducing agent lithium aluminum hydride ($Li^+ \ AlH_4^-$). This reagent also reduces other carbonyl groups such as keto groups but does not usually reduce carbon–carbon double bonds (see Section 10.9).

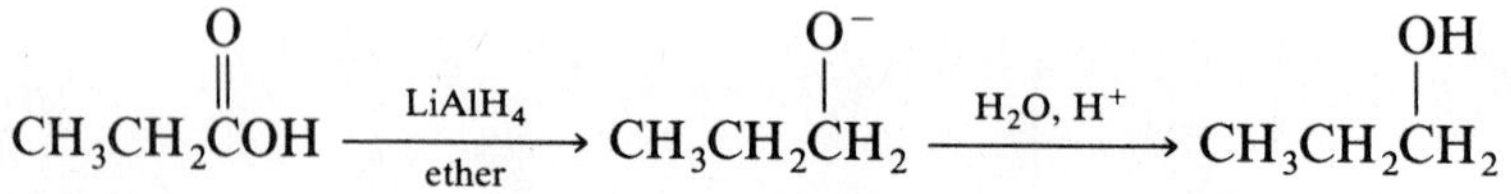

$$CH_3CH_2COH \xrightarrow[\text{ether}]{LiAlH_4} CH_3CH_2CH_2 \xrightarrow{H_2O, \ H^+} CH_3CH_2CH_2$$

Propanoic acid   Propoxide ion   1-Propanol

PROBLEM 11.15.   Write flow equations showing how you would prepare the following alcohols by the reduction of carboxylic acids:

(a) 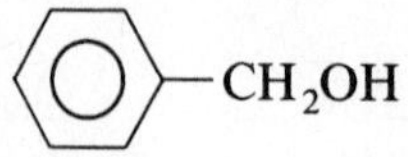 ⟨O⟩—CH₂OH   (b) CH₃CHCH₂CH₂OH   (two routes)

with OH above the CH in (b)

B. Esterification

The reaction of a carboxylic acid with an alcohol and a trace of strong acid catalyst (usually H_2SO_4) yields an **ester**, RCO_2R'. The reaction is called an **esterification reaction**.

General Equation for Esterification:

$$\underset{\text{A carboxylic acid}}{\overset{O}{\overset{\|}{RCOH}}} + \underset{\text{An alcohol}}{R'OH} \underset{}{\overset{H^+,\text{ heat}}{\rightleftharpoons}} \underset{\text{An ester}}{\overset{O}{\overset{\|}{RCOR'}}} + \underset{\text{Water}}{HOH}$$

In this reaction, the hydroxyl group of the carboxylic acid is replaced by the alkoxyl group ($-OR'$) of the alcohol.

Esterification reactions are *reversible*; therefore, the reaction mixture is an equilibrium mixture of reactants and products. To make this reaction useful for the synthesis of esters, we must drive the equilibrium to the ester side of the equation. We do this by adding an excess of one reactant or by removing one or both products as they are formed (by distillation, for example).

If the carboxylic acid or the alcohol is sterically hindered or if a phenol is used instead of an alcohol, the equilibrium favors the reactant side of the equation. Consequently, hindered esters or phenyl esters cannot be prepared by this direct esterification reaction. In Chapter 12 we will discuss some alternative ways of preparing these esters.

$$\underset{\text{Hindered reactants}}{(CH_3)_2CH\overset{O}{\overset{\|}{C}}OH + (CH_3)_3COH} \xrightarrow{H^+,\text{ heat}} \text{no appreciable ester formation}$$

$$CH_3\overset{O}{\overset{\|}{C}}OH + \langle\bigcirc\rangle\text{—OH} \xrightarrow{H^+,\text{ heat}} \text{no appreciable ester formation}$$

Phenol

C. Mechanism of Esterification

The mechanism of ester formation, shown in Figure 11.2, is a lengthy series of reaction steps beginning with a protonation step. Because protonation adds a positive charge to a carbonyl group, this group's reactivity toward weak nucleophiles (an alcohol, in this reaction) is enhanced.

Step 2 in the mechanism is the addition of the nucleophilic alcohol to the carbonyl group. The product of this step contains a protonated $-OR'$ group.

Step 3 is the loss of a proton from the $-OR'$ group, and step 4 is protonation of one of the $-OH$ groups (either one) to form a protonated hydroxyl group, $-OH_2^+$. Step 5 is the loss of this protonated hydroxyl group as the excellent leaving group H_2O.

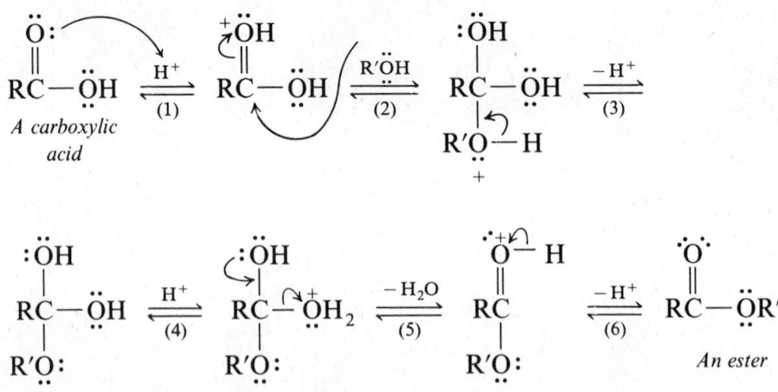

$$
\overset{\overset{\displaystyle \ddot{O}:}{\|}}{\underset{\substack{\ddot{} \\ \text{A carboxylic} \\ \text{acid}}}{RC-\ddot{O}H}}
\;\underset{(1)}{\overset{H^+}{\rightleftharpoons}}\;
\overset{\overset{\displaystyle {}^+\ddot{O}H}{\|}}{RC-\ddot{O}H}
\;\underset{(2)}{\overset{R'OH}{\rightleftharpoons}}\;
\overset{:\ddot{O}H}{\underset{\overset{|}{R'\ddot{O}-H}}{RC-\ddot{O}H}}
\;\underset{(3)}{\overset{-H^+}{\rightleftharpoons}}
$$

$$
\overset{:\ddot{O}H}{\underset{\overset{|}{R'\ddot{O}:}}{RC-\ddot{O}H}}
\;\underset{(4)}{\overset{H^+}{\rightleftharpoons}}\;
\overset{\ddot{O}H}{\underset{\overset{|}{R'\ddot{O}:}}{RC-{}^+\ddot{O}H_2}}
\;\underset{(5)}{\overset{-H_2O}{\rightleftharpoons}}\;
\overset{\overset{\displaystyle {}^+\ddot{O}-H}{\|}}{\underset{\overset{|}{R'\ddot{O}:}}{RC}}
\;\underset{(6)}{\overset{-H^+}{\rightleftharpoons}}\;
\underset{\text{An ester}}{\overset{\overset{\displaystyle \ddot{O}}{\|}}{RC-\ddot{O}R'}}
$$

Figure 11.2 Mechanism of acid-catalyzed esterification of a carboxylic acid with an alcohol.

The product of step 5 is a protonated ester, which loses its proton in step 6 to yield the product ester. All the protonation and deprotonation steps in the sequence are simply acid-base equilibria that occur in an acidic medium.

Let us summarize the mechanism to show the key intermediate and to emphasize the replacement of the —OH of the carboxylic acid by the —OR' of the alcohol.

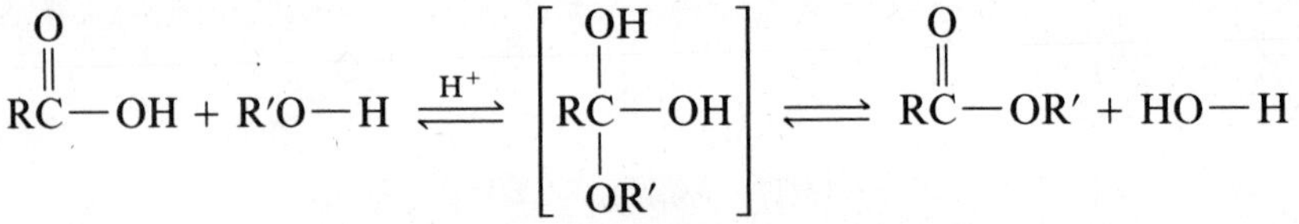

$$
\overset{\overset{\displaystyle O}{\|}}{RC-OH} + R'O-H
\;\overset{H^+}{\rightleftharpoons}\;
\left[\overset{OH}{\underset{\overset{|}{OR'}}{RC-OH}} \right]
\;\rightleftharpoons\;
\overset{\overset{\displaystyle O}{\|}}{RC-OR'} + HO-H
$$

Principal intermediate

Table 11.6 summarizes the reactions of carboxylic acids.

PROBLEM 11.16. Write a flow equation showing all the steps in the mechanism of the esterification of acetic acid with ethanol, as we have done for the general case in Figure 11.2.

PROBLEM 11.17. A pair of isotopes of an element are atoms containing the same number of protons, but a different number of neutrons. Two isotopes of oxygen are ^{16}O (containing 8 protons and 8 neutrons, atomic mass 16), which is the common naturally occurring isotope, and ^{18}O (containing 8 protons and 10 neutrons, atomic mass 18). When benzoic acid is esterified with methanol containing the isotope ^{18}O instead of the usual isotope ^{16}O, all the ^{18}O is found in the product ester and none is found in the water. Using the mechanism of esterification, explain why this is so.

$$\langle\bigcirc\rangle-\overset{\overset{\displaystyle O}{\|}}{C}OH + CH_3{}^{18}OH \underset{}{\overset{H^+}{\rightleftharpoons}} \langle\bigcirc\rangle-\overset{\overset{\displaystyle O}{\|}}{C}-{}^{18}OCH_3 + H_2O$$

TABLE 11.6   SUMMARY OF THE REACTIONS OF CARBOXYLIC ACIDS

Reaction	Product
Neutralization	
$RCOH + :B^{-a} \xrightarrow{-HB} RCO^-$ (with $\overset{O}{\|}$ carbonyls)	*A carboxylate*
Reduction	
$RCOH \xrightarrow[(2)\ H_2O,\ H^+]{(1)\ LiAlH_4} RCH_2$ (with $\overset{O}{\|}$ and OH)	*A 1° alcohol*
Esterification	
$RCOH + HOR' \underset{}{\overset{H^+,\ heat,\ -H_2O}{\rightleftharpoons}} RCOR'$ (with $\overset{O}{\|}$ carbonyls)	*An ester*

^{a} : B^- is a base, such as ^-OH, $HCO_3{}^-$, or RNH_2.

SUMMARY

Carboxylic acids can form strong multiple hydrogen bonds in their pure state or in water. In non-hydrogen-bonding solvents, carboxylic acid molecules form hydrogen-bonded dimers. The melting points and boiling points of carboxylic acids are relatively high.

In the IUPAC system, carboxylic acids are named as **-anoic acids**. Trivial names of low-formula-weight carboxylic acids are common. Examples are formic acid and acetic acid. In the trivial names, Greek letters can be used to designate a position of substitution—for example, α-chlorobutyric acid.

Carboxylic acids can be prepared by the **hydrolysis** of their derivatives: acid halides, acid anhydrides, esters, amides, or nitriles. Hydrolysis can occur with aqueous acid or base. **Oxidation reactions** of primary alcohols, aldehydes, alkenes containing an RCH= group, and alkyl aromatic compounds also yield carboxylic acids. **Grignard reactions** with carbon dioxide can yield a wide variety of types of carboxylic acids. Table 11.4 summarizes the synthesis of carboxylic acids.

The acidity of weak acids is measured by K_a or pK_a.

$$K_a = \frac{[RCO_2^-][H^+]}{[RCO_2H]} \qquad pK_a = -\log K_a$$

Increasing K_a or decreasing pK_a indicates increasing acid strength.

The principal reason for the acidity of carboxylic acids is the resonance stabilization of the carboxylate ion. A nearby substituent that withdraws electron density by the inductive effect also stabilizes the carboxylate ion and increases the strength of the carboxylic acid.

Stabilization relative to
the acid increases the
ionization and thus the
strength of the acid.

$$\underset{RCOH}{\overset{O}{\parallel}} \ \underset{H_2O}{\overset{}{\rightleftharpoons}} \ \underset{RCO^-}{\overset{O}{\parallel}} + H^+$$

Treatment of a carboxylic acid with relatively strong base yields carboxylates (RCO_2^-).

A **buffer solution** is a solution of a weak acid and its anion. A buffer solution can react with either added acid or added base and thus resists changes in pH.

Carboxylic acids can undergo *reduction* with lithium aluminum hydride or *esterification* with nonhindered alcohols. See Table 11.6 for equations for these reactions.

STUDY PROBLEMS

11.18. Draw formulas for the following compounds:
  (a) acetic acid
  (b) bromoacetic acid
  (c) β-bromopropionic acid
  (d) 3-bromopropanoic acid
  (e) *m*-methylbenzoic acid
  (f) 2,3-dichlorobutanedioic acid
  (g) *trans*-3-isopropylcyclohexanecarboxylic acid
  (h) *cis*-1,4-cyclohexanedicarboxylic acid
  (i) (*S*)-2-hydroxypropanoic acid
  (j) 3-methyl-4-hexenoic acid

11.19. Name the following compounds by the IUPAC system:

(a)
$$\underset{\underset{CH_3}{|}}{\overset{\overset{CH_3}{|}}{CH_3C}} - \underset{\underset{CH_3}{|}}{\overset{\overset{NO_2}{|}}{CH_2C}} \overset{\overset{O}{\parallel}}{CH_2COH}$$

(b) a benzene ring with CO_2H and Cl substituents

(c) $\underset{\displaystyle \underset{Cl}{|}}{HOC}CHCH_2\underset{\displaystyle \underset{Cl}{|}}{CH}COH$ (with two C=O groups)

(d) $\langle \bigcirc \rangle - CH_2\overset{\displaystyle O}{\overset{\|}{C}}OH$

(e) $(CH_3)_2CHCH_2\overset{\displaystyle O}{\overset{\|}{C}}O^- \ K^+$

(f) $\langle \bigcirc \rangle - \overset{\displaystyle O}{\overset{\|}{C}}O^- \ Na^+$

11.20. Complete the following equations, showing only the major organic products:

(a) $CH_3CH_2O\overset{\displaystyle O}{\overset{\|}{C}} - \langle \bigcirc \rangle + H_2O \ \xrightarrow[heat]{H^+}$

(b) $\langle \bigcirc \rangle - C \equiv N + H_2O \ \xrightarrow[heat]{H^+}$

(c) $CH_3CH_2CH = C(CH_3)_2 \ \xrightarrow[heat]{CrO_3 + H_2SO_4}$

(d) $CH_3CH_2MgCl \ \xrightarrow[(2)\ H_2O,\ H^+]{(1)\ CO_2}$

(e) $HO - \langle \bigcirc \rangle - CH_2OH \ \xrightarrow[heat]{CrO_3 + H_2SO_4}$

(f) $CH_3CH_2\overset{\displaystyle O}{\overset{\|}{C}}NH_2 \ \xrightarrow[heat]{H_2O,\ H^+}$

(g) $\langle \overset{O}{\underset{O}{\bigcirc}} {=} O \ \xrightarrow[(2)\ H^+]{(1)\ Na^+ \ ^-OH,\ H_2O,\ heat}$

(h) $CH_3 - \langle \bigcirc \rangle - CH_2CH_2 - \langle \bigcirc \rangle \ \xrightarrow[(2)\ H^+]{(1)\ K^+ \ MnO_4^-,\ heat}$

11.21. Write flow equations for the following conversions:

(a) $(CH_3)_2CHCH_2CH_2Cl \longrightarrow (CH_3)_2CHCH_2CH_2CO_2H$ (two routes)

(b) $(CH_3)_2CHCH_2CH_2Cl \longrightarrow (CH_3)_2CHCH_2CO_2H$

(c) $(CH_3)_2CHCH_2CH_2Cl \longrightarrow (CH_3)_2CHCO_2H$

11.22. Write an equation for the equilibrium between propanoic acid and water.

11.23. Write the equation for the definition of the acidity constant for propanoic acid in water.

11.24. The acidity constant of propanoic acid is 1.3×10^{-5}. In a $0.00100\ M$ solution of propanoic acid:

(a) What is the hydrogen ion concentration?

(b) What is the concentration of propanoate ion?

11.25. What is the pK_a corresponding to each of the following K_a values?

(a) 1.0×10^{-8} (b) 10.0×10^{-9}

(c) 0.010×10^{-5} (d) $1.0 \times 10^{-7.5}$

11.26. Convert each pK_a to K_a: **(a)** 6; **(b)** -6; **(c)** 0.

11.27. Which acid of each pair would be more acidic? Explain your answers:

(a) CH_3CO_2H or $BrCH_2CO_2H$

(b) $BrCH_2CH_2CO_2H$ or CH_3CHCO_2H
$\qquad\qquad\qquad\qquad\qquad\quad |$
$\qquad\qquad\qquad\qquad\qquad\quad Br$

(c) $BrCH_2CO_2H$ or Br_2CHCO_2H

(d) $BrCH_2CO_2H$ or FCH_2CO_2H

11.28. Using the K_a values, tell which is the strongest acid and which is the weakest.

(a) 1.8×10^{-5} **(b)** 2.5×10^{-5} **(c)** 3.8×10^{-6}

11.29. Using the pK_a values, tell which is the strongest acid and which is the weakest.

(a) 3.29   **(b)** 7.85   **(c)** 2.13

11.30. Complete the following equations. (If no appreciable reaction occurs, write *no reaction*.)

(a) $CH_3CH_2CO_2H + Na^+\ Cl^- \xrightarrow{H_2O}$

(b) $CH_3CH_2CO_2H + K^+\ {}^-OH \xrightarrow{H_2O}$

(c) $CH_3CH_2CO_2H + K^+\ HCO_3^- \xrightarrow{H_2O}$

(d) $CH_3CH_2CO_2H +$ ⟨◯⟩$-O^-\ K^+ \xrightarrow{H_2O}$

(e) $CH_3CH_2CO_2^- + CH_3CH_2NH_2 \xrightarrow{H_2O}$

(f) $CH_3CH_2CO_2^-\ CH_3\overset{+}{N}H_3 + K^+\ {}^-OH \xrightarrow{H_2O}$

(g) $CH_3CH_2CO_2^-\ CH_3\overset{+}{N}H_3 + HCl \xrightarrow{H_2O}$

(h) $CH_3CH_2CO_2H + CH_3CH_2NHCH_3 \xrightarrow{H_2O}$

(i) ⟨◯⟩$-\overset{\overset{\textstyle O}{\|}}{C}OH + H_2 \xrightarrow[25°C]{Pt}$

(j) ⟨◯⟩$-\overset{\overset{\textstyle O}{\|}}{C}OH \xrightarrow[\text{(2) } H_2O,\ H^+]{\text{(1) LiAlH}_4}$

(k) $CH_3\overset{\overset{\textstyle O}{\|}}{C}CH_2CH_2\overset{\overset{\textstyle O}{\|}}{C}OH \xrightarrow[\text{(2) } H_2O,\ H^+]{\text{(1) LiAlH}_4}$

(l) ⟨◯⟩$-\overset{\overset{\textstyle O}{\|}}{C}OH + CH_3CH_2CH_2OH \xrightarrow[\text{heat}]{H^+}$

(m) ⟨◯⟩$-OH + CH_3CH_2CO_2H \xrightarrow[\text{heat}]{H^+}$

11.31. How would you distinguish between the following pairs of compounds by simple chemical tests? Use equations in your answers.
(a) propanoic acid and 1-propanol
(b) propanoic acid and propanal
(c) propanoic acid and propanone (acetone)
(d) benzoic acid and phenol
(e) phenol and cyclohexanol
(f) acetic acid and ethyl acetate, $CH_3CO_2CH_2CH_3$
(g) acetic acid and sodium acetate

11.32. How would you make the following conversions?

(a) $HO_2CCH_2CO_2H \longrightarrow HOCH_2CH_2CH_2OH$

(b) $\longrightarrow HOCH_2(CH_2)_4CH_2OH$

(c) $\longrightarrow$ $-CO_2H$

(d) $CH_3\overset{O}{\overset{\|}{C}}OH \longrightarrow CH_3\overset{O}{\overset{\|}{C}}OCH_2CH_2O\overset{O}{\overset{\|}{C}}CH_3$

(e) $CH_3CH_2CH_2CH_2CO_2H \longrightarrow CH_3CH_2CH_2CH_2CH_2OH$

(f) $CH_3CH_2CH_2CH_2OH \longrightarrow CH_3CH_2CH_2CH_2CO_2H$

(g) $-CH_2OH$ as the only organic reactant $\longrightarrow$ $-\overset{O}{\overset{\|}{C}}OCH_2-$

(h) $(CH_3)_2C{=}CH_2 \longrightarrow CH_3\overset{\overset{\displaystyle CH_3}{|}}{\underset{\underset{\displaystyle CH_3}{|}}{C}}-CO_2H$

11.33. What is the acidity constant (K_a) of a compound if a 0.0200 M solution of the compound has a pH of 2.00?

11.34. Using formulas in your answer, explain why benzoic acid is a stronger acid than is phenol.

11.35. Write a flow equation for the mechanism of the esterification of acetic acid with (S)-2-butanol.

11.36. Explain the following observation: When 7-bromoheptanoic acid is treated with potassium hydroxide in ethanol and then heated with aqueous acid, the following compound is formed.

POINT OF INTEREST 11

Perfumes

A **perfume** is a liquid used to impart a fragrant or pleasing aroma. The term arises from the Latin *per* and *fumus*, meaning "through smoke," probably because the first perfumes were generated by burning plant gums, such as myrrh and frankincense. Today's perfumes are solutions of 10–25% perfume essences in ethanol.

The art of perfumery can be traced to the early Egyptians. The Romans brought the art to Gaul (now France), where it has flourished ever since. Perfume was used by the ancients for religious ceremonies and embalming. Of course, a major use was, and still is, enhancing the elegant images of both men and women.

A good perfume is more than just a fragrant liquid. It is an artistic creation capable of arousing emotions, affecting moods, and suggesting pleasure. The creator of a fine perfume must be able to discriminate among the odors of 500 to 2000 different chemicals of the about 5000 used in perfumery. The creator also must have a sense of odor harmony so that he or she can blend the components to achieve a theme, such as an aroma suggestive of forest earth or of spice.

The principal natural components of perfumes are **essential oils**, obtained from plants by steam distillation; **flower oils**, obtained from flowers by extraction with a solvent; **natural extracts**, obtained from plant gums, resins, balsams, mosses, and so forth; and **animal fixatives**, which are products such as ambergris (a rare secretion from sperm whales), castoreum (from beavers), civet (from the Ethiopian civet cat), and musk (from the Asian male musk deer). Today, many of these natural products can be mimicked by skillful blends of synthetic mixtures.

Each perfume contains three groups of components, called the **top note**, the **middle note**, and the **end note**. The combination of these three groups provides the theme and the harmony of the perfume.

The top-note components of a perfume are the most volatile. Their odors are the first perceived when a perfume is used. The top notes of perfumes may arise from esters of carboxylic acids, which often have sweet fruity odors.

γ-Undecalactone

*a cyclic ester with
jasmine odor*

p-Cresyl acetate

heavy floral odor

Geranyl acetate

fresh leafy odor

The fragrance of the top note disappears quickly, unveiling the aroma of the less volatile middle note. A middle note is often a flowery enhancer of the end note. Typical components used to formulate a middle note are 2-phenylethanol and rhodinol. Natural oils such as oil of roses (attar of rose) are also used.

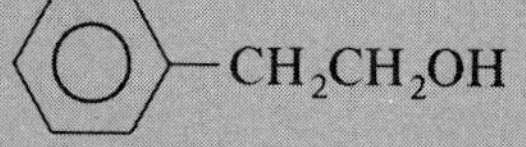

2-Phenylethanol 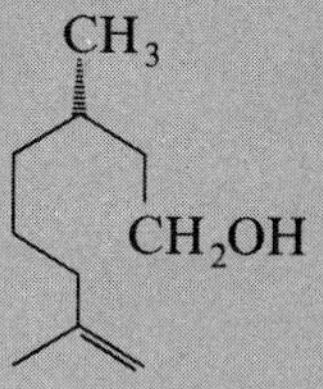   Rhodinol

The end note of a perfume is the least volatile part and thus the long-lasting odor that stays with the user. Natural plant products, such as the heavy essential oils of sandalwood or patchouli, and animal products, such as castoreum, civet, or natural musk, are used. In less expensive perfumes, synthetic products, such as synthetic civetone or musk xylene, are used.

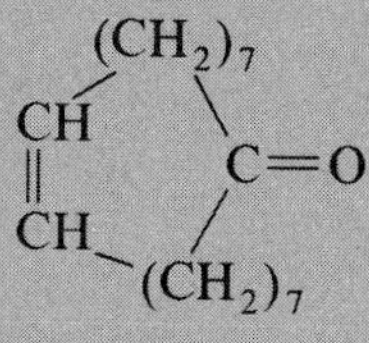

Civetone

*the principal component
of civet*

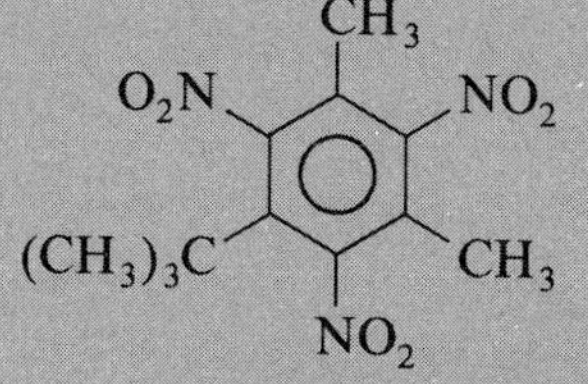

Musk xylene

a synthetic musk

Chapter 12

Derivatives of Carboxylic Acids

A derivative of a carboxylic acid is a compound that yields a carboxylic acid when it is hydrolyzed. In this chapter we discuss carboxylic acid halides, anhydrides, esters, and amides.

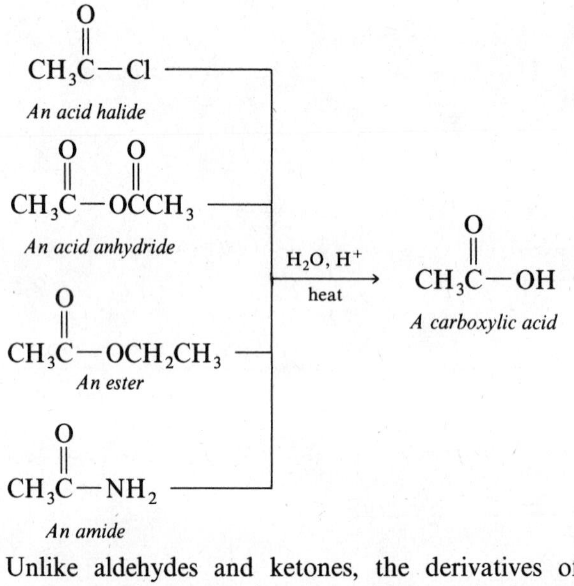

$$CH_3\overset{\displaystyle O}{\overset{\displaystyle \|}{C}}-Cl$$

An acid halide

$$CH_3\overset{\displaystyle O}{\overset{\displaystyle \|}{C}}-O\overset{\displaystyle O}{\overset{\displaystyle \|}{C}}CH_3$$

An acid anhydride

$$CH_3\overset{\displaystyle O}{\overset{\displaystyle \|}{C}}-OCH_2CH_3$$

An ester

$$CH_3\overset{\displaystyle O}{\overset{\displaystyle \|}{C}}-NH_2$$

An amide

$$\xrightarrow[\text{heat}]{H_2O,\ H^+}$$

$$CH_3\overset{\displaystyle O}{\overset{\displaystyle \|}{C}}-OH$$

A carboxylic acid

Unlike aldehydes and ketones, the derivatives of carboxylic acids contain *leaving groups*, electronegative groups that can be lost as anions (X^- or RCO_2^-) or as protonated anions (ROH or R_2NH). Recall that weak bases are better leaving groups than strong bases.

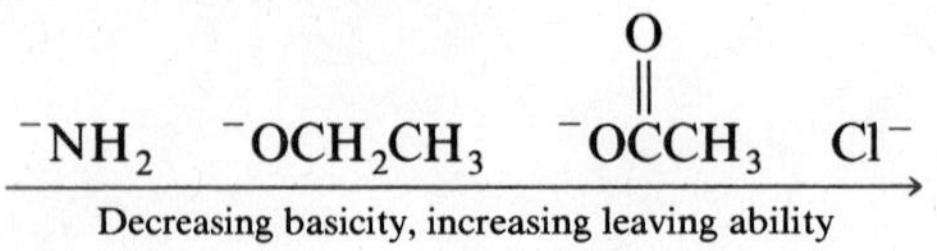

$$\xrightarrow{}$$

$$^-NH_2 \qquad ^-OCH_2CH_3 \qquad ^-O\overset{\displaystyle O}{\overset{\displaystyle \|}{C}}CH_3 \qquad Cl^-$$

Decreasing basicity, increasing leaving ability

The reactivity of the carboxylic acid derivatives directly parallels the leaving ability of the group bonded to the carbonyl carbon.

$$\underset{\text{Increasing reactivity toward water and other nucleophiles} \longrightarrow}{\underset{\displaystyle RC-NH_2 \quad\quad RC-OR' \quad\quad RC-OCR' \quad\quad RC-X}{\overset{\displaystyle O}{\overset{\displaystyle \|}{}} \quad\quad \overset{\displaystyle O}{\overset{\displaystyle \|}{}} \quad\quad \overset{\displaystyle O \quad O}{\overset{\displaystyle \| \quad \|}{}} \quad\quad \overset{\displaystyle O}{\overset{\displaystyle \|}{}}}}$$

Acid halides are sufficiently reactive toward water that they are never found in nature. Anhydrides are only rarely found. Their reactivity, however, makes both these derivatives valuable laboratory reagents.

Esters and amides are less reactive and are common in nature. Proteins are polyamides. Fats and waxes are high-formula-weight esters. Low-formula-weight esters occur in many fruits and are, to a large extent, responsible for flavors and odors of fruit.

$$\overset{\displaystyle O}{\overset{\displaystyle \|}{CH_3COCH_2CH_2CH_3}} \qquad\qquad \overset{\displaystyle O}{\overset{\displaystyle \|}{CH_3CH_2CH_2COCH_2CH_3}}$$

n-Propyl acetate   Ethyl butyrate

odor of pears   *odor of pineapple*

We will begin our discussion of carboxylic acid derivatives with the most reactive acid halides and then progress to the acid anhydrides, esters, and amides.

12.1 CARBOXYLIC ACID HALIDES

A. Nomenclature

Acid halides are also called **acyl halides** because they have the structure of an acyl group bonded to a halide.

Acyl groups   *Acid halides, or acyl halides*

In either the IUPAC system or trivial nomenclature systems, acid halides are named after their parent carboxylic acids with the *-ic acid* ending changed to *-yl halide*.

$$CH_3\overset{\overset{\displaystyle O}{\|}}{C}-OH \qquad\qquad CH_3\overset{\overset{\displaystyle O}{\|}}{C}-Cl$$

IUPAC:	Ethanoic acid	Ethanoyl chloride
Trivial:	Acetic acid	Acetyl chloride

$$CH_3CH_2CH_2\overset{\overset{\displaystyle O}{\|}}{C}-OH \qquad\qquad CH_3CH_2CH_2\overset{\overset{\displaystyle O}{\|}}{C}-Br$$

IUPAC:	Butanoic acid	Butanoyl bromide
Trivial:	Butyric acid	Butyryl bromide

PROBLEM 12.1.   Name the following acid halides:

(a) 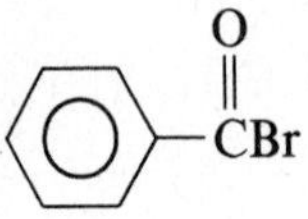 $\langle\bigcirc\rangle-\overset{\overset{\displaystyle O}{\|}}{C}Br$ **(b)** $CH_3CH_2CH_2CH_2\overset{\overset{\displaystyle O}{\|}}{C}Cl$

B. Preparation of Acid Halides

Because of their reactivity and because of their easy preparation in the laboratory, acid chlorides and acid bromides are useful reactants in the organic laboratory. These are prepared by treatment of a carboxylic acid with a commercially available active halogenating agent, such as $SOCl_2$, PCl_5, or PBr_3.

General Equations for Acid Halide Formation:

$$3R\overset{\overset{\displaystyle O}{\|}}{C}OH + \underset{\substack{\textit{A phosphorus} \\ \textit{trihalide}}}{PX_3} \longrightarrow 3R\overset{\overset{\displaystyle O}{\|}}{C}X + H_3PO_3$$

$$R\overset{\overset{\displaystyle O}{\|}}{C}OH + \underset{\textit{Thionyl chloride}}{Cl\overset{\overset{\displaystyle O}{\|}}{S}Cl} \longrightarrow R\overset{\overset{\displaystyle O}{\|}}{C}Cl + SO_2 + HCl$$

PROBLEM 12.2. Complete the following equations for specific examples of acid halide formation.

(a) $CH_3\overset{\displaystyle O}{\overset{\displaystyle \|}{C}}OH + PCl_3 \longrightarrow$

(b) $\overset{\displaystyle O}{\overset{\displaystyle \|}{C}}OH + SOCl_2 \longrightarrow$

(c) $HO\overset{\displaystyle O}{\overset{\displaystyle \|}{C}}-\overset{\displaystyle O}{\overset{\displaystyle \|}{C}}OH + PBr_3 \longrightarrow$

C. Reactions of Acid Halides

The carbonyl carbon is a partially positive site in a molecule, and halide ions are excellent leaving groups. Therefore, acid halides can be attacked by weak nucleophiles such as water and alcohols in a type of reaction called a **nucleophilic acyl substitution reaction**. These reactions are addition-elimination reactions, similar to reactions you have already encountered.

The mechanism for nucleophilic acyl substitution reactions of acid halides is general.

Reaction with Water   The reaction of an acid halide with water proceeds by the general addition-elimination mechanism shown. An additional step in the mechanism of the hydrolysis is the final loss of a proton from the $-\overset{+}{O}H_2$ group.

Reaction with Alcohols The reaction of an acid halide with an alcohol or a phenol proceeds by a path similar to that of hydrolysis and yields an ester. This reaction is an excellent route to all kinds of esters including sterically hindered esters and phenyl esters. These hindered esters and phenyl esters cannot be prepared by direct esterification of carboxylic acids (see Section 11.7B).

Synthesis of a Hindered Ester:

$$
\underset{\text{Acetyl chloride}}{CH_3\overset{\displaystyle O}{\overset{\|}{C}}-Cl} + \underset{\text{t-Butyl alcohol}}{(CH_3)_3C-OH} \longrightarrow \underset{\text{t-Butyl acetate}}{CH_3\overset{\displaystyle O}{\overset{\|}{C}}-OC(CH_3)_3} + HCl
$$

Synthesis of a Phenyl Ester:

$$
\underset{\text{Acetyl chloride}}{CH_3\overset{\displaystyle O}{\overset{\|}{C}}-Cl} + \underset{\text{Phenol}}{\langle \bigcirc \rangle-OH} \longrightarrow \underset{\text{Phenyl acetate}}{CH_3\overset{\displaystyle O}{\overset{\|}{C}}O-\langle \bigcirc \rangle} + HCl
$$

PROBLEM 12.3. Write a flow equation to show the mechanism of the reaction of acetyl chloride with methanol. (*Hint:* Refer to the general mechanism at the start of Section 12.1C and to the mechanism for the reaction with water.)

In these esterification reactions, a tertiary amine or pyridine is often added to the reaction mixture to remove the product hydrogen halide. The amine undergoes a simple acid-base reaction with the hydrogen halide to yield an amine salt. An amine used to remove an acid from a reaction mixture is sometimes referred to as an **acid scavenger**.

Removal of HX:

$$
\underset{\text{A 3° amine}}{R_3N\colon} + H-\overset{..}{\underset{..}{X}}\colon \longrightarrow \underset{\text{An amine salt}}{R_3\overset{+}{N}-H \ \colon\overset{..}{\underset{..}{X}}\colon^-}
$$

$$
\underset{\text{Pyridine}}{\langle \text{N} \rangle} + H-\overset{..}{\underset{..}{Cl}}\colon \longrightarrow \underset{\text{Pyridinium chloride}}{\langle \overset{+}{\underset{H}{N}} \rangle \ \colon\overset{..}{\underset{..}{Cl}}\colon^-}
$$

Reaction with Ammonia and Amines Acid chlorides react with ammonia, with primary amines (RNH_2), and with secondary amines (R_2NH) to yield amides. Tertiary amines and pyridine do not yield amides when treated with an acid halide because they do not have an NH group. This is the reason that these amines can be used as acid scavengers in esterification reactions of acid halides. In the reaction leading to an amide, excess ammonia or amine is used to react with the product hydrogen halide.

Reaction with NH_3 Yields Amides of the Type $\overset{\displaystyle O}{\overset{\|}{RCNH_2}}$:

$$CH_3\overset{\ddot{O}:}{\overset{\|}{C}}-Cl \ + \ :NH_3 \longrightarrow CH_3\overset{:\ddot{O}:^-}{\overset{|}{C}}-\overset{..}{C}l: \longrightarrow CH_3\overset{\ddot{O}:}{\overset{\|}{C}}-\overset{+}{N}H_3 + :\overset{..}{\underset{..}{C}}l:^-$$

$$\qquad\qquad\qquad\qquad\qquad\qquad\overset{|}{{}^+NH_3} \qquad\qquad\qquad \Big\downarrow {}_{NH_3}$$

$$\qquad\qquad\qquad\qquad\qquad\qquad\qquad\qquad CH_3\overset{\ddot{O}\cdot}{\overset{\|}{C}}NH_2 + \overset{+}{N}H_4$$

Reaction with 1° Amines Yields $\overset{\displaystyle O}{\overset{\|}{RCNHR'}}$:

$$\qquad\qquad\qquad\qquad\qquad\qquad A\ 1°\ amine$$

Reaction with a 2° Amine Yields $\overset{\displaystyle O}{\overset{\|}{RCNR'_2}}$:

$$CH_3\overset{O}{\overset{\|}{C}}Br \ + \ 2(CH_3)_2NH \longrightarrow CH_3\overset{O}{\overset{\|}{C}}N(CH_3)_2 + (CH_3)_2\overset{+}{N}H_2\ Cl^-$$

Table 12.1 summarizes the reactions of acid halides.

PROBLEM 12.4. Complete the following equations:

(a) $CH_3CH_2\overset{O}{\overset{\|}{C}}Cl +$ $-OH \longrightarrow$

Step 1, Saponification:

$$
\underset{\text{An ester}}{\overset{\overset{\displaystyle O}{\|}}{RCOR'}} + {}^{-}OH \xrightarrow[\text{heat}]{H_2O} \underset{\text{A carboxylate}}{\overset{\overset{\displaystyle O}{\|}}{RCO^{-}}} + \underset{\text{An alcohol}}{HOR'}
$$

Step 2, Acidification:

$$
\overset{\overset{\displaystyle O}{\|}}{RCO^{-}} + H^{+} \longrightarrow \underset{\text{A carboxylic acid}}{\overset{\overset{\displaystyle O}{\|}}{RCOH}}
$$

The mechanism for saponification is similar to those for other nucleophilic acyl substitutions—addition followed by elimination. A proton transfer step converts the elimination products into the observed products—carboxylate and alcohol.

$$
\underset{\ }{\overset{\overset{\displaystyle \ddot{O}:}{\|}}{RC{-}OR'}} + {}^{-}:\ddot{O}H \underset{}{\overset{\text{addition}}{\rightleftharpoons}} \underset{:OH}{RC{-}\overset{:\ddot{O}:{}^{-}}{|}OR'} \xrightarrow{\text{elimination}}
$$

$$
\underset{\ }{\overset{\overset{\displaystyle \ddot{O}:}{\|}}{RC{-}\ddot{O}H}} + {}^{-}:\ddot{O}R' \xrightarrow[\text{(not reversible)}]{\text{proton transfer}} \underset{\ }{\overset{\overset{\displaystyle \dot{\ddot{O}}}{\|}}{RC{-}\ddot{O}:{}^{-}}} + HOR'
$$

The irreversibility of the final proton transfer step is the reason that saponification is an irreversible reaction. The product carboxylate ion is far too weak a base to remove a proton from an alcohol (the reverse of the last step).

Specific Example of Saponification:

Methyl benzoate → Sodium benzoate + Methanol

Benzoic acid

PROBLEM 12.11. Predict the organic products:

$$
\textbf{(a)}\quad CH_3CH_2\overset{\overset{\displaystyle O}{\|}}{C}OCH(CH_3)_2 \xrightarrow[\text{(2) } H_2O, H^{+}]{\text{(1) } {}^{-}OH, H_2O, \text{ heat}}
$$

(b) $\underset{\parallel}{\overset{O}{C}}H_3CH_2O\overset{O}{\overset{\parallel}{C}}CH_2\overset{O}{\overset{\parallel}{C}}OCH_2CH_3 \xrightarrow[\text{heat}]{^-OH, H_2O}$

(c) $\langle\bigcirc\rangle-\overset{O}{\overset{\parallel}{C}}O-\underset{\overset{|}{CH_3}}{\overset{(R)}{\overset{\overset{CH_2CH_3}{\diagup}}{C}}}\overset{\diagdown}{\underset{H}{\text{\tiny ...}}} \xrightarrow[(2)\ H_2O,\ H^+]{(1)\ ^-OH,\ H_2O,\ \text{heat}}$

[*Hint:* In (c), consider the mechanism and how this will affect the stereochemistry of the product alcohol.]

Ammonolysis   Esters undergo reaction with ammonia or amines to yield amides by a reaction called *ammonolysis*. The mechanism is very similar to that for saponification.

General Equation for Ammonolysis:

$$RC\overset{\ddot{O}:}{\overset{\parallel}{-}}\ddot{O}R' \ + \ :NH_3 \xrightarrow{\text{addition}} RC\overset{:\ddot{O}:^-}{\underset{^+NH_3}{\overset{|}{-}}}\ddot{O}R' \xrightarrow{\text{elimination}}$$

$$RC\overset{\ddot{O}:}{\overset{\parallel}{-}}\overset{+}{N}H_3 \ + \ ^-:\ddot{O}R' \xrightarrow[-R'OH]{\text{proton transfer}} RC\overset{\dot{\overset{\cdot\cdot}{O}}}{\overset{\parallel}{-}}\ddot{N}H_2$$

An amide

Specific Example of Ammonolysis:

$$ClCH_2\overset{O}{\overset{\parallel}{C}}OCH_2CH_3 + NH_3 \xrightarrow[1h]{0°} ClCH_2\overset{O}{\overset{\parallel}{C}}NH_2 + CH_3CH_2OH$$

Ethyl 2-chloroacetate   2-Chloroacetamide   Ethanol

Grignard Reactions   An ester contains a carbonyl group that can be attacked by a Grignard reagent. The intermediate product is a ketone, which undergoes further reaction with the Grignard reagent. The final product, after hydrolysis, is a tertiary alcohol.

Step 1, Attack of First Mole of Grignard Reagent:

$$\overset{\ddot{\text{O}}:}{\underset{\displaystyle \text{RC} - \text{OR}'}{\|}} + \text{R}'' - \text{MgX} \xrightarrow{\text{addition}} \underset{\displaystyle \underset{\text{R}''}{|}}{\overset{\displaystyle :\ddot{\text{O}}:^{-}}{\underset{\displaystyle \text{RC} - \ddot{\text{O}}\text{R}'}{|}}} \xrightarrow[\;\; -\;^{-}\text{OR}' \;\;]{\text{elimination}} \overset{\ddot{\text{O}}:}{\underset{\displaystyle \text{RC} - \text{R}''}{\|}}$$

Steps 2 and 3, Attack of Second Mole and Hydrolysis:

$$\overset{\ddot{\text{O}}:}{\underset{\displaystyle \text{RC} - \text{R}''}{\|}} + \text{R}'' - \text{MgX} \longrightarrow \underset{\displaystyle \underset{\text{R}''}{|}}{\overset{\displaystyle :\ddot{\text{O}}:^{-}}{\underset{\displaystyle \text{RC} - \text{R}''}{|}}} \xrightarrow{\text{H}_2\text{O, H}^+} \underset{\displaystyle \underset{\text{R}''}{|}}{\overset{\displaystyle \text{OH}}{\underset{\displaystyle \text{RC} - \text{R}''}{|}}}$$

Two groups the same—
from R''MgX

Overall Reaction:

$$\overset{\text{O}}{\underset{\displaystyle \text{RCOR}'}{\|}} \xrightarrow[\text{(2) H}_2\text{O, H}^+]{\text{(1) 2R''MgX}} \underset{\displaystyle \underset{\text{R}''}{|}}{\overset{\displaystyle \text{OH}}{\underset{\displaystyle \text{R} - \text{C} - \text{R}''}{|}}} + \text{R}'\text{OH}$$

An ester   *A 3° alcohol*

The product tertiary alcohol contains two identical groups (R'') bonded to the hydroxyl's carbon; both these R'' groups came from the Grignard reagent (R''MgX).

EXAMPLE

How would you synthesize 2-methyl-2-butanol from an ester?

Solution:

1. Write the formula, and determine whether the hydroxyl carbon is bonded to two groups that are the same. Here, the two groups are $-\text{CH}_3$; therefore, the Grignard reagent must be CH_3MgX.

$$\underset{\displaystyle \underset{\text{CH}_3}{|}}{\overset{\displaystyle \text{OH}}{\underset{\displaystyle \text{CH}_3\text{CH}_2 - \text{C} - \text{CH}_3}{|}}} \xleftarrow{\text{From CH}_3\text{MgX}}$$

2. Use the rest of the formula to determine the ester needed.

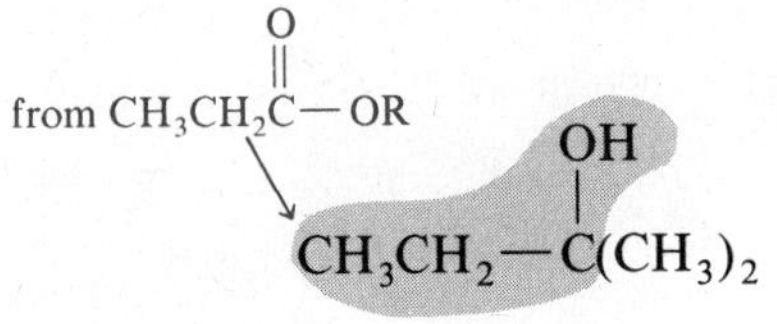

$$\text{from } CH_3CH_2\overset{\displaystyle O}{\overset{\displaystyle \|}{C}}-OR$$

$$\underset{\displaystyle CH_3CH_2-\overset{\displaystyle OH}{\overset{\displaystyle |}{C}}(CH_3)_2}{}$$

3. Write a flow equation.

$$CH_3CH_2\overset{\displaystyle O}{\overset{\displaystyle \|}{C}}OCH_2CH_3 \xrightarrow[\text{(2) } H_2O, H^+]{\text{(1) } 2CH_3MgI} CH_3CH_2\overset{\displaystyle OH}{\underset{\displaystyle CH_3}{\overset{\displaystyle |}{C}}}-CH_3$$

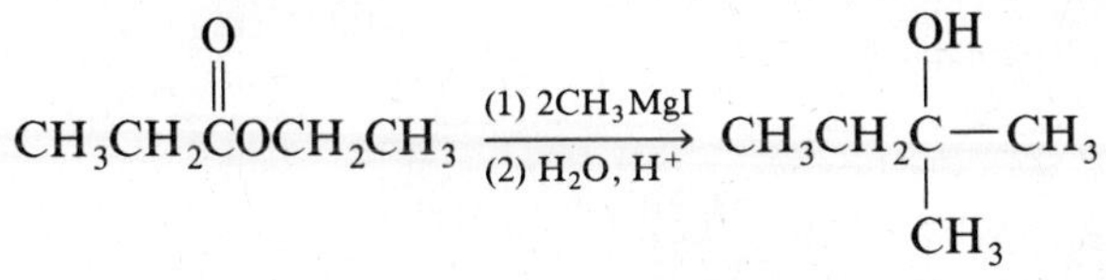

This tertiary alcohol could also have been synthesized from a ketone.

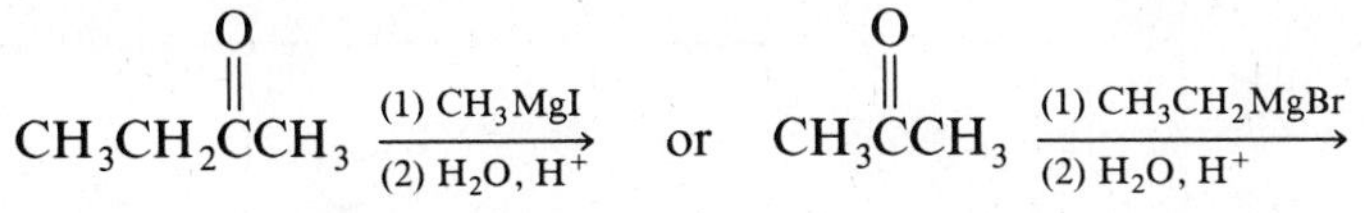

$$CH_3CH_2\overset{\displaystyle O}{\overset{\displaystyle \|}{C}}CH_3 \xrightarrow[\text{(2) } H_2O, H^+]{\text{(1) } CH_3MgI} \quad \text{or} \quad CH_3\overset{\displaystyle O}{\overset{\displaystyle \|}{C}}CH_3 \xrightarrow[\text{(2) } H_2O, H^+]{\text{(1) } CH_3CH_2MgBr}$$

In summary, two types of Grignard reactions can be used to prepare tertiary alcohols: (1) reaction of Grignard reagents with ketones and (2) reaction of Grignard reagents with esters.

PROBLEM 12.12.   Write flow equations showing how you would synthesize the following compounds from esters:

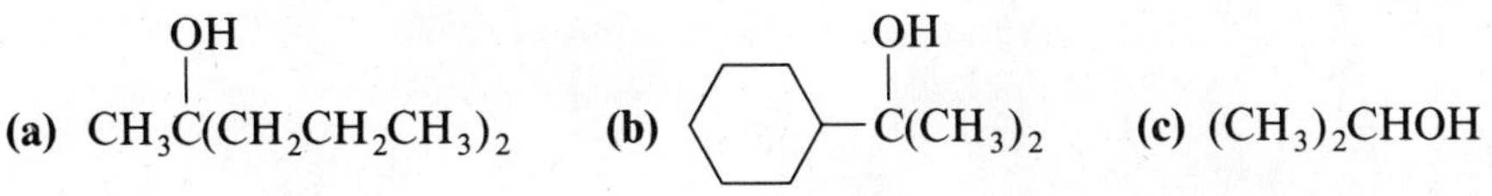

(a) $CH_3\overset{\displaystyle OH}{\overset{\displaystyle |}{C}}(CH_2CH_2CH_3)_2$ (b) ⟨hexagon⟩$-\overset{\displaystyle OH}{\overset{\displaystyle |}{C}}(CH_3)_2$ (c) $(CH_3)_2CHOH$

Reduction   Esters can be reduced by catalytic hydrogenation using hydrogen gas, an appropriate catalyst, heat, and pressure or by reaction with lithium aluminum hydride followed by hydrolysis. In either type of reduction, the organic products are *two alcohols*. The alcohol from the carbonyl portion of the ester must be a *primary alcohol*.

General Equation for Ester Reduction:

$$\underset{\substack{\uparrow \\ \text{To a 1° alcohol}}}{RC-OR'} \xrightarrow{[H]} RCH_2OH + HOR'$$

$$\underset{A\ 1°\ alcohol}{} \qquad \underset{1°,\ 2°,\ or\ 3°}{}$$

Specific Examples:

$$CH_3\overset{O}{\overset{\|}{C}}OCH_2CH_3 \xrightarrow[\text{heat, pressure}]{H_2,\ \text{catalyst}} CH_3CH_2OH + HOCH_2CH_3$$

$$CH_3CH_2O\overset{O}{\overset{\|}{C}}CH_2\overset{O}{\overset{\|}{C}}OCH_2CH_3 \xrightarrow[\text{(2) H}_2\text{O, H}^+]{\text{(1) LiAlH}_4}$$

$$CH_3CH_2OH + HOCH_2CH_2CH_2OH + HOCH_2CH_3$$

PROBLEM 12.13. Predict the organic products:

(a) $CH_3CH_2\overset{O}{\overset{\|}{C}}OCH_2(CH_2)_4CH_3 \xrightarrow[\text{(2) H}_2\text{O, H}^+]{\text{(1) LiAlH}_4}$

(b) $(CH_3)_2CHO\overset{O}{\overset{\|}{C}}CH_2CH_3 \xrightarrow[\text{(2) H}_2\text{O, H}^+]{\text{(1) LiAlH}_4}$

(c) 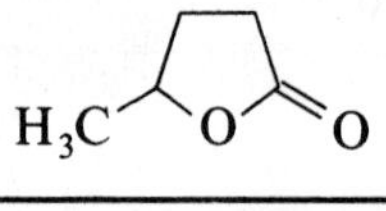 $\xrightarrow[\text{(2) H}_2\text{O, H}^+]{\text{(1) LiAlH}_4}$

Table 12.3 summarizes the reactions of esters.

D. Polyesters

A **polyester** is a polymer containing repetitive ester units. A polyester is synthesized by an esterification reaction between a carboxylic acid (or its derivative) containing more than one acyl group and an alcohol containing more than one hydroxyl group.

Dacron is a common polyester that is used as a fiber and to make Mylar sheeting. Dacron is made by a **transesterification reaction** (a reaction in which

TABLE 12.3   SUMMARY OF ESTER REACTIONS

Reactants[a]				Products		

$$\underset{\text{RC}-\text{OR}'}{\overset{\text{O}}{\|}} \quad + \quad \underset{\text{Water}}{\text{H}-\text{OH}} \quad \overset{\text{H}^+,\ \text{heat}}{\underset{\longleftarrow}{\longrightarrow}} \quad \underset{\substack{\textit{A carboxylic} \\ \textit{acid}}}{\overset{\text{O}}{\|}}\text{RC}-\text{OH} \quad + \quad \underset{\textit{An alcohol}}{\text{HOR}'}$$

$$\underset{\text{RC}-\text{OR}'}{\overset{\text{O}}{\|}} \quad + \quad \underset{\substack{ \\ \text{Hydroxide ion}}}{{}^-\text{OH}} \quad \overset{\text{heat}}{\longrightarrow} \quad \underset{\textit{A carboxylate}}{\overset{\text{O}}{\|}}\text{RC}-\text{O}^- \quad + \quad \text{HOR}'$$

$$\underset{\text{RC}-\text{OR}'}{\overset{\text{O}}{\|}} \quad + \quad \underset{\substack{\textit{Ammonia or} \\ \textit{an amine}}}{\text{NH}_3} \quad \longrightarrow \quad \underset{\textit{An amide}}{\overset{\text{O}}{\|}}\text{RC}-\text{NH}_2 \quad + \quad \text{HOR}'$$

$$\underset{\text{RC}-\text{OR}'}{\overset{\text{O}}{\|}} \quad + \quad \underset{\substack{\textit{A Grignard} \\ \textit{reagent}}}{2\text{R}''\text{MgX}^b} \quad \longrightarrow \overset{\text{H}_2\text{O},\ \text{H}^+}{\longrightarrow} \quad \underset{\textit{A 3° alcohol}}{\overset{\text{OH}}{\underset{|}{\text{RCR}''_2}}} \quad + \quad \text{HOR}'$$

$$\underset{\text{HC}-\text{OR}'}{\overset{\text{O}}{\|}} \quad + \quad \underset{\substack{\textit{A Grignard} \\ \textit{reagent}}}{2\text{R}''\text{MgX}^b} \quad \longrightarrow \overset{\text{H}_2\text{O},\ \text{H}^+}{\longrightarrow} \quad \underset{\textit{A 2° alcohol}}{\overset{\text{OH}}{\underset{|}{\text{HCR}''_2}}} \quad + \quad \text{HOR}'$$

$$\underset{\text{RC}-\text{OR}'}{\overset{\text{O}}{\|}} \quad + \quad \underset{\substack{\textit{A reducing} \\ \textit{agent}}}{\text{LiAlH}_4^b} \quad \longrightarrow \overset{\text{H}_2\text{O},\ \text{H}^+}{\longrightarrow} \quad \underset{\textit{Alcohols}}{\overset{\text{OH}}{\underset{|}{\text{RCH}_2}}} \quad + \quad \text{HOR}'$$

[a] R, R′, and R″ can be alkyl, aryl, etc.

[b] The second reaction arrow indicates that hydrolysis is a separate reaction.

alcohols are exchanged in an ester). In this reaction, a methyl diester is treated with 1,2-ethanediol.

$$CH_3O-\overset{\overset{\textstyle O}{\|}}{C}-\underset{}{\bigcirc}-\overset{\overset{\textstyle O}{\|}}{C}-OCH_3 + 2HOCH_2CH_2OH \xrightarrow[-2CH_3OH]{H^+}$$

Dimethyl terephthalate   1,2-Ethanediol
(ethylene glycol)

$$HOCH_2CH_2O-\overset{\overset{\textstyle O}{\|}}{C}-\underset{}{\bigcirc}-\overset{\overset{\textstyle O}{\|}}{C}-OCH_2CH_2OH \xrightarrow{many\ steps}$$

Can react with more
dimethyl terephthalate

$$\left\{-OC-\bigcirc-COCH_2CH_2\left[\overset{O}{OC}-\bigcirc-\overset{O}{COCH_2CH_2}\right]_n OC-\bigcirc-C-\right\}$$

A Dacron chain contains about 100 repeating
ethylene-terephthalate units.

12.4 CARBOXYLIC ACID AMIDES

Amides are structurally similar to esters, but they contain a nitrogen group instead of an oxygen group bonded to the carbonyl carbon. Amides undergo reactions similar to those of esters, but they are slightly less reactive. One reason for the lesser reactivity is the importance of resonance stabilization of the amide group.

Resonance Structures for an Amide:

$$\left[R-\overset{\overset{\textstyle :\ddot{O}:}{\diagup}}{C}\underset{\overset{\textstyle :}{NR'_2}}{} \longleftrightarrow R-\overset{\overset{\textstyle :\ddot{O}:^-}{\diagup}}{C}\underset{\overset{\textstyle +}{NR'_2}}{}\right] \quad or \quad R-\overset{\overset{\textstyle O^{\delta-}}{\diagup}}{C}\underset{\overset{\textstyle NR'_2}{\delta+}}{}$$

When treated with dilute acid, amides do not form salts as amines do. The reason is that the "unshared" valence electrons of an amide nitrogen are used in the partial double bond and are thus not available for donation.

Electrons available
for donation

An amine: $R_3N:$ $+ \ H{-}Cl \longrightarrow R_3\overset{+}{N}{-}H \quad Cl^-$

An amide: $RC\underset{\cdot\cdot}{N}R'_2$ $+ \ H{-}Cl \ \xrightarrow{H_2O} \ $ no salt formation

Electrons not available
for donation

Amides containing an N—H group can undergo hydrogen bonding. These amides have higher melting points and boiling points than amides containing an —NR$_2$ group.

No N—H to form
hydrogen bonds

$CH_3\overset{O}{\overset{\|}{C}}NH_2$ $CH_3\overset{O}{\overset{\|}{C}}NHCH_3$ $CH_3\overset{O}{\overset{\|}{C}}N(CH_3)_2$

Acetamide   N-Methylacetamide   N,N-Dimethylacetamide
(mp 82°C, bp 221°C)   (mp 28°C, bp 204°C)   (mp −20°C, bp 165°C)

PROBLEM 12.14. Write formulas for two molecules of acetamide to show a hydrogen bond between them.

A. Nomenclature of Amides

The name of an amide with a —NH$_2$ group is derived from the name of the parent carboxylic acid with the *-ic acid* or *-oic acid* ending changed to *-amide*.

$CH_3\overset{O}{\overset{\|}{C}}{-}OH$ $CH_3\overset{O}{\overset{\|}{C}}{-}NH_2$ $\langle\bigcirc\rangle{-}\overset{O}{\overset{\|}{C}}{-}OH$ $\langle\bigcirc\rangle{-}\overset{O}{\overset{\|}{C}}{-}NH_2$

Acetic acid   Acetamide   Benzoic acid   Benzamide

An amide can have one or two substituents besides the carbonyl group bonded to the N (—NHR or —NR$_2$). If the nitrogen is bonded to one alkyl group, the amide name is preceded by an *N-alkyl-* to show this. If the nitrogen is bonded to two groups, the amide name is preceded by *N, N*-dialkyl-. (*N*-Aryl groups are also possible.) The following examples show the naming of an *N*-methyl amide and an *N, N*-diethyl amide.

From acetamide

Methyl group bonded to N

$$CH_3\overset{\overset{\displaystyle O}{\|}}{C}NH{-}CH_3$$

N-Methylacetamide

$$\underset{\displaystyle}{\bigcirc}{-}\overset{\overset{\displaystyle O}{\|}}{C}N(CH_2CH_3)_2$$

N, N-Diethylbenzamide

B. Preparation of Amides

Amides are synthesized from other carboxylic acid derivatives and ammonia or amines. These reactions are discussed in Sections 12.1C, 12.2C, and 12.3C.

$$CH_3\overset{\overset{\displaystyle O}{\|}}{C}Cl \;+\; HN(CH_3)_2 \xrightarrow{-HCl} CH_3\overset{\overset{\displaystyle O}{\|}}{C}{-}\underset{\underbrace{\qquad}}{N(CH_3)_2} \;\;{-}NR_2$$

Acetyl chloride   *A 2° amine*   *N, N*-Dimethylacetamide

$$CH_3\overset{\overset{\displaystyle O}{\|}}{C}{-}O\overset{\overset{\displaystyle O}{\|}}{C}CH_3 \;+\; H_2NCH_3 \xrightarrow{-CH_3CO_2H} CH_3\overset{\overset{\displaystyle O}{\|}}{C}{-}\underset{\underbrace{\qquad}}{NHCH_3} \;\;{-}NHR$$

Acetic anhydride   *A 1° amine*   *N*-Methylacetamide

$$CH_3\overset{\overset{\displaystyle O}{\|}}{C}{-}OCH_2CH_3 \;+\; NH_3 \xrightarrow{-CH_3CH_2OH} CH_3\overset{\overset{\displaystyle O}{\|}}{C}{-}\underset{\underbrace{\qquad}}{NH_2} \;\;{-}NH_2$$

Ethyl acetate   Ammonia   Acetamide

C. Reactions of Amides

Hydrolysis   Amides, like esters, undergo hydrolysis when heated with aqueous acid or base. A fairly concentrated solution of acid or base and a long heating period may be necessary.

Hydrolysis in Aqueous Acid:

Product amine forms
a salt in acid

$$\underset{RC-NR_2' + H_2O + H^+}{\overset{O}{\parallel}} \xrightarrow{\text{heat}} \underset{RC-OH}{\overset{O}{\parallel}} + H_2\overset{+}{N}R_2'$$

Hydrolysis in Aqueous Base:

Carboxylate in base

$$\underset{RC-NR_2'}{\overset{O}{\parallel}} + {}^-OH \xrightarrow{\text{heat}} \underset{RCO^-}{\overset{O}{\parallel}} + HNR_2'$$

$$\xrightarrow{H^+} \underset{RCOH}{\overset{O}{\parallel}}$$

Specific Example of Amide Hydrolysis:

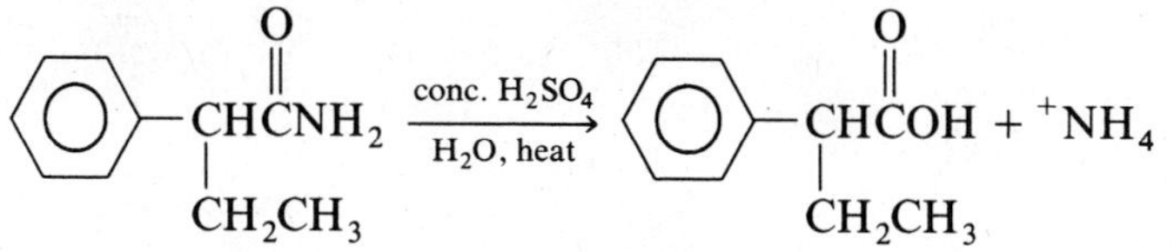

$$\bigcirc\!\!\!\!\bigcirc - \underset{CH_2CH_3}{\underset{|}{CH}}\overset{O}{\overset{\parallel}{C}}NH_2 \xrightarrow[H_2O, \text{ heat}]{\text{conc. } H_2SO_4} \bigcirc\!\!\!\!\bigcirc - \underset{CH_2CH_3}{\underset{|}{CH}}\overset{O}{\overset{\parallel}{C}}OH + {}^+NH_4$$

Reduction Amides undergo reduction with lithium aluminum hydride to yield amines. The type of amine product (primary, secondary, or tertiary) depends on the nitrogen substituents in the amide. In all these reductions, the carbonyl group is reduced to $-CH_2-$.

$$\underset{RCNH_2}{\overset{O}{\parallel}} \xrightarrow{[H]} RCH_2NH_2$$

A 1° amine

$$\underset{RCNHR'}{\overset{O}{\parallel}} \xrightarrow{[H]} RCH_2NHR'$$

A 2° amine

$$\underset{RCNR_2'}{\overset{O}{\parallel}} \xrightarrow{[H]} RCH_2NR_2'$$

A 3° amine

Table 12.4 summarizes the reactions of amides.

TABLE 12.4   SUMMARY OF AMIDE REACTIONS

Reactants[a]			Products		
$\underset{\displaystyle RC-NR'_2}{\overset{\displaystyle O \atop \displaystyle \|}{}}$	$+$	$\underset{\text{Water}}{H_2O}$	$\xrightarrow{\;H^+ \text{ or } {}^-OH\;}$	$\underset{\text{A carboxylic acid}}{\overset{\displaystyle O \atop \displaystyle \|}{RCOH}}$	$+\;\;\underset{\text{An amine}}{HNR'_2}{}^{\,b}$
$\underset{\displaystyle RCNR'_2}{\overset{\displaystyle O \atop \displaystyle \|}{}}$	$+$	$\underset{\text{A reducing agent}}{LiAlH_4}$	$\longrightarrow$	$\underset{\text{An amine}}{RCH_2NR'_2}$	

[a] R and R' can be alkyl, aryl, H, etc.

[b] In acidic solution, the amine is protonated. In alkaline solution, the carboxylate is formed instead of the carboxylic acid.

D. Some Interesting Amides

Because they are the least reactive of the carboxylic acid derivatives, amides are widely distributed in nature. They are also important compounds in commerce. **Proteins** are polyamides, as are the synthetic **nylons**.

Portion of a protein molecule   Portion of a nylon molecule

The **barbiturates**, commonly used as prescription sedatives, are cyclic amides that vary in the substituents on one carbon.

Phenobarbital

Urea is a compound excreted in the urine by mammals to rid the system of excess nitrogen from ingested protein. Urea is also used as a fertilizer and as a starting material for synthetic polymers and drugs, including the barbiturates. The closely related **carbamates**, compounds containing amide-ester groups, are found in drugs and insecticides.

$$H_2N-\overset{\overset{\displaystyle O}{\|}}{C}-NH_2$$

Urea

$$-O-\overset{\overset{\displaystyle O}{\|}}{C}-N\overset{\diagup}{\diagdown}$$

Carbamate group

$$\overset{\overset{\displaystyle O}{\|}}{O\,C\,NHCH_3}$$

Sevin

an insecticide

PROBLEM 12.15.   The following formula represents a dipeptide, a type of compound that can be obtained by the partial hydrolysis of proteins.

$$\overset{+}{H_3}NCH_2\overset{\overset{\displaystyle O}{\|}}{C}NH\underset{\underset{\displaystyle CH_3}{|}}{CH}\overset{\overset{\displaystyle O}{\|}}{C}O^-$$

Predict the organic products when this dipeptide is treated with each of the following sets of reagents:

(a) NaOH, H_2O, heat   **(b)** H_2SO_4, H_2O, heat
(c) (1) $LiAlH_4$, (2) H_2O, H^+

SUMMARY

Derivatives of carboxylic acids yield carboxylic acids when they are hydrolyzed.

Acid halides are named after the carboxylic acid with the ending *-yl halide* (chloride, bromide, iodide) replacing *-ic acid*. Anhydrides are named with the word *anhydride*.

$$CH_3\overset{\overset{\displaystyle O}{\|}}{C}-OH \qquad CH_3\overset{\overset{\displaystyle O}{\|}}{C}-Cl \qquad CH_3\overset{\overset{\displaystyle O}{\|}}{C}-O\overset{\overset{\displaystyle O}{\|}}{C}CH_3$$

Acetic acid   **Acetyl chloride**   **Acetic anhydride**

Esters are named as *alkyl* or *aryl -ates*, while amides are named with the ending *-amide* replacing *-ic acid* or *-oic acid*.

$$CH_3\overset{\overset{\displaystyle O}{\|}}{C}-OCH_2CH_3 \qquad CH_3\overset{\overset{\displaystyle O}{\|}}{C}-NH_2 \qquad CH_3\overset{\overset{\displaystyle O}{\|}}{C}-N(CH_3)_2$$

Ethyl **acetate**   **Acetamide**   *N,N*-Dimethyl**acet**amide

TABLE 12.5 SUMMARY OF PREPARATIONS OF CARBOXYLIC ACID DERIVATIVES

Reaction	Section reference

Acid halides

$$RCOOH + XSX,\ PX_3,\ \text{or}\ PX_5 \longrightarrow RCOX \qquad\qquad 12.1B$$

Acid anhydrides

$$RCOX + {}^-OCOR' \longrightarrow RC(O)-OCOR' \qquad\qquad 12.2B$$

$$2RCOOH + CH_3COCCH_3 \xrightarrow[-2CH_3CO_2H]{heat} RC(O)-OCOR \qquad\qquad 12.2B$$

Esters

$$RCOOH + H-OR' \overset{H^+,\ heat,\ -H_2O}{\rightleftharpoons} RCOOR'^{\,a} \qquad\qquad 11.7B,\ 12.3B$$

$$RCOX + H-OR' \xrightarrow{-HX} RCOOR' \qquad\qquad 12.1C,\ 12.3B$$

$$RC(O)-OCOR + H-OR' \xrightarrow{-RCO_2H} RCOOR' \qquad\qquad 12.2C,\ 12.3B$$

Amides

$$RCOX + 2HNR'_2 \xrightarrow{-R'_2\overset{+}{N}H_2\ X^-} RCONR'_2 \qquad\qquad 12.1C,\ 12.4B$$

$$RC(O)-OCOR + 2HNR'_2 \xrightarrow{-RCO_2^-\ H_2\overset{+}{N}R'_2} RCONR'_2 \qquad\qquad 12.2C,\ 12.4B$$

$$RCOOR'' + HNR'_2 \xrightarrow{-R''OH} RCONR'_2 \qquad\qquad 12.3C,\ 12.4B$$

[a] This reaction is not practical for hindered esters or phenyl esters.

The preparations of carboxylic acid derivatives are summarized in Table 12.5.

All the carboxylic acid derivatives can be hydrolyzed in aqueous acid or base to yield *carboxylic acids*. Alkaline hydrolysis of esters is called **saponification**.

$$
\begin{array}{c}
\text{O} \\
\parallel \\
\text{RC}-\text{X} \\[6pt]
\text{O}\quad\text{O} \\
\parallel\quad\parallel \\
\text{RC}-\text{OCR} \\[6pt]
\text{O} \\
\parallel \\
\text{RC}-\text{OR}' \\[6pt]
\text{O} \\
\parallel \\
\text{RC}-\text{NR}'_2
\end{array}
\quad
\xrightarrow[\text{H}^+ \text{ or }^-\text{OH}]{\text{H}_2\text{O}}
\quad
\begin{array}{c}
\text{O}\qquad\quad\text{O} \\
\parallel\qquad\quad\parallel \\
\text{RCOH (or RCO}^-\text{ in base)} \\[4pt]
\textit{A carboxylic} \\
\textit{acid}
\end{array}
$$

Increasing reactivity (arrow pointing up along the left)

Acid halides, anhydrides, or esters react with ammonia or amines to yield *amides*.

$$
\begin{array}{c}
\text{O} \\
\parallel \\
\text{RC}-\text{X} \\[6pt]
\text{O}\quad\text{O} \\
\parallel\quad\parallel \\
\text{RC}-\text{OCR} \\[6pt]
\text{O} \\
\parallel \\
\text{RC}-\text{OR}'
\end{array}
\quad
\xrightarrow{\text{R}''_2\text{NH}}
\quad
\begin{array}{c}
\text{O} \\
\parallel \\
\text{RCNR}''_2 \\[4pt]
\textit{An amide}
\end{array}
$$

R″ can be H

Increasing reactivity (arrow pointing up along the left)

Acid halides and anhydrides react with alcohols to yield *esters*.

$$
\begin{array}{c}
\text{O} \\
\parallel \\
\text{RC}-\text{X} \\[12pt]
\text{O}\quad\text{O} \\
\parallel\quad\parallel \\
\text{RC}-\text{OCR}
\end{array}
\quad
\xrightarrow{\text{R}'\text{OH}}
\quad
\begin{array}{c}
\text{O} \\
\parallel \\
\text{RCOR}' \\[4pt]
\textit{An ester}
\end{array}
$$

Other reactions discussed in this chapter are *Grignard reactions of esters* to yield secondary and tertiary alcohols (Section 12.3C); *reduction of esters* to yield two alcohols, one of them primary (Section 12.3C); and *reduction of amides* to yield amines (Section 12.4C).

STUDY PROBLEMS

12.16. Identify the acid halide, acid anhydride, ester, or amide groups in each of the following structures. (Each structure may contain none or more than one of the functional groups.)

(a)

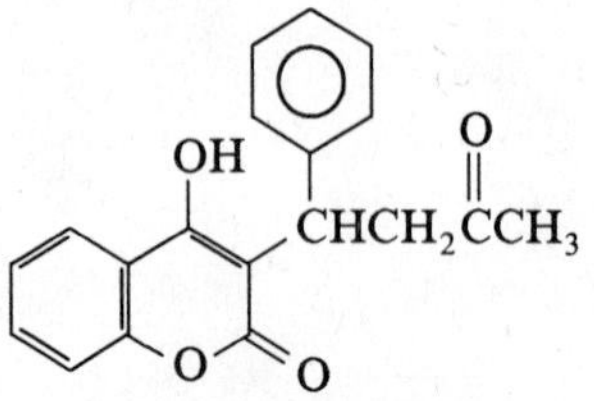

Warfarin
a rat poison

(b)

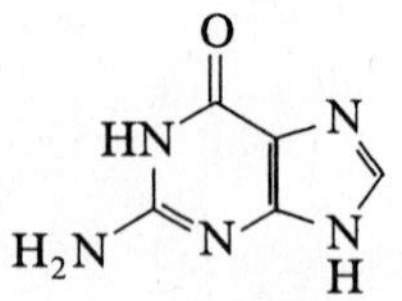

Guanine
a component of nucleic acids

(c)

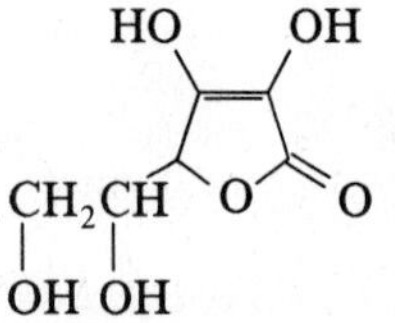

Ascorbic acid
(vitamin C)

(d)

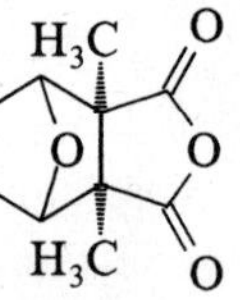

Cantharidin
reputed to be an aphrodisiac and a wart remover

12.17. Name the following compounds by the IUPAC system:

(a) $CH_3CH=CHCCl$ with O double bond

(b) CH_3CH_2-⟨benzene ring⟩$-CCl$ with O double bond

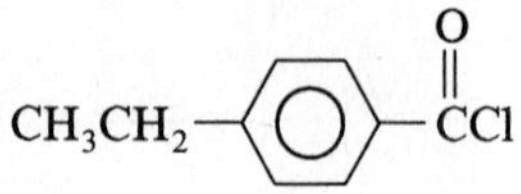

(c) $CH_3CH_2CH_2CCl$ with O double bond

(d) $(CH_3)_3CCCl$ with O double bond

12.18. Write equations showing how each of the compounds in the preceding problem can be prepared from a carboxylic acid.

12.19. Name the following compounds:

(a) $CH_3CH_2CH_2CH_2CH_2COCCH_2CH_2CH_2CH_3$ with two O double bonds

(b) CH_3CH_2COC-⟨benzene ring⟩$-OH$ with two O double bonds

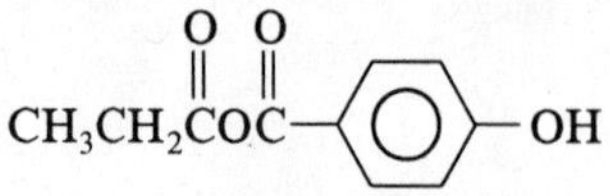

(c) $Br-$⟨benzene ring⟩$-COCH(CH_3)_2$ with O double bond

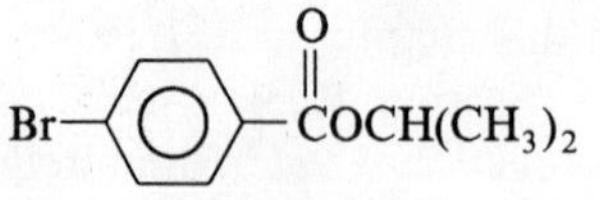

(d) 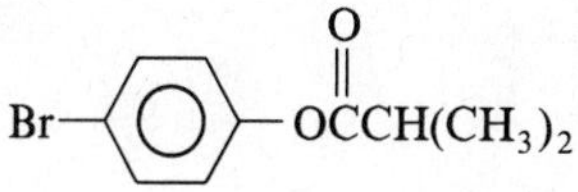 $Br-\!\!\left\langle\!\!\bigcirc\!\!\right\rangle\!\!-\overset{\overset{\displaystyle O}{\parallel}}{O}CCH(CH_3)_2$

(e) $CH_3CH_2\overset{\overset{\displaystyle O}{\parallel}}{C}NH_2$

(f) $H\overset{\overset{\displaystyle O}{\parallel}}{C}NHCH_2CH_3$

(g) 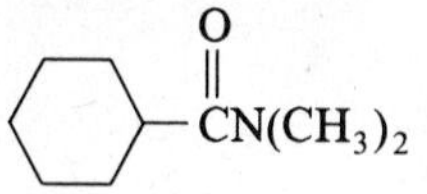 $\left\langle\!\!\bigcirc\!\!\right\rangle\!\!-\overset{\overset{\displaystyle O}{\parallel}}{C}N(CH_3)_2$

(h) $CH_3O\overset{\overset{\displaystyle O}{\parallel}}{C}CH_2CH_2CH_2\overset{\overset{\displaystyle O}{\parallel}}{C}OCH_3$

12.20. Write formulas for the following compounds:
  (a) benzoic anhydride
  (b) benzamide
  (c) phenyl *m*-hydroxybenzoate
  (d) *N*-ethylacetamide
  (e) ethyl formate
  (f) *t*-butyl acetate
  (g) 2,4-dinitrobenzoyl chloride
  (h) methyl propenoate
  (i) dimethyl oxalate
  (j) *N,N'*-dimethylurea
  (k) 3-methylcyclopentyl 2-chloropentanoate

12.21. Write equations for the acidic hydrolyses of the following compounds:

(a) $ClCH_2\overset{\overset{\displaystyle O}{\parallel}}{C}\overset{\overset{\displaystyle O}{\parallel}}{O}CCH_3$ **(b)** $ClCH_2\overset{\overset{\displaystyle O}{\parallel}}{C}Cl$

(c) $CH_3CH_2CH_2\overset{\overset{\displaystyle O}{\parallel}}{C}NHCH_2CH_3$ **(d)** 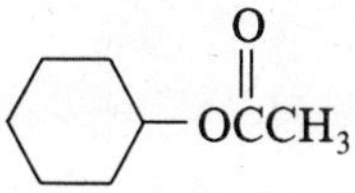 $\left\langle\!\!\bigcirc\!\!\right\rangle\!\!-O\overset{\overset{\displaystyle O}{\parallel}}{C}CH_3$

(e) 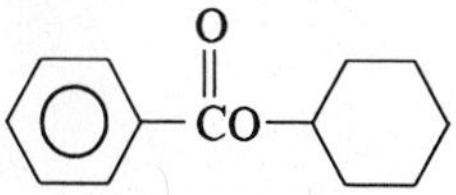 $\left\langle\!\!\bigcirc\!\!\right\rangle\!\!-\overset{\overset{\displaystyle O}{\parallel}}{C}O-\left\langle\!\!\bigcirc\!\!\right\rangle$ **(f)** 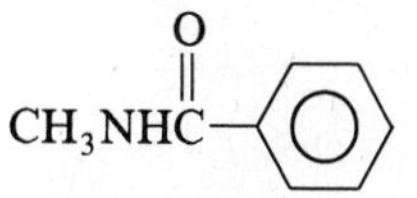 $CH_3NH\overset{\overset{\displaystyle O}{\parallel}}{C}-\left\langle\!\!\bigcirc\!\!\right\rangle$

12.22. List the following compounds in order of increasing reactivity toward water (least reactive first):

(a) $CH_3C\overset{\overset{\displaystyle O}{\parallel}}{}O\overset{\overset{\displaystyle O}{\parallel}}{C}CH_3$ **(b)** $CH_3\overset{\overset{\displaystyle O}{\parallel}}{C}CH_3$ **(c)** $CH_3\overset{\overset{\displaystyle O}{\parallel}}{C}Cl$

(d) $CH_3CH_2O\overset{\overset{\displaystyle O}{\parallel}}{C}CH_3$ **(e)** $CH_3CH_2NH\overset{\overset{\displaystyle O}{\parallel}}{C}CH_3$ **(f)** $CH_3CH_2CH_3$

12.23. Complete the following equations:

(a) $CH_2=CHC(CH_3)_2\overset{\overset{\displaystyle O}{\displaystyle \|}}{C}Cl + H_2O \longrightarrow$

(b)

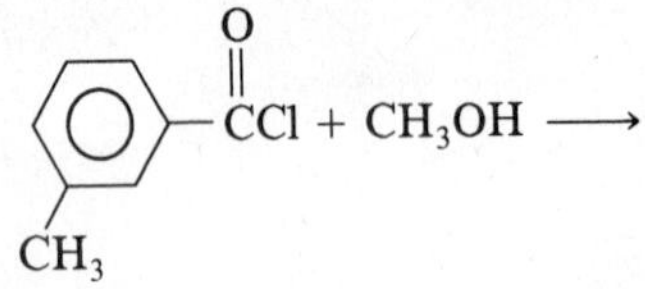

(c)

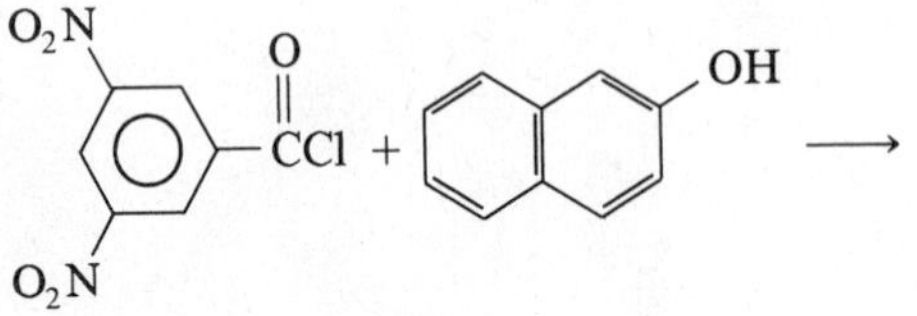

(d) 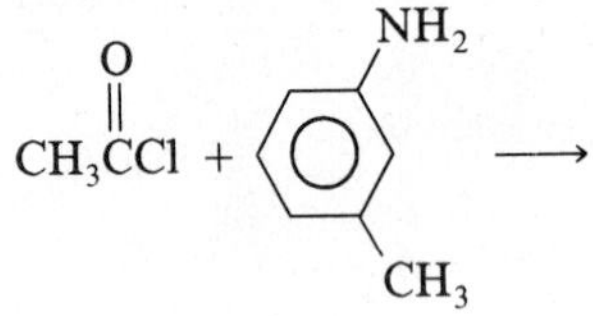 $CH_3\overset{\overset{\displaystyle O}{\displaystyle \|}}{C}Cl +$

12.24. Write equations to show how acetyl chloride could be converted to:

(a) $CH_3\overset{\overset{\displaystyle O}{\displaystyle \|}}{C}NH_2$ (b) $CH_3\overset{\overset{\displaystyle O}{\displaystyle \|}}{C}NHCH_3$ (c) $CH_3\overset{\overset{\displaystyle O}{\displaystyle \|}}{C}N(CH_3)_2$

12.25. Write equations to show how acetic anhydride could be converted to the compounds in the preceding problem.

12.26. Write equations showing two procedures for the preparation of each of the following compounds:

(a)

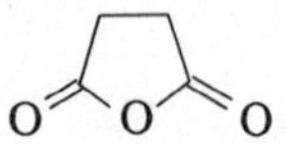

(b)

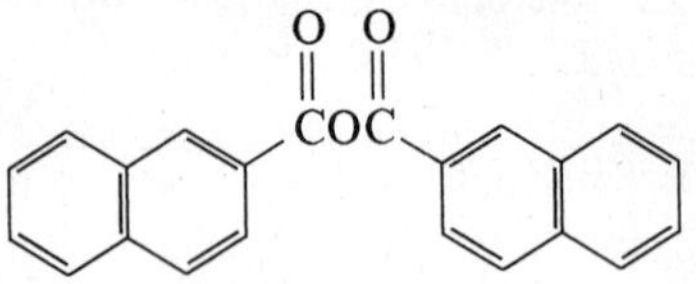

12.27. Write equations showing how the following compounds can be prepared from acetic anhydride and other appropriate reagents:

(a) $CH_3CH_2CH_2O\overset{\overset{\displaystyle O}{\displaystyle \|}}{C}CH_3$ (b)

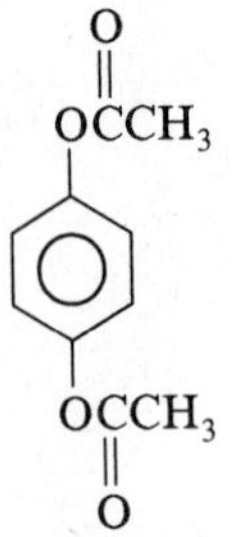

(c)

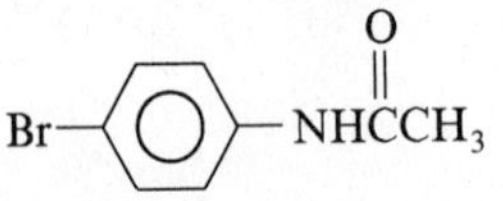

12.28. Write equations showing how *n*-propyl acetate can be prepared from (a) an acid halide and (b) a carboxylic acid.

12.29. Write equations showing how each of the following compounds can be prepared starting with a carboxylic acid.

(a)

$$\text{C}_6\text{H}_5-\overset{\displaystyle O}{\overset{\displaystyle \|}{\text{C}}}\text{NH}-\text{C}_6\text{H}_5$$

(b) O_2N ... $-\overset{\displaystyle O}{\overset{\displaystyle \|}{\text{C}}}\text{OCH}_2\text{CH(CH}_3)_2$ with O_2N

12.30. Write equations for the reactions of cyclohexyl isobutyrate with the following reagents:
   (a) aqueous HCl, heat
   (b) aqueous NaOH, heat
   (c) methylamine
   (d) $CH_2{=}CHMgI$ in diethyl ether, followed by treatment with cold dilute HCl
   (e) $LiAlH_4$, followed by treatment with cold dilute HCl

12.31. Write equations for three routes to N-methylbutanamide from butanoic acid.

12.32. Write equations for the acidic hydrolysis reactions of the following compounds:

(a) $\overset{\displaystyle O}{\overset{\displaystyle \|}{\text{ClC}}}-\overset{\displaystyle O}{\overset{\displaystyle \|}{\text{CCl}}}$

(b) $CH_3\overset{\displaystyle O}{\overset{\displaystyle \|}{\text{OC}}}OCH_3$

(c) $\overset{\displaystyle O}{\overset{\displaystyle \|}{\text{ClCCl}}}$

(d) phthalic anhydride

(e) $Cl\overset{\displaystyle O}{\overset{\displaystyle \|}{\text{C}}}OCH_2CH_3$

(f) $H_2N\overset{\displaystyle O}{\overset{\displaystyle \|}{\text{C}}}NH_2$

12.33. Write equations for the alkaline hydrolysis of the compounds in the preceding problem.

12.34. Complete the following equations:

(a) $HO-\text{C}_6\text{H}_4-\overset{\displaystyle O}{\overset{\displaystyle \|}{\text{C}}}OCH_3$ + excess NaOH $\xrightarrow[\text{heat}]{H_2O}$

(b) benzene ring with $\overset{\displaystyle O}{\overset{\displaystyle \|}{\text{C}}}OCH_3$ and $\overset{\displaystyle }{\underset{\displaystyle O}{\overset{\displaystyle \|}{\text{C}}}}OCH_3$ + H_2O $\xrightarrow[\text{heat}]{H^+}$

(c) $(CH_3)_2CHCH_2\overset{\displaystyle O}{\overset{\displaystyle \|}{\text{C}}}OCH_2CH_3$ $\xrightarrow[\text{(2) } H_2O, H^+]{\text{(1) excess } CH_3CH_2CH_2MgBr}$

(d) $\text{C}_6\text{H}_5-O\overset{\displaystyle O}{\overset{\displaystyle \|}{\text{C}}}O-\text{C}_6\text{H}_5$ $\xrightarrow[\text{(2) } H_2O, H^+]{\text{(1) } LiAlH_4}$

(e) $HO\overset{\displaystyle O}{\overset{\displaystyle \|}{\text{C}}}CH_2CH_2\overset{\displaystyle Cl}{\overset{\displaystyle |}{\text{CH}}}-\overset{\displaystyle O}{\overset{\displaystyle \|}{\text{C}}}OH$ + excess $SOCl_2$ $\longrightarrow$

(f) $\underset{\text{HOC(CH}_2)_3\text{COH}}{\overset{\text{O} \quad \cdot \quad \text{O}}{\|\qquad\|}} + \underset{\text{CH}_3\text{COCCH}_3}{\overset{\text{O}\;\;\text{O}}{\|\;\;\|}} \xrightarrow{\text{warm}}$

(g) $\underset{\text{CH}_3\text{CH}_2\text{OCH}}{\overset{\text{O}}{\|}} \xrightarrow[\text{(2) H}_2\text{O, H}^+]{\text{(1) excess (CH}_3)_2\text{CHMgBr}}$

(h) ⟨◯⟩$-\underset{\text{CNHCH}_2\text{CH}_2\text{CH}_3}{\overset{\text{O}}{\|}}$ $\xrightarrow[\text{(2) H}_2\text{O, H}^+]{\text{(1) LiAlH}_4}$

(i) ⟨◯⟩$-\underset{\text{CCl}}{\overset{\text{O}}{\|}} + (\text{CH}_3)_3\text{CNH}_2 \longrightarrow$

(j) (structure of phthalic anhydride) $+ \text{ excess NH}_3 \longrightarrow$

12.35. Write flow equations for the conversion of 2-methylpropanoic acid to the following compounds. (More than one step may be required. More than one correct answer may be possible.)

(a) $(\text{CH}_3)_2\text{CH}\overset{\text{O}}{\overset{\|}{\text{C}}}\text{NH}_2$

(b) $(\text{CH}_3)_2\text{CH}\overset{\text{O}\;\text{O}}{\overset{\|\;\;\|}{\text{CO}}}\text{CCH(CH}_3)_2$

(c) $(\text{CH}_3)_2\text{CH}\overset{\text{O}}{\overset{\|}{\text{C}}}\text{OC(CH}_3)_3$

(d) $(\text{CH}_3)_2\text{CH}\overset{\text{O}}{\overset{\|}{\text{C}}}\text{OCH}_3$

(e) ⟨◯⟩$-\text{O}\overset{\text{O}}{\overset{\|}{\text{C}}}\text{CH(CH}_3)_2$

(f) $(\text{CH}_3)_2\text{CH}\overset{\text{O}}{\overset{\|}{\text{C}}}-\text{N}⟨◯⟩$

(g) $(\text{CH}_3)_2\text{CH}\overset{\text{OH}}{\overset{|}{\text{C}}}(\text{CH}_3)_2$

12.36. The pain reliever *phenacetin* can be synthesized from acetic anhydride and an amine. Write an equation for this reaction.

$\text{CH}_3\text{CH}_2\text{O}-⟨◯⟩-\text{NH}\overset{\text{O}}{\overset{\|}{\text{C}}}\text{CH}_3$

Phenacetin

12.37. Predict the products of the alkaline saponification of the pain reliever lidocaine (Xylocaine).

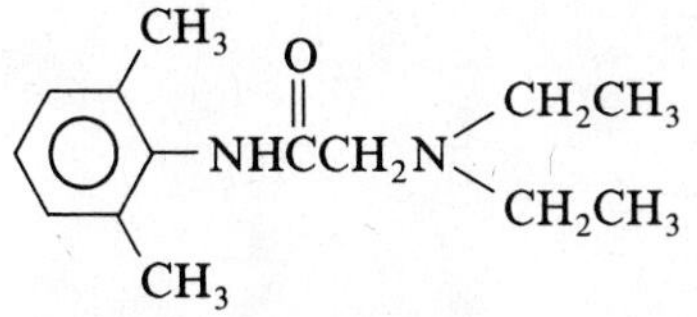

Lidocaine

12.38. When salicylic acid (*o*-hydroxybenzoic acid) is heated with methanol and a trace of acid, the product is *oil of wintergreen*. What is the structure of oil of wintergreen?

12.39. Acetylcholine, a neurotransmitter, is deactivated by an ester hydrolysis reaction catalyzed by the enzyme acetylcholinesterase. Write an equation for this reaction.

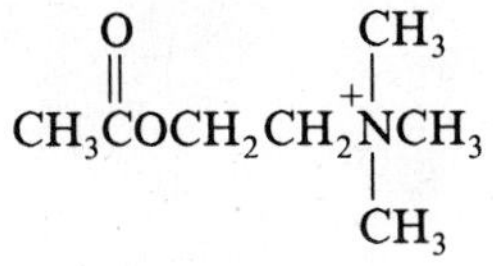

Acetylcholine

12.40. Nicotinamide, a component of oxidized nicotinamide-adenine dinucleotide (NAD^+), can be prepared by the reaction of nicotinic acid with thionyl chloride and then ammonia. (Nicotinic acid is used to cure the dietary disease pellagra.) Write equations for the preparation of nicotinamide.

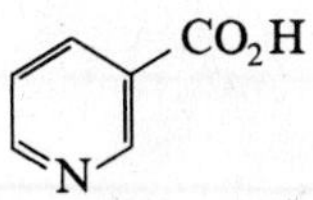

Nicotinic acid

12.41. How would you make the following conversions?

(a)

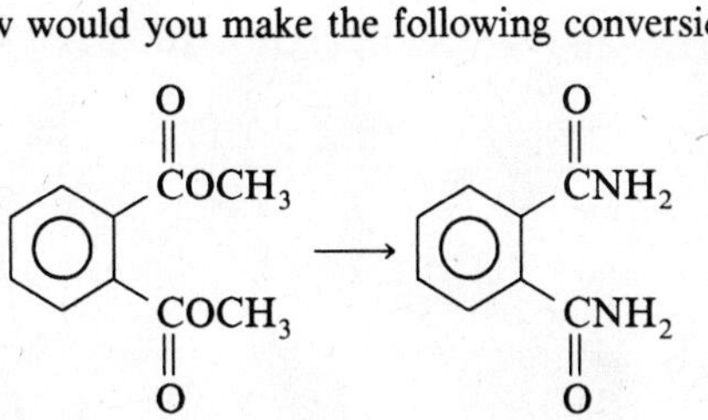

(b)

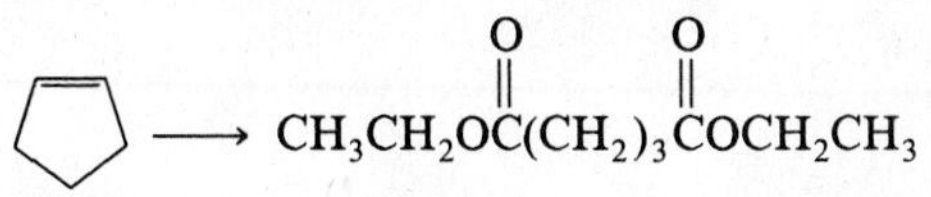

(c) $CH_3CH_2CH_2COCH_3 \longrightarrow CH_3CH_2CH_2CH_2Br$

(d)

$$\text{(benzene-1,2-dicarboxylic acid)} \longrightarrow \text{(phthalic anhydride)}$$

(e)

$$\begin{array}{c} CH_2OH \\ | \\ CHOH \\ | \\ CH_2OH \end{array} \longrightarrow \begin{array}{c} O \\ \| \\ CH_2OCCH_2CH_2CH_3 \\ | \quad O \\ \quad \| \\ CHOCCH_2CH_2CH_3 \\ | \quad O \\ \quad \| \\ CH_2OCCH_2CH_2CH_3 \end{array}$$

(f) $\quad ClCH_2CH_2CH_2Cl \longrightarrow \overset{O}{\overset{\|}{CH_3OC}}CH_2CH_2CH_2\overset{O}{\overset{\|}{COCH_3}}$

(g) $\quad CH_3CH_2CH_2\overset{O}{\overset{\|}{COH}} \longrightarrow CH_3CH_2CH_2\overset{O}{\overset{\|}{CN}}(CH_3)_2$

POINT OF INTEREST 12

Soaps and Detergents

The verb form *to saponify* means "to make soap" (Latin *sapon*, "soap," and *-fy*, a combining suffix form meaning "make"). Soap is indeed made by the saponification of fats, which are triesters of the triol glycerol. The carboxylic acid parents of fats, called **fatty acids**, contain long continuous hydrocarbon chains bonded to their carboxyl groups.

Fatty acid chains can vary in length and may contain double bonds.

$$\begin{array}{c} O \\ \| \\ CH_2OC(CH_2)_{14}CH_3 \\ | \quad O \\ \quad \| \\ CHOC(CH_2)_{14}CH_3 \; + \; 3\,Na^+\ {}^-OH \\ | \\ CH_2OC(CH_2)_{14}CH_3 \\ \quad \| \\ \quad O \end{array} \xrightarrow[\text{heat}]{H_2O} \begin{array}{c} CH_2OH \\ | \\ CHOH \\ | \\ CH_2OH \end{array} + \; 3\,Na^+\ {}^-\overset{O}{\overset{\|}{OC}}(CH_2)_{14}CH_3$$

Glyceryl tripalmitate

a typical fat

Glycerol (glycerin)

Sodium palmitate

a typical soap

The historical origin of soap is unknown. The Babylonians are reputed to have used soap in 2800 B.C. Soap was used as an article of commerce by the Phoenicians and the Romans. During the Dark Ages, the use of soap declined, and the art of making it was lost in many parts of Europe. Soap reappeared during the Middle Ages. At this time, it was regarded by many as a curiosity. As late as 1672, in one particular circumstance, a gift of soap to a lady of the nobility was accompanied by a detailed description of its use.

Soap was brought to America in the 1600s by German and Polish immigrants. Until the 1800s, soapmaking in America remained a household art, accomplished by the same process the ancients used—heating surplus grease and cooking oil with alkali obtained by leaching wood ashes. Soap is manufactured today by similar chemical reactions starting with refined vegetable oils and fats. Today's alkali is either sodium or potassium hydroxide.

Fats and oils with long-chain (C_{16}–C_{18}) saturated fatty acids yield hard soaps, while fats or oils with shorter-chain (C_{12}–C_{14}) unsaturated fatty acids yield softer, more soluble soaps. Soaps formed with sodium hydroxide are less soluble than soaps formed with potassium hydroxide.

Today's soaps are blended to attain the desired properties. Toilet soaps contain perfumes, dyes, and medicinal agents. Shampoos contain soap along with agents such as diethanolamine ($HOCH_2CH_2NHCH_2CH_2OH$) that adjust the pH, lanolin as a conditioner, and proteins to give the hair body.

Soaps are one type of *surfactants* (surface-active agents), compounds that lower the surface tension of water. This property allows soapy water to penetrate fabric, where it can loosen and dislodge dirt and oil. Once the oily grime has been removed from the fabric surface, soaps aid in washing it away because of their chemical structure—a hydrophilic (water-loving) ionic end and a hydrophobic (water-hating) hydrocarbon chain. The hydrocarbon chain dissolves in the water-insoluble oily particle. The ionic end dissolves in water. Thus, the soap molecules keep oil droplets dispersed, or emulsified, in water so that they can be washed away.

The negative ionic charges of the soap ions also cause the soap-oil droplets to repel one another so that the emulsified oils cannot coagulate. Figure 12.1 illustrates the emulsification of oils by soaps.

One disadvantage of soaps as cleansing agents is that they precipitate with calcium ions and magnesium ions, which are common cations in hard water.

$$2CH_3(CH_2)_{14}\overset{\overset{\textstyle O}{\|}}{C}O^-\,Na^+ + Ca^{2+} \longrightarrow \left[CH_3(CH_2)_{14}\overset{\overset{\textstyle O}{\|}}{C}O\right]_2 Ca\downarrow + 2Na^+$$

A soluble soap   *A precipitated soap*

A precipitated soap cannot act to remove dirt, but instead forms a scum ("bathtub ring"). One way to prevent scum formation is to use naturally soft or artificially softened water, which (by definition) contains no calcium or magnesium ions.

Detergents, which are similar to soaps in structure, also act as surfactants. The first detergents were patented in 1925 in Germany and were introduced into the United States in the 1930s. Like soaps, these synthetic compounds contain long

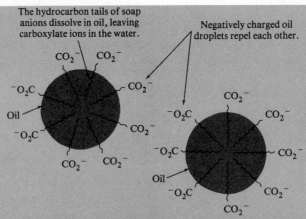

Figure 12.1 Soap molecules can keep oil droplets suspended in water.

hydrocarbon chains, but the ionic end is typically a *sulfate* or *sulfonate* group instead of a carboxylate group.

Two Typical Detergents:

$$CH_3(CH_2)_{11}OSO_3^- \ Na^+ \qquad CH_3(CH_2)_{11} - \langle\bigcirc\rangle - SO_3^- \ Na^+$$

Sodium lauryl sulfate   Sodium *p*-dodecylbenzenesulfonate

Sulfates and sulfonates do not precipitate with metal ions and are thus effective cleaning agents even in hard water.

Modern synthetic detergents, like the natural soaps, have only continuous hydrocarbon chains as their hydrophobic groups. Continuous-chain hydrocarbon groups are *biodegradable*, able to be metabolized by bacteria in waste-treatment facilities. Detergents that contain branched-chain alkyl groups pass through the waste-treatment plants and contaminate lakes and streams. These detergents are no longer manufactured.

Detergents often contain additives to enhance their usefulness. For example, detergents may contain enzymes (proteases or amylases) to break down proteins and starches. They may also contain builders, softeners, antistatic agents, and optical bleaches.

Builders are compounds like sodium tripolyphosphate (STP, $Na_5P_3O_{10}$) that increase the alkalinity of and soften the wash water to increase the efficiency of the detergent. Use of the phosphate builders is decreasing because they contaminate waste water and cause excessive growth of algae, thereby depleting the oxygen supply in the water.

Softeners and antistatic agents are quaternary ammonium compounds, such as distearyldimethylammonium chloride. Optical bleaches are compounds that absorb ultraviolet light and fluoresce to emit blue light, thus masking the natural yellowing of white fabrics.

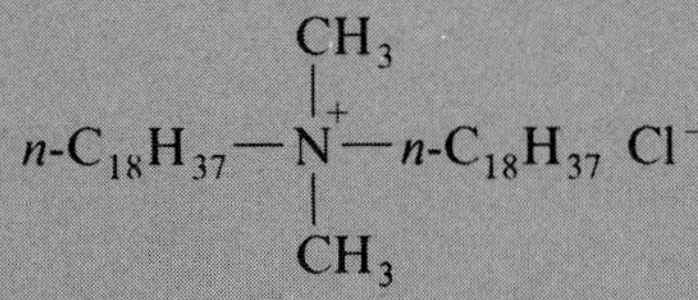

$$n\text{-}C_{18}H_{37}\overset{\overset{\displaystyle CH_3}{\displaystyle |}}{\underset{\underset{\displaystyle CH_3}{\displaystyle |}}{\overset{+}{N}}}\text{-}n\text{-}C_{18}H_{37}\;\;Cl^-$$

Distearyldimethylammonium
chloride

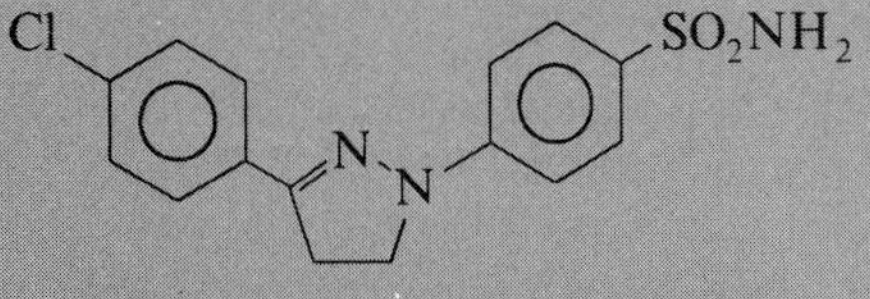

*A pyrazoline optical bleach
added to detergents to whiten
synthetic fabrics*

Nonionic detergents are gaining in popularity because they are superior for washing modern synthetic fabrics. These detergents contain polar groups that are nonionic, such as the oxygen functional groups in the following formula.

Long chain

$$\overset{\displaystyle O}{\underset{\displaystyle \|}{}}$$
$$R C O - CH_2CH_2O - (CH_2CH_2O)_2 - CH_2CH_2OH$$

One type of nonionic detergent

Chapter 13

Addition and Condensation Reactions

Reactions that result in new carbon–carbon bonds are exceptionally useful to organic chemists, medicinal chemists, biochemists, and anyone else who needs to synthesize large molecules from smaller ones. Grignard reactions are one type of reaction used to build carbon skeletons. In this chapter we discuss two other types of reactions used to build molecules: the aldol addition and the ester condensation.

13.1 THE ALDOL ADDITION

When treated with dilute base, an aldehyde with at least one alpha hydrogen undergoes a self-addition reaction. Two molecules of the aldehyde combine to form a 3-hydroxy aldehyde called an **aldol** (from *ald*ehyde and alco*hol*). The reaction itself is called an **aldol addition** or **aldol condensation**. A *condensation reaction* is a reaction in which two organic molecules combine, usually (but not here) with the loss of water or other small molecule.

An Aldol Addition:

$$
\underset{\text{Acetaldehyde}}{\overset{\overset{\displaystyle O}{\|}}{CH_3CH}} + \underset{}{\overset{\overset{\displaystyle O}{\|}}{CH_3CH}} \xrightleftharpoons{^-OH} \underset{\substack{\text{3-Hydroxybutanal}\\(\textit{also called}\ \text{aldol})}}{\overset{\overset{\displaystyle OH}{|}}{CH_3CH} - \overset{\overset{\displaystyle O}{\|}}{CH_2CH}}
$$

From one aldehyde molecule From the other

We will first discuss the mechanism of an aldol addition and then show some variations of this reaction.

A. Acidity of the Alpha Hydrogen

The alpha hydrogen of an aldehyde or a ketone is slightly acidic—more acidic than a hydrogen of an alkane but less acidic than the hydrogen of an alcohol or water.

$$
\underset{\sim\,50}{CH_3CH_2} \qquad \underset{\sim\,20\text{-}25}{\overset{\displaystyle H}{\underset{\displaystyle |}{-}}\overset{\displaystyle O}{\overset{\displaystyle \|}{C}}\overset{}{-}C-} \qquad \underset{15\text{-}19}{RO-H} \qquad \underset{15.7}{HO-H} \qquad \underset{3\text{-}6}{R\overset{\displaystyle O}{\overset{\displaystyle \|}{C}}O-H}
$$

pK$_a$:

→ Increasing acidity

When the aldehyde or ketone with an alpha hydrogen is treated with a strong base ($^-$OH or $^-$OR), a small percentage of alpha protons are abstracted in an acid-base reaction.

$$
\underset{H}{\overset{\displaystyle O}{\overset{\displaystyle \|}{RCHCH}}} + \quad ^-\!:\!\ddot{O}H \;\rightleftharpoons\; R\bar{C}H\overset{\displaystyle O}{\overset{\displaystyle \|}{CH}} + H-\ddot{O}H
$$

One reason for the acidity of the alpha hydrogen is that the carbonyl carbon withdraws electron density from this hydrogen by the inductive effect. The electron withdrawal causes the alpha hydrogen to be more positive than an ordinary alkyl hydrogen is.

$$
\overset{\displaystyle H}{\overset{\downarrow}{R-C}} \overset{\displaystyle O}{\overset{\displaystyle \|}{\;\rightarrow CH}}
$$

$\uparrow$
H ←——— Becomes $\delta+$ by the inductive effect

Because this alpha hydrogen is partially positive, it is more attractive to a base than is an ordinary alkyl hydrogen and is more easily removed.

$$
\underset{\underset{\delta+}{H}}{\overset{\displaystyle H}{\overset{\displaystyle |}{R-C}}}\overset{\displaystyle O}{\overset{\displaystyle \|}{-CH}}
$$

$\qquad ^-\!:\!\ddot{O}H$

Another reason for the increased acidity of a proton alpha to a carbonyl group is that the product anion is resonance stabilized. This anion is called an **enolate ion** from the structure of one of the resonance formulas.

Resonance Structures for the Anion:

$$
\underset{}{\bar{R}CH-}\overset{\displaystyle :\ddot{O}}{\overset{\displaystyle \|}{CH}} \longleftrightarrow RCH=\overset{\displaystyle :\ddot{O}:^-}{\overset{\displaystyle |}{CH}}
$$

An enolate ion

The resonance stabilization of the enolate ion is the main reason that the alpha hydrogen can be removed by base.

$$RCH_2\overset{\displaystyle O}{\overset{\|}{C}}H + {}^-OH \rightleftharpoons R\overset{..}{\underset{..}{C}}H\overset{\displaystyle O}{\overset{\|}{C}}H + H_2O$$

Resonance stabilized

$$RCH_2CH_3 + {}^-OH \longrightarrow \text{no reaction because there is no product stabilization}$$

B. Mechanism of the Addition

In an alkaline solution, a small percentage of aldehyde molecules are in the anionic form. The negative carbon of an enolate ion attacks the partially positive carbonyl carbon of an aldehyde molecule in a typical nucleophilic addition reaction. A final acid-base reaction with the solvent yields the aldol product.

Step 1, Formation of Enolate Ion:

$$RCH_2\overset{\displaystyle O}{\overset{\|}{C}}H + {}^-:\overset{..}{\underset{..}{O}}H \rightleftharpoons R\overset{..}{\underset{..}{C}}H\overset{\displaystyle O}{\overset{\|}{C}}H + H_2\overset{..}{\underset{..}{O}}:$$

Step 2, Nucleophilic Addition:

$$RCH_2\overset{\displaystyle \overset{..}{\underset{..}{O}}:}{\overset{\|}{C}}H \;\; + \;\; :\overset{\displaystyle O}{\underset{\underset{R}{|}}{\overset{\|}{C}}}HCH \;\; \rightleftharpoons \;\; RCH_2\overset{\displaystyle :\overset{..}{\underset{..}{O}}:^-}{\underset{}{\overset{|}{C}}}H - \overset{\displaystyle O}{\underset{\underset{R}{|}}{\overset{\|}{C}}}HCH$$

An alkoxide ion

Step 3, Acid-Base Reaction with Water:

$$RCH_2\overset{\displaystyle :\overset{..}{\underset{..}{O}}:^-}{\overset{|}{C}}H - \overset{\displaystyle O}{\underset{\underset{R}{|}}{\overset{\|}{C}}}HCH \;\; \xrightarrow{H-\overset{..}{O}H} \;\; RCH_2\overset{\displaystyle :\overset{..}{O}H}{\overset{|}{C}}H - \overset{\displaystyle O}{\underset{\underset{R}{|}}{\overset{\|}{C}}}HCH + {}^-:\overset{..}{\underset{..}{O}}H$$

An aldol

> **PROBLEM 13.1.** Write equations for the three steps in the mechanism of the aldol condensation of propanal, $CH_3CH_2CH{=}O$.

EXAMPLE

Write an equation that shows the synthesis of the following compound by an aldol addition.

Solution:

1. Determine which bond is the new carbon–carbon bond. *This bond must join the hydroxyl carbon and the alpha carbon.*

2. Write the equation:

PROBLEM 13.2. Write an equation showing the aldol addition leading to the following product:

$$\underset{\substack{|\\ \mathrm{CH_2CH_3}}}{\mathrm{CH_3CH_2CH_2}\overset{\overset{\displaystyle \mathrm{OH}}{|}}{\mathrm{CH}}\overset{\overset{\displaystyle \mathrm{O}}{\|}}{\mathrm{CH}}\mathrm{CH}}$$

C. Scope of the Aldol Addition

Although some ketones can undergo an aldol-type addition to yield a 3-hydroxy ketone, the reaction is not general. The reason is probably that the transition state leading to the addition product from the ketone has a large amount of steric hindrance.

An aldehyde without an alpha hydrogen cannot undergo self-addition.

$$\mathrm{H{-}\overset{\overset{\displaystyle O}{\|}}{C}{-}H} \xrightarrow{\ ^-\mathrm{OH}\ } \text{no addition}$$

no α hydrogen

$$\text{(phenyl)}{-}\overset{\overset{\displaystyle O}{\|}}{\mathrm{CH}} \xrightarrow{\ ^-\mathrm{OH}\ } \text{no addition}$$

An important variation of the aldol condensation, called a **crossed aldol addition** or **mixed aldol addition**, involves the addition of an enolate to an aldehyde that has no alpha hydrogen.

Formation of the Enolate:

$$\mathrm{CH_3}\overset{\overset{\displaystyle O}{\|}}{\mathrm{CH}} \underset{\ ^-\mathrm{OH}\ }{\rightleftharpoons} {^-}{:}\mathrm{CH_2}\overset{\overset{\displaystyle O}{\|}}{\mathrm{CH}} + \mathrm{H_2O}$$

Crossed Aldol Addition:

$$\mathrm{H}\overset{\overset{\displaystyle O}{\|}}{\mathrm{CH}} + {^-}{:}\mathrm{CH_2}\overset{\overset{\displaystyle O}{\|}}{\mathrm{CH}} \rightleftharpoons \underset{\substack{\uparrow\\ \text{From}\\ \mathrm{H_2C{=}O}}}{\mathrm{CH_2}}\overset{\overset{\displaystyle {:}\ddot{O}{:}^-}{|}}{\text{---}}\underset{\substack{\uparrow\\ \text{From}\\ \mathrm{CH_3CH{=}O}}}{\mathrm{CH_2}}\overset{\overset{\displaystyle O}{\|}}{\mathrm{CH}} \xrightarrow[\ \ ]{\mathrm{H_2O}} \mathrm{CH_2}\overset{\overset{\displaystyle OH}{|}}{\text{---}}\mathrm{CH_2}\overset{\overset{\displaystyle O}{\|}}{\mathrm{CH}}$$

From H$_2$C=O   From CH$_3$CH=O

When two aldehydes are used, as in the preceding example, a mixture of crossed addition and self-addition products would be formed. The self-addition reaction that would occur in this example follows.

Self-Addition:

$$\underset{\text{O}}{\overset{\text{O}}{CH_3CH}} + \; ^-:CH_2\overset{\text{O}}{CH} \rightleftharpoons CH_3\overset{:\overset{..}{O}:^-}{CH}-CH_2\overset{\text{O}}{CH} \overset{H_2O}{\rightleftharpoons} CH_3\overset{OH}{CH}-CH_2\overset{\text{O}}{CH}$$

When a ketone is used to form the enolate, good yields of a single product can result because the ketone does not readily undergo self-addition.

EXAMPLE

Predict the organic product of the reaction of benzaldehyde with acetone in the presence of base.

Solution:

1. Write the formulas and determine which reactant is more likely to lose an alpha hydrogen to base.

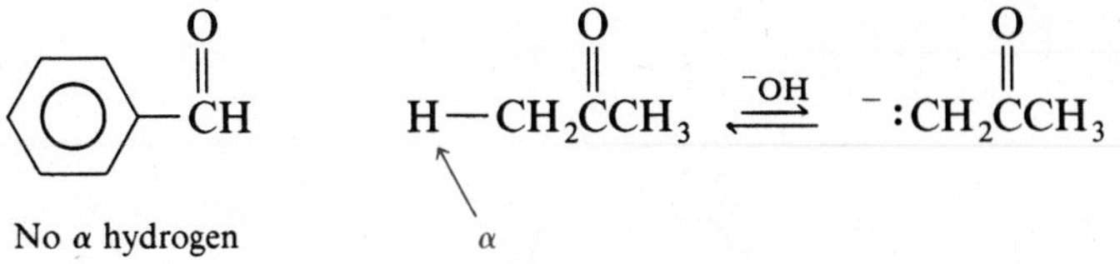

No α hydrogen

2. Write an equation that shows attack of the negative carbon on the carbonyl group.

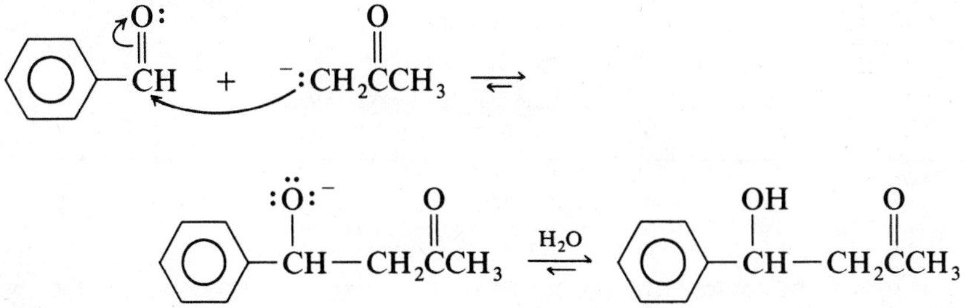

PROBLEM 13.3. Complete the following equations:

(a) $\langle O \rangle - \overset{\overset{\text{O}}{\|}}{\text{CH}} + (CH_3)_3C\overset{\overset{\text{O}}{\|}}{C}CH_3 \overset{^-OH}{\rightleftharpoons}$

(b) $(CH_3)_3C\overset{\overset{\text{O}}{\|}}{C}H + CH_3\overset{\overset{\text{O}}{\|}}{C}H \overset{^-OH}{\rightleftharpoons}$

PROBLEM 13.4. Write an equation showing the aldol addition leading to the following product:

$$\overset{\overset{\text{OH}}{|}}{CH_2}CH_2\overset{\overset{\text{O}}{\|}}{C}CH_3$$

D. Dehydration of Aldol Products

Simple alcohols undergo dehydration when treated with a strong acid but not when treated with base.

$$\overset{\overset{\text{H}}{|}}{R_2C}-\overset{\overset{\text{OH}}{|}}{CR_2} \xrightarrow[\text{heat}]{H^+} R_2C{=}CR_2 + H_2O$$

$$\overset{\overset{\text{H}}{|}}{R_2C}-\overset{\overset{\text{OH}}{|}}{CR_2} \xrightarrow[\text{heat}]{^-OH} \text{no reaction}$$

3-Hydroxy carbonyl compounds can undergo dehydration in either acid or base, however, because the dehydrated product contains a lower-energy conjugated double bond.

$$\overset{\overset{\text{OH}}{|}}{RCH}-\overset{\overset{\text{H}}{|}}{CH}-\overset{\overset{\text{O}}{\|}}{CR} \xrightarrow[\text{heat}]{^-OH \text{ or } H^+} RCH{=}CH-\overset{\overset{\text{O}}{\|}}{CR} + H_2O$$

Conjugated with C=O

When the potential carbon–carbon double bond is conjugated with other unsaturation besides the carbonyl group, such as that of a benzene ring, it may be difficult to isolate the hydroxy compound. Dehydration can be spontaneous under the conditions of the aldol addition.

$$\text{C}_6\text{H}_5\overset{\displaystyle O}{\overset{\|}{\text{C}}}\text{H} + \text{CH}_3\overset{\displaystyle O}{\overset{\|}{\text{C}}}\text{H} \underset{}{\overset{^-\text{OH}}{\rightleftharpoons}} \text{C}_6\text{H}_5\underset{\overset{\displaystyle |}{\text{OH}}}{\overset{}{\text{C}}}\text{H}-\text{CH}_2\overset{\displaystyle O}{\overset{\|}{\text{C}}}\text{H}$$

Easily dehydrated

Conjugated with C=O
and benzene ring

$$\xrightarrow{-\text{H}_2\text{O}} \text{C}_6\text{H}_5-\text{CH}=\text{CH}\overset{\displaystyle O}{\overset{\|}{\text{C}}}\text{H}$$

Cinnamaldehyde

*a flavoring agent
in cinnamon*

PROBLEM 13.5. Write equations showing how you would prepare the following compounds by aldol additions followed by dehydration.

(a) $\text{C}_6\text{H}_5-\text{CH}_2\text{CH}=\text{C}\overset{\displaystyle O}{\overset{\|}{\text{C}}}\text{H}$ (with phenyl substituent on the C)   (b) $\text{C}_6\text{H}_5-\text{CH}=\text{CH}\overset{\displaystyle O}{\overset{\|}{\text{C}}}\text{CH}_3$

E. A Biological Reverse Aldol Addition

The aldol addition is a reversible reaction commonly encountered in biochemical metabolic pathways where large molecules are synthesized from small ones or where large molecules are fragmented into smaller ones. For example, an intermediate in the metabolism of glucose, fructose 1,6-diphosphate, is broken down into two three-carbon fragments by a reverse aldol reaction. The two three-carbon units can then be metabolized further, eventually yielding CO_2 and biological energy.

$$\begin{array}{c} \text{CH}_2\text{OPO}_3{}^{2-} \\ | \\ \text{C}=\text{O} \\ | \\ \text{HO}-\text{C}-\text{H} \\ | \\ \text{H}-\text{C}-\text{OH} \\ | \\ \text{H}-\text{C}-\text{OH} \\ | \\ \text{CH}_2\text{OPO}_3{}^{2-} \end{array} \xrightarrow{\text{aldolase}} \begin{array}{c} \text{CH}_2\text{OPO}_3{}^{2-} \\ | \\ \text{C}=\text{O} \\ | \\ \text{HOCH}_2 \end{array} \quad \begin{array}{c} \text{HC}=\text{O} \\ | \\ \text{H}-\text{C}-\text{OH} \\ | \\ \text{CH}_2\text{OPO}_3{}^{2-} \end{array}$$

Bond joining the
α carbon to
the C=O carbon

The α carbon

The carbonyl group

Fructose
1,6-diphosphate

PROBLEM 13.6. What would be the products of a reverse aldol reaction of the following compound?

$$CH_3\overset{\overset{\displaystyle O}{\|}}{C}CH_2\overset{\overset{\displaystyle OH}{|}}{C}HCH_3$$

13.2 ESTER CONDENSATION REACTIONS

Esters containing alpha hydrogens can undergo **ester condensation reactions** to yield β-keto esters. Ester condensations are also called **Claisen condensations** because they were first reported in 1887 by the German chemist Ludwig Claisen.

In an ester condensation, the starting ester is treated with a very strong base, such as an alkoxide or sodium hydride ($Na^+ H^-$). After the reaction is complete, acidification of the product mixture yields the keto ester.

An Ester Condensation, or Claisen Condensation:

$$CH_3\overset{\overset{\displaystyle O}{\|}}{C}-OCH_2CH_3 + CH_3\overset{\overset{\displaystyle O}{\|}}{C}OCH_2CH_3 \xrightarrow[\text{(2) } H^+]{\text{(1) } Na^+ \text{ }^-OCH_2CH_3}$$

Ethyl acetate

A new C—C bond

$$CH_3\overset{\overset{\displaystyle O}{\|}}{C}-CH_2\overset{\overset{\displaystyle O}{\|}}{C}OCH_2CH_3 + CH_3CH_2OH$$

Ethyl acetoacetate   Ethanol

A β-keto ester

In an ester condensation, the carbonyl portion of one ester molecule forms a bond with the alpha carbon of the other ester molecule. The product, therefore, contains a keto group beta to the ester carbonyl group.

$$\overset{\overset{\displaystyle O}{\|}}{R}C-OR'' + \overset{\overset{\displaystyle O}{\|}}{CH_2}COR'' \longrightarrow R\overset{\overset{\displaystyle O}{\|}}{C}-\overset{}{CH}COR''$$

α to ester carbonyl group

β to ester
α to ester

R'

new C—C bond

EXAMPLE

What is the ester condensation product from methyl propanoate?

Solution: Write the formula for the starting ester twice. Identify the carbonyl carbon of one ester and the alpha carbon of the other ester. Here we have written the second formula differently from the first to emphasize the alpha carbon.

$$
\underset{\text{O}}{\overset{\text{O}}{\underset{\|}{}}}
$$

α carbon

$$
\mathrm{CH_3CH_2\overset{\overset{\displaystyle O}{\|}}{C}-OCH_3} \;+\; \mathrm{CH_2\overset{\overset{\displaystyle O}{\|}}{C}OCH_3} \quad\xrightarrow[\text{(2) H}^+]{\text{(1) }^-\text{OCH}_3}
$$

$$
\underset{\displaystyle \mathrm{CH_3}}{}
$$

Write formulas for the structures of the products.

$$
\mathrm{CH_3CH_2\overset{\overset{\displaystyle O}{\|}}{C}-\underset{\underset{\displaystyle CH_3}{|}}{CH}\overset{\overset{\displaystyle O}{\|}}{C}OCH_3} \;+\; \mathrm{HOCH_3}
$$

From the
first molecule

From the second

PROBLEM 13.7. Complete the following equations for ester condensations.

(a) $2\,\mathrm{CH_3\overset{\overset{\displaystyle O}{\|}}{C}OCH_2CH_2CH_3}$ $\xrightarrow[\text{(2) H}^+]{\text{(1) }^-\text{OCH}_2\text{CH}_2\text{CH}_3}$

(b) $2\,\mathrm{CH_3CH_2CH_2\overset{\overset{\displaystyle O}{\|}}{C}OCH_3}$ $\xrightarrow[\text{(2) H}^+]{\text{(1) }^-\text{OCH}_3}$

(c) 2 $\mathrm{\overset{\overset{\displaystyle O}{\|}}{C}OCH_3}$ $\xrightarrow[\text{(2) H}^+]{\text{(1) }^-\text{OCH}_3}$

A. Mechanism of the Ester Condensation

Unlike aldol addition reactions, ester condensations are nucleophilic acyl substitution (addition-elimination) reactions. The first step in the reaction sequence is the generation of a nucleophile—the enolate ion of an ester molecule. This step is a simple acid-base reaction initiated by the strongly basic alkoxide ion and is directly analogous to the first step in an aldol addition.

Step 1, Formation of the Enolate:

$$CH_3CH_2\ddot{O}:^- \ Na^+ \ + \ CH_2\overset{\overset{O}{\|}}{C}OCH_2CH_3 \ \rightleftharpoons$$

$$pK_a \sim 25$$

$$CH_3CH_2\ddot{O}H \ + \ Na^+ \ ^-:CH_2\overset{\overset{O}{\|}}{C}OCH_2CH_3$$

Enolate of the ester

The second step is substitution—the addition of the enolate ion to an ester molecule, followed by elimination of an alkoxide ion. A proton transfer step generates a new enolate, the enolate of the β-keto ester.

Step 2, The Substitution Reaction:

$$CH_3\overset{\overset{\ddot{O}:}{\|}}{C}-OCH_2CH_3 \ + \ Na^+ \ ^-:CH_2\overset{\overset{O}{\|}}{C}OCH_2CH_3 \ \xrightarrow{\text{addition}}$$

$$CH_3\overset{\overset{:\ddot{O}:^- \ Na^+}{|}}{C}\text{—}\ddot{O}CH_2CH_3 \ \xrightarrow{\text{elimination}} \ CH_3\overset{\overset{\ddot{O}:}{\|}}{C} \ \quad + \ Na^+ \ ^-:\ddot{O}CH_2CH_3$$

$$\overset{|}{CH_2}\overset{\overset{}{C}OCH_2CH_3}{} \qquad\qquad \overset{|}{CH_2}COCH_2CH_3$$

$$\overset{\overset{}{\|}}{O} \qquad\qquad\qquad \overset{\overset{}{\|}}{O}$$

$$pK_a \sim 13$$

$$\xrightarrow{\text{proton transfer}} \ CH_3\overset{\overset{O}{\|}}{C}-\overset{-}{\underset{\cdot\cdot}{C}}H\overset{\overset{O}{\|}}{C}OCH_2CH_3 + Na^+ \ + H\ddot{O}CH_2CH_3$$

A new enolate

The enolate of a β-keto ester is relatively stable because the negative charge is delocalized by *two* carbonyl groups. The enolate ion of a β-keto ester is too weak a base to remove a proton from an alcohol, a proton transfer that would be required for the reverse of the last step. Therefore, the last step in the sequence is irreversible; this step drives the entire sequence to completion.

Resonance Formulas for the Enolate Ion of a β-Keto Ester:

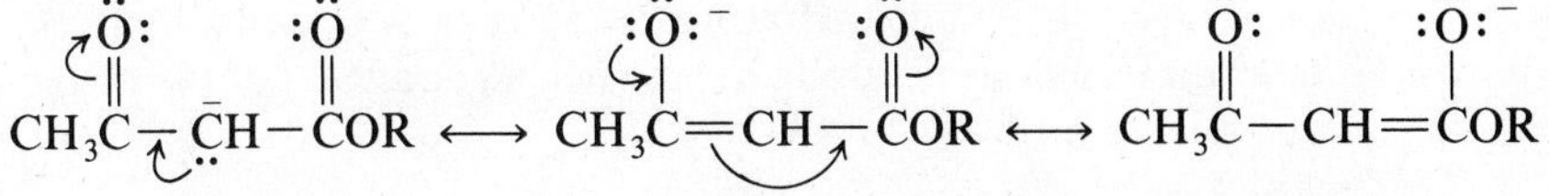

$$CH_3C-\overset{..}{C}H-COR \longleftrightarrow CH_3C=CH-COR \longleftrightarrow CH_3C-CH=COR$$

The product of the nucleophilic acyl substitution is an enolate of a β-keto ester. The β-keto ester itself is formed when the reaction mixture is acidified.

Step 3, Acidification:

$$CH_3\overset{O}{\overset{||}{C}}-\overset{-}{\underset{..}{C}}HCOR \xrightarrow{\;H^+\;} CH_3\overset{O}{\overset{||}{C}}-CH_2\overset{O}{\overset{||}{C}}OR$$

EXAMPLE

How would you prepare the following keto ester from a simple ester?

$$(CH_3)_2CHCH_2\overset{O}{\overset{||}{C}}\underset{\underset{CH(CH_3)_2}{|}}{CH}\overset{O}{\overset{||}{C}}OCH_2CH_3$$

Solution:

1. Circle the ketone portion of the formula to indicate the starting esters.

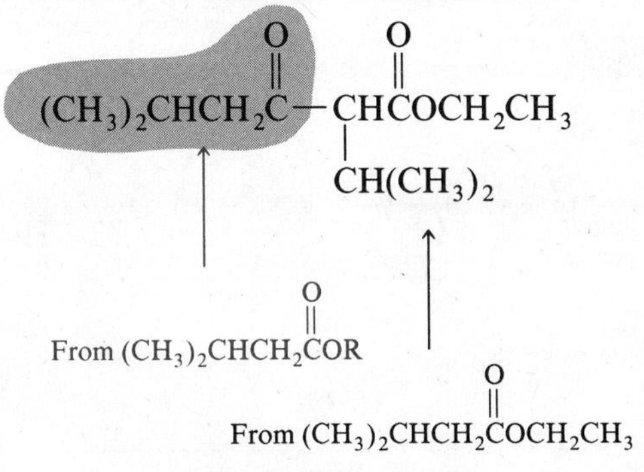

$$(CH_3)_2CHCH_2\overset{O}{\overset{||}{C}}-\underset{\underset{CH(CH_3)_2}{|}}{CH}\overset{O}{\overset{||}{C}}OCH_2CH_3$$

From $(CH_3)_2CHCH_2\overset{O}{\overset{||}{C}}OR$

From $(CH_3)_2CHCH_2\overset{O}{\overset{||}{C}}OCH_2CH_3$

2. Write a flow equation for the preparation:

$$2(CH_3)_2CHCH_2\overset{O}{\overset{||}{C}}OCH_2CH_3 \xrightarrow[\text{(2) } H^+]{\text{(1) } Na^+ \; {}^-OCH_2CH_3} \text{product}$$

B. Mixed Ester Condensations

If a mixture of esters with alpha hydrogens is treated with a strong base, a mixture of products is obtained. However, if an ester with alpha hydrogens is mixed with an ester that contains no alpha hydrogens, a good yield of the mixed or crossed product can be obtained.

Examples of Esters Containing No Alpha Hydrogens:

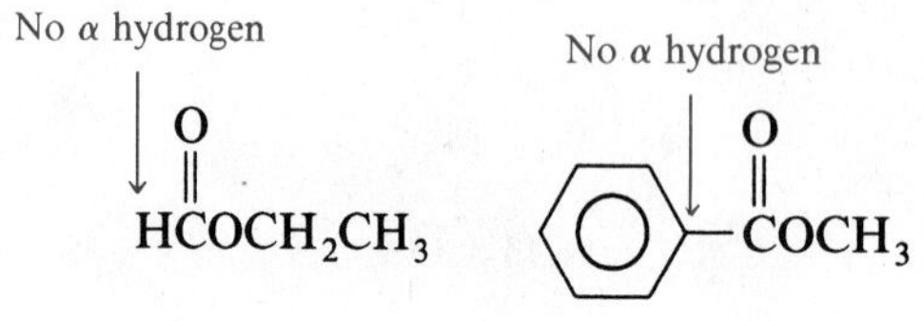

No α hydrogen

$$HCOCH_2CH_3$$

Ethyl formate

No α hydrogen

$$\langle\bigcirc\rangle - COCH_3$$

Methyl benzoate

Examples of Mixed Ester Condensations:

$$HCOCH_2CH_3 + CH_3COCH_2CH_3 \xrightarrow[\text{(2) H}^+]{\text{(1) Na}^+ \ ^-OCH_2CH_3}$$

$$HC-CH_2COCH_2CH_3 + CH_3CH_2OH$$

$$\langle\bigcirc\rangle - C-OCH_2CH_3 + CH_3COCH_2CH_3 \xrightarrow[\text{(2) H}^+]{\text{(1) Na}^+ \ ^-OCH_2CH_3}$$

$$\langle\bigcirc\rangle - C-CH_2COCH_2CH_3 + CH_3CH_2OH$$

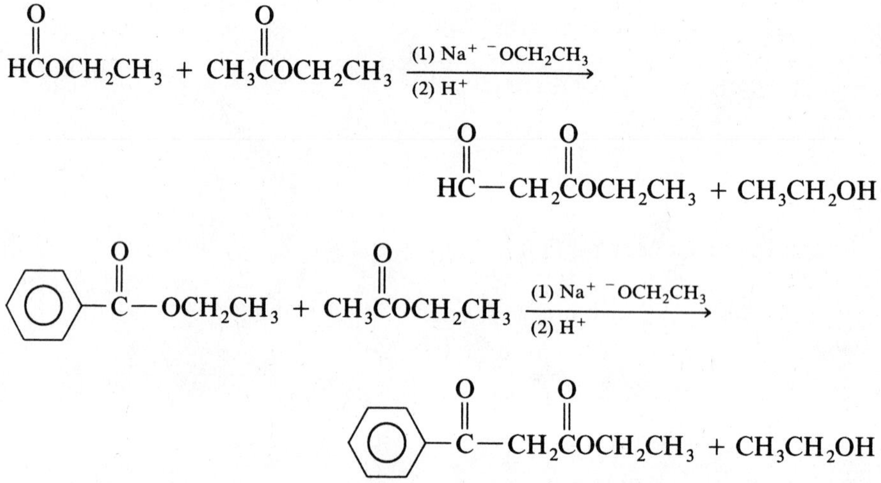

EXAMPLE

How would you prepare the following keto ester from one or more simple esters?

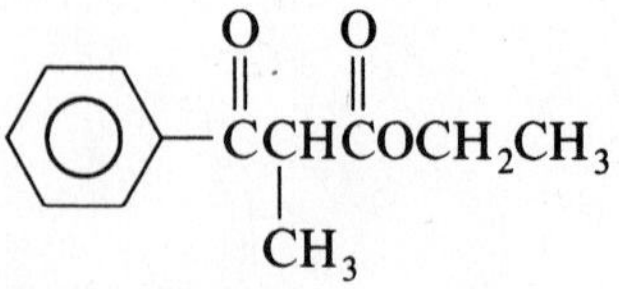

$$\langle\bigcirc\rangle - CCHCOCH_2CH_3$$
$$\qquad\quad |$$
$$\qquad\quad CH_3$$

Solution:

1. Circle the ketone portion of the formula to indicate the starting esters.

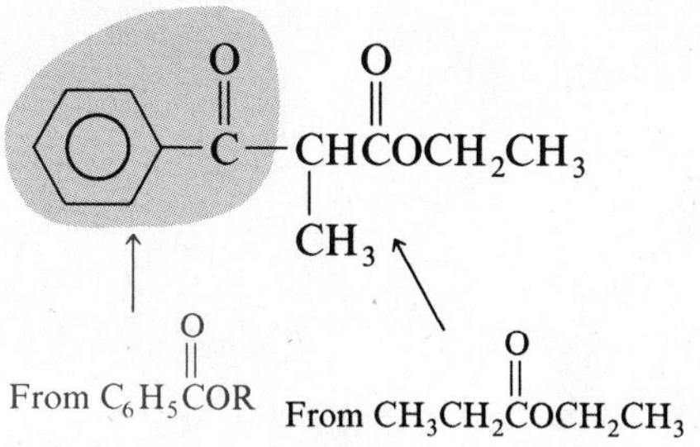

From C_6H_5COR From $CH_3CH_2COCH_2CH_3$

2. Write a flow equation for the preparation:

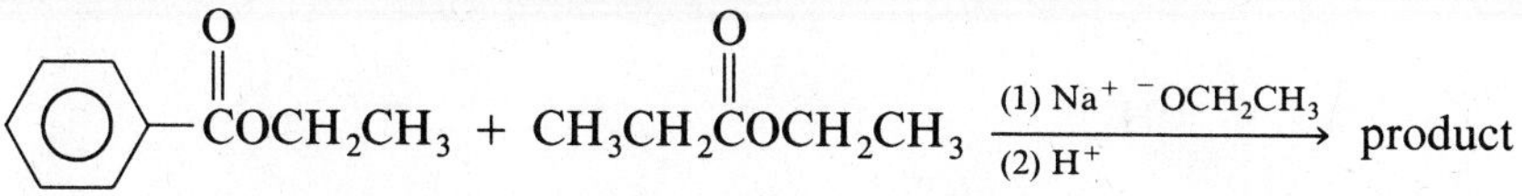

PROBLEM 13.8. Write flow equations for the preparation of the following esters:

(a) 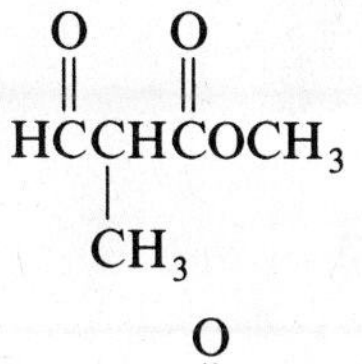 $HCCHCOCH_3$
       CH_3

(b) 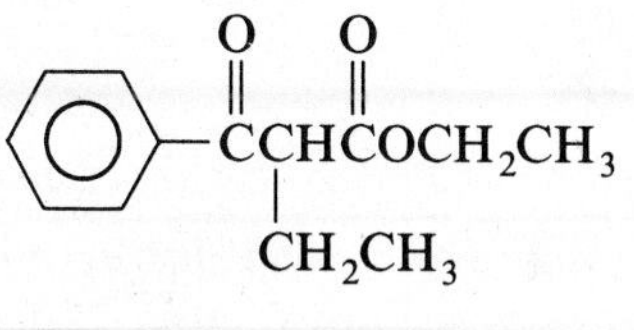 C_6H_5—$CCHCOCH_2CH_3$
           CH_2CH_3

(c)

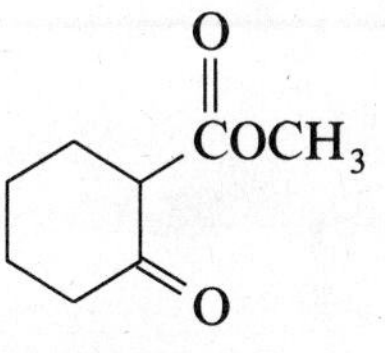

C. Hydrolysis and Decarboxylation of β-Keto Esters

A β-keto ester can be hydrolyzed or saponified to a β-keto carboxylic acid. However, a carboxylic acid containing another carbonyl group beta to the carboxyl group can lose carbon dioxide gas when it is heated. This reaction is called **decarboxylation** (loss of a carboxyl group). Decarboxylation often occurs under acidic hydrolysis conditions, or it may occur when the keto acid is distilled. The product of the decarboxylation is a ketone.

Step 1, Hydrolysis:

$$CH_3CCH_2COCH_2CH_3 + H_2O \xrightarrow[-CH_3CH_2OH]{H^+, \text{ heat}} CH_3CCH_2COH$$

A β-keto ester   *A β-keto acid*

Step 2, Decarboxylation:

$$CH_3CCH_2-C-OH \xrightarrow{\text{heat}} CH_3CCH_3 + CO_2$$

A β-keto acid   *A ketone*   Carbon dioxide

Decarboxylation occurs by way of a cyclic intermediate.

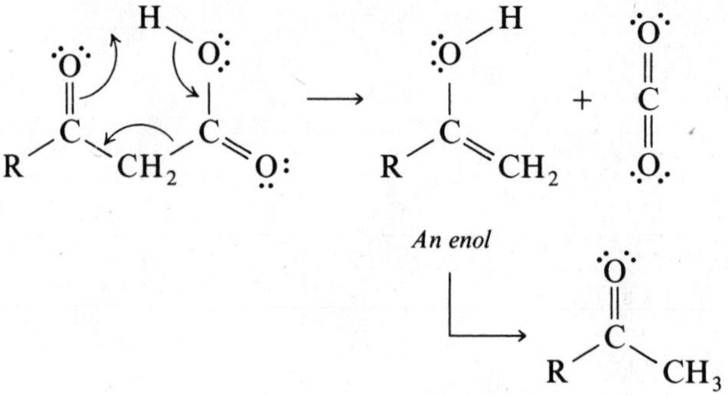

An enol

A ketone

Decarboxylation is general for carboxylic acids with beta carbonyl groups, not only for β-keto acids. For example, β-diacids also can undergo decarboxylation.

$$HOC-CH_2-COH \xrightarrow{\text{heat}} CH_3COH + CO_2$$

Malonic acid   Acetic acid

In biochemical reactions, which are catalyzed by enzymes, decarboxylations of both α- and β-keto acids are common.

$$HOCCH_2CHC-COH \xrightarrow{\text{enzyme}} HOCCH_2CH_2C-COH + CO_2$$

$$\underset{\underset{O}{\overset{\|}{}}}{COH}$$

α-Ketoglutaric acid   Carbon dioxide

Oxalosuccinic acid

an intermediate in the metabolism of glucose

Saponification under alkaline conditions does not cause decarboxylation because the salt of the keto acid, not the acid itself, is heated during the reaction. Thus, a chemist can avoid decarboxylation and prepare the keto acid if that is the desired product.

Step 1, Saponification (No Decarboxylation):

$$CH_3CCH_2COCH_2CH_3 + {}^-OH \xrightarrow[-CH_3CH_2OH]{heat} CH_3CCH_2CO^-$$

A β-keto ester   *A β-keto carboxylate*

Step 2, Acidification (No Decarboxylation):

$$CH_3CCH_2CO^- + H^+ \xrightarrow{cold} CH_3CCH_2COH$$

A β-keto carboxylate   *A β-keto acid*

PROBLEM 13.9.   Complete the following equations:

(a)
$$CH_3CH_2CCHCOCH_3 + H_2O \xrightarrow[heat]{H^+}$$
$$\qquad\qquad |$$
$$\qquad\qquad CH_3$$

(b)
a cyclohexane ring bearing $-COH$ (C=O) and a ring ketone (=O), $\xrightarrow{heat}$

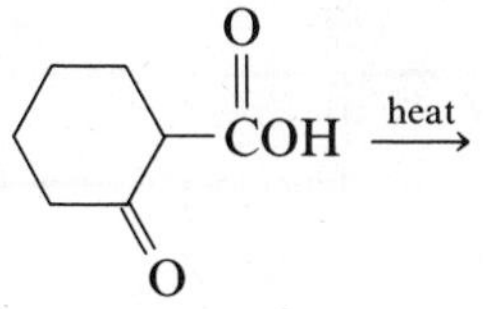

(c)
$$HOCCHCOH \xrightarrow{heat}$$
$$\quad\;\; |$$
$$\quad\;\; CH_3$$

13.3 SYNTHESES USING ALDOL ADDITION REACTIONS AND ESTER CONDENSATION REACTIONS

Examination of Figures 13.1 and 13.2 reveals that a wide variety of products can be synthesized by aldol additions and ester condensations. To solve a problem involving the synthesis of a compound by one of these routes, you must first analyze the formula of the desired product as shown in the figures.

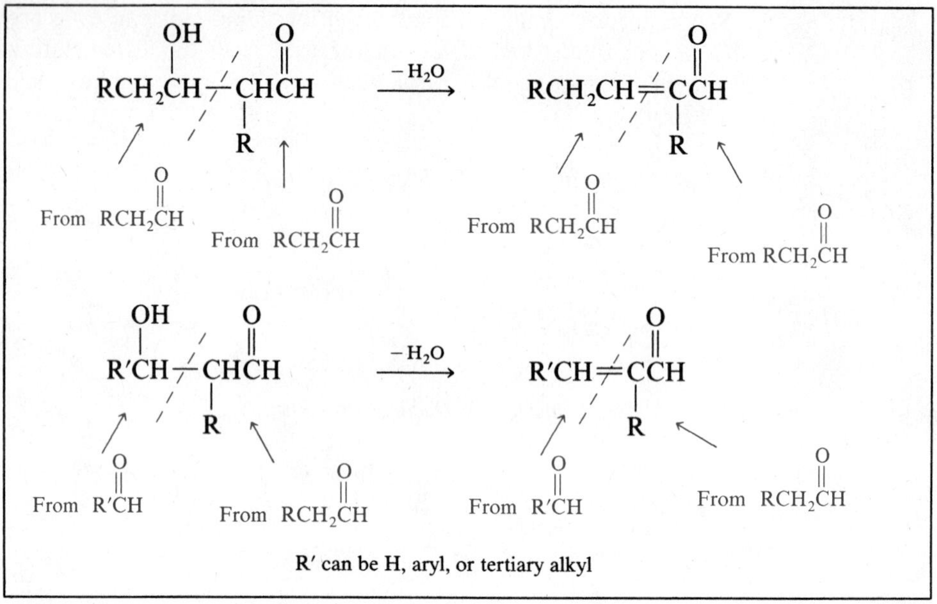

Figure 13.1 A β-hydroxy aldehyde or an α,β-unsaturated aldehyde is synthesized by an aldol addition.

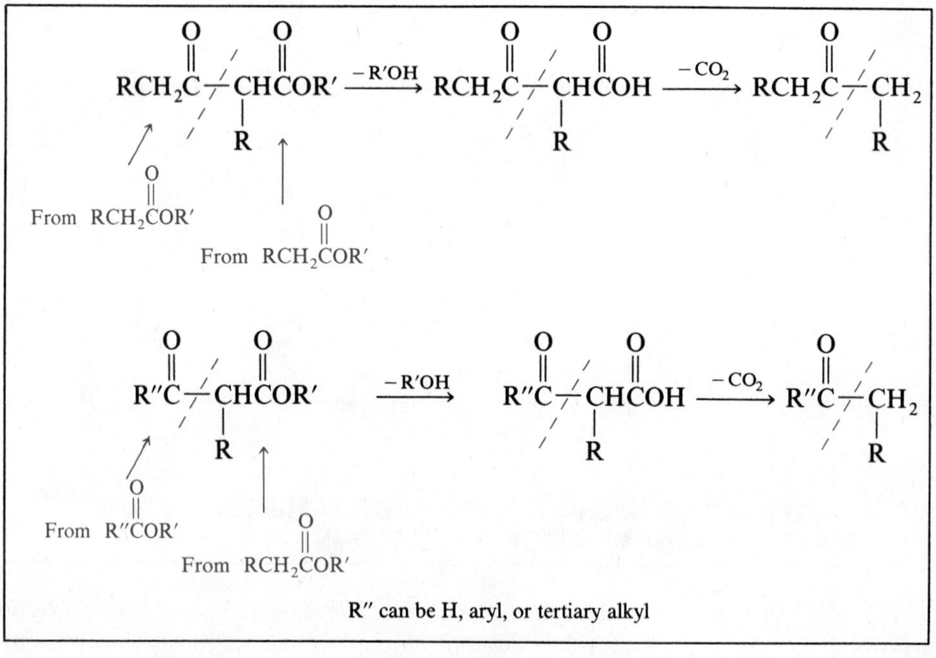

Figure 13.2 A β-keto ester, a β-keto acid, or a ketone is synthesized by an ester condensation.

EXAMPLE

Analyze the following formulas as we have done the general formulas in Figures 13.1 and 13.2. Be sure to show the new carbon–carbon bond and how its position determines the starting materials.

(a)

$$
\underset{\underset{\displaystyle OH}{\displaystyle |}}{(CH_3)_2CHCH\overset{\displaystyle \overset{O}{\|}}{\underset{\displaystyle |}{C}}(CH_3)_2}
$$

with the CH above the carbonyl carbon.

(b)

Phenyl–$\overset{O}{\overset{\|}{C}}\overset{O}{\underset{\underset{\displaystyle CH(CH_3)_2}{\displaystyle |}}{\overset{\|}{C}}HC}OH$

(c)

Phenyl–$\overset{O}{\overset{\|}{C}}CH_2CH_2CH_3$

(d)

Cyclohexenyl–$\overset{O}{\overset{\|}{C}}H$

Solution:

(a) A β-hydroxy aldehyde is synthesized by an aldol addition. The new carbon–carbon bond joins the alpha carbon and the hydroxyl carbon.

$$
(CH_3)_2CHCH\underset{\underset{\displaystyle OH}{\displaystyle |}}{\overset{\overset{\displaystyle CHO}{\displaystyle |}}{-}}C(CH_3)_2
$$

From $(CH_3)_2CHCH\overset{O}{\overset{\|}{}}$ From $(CH_3)_2CHCH\overset{O}{\overset{\|}{}}$

(b) A β-keto acid is synthesized by an ester condensation. The new carbon–carbon bond joins the alpha carbon to the carbonyl carbon.

Phenyl–$\overset{O}{\overset{\|}{C}}\diagup\overset{O}{\underset{\underset{\displaystyle CH(CH_3)_2}{\displaystyle |}}{\overset{\|}{C}}HC}OH$

From Phenyl–$\overset{O}{\overset{\|}{C}}OR$

From $(CH_3)_2CHCH_2\overset{O}{\overset{\|}{C}}OR$
and subsequent saponification
and acidification

(c) A ketone is synthesized by an ester condensation, followed by hydrolysis and decarboxylation. The new carbon–carbon bond joins the ketone carbonyl carbon to the former alpha carbon, which must be a $-CH_2-$ group.

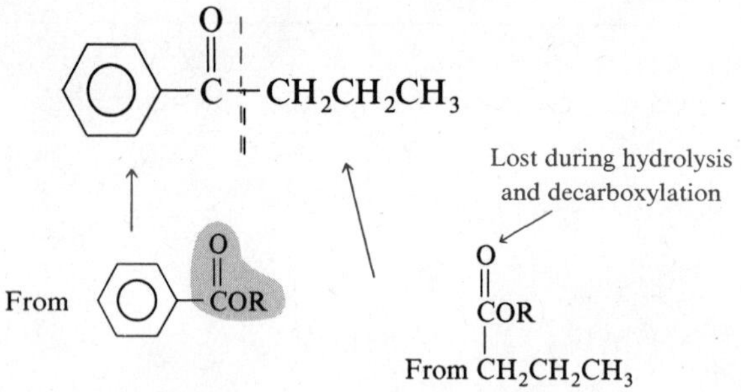

(d) An α,β-unsaturated aldehyde is synthesized by an aldol addition followed by dehydration. The new carbon–carbon bond is the double bond to the alpha carbon. In this cyclic product, both starting aldehyde groups must be in the same chain.

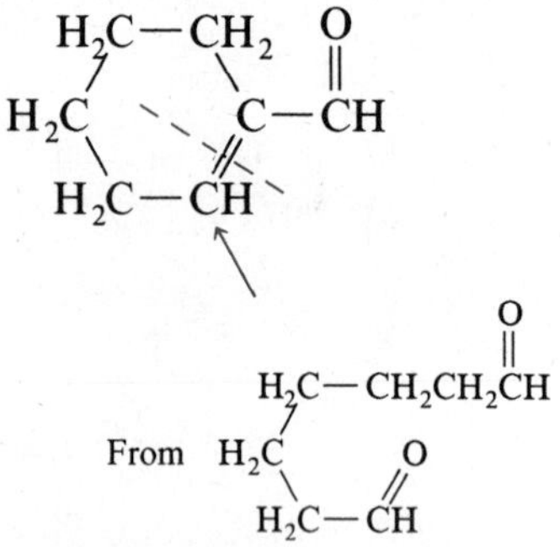

PROBLEM 13.10. Write flow equations for the syntheses of compounds **(a)–(d)** in the preceding example.

PROBLEM 13.11. Write flow equations for the syntheses of the following compounds by aldol additions or ester condensations.

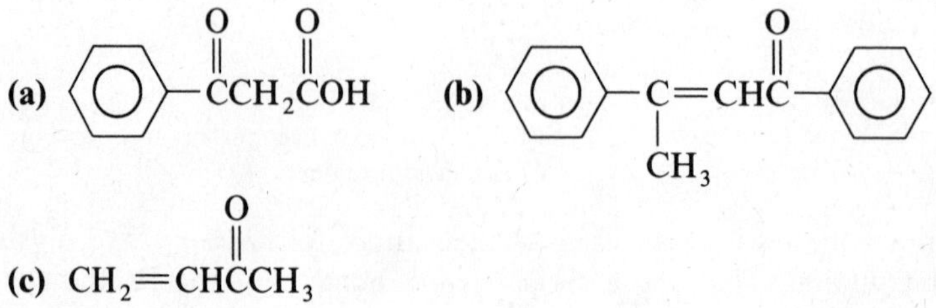

SUMMARY

A hydrogen alpha to a carbonyl group is somewhat acidic because the product **enolate ion** is resonance stabilized. An aldehyde with at least one alpha hydrogen can undergo an **aldol addition** when treated with base.

$$R_2CCH \overset{^-OH}{\rightleftharpoons} \left[R_2\ddot{C}{-}CH \longleftrightarrow R_2C{=}CH \right] \overset{R_2CHCH}{\rightleftharpoons} R_2CHCH{-}\underset{R}{C}{-}CH$$

An aldol

A **mixed aldol addition** is the aldol addition of two different carbonyl compounds. This reaction is useful only with an aldehyde with no alpha hydrogens and another aldehyde or ketone.

$$R'CH + R_2CCR \overset{^-OH}{\rightleftharpoons} R'CH{-}CR_2CR$$

No α hydrogen

An aldol that still contains a hydrogen alpha to the carbonyl group can undergo *dehydration*. Therefore, the observed product of an aldol addition can be an α,β-unsaturated carbonyl compound.

$$R'CH{-}CRCR \overset{-H_2O}{\longrightarrow} R'CH{=}CRCR$$

An ester with an alpha hydrogen can undergo an **ester condensation** when treated with a strong base. Because a hydrogen alpha to *two* carbonyl groups is more acidic than the ROH proton, the initial product is an enolate ion.

$$RCH_2C{-}OR' + CHRCOR' \overset{^-OR'}{\underset{-2R'OH}{\longrightarrow}} RCH_2C{-}\ddot{C}RCOR'$$

An enolate ion

$$\overset{H^+}{\longrightarrow} RCH_2C{-}CHRCOR'$$

A β-keto ester

A **mixed ester condensation** is the condensation reaction of an ester with no alpha hydrogens and another ester.

$$R''C{-}OR' + CHRCOR' \overset{(1) \ ^-OR'}{\underset{(2) \ H^+}{\longrightarrow}} R''C{-}CHRCOR'$$

No α hydrogen

β-Keto esters can be saponified and acidified to β-keto acids.

$$\underset{RCCHRC}{\overset{O\quad\;\; O}{\parallel\quad\;\; \parallel}}{RCCHRC}—OR' \xrightarrow[\text{(2) H}^+\text{, cold}]{\text{(1) }^-\text{OH, heat}} \underset{RCCHRCOH}{\overset{O\quad\;\; O}{\parallel\quad\;\; \parallel}}{RCCHRCOH}$$

A β-keto acid

If an ester with a beta carbonyl group is hydrolyzed in acidic solution, **decarboxylation** can occur.

$$\underset{RCCHR}{\overset{O\qquad O}{\parallel\qquad \parallel}}{RCCHR}—COR' \xrightarrow[\text{heat}]{\text{H}_2\text{O, H}^+} \underset{RCCHR}{\overset{O\qquad O}{\parallel\qquad \parallel}}{RCCHR}—COH \xrightarrow[-\text{CO}_2]{\text{heat}} \underset{RCCH_2R}{\overset{O}{\parallel}}{RCCH_2R}$$

A β-keto ester　　　　　　　*A β-keto acid*　　　　　　*A ketone*

STUDY PROBLEMS

13.12. Circle the most acidic hydrogen(s) in each of the following structures. (You may find it helpful to redraw the cyclic structures to show the carbons and hydrogens.)

(a) $\underset{CH_3CH_2CH}{\overset{O}{\parallel}}{CH_3CH_2CH}$

(b) 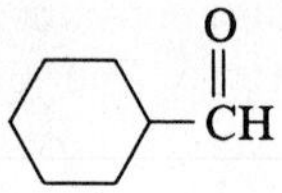 (cyclohexyl)$\overset{O}{\overset{\parallel}{-CH}}$

(c) 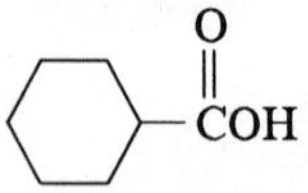 (cyclohexyl)$\overset{O}{\overset{\parallel}{-COH}}$

(d) $\underset{CH_3CH_2COCH_3}{\overset{O}{\parallel}}{CH_3CH_2COCH_3}$

(e) 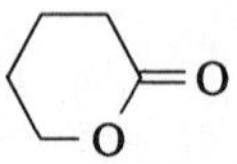 (cyclic structure)=O with O

(f) $\underset{HOCH_2CH_2CH}{\overset{O}{\parallel}}{HOCH_2CH_2CH}$

(g) $\underset{CH_3CCH_2COCH_2CH_3}{\overset{O\quad\; O}{\parallel\quad\; \parallel}}{CH_3CCH_2COCH_2CH_3}$

(h) 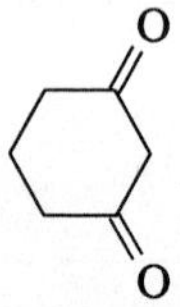 (cyclohexane-1,3-dione structure)

13.13. For each of the above compounds, show the structure of the conjugate base (the anion that would result from an acid-base reaction) and its resonance structure, if any.

13.14. Complete the following equations showing the product of the aldol addition.

(a) 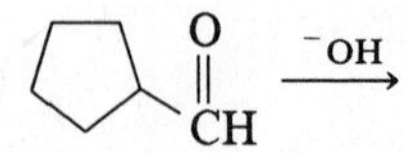 2 (cyclopentyl)$\overset{O}{\overset{\parallel}{CH}}$ $\xrightarrow{^-\text{OH}}$

(b) 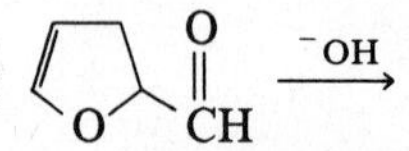 2 (furanyl)$\overset{O}{\overset{\parallel}{CH}}$ $\xrightarrow{^-\text{OH}}$

(c) C_6H_5—$\overset{\overset{\text{O}}{\|}}{CH}$ + $CH_3CH_2\overset{\overset{\text{O}}{\|}}{CH}$ $\xrightarrow{\ ^-OH\ }$

(d) C_6H_{11}—$\overset{\overset{\text{O}}{\|}}{CH}$ + $H\overset{\overset{\text{O}}{\|}}{CH}$ $\xrightarrow{\ ^-OH\ }$

(e) C_6H_5—$CH_2\overset{\overset{\text{O}}{\|}}{CH}$ + C_6H_5—$\overset{\overset{\text{O}}{\|}}{CH}$ $\xrightarrow{\ ^-OH\ }$

(f) C_6H_{10}=O + $H\overset{\overset{\text{O}}{\|}}{CH}$ $\xrightarrow{\ ^-OH\ }$

13.15. Write the formula of the aldehyde that was used to prepare each of the following hydroxy aldehydes by an aldol condensation.

(a) $(CH_3)_2CHCH_2\overset{\overset{\text{OH}}{|}}{CH}\underset{\underset{CH(CH_3)_2}{|}}{CH}\overset{\overset{\text{O}}{\|}}{CH}$

(b) $CH_3CH_2CH_2CH_2\overset{\overset{\text{OH}}{|}}{CH}\underset{\underset{CH_3CH_2CH_2}{|}}{CH}\overset{\overset{\text{O}}{\|}}{CH}$

13.16. The following compounds were prepared by mixed aldol additions. What were the reactants?

(a) $(CH_3)_3C\overset{\overset{\text{OH}}{|}}{CH}CH_2\overset{\overset{\text{O}}{\|}}{C}CH_3$

(b) $HOCH_2CH_2\overset{\overset{\text{O}}{\|}}{CH}$

(c) C_6H_5—$\overset{\overset{\text{OH}}{|}}{CH}\underset{\underset{CH_3}{|}}{CH}\overset{\overset{\text{O}}{\|}}{CH}$

(d) cyclohexane ring with $\overset{\overset{\text{O}}{\|}}{CH}$ and OH substituents

13.17. Each of the following compounds arose from the dehydration of a mixed aldol addition product. Write flow equations for the formation of the aldol products and their dehydration.

(a) C_6H_5—CH=$\underset{\underset{C_6H_5}{|}}{C}\overset{\overset{\text{O}}{\|}}{CH}$

(b) cyclohexanone ring =CH—C_6H_5

(c) $(CH_3)_3CCH$=$CH\overset{\overset{\text{O}}{\|}}{CH}$

(d) cyclopentene ring with CH_3 and $\overset{\overset{\text{O}}{\|}}{C}CH_3$ substituents

13.18. Write equations for acid-base reactions of the following compounds with sodium methoxide (Na^+ $^-OCH_3$). If the product is resonance stabilized, show the resonance structures.

(a) $C_6H_5-CH_2\overset{\displaystyle O}{\overset{\|}{C}}OCH_3$

(b) $CH_3\overset{\displaystyle O}{\overset{\|}{C}}CH_2\overset{\displaystyle O}{\overset{\|}{C}}OCH_3$

(c) $CH_3CH_2O\overset{\displaystyle O}{\overset{\|}{C}}CH_2CH_3$

(d) cyclohexane with $\overset{\displaystyle O}{\overset{\|}{C}}OCH_3$ substituent and ketone =O

13.19. Write equations for the ester condensations of the following esters:

(a) $CH_3CH_2\underset{\underset{\displaystyle CH_3}{|}}{CH}\overset{\displaystyle O}{\overset{\|}{C}}OCH_3$

(b) $CH_3\overset{\displaystyle O}{\overset{\|}{C}}OCH_2CH_2CH(CH_3)_2$

(c) $C_6H_5-CH_2CH_2\overset{\displaystyle O}{\overset{\|}{C}}OCH_2CH_3$

(d) $3,5-(CH_3)_2C_6H_3-CH_2\overset{\displaystyle O}{\overset{\|}{C}}OCH_3$

(e)
$$H_3C-\underset{\underset{\displaystyle H_3C-\underset{\displaystyle CH_2}{|}}{|}}{\overset{\displaystyle \overset{O}{\overset{\|}{C}}OCH_2CH_3}{}} \quad CH_2-\overset{\underset{\displaystyle O}{\overset{\|}{}}}{C}OCH_2CH_3$$

13.20. Complete the following equations showing the mixed ester condensation product.

(a) $C_6H_5-\overset{\displaystyle O}{\overset{\|}{C}}OCH_3 + CH_3\overset{\displaystyle O}{\overset{\|}{C}}OCH_3 \xrightarrow[\text{(2) } H^+]{\text{(1) } ^-OCH_3}$

(b) $CH_3CH_2CH_2\overset{\displaystyle O}{\overset{\|}{C}}OCH_2CH_3 + H\overset{\displaystyle O}{\overset{\|}{C}}OCH_2CH_3 \xrightarrow[\text{(2) } H^+]{\text{(1) } ^-OCH_2CH_3}$

(c) $C_6H_5-\overset{\displaystyle O}{\overset{\|}{C}}OCH_3 + CH_3\overset{\displaystyle O}{\overset{\|}{C}}CH_3 \xrightarrow[\text{(2) } H^+]{\text{(1) } ^-OCH_3}$

(d) $C_6H_5-CH_2\overset{\displaystyle O}{\overset{\|}{C}}OCH_2CH_3 + CH_3CH_2O\overset{\displaystyle O}{\overset{\|}{C}}OCH_2CH_3 \xrightarrow[\text{(2) } H^+]{\text{(1) } ^-OCH_2CH_3}$

13.21. (1) Which of the following compounds can undergo decarboxylation upon acidic hydrolysis and heating? (2) Circle the atoms lost as CO_2 in each.

(a) $CH_3\overset{O}{\underset{}{\overset{\|}{C}}}\overset{}{\underset{CH_3}{\overset{|}{C}H}}\overset{O}{\overset{\|}{C}}OCH_2CH_3$

(b) $CH_3\overset{O}{\overset{\|}{C}}CH_2\overset{}{\underset{CH_3}{\overset{|}{C}H}}\overset{O}{\overset{\|}{C}}OCH_3$

(c) cyclohexanone with $\overset{O}{\overset{\|}{C}}OCH_3$ substituent

(d) cyclohexanone with $CH_2\overset{O}{\overset{\|}{C}}OH$ substituent

13.22. For the compounds in the preceding problem that can undergo decarboxylation, write the formula for the decarboxylation product.

13.23. Write equations for the acidic hydrolysis and decarboxylation of the following esters:

(a) $C_6H_5-\overset{O}{\overset{\|}{C}}CH_2\overset{O}{\overset{\|}{C}}OCH_2CH_3$

(b) $CH_3CH_2\overset{O}{\overset{\|}{C}}\overset{}{\underset{CH_2CH_3}{\overset{|}{C}H}}\overset{O}{\overset{\|}{C}}OCH_2CH_3$

13.24. Write equations for the saponification of the esters in the preceding problem, followed by acidification.

13.25. Write flow equations for the mechanisms of the following aldol addition reactions:

(a) $CH_3OCH_2CH_2\overset{O}{\overset{\|}{C}}H \xrightarrow{\ ^-OH\ }$

(b) $C_6H_5-CH_2CH_2\overset{O}{\overset{\|}{C}}H \xrightarrow{\ ^-OH\ }$

13.26. Write a flow equation for the mechanism of the ester condensation of the following ester:

$C_6H_5-CH_2\overset{O}{\overset{\|}{C}}OCH_3 \xrightarrow{\ ^-OCH_3\ }$

13.27. Write an equation for the mechanism of the decarboxylation of malonic acid.

$HO\overset{O}{\overset{\|}{C}}CH_2\overset{O}{\overset{\|}{C}}OH \xrightarrow{\ heat\ }$

Malonic acid

13.28. Write flow equations for the following conversions:

(a) $C_6H_5-\overset{O}{\overset{\|}{C}}OH \longrightarrow C_6H_5-\overset{O}{\overset{\|}{C}}CH(\overset{O}{\overset{\|}{C}}OCH_2CH_3)_2$

(b)

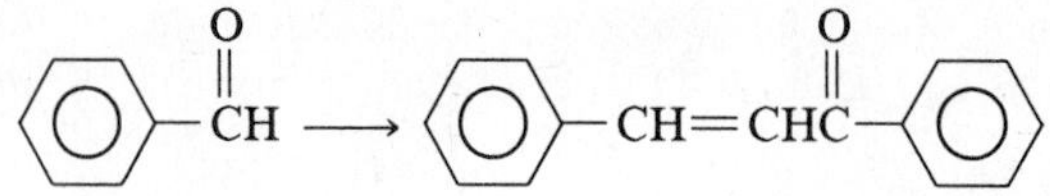

(c)

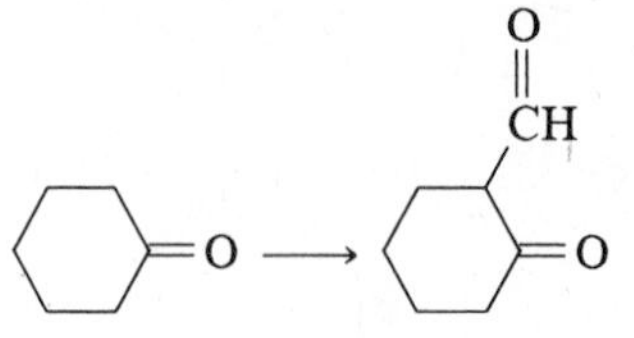

(d)

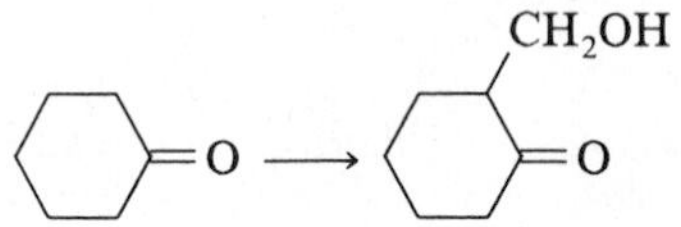

(e) 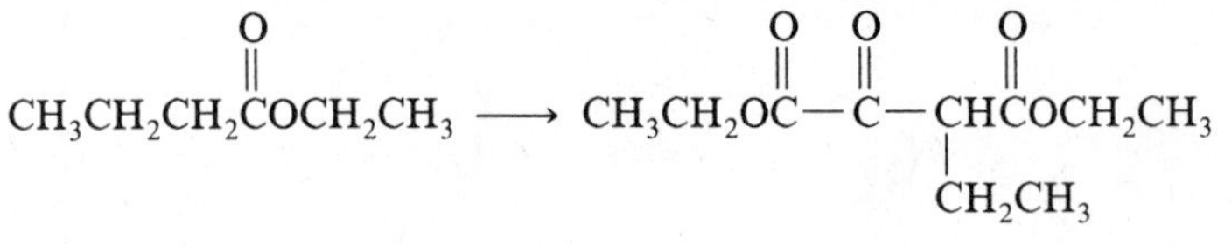 $CH_3CH_2CH_2\overset{\overset{\displaystyle O}{\|}}{C}OCH_2CH_3 \longrightarrow CH_3CH_2O\overset{\overset{\displaystyle O}{\|}}{C}-\overset{\overset{\displaystyle O}{\|}}{C}-\underset{\underset{\displaystyle CH_2CH_3}{|}}{CH}C\overset{\overset{\displaystyle O}{\|}}{O}CH_2CH_3$

(f)

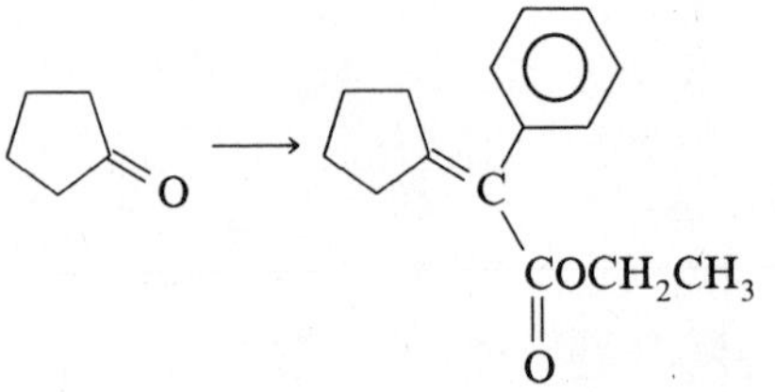

POINT OF INTEREST 13

Ketone Bodies and Diabetes

Diabetes mellitus is a condition in which glucose cannot be properly utilized by the body's cells. In juvenile-onset diabetes, the cause is a deficiency of insulin, a hormone that promotes glucose metabolism. About 0.04% of the population develop juvenile-onset diabetes by the time they are 8 to 12 years old. In adult-onset diabetes, which affects about 3% of people over 40, the problem is decreasing sensitivity of the insulin receptor sites.

Because untreated diabetics cannot utilize glucose, they depend on the metabolism of fats for energy. A similar situation is encountered in cases of starvation. If no carbohydrates are present in the diet, the body's supply of glucose is depleted and, again, the body must depend on fat metabolism for energy.

One of the degradation products of nutrients of all sorts is the acetyl group. This group forms a thioester with a complex molecule called *coenzyme A* and abbreviated HS—CoA. The thioester is called *acetylcoenzyme A* and is abbreviated acetyl CoA or $CH_3CO-SCoA$.

$$H-SCoA$$

Coenzyme A

a thiol (HSR)

$$CH_3\overset{\overset{\displaystyle O}{\|}}{C}-SCoA$$

S instead of O
in a thioester

Acetyl CoA

a thioester $(RC-SR)$
$\overset{}{\underset{\displaystyle O}{\|}}$

In one biochemical pathway, acetylcoenzyme A undergoes an ester condensation to yield a β-keto carboxylate, the acetoacetate ion. The acetoacetate ion formed in this condensation can be either reduced to β-hydroxybutyrate or decarboxylated to acetone. As a group, these three compounds (the acetoacetate, the β-hydroxybutyrate, and acetone) are called **ketone bodies**.

Ketone Body Formation:

$$2CH_3\overset{\overset{\displaystyle O}{\|}}{C}SCoA \xrightarrow[-2HSCoA]{\text{many steps}} CH_3\overset{\overset{\displaystyle O}{\|}}{C}CH_2\overset{\overset{\displaystyle O}{\|}}{C}O^-$$

Acetyl CoA

Ketone bodies

Acetoacetate ion

[H] $-CO_2$

$$CH_3\overset{\overset{\displaystyle OH}{|}}{C}HCH_2\overset{\overset{\displaystyle O}{\|}}{C}O^- \qquad CH_3\overset{\overset{\displaystyle O}{\|}}{C}CH_3$$

β-Hydroxybutyrate ion   Acetone

Ketone bodies are normal metabolites, synthesized in liver and kidney tissue. Once synthesized, ketone bodies diffuse into the bloodstream and are carried to the muscles and other tissues, where they are oxidized to carbon dioxide and water to provide energy.

When a person has diabetes or is starving, the metabolism of large quantities of fatty acids leads to an excess of acetylcoenzyme A, and the level of ketone bodies rises from their normal level of less than 3 mg per 100 mL of blood to values as high as 90 mg per 100 mL. At these high concentrations, the ketone bodies are removed by the kidneys and excreted in the urine. A diabetic individual may excrete as much as 5000 mg of ketone bodies in 24 hours. The odor of acetone may be evident in the breath. This excess of ketone bodies in the system is called **ketosis**.

The ester condensation of acetyl CoA is catalyzed by the enzyme acetoacetyl-CoA thiolase. In the reaction, the acetyl CoA first forms an enolate ion. The sulfur atom of acetyl CoA aids in the formation of the enolate by increasing the acidity of the alpha hydrogens of the acetyl group compared to the acidity of an ordinary ester's alpha hydrogens. The enolate (probably enzyme bound) attacks the carbonyl group of a second acetyl CoA. Loss of HSCoA results in the condensation product.

Enzyme-Catalyzed Thioester Condensation:

$$H \overset{\curvearrowleft}{-} CH_2CSCoA \xrightarrow{-base-H^+} \left[{}^-\overset{\cdot\cdot}{C}H_2 \overset{\curvearrowright}{-} CSCoA \longleftrightarrow CH_2 = CSCoA \right]$$

Base: Thioenolate ion

$$CH_3CSCoA \; + \; :\overset{-}{C}H_2CSCoA \longrightarrow CH_3\overset{\curvearrowleft}{C} - CH_2CSCoA$$
$$\overset{|}{SCoA}$$

$$\xrightarrow[-HSCoA]{H^+} CH_3CCH_2CSCoA$$

Acetoacetyl CoA

Figure 13.3 outlines this reaction (Reaction 1) and the remaining reactions leading to the ketone bodies.

The next step in this sequence is either the direct hydrolysis of the acetoacetyl CoA to acetoacetate ion or, more commonly, condensation of the acetoacetyl CoA with another molecule of acetyl CoA to yield β-hydroxy-β-methylglutaryl CoA (abbreviated HMG CoA). This condensation reaction is Reaction 2 in Figure 13.3. In Reaction 3, HMG CoA is enzymatically cleaved to the acetoacetate ion and acetylcoenzyme A.

In the cells of the liver and kidney, a portion of the acetoacetate is reduced to (R)-β-hydroxybutyrate (reaction 4). The formation of acetone in the liver and kidney (reaction 5) is a decarboxylation reaction (a reaction in which carbon dioxide is lost). This reaction is of minor importance in higher animals. However, in one species of bacteria, *Clostridium acetobutylicum*, the decarboxylation of acetoacetate is a major reaction. In this bacterial system, the decarboxylation is catalyzed by an enzyme involving an imine intermediate.

$$CH_3CCH_2CO^- \; + \; enzyme-NH_2 \xrightarrow{-H_2O} CH_3C \overset{\curvearrowleft}{-} CH_2 \overset{\curvearrowright}{-} C \overset{\curvearrowright}{-} \overset{\cdot\cdot}{O}:^- \xrightarrow{-CO_2}$$

Acetoacetate ion   N—enzyme
                                        H^+   *An imine*

$$CH_3C = CH_2 \longrightarrow CH_3CCH_3 \xrightarrow{H_2O} CH_3CCH_3 \; + \; enzyme-NH_2$$
$$H \overset{\curvearrowleft}{-} N-enzyme \qquad N-enzyme \qquad \qquad O$$

Acetone

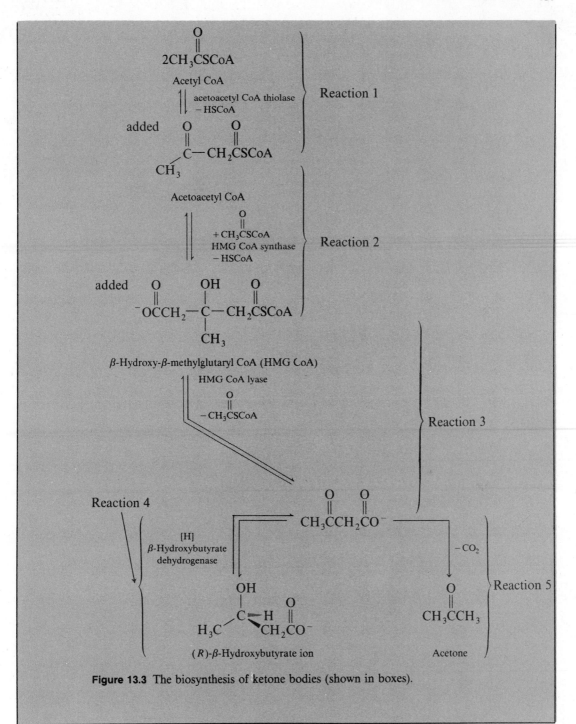

Figure 13.3 The biosynthesis of ketone bodies (shown in boxes).

Chapter 14

Amines

An **amine** is a compound that contains an **amino group** ($-NH_2$, $-NHR$, or $-NR_2$). The amino group contains a nitrogen bonded to one to three carbons (but not a carbonyl group) and an appropriate number of hydrogens (none, one, or two). When one of the carbons bonded to the nitrogen is a carbonyl carbon, the compound is an amide, not an amine (see Section 12.4).

General Formulas for Amines:

$$R\ddot{N}H_2 \qquad R_2\ddot{N}H \qquad R_3N:$$

where R can be alkyl or aryl

Amines are exceptionally important in biochemistry. For example, *serotonin*, a compound found in the central nervous system, transmits nerve impulses and constricts blood vessels. *Histamine* is a compound that is in part responsible for allergic symptoms.

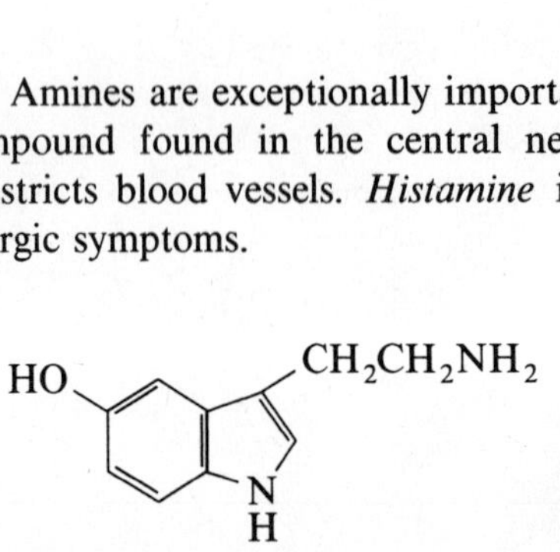

Serotonin

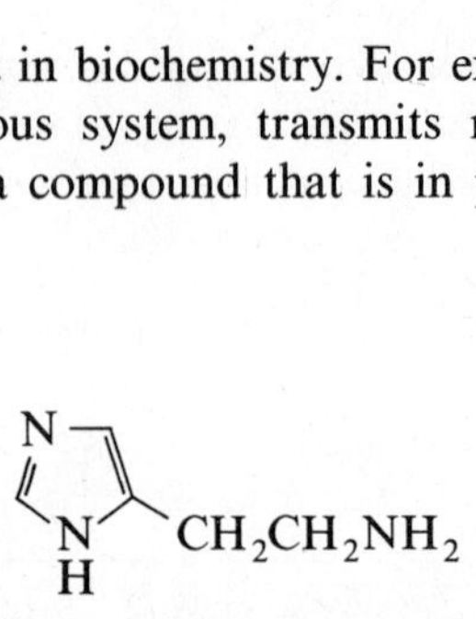

Histamine

14.1 CLASSIFICATION AND BONDING OF AMINES

Amines are classed as *primary* (RNH_2), *secondary* (R_2NH), or *tertiary* (R_3N), depending on the number of carbons bonded to the nitrogen (not to a carbon, as in alcohols).

Some Primary (1°) Amines (One Carbon Bonded to N):

$$CH_3NH_2 \qquad CH_3-\underset{\underset{CH_3}{|}}{\overset{\overset{CH_3}{|}}{C}}-NH_2 \qquad \text{⬡}-NH_2$$

Some Secondary (2°) Amines (Two Carbons Bonded to N):

$$CH_3-NH-CH_3 \qquad \text{⬡}-NHCH_3 \qquad$$

Some Tertiary (3°) Amines (Three Carbons Bonded to N):

$$CH_3-\underset{\underset{CH_3}{|}}{N}-CH_3 \qquad \text{⬡}-\underset{\overset{|}{CH_3}}{N}-\text{⬡} \qquad$$

PROBLEM 14.1. Classify each of the following amines as primary, secondary, or tertiary:

(a) $(CH_3CH_2)_3N$ **(b)** $(CH_3CH_2)_3CNH_2$

(c) ⬠ $\overset{\overset{O}{\|}}{COH}$ **(d)** ⬡$-NHCH_3$

The nitrogen in an amine is sp^3 hybridized, just as the nitrogen in ammonia is. In both ammonia and amines, the bonds are pyramidal—nearly tetrahedral if we include the unshared pair of valence electrons. Figure 14.1 shows the sp^3-hybrid orbitals of an amine.

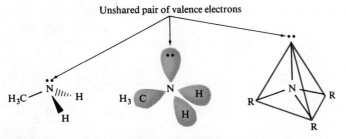

Unshared pair of valence electrons

Figure 14.1 The bonding in an amine: RNH_2, R_2NH, or R_3N. We do not always show the unshared pair of valence electrons in formulas.

We might expect an amine with three different substituents on the nitrogen to be chiral.

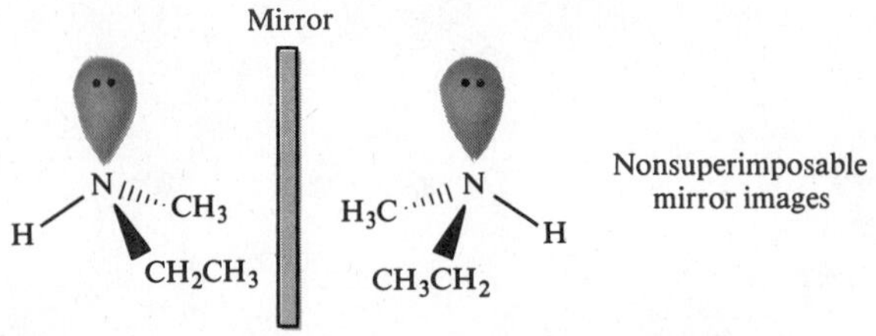

Such structures are indeed chiral because the mirror images are not superimposable. However, amine enantiomers such as those shown above cannot be isolated because they are interconvertible at room temperature.

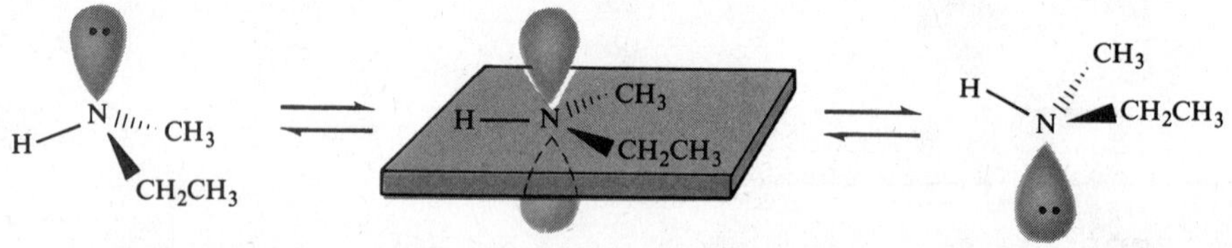

Left-hand structure   Transition state (planar)   Right-hand structure

14.2 NOMENCLATURE OF AMINES

Simple aliphatic amines are usually named by listing the alkyl or aryl substituents and adding the ending *-amine*. The parts of an amine name are combined into one word.

$$CH_3NH_2 \qquad (CH_3)_2NH \qquad CH_3CH_2NHCH(CH_3)_2$$

Methylamine   Dimethylamine   Ethylisopropylamine

Aryl amines are named as derivatives of the parent aromatic amine, such as aniline. Substituents on the nitrogen are named with a prefix preceded by *N-* to show that these substituents are bonded to the nitrogen, not on the ring. (The *N-* prefix is used in some alkyl amine names as well.)

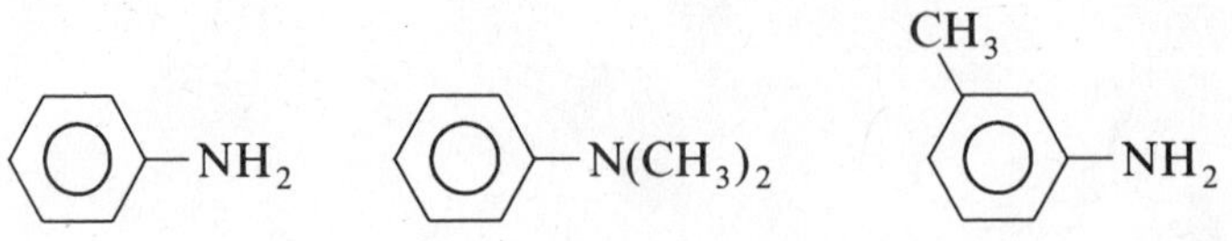

Aniline   *N,N*-Dimethylaniline   *m*-Methylaniline

Heterocyclic amines, with the nitrogen in the ring, have individual names. A few examples follow:

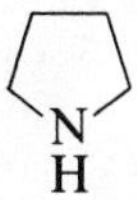

Pyrrolidine 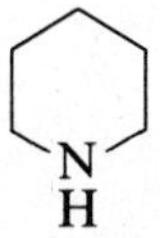   Piperidine 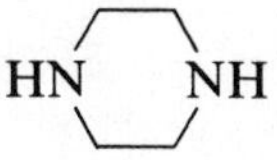   Piperazine

If it is not possible to name the compound as an alkyl amine or an aryl amine, we use the prefix *amino-* for the amino group and designate its position in the parent with a number, if necessary.

$H_2NCH_2CH_2OH$

2-Aminoethanol
(ethanolamine)

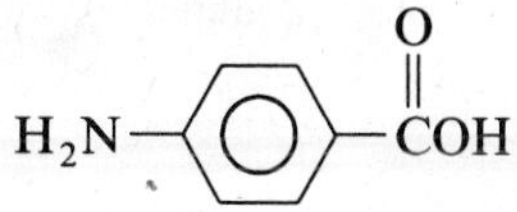

p-Aminobenzoic acid
(PABA, *a sunscreen*)

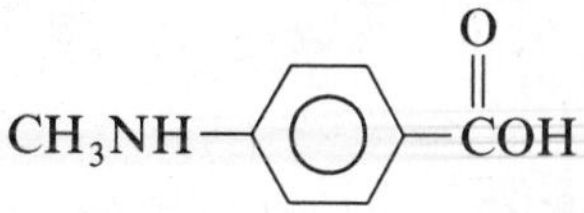

N-Methyl-*p*-amino-
benzoic acid

PROBLEM 14.2.   Name the following amines:

(a) 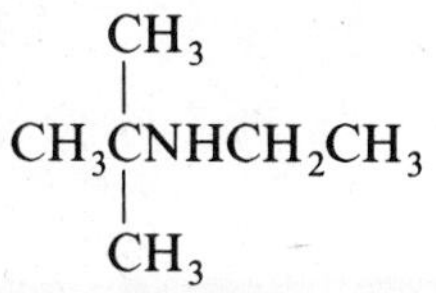 CH_3 — $CH_3CNHCH_2CH_3$ — CH_3

(b) 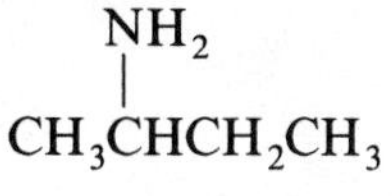 NH_2 — $CH_3CHCH_2CH_3$

(c) 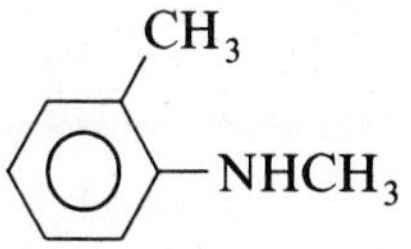 CH_3 — $NHCH_3$

(d) 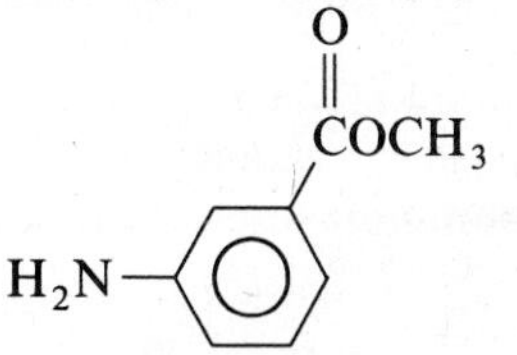 H_2N — $COCH_3$ with O

14.3 PHYSICAL PROPERTIES OF AMINES

Table 14.1 lists the boiling points, water solubilities, and basicity constants of some amines. (We will discuss the basicities in Section 14.4.)

An amine containing an N—H bond can form hydrogen bonds with the unshared valence electrons of oxygen or another nitrogen. Of the two types of hydrogen bonds, the NH-----N bond is much weaker than the OH-----O bond. The reason for the difference in hydrogen bond strength (as measured by dissociation energies) is that the nitrogen is less electronegative than is the oxygen, and the

TABLE 14.1 PROPERTIES OF AMMONIA AND SOME COMMON AMINES

Formula	Name	Bp (°C)	K_b	pK_b
NH_3	Ammonia	-33	1.79×10^{-5}	4.75
CH_3NH_2	Methylamine	-6	45×10^{-5}	3.35
$(CH_3)_2NH$	Dimethylamine	7	54×10^{-5}	3.27
$(CH_3)_3N$	Trimethylamine	3	6.5×10^{-5}	4.19
⬡—NH_2	Cyclohexylamine	134	45×10^{-5}	3.35
⬡—NH_2	Aniline	184	4.2×10^{-10}	9.38
Pyridine structure	Pyridine	116	18×10^{-10}	8.75

$N-H$ bond is thus less polar.

$$ROH\text{-----}:\underset{\cdot\cdot}{O}R \qquad R_2NH\text{-----}:NR_2$$

with H atoms above each O and N

Dissociation energy: 5 kcal/mol   3 kcal/mol

The boiling points of amines containing an $N-H$ bond are intermediate between those of alkanes (no hydrogen bonding) and those of alcohols (strong hydrogen bonding).

	$CH_3CH_2CH_3$	$CH_3CH_2NH_2$	CH_3CH_2OH
	Propane	Ethylamine	Ethanol
Formula weight:	44	45	46
Boiling point (°C):	-42	17	78.5

The boiling points of amines containing no $N-H$ bonds, and thus no hydrogen bonds, are lower than those of amines that do contain hydrogen bonds. Trimethylamine boils at a lower temperature than ethylmethylamine.

	CH_3CHCH_3 (CH_3)	CH_3NCH_3 (CH_3)	$CH_3NHCH_2CH_3$
	Methylpropane	Trimethylamine	Ethylmethylamine
Formula weight:	58	59	59
Boiling point (°C):	-10	3	35

Because amines can form strong hydrogen bonds with the hydrogens in water, amines of low formula weight are water soluble just as alcohols are.

Because N is less electronegative,
its electrons are more readily
donated than those of O.

$$HO-H \dashrightarrow :NR_2$$

Dissociation energy: 7 kcal/mol

PROBLEM 14.3.   Explain why trimethylamine, which has no N—H bond, is as water soluble as methylamine, which contains a hydrogen-bonding —NH$_2$ group.

14.4 BASICITY OF AMINES

Like ammonia, amines are weak bases, far weaker than the hydroxide ion. They can donate a pair of unshared valence electrons from the nitrogen and form a bond with a proton. Water-soluble amines undergo a reversible reaction with water, which releases hydroxide ions.

$$\ddot{N}H_3 + H-\overset{..}{\underset{..}{O}}H \rightleftharpoons {}^+NH_4 + {}^-:\overset{..}{\underset{..}{O}}H$$

$$CH_3\ddot{N}H_2 + H-\overset{..}{\underset{..}{O}}H \rightleftharpoons CH_3\overset{+}{N}H_3 + {}^-:\overset{..}{\underset{..}{O}}H$$

A. Basicity Constants

The basicity of a compound, such as an amine, is specified by the **basicity constant** (K_b), which is the equilibrium constant for the compound's reaction with water.

$$RNH_2 + H_2O \rightleftharpoons R\overset{+}{N}H_3 + {}^-OH$$

$$K_b = \frac{[R\overset{+}{N}H_3][{}^-OH]}{[RNH_2]}$$

where $[R\overset{+}{N}H_3]$ = molarity of $R\overset{+}{N}H_3$
        $[{}^-OH]$ = molarity of ^-OH
        $[RNH_2]$ = molarity of RNH_2
        and the concentration of H$_2$O is included in K_b

The term pK_b, which is analogous to pK_a, is often used to designate the base strength of a compound.

$$pK_b = -\log K_b$$

pK_b value

$\downarrow$

If $K_b = 1.0 \times 10^{-5}$, $pK_b = 5$

As the base strengths of a series of compounds increase, K_b values become larger and pK_b values *decrease*. Table 14.1 lists some K_b and pK_b values.

	NH_3	CH_3NH_2	CH_3NHCH_3
K_b:	1.79×10^{-5}	45×10^{-5}	54×10^{-5}
pK_b:	4.75	3.35	3.27

Increasing base strength
(increasing K_b; decreasing pK_b)

PROBLEM 14.4. On the basis of K_b values, which of the following compounds is the strongest base?

(a) $(CH_3)_3CNH_2$ (b) $(CH_3)_3N$ (c) NH_3

K_b: 6.8×10^{-4} 6.5×10^{-5} 1.79×10^{-5}

PROBLEM 14.5. On the basis of pK_b values, which of the following compounds is the strongest base?

(a) ⬡—NH_2 (b) ⬡—NH_2 (c) $(CH_3)_2NH$

pK_b: 3.35   9.38   3.27

B. Factors Affecting Base Strength

An acid-base reaction is an equilibrium that can be shifted to one side of the equation or the other by the relative stability of reactants or products. Any feature of the structure or environment that stabilizes the protonated amine relative to the free or unprotonated amine increases the base strength of the amine.

Increasing stabilization
relative to products
shifts equilibrium
to this side.

Increasing stabilization
relative to reactants
shifts equilibrium
to this side.

$\downarrow$ $\downarrow$

$$RNH_2 + H_2O \rightleftharpoons \overset{+}{R}NH_3 + {}^-OH$$

Alkylamines, dialkylamines, and trialkylamines have basicity constants some-what larger than that of ammonia (see Table 14.1). This increase in base strength is attributed, in part, to the *electron-releasing inductive effect of alkyl groups*, which helps stabilize the positive charge of the product and shifts the equilibrium to the right.

$$H_3C \rightarrow \overset{\overset{\displaystyle H}{|}}{N}-H + H_2O \rightleftharpoons H_3C \rightarrow \overset{\overset{\displaystyle H}{|}}{\underset{\underset{\displaystyle H}{|}}{\overset{+}{N}}}-H + \; ^-OH$$

More stabilized with
respect to reactants
than is $^+NH_4$

Another factor contributing to the base strength is the *availability of the unshared pair of valence electrons for bonding to a proton*. Compare the basicity constants of cyclohexylamine (45×10^{-5}; pK_b, 3.35) and aniline (4.2×10^{-10}; pK_b, 9.37). Cyclohexylamine is much more basic.

Cyclohexylamine is more basic than aniline because its unshared valence electrons are localized on the nitrogen and are more available for bonding. The unshared valence electrons on the nitrogen in aniline are delocalized by resonance and are less available for bonding to a proton.

Resonance Structures of Aniline:

Delocalized electrons

C. Amine Salts

Alkyl amines undergo complete reaction with strong acids to yield stable com-pounds called **amine salts**, usually named as **alkylammonium salts**. Aryl amines and some heterocyclic amines also react with strong acids, as shown in the specific examples that follow.

General Equation for Amine Salt Formation:

$$R_3N\!: \; + \; H-X \longrightarrow R_3\overset{+}{N}-H \; \; X^-$$

An amine salt

Specific Examples:

$$CH_3NH_2 + HCl \longrightarrow CH_3\overset{+}{N}H_3 \ Cl^-$$

Methylamine   Methylammonium chloride

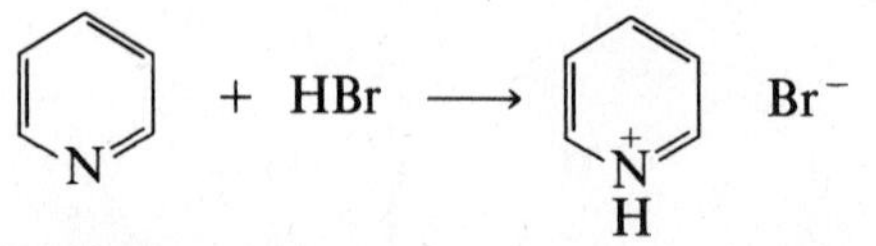

Pyridine   Pyridinium bromide

Amino acids, which can be obtained by hydrolysis of proteins, are saltlike (high melting, water soluble) organic compounds because they undergo an internal acid-base reaction to yield a **dipolar ion**, an ionic structure with positive and negative charges in the same molecule.

Amino Acids, Such as Alanine, Form Dipolar Ions:

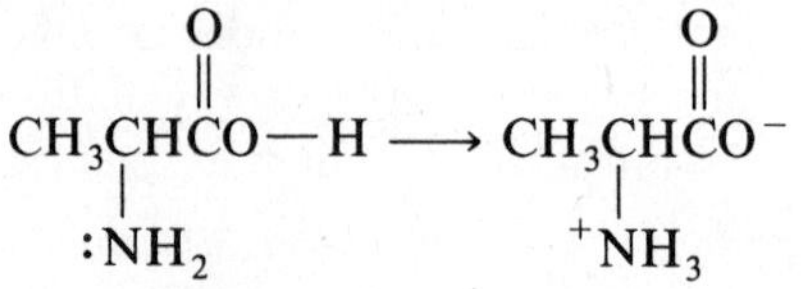

Alanine   *A dipolar ion*

EXAMPLE

(1) Identify each of the following salts as the salt of a primary, secondary, or tertiary amine. (2) Write an equation for the preparation of each.

(a) 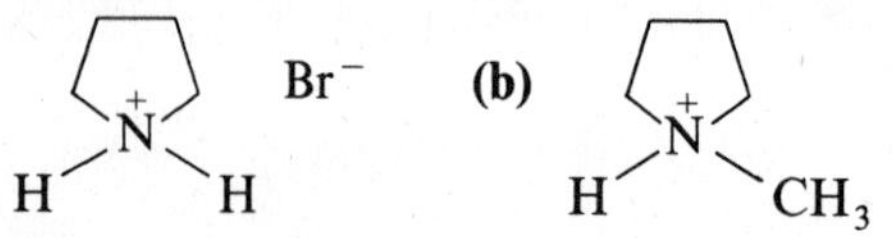   Br⁻   **(b)**   Cl⁻   **(c)** $(CH_3)_3C\overset{+}{N}H_3 \ I^-$

Solution:

1. The definitions for primary, secondary, and tertiary amines also apply to amine salts. In **(a)** the nitrogen is bonded to *two* carbons; this is the salt of a secondary amine. In **(b)** the nitrogen is bonded to *three* carbons; this is the salt of a tertiary amine. In **(c)** the nitrogen is bonded to only *one* carbon; this is the salt of a primary amine.

2. **(a)** 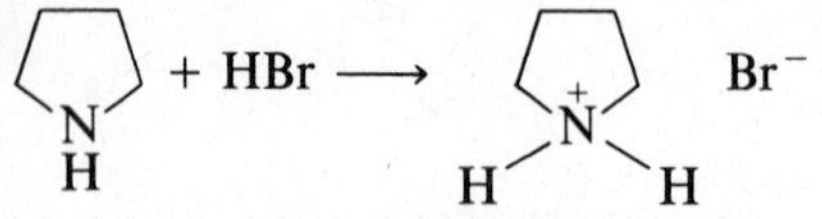 + HBr $\longrightarrow$ Br⁻

(b) ⬡N(ring) + HCl ⟶ ⬡N⁺(ring) Cl⁻
with CH₃ on N; product has H and CH₃ on N⁺

(c) $(CH_3)_3CNH_2 + HI \longrightarrow (CH_3)_3C\overset{+}{N}H_3\ I^-$

An amine can be regenerated from an amine salt by treatment with a base that is stronger than the amine.

$$R_3\overset{+}{N}\text{---}H \ + \ {}^-{:}\overset{\cdot\cdot}{O}H \longrightarrow R_3N\!: \ + \ H\text{---}\overset{\cdot\cdot}{O}H$$

PROBLEM 14.6. Complete the following flow equations by filling in the blanks:

(a) $CH_3CH_2NHCH_2CH_3 \xrightarrow{\ NaH_2PO_4\ }$ _____ $\xrightarrow{\ ^-OH\ }$ _____

(b) ⬡NH (pyrrolidine) $\xrightarrow{\ H_3PO_4\ }$ _____ $\xrightarrow{\ ^-OH\ }$ _____

(c) ⬡N—CH₂COH (with C=O) $\longrightarrow$ _____ $\xrightarrow{\ ^-OH\ }$ _____

An amine can be separated from a mixture of water-insoluble organic compounds by (1) dissolving the mixture in a water-insoluble organic solvent, such as diethyl ether; (2) treating the solution with aqueous acid, which converts the amine to a water-soluble, ether-insoluble salt; (3) separating the aqueous layer from the ether layer; and (4) adding base to the aqueous layer to regenerate the amine. Figure 14.2 shows an example of this type of separation.

PROBLEM 14.7. Acetanilide is prepared from aniline and acetic anhydride. If excess aniline were used, how could it be separated from the product acetanilide? (Use both words and equations in your answer.)

$$\langle\!\bigcirc\!\rangle\text{---}NH\overset{O}{\overset{\|}{C}}CH_3$$

Acetanilide

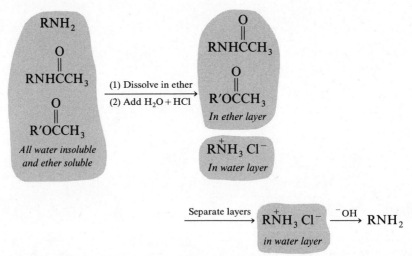

Figure 14.2 Separation of a water-insoluble amine from a mixture of other water-insoluble compounds.

D. Quaternary Ammonium Salts

A **quaternary ammonium salt** is an amine salt in which the nitrogen is bonded to four carbons (Latin *quaterni*, "four each"). Because the unshared pair of electrons on the nitrogen are used in bond formation, the nitrogen has a positive ionic charge. The bonding around the nitrogen of a quaternary ammonium salt is analogous to that around an sp^3-hybridized carbon in an alkane or a substituted alkane. If the nitrogen is bonded to four *different* groups, the cations are chiral and the quaternary ammonium salt exists as a pair of resolvable enantiomers.

Tetramethylammonium chloride
a quaternary ammonium salt

A chiral quaternary ammonium salt

Unlike salts of primary, secondary, or tertiary amines, quaternary ammonium salts are not converted to amines when treated with base because they do not contain an acidic hydrogen.

$$CH_3-\overset{\overset{\displaystyle CH_3}{|+}}{\underset{\underset{\displaystyle CH_3}{|}}{N}}-CH_3 \ + \ {}^-OH \longrightarrow \text{no reaction}$$

TABLE 14.2   SUMMARY OF PREPARATIONS OF AMINES

Reaction	Section reference
Substitution	
$RX^a + NH_3 \longrightarrow R\overset{+}{N}H_3\ X^- \xrightarrow{\ ^-OH\ } RNH_2$	7.6G, 14.5A
Reduction	
$\underset{\displaystyle RCNR'_2}{\overset{\displaystyle \overset{O}{\|}}{}}$ or $RC\equiv N \xrightarrow{\ [H]\ } RCH_2NR'_2$ or RCH_2NH_2	12.4C
$ArNO_2 \xrightarrow{\ [H]\ } ArNH_2$	14.5B

^{a}R must be methyl, primary alkyl, benzylic, allylic, or, in some cases, secondary alkyl.

14.5 PREPARATION OF AMINES

Two general paths leading to amines are substitution and reduction, as shown in Table 14.2.

A. Substitution Reactions of Alkyl Halides

Ammonia and amines contain an unshared pair of valence electrons on the nitrogen; therefore, these compounds can act as nucleophiles in nucleophilic substitution reactions of alkyl halides. The reaction with ammonia yields the salt of a primary amine. Treatment of this amine salt with base liberates the free amine.

$$CH_3CH_2Br \xrightarrow[S_N2]{:NH_3} CH_3CH_2\overset{+}{N}H_3\ Br^- \xrightarrow[-H_2O]{\ ^-OH\ } CH_3CH_2NH_2 + Br^-$$

$\uparrow$
CH_3X or $1°RX$

A 1° amine

Reaction of an alkyl halide with an amine instead of ammonia yields a secondary amine, a tertiary amine, or a quaternary ammonium salt, depending on the amine used.

$$CH_3CH_2Br + CH_3NH_2 \longrightarrow CH_3CH_2\overset{+}{N}H_2CH_3\ Br^- \xrightarrow{\ ^-OH\ } CH_3CH_2NHCH_3$$

A 1° amine   *A 2° amine*

$$CH_3CH_2Br + (CH_3)_2NH \longrightarrow CH_3CH_2\overset{+}{N}H(CH_3)_2\ Br^- \xrightarrow{\ ^-OH\ } CH_3CH_2N(CH_3)_2$$

A 2° amine   *A 3° amine*

$$CH_3CH_2Br + (CH_3)_3N \longrightarrow CH_3CH_2\overset{+}{N}(CH_3)_3\ Br^-$$

A 3° amine   *A quaternary ammonium salt*

Although reasonable yields can be obtained in some sequences, the yields of this type of reaction are often low because the product amine (present in small quantities in the equilibrium mixture) can also react with the alkyl halide to produce an overalkylated product.

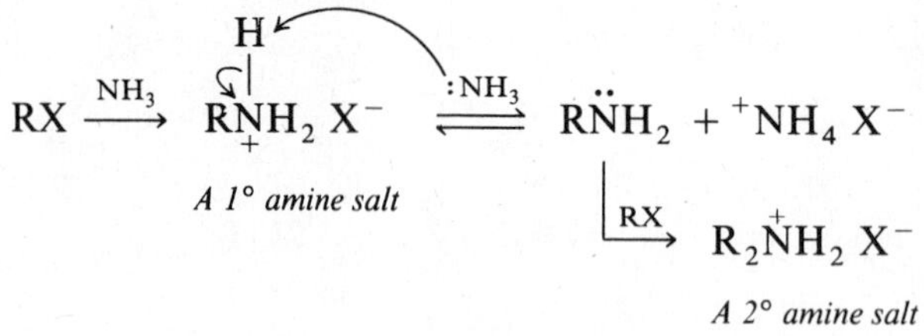

$$RX \xrightarrow{NH_3} R\overset{+}{N}H_2 \ X^- \xrightleftharpoons[\quad]{:NH_3} R\ddot{N}H_2 \ + \ ^+NH_4 \ X^-$$

A 1° amine salt

$$\downarrow RX$$

$$R_2\overset{+}{N}H_2 \ X^-$$

A 2° amine salt

PROBLEM 14.8. One way to minimize product mixtures is to use an excess of one reagent. Which of the following reactions would yield $CH_3CH_2CH_2NH_2$ containing minimal amounts of secondary and tertiary amines? Explain your answer.

(a) excess $CH_3CH_2CH_2Br + NH_3$ **(b)** $CH_3CH_2CH_2Br + $ excess NH_3

B. Reduction of Other Nitrogen Compounds

Reduction of amides or nitriles with lithium aluminum hydride or with hydrogen gas (catalytic hydrogenation) yields amines. With amides, primary, secondary, or tertiary amines can be obtained, depending on the amount of substitution on the amide nitrogen. With nitriles, only primary amines of the type RCH_2NH_2 can be obtained because the carbon bonded to the nitrogen can have only one substituent (R) in the nitrile.

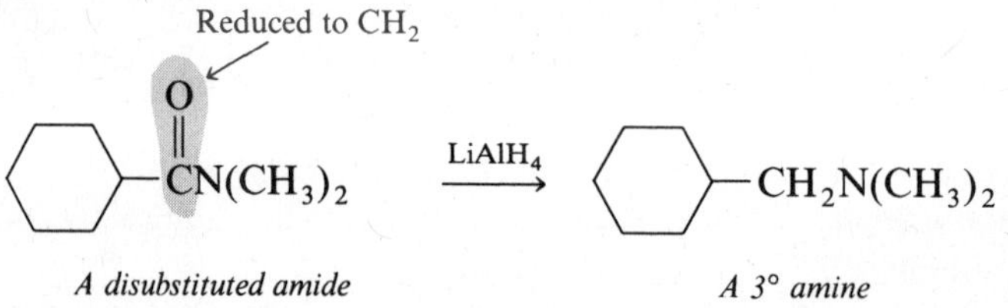

Reduced to CH_2

A disubstituted amide   *A 3° amine*

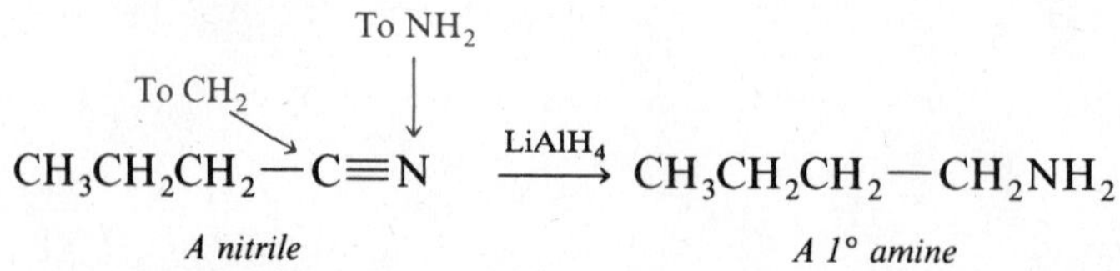

To NH_2

To CH_2

$$CH_3CH_2CH_2 \overset{}{-} C \equiv N \xrightarrow{LiAlH_4} CH_3CH_2CH_2 - CH_2NH_2$$

A nitrile   *A 1° amine*

Nitro groups can also be reduced to primary amines. Aromatic nitro compounds are often used because they are easily obtained from aromatic hydrocarbons by aromatic nitration (see Section 5.3B). Nitro compounds can be reduced by catalytic hydrogenation or by metal reducing agents, such as metallic iron with hydrochloric acid.

$$\text{(benzene)} \xrightarrow[\text{H}_2\text{SO}_4]{\text{HNO}_3} \text{(benzene)}-\text{NO}_2 \xrightarrow[\text{heat}]{\text{Fe, HCl}} \text{(benzene)}-\overset{+}{\text{N}}\text{H}_3 \text{ Cl}^- \xrightarrow{^-\text{OH}} \text{(benzene)}-\text{NH}_2$$

Anilinium chloride

an amine salt

Aniline

PROBLEM 14.9. Write flow equations for the synthesis of the following amines from organic compounds containing no nitrogen.

(a) (cyclopentane)$\text{CH}_2\text{CH}_2\text{NH}_2$

(b) $\text{H}_2\text{N}-$(benzene, with NH_2 top and CH_3 right)$-\text{CH}_3$

PROBLEM 14.10. Complete the following equations:

(a) $\text{N}\equiv\text{C(CH}_2)_4\text{C}\equiv\text{N} \xrightarrow{\text{LiAlH}_4}$

(b) $\text{H}_2\overset{\displaystyle O}{\overset{\displaystyle \|}{\text{NC}}}\text{(CH}_2)_4\overset{\displaystyle O}{\overset{\displaystyle \|}{\text{C}}}\text{NH}_2 \xrightarrow[\text{heat, pressure}]{\text{excess H}_2\text{, catalyst}}$

(c) $\text{O}_2\text{N}-$(benzene)$-\overset{\displaystyle O}{\overset{\displaystyle \|}{\text{C}}}\text{OH} \xrightarrow[\text{heat}]{\text{Fe, HCl}}$

(d) (pyridine ring with N)$-\underset{\displaystyle \text{Br}}{\text{CH}}\text{CH}_2\text{CH}_2\text{CH}_2\text{NHCH}_3 \longrightarrow$ nicotine

14.6 REACTIONS OF AMINES

We have already discussed two reactions of amines in this chapter: (1) reaction with acids to yield amine salts (Section 14.4C) and (2) alkylation with alkyl halides to yield more-substituted amines (Section 14.5A). Table 14.3 summarizes these reactions and others that we will discuss in this section.

A. Acylation

Recall that the term *acylation* means substitution with an acyl group. The nucleophilic acylation of an amine, in which the amine loses a proton and gains an acyl group, yields an amide.

TABLE 14.3 SUMMARY OF REACTIONS OF AMINES

Reaction	Section reference
With acid	
$R_3N^a + HX \longrightarrow R_3\overset{+}{N}H \ X^-$	14.4
Alkylation	
$R_3N^a + R'X^b \longrightarrow R_3\overset{+}{N}R' \ X^-$	14.5A
Acylation	
$R_2NH + Y-\overset{\overset{\displaystyle O}{\|}}{C}R'^c \longrightarrow R_2N-\overset{\overset{\displaystyle O}{\|}}{C}R'$	12.1C, 12.2C, 12.3C
With nitrous acid: see Table 14.4	

[a] R can be H, alkyl, or aryl.

[b] The alkyl halide can be methyl, primary, allylic, benzylic, or, in some cases, secondary.

[c] $R\overset{\overset{\displaystyle O}{\|}}{C}-Y$ can be a carboxylic acid halide, anhydride, or ester.

An acyl group

$$R'_2N-H \ + \ \overset{\overset{\displaystyle O}{\|}}{RC}-Y \longrightarrow R'_2N-\overset{\overset{\displaystyle O}{\|}}{C}R \ + \ H-Y$$

An amide

The compounds that react with amines to yield amides are esters, acid anhydrides, and acid halides. The acid halides are the most reactive of these three because they contain the best leaving groups, while the esters are the least reactive (see the introduction to Chapter 12).

Reaction with Esters:

$$CH_3\overset{\overset{\displaystyle O}{\|}}{C}-OCH_2CH_3 + H\overset{\overset{\displaystyle H}{|}}{N}CH_3 \longrightarrow CH_3\overset{\overset{\displaystyle O}{\|}}{C}-NHCH_3 + CH_3CH_2OH$$

Reaction with Acid Anhydrides:

$$CH_3\overset{\overset{\displaystyle O}{\|}}{C}-O\overset{\overset{\displaystyle O}{\|}}{C}CH_3 + H\overset{\overset{\displaystyle H}{|}}{N}CH_3 \longrightarrow CH_3\overset{\overset{\displaystyle O}{\|}}{C}-NHCH_3 + HO\overset{\overset{\displaystyle O}{\|}}{C}CH_3$$

$$\Big\downarrow^{CH_3NH_2}$$

$$CH_3\overset{+}{N}H_3 \quad \ ^-O\overset{\overset{\displaystyle O}{\|}}{C}CH_3$$

Reaction with Acid Halides:

$$\underset{\displaystyle CH_3C-Cl}{\overset{\displaystyle O \atop \displaystyle \|}{}} + \underset{\displaystyle HNCH_3}{\overset{\displaystyle H \atop \displaystyle |}{}} \longrightarrow \underset{\displaystyle CH_3C-NHCH_3}{\overset{\displaystyle O \atop \displaystyle \|}{}} + \quad HCl$$

$$\Big\downarrow CH_3NH_2$$

$$CH_3\overset{+}{N}H_3 \quad Cl^-$$

B. Reaction with Nitrous Acid

Nitrous acid (HONO) is prepared from sodium nitrite ($Na^+\ ^-NO_2$) and ice-cold hydrochloric acid.

$$Na^+\ ^-:\!\ddot{O}\!-\!\ddot{N}\!\!=\!\!\ddot{O}\!: + H^+ \xrightarrow{0°C} H\!-\!\ddot{O}\!-\!\ddot{N}\!\!=\!\!\ddot{O}\!: + Na^+$$

Nitrous acid undergoes different reactions with different types of amines. We will describe these reactions and then summarize them in Table 14.4 at the end of this section.

Tertiary alkyl amines react with nitrous acid to yield amine salts, just as they do with any acid.

$$R_3N\!: + HONO \longrightarrow R_3\overset{+}{N}H\ ^-ONO$$

Tertiary aryl amines, however, undergo electrophilic aromatic substitution reactions to yield a *nitroso compound*, a compound containing the nitroso group, $-N\!=\!O$.

$$\langle\!\bigcirc\!\rangle\!-\!N(CH_3)_2 + HONO \longrightarrow O\!=\!N\!-\!\langle\!\bigcirc\!\rangle\!-\!N(CH_3)_2$$

Secondary amines react with nitrous acid to yield **N-nitrosoamines**, usually called **nitrosamines**, compounds with the nitroso group bonded to the amine nitrogen. Many compounds containing N-nitroso groups have proved to be carcinogenic (cancer-causing) in laboratory animals.

$$\underset{\displaystyle CH_3NCH_3}{\overset{\displaystyle H \atop \displaystyle |}{}} + HONO \longrightarrow \underset{\displaystyle CH_3NCH_3}{\overset{\displaystyle N\!=\!O \atop \displaystyle |}{}} + H_2O$$

N-Nitrosodimethylamine
(dimethylnitrosamine)

Primary alkyl amines react with nitrous acid to form **alkyl diazonium salts**, $R\!-\!\overset{+}{N}\!\!\equiv\!\!N\ Cl^-$. These salts are unstable, lose nitrogen gas (N_2, an excellent leaving group), and yield carbocations. These unstable carbocations then undergo substitution and elimination reactions to yield mixtures of products.

$$\underset{\substack{|\\CH_3CHCH_3}}{NH_3} \xrightarrow[0°C]{HONO,\ HCl} \underset{\substack{|\\CH_3CHCH_3}}{\overset{+}{\overset{\displaystyle N\equiv N:}{\curvearrowleft}}} \xrightarrow{-:N\equiv N:} \underset{\substack{|\\CH_3\overset{+}{C}HCH_3}}{} \xrightarrow[-H^+]{Cl^-,\ H_2O}$$

$$\underbrace{CH_2{=}CHCH_3 \quad + \quad \underset{\substack{Cl\\|\\CH_3CHCH_3}}{} \quad + \quad \underset{\substack{OH\\|\\CH_3CHCH_3}}{}}$$

A mixture of products

Primary aryl amines also react with nitrous acid to yield diazonium salts. This reaction is the most important of the nitrous acid reactions. Unlike alkyl diazonium salts, aryl diazonium salts are relatively stable if they are kept cold. If the aqueous reaction mixture is allowed to warm, the diazonium ion reacts with water to form a phenol.

$$\langle\!\!\!\bigcirc\!\!\!\rangle{-}NH_2 \xrightarrow[0°C]{HONO,\ HCl} \langle\!\!\!\bigcirc\!\!\!\rangle{-}N_2^+\ Cl^- \xrightarrow[warm]{H_2O} \langle\!\!\!\bigcirc\!\!\!\rangle{-}OH + N_2$$

Aniline $\qquad\qquad$ Benzenediazonium $\qquad\qquad$ Phenol
$\qquad\qquad\qquad\quad$ chloride

Syntheses Using Aryl Diazonium Salts A variety of other reagents can undergo reaction with aryl diazonium salts to yield a wide variety of substituted products. For example, copper(I) chloride and aryl diazonium ions yield aryl chlorides. Many of the products of diazonium reactions, such as phenols (ArOH), aryl iodides (ArI), and aryl nitriles (ArCN), are difficult or impossible to prepare by any other route.

$$\langle\!\!\!\bigcirc\!\!\!\rangle{-}N_2^+\ Cl^- \begin{cases} \xrightarrow[-N_2]{CuCl} \langle\!\!\!\bigcirc\!\!\!\rangle{-}Cl \\[1em] \xrightarrow[-N_2]{CuBr} \langle\!\!\!\bigcirc\!\!\!\rangle{-}Br \\[1em] \xrightarrow[-N_2]{KI} \langle\!\!\!\bigcirc\!\!\!\rangle{-}I \\[1em] \xrightarrow[-N_2]{CuCN} \langle\!\!\!\bigcirc\!\!\!\rangle{-}CN \end{cases}$$

Another type of reaction that aryl diazonium salts undergo is a **coupling reaction** with aromatic rings activated by an *o,p*-directing group. These coupling reactions are examples of electrophilic aromatic substitution reactions and occur primarily at the *para* position of the activated ring. The products, called **azo compounds**, are colored and some are used as dyes.

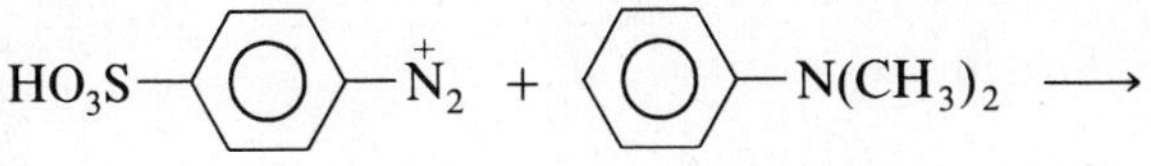

$$HO_3S-\!\!\bigcirc\!\!-\overset{+}{N}_2 \ + \ \bigcirc\!\!-N(CH_3)_2 \ \longrightarrow$$

Activated toward

electrophilic substitution

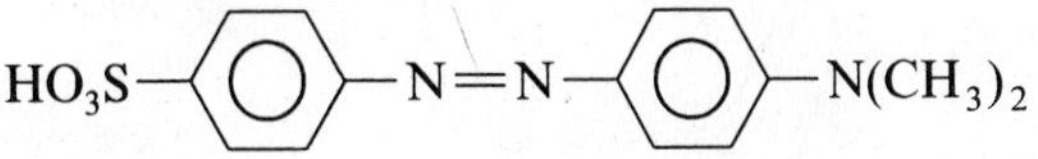

$$HO_3S-\!\!\bigcirc\!\!-N\!=\!N-\!\!\bigcirc\!\!-N(CH_3)_2$$

Methyl orange

an acid-base indicator

PROBLEM 14.11. Complete the following equations, showing the major organic products. (*Hint:* First, ask yourself if the amine reactant is primary, secondary, or tertiary, and alkyl or aryl.)

(a) 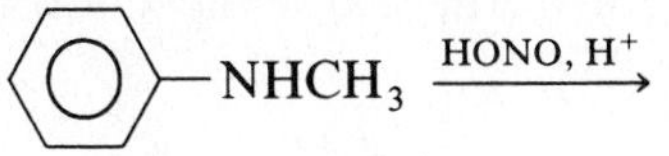 $\bigcirc\!\!-NHCH_3 \ \xrightarrow{\text{HONO, H}^+}$

(b) 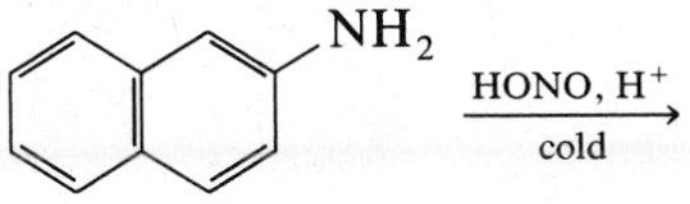 naphthalene$-NH_2 \ \xrightarrow[\text{cold}]{\text{HONO, H}^+}$

(c) 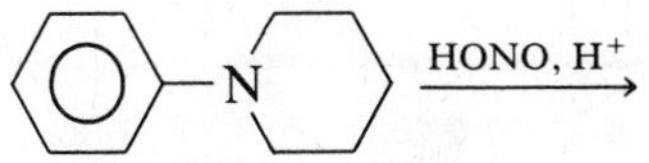 $\bigcirc\!\!-N\!\!\bigcirc \ \xrightarrow{\text{HONO, H}^+}$

(d) 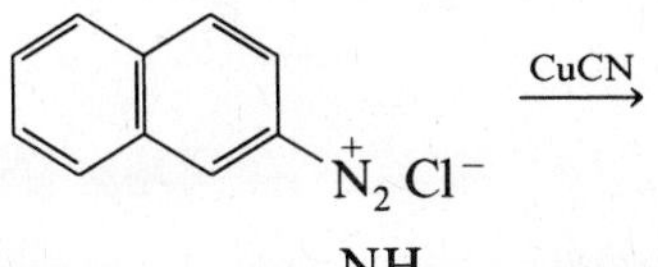 naphthalene$-\overset{+}{N}_2 \ Cl^- \ \xrightarrow{\text{CuCN}}$

(e) 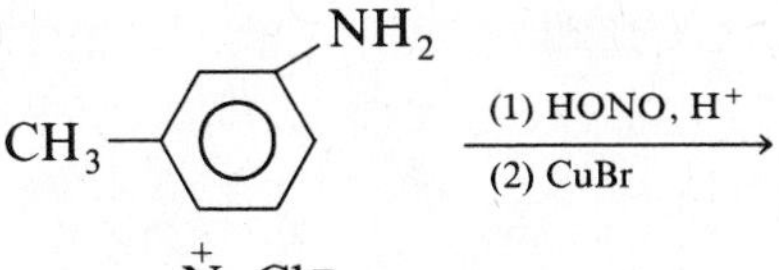 $CH_3-\!\!\bigcirc\!\!-NH_2 \ \xrightarrow[\text{(2) CuBr}]{\text{(1) HONO, H}^+}$

(f) 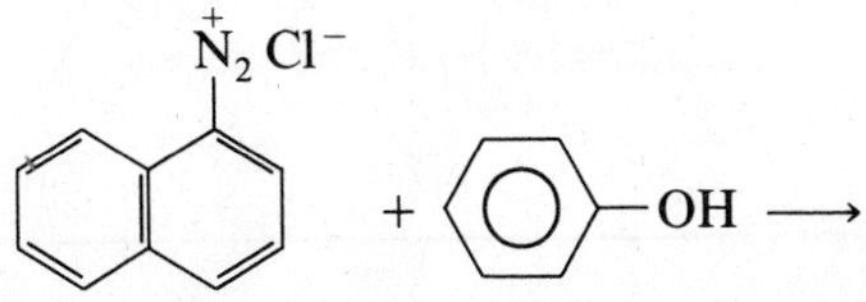 naphthalene$-\overset{+}{N}_2 \ Cl^- \ + \ \bigcirc\!\!-OH \ \longrightarrow$

In synthesis problems using aryl diazonium salts, you may be asked to start with benzene or a substituted benzene. In this type of problem you would first nitrate the ring, then reduce the nitro group to an amino group, and finally treat the arylamine with nitrous acid.

Synthesis of an Aryl Diazonium Salt:

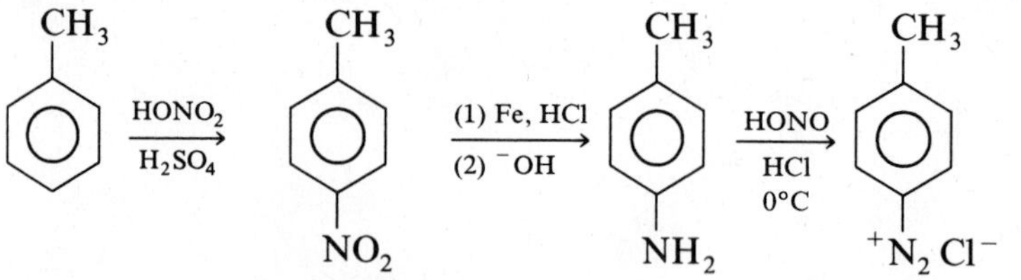

(along with the *o* isomer)

EXAMPLE

How would you prepare *m*-dibromobenzene from benzene?

Solution: Simple dibromination of benzene would yield *o*- and *p*-dibromobenzene, not *m*-dibromobenzene. However, we could obtain the desired compound from a diazonium salt.

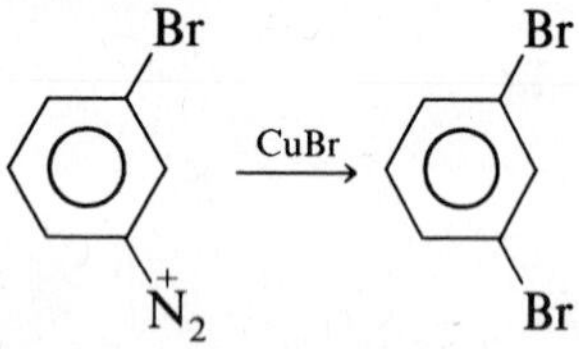

We must prepare the diazonium salt from *m*-bromoaniline. This reagent, in turn, can be obtained from *m*-bromonitrobenzene.

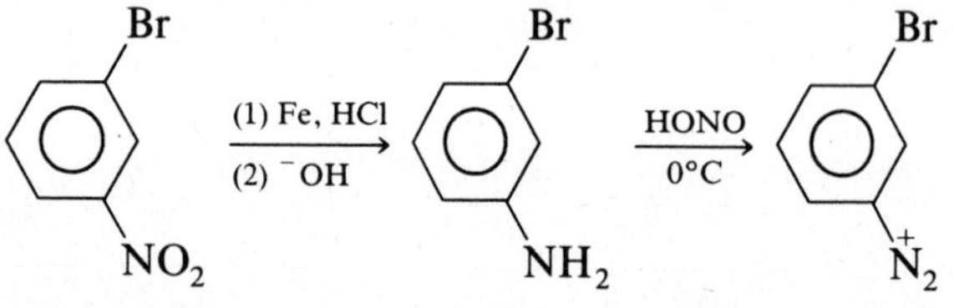

The synthesis of *m*-bromonitrobenzene is a straightforward nitration, which places the *m*-directing nitro group on the ring, followed by bromination.

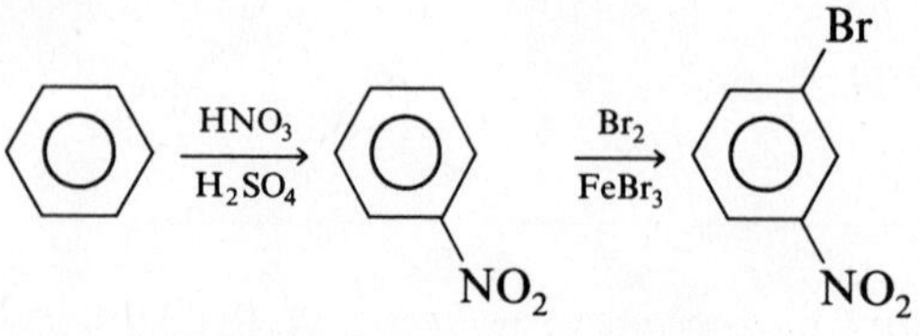

Table 14.4 summarizes the reactions of amines with nitrous acid.

TABLE 14.4   REACTIONS OF AMINES WITH NITROUS ACID

Tertiary amine[a]

$$R_3N + H—ONO \longrightarrow R_3\overset{+}{N}H \; {}^-ONO \quad \text{Amine salt}$$

Secondary amine

$$R_2NH + HO—NO \longrightarrow R_2N—N{=}O \quad \text{Nitrosamine}$$

Primary amine

$$RNH_2 + HONO \xrightarrow{\text{cold}} R—\overset{+}{N}_2 \xrightarrow{-N_2} R^+ \quad \text{(unstable)}$$

$$ArNH_2 + HONO \xrightarrow[\text{HCl}]{\text{cold}} Ar—\overset{+}{N}_2 \; Cl^- \quad \text{Aryl diazonium salt}^b$$

[a] Tertiary aryl amines undergo ring substitution by —NO.

[b] Aryl diazonium salts undergo reaction with CuCl, CuBr, or KI to yield ArX; with CuCN to yield ArCN; with H_2O to yield ArOH; or with activated aromatic rings to yield azo coupling products.

SUMMARY

Amines are classified as **primary** (RNH_2), **secondary** (R_2NH), or **tertiary** (R_3N). Simple amines are named by listing the alkyl groups and adding the ending *-amine*. Aryl amines and heterocyclic amines generally have their own individual names.

Primary and secondary amines form hydrogen bonds in their pure states. All amines can form hydrogen bonds with water or alcohols.

Amines are weak bases. Their basicity is measured by a K_b or pK_b value.

$$K_b = \frac{[R_3\overset{+}{N}H][{}^-OH]}{[R_3N]} \qquad pK_b = -\log K_b$$

Increasing K_b or decreasing pK_b indicates increasing base strength.

Factors that increase the stabilization of the amine salt relative to the amine cause an increase in the base strength of the amine. Examples of factors affecting basicity are the inductive effect of substituents, solvent stabilization of the cation, and resonance stabilization of either the amine or its cation.

Amines react with acids to yield **amine salts**. The amine can be regenerated by treatment with base.

$$R_3N: \xrightarrow{HX} R_3\overset{+}{N}H \; X^- \xrightarrow[-HX]{{}^-OH} R_3N:$$

An amine salt

A **quaternary ammonium salt** ($R_4N^+ \; X^-$) cannot be converted to an amine by treatment with base.

An amine can be prepared by the reaction of ammonia or an amine with an alkyl halide or by reduction of an amide, a nitrile, or a nitro compound.

Amines undergo *substitution reactions* with alkyl halides to yield salts of more-substituted amines. Amines undergo *acylation reactions* with esters, acid anhydrides, or acid halides to yield amides, as shown in Table 14.3.

The most important reaction of amines with nitrous acid is the reaction of a primary aryl amine to yield a relatively stable **aryl diazonium salt**. These salts are used to synthesize a variety of substituted aromatic compounds (see Section 14.6B).

STUDY PROBLEMS

14.12. Classify each of the following amines as primary, secondary, or tertiary:

(a)

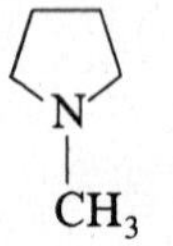

(b)

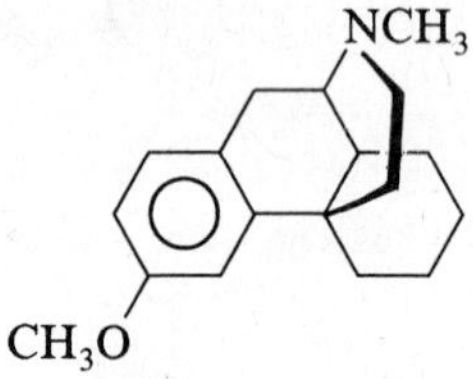

Dextromethorphan

a cough suppressant

(c) $CH_3CHCH_2NH_2$

(d) $(CH_3CHCH_2)_2NH$

(e)

(f)

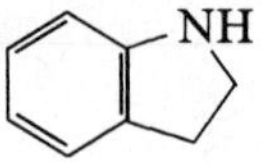

14.13. Name the following amines:

(a) $CH_3CH_2CH_2CH_2NH_2$

(b) $CH_3N(CH_2CH_3)_2$

(c) 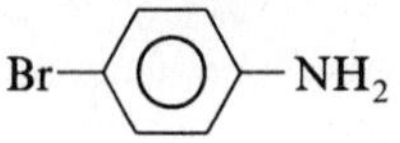 Br—⟨O⟩—NH_2

(d) $CH_3CH_2CH_2NHCH_2CH_2CH_3$

(e) $H_2NCH_2\overset{\overset{\displaystyle O}{\|}}{C}CH_2CH_3$

(f) 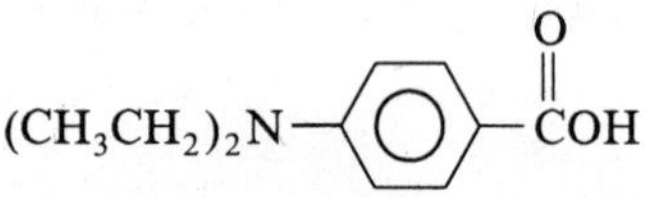 $(CH_3CH_2)_2N$—⟨O⟩—$\overset{\overset{\displaystyle O}{\|}}{C}OH$

(g) 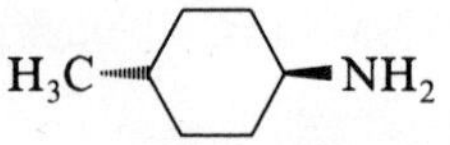 H_3C—⟨ ⟩—NH_2

(h)

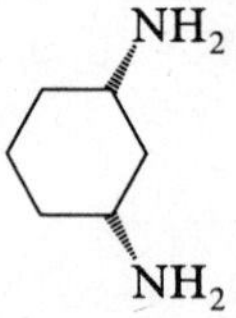

14.14. Draw formulas for the following amines:
(a) *N*-isopropylaniline
(b) benzylamine
(c) trimethylamine
(d) *N*-methylcyclohexylamine
(e) ethyl 2-aminopropanoate
(f) *N,N*-dimethyl-4-penten-1-amine

14.15. Draw formulas showing the hydrogen bonding between methanol and trimethylamine.

14.16. Arrange the following compounds in order of increasing boiling point (lowest boiling point first).
(a) *n*-pentylamine   (b) hexane
(c) 1-pentanol   (d) ethyl *n*-propyl ether
(e) butanoic acid

14.17. Convert the following K_b values to pK_b values: (a) 1.0×10^{-4}; (b) 10.0×10^{-6}; (c) $10.0 \times 10^{-5.5}$.

14.18. Convert the following pK_b values to K_b values: (a) 3.00; (b) 10.00; (c) 9.20.

14.19. Which amine in each pair is the stronger base?
(a) pyridine ($K_b = 18 \times 10^{-10}$) or aniline ($K_b = 4.2 \times 10^{-10}$)
(b) aniline ($pK_b = 9.37$) or *p*-methylaniline ($pK_b = 8.9$)
(c) pyridine ($pK_b = 8.75$) or piperidine ($pK_b = 2.88$)

14.20. What is the ^-OH concentration in a 0.00100 *M* aqueous solution of dimethylamine ($K_b = 5.4 \times 10^{-4}$)?

14.21. Calculate the K_b of a 0.00100 *M* aqueous solution of an amine that has a hydroxide ion concentration of 0.00025 *M*.

14.22. On the basis of structure, predict which compound in each pair would be more basic. Explain your answers.

(a) $CH_3CH_2NH_2$ or $HOCH_2CH_2NH_2$

(b) CH_3—⟨O⟩—NH_2 or Cl_3C—⟨O⟩—NH_2

(c) $CH_3\overset{O}{\overset{\|}{C}}CH_2NH_2$ or $CH_3CH_2\overset{O}{\overset{\|}{C}}NH_2$

14.23. Complete the following equations. If no reaction occurs, write "no reaction."

(a) ⟨N–CH₃⟩ + HBr ⟶

(b) $H_2N\overset{O}{\overset{\|}{C}}(CH_2)_2NH_2$ + excess H^+ $\xrightarrow[H_2O]{\text{cold}}$

(c) $(CH_3)_4N^+\ Cl^- + H_2SO_4$ $\xrightarrow{\text{cold}}$

(d) ⟨quinoline⟩ + HCl ⟶

(e) ⟨O⟩—NH_2 + ⟨⟩—$\overset{+}{N}H_3\ Cl^-$ ⟶

(f) ⟨O⟩—$\overset{+}{N}H_3\ Cl^-$ + ⟨⟩—NH_2 ⟶

(g) $(CH_3)_3\overset{+}{N}H\ Br^- + {}^-OH$ ⟶

(h) 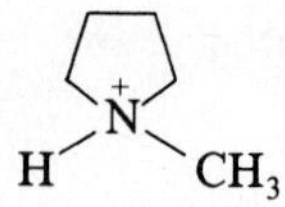 $Br^- + {}^-OH \longrightarrow$

(i) $(CH_3)_4N^+ \ Br^- + {}^-OH \longrightarrow$

(j) 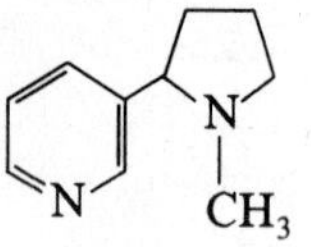 $+ \ 1HBr \longrightarrow$

Nicotine

14.24. Draw the dipolar ion for each of the following amino acids:

(a) $H_2NCH_2\overset{\overset{\displaystyle O}{\|}}{C}OH$

Glycine

(b) $HOCH_2\underset{\underset{\displaystyle NH_2}{|}}{C}H\overset{\overset{\displaystyle O}{\|}}{C}OH$

Serine

(c)

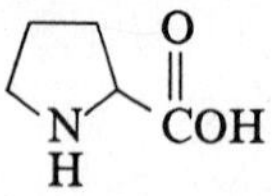

Proline

(d) $HO\overset{\overset{\displaystyle O}{\|}}{C}CH_2CH_2\underset{\underset{\displaystyle NH_2}{|}}{C}H\overset{\overset{\displaystyle O}{\|}}{C}OH$

Glutamic acid

14.25. Write equations showing how each of the following amines can be prepared (1) by reduction of an amide and (2) from an alkyl halide.

(a) 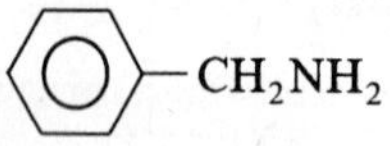 $-CH_2NH_2$ **(b)** $CH_3CH_2CH_2CH_2NH_2$

14.26. Write flow equations for the following conversions. (Assume you can separate *o*- and *p*-isomers.)

(a)

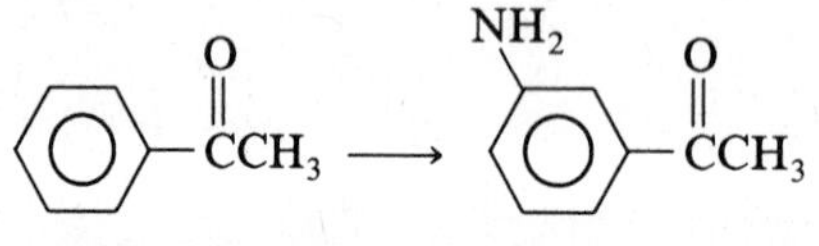

(b) $(CH_3)_3C\overset{\overset{\displaystyle O}{\|}}{C}N(CH_3)_2 \longrightarrow (CH_3)_3CCH_2N(CH_3)_2$

(c) $CH_3CH_2CH_2Br \longrightarrow CH_3CH_2CH_2CH_2NH_2$

(d) $CH_3CH_2CH_2\overset{\overset{\displaystyle O}{\|}}{C}H \longrightarrow CH_3CH_2CH_2\underset{\underset{\displaystyle OH}{|}}{C}HCH_2NH_2$

14.27. Complete the following equations.

(a) $CH_3\overset{\overset{\displaystyle O}{\|}}{C}Cl + 2$ 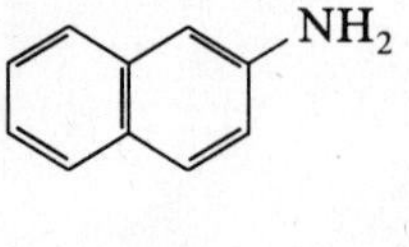 $\longrightarrow$

(b) $\underset{\substack{O \quad O \\ \| \quad \|}}{CH_3COCCH_3}$ + $O_2N\!-\!\langle\!\bigcirc\!\rangle\!-\!NH_2 \longrightarrow$

(c) $CH_3CH_2CH_2\overset{\substack{O \\ \|}}{C}OCH_3$ + $CH_3CH_2CH_2NH_2 \longrightarrow$

(d) $\langle\!\bigcirc\!\rangle$ (with NH_2 top and CH_3 bottom)   + HONO $\xrightarrow{H^+}$

(e) $\langle\!\bigcirc\!\rangle\!-\!CH_2\overset{\substack{NHCH_3 \\ |}}{C}HCH_3$ + HONO $\longrightarrow$

(f) $\langle\!\bigcirc\!\rangle\!-\!N\langle\!\bigcirc\!\rangle$ + HONO $\longrightarrow$

(g)   (piperidine ring with $N\!-\!CH_3$)   + HONO $\longrightarrow$

14.28. Show how the following compounds can be prepared from benzene as the only organic reactant. (Assume that you can separate o- and p-isomers.)

(a) $\langle\!\bigcirc\!\rangle$ (with Cl top and OH bottom)

(b) $\langle\!\bigcirc\!\rangle\!-\!N\!=\!N\!-\!\langle\!\bigcirc\!\rangle\!-\!OH$

(c) $\langle\!\bigcirc\!\rangle\!-\!Br$ (with I top)

(d) $Br\!-\!\langle\!\bigcirc\!\rangle\!-\!CN$

14.29. How would you make the following conversions?

(a) $CH_3CH_2NH_2 \longrightarrow CH_3CH_2NHCH_3$

(b) $CH_3CH_2NH_2 \longrightarrow CH_3CH_2NH\overset{\substack{O \\ \|}}{C}CH_2CH_2CH_3$

(c) $\langle\!\bigcirc\!\rangle\!-\!NO_2 \longrightarrow \langle\!\bigcirc\!\rangle\!-\!\overset{\substack{O \\ \|}}{C}OH$

(d) $\langle \bigcirc \rangle \longrightarrow$ Br$-\langle \bigcirc \rangle-$OH with Br substituents (ortho positions)

(e) $CH_3CH_2CH_2NH_2 \longrightarrow CH_3CH_2CH_2\overset{+}{N}(CH_3)_3 \ Br^-$

(f) $\underset{\displaystyle CH_3CH_2CH}{\overset{\displaystyle \overset{O}{\|}}{}} \longrightarrow \underset{\displaystyle CH_3CH_2CHCOH}{\overset{\displaystyle \overset{O}{\|}}{}}$
        $\underset{OH}{}$

14.30. *Guanidine* is a very alkaline compound found in turnip juice, mushrooms, mussels, and earthworms. Using formulas in your answer, explain why guanidine is so basic.

$$\overset{\displaystyle \overset{NH}{\|}}{H_2NCNH_2}$$

Guanidine

POINT OF INTEREST 14

Quaternary Ammonium Salts

A quaternary ammonium salt with sufficient hydrocarbon character can act as a detergent or an emulsifying agent (an agent that can disperse an oil in water). (Detergents are discussed in Point of Interest 12.) The properties that allow a compound to act as an emulsifying agent arise from a dual structure that contains both hydrocarbon and ionic structural components. The hydrophobic hydrocarbon portion of a quaternary ammonium salt can dissolve in nonpolar organic material, such as grease or oil, while the ionic portion is attracted to water.

A Benzalkonium Chloride:

Attracted to H_2O

$$\langle \bigcirc \rangle-CH_2\overset{\overset{\displaystyle CH_3}{|}}{\underset{\underset{\displaystyle CH_3}{|}}{\overset{+}{N}}}-(CH_2)_{11}CH_3 \qquad Cl^-$$

Attracted to nonpolar
organic material

Cetylpyridinium chloride and tetradecyltrimethylammonium bromide are quaternary ammonium cationic antiseptics, commonly called *quats*. These agents are active against both Gram-positive and Gram-negative bacteria, although a high concentration is needed for the latter class of bacteria. Soaplike solutions of these compounds are used for preoperative skin cleaning, for cleaning wounds and burns, and for removing scabs and skin crusts. Cetylpyridinium chloride is also used in mouthwash formulations.

Common Cationic Antiseptics:

$$CH_3(CH_2)_{14}CH_2 \overset{+}{-N} \text{(pyridinium ring)} \quad Cl^-$$

$$CH_3(CH_2)_{12}CH_2 \overset{CH_3}{\underset{CH_3}{\overset{|}{\underset{|}{+NCH_3}}}} \quad Br^-$$

Cetylpyridinium chloride   Tetradecyltrimethylammonium bromide

How the quats act as antiseptics is not fully understood, but it is thought that they denature the lipoproteins found in the cell membrane.

Choline chloride, $HOCH_2CH_2\overset{+}{N}(CH_3)_3$ Cl^-, which is involved in the transmission of nerve impulses, is a quaternary ammonium salt. **Lecithins** are substituted cholines. These compounds are natural emulsifying agents found in the brain, nerve cells, and the liver. They are also found in egg yolk, soybeans, and other foods.

R and R′ are long hydrophobic hydrocarbon chains.

$$R'CO \overset{O}{\overset{\|}{-}} \overset{CH_2OCR (O\|)}{\underset{CH_2-O-P-OCH_2CH_2\overset{+}{N}(CH_3)_3}{\overset{|}{\underset{\|}{\underset{O}{C-H}}}}}$$

Hydrophilic ionic end

From choline

Egg yolks are one ingredient in mayonnaise. It is the lecithins from the egg yolks that keep the salad oils, with their long hydrocarbon chains, dispersed in the water and vinegar.

One important laboratory use of quaternary ammonium salts is as **phase-transfer agents**. Like crown ethers (Section 9.6), quaternary ammonium salts can carry naked unsolvated inorganic ions into nonpolar organic solvents. In the organic layer, these ions can react with water-insoluble organic compounds. The following nucleophilic substitution reaction of a water-insoluble alkyl halide proceeds rapidly in the presence of a quaternary ammonium salt to yield the product nonanenitrile.

These groups dissolve in the organic solvent— carrying the anion along.

$$(CH_3CH_2CH_2CH_2)_4 N^+ \quad {}^-C \equiv N$$

$$CH_3(CH_2)_7Cl + \underbrace{(CH_3CH_2CH_2CH_2)_4N^+\ ^-C\equiv N} \xrightarrow{S_N2}$$

Fast reaction in the organic solvent

$$CH_3(CH_2)_7C\equiv N + (CH_3CH_2CH_2CH_2)_4N^+\ \ Cl^-$$

Nonanenitrile

Without the phase-transfer agent, the reaction is very slow because reaction between the cyanide ion and the alkyl halide can occur only at the interface between the aqueous phase and the organic phase.

$$CH_3(CH_2)_7Cl \qquad\qquad Na^+\ ^-C\equiv N$$

Soluble in organic phase,   *Soluble in water,*
not in water   *not in organic solvents*

Chapter 15

Spectroscopy

Spectroscopy is the measurement of the interactions of electromagnetic radiation with substances. Spectroscopy is important to organic chemists because the absorption of energy from the electromagnetic spectrum can be correlated with the structure of a compound. This information can be used to provide further insight into bonding of a compound, to determine the structures of unknown compounds, or to measure the quantities of known compounds. In this chapter we emphasize structure determination.

Determination of the structure of organic compounds is important to organic chemists, biochemists, medicinal chemists, and anyone else involved in a chemical or biological area of study. With knowledge of the structure of a compound, a chemist can predict the chemical, physical, and, to a limited extent, biological properties of a compound.

Until around 1950, chemists relied almost entirely on elemental analysis and chemical reactions of compounds (along with logic, skill, intuition, and patience) to determine molecular structures. With the development of spectroscopic instrumentation and computers, the task of structure determination has become much more efficient. With modern spectral instruments, a chemist can rapidly obtain a large amount of information about the structure of a compound using a very small sample of material.

In this chapter we briefly study three spectral tools that are extensively used in structure determinations: infrared spectroscopy, ultraviolet-visible spectroscopy, and nuclear magnetic resonance spectroscopy. Infrared spectra help us identify the functional groups in a molecule. Ultraviolet and visible spectra help us determine the extent of pi-bond conjugation in a compound and are very useful in quantitative analysis. Nuclear magnetic resonance spectra help us determine the carbon skeletons and the hydrocarbon groups of the structure. Individually, each tool provides valuable information. However, when all three tools are used in conjunction, the resulting data can sometimes provide sufficient information that the complete structure of a compound can be deduced without chemical testing.

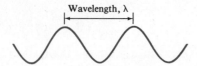

Figure 15.1 The wavelength is the distance from crest to crest in a continuous waveform.

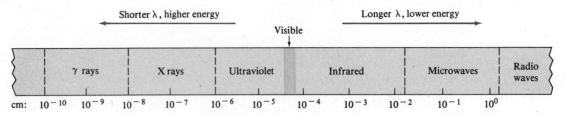

Figure 15.2 The electromagnetic spectrum.

15.1 SOME FEATURES OF ELECTROMAGNETIC RADIATION

Electromagnetic radiation is energy emitted from a source and transmitted through space or matter in the form of waves. The most familiar example of electromagnetic radiation is visible light.

Different types of electromagnetic radiation are characterized by their **wavelength**, symbolized *lambda* (λ), the distance from crest to crest in the waveform, as shown in Figure 15.1.

Figure 15.2 shows the electromagnetic spectrum—the various forms of radiant energy and their range of wavelengths. At one extreme of the spectrum are radio waves, radiation with very long wavelengths. At the other end of the spectrum are x rays and gamma rays—radiation with very short wavelengths. Intermediate are microwaves, infrared rays, visible light, and ultraviolet rays. The wavelengths of these rays range from about 0.01 cm to 0.000001 cm.

The **frequency** of radiation, symbolized *nu* (ν), is also used to characterize electromagnetic radiation. Frequency is defined as the number of cycles (complete waves) per second, or *hertz* (Hz). Frequency is inversely proportional to wavelength —the shorter the wavelength, the greater the frequency.

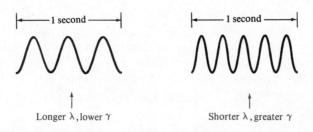

In addition to being a collection of waves, a beam of light can be thought of as a collection of particles, or discrete packets of energy, called **photons**. The energy of a photon is inversely proportional to the wavelength of the light beam—the longer the wavelength, the lower the energy of the photon. In Figure 15.2 we have indicated the lower-energy and higher-energy ends of the electromagnetic spectrum. Radio waves (long wavelength) contain low-energy photons, while gamma rays (short wavelength) contain high-energy photons.

When a compound absorbs energy, it undergoes transitions to different energy states. For example, all covalent bonds of a compound are continuously vibrating and oscillating. A vibrating bond can absorb energy and vibrate at a higher-energy vibrational state. The vibrational states of a bond are *quantized*—the bond can vibrate at certain energy levels only. Therefore, a particular bond can absorb a photon of only the precise energy to excite the bond to a higher vibrational state.

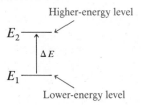

Because photons of different energies (different wavelengths or frequencies) cause vibrational changes in different types of bonds, it is possible to correlate the wavelength or frequency of absorption with the types of bonds in the structure. We measure the wavelength or frequency of the energy absorbed by a compound in an instrument called a **spectrometer** or **spectrophotometer**. Figure 15.3 shows a generalized diagram of a spectrometer. Electromagnetic radiation from a radiation source is passed through a sample. The detector determines whether radiation of a particular wavelength is absorbed or not.

Table 15.1 lists the three types of spectroscopy that we discuss in this chapter, the type of electromagnetic radiation needed for each, and the changes in the molecule caused by the energy absorption.

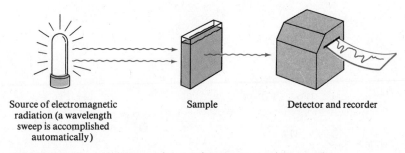

Source of electromagnetic radiation (a wavelength sweep is accomplished automatically) Sample Detector and recorder

Figure 15.3 General diagram showing how a spectrometer works.

TABLE 15.1 SOME TYPES OF SPECTROSCOPY

Type of spectroscopy	Type of radiation needed	Changes caused in molecules
Ultraviolet (uv)-visible	Uv-visible (λ = 200–750 nm)	Promotion of an electron to a higher-energy orbital[a]
Infrared (ir)	Infrared (λ = 2.5–16 μm)	Vibrational excitation
Nuclear magnetic resonance (nmr)	Radio[b]	Change in nuclear spin state in magnetic field

[a] Ultraviolet and visible radiation can also cause some chemical reactions to occur, such as the splitting of halogen molecules into radicals.

[b] The particular wavelength depends on the nuclei being studied and on the strength of the applied magnetic field.

15.2 INFRARED SPECTROSCOPY

Infrared spectroscopy aids in the identification of the bond types present in a compound. With the knowledge of which types of covalent bonds are present and which are absent, we can deduce the types of functional groups that are present or absent in a structure. For example, if a compound has an O—H bond, then the compound could be a carboxylic acid (RCO_2H), an alcohol (ROH), or a phenol (ArOH).

A. Vibrational Modes

The atoms bonded by a covalent bond in a molecule vibrate, or oscillate, much like two balls connected by a spring. Bonds can stretch or shrink. The angles between pairs of bonds can expand or contract. Examples of three different types or modes of vibration of covalent bonds follow:

Stretching   Bending

The vibrational modes of bonds in molecules are restricted to specific energy levels. Each type of bond, such as C—H, O—H, or C=O, has different vibrational energy levels. Therefore, changing the stretching or bending modes of an O—H bond, a C—H bond, or a C=O bond requires different amounts of energy.

Chemists have studied an enormous number of organic compounds and have correlated the bond types with their wavelengths or frequencies of infrared absorption. Therefore, to determine if a compound contains a specific type of bond ($O—H$, for example), we measure the absorption of energy by that compound at the wavelength or frequency known to be characteristic for that bond. If the compound absorbs energy, that type of bond is present in the structure. If energy is not absorbed, that bond is absent.

B. The Infrared Spectrum

In an infrared spectrophotometer, infrared radiation of successively increasing wavelength is passed through the sample and the **percent transmittance** measured. The percent transmittance is an inverse measure of the absorption of infrared radiation. Theoretically, if a compound absorbs no radiation at a particular wavelength, the transmittance is 100%. If all the radiation is absorbed, the transmittance is 0%.

A **spectrum** is a graph of wavelength or frequency versus energy absorbed by a compound. An infrared spectrum is the graph of percent transmittance versus either increasing wavelength or decreasing frequency. Figure 15.4 is the infrared spectrum of 1-hexanol. This particular infrared spectrum and others in this text show percent transmittance versus frequency expressed as **wavenumbers**, which have the units of *reciprocal centimeters* (cm^{-1}). Spectra obtained with other models of infrared spectrophotometers may show percent transmittance versus wavelength in microns (μ) or micrometers (μm), where $1\ \mu = 1\ \mu m = 10^{-6}$ m. In this text, the wavelength units are shown at the *top* of the spectra.

Examine the infrared spectrum in Figure 15.4. Each dip, called a **band** or **peak**, represents absorption of infrared radiation at that frequency by the sample. The

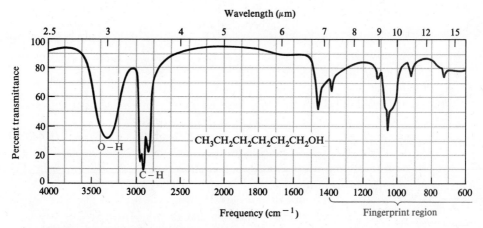

Figure 15.4 Infrared spectrum of 1-hexanol.

frequency at which each absorption occurs is measured at the low point of the dip. The relative intensity of each band is characteristic of the particular bond, such as C—H or C=O.

The entire infrared spectrum, such as the one for 1-hexanol, represents a series of physical constants for the compound. Under the same conditions (solvent, if any; concentration; temperature; instrument type; and so forth), 1-hexanol will always show the same absorption pattern, a unique pattern that will be different from that of any other compound.

C. Functional Groups in Infrared Spectra

The region of an infrared spectrum to the *left* of about 1400–1500 cm^{-1} is called the **functional-group region**. This portion of the spectrum shows absorption arising from distinctive stretching vibrational changes of various bonds and groups. Most absorption peaks in this region of the spectrum are readily identified as arising from one or more specific functional groups. Table 15.2 lists absorption frequencies of many common functional groups. A chart that correlates functional groups with their infrared absorption frequencies is located inside the back cover of this text.

The region of an infrared spectrum to the *right* of 1400–1500 cm^{-1} is called the **fingerprint region**. Absorptions from many types of vibrational changes, both stretching and bending, cause this region of a spectrum to be complex and generally difficult to interpret. However, we can use the fingerprint region to identify a compound by comparing its spectrum with the spectra of known compounds located in a library of infrared spectra. If the spectrum of the unknown compound and the spectrum of a known compound show the same absorption pattern in *both* the functional-group region and the fingerprint region, the compounds are identical.

Carbon–Carbon Absorption Carbon–carbon single bonds in alkanes and alkyl groups show absorption in the infrared spectrum that is too weak to be useful. Carbon–carbon double and triple bonds show weak to strong absorption, depending on the compound. Bonds formed from sp^2-hybridized carbon atoms (C=C bonds and aromatic carbon–carbon bonds) show absorption in the 1450–1700 cm^{-1} range (see Table 15.2). Carbon–carbon triple bonds (C≡C) show weak, but characteristic, absorption between 2100 and 2250 cm^{-1}. The only other bonds that absorb in this general region are C≡N and Si—H.

Carbon–Hydrogen Absorption Carbon–hydrogen absorption occurs at 2700–3300 cm^{-1}, toward the left side of an infrared spectrum.

Figure 15.4 clearly shows strong absorption at about 2750–2800 cm^{-1} arising from bonds between sp^3-hybridized carbons and hydrogens. All organic compounds that contain alkyl groups in their structure show absorption in this region of the infrared spectrum.

Absorption by Alcohols, Amines, and Ethers Infrared absorption by O—H and N—H single bonds is very distinctive because it occurs to the *left* of C—H absorption. The O—H bond usually shows strong, broad absorption between 3200

TABLE 15.2 COMMON INFRARED ABSORPTION FREQUENCIES

Bond type	Absorption frequency (cm^{-1})[a]
Carbon–carbon	
C$=$C (alkenyl)	1600–1700 (5.9–6.2 μm)
C$-$C (aryl)	1450–1600 (6.2–6.9 μm)
C$\equiv$C (alkynyl)	2100–2250 (4.4–4.8 μm)
Carbon–hydrogen	
sp^3 C$-$H	2800–3000 (3.3–3.6 μm)
sp^2 C$-$H	3000–3300 (3.0–3.3 μm)
sp C$-$H	$\sim$ 3300 (3.0 μm)
Aldehyde C$-$H	2820–2900 (3.4–3.5 μm)
	2700–2780 (3.6–3.7 μm)
Alcohols, ethers, phenols, and amines	
O$-$H[b]	3000–3700 (2.7–3.3 μm)
N$-$H[c]	3000–3700 (2.7–3.3 μm)
Alcohol C$-$O	900–1300 (8–11 μm)
Amine C$-$N	900–1300 (8–11 μm)
Ether C$-$O	1050–1260 (7.9–9.5 μm)
Carbonyl compounds[d]	
Aldehyde C$=$O	1720–1740 (5.75–5.81 μm)
Ketone C$=$O	1705–1750 (5.71–5.87 μm)
Carboxyl C$=$O	1700–1725 (5.80–5.88 μm)
Ester C$=$O	1735–1750 (5.71–5.76 μm)
Nitrile	
C$\equiv$N	2200–2400 (4.2–4.5 μm)

[a] Values for the corresponding wavelengths are shown in parentheses.

[b] Hydrogen-bonded.

[c] Hydrogen-bonded; double peak for $-$NH$_2$.

[d] Conjugation shifts carbonyl absorption slightly to the right (lower frequency) of the value given.

and 3500 cm^{-1}. (Again, refer to Figure 15.4, the infrared spectrum of 1-hexanol.) The N$-$H absorption band is found in the same region but is sharper and weaker. The primary amino group in RNH$_2$ (Figure 15.5) shows a double peak in this region (one peak for each H), while the secondary amino group in R$_2$NH shows a single peak.

Hydrogen bonding has a definite effect on the broadness of the infrared absorption band arising from the OH or NH bonds. Amine (NH) absorption is weaker than alcohol (OH) absorption because amines form weaker hydrogen bonds than do alcohols. In the vapor state, where hydrogen bonding does not occur, the O$-$H absorption of an alcohol becomes weaker and is shifted to higher frequen-

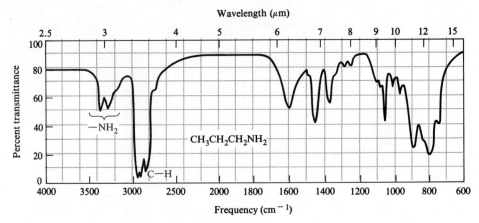

Figure 15.5 Infrared spectrum of *n*-propylamine.

cies. In carboxylic acids, the strong hydrogen bonding between carboxyl groups leads to strong and very broad OH absorption, stronger and broader than even alcohol absorption.

Compounds containing C—O and C—N bonds, including alcohols, amines, ethers, esters, and amides, also show stretching absorption in the fingerprint region between 900 and 1300 cm^{-1}. The absorption is strong if it is due to C—O and weak if it is due to C—N. In either case, the peak may be difficult to identify with certainty because other peaks of similar intensity frequently appear in this portion of the fingerprint region.

Absorption by Carbonyl Compounds Compounds containing carbonyl groups (C=O) show strong and distinctive infrared absorption about 1650–1800 cm^{-1}. In addition to this absorption by the carbonyl group, aldehydes show unusual aldehyde C—H absorption (2820–2900 cm^{-1}; 2700–2780 cm^{-1}). Carboxylic acids also show strong O—H absorption (beginning at 3330 cm^{-1}). Esters also show C—O absorption (1100–1300 cm^{-1}). Amides containing an N—H bond show N—H stretching absorption at 3000–3500 cm^{-1} and N—H bending absorption (1515–1670 cm^{-1}) just to the right of the C=O absorption.

$$
\begin{array}{ccccc}
\overset{\displaystyle O}{\overset{\displaystyle \|}{\text{RCR}'}} &
\overset{\displaystyle O}{\overset{\displaystyle \|}{\text{RC—H}}} &
\overset{\displaystyle O}{\overset{\displaystyle \|}{\text{RC—OH}}} &
\overset{\displaystyle O}{\overset{\displaystyle \|}{\text{RC—OR}'}} &
\overset{\displaystyle O}{\overset{\displaystyle \|}{\text{RC—NHR}'}} \\
\uparrow & \uparrow & \uparrow & \uparrow & \uparrow \\
\text{C=O only} & \text{C=O and} & \text{C=O} & \text{C=O} & \text{C=O} \\
 & \text{aldehyde C—H} & \text{and O—H} & \text{and C—O} & \text{and N—H}
\end{array}
$$

Figure 15.6 shows the spectrum of an ester, illustrating the absorption by alkyl C—H bonds, the C=O group, and the C—O single bond.

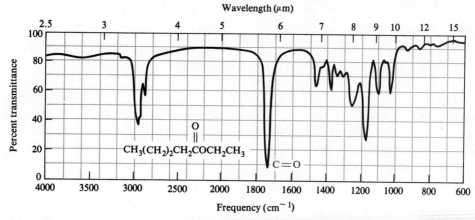

Figure 15.6 Infrared spectrum of ethyl pentanoate.

EXAMPLE

Examine Figure 15.7. (1) Point out each pertinent absorption peak. (2) Is this compound **(a)** butanone (methyl ethyl ketone), **(b)** butanal (butyraldehyde), or **(c)** butanoic acid?

Solution: (1) Beginning from the left side of the spectrum, we see sp^3 C—H absorption at 3000–2800 cm^{-1}, aldehyde C—H absorption at about 2800 cm^{-1} and 2700 cm^{-1}, and C=O absorption at 1720 cm^{-1}. The remaining absorption is in the fingerprint region; we do not usually try to assign these peaks. (2) The

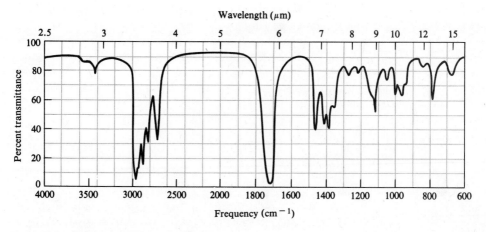

Figure 15.7 Infrared spectrum for the example.

compound is an aldehyde—**(b)**, butanal. It could not be **(c)**, butanoic acid, because the spectrum shows no O—H absorption. The aldehyde C—H absorption distinguishes **(b)** from **(a)**, the ketone.

PROBLEM 15.1.   How would you distinguish between the following pairs of compounds by infrared spectroscopy? In your answer, discuss specific absorption peaks by bond type and by frequency (or wavelength).

(a) 1-propanol and *n*-propylamine

(b) *n*-propylamine and dimethylamine

(c) benzoic acid and benzaldehyde

15.3 ULTRAVIOLET-VISIBLE SPECTROSCOPY

Both ultraviolet (uv) radiation and visible radiation contain photons of higher energy than the energy of infrared photons. When either ultraviolet or visible radiation is absorbed by a compound, the result is the transition of an electron from the ground state (lowest-energy state) of the compound to an excited higher-energy state. These electronic transitions can be correlated with the extent of pi-bond conjugation in a compound. However, ultraviolet and visible spectroscopy finds only minor use in structure identification. The principal use of this type of spectroscopy is in quantitative analysis.

A. The Ultraviolet-Visible Spectrum

Wavelengths of ultraviolet and visible light are measured in *nanometers* (nm), where $1 \text{ nm} = 10^{-9} \text{ m}$. You may also encounter the unit *angstrom*, where $1 \text{ Å} = 10^{-10} \text{ m}$, or the unit *millimicron* (mμ), where $1 \text{ m}\mu = 1 \text{ nm}$.

The wavelengths of visible light—radiation that we can see—range from about 400 nm (violet light) to 750 nm (red light). Intermediate wavelengths give rise to the other colors—blue, green, yellow, orange, and intermediate colors. Ultraviolet radiation is invisible but can cause burns (sunburn, for example); it has wavelengths from 100 nm to 400 nm.

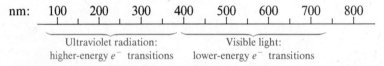

Figure 15.8 shows the ultraviolet spectrum of isoprene, 2-methyl-1,3-butadiene. In an ultraviolet or visible spectrum, the wavelength is plotted versus **absorbance**, which is the logarithm of the ratio of the intensity of radiation entering the sample

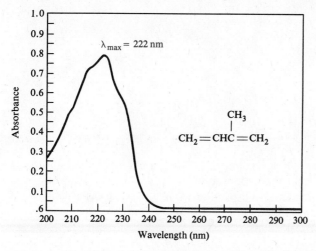

Figure 15.8 Ultraviolet spectrum of isoprene (2-methyl-1,3-butadiene, $CH_2 = CH - \underset{\underset{CH_3}{\displaystyle |}}{C} = CH_2$) in methanol.

(I_0) to the intensity of the radiation leaving the sample (I).

$$\text{Absorbance,} \quad A = \log \frac{I_0}{I}$$

The intensity of a beam of light is proportional to the number of photons—more photons, greater intensity. A low value of intensity of a light beam leaving the sample (I) compared to that entering the sample (I_0) means photons were absorbed by the sample. The lower the value, the greater the number that were absorbed and the greater will be the absorbance.

The absorbance recorded by the instrument is called the *observed absorbance*. Because absorbance depends on the number of molecules in the path of the light, the observed absorbance is usually converted to **molar absorptivity** (ϵ), a calculation that corrects the observed value for concentration and for length of the sample cell in the light path.

$$\text{Molar absorptivity,} \quad \epsilon = \frac{A}{cl}$$

where A = observed absorbance
c = concentration (M)
l = cell length (cm)

EXAMPLE

Calculate the molar absorptivity for a compound with $A = 1.2$ and $c = 9.2 \times 10^{-5}\ M$ in a 1.0-cm cell.

Solution:

$$\epsilon = \frac{A}{cl}$$

$$= \frac{1.2}{(9.2 \times 10^{-5})(1.0)}$$

$$= 13{,}000$$

In an ultraviolet spectrum, the absorptions of photons appear as increases, or peaks, of the traced curve, not as dips as observed in infrared spectra. The wavelength of maximum absorption in an ultraviolet or visible spectrum is called the λ_{max} ("lambda max"). For isoprene, the λ_{max} is 222 nm, and its molar absorptivity is 10,800.

B. Ultraviolet-Visible Absorption and Structure

Ultraviolet and visible radiation do not affect the vibrational modes of covalent bonds. Instead, electrons absorb the photons and are promoted from a filled molecular orbital to the higher-energy unoccupied molecular orbital.

One electron has been promoted to higher-energy orbital.

The vertical arrows represent two pairs of electrons in the orbitals of a double bond.

The wavelength of ultraviolet or visible radiation absorbed by a compound depends on how much energy is required to promote an electron in the compound. Molecules whose electrons all require high energy for promotion absorb only short-wavelength radiation. However, if the molecule has some electrons that require less energy for promotion, then longer wavelengths of radiation are absorbed.

An ultraviolet spectrum is usually run from 200 to 400 nm. The types of electrons that can be promoted at these wavelengths are primarily pi-bond electrons in conjugated systems or in aromatic rings. The greater the amount of conjugation or aromaticity in a compound, the easier it is to promote an electron. (Figure 15.9 shows the wavelengths absorbed by some compounds.) Sufficient conjugation leads to colored compounds—compounds that absorb some wavelengths of visible

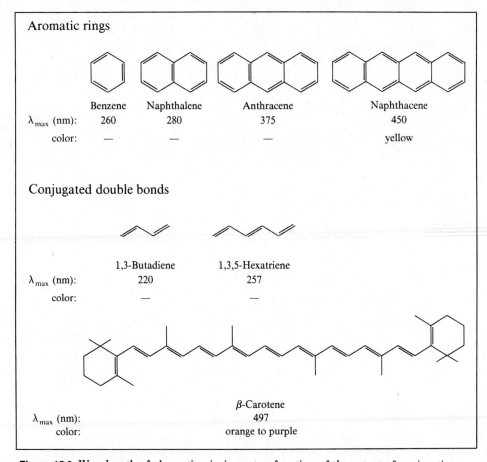

Figure 15.9 Wavelength of absorption is, in part, a function of the extent of conjugation.

light and reflect the remaining wavelengths of visible light back to our eyes (see Figure 15.10).

Increasing conjugation
⟶
Easier electron promotion;
longer wavelengths absorbed

PROBLEM 15.2. A chemist obtained ultraviolet spectra of a series of unsaturated aldehydes. Match each λ_{max} given to a structure:

$$\text{O}$$
$$\|$$
(a) $CH_3CH{=}CH{-}CH$ (1) 270 nm

(b) $CH_3(CH=CH)_2-\overset{\displaystyle O}{\overset{\displaystyle \|}{C}}H$ (2) 343 nm

(c) $CH_3(CH=CH)_3-\overset{\displaystyle O}{\overset{\displaystyle \|}{C}}H$ (3) 312 nm

(d) $CH_3(CH=CH)_4-\overset{\displaystyle O}{\overset{\displaystyle \|}{C}}H$ (4) 217 nm

C. Use of Ultraviolet-Visible Spectroscopy in Quantitative Analysis

Reconsider the equation that relates molar absorptivity (ϵ) with observed absorbance (A), concentration (c, in moles per liter), and cell length (l).

$$\epsilon = \frac{A}{cl}$$

In this equation, ϵ is a physical constant for a specific compound under a specified set of experimental conditions (such as solvent), and these can be held constant by the investigator. On rearranging the equation, we can see that, for a single compound in cells of the same path length, absorbance A is directly proportional to the concentration and that (ϵl) is the proportionality constant.

$$A = (\epsilon l)c$$

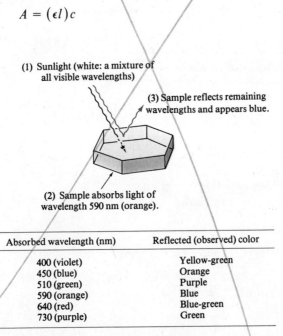

(1) Sunlight (white: a mixture of all visible wavelengths)

(3) Sample reflects remaining wavelengths and appears blue.

(2) Sample absorbs light of wavelength 590 nm (orange).

Absorbed wavelength (nm)	Reflected (observed) color
400 (violet)	Yellow-green
450 (blue)	Orange
510 (green)	Purple
590 (orange)	Blue
640 (red)	Blue-green
730 (purple)	Green

Figure 15.10 An object appears colored to us when it absorbs some wavelengths of white light and reflects the other wavelengths to our eyes.

The direct relationship between concentration and absorbance is called the **Beer-Lambert law** or, more commonly, **Beer's law**. This relationship provides the basis of numerous analytical procedures for quantitative measurements of substances and is of the utmost importance in many phases of chemistry, such as environmental work and clinical analyses (see Point of Interest 15). Because ϵ for many compounds is quite large (10,000 to 100,000) and because absorbance can be measured at a specific wavelength, procedures based on Beer's law are both specific and sensitive.

EXAMPLE

If the ϵ for a compound at a specific wavelength is 30,000, what is the concentration (in M) of that compound in a solution with an absorbance of 0.80 in a 1.0-cm cell?

Solution:

$$A = \epsilon lc$$
$$0.80 = (30{,}000)(1.0)(c)$$
$$c = 2.7 \times 10^{-5}\ M$$

Unfortunately, the direct relationship between absorbance and concentration is not as direct as theory would imply. The problem can be traced in part to the nonlinear response of instrument detectors over large ranges of input. Consequently, in practice, an analyst first obtains a *standard curve* for the instrument using the exact procedure used in the analysis. This is done by measuring and plotting the absorbance of samples containing known amounts of the compound. The analysis of the unknown is then done using the same instrument and procedure, and the unknown's concentration is determined using a standard curve as shown in Figure 15.11.

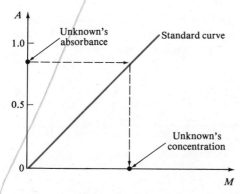

Figure 15.11 Typical standard curve for determination of the concentration of a sample from its absorbance.

15.4 NUCLEAR MAGNETIC RESONANCE SPECTROSCOPY

Nuclear magnetic resonance (nmr) spectroscopy involves the change of the spin state of a nuclear magnetic moment when the nucleus absorbs electromagnetic radiation in a strong magnetic field. Two types of nmr spectroscopy in common use today are ^{1}H (proton) and ^{13}C (carbon-13) nmr. Proton nmr spectra are useful in determination of the hydrocarbon portion of a compound. In recent years, proton nmr spectroscopy has become a standard tool in medical practice, where it is used to measure the density of tissue and thus to show the location of tumors in the tissues. ^{13}C nmr spectroscopy, a fairly new tool, is used to identify the different types of carbons in a compound.

Before discussing the interpretation of nmr spectra, we must first consider the theory of nmr spectroscopy and show why energy can be absorbed by nuclei of atoms.

A. Theory of Nmr Spectroscopy

Some elements have isotopes with nuclei that behave as though they were spinning about an axis, much as the Earth or a child's top does.

Axis of spin

To organic chemists, the two most important isotopes that have nuclear spin are ^{1}H, the common isotope of hydrogen, and ^{13}C, a minor isotope of carbon. The common isotope of carbon (^{12}C) and the common isotopes of nitrogen (^{14}N) and oxygen (^{16}O) do *not* have nuclear spin. The isotope ^{1}H accounts for 99.98% of naturally occurring hydrogen atoms, while ^{13}C accounts for only 1.1% of naturally occurring carbon.

The spinning of a charged particle creates a magnetic field. Isotopes with nuclear spin therefore have small magnetic fields, whose magnitude and direction can be described by a vector called a **magnetic moment**.

Magnetic field

The magnetic moment is a vector used to describe the magnitude and direction of the magnetic field.

The magnetic moment of ^1H or ^{13}C can be likened to a tiny bar magnet, which has two poles, a north pole and a south pole.

When a compound containing ^1H or ^{13}C is placed in a strong applied magnetic field, which we designate as H_0, the magnetic moments of the spinning nuclei become aligned in one of two directions in the applied field. They become aligned either with the direction of the applied field (**parallel**) or opposed to the direction of applied field (**antiparallel**). In nmr spectroscopy, the applied magnetic field is generated by a strong permanent magnet or by an electromagnet; the sample is placed between the poles of the magnet.

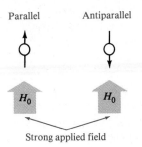

The nuclei with parallel alignment have slightly lower energy than those with antiparallel alignment. For this reason, slightly more than half the nuclei have their magnetic moments in the parallel state, while slightly fewer than half have their moments in the antiparallel state. Figure 15.12 illustrates the alignment of the spin states of the nuclei. Note that only the magnetic moments of the nuclei—*not* the molecule as a whole—become aligned with the field.

When a compound in the applied field is irradiated with electromagnetic radiation of the proper frequency, a few of the lower-energy parallel nuclei absorb the energy and switch their spin states from the lower-energy parallel state to the higher-energy antiparallel state. This conversion from parallel spin state to antiparallel spin state is called a "flip." The higher-energy antiparallel nuclei can lose

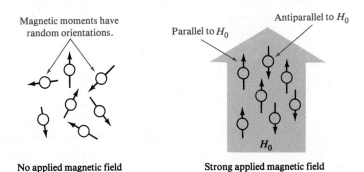

Figure 15.12 Nuclei with spin align their spin states either parallel or antiparallel to a strong applied magnetic field (H_0).

energy to their surroundings (the sample gets warmer) and return to the parallel state. Therefore, energy can be absorbed continuously while a spectrum is being run.

At the frequency that causes flips, the nuclei are said to be in *resonance* with the radiation. This terminology is the origin of the name "nuclear magnetic resonance." (The word resonance used here has no relationship to the word resonance used to describe the delocalization of pi electrons in benzene.)

The exact frequency of energy required for resonance depends on the strength of the applied field and on the isotope being brought into resonance. In a typical nmr spectrometer, a magnetic field of 14,100 gauss is applied. At this field strength, electromagnetic radiation of 60 MHz (60 megahertz, or 60 million hertz) brings ^{1}H nuclei (protons) into resonance. At the same field strength, 15 MHz brings ^{13}C nuclei into resonance. These frequencies are in the radio-frequency range of the electromagnetic spectrum.

B. Molecular Magnetic Fields

If all ^{1}H nuclei or all ^{13}C nuclei came into resonance at the same combination of applied magnetic field and radio frequency, we could not obtain a useful spectrum. However, the nuclei in an organic molecule are surrounded by electrons. Under the influence of the strong magnetic field (H_0), these electrons circulate within their orbitals to create their own tiny magnetic fields, called **molecular fields**, that oppose the field generated by the magnet of the instrument. Thus, the actual magnetic field strength at the location of the ^{1}H or ^{13}C nucleus in the compound is actually slightly less than the magnetic field strength applied to the sample by the instrument. We say that the nucleus is **shielded** from the applied field by the molecular field of the molecule.

$$H_{actual} = H_0 - H_{molecular}$$

The molecular magnetic field of a compound varies within its structure because electron density varies from one part of the molecule to the next. Depending on their location in the molecule, different ^{1}H or ^{13}C nuclei see slightly different magnetic fields. Therefore, these different nuclei come into resonance at slightly different combinations of applied magnetic field strength and radio frequency. By varying the applied magnetic field strength (or the applied radio frequency) over the appropriate range, we can obtain a spectrum that shows energy absorption as a function of the applied magnetic field strength. Therefore, by holding the radio frequency constant and varying the magnetic field—for example, from low field strength to high—while recording the energy absorbed by the sample, we can obtain a spectrum that shows absorption by the nuclei in the sample in increasing order of molecular shielding.

C. The Proton Nmr Spectrum

Figure 15.13 shows the proton (^{1}H) nmr spectrum of *p*-dimethoxybenzene. The spectrum is a graph of the **intensity** of the absorption of radio waves versus **delta values** (δ), where 1 δ value = 1 part per million of the instrument's radio frequency.

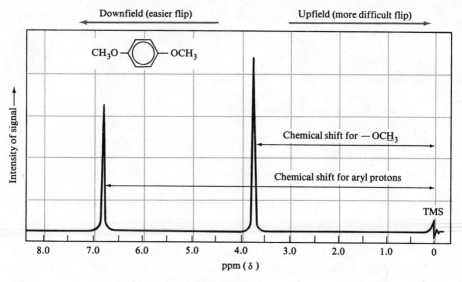

Figure 15.13 Proton nmr spectrum of *p*-dimethoxybenzene.

For example, with an instrument operating at 60 MHz, 1 δ (one millionth of 60 MHz) is equal to 60 Hz. Although δ is calibrated in units of radio frequency, it is more convenient to think of the instrument sweeping from a high δ value to δ value of zero (from the spectrum's left to its right) as representing a change from low applied magnetic field strength to high field strength.

The value for $\delta = 0$ is set where the protons in tetramethylsilane (TMS) produce a signal in the nmr spectrum.

$$CH_3 - \underset{\underset{CH_3}{|}}{\overset{\overset{CH_3}{|}}{Si}} - CH_3$$

Tetramethylsilane (TMS)

*an internal standard used
in nmr spectroscopy*

To determine where this position is, the instrument operator adds a small amount of TMS directly to the solution of sample whose spectrum is to be run; therefore, TMS acts as an internal standard for that sample. Before running the sample spectrum, the operator adjusts the instrument settings so that the absorption by the TMS protons is set at zero on the spectrum paper. An internal standard in a sample allows us to compare nmr spectra obtained using different instruments, even instruments with different combinations of field strength and radio frequency.

Most protons in organic compounds absorb *to the left of*, or **downfield from**, TMS in the nmr spectrum. The distance in δ values from TMS to each signal (absorption) in the spectrum is called the **chemical shift** for the proton or protons giving rise to the signal. In Figure 15.13, two types of protons give two signals with chemical shifts of $\delta 3.75$ and 6.80.

D. Chemical Shifts in Proton Nmr Spectra

Sigma-Electron-Induced Fields   In an organic compound, a proton is covalently bonded to carbon, nitrogen, oxygen, or other atoms by a sigma bond. In a strong magnetic field, the electrons of the sigma bond circulate and create a small magnetic field that opposes the applied field (see Figure 15.14).

The electron density of the sigma bond near the proton can be changed by the presence of other atoms in the structure. A nearby electronegative atom withdraws electron density from the neighborhood of the proton. The decrease in electron density means that a smaller molecular magnetic field opposing the applied field is induced. Consequently, a smaller applied field is needed to cause the spin state of the proton to flip. The signal for a deshielded proton (one surrounded by less electron density) is observed downfield from the signals for protons that are not deshielded by electronegative atoms.

To illustrate this effect, let us consider a series of specific compounds—the series of halomethanes. The electronegativity of the halogens increases from iodine to fluorine: I < Br < Cl < F. The chemical shift for the protons of a halomethane is directly related to the electronegativity of the halogen atom. Of the halogens, fluorine is the most effective at drawing electron density toward itself. Therefore, we find the protons in fluoromethane to be the most deshielded of any of the protons in the series; their signal is found farthest downfield (most distant from TMS).

Iodine is the least electronegative of these halogens. Consequently, the signal from the iodomethane protons are found the farthest upfield (closest to TMS). The

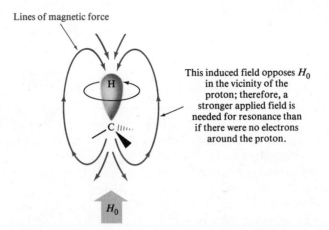

Lines of magnetic force

This induced field opposes H_0 in the vicinity of the proton; therefore, a stronger applied field is needed for resonance than if there were no electrons around the proton.

H_0

Figure 15.14 The electrons in a sigma bond rotate under the influence of a strong applied magnetic field (H_0). The movement of charged species induces a small magnetic field that opposes the applied field.

signals of the protons of chloromethane and bromomethane are intermediate.

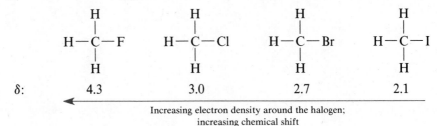

$$\delta: \qquad\qquad 4.3 \qquad\qquad 3.0 \qquad\qquad 2.7 \qquad\qquad 2.1$$

Increasing electron density around the halogen;
increasing chemical shift

PROBLEM 15.3. Suggest a reason why the nmr signal for tetramethylsilane is upfield (to the right of) the signals for most protons in organic molecules.

Pi-Electron-Induced Fields Pi electrons, like sigma electrons, can be induced to circulate and to generate a molecular field. In general, the strength of the magnetic field generated by circulating pi electrons is stronger than that generated by sigma electrons. Consequently, the presence of pi electrons can have a pronounced effect on chemical shifts of nearby protons. Figure 15.15 shows the types of molecular fields induced by pi electrons.

For benzene, alkenes, and aldehydes, the molecular field generated by the moving pi electrons *augments the applied field in the vicinity of the protons*. For these protons, the directions of the molecular field and the applied field are the same. The

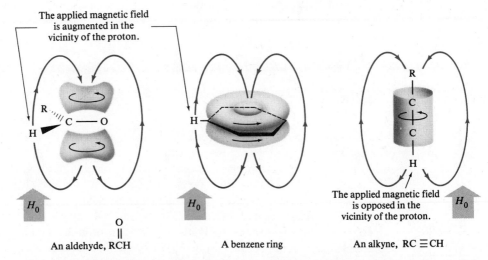

Figure 15.15 The molecular magnetic field induced by pi electrons opposes the applied magnetic field (H_0), but may either augment or oppose H_0 in the vicinity of the proton.

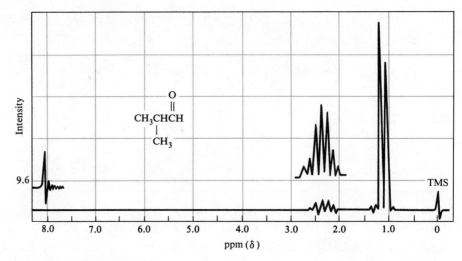

Figure 15.16 Proton nmr spectrum of 2-methylpropanal. The peaks between $\delta 2.0$ and 2.9 have been retraced above the spectrum. The aldehyde proton signal at $\delta 9.6$ has also been traced above the spectrum.

protons on sp^2-hybridized carbons all absorb relatively far downfield because less external field strength is needed to bring them into resonance.

The molecular field surrounding an aldehyde proton is affected by the combined effects of an electronegative oxygen and the carbonyl pi electrons. This proton is sufficiently deshielded that its signal is found at $\delta 9$–10, which is off the spectrum paper of some instruments. Figure 15.16 shows the nmr spectrum of an aldehyde; the operator has traced the off-paper downfield portion of the spectrum ($\delta 8$–12) above the tracing for the rest of the compound.

Table 15.3 is a listing of the expected chemical shifts for different types of protons.

E. Equivalent and Nonequivalent Protons

Protons that are in identical molecular fields in a molecule are said to be **magnetically equivalent**. If protons are magnetically equivalent, they have the same chemical shift—that is, their absorptions occur at the same location in an nmr spectrum.

In general, protons that are magnetically equivalent are also chemically equivalent. It is relatively easy to identify chemically equivalent protons by inspection of a structure. To determine if protons are equivalent, we carry out a simple test. We inspect the structure and mentally replace each proton by another atom. If the replacement results in only one product (not isomers), the protons are equivalent. If, however, the replacement results in isomers, the protons are not equivalent.

TABLE 15.3   SOME TYPICAL PROTON CHEMICAL SHIFTS

Type of proton	δ (ppm)
sp^3 C—H:	
$RC\underline{H}_3$	0.8–1.2
$R_2C\underline{H}_2$	1.1–1.5
$R_3C\underline{H}$	~ 1.5
$ArC\underline{H}_3$	2.2–2.5
$R_2NC\underline{H}_3$	2.2–2.6
$R_2C\underline{H}OR$	3.2–4.3
$RC\underline{H}_2Cl$	3.4–3.5

$$R_2C=C\overset{\displaystyle CH_3}{\underset{\displaystyle R}{\Big\langle}} \qquad 1.6\text{–}1.9$$

$$\underset{\displaystyle RC\underline{CH}_2R}{\overset{\displaystyle O}{\overset{\displaystyle \|}{}}} \qquad 2.0\text{–}2.7$$

sp and sp^2 C—H:

$$R_2C=C\overset{\displaystyle H}{\underset{\displaystyle R}{\Big\langle}} \qquad 4.9\text{–}5.9$$

$Ar\underline{H}$	6.0–8.0

$$\underset{\displaystyle RC\underline{H}}{\overset{\displaystyle O}{\overset{\displaystyle \|}{}}} \qquad 9.4\text{–}10.4$$

$RC\equiv C\underline{H}$	2.3–2.9

—OH and —NH:

$RO\underline{H}$	1–6
$R_2N\underline{H}$	2–4
$ArO\underline{H}$	6–8

$$\underset{\displaystyle RCO\underline{H}}{\overset{\displaystyle O}{\overset{\displaystyle \|}{}}} \qquad 10\text{–}12$$

Example of a Compound Containing Chemically Equivalent Protons:

$$\underset{\displaystyle CH_3CCH_3}{\overset{\displaystyle O}{\overset{\displaystyle \|}{}}} \quad \xrightarrow[-HCl]{Cl_2} \quad \underset{\displaystyle CH_3CCH_2Cl}{\overset{\displaystyle O}{\overset{\displaystyle \|}{}}}$$

Replacement of any one of six H's with Cl yields the same product— chloroacetone; therefore, the six H's are equivalent.

Example of a Compound with Chemically Nonequivalent Protons:

$$\underset{\substack{\uparrow \quad \nearrow \\ \text{These protons are}}}{Br_2CHCCHI_2} \xrightarrow[-HCl]{Cl_2} \underset{\substack{| \\ Cl}}{Br_2CCCHI_2} + \underset{\substack{| \\ Cl}}{Br_2CHCCl_2}$$

These protons are
not equivalent.
Replacement of each
results in isomers.

Isomers

Most organic compounds contain groups of equivalent protons that are nonequivalent to other groups of protons in the structure. For example, chloroethane contains two types of protons—the three $-CH_3$ protons and the two $-CH_2-$ protons. The three $-CH_3$ protons are equivalent to one another but are not equivalent to the CH_2 protons. The two CH_2 protons are equivalent to each other but are not equivalent to the $-CH_3$ protons.

Chloroethane Contains Two Sets of Protons:

$$CH_3CH_2Cl$$

One set contains three   Another set contains
equivalent protons   two equivalent protons

EXAMPLE

Identify the equivalent protons and the nonequivalent protons in the following structures. How many sets or groups of protons does each compound have?

$$\underset{\substack{| \\ CH_3}}{\overset{\substack{CH_3 \\ |}}{CH_3SiCH_3}}$$ ⬡ $$\overset{\substack{O \\ ||}}{CH_3CCH_2Cl}$$

TMS   Benzene   Chloroacetone

Solution:

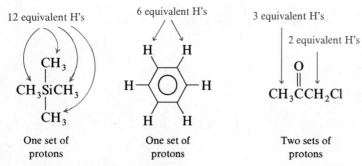

One set of   One set of   Two sets of
protons   protons   protons

PROBLEM 15.4. Show the nonequivalence of the two types of protons in chloroethane by writing two equations: **(a)** an equation for the replacement of a $-CH_3$ proton by Cl and **(b)** an equation for the replacement of a CH_2 proton by Cl. Explain how these equations show the equivalence or nonequivalence of the protons.

Protons that are magnetically nonequivalent have different chemical shifts. Therefore, in an nmr spectrum, we can count the number of different types of protons by counting the number of principal absorption signals. Figure 15.13, the proton nmr spectrum of *p*-dimethoxybenzene, shows two absorption signals, indicating that this compound has two types of magnetically nonequivalent protons in its structure.

F. Counting the Protons in an Nmr Spectrum

In a proton nmr spectrum, *the area under each principal signal is proportional to the number of protons giving rise to that signal.* In CH_3CH_2Cl, we observe that two proton signals have an area ratio of 3 (for $-CH_3$) to 2 (for $-CH_2-$).

EXAMPLE

For the following compounds, list (1) the number of principal nmr signals (the *types* of protons) and (2) the relative areas of the signals.

        O   O
        ‖   ‖

(a) CH_3COCH_3 **(b)** $CH_3CH_2CCH_2CH_3$ **(c)** $Cl-\!\!\left\langle\!\bigcirc\!\right\rangle\!\!-OCH_3$

Solution:
(a) Methyl acetate contains two types of protons with an area ratio of 3 : 3. A ratio is generally expressed as the smallest possible whole numbers—in this example, 1 : 1.

            O
            ‖
       CH_3C-OCH_3

    3 H's on an   3 H's in a
      α carbon $-OCH_3$ group

(b) 3-Pentanone contains two types of protons with an area ratio of 6 : 4, or 3 : 2.

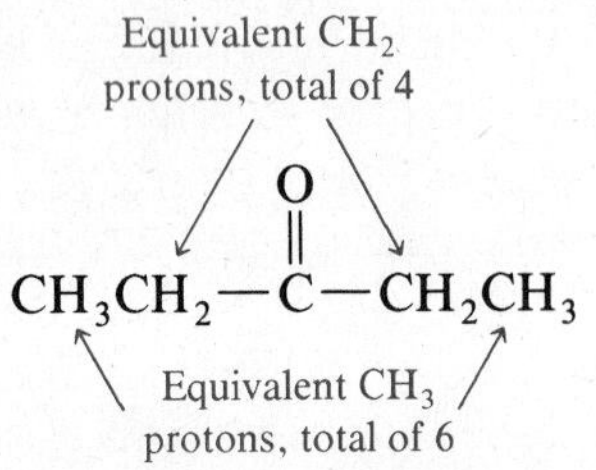

Equivalent CH_2
protons, total of 4

$$CH_3CH_2 - \overset{\overset{\textstyle O}{\|}}{C} - CH_2CH_3$$

Equivalent CH_3
protons, total of 6

(c) *p*-Chloroanisole contains three types of protons with an area ratio of 3 : 2 : 2.

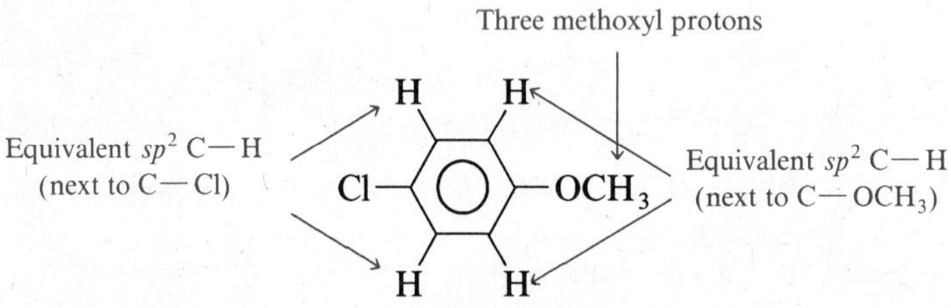

Three methoxyl protons

Equivalent sp^2 C—H
(next to C—Cl)

Cl—⟨O⟩—OCH_3

Equivalent sp^2 C—H
(next to C—OCH_3)

PROBLEM 15.5. Analyze the following structures as we have done in the example.

(a) $ClCH_2CH_2OCH_2CH_2Cl$ **(b)** $ClCH_2CH_2OCH_2CHCl_2$

(c)  CH_3—⟨O⟩—CH_3 **(d)** 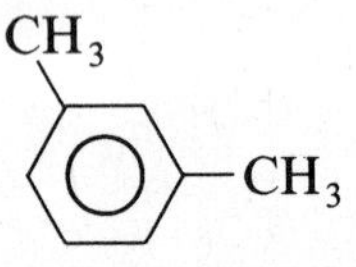 ⟨O⟩ with CH_3 (top) —CH_3

PROBLEM 15.6. A proton nmr spectrum showed three signals with the area ratio of 3 : 3 : 2. How many protons does a molecule of the compound contain? Explain.

The relative areas of the peaks in a spectrum are determined instrumentally and recorded as an **integration curve** superimposed on the spectrum, as shown in Figure 15.17. The height of each step in the integration curve is proportional to the area under the nmr signal.

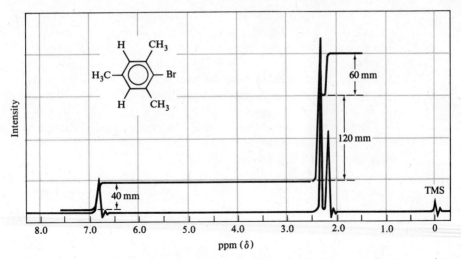

Figure 15.17 Proton nmr spectrum of 1-bromo-2,4,6-trimethylbenzene, showing an integration curve.

G. Splitting Patterns in Proton Nmr Spectra

The $n + 1$ rule   Figure 15.18 is the proton nmr spectrum of chloroethane. As we have predicted, the spectrum shows two principal signals—the $-\underline{C}H_3$ signal (upfield, area 3) and the $-\underline{C}H_2Cl$ signal (downfield because of the electron-withdrawing effect of the electronegative Cl, area 2). (The relative areas are shown above the signals rather than as an integration curve.)

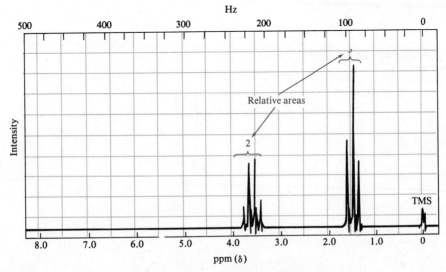

Figure 15.18 Proton nmr spectrum of chloroethane, CH_3CH_2Cl.

As you can see in Figure 15.18, the two principal signals for chloroethane are a composite of a number of smaller signals. The $—CH_3$ signal is split into a *triplet* (three peaks), while the $—CH_2Cl$ signal is split into a *quartet* (four peaks). To measure the area of a principal signal, we sum all the small peak areas. Thus to determine the ratio of protons in chloroethane, we add the areas under the three peaks from the $—CH_3$ signal and compare that area with the sum of the areas of the four peaks arising from the CH_2 signal.

The splitting of a principal signal into smaller peaks is called **spin–spin splitting**. Spin–spin splitting is caused by the parallel and antiparallel spin states of nonequivalent protons on neighboring (adjacent) carbon atoms; we will discuss this aspect shortly. Protons that split one another's signals are said to be **coupled protons**. Equivalent protons do not split one another's signals; only nonequivalent protons cause splitting.

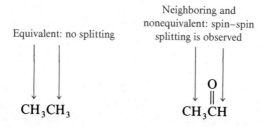

The type of splitting pattern observed (doublet, triplet, quartet, and so forth) depends on the *number* of neighboring nonequivalent protons. For many simple compounds, we can predict the splitting pattern with the ***n* + 1 rule**, where *n* is the number of neighboring protons.

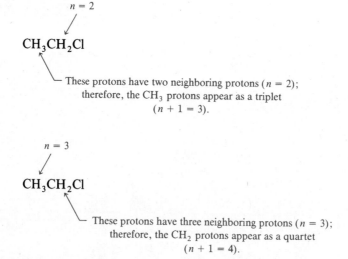

EXAMPLE

Use the $n + 1$ rule to predict the splitting pattern for the following compound. What would be the ratio of the areas of the principal signals?

$$Cl_2CH—CH_2Cl$$

Solution:

This proton has two neighboring H's; $n + 1 = 3$.
Its signal appears in the spectrum as a triplet.

$$Cl_2CH—CH_2Cl$$

These two protons have one neighboring H; $n + 1 = 2$.
They appear in the spectrum as a doublet.

The relative area for the left-hand single proton is 1, while that for the two right-hand protons is 2.

Spin–spin splitting occurs because a neighboring proton has two spin states (parallel and antiparallel). The signal for the absorbing proton is split because half its neighbors have parallel spin states and the other half have antiparallel spin states. Parallel neighbors deshield the absorbing proton, whereas antiparallel neighbors shield the absorbing proton. Thus, there are really two different types of protons absorbing the electromagnetic radiation—those with parallel neighbors and those with antiparallel neighbors.

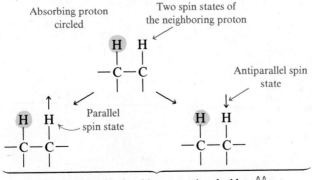

The signal for the absorbing proton is a doublet, ⋀⋀

Two or more neighboring protons lead to more possible combinations of neighboring spin states. The situation is simplified somewhat because some combinations are equivalent and therefore are additive.

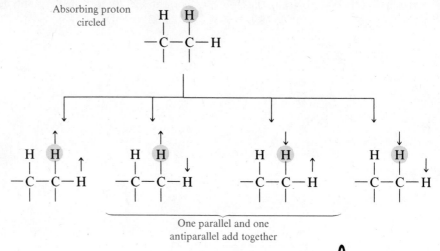

One parallel and one
antiparallel add together

The signal for the absorbing proton is a triplet, ⠀⠀⠀⠀.

Figure 15.19 summarizes the $n + 1$ splitting patterns.

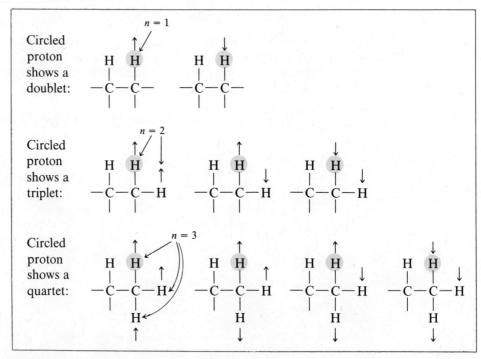

Figure 15.19 The $n + 1$ rule used to predict proton nmr splitting patterns.

Because of the great variety of possible organic structures, not all proton nmr spectra follow simple splitting rules. Nmr spectra can become quite complex for a number of reasons. First, two or more different signals may overlap and mask the expected patterns. Second, a proton signal can be split by two or more nonequivalent neighboring protons that are nonequivalent to one another. This type of splitting does not generally follow the $n + 1$ rule.

$$\text{ClCH}_2\text{—}\overset{\displaystyle \overset{\textstyle \text{H}}{|}}{\underset{\displaystyle \underset{\textstyle \text{I}}{|}}{\text{C}}}\text{—CH}_2\text{OCH}_3$$

These nonequivalent sets of protons independently
split the signal for the circled H.

PROBLEM 15.7. Using the $n + 1$ rule, predict the proton nmr splitting pattern of each proton in the following compounds:

(a) $\underset{\displaystyle }{\text{CH}_3}\overset{\displaystyle \overset{\textstyle \text{O}}{\|}}{\text{CH}}$

(b) ⬡—OCH_3

(c) Cl—⬡—$\overset{\displaystyle \overset{\textstyle \text{O}}{\|}}{\text{CH}}$

(d) $\text{CH}_3\text{CH}_2\text{O}\overset{\displaystyle \overset{\textstyle \text{CH}_3}{|}}{\text{CH}}\text{CH}_3$

Area Ratios Within Splitting Patterns   Splitting patterns have characteristic area ratios within the patterns themselves. The two peaks in a doublet have the area ratio of $1:1$. The three peaks in a triplet have an area ratio of $1:2:1$. The four peaks in a quartet have an area ratio of $1:3:3:1$. These area ratios are idealized; in most spectra, the splitting patterns are distorted.

$1:1$ $\qquad$ $1:2:1$ $\qquad$ $1:3:3:1$

The important features of proton nmr spectroscopy are summarized in Figure 15.20.

H. Carbon-13 Nmr Spectra

Because ^{13}C is only 1.1% of naturally occurring carbon atoms, only 1.1% of carbon nuclei have spin and can absorb radio frequency in an nmr instrument. For this reason, an nmr spectrometer for ^{13}C nmr spectroscopy must be very sensitive. In practice, a computer is used to assist in gathering and storing the data; only in recent years have such spectrometers been available for general use.

Proton nmr spectroscopy

1. The *number of principal signals* can sometimes be used to determine the number of types of protons.
2. The *chemical shifts* provide clues to the chemical environments of the types of protons.
3. The *areas under the signals* are proportional to the numbers of protons giving rise to the signals.
4. The *splitting patterns* provide information about the neighboring protons (protons on adjacent carbons).

Figure 15.20 Summary of proton nmr spectroscopy.

In many respects, ^{13}C nmr spectra are more easily interpreted than proton nmr spectra. (1) The common range of energy absorption for ^{13}C is $\delta 0$–200 relative to TMS, contrasted with $\delta 0$–15 for proton nmr. This wide range means that fewer peaks overlap in ^{13}C spectra. (2) Because only 1.1% of carbon in a compound is ^{13}C, ^{13}C-^{13}C coupling is negligible and thus is not observed. Therefore, in one type of ^{13}C nmr spectrum, each magnetically nonequivalent carbon is seen as a single unsplit peak. (3) The areas under the peaks in ^{13}C nmr spectra are not necessarily proportional to the number of carbons giving rise to the signal; therefore, we do not consider the area ratios.

Figure 15.21 shows two ^{13}C nmr spectra for methyl (E)-2-methyl-2-butenoate. The lower spectrum is a **proton-decoupled spectrum**, in which each carbon or group of equivalent carbons appears as a singlet. Note how far downfield the carbonyl signal appears. In general, the relative relationships of chemical shifts of carbons are similar to those observed for protons in proton nmr spectra. Figure 15.22 is a correlation chart of ^{13}C nmr spectroscopy showing the usual chemical shifts for various types of carbons.

The upper spectrum in Figure 15.21 is a **proton-coupled spectrum** for methyl (E)-2-methyl-2-butenoate, in which the signal for each carbon-13 is split by protons bonded directly to it. (Instrument design allows both the coupled spectrum and the decoupled one to be carried out, one after the other, on the same grid of recorder paper.) The $n + 1$ rule is followed in ^{13}C nmr spectra. The n in this case is the number of protons bonded *to* the carbon giving the signal. The number of peaks in each signal equals the number of protons bonded to that carbon plus 1.

$$-CH_3 \qquad\qquad -CH_2- \quad \text{or} \quad =CH_2$$

Three H's: a quartet   Two H's: a triplet

$$-\overset{|}{\underset{}{CH}}- \quad \text{or} \quad =CHR \qquad -\overset{|}{\underset{|}{C}}- \quad \text{or} \quad -\overset{O}{\overset{\|}{C}}-$$

One H: a doublet   No H's: a singlet

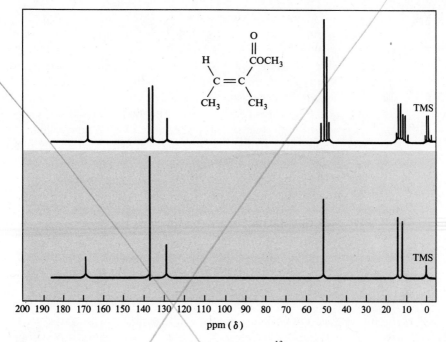

Figure 15.21 Coupled (top) and decoupled (bottom) ^{13}C nmr spectra of methyl (E)-2-methyl-2-butenoate.

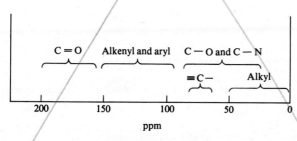

Figure 15.22 Correlation chart for ^{13}C nmr spectroscopy.

Again, note the carbonyl signal in Figure 15.21. Besides being far downfield, this signal remains a singlet in the coupled spectrum because it has no proton bonded to the carbon.

SUMMARY

Spectroscopy, the measurement of absorption of electromagnetic radiation by substances, is used in determining the structures of organic compounds.

Infrared radiation is absorbed by bonds and promotes the bonds to excited (higher-energy) vibrational states. Different types of bonds absorb infrared radiation at different frequencies, as summarized in Table 15.2. An infrared spectrum is an instrument-generated graph of percent transmittance versus either frequency in wavenumbers (cm^{-1}) or wavelength (μm).

Ultraviolet radiation and visible radiation are higher-energy radiation than infrared. These types of radiation cause promotion of electrons to excited electronic states (higher-energy orbitals). Of special interest are the ultraviolet spectra of compounds with conjugated pi bonds. The more extensive the conjugation, the easier it is to promote electrons and the more likely a compound is to be colored.

In ultraviolet and visible spectra, the wavelength of maximum absorption is called the λ_{max}. The **molar absorptivity** ϵ is a measure of the quantity of radiation absorbed.

Nmr spectroscopy is based on the ability of nuclei with spin (^1H and ^{13}C) to undergo a change in spin state ("flip") in an applied magnetic field when subjected to electromagnetic radiation.

The **chemical shift (δ)** is the distance from the TMS signal, set at 0, and the signal in question. Nmr absorption by most organic compounds is *downfield* from, or to the left of, the peak for TMS. Nearby electronegative groups increase the chemical shift of a proton by *deshielding*. Pi electrons also affect the chemical shift, usually also increasing it. Table 15.3 lists some typical chemical shifts.

In proton nmr spectra, magnetically nonequivalent protons exhibit different chemical shifts. Neighboring protons that are nonequivalent to one another also split one another's signals into groups of smaller peaks. In simple cases, the ***n* + 1 rule** is followed—the splitting pattern is composed of $n + 1$ peaks, where n is the number of neighboring protons.

In proton nmr spectra, the relative areas under the principal signals are proportional to the numbers of protons giving rise to the signal.

Carbon-13 nmr spectra are of two types: **proton-decoupled spectra** and **proton-coupled spectra**. In proton-decoupled spectra, the signal for each carbon (or each group of magnetically equivalent carbons) appears as a singlet, unsplit by any other portions of the molecule. In proton-coupled spectra, the signal for each carbon (or group of magnetically equivalent carbons) is split by the protons bonded directly to that carbon. Again, the $n + 1$ rule is followed.

STUDY PROBLEMS

15.8. Which radiation contains the higher or highest energy:
   (a) radio waves, ultraviolet radiation, or infrared radiation?
   (b) radiation with $\nu = 5$ Hz or 10 Hz?
   (c) radiation with $\lambda = 0.01$ cm or 0.0001 cm?

15.9. Make the following conversions:
   (a) 15 μm to m   **(b)** 400 Å to m   **(c)** 200 nm to m

15.10. How would you distinguish between each of the following pairs of compounds by their infrared spectra? (Use specific wavelengths or frequencies in your answers.)

$$\text{(a)} \quad \underset{\overset{|}{\text{OH}}}{CH_3CH_2CHCH_3} \quad \text{and} \quad \underset{\overset{\parallel}{\text{O}}}{CH_3CH_2CCH_3}$$

(b) $CH_3CH_2\overset{\overset{\displaystyle O}{\|}}{C}OCH_3$ and $CH_3CH_2\overset{\overset{\displaystyle O}{\|}}{C}OH$

(c) $CH_3OCH_2CH_2CH_3$ and $CH_3\overset{\overset{\displaystyle O}{\|}}{C}CH_2CH_2CH_3$

(d) $CH_3CH_2CH_2NHCH_3$ and $CH_3CH_2CH_2N(CH_3)_2$

(e) $(CH_3)_2CHCH_2CH_3$ and $(CH_3)_2CHCH=CH_2$

(f) $CH_3CH_2CH=CH_2$ and $CH_3CH_2C\equiv CH$

15.11. How would you use infrared spectroscopy to determine when the following reaction is completed?

$$CH_3\overset{\overset{\displaystyle O}{\|}}{C}CH_2CH_3 \xrightarrow[CH_3OH]{NaBH_4} CH_3\overset{\overset{\displaystyle OH}{|}}{C}HCH_2CH_3$$

15.12. A compound shows its principal infrared absorption at 2950, 1730, and 1170 cm^{-1}. Which of the following structures could the compound be? Explain your answer.

(a) $CH_3CH_2CH_2CH_2OH$ (b) $CH_3(CH_2)_3CO_2H$

(c) $CH_3CH_2O\overset{\overset{\displaystyle O}{\|}}{C}(CH_2)_3CH_3$

15.13. A compound with the formula C_4H_8O shows no significant absorption in the infrared spectrum at 3000–3700 cm^{-1} or at 1700–1750 cm^{-1}. What are the possible structures for this compound?

15.14. How would you distinguish between the following pairs of compounds by ultraviolet spectroscopy?

(a)

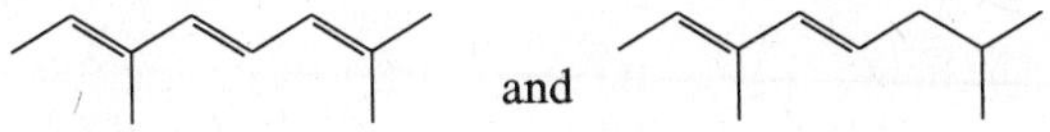

and

(b) $CH_3CH=CH\overset{\overset{\displaystyle O}{\|}}{C}H$ and $CH_2=CHCH_2\overset{\overset{\displaystyle O}{\|}}{C}H$

(c)

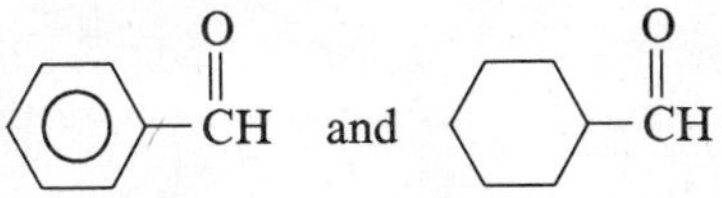

15.15. Calculate the molar absorptivity ϵ of a compound if the observed absorbance is 8.60, the concentration of the sample is 0.010 M, and the cell length is 0.10 cm.

15.16. Which form of the acid-base indicator phenolphthalein is more likely to be colored? Explain.

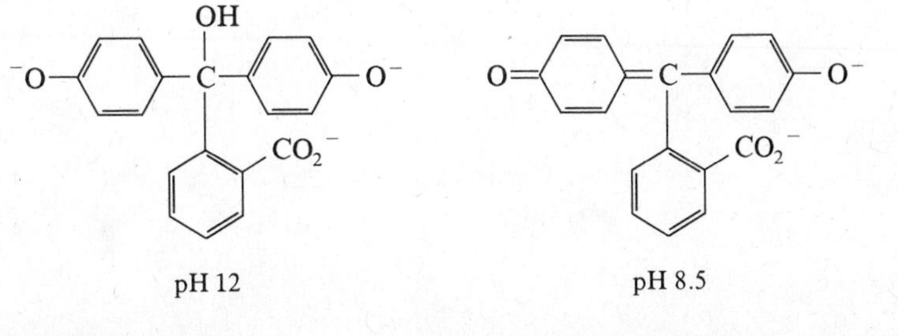

pH 12   pH 8.5

15.17. Which indicated proton in each pair would exhibit the larger chemical shift in a proton nmr spectrum? Explain your answers.

(a) $CH_3CH_2C\underline{H}_2I$ or $CH_3CH_2C\underline{H}_2Cl$

(b) $CH_3CH_2CH_2Cl$ or $CH_3C\underline{H}_2CH_2Cl$

(c) $CH_3CH=C\underline{H}_2$ or   ⬡—$\underline{H}$

(d) $CH_3CH_2CC\underline{H}_3$ or $CH_3CH_2C\underline{H}$
(with $\overset{O}{\underset{||}{}}$ above each carbonyl)

(e) ⬡—$C\underline{H}_3$ or   ⬡—$\underline{H}$

15.18. Indicate the groups of chemically equivalent protons in each structure:

(a) $CH_3CH_2\overset{O}{\overset{||}{C}}OCH_3$

(b) CH_3O—⬡—CH_3

(c) $CH_3CH_2\overset{Cl}{\overset{|}{C}}HCH_3$

(d) $\overset{H}{\underset{Cl}{}}C=C\overset{CH_3}{\underset{H}{}}$

(e) $CH_3CH_2OCH_2CH_2CH_3$

(f) $CH_3\overset{CH_3}{\overset{|}{C}}HCH_2\overset{CH_3}{\overset{|}{C}}HCH_2CH_3$

15.19. In Problem 15.18, tell (1) the numbers of *principal* peaks in each proton nmr spectrum and (2) the relative areas under these peaks.

15.20. Predict the splitting pattern (singlet, doublet, quartet) of each type of proton in the proton nmr spectrum of the local anesthetic lidocaine (Problem 12.37, page 425).

15.21. Indicate the different nonequivalent carbon atoms that would be seen in the ^{13}C nmr spectrum of lidocaine.

15.22. (a) How many peaks does tetramethylsilane, $(CH_3)_4Si$, show in the proton nmr spectrum? **(b)** How many peaks would it show in the proton-decoupled ^{13}C nmr spectrum? **(c)** How many peaks would it show in the proton-coupled ^{13}C nmr spectrum?

15.23. A compound with the molecular formula C_5H_{12} shows a single peak in its proton nmr spectrum. What is the structure of this hydrocarbon?

15.24. Sketch the expected proton nmr spectra of the following compounds. Be sure to consider chemical shift (Table 15.3), splitting patterns, and approximate areas.

(a) $CH_3\overset{Br}{\overset{|}{C}}H-\overset{Br}{\overset{|}{C}}HCH_3$

(b) $CH_3\overset{O}{\overset{||}{C}}OCH_2CH_3$

(c) $(CH_3)_2CHCl$

(d) $\overset{CH_3CH_2}{\underset{CH_3CH_2}{}}NCH_2CH_2O\overset{O}{\overset{||}{C}}$—⬡—$NH_2$

Novocaine

15.25. A compound C_4H_8O shows four peaks in the proton-decoupled ^{13}C nmr spectrum at $\delta 8$, 30, 37, and 208. The proton-coupled spectrum shows these peaks as a quartet, another quartet, a triplet, and a singlet, respectively. Assign each peak to a group, and write the condensed structural formula for the compound.

15.26. A compound $C_8H_{10}O$ shows the following absorption in the proton nmr spectrum: singlets at $\delta 2.15$ and 3.45 and four peaks (not necessarily a quartet) at about $\delta 6.7$. The area ratios are $3:3:4$. What is the structure of the compound?

15.27. The proton nmr spectrum of a compound $C_3H_6Br_2$ shows a quintet (splitting pattern of five peaks) at $\delta 2.3$ and a triplet at $\delta 3.7$. What is the structure of this compound?

15.28. A compound $C_5H_{10}O$ shows its principal infrared absorption at about 2800–3000 and 1710 cm^{-1}. Its proton nmr spectrum shows a doublet at $\delta 0.95$, a singlet at $\delta 2.10$, and a multiplet (signal split into several peaks) at $\delta 2.43$. The area ratios are $6:3:1$. What is the structure of the compound?

POINT OF INTEREST 15

Clinical Chemistry. Colorimetric Measurement of Protein in the Urine

The ability of physicians to diagnose disease and dystrophies (faulty development or metabolism) has been enormously advanced by clinical chemical produces—analyses of body fluids and tissues for specific compounds and ions.

One valuable clinical procedure is the analysis for protein in urine. As is the case with most clinical tests, a number of different procedures have been developed. For urinalysis, most clinical laboratories carry out two routine produces—a qualitative test to determine if protein is present and a more elaborate quantitative analysis if the qualitative test is positive.

The medical importance of these tests stems from the fact that protein in the urine (proteinuria) is a symptom of kidney malfunction. Although protein can appear in the urine because of a number of temporary conditions, such as muscular exertion, persistent proteinuria almost always signifies disease of the kidneys. Because proteinuria causes no outward symptoms (such as pain), clinical analysis is imperative.

The Biuret Procedure for Protein Analysis   If the qualitative test for protein in the urine is positive, then a quantitative test (usually the *biuret procedure*) is usually performed to determine the exact amount of protein present.

For the biuret procedure, a 24-hour urine sample is collected and the protein of an aliquot (sample) is precipitated by trichloroacetic acid. (The acid disrupts the hydrogen bonding of the protein, causing it to coagulate.) The precipitated protein is separated from the urine by centrifugation and then is redissolved in an alkaline solution. These steps concentrate the protein and thus allow a more accurate analysis.

The redissolved protein is treated with copper(II) ions in an alkaline solution. A protein–copper(II) complex, whose structure is similar to that of the biuret-copper(II) complexes, is formed. Because biuret is the simplest amide that undergoes this type of reaction, the reaction is called the *biuret reaction*.

The Biuret Reaction:

$$Cu^{2+} + 2H_2N\overset{\overset{O}{\|}}{C}NH\overset{\overset{O}{\|}}{C}NH_2 + 4NaOH \longrightarrow$$

Biuret

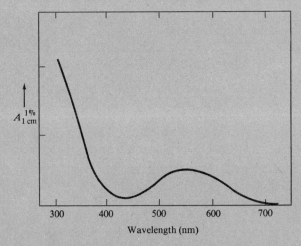

+ $4Na^+$ + $4H_2O$

Violet-pink

The protein–copper(II) complex absorbs light in the 550-nm region (green) of the spectrum and thus appears pink or reddish (see Figure 15.23).

If excess copper(II) ions are present, the intensity of the pink color is proportional to the concentration of the protein–copper(II) complex and, thus, also to the original protein concentration (Beer's law). With the aid of a standard curve, a clinician can determine the amount of protein in the urine sample.

Figure 15.23 Visible spectrum of the protein–copper(II) complex formed in the biuret reaction. (The term $A_{1\,cm}^{1\%}$ is the absorbance of a 1% solution with a sample path length of 1 cm.)

Chapter 16

Carbohydrates

The term **carbohydrate** arose from an early misconception of the structure of sugars. In the early days of chemistry, analysis of different sugars gave an empirical formula of CH_2O and a molecular formula that could be interpreted as $C_x(H_2O)_y$. For example, the molecular formula for glucose ($C_6H_{12}O_6$) was interpreted as $C_6(H_2O)_6$. These formulas led to early chemists to the conclusion that sugars were hydrates of carbon—hence the term "carbohydrates."

When the structural formulas of the carbohydrates were determined, it became apparent that these compounds are more complex than simple hydrates of carbon. Based on the knowledge of their structures, we now define **carbohydrates** as a class of compounds that consist of, or can be hydrolyzed to, polyhydroxy aldehydes and ketones. In this context, polyhydroxy means two or more hydroxyl groups.

Carbohydrates are the main energy source for human beings. Most of the carbohydrate we eat is starch, which occurs in wheat, corn, rice, potatoes, and other grains, fruits, and vegetables.

The word **sugar** is part of our everyday language and refers to a sweet crystalline carbohydrate, usually table sugar. Sugar is an ancient word that can be traced to the Sanskrit word *sarkara*, which means "sugar." The term **saccharide** (Latin *saccharum*, "sugar") is also used to signify a sugar.

The different classes of carbohydrates can be related to one another by hydrolysis. The simple sugars, or **monosaccharides**, are the polyhydroxy aldehydes and ketones that cannot be hydrolyzed to smaller carbohydrate fragments. The monosaccharides, therefore, are monomers, the building blocks for all other carbohydrates. A structure containing two monosaccharides bonded together is a **disaccharide** (from *di*, "two"). A structure containing three monosaccharides bonded together is a **trisaccharide**.

Carbohydrates consisting of 2 to 10 saccharide units are classified as **oligosaccharides** (Greek *oligo-*, "a few"). A structure containing more saccharide units is classified as a **polysaccharide**. There is no clear-cut dividing line between oligosaccharides and polysaccharides because the properties of the higher oligosaccharides merge with those of the lower polysaccharides.

Polysaccharides
(> 10 saccharide units)

The carbohydrates
↓ H$_2$O

Oligosaccharides
(2–10 saccharide units)

↓ H$_2$O

Monosaccharides
(one saccharide unit)

↓ H$_2$O

No hydrolytic cleavage

All naturally occurring monosaccharides are continuous-chain aldehydes or ketones. In a ketone monosaccharide, the carbonyl group is always located at the second carbon of the chain. The naturally occurring monosaccharides generally have a hydroxyl group bonded to each tetrahedral carbon of the chain. The principal exceptions are the deoxy sugars, the amino sugars, and the glucuronic acids. We will discuss these carbohydrates later in this chapter.

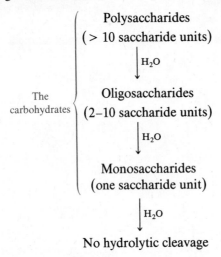

Glucose

a common monosaccharide

16.1 CLASSIFICATION OF THE MONOSACCHARIDES

The suffix designating a sugar is *-ose*, which is used for most, but not all, saccharides. For example, glucose, sucrose, and cellulose are saccharides with names that end in -ose, while glyceraldehyde, starch, and glycogen are saccharides with names that do not.

TABLE 16.1 CLASSIFICATION SYSTEM FOR THE MONOSACCHARIDES

Number of carbons	Aldoses	Ketoses
3 (triose)	Aldotriose	Ketotriose or triulose
4 (tetrose)	Aldotetrose	Ketotetrose or tetrulose
5 (pentose)	Aldopentose	Ketopentose or pentulose
6 (hexose)	Aldohexose	Ketohexose or hexulose
7 (heptose)	Aldoheptose	Ketoheptose or heptulose

In general, the carbohydrates and their derivatives are named using trivial, rather than IUPAC, nomenclature. Each carbohydrate has its own unique name. We will present carbohydrate nomenclature as we discuss the structures of the carbohydrates.

The monosaccharides can be classified by their principal functional group (aldehyde or ketone) and by the number of carbons in the carbon chain. If a monosaccharide is an aldehyde, it is classified as an **aldose**; if it is a ketone, a **ketose**.

A monosaccharide with three carbons is classified as a **triose**; with four carbons, a **tetrose**; and so forth. These two classifications can be combined. For example, an **aldohexose** is a six-carbon aldehyde monosaccharide and a **ketohexose** is a six-carbon ketone monosaccharide. When ketoses are classified by both functional group and the number of carbons they contain, the suffix *-ulose* is often used. For example, a ketohexose could also be classified as a *hexulose*.

Table 16.1 summarizes the conventions for classifying the monosaccharides.

EXAMPLE

Classify each of the following monosaccharides as to (1) its carbon content, (2) its principal functional group, and (3) its combination of carbon content and principal functional group.

$$
\begin{array}{c}
\text{O} \\
\parallel \\
\text{CH} \\
\mid \\
\text{(a)} \quad \text{HO} \mathbin{\blacktriangleright} \text{C} \mathbin{\blacktriangleleft} \text{H} \\
\mid \\
\text{HO} \mathbin{\blacktriangleright} \text{C} \mathbin{\blacktriangleleft} \text{H} \\
\mid \\
\text{CH}_2\text{OH}
\end{array}
\qquad
\begin{array}{c}
\text{CH}_2\text{OH} \\
\mid \\
\text{C}\!=\!\text{O} \\
\mid \\
\text{(b)} \quad \text{H} \mathbin{\blacktriangleright} \text{C} \mathbin{\blacktriangleleft} \text{OH} \\
\mid \\
\text{H} \mathbin{\blacktriangleright} \text{C} \mathbin{\blacktriangleleft} \text{OH} \\
\mid \\
\text{CH}_2\text{OH}
\end{array}
$$

Solution: Compound **(a)** contains four carbon atoms (a *tetrose*) and an aldehyde group (an *aldose*); therefore, **(a)** is an *aldotetrose*. Compound **(b)** contains five carbon atoms (a *pentose*) and a keto group (a *ketose*); therefore, it is a *ketopentose* or a *pentulose*.

16.2 THE TRIOSES AND TETROSES

A. Fischer Projections of the Monosaccharides

The Fischer projection is used to emphasize the (R) and (S) configurations of chiral carbons (see Sections 6.1C and 6.3A). In a Fischer projection of a carbohydrate, the carbon chain is drawn vertically with the aldehyde or keto group located toward the top of the formula. The horizontal bonds to the chiral carbons are projected toward the viewer and the vertical bonds, including those of the carbon chain, are projected away from the viewer.

Figure 16.1 shows Fischer projections for glyceraldehyde. Carbon-2 of glyceraldehyde is chiral; consequently, glyceraldehyde exists as a pair of enantiomers (nonsuperimposable mirror images; see Section 6.1B). These enantiomers are named (R)-2,3-dihydroxypropanal and (S)-2,3-dihydroxypropanal. However, they are usually referred to by their classical names, D-glyceraldehyde and L-glyceraldehyde.

The Fischer projection for D-glyceraldehyde is drawn with the hydroxyl group on the chiral carbon projected to the right, while that for the L enantiomer is drawn with this hydroxyl group projected to the left.

B. Diastereomers

When a compound has one chiral carbon, there can be only two stereoisomers, which are enantiomers of each other. Glyceraldehyde is an example of a compound that has one chiral carbon and exists as a pair of enantiomers. When a compound has two chiral carbons, there can be up to four stereoisomers (2^n isomers, where n is the number of chiral carbons; see Section 6.4). The aldotetroses contain two chiral carbons; as a consequence, a total of four stereoisomeric monosaccharides are possible.

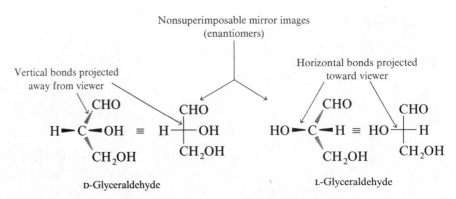

Figure 16.1 Fischer projections of D- and L-glyceraldehyde.

The Four Stereoisomeric Aldotetroses:

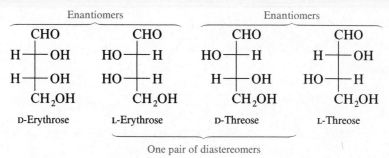

D-Erythrose and L-erythrose are enantiomers, as are D-threose and L-threose. (The significance of the D- and L- in these names will be discussed later in this section.) D-Erythrose and D-threose are not enantiomers; that is, they are not mirror images. A pair of the stereoisomers that are not enantiomers are called **diastereomers**.

CHO   CHO   CHO   CHO
H——OH   HO——H   HO——H   H——OH
H——OH   HO——H   H——OH   HO——H
CH₂OH   CH₂OH   CH₂OH   CH₂OH
D-Erythrose L-Erythrose D-Threose   L-Threose

Enantiomers — D-Erythrose, L-Erythrose; Enantiomers — D-Threose, L-Threose; One pair of diastereomers — L-Erythrose, D-Threose

---

PROBLEM 16.1. We have shown one diastereomeric pair for the aldotetroses. Draw Fischer projections for three other diastereomeric pairings.

A pair of enantiomers are isomers with almost all the same chemical and physical properties. Enantiomers differ only in their interaction with polarized light (equal in magnitude and opposite in direction) and in their interaction with other chiral compounds. Diastereomers are different compounds in the usual chemical sense. They have different solubilities, different melting points, different chemical reactivities, etc.

Of the four tetroses, only D-erythrose is a natural product. It has been isolated from red algae, and D-erythrose 4-phosphate is a key intermediate in photosynthesis.

C. The D and L Families of Monosaccharides

Two of the aldotetroses, D-erythrose and D-threose, have their last chiral hydroxyl group (the hydroxyl group on carbon 3) projected to the right. This chiral carbon has the same configuration as the chiral carbon in D-glyceraldehyde.

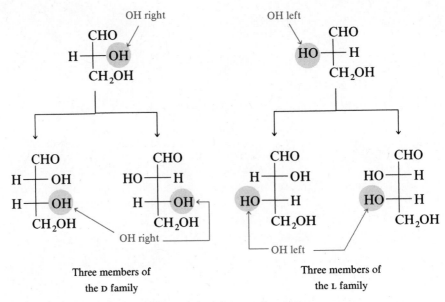

Figure 16.2 The D and L families of the trioses and tetroses.

The other two aldotetroses have the hydroxyl group at carbon 3 projected to the left, the same configuration as that in L-glyceraldehyde, as shown in Figure 16.2.

On the basis of the configuration of the last chiral carbon, all carbohydrates can be classified into one of two major subdivisions or families, the **D series** or the **L series.** The D-monosaccharides all have the hydroxyl group of the lowermost chiral carbon projected to the right in the Fischer projection. The L sugars are just the opposite; the hydroxyl group on their lowermost chiral carbon is projected to the left.

Figure 16.3 shows the D-aldose family of monosaccharides. Similar charts can be drawn for the L-aldoses and for the D- and L-ketoses.

Most of the abundant naturally occurring monosaccharides belong to the D series.

Some Naturally Occurring Sugars That Belong to the D *Series:*

CHO H—OH CH₂OH	CHO H—OH H—OH CH₂OH	CHO H—OH H—OH H—OH CH₂OH	CHO H—OH HO—H H—OH H—OH CH₂OH
D-Glyceraldehyde	D-Erythrose	D-Ribose	D-Glucose
a triose	*a tetrose*	*a pentose*	*a hexose*

To the right in the D series

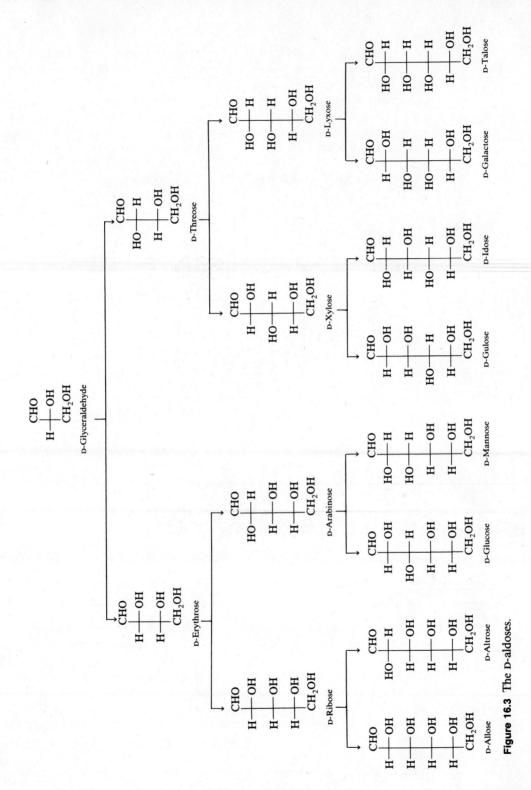

Figure 16.3 The D-aldoses.

Some L sugars are found in nature. For example, L-arabinose, an aldopentose, is widely distributed, whereas D-arabinose has been found in only a few bacteria and sponges. L-Fucose, an aldohexose, is one of the monosaccharides found in the oligosaccharides that determine blood type (A, B, AB, or O). These oligosaccharides are bonded to the surface protein of erythrocytes (red blood cells). L-Fucose is 6-deoxy-L-galactose. (The term 6-deoxy in the name signifies that the hydroxyl group at position 6 has been replaced by a hydrogen.)

Some Naturally Occurring Sugars That Belong to the L Series:

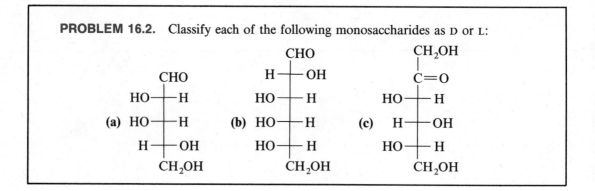

L-Arabinose   L-Fucose   L-Galactose

PROBLEM 16.2.   Classify each of the following monosaccharides as D or L:

(a)
```
   CHO
HO—H
HO—H
H —OH
  CH₂OH
```

(b)
```
   CHO
H —OH
HO—H
HO—H
HO—H
  CH₂OH
```

(c)
```
  CH₂OH
   C=O
HO—H
H —OH
HO—H
  CH₂OH
```

16.3 THE HEXOSES

The aldohexoses are the six-carbon aldehyde sugars, while the ketohexoses, or hexuloses, are the six-carbon ketone sugars. The aldohexoses contain four chiral carbons and consequently there are 2^4, or 16, stereoisomeric aldohexoses. Eight of these sixteen isomers belong to the L series and eight belong to the D series. Of the eight belonging to the D series, three (D-glucose, D-galactose, and D-mannose) are found in the greatest abundance in nature.

D-Glucose (dextrose, blood sugar) is found in the blood serum of animals. Its constant concentration is imperative for the well-being of the organism. D-Glucose is the monomeric unit of the starches and cellulose. Monomeric D-glucose and D-fructose (levulose, fruit sugar), a ketohexose, are found in the juices of fruits and in honey. D-Glucose and D-fructose are the only monosaccharides that can be utilized directly by animals; consequently, they can be administered intravenously.

D-Galactose is one of the saccharides of the disaccharide lactose (milk sugar). D-Mannose is the monomer of the polysaccharide mannan, found in bacteria, yeasts, fungi, and higher plants. Both galactose and mannose are utilized indirectly as food. By metabolic isomerization of these sugars' derivatives, cells can convert these two monosaccharides to glucose, which is then metabolized for energy.

The ketohexoses, the six-carbon keto sugars, have only three chiral carbons; therefore, there are eight stereoisomeric ketohexoses—four in the D series and four in the L series. Of the D-ketohexoses, only D-fructose is of major importance. D-Fructose, besides occurring in fruit juice and honey, is one of the monosaccharide units in the disaccharide sucrose (table sugar). Fructose is the sweetest of the sugars. In terms of relative sweetness, sucrose (the reference compound) has a rating of 100; fructose, 173; glucose 74; galactose, 73; and lactose (milk sugar), 16. Because of the low sweetness of lactose, sucrose is often added to milk products as a sweetener.

The Most Common Aldohexoses and Ketohexoses:

D-Glucose   D-Galactose   D-Mannose   D-Fructose

A. Epimers

Diastereomers that differ in the configuration of only one chiral carbon are called **epimers**. D-Glucose and D-galactose are diastereomers that are epimers because they differ in the projection of the hydroxyl group at carbon 4.

A Pair of Epimers (Diastereomers That Differ in the Configuration of Only One Chiral Center):

D-Glucose   D-Galactose

D-Galactose and D-mannose are diastereomers, but not epimers, because they differ in the configuration of two chiral carbons, carbons 2 and 4. D-Fructose and any one of the aldohexoses are simply structural isomers of each other, having the same molecular formula but differing in the sequence of atoms in their structures.

B. Cyclization of the Monosaccharides

In Section 10.6B, we discussed the reaction of an aldehyde with an alcohol to yield a hemiacetal. The mechanism of the acid-catalyzed equilibration involves a series of steps. The result is the addition of the alcohol across the carbonyl double bond.

An aldehyde   *An alcohol*   *A hemiacetal*

If the hydroxyl and aldehyde groups are located in the same molecule, intramolecular hemiacetal formation yields a cyclic hemiacetal.

A hydroxy aldehyde   A cyclic hemiacetal

Monosaccharides, such as glucose, contain an aldehyde group and a hydroxyl group. These sugars can undergo cyclization in water to yield both five- and six-membered-ring hemiacetals; however, we will show only the six-membered ring of glucose to simplify the discussion.

The reaction takes place by the attack of the carbon-5 hydroxyl group (the same hydroxyl group used for the D series designation) on the carbonyl carbon. The reaction generates a new chiral carbon at carbon 1. Both configurations of this new chiral carbon are generated—one in which the new hydroxyl group is projected to the right in the Fischer projection (α-D-glucose) and one in which the new hydroxyl group is projected to the left (β-D-glucose).

Fischer Projections for the Cyclization of Glucose:

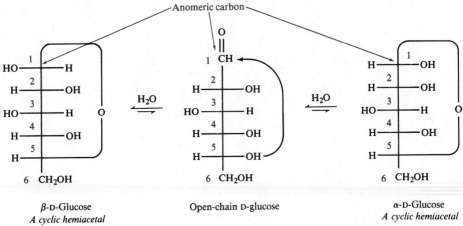

β-D-Glucose
A cyclic hemiacetal

Open-chain D-glucose

α-D-Glucose
A cyclic hemiacetal

In carbohydrate chemistry, the chiral carbon generated by the cyclization reaction is called the **anomeric carbon**. In glucose, the anomeric carbon is carbon 1. In fructose, it is carbon 2.

The two newly generated cyclic structures, α-D-glucose and β-D-glucose, are diastereomers and not enantiomers. These two cyclic structures differ in the configuration of only carbon 1, not in configurations of the other chiral carbons in the structure. The enantiomer of β-D-glucose is β-L-glucose (shown below), an L sugar.

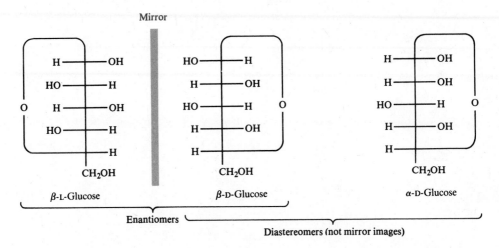

β-L-Glucose

β-D-Glucose

α-D-Glucose

Enantiomers

Diastereomers (not mirror images)

C. Anomers

In carbohydrate chemistry, diastereomers generated by a cyclization reaction at the anomeric carbon are called **anomers**. Because anomers differ in the configuration of only one carbon, the anomeric carbon, they are also epimers. Thus, α-D-glucose and β-D-glucose are epimers as well as anomers.

> *Structural isomers:* compounds with the same molecular formula but different structures.
>
> *Diastereomers:* stereoisomers that are not enantiomers, but differ in the configuration of one or more chiral centers.
>
> *Epimers:* diastereomers differing in the configuration of only one chiral center.
>
> *Anomers:* carbohydrate epimers differing in the configuration of only the hemiacetal or hemiketal carbon, the anomeric carbon.

D. Haworth Formulas

The cyclization of glucose forms a six-membered ring consisting of five carbon atoms and one oxygen. To emphasize this fact and the fact that the groups bonded on the ring are cis and trans to one another, a **Haworth formula** is used instead of a Fischer projection. In the Haworth formula, the ring is drawn as a flat six-membered ring. For the aldohexoses, the oxygen of the ring is positioned in the upper right-hand corner. For the D series of sugars, the CH_2OH (carbon 6 and its hydroxyl group) are projected *above the plane of the ring*. If the sugar belongs to the L series, the CH_2OH is projected *below the plane of the ring*. Carbon 1, the hemiacetal carbon, is positioned to the right in the formula.

Conventions Used in a Haworth Formula:

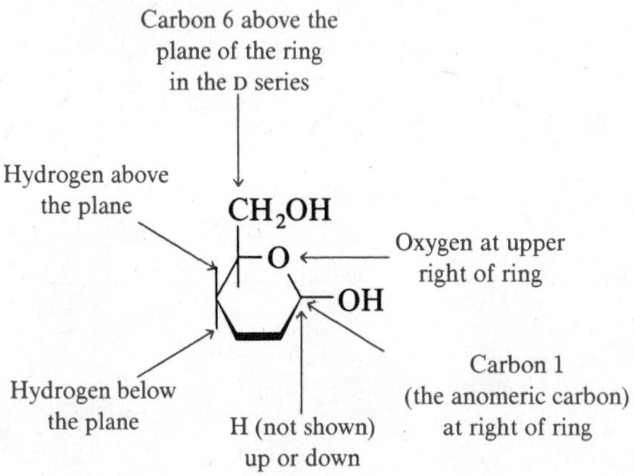

To convert the formula of a sugar drawn in a Fischer projection to a Haworth formula, draw each group that is *projected to the right in the Fischer projection below the plane of the ring in the Haworth formula.*

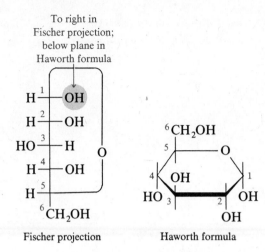

To right in
Fischer projection;
below plane in
Haworth formula

Fischer projection   Haworth formula

The equation for the cyclization of glucose can be rewritten with Haworth formulas to emphasize the six-membered cyclic hemiacetals.

α: down

H_2O

or

α-D-Glucose   Open-chain D-glucose

β: up

H_2O

β-D-Glucose

E. Classification of the Monosaccharides by Ring Size

D-Fructose also exists as cyclic structures in solution. Because fructose is a ketone, the rings are cyclic hemiketals, rather than cyclic hemiacetals as are formed by the aldoses. When fructose is dissolved in water, an equilibrium is established among the open-chain form, the five-membered ring, and the six-membered ring. The alpha and beta anomers of each size of ring are formed. (We will ignore the α- and β-anomers of each sized ring to simplify the discussion.)

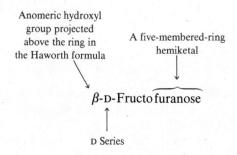

β-D-Fructose
six-membered ring (80%)   Open-chain fructose β-D-Fructose
five-membered ring (20%)

To distinguish the two different ring sizes, we add the terms **pyran** and **furan** to the name of the sugar. Pyran is the name of the six-membered ring containing one oxygen and two double bonds. Furan is the name of the corresponding five-membered ring.

Pyran   Furan

Six-membered ring saccharides are called **pyranoses**, and five-membered ring saccharides are called **furanoses**. These two terms are incorporated into the name to specify the size of the ring of the saccharide. For example, α-D-glucopyranose is a six-membered-ring hemiacetal and β-D-fructofuranose is a five-membered-ring hemiketal.

Anomeric hydroxyl
group projected
above the ring in
the Haworth formula

A five-membered-ring
hemiketal

β-D-Fructofuranose

D Series

PROBLEM 16.3. Identify the following monosaccharides as pyranoses or furanoses.

(a)

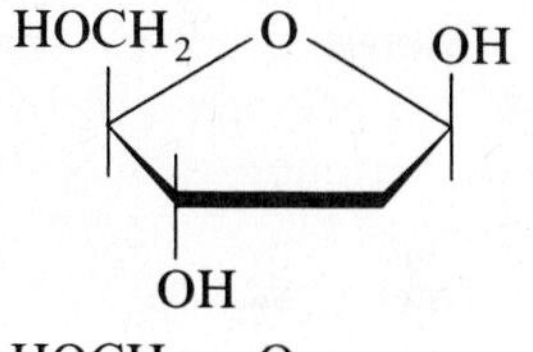

(b)

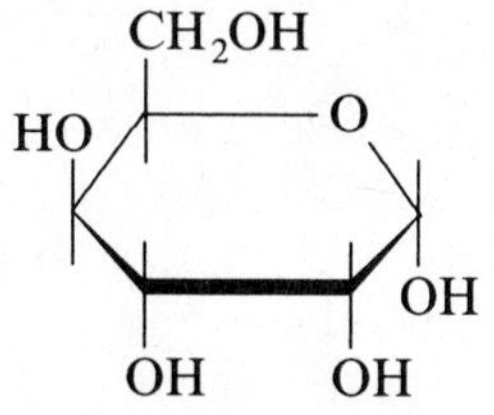

(c)

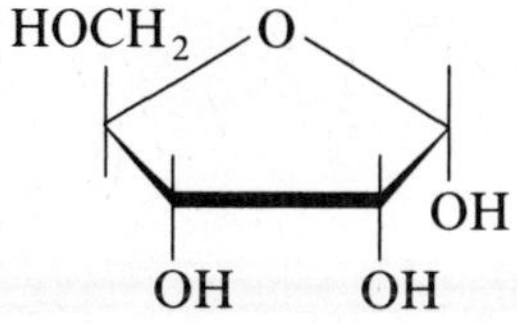

F. Conformation of the Monosaccharides

In Section 2.7C, we discussed the fact that the cyclohexane ring is not flat, but is puckered and assumes primarily the most stable chair conformations. Substituents on a cyclohexane ring are projected as either equatorial (*e*) or axial (*a*). The ring system assumes the chair form that places the largest of the groups in the less sterically hindered equatorial position.

The six-membered cyclic hemiacetals also exist in chair conformations.

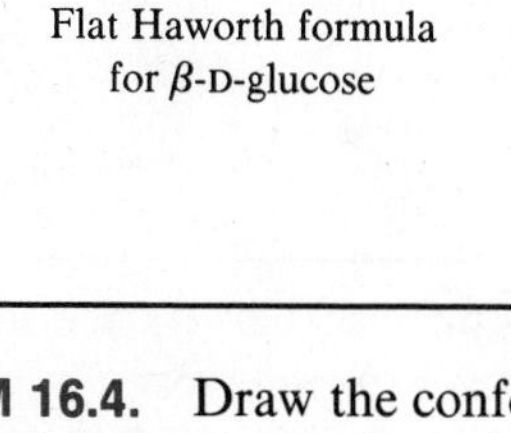

Flat Haworth formula
for β-D-glucose

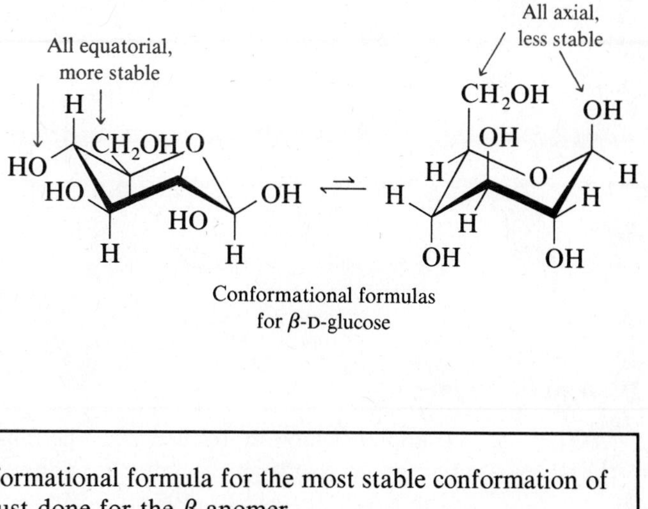

Conformational formulas
for β-D-glucose

PROBLEM 16.4. Draw the conformational formula for the most stable conformation of α-D-glucose, as we have just done for the β-anomer.

16.4 OTHER MONOSACCHARIDES

A. Aldopentoses

Two aldopentoses, D-ribose and 2-deoxy-D-ribose, are very abundant in nature. Ribose, as a furanose, is the sugar found in the polymeric chain of RNA (ribonucleic acid). Deoxyribose, also as a furanose, is found in DNA (deoxyribonucleic acid). DNA is the carrier of the genetic code, while different types of RNA are involved in the translation of the code into proteins. We will discuss nucleic acids in Chapter 19.

Formulas for D-*Ribose:*

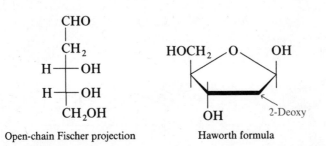

Open-chain Fischer projection   Haworth formula

*Formulas for 2-Deoxy-*D-*Ribose:*

Open-chain Fischer projection   Haworth formula

B. Amino Sugars

Two amino sugars are formed when certain polysaccharides are hydrolyzed. These amino sugars are D-glucosamine (2-amino-2-deoxy-D-glucose) and D-galactosamine (2-amino-2-deoxy-D-galactose). In their parent polysaccharides, these amino sugars contain an *N*-acetyl group ($N{-}COCH_3$) or an *N*-sulfonic acid group ($N{-}SO_3H$).

These groups are lost during the hydrolysis of the polysaccharides.

$$
\begin{array}{cc}
\text{CHO} & \text{CHO} \\
\text{H}-\text{NH}_2 & \text{H}_2\text{N}-\text{H} \\
\text{HO}-\text{H} & \text{HO}-\text{H} \\
\text{H}-\text{OH} & \text{H}-\text{OH} \\
\text{H}-\text{OH} & \text{H}-\text{OH} \\
\text{CH}_2\text{OH} & \text{CH}_2\text{OH}
\end{array}
$$

D-Glucosamine   D-Galactosamine

the 2-amino derivative   *the 2-amino derivative*
of glucose   *of galactose*

16.5 REACTIONS OF THE MONOSACCHARIDES

A. Mutarotation

α-D-Glucose and β-D-glucose are diastereomers. Each can be isolated as a pure compound and its physical constants, including optical rotation (see Section 6.2B), can be measured. The specific rotation of a freshly prepared solution of pure α-D-glucose is 112.2° and that of a freshly prepared solution of β-D-glucose is 18.7°. When either compound is placed in water, it is converted to an equilibrium mixture consisting of 36% of the α-isomer, 64% of the β-isomer, and a trace (0.02%) of the open-chain aldehyde. Because an equilibrium is established starting with either anomer, the initial optical rotation of the solution changes to that of the equilibrium mixture, 52.7°. The spontaneous change in optical rotation is called **mutarotation**. Mutarotation is very slow in neutral solution, but is fast, requiring only a few seconds, if either acid or base is present.

Mutarotation is due to the equilibration of the two anomers of a sugar. All sugars that have a hemiacetal (or hemiketal) group exhibit mutarotation. If the hemiacetal (or hemiketal) hydroxyl group is converted to an acetal (or ketal), mutarotation is not observed.

Undergoes mutarotation in solution   Cannot undergo mutarotation

B. Oxidation

Aldehydes and primary alcohols can both be oxidized to carboxylic acids.

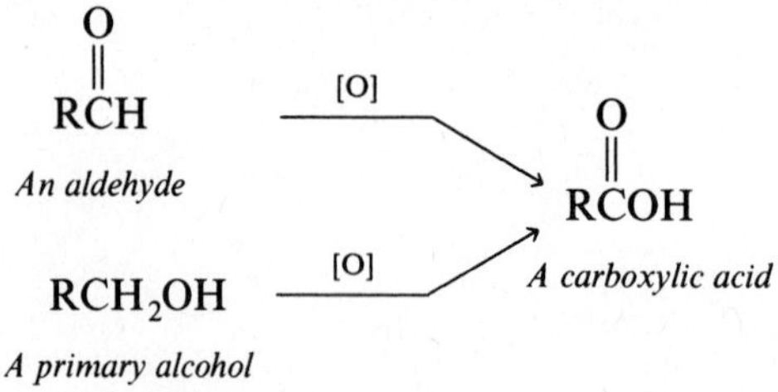

$$\underset{\text{An aldehyde}}{\overset{\overset{\displaystyle O}{\|}}{RCH}} \xrightarrow{\ [O]\ } \underset{\text{A carboxylic acid}}{\overset{\overset{\displaystyle O}{\|}}{RCOH}}$$

$$\underset{\text{A primary alcohol}}{RCH_2OH} \xrightarrow{\ [O]\ }$$

Aldoses contain both aldehyde groups and primary alcohol groups; under the proper reaction conditions, both groups can be oxidized to carboxyl groups.

Aldonic Acids   Aldehydes are easier to oxidize than alcohols. Consequently, we can oxidize the aldehyde group of a carbohydrate without oxidizing any of the alcohol hydroxyl groups. Several reagents can be used to oxidize the aldehyde group: Tollens reagent [$Ag(NH_3)_2{}^+$], Fehling solution (Cu^{2+}-tartrate complex), and Benedict solution (Cu^{2+}-citrate complex). In their reactions with the sugars, these oxidizing agents are reduced by the sugar and, therefore, the sugar is called a **reducing sugar**.

The product of the oxidation is a carboxylate, which can be converted to a carboxylic acid. A monosaccharide with a carboxyl group at carbon 1 is called an **aldonic acid**.

Step 1, Oxidation:

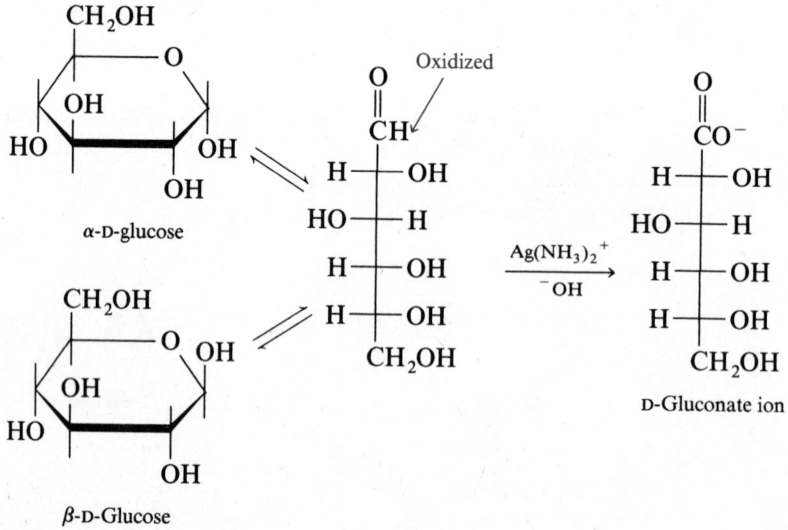

α-D-glucose

β-D-Glucose

D-Gluconate ion

Step 2, Acidification:

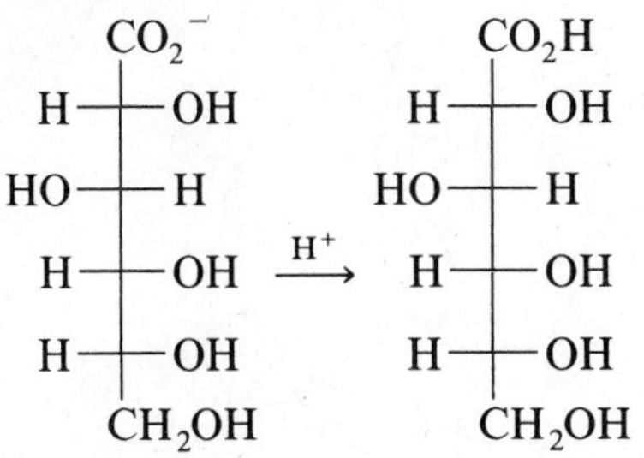

D-Gluconate ion   D-Gluconic acid

an aldonic acid

To be a reducing sugar, the anomeric carbon must have a hydroxyl group—that is, the sugar must be in the hemiacetal form. An acetal is not in equilibrium with an aldehyde and, consequently, oxidation does not occur.

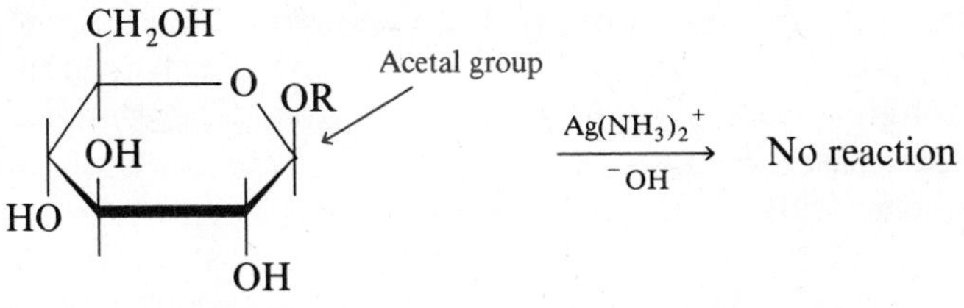

Not a reducing sugar

When acidified, the carboxylate of an aldonic acid initially forms an aldonic acid, which then may cyclize to a cyclic ester, called a *lactone*. Although both five- and six-membered-ring lactones can be formed by the cyclization, D-gluconic acid preferentially forms the five-membered-ring lactone.

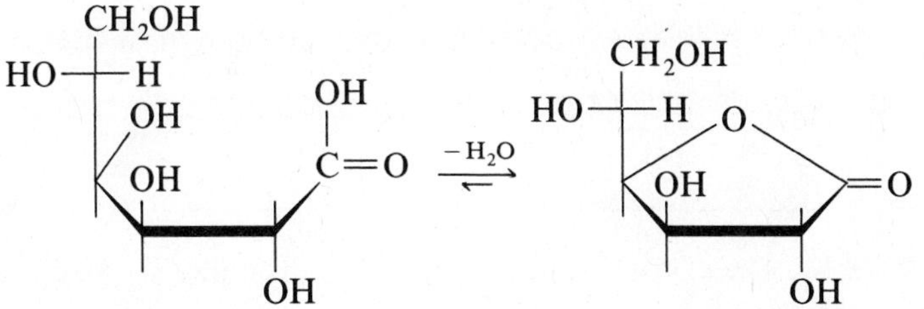

D-Gluconic acid   A lactone (cyclic ester)
                          of D-gluconic acid

Only a carbohydrate with a hemiacetal group is a reducing sugar. An apparent exception is fructose, a ketohexose that readily reacts with oxidizing agents. Fructose is not oxidized directly. Rather, the alkaline conditions of the reaction mixture convert fructose to an aldose (glucose or mannose), which is oxidized to the aldonic acid. The intermediate in the conversion is an enediol, a tautomer of fructose.

The enediol group

$$
\begin{array}{c}
\text{CH}_2\text{OH} \\
| \\
\text{C}=\text{O} \\
\text{HO}\!-\!\!-\!\text{H} \\
\text{H}\!-\!\!-\!\text{OH} \\
\text{H}\!-\!\!-\!\text{OH} \\
\text{CH}_2\text{OH}
\end{array}
\quad \underset{}{\overset{-\text{OH}}{\rightleftharpoons}} \quad
\begin{array}{c}
\text{CHOH} \\
\| \\
\text{C}\!-\!\text{OH} \\
\text{HO}\!-\!\!-\!\text{H} \\
\text{H}\!-\!\!-\!\text{OH} \\
\text{H}\!-\!\!-\!\text{OH} \\
\text{CH}_2\text{OH}
\end{array}
\quad \rightleftharpoons \quad
\begin{array}{c}
\overset{\text{O}}{\overset{\|}{\text{CH}}} \\
\text{CHOH} \\
\text{HO}\!-\!\!-\!\text{H} \\
\text{H}\!-\!\!-\!\text{OH} \\
\text{H}\!-\!\!-\!\text{OH} \\
\text{CH}_2\text{OH}
\end{array}
\quad \overset{(1)\,[\text{O}]}{\underset{(2)\,\text{H}^+}{\longrightarrow}} \quad
\begin{array}{c}
\overset{\text{O}}{\overset{\|}{\text{COH}}} \\
\text{CHOH} \\
\text{HO}\!-\!\!-\!\text{H} \\
\text{H}\!-\!\!-\!\text{OH} \\
\text{H}\!-\!\!-\!\text{OH} \\
\text{CH}_2\text{OH}
\end{array}
$$

D-Fructose   *An enediol*   *An aldose*   *An aldonic acid*

Aldaric Acids Treatment of an aldohexose with a more vigorous oxidizing agent, such as hot dilute nitric acid, causes both the aldehyde group and the terminal CH_2OH group to be oxidized to carboxyl groups. The resulting dicarboxylic acids are called **aldaric acids**. D-Glucaric acid is the aldaric acid of glucose. Aldaric acids, like aldonic acids, are in equilibrium with their lactones.

$$
\begin{array}{c}
\text{CHO} \\
\text{H}\!-\!\!-\!\text{OH} \\
\text{HO}\!-\!\!-\!\text{H} \\
\text{H}\!-\!\!-\!\text{OH} \\
\text{H}\!-\!\!-\!\text{OH} \\
\text{CH}_2\text{OH}
\end{array}
\quad \overset{\text{HNO}_3}{\underset{\text{heat}}{\longrightarrow}} \quad
\begin{array}{c}
\text{CO}_2\text{H} \\
\text{H}\!-\!\!-\!\text{OH} \\
\text{HO}\!-\!\!-\!\text{H} \\
\text{H}\!-\!\!-\!\text{OH} \\
\text{H}\!-\!\!-\!\text{OH} \\
\text{CO}_2\text{H}
\end{array}
$$

Oxidized to carboxyl groups

D-Glucose   D-Glucaric acid

PROBLEM 16.5. Draw Fischer projections for all five- and six-membered-ring lactones of D-glucaric acid.

Uronic Acids The terminal CH_2OH group can be enzymatically oxidized in a living system to a carboxylic acid without the concurrent oxidation of the aldehyde group. This reaction cannot be carried out by a direct laboratory oxidation reaction —the aldehyde group would be preferentially oxidized. The product carboxylic acid of the oxidation of the CH_2OH group is called a **uronic acid**. The uronic acid of

D-glucose is D-glucuronic acid.

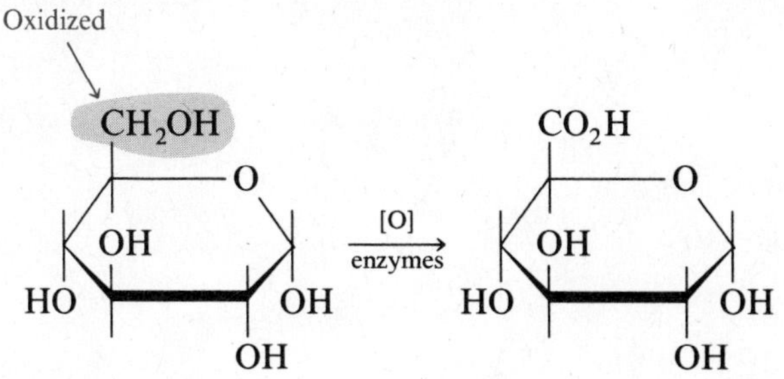

Oxidized

α-D-Glucose

$\xrightarrow[\text{enzymes}]{[O]}$

α-D-Glucuronic acid

a uronic acid

Animals eliminate toxic phenols and other waste products as glucuronides, esters of glucuronic acid.

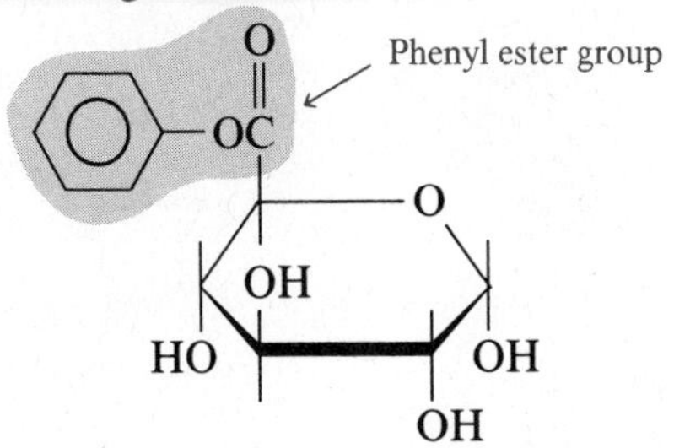

Phenyl ester group

A phenyl glucuronide

Glucuronic acid is also one of the saccharides found in hyaluronic acid, the viscous polysaccharide found in joints, in the vitreous humor of the eye, and in umbilical cord jelly.

Figure 16.4 summarizes the oxidation reactions of the monosaccharides.

C. Reduction

The aldehyde and keto groups can be reduced to alcohols. The product polyhydroxy compounds are called **alditols**. The reduction of D-glucose forms D-glucitol, also called sorbitol.

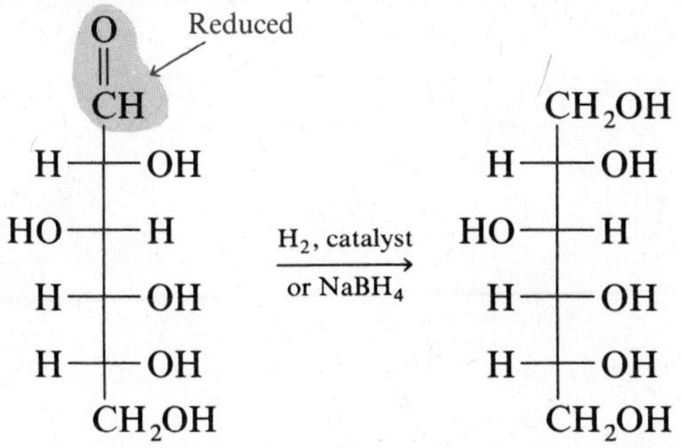

Reduced

$\xrightarrow[\text{or NaBH}_4]{\text{H}_2, \text{ catalyst}}$

D-Glucose

D-Glucitol
(sorbitol)

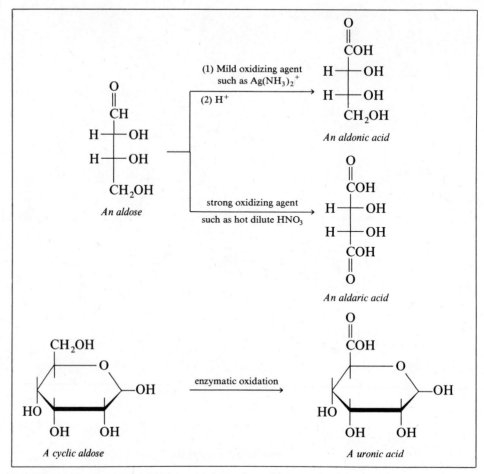

Figure 16.4 Summary of oxidation reactions of an aldose.

Synthetic sorbitol is used as an artificial sweetener. Mannitol, the reduction product of mannose, is obtained from seaweed and used as a nutrient in bacteriology. Xylitol, the reduction product of xylose, is used as a sweetener and an anticavity agent in some chewing gums.

PROBLEM 16.6. Write equations for the $NaBH_4$-reduction reactions leading to (a) mannitol and (b) xylitol.

D. Ester Formation

The hydroxyl groups of the monosaccharides can undergo typical alcohol reactions. For example, they can form esters with organic acids and inorganic acids. When

β-D-glucose is treated with acetic anhydride, a reagent used to form acetate esters (see Section 12.2C), a pentacetate is formed.

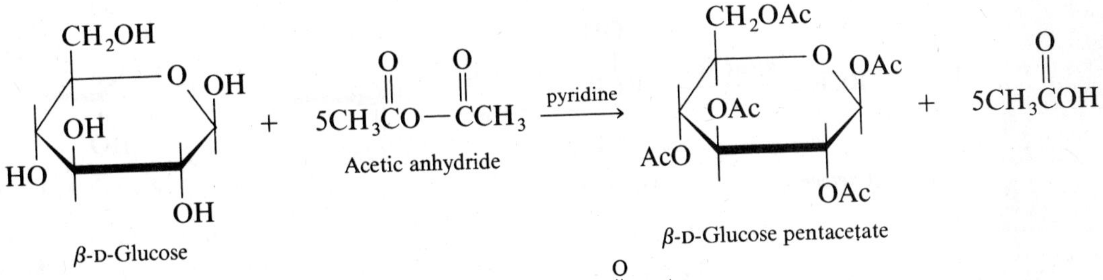

β-D-Glucose + $5CH_3CO-CCH_3$ $\xrightarrow{\text{pyridine}}$ β-D-Glucose pentacetate + $5CH_3COH$

Acetic anhydride

$\left(\text{where} - OAc = -O\overset{O}{\overset{\|}{C}}CH_3 \right)$

The most important inorganic esters of sugars are the phosphates, which are key intermediates in the metabolism of carbohydrates. The formation of a typical phosphate derivative is the first step in the biological oxidation of glucose. In this reaction step, glucose is phosphorylated with ATP (adenosine triphosphate). The reaction is catalyzed by the enzyme glucokinase. (The suffix *kinase* indicates that the enzyme transfers a phosphate from ATP to an acceptor, in this case, glucose.)

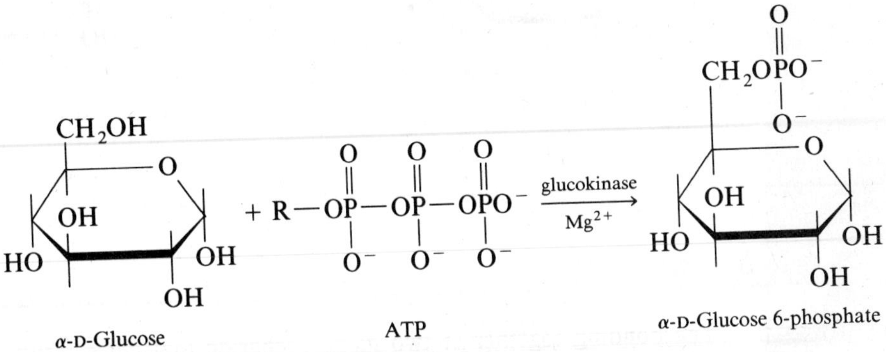

α-D-Glucose   + R—OP—OP—OPO⁻ $\xrightarrow[\text{Mg}^{2+}]{\text{glucokinase}}$ α-D-Glucose 6-phosphate

ATP

E. Glycoside Formation

Hemiacetals react with alcohols to form acetals (see Section 10.6B). In the reaction, a molecule of water is lost.

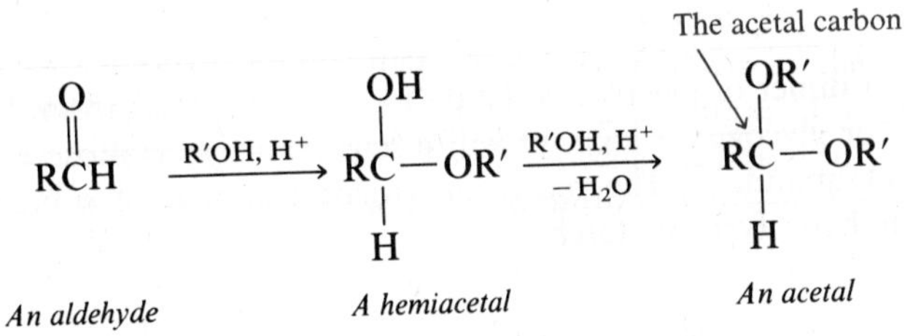

RCH $\xrightarrow{\text{R'OH, H}^+}$ RC—OR' $\xrightarrow[-\text{H}_2\text{O}]{\text{R'OH, H}^+}$ RC—OR'

An aldehyde   *A hemiacetal*   *An acetal*

16.27. The crystalline form of maltose is β-maltose. When this compound is dissolved in water, the specific rotation slowly changes from an initial value of +111.7° to +130.4°. Explain this change in optical rotation.

16.28. In aqueous solution, the open-chain form of D-ribose is in equilibrum with α- and β-pyranoses (20 and 56%, respectively) and α- and β-furanoses (6 and 18%). Write equations that show these compounds. (Use Haworth formulas where appropriate.)

16.29. The fabric cellulose acetate is made by heating cellulose with acetic anhydride. For conversion to threads, the crude cellulose acetate product is dissolved in a volatile solvent and passed through orifices as the solvent is evaporated. Complete the following equation that shows the initial formation of cellulose acetate. Use Ac— to represent the acetyl group, $CH_3\overset{\underset{\|}{O}}{C}{-}$.

16.30. Oxidation of D-erythrose and D-threose (Figure 16.3) yields two diacids called tartaric acids—one is optically active and the other is the optically inactive *meso* form. Write equations that show which monosaccharide yields which tartaric acid.

16.31. *Heparin* is an anticoagulant (a compound that prevents blood clots from forming) found in various body tissues. This compound is a series of sulfated polysaccharides. The formula for a fragment follows:

Predict the products of complete hydrolysis of this heparin fragment.

16.32. *Trehalose* is a disaccharide found in the poisonous mushroom *Amanita muscaria* and in the cocoons of parasitic beetles of the *Larinus* species. Trehalose is a nonreducing sugar. When hydrolyzed, it yields only D-glucose. What are the possible structures of trehalose?

16.6 D

A. Malto

POINT OF INTEREST 16

Naturally Occurring Glycosides

Some fairly simple glycosides are common in nature. Generally, the carbohydrate portion of the glycoside is a monosaccharide or a higher saccharide, sometimes altered from the true sugar, as in the glucuronic acid portion of laetrile, the structure of which follows. The nonsugar portion of the glycoside is called an *aglycone*.

Plants of the *Prunus* genus, which includes apricots, peaches, and cherries, contain the glycosides *amygdalin* and *laetrile* in their pits. In these compounds, mandelonitrile (the cyanohydrin of benzaldehyde) is the aglycone. (Several years ago, these compounds were touted as cancer cures, a claim that has been refuted.) Both amygdalin and laetrile are hydrolyzed in the stomach to the sugar, benzaldehyde, and HCN, which is toxic.

Amygdalin

Laetrile

Cardiac glycosides, more complex in structure than amygdalin or laetrile, are heart stimulants used in the treatment of congestive heart failure, a condition in which the heart does not beat sufficiently strongly. Cardiac stimulants are highly toxic in doses over the therapeutic levels. These compounds are found in certain plants, such as the dogbane family (Apocynaceae), foxglove (*Digitalis purpurea*), and others. Similar cardiac stimulants, but not as glycosides, are found in poisonous toads. (The ancient Chinese used dried toad skins medicinally for centuries.)

Certain cardiac glycosides from plants are mentioned in ancient Egyptian papyruses. The Romans used cardiac glycosides as heart tonics and as rat poisons. The African Zulu arrowhead poisons were cardiac glycosides. In 1785 the English physician William Withering reported on the use and limitations of digitalis, the extract from foxglove leaves, in the treatment of dropsy (fluid accumulation). Its profound effect on the heart muscle was noted shortly thereafter.

The cardiac glycosides are all glycosides of *steroids*, compounds with distinctive ring systems that will be discussed in Chapter 17. The most important glycoside in digitalis is digitoxin, a glycoside formed from a trisaccharide of 2-deoxy-D-ribose and the steroid aglycone *digitoxigenin*.

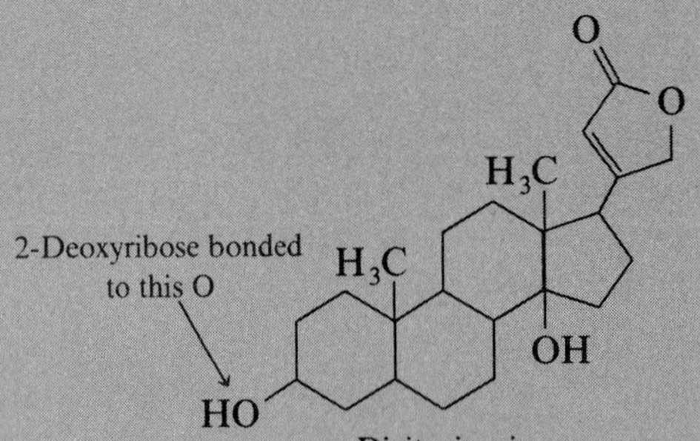
2-Deoxyribose bonded
to this O
HO
H$_3$C
H$_3$C
OH
O
O

Digitoxigenin
aglycone of digitoxin

Chapter 17

Lipids

Lipids (Greek *lipos*, "fat") are water-insoluble compounds that can be isolated from cells and tissues by extraction with a relatively nonpolar organic solvent, such as diethyl ether or chloroform. This class of compounds is, therefore, defined by physical properties—solubility in nonpolar solvents and insolubility in water—and not by structure. It is not surprising, then, that the class of lipids encompasses a variety of different compound types. Animal fats, vegetable oils, phosphate-containing fatlike compounds, waxes, terpenes, steroids, and prostaglandins are all lipids. In this chapter, we will be able to survey these various types of lipids only briefly. (Terpenes were discussed in Point of Interest 4.) We will concentrate on the structures, the properties, and, to a limited extent, the reactions of the lipids.

Although the structures of lipids vary, all lipids are typified by a structure that has a large hydrophobic hydrocarbon group or groups and only a few, if any, hydrophilic groups. These are the structural features that make lipids insoluble in water and soluble in nonpolar solvents.

17.1 TRIGLYCERIDES. FATS AND OILS

There is no major structural difference between fats, such as beef fat, and fatty oils, such as corn oil. Both classes of compounds are triesters formed from the triol glycerol and three long continuous-chain carboxylic acids called **fatty acids**. These triesters are called **triacylglycerols** or **triglycerides** regardless of whether they are isolated from fats or oils.

Structure of the Triglycerides:

$$
\begin{matrix}
CH_2OH \\
| \\
CHOH \\
| \\
CH_2OH
\end{matrix}
\;+\;
\left\{
\begin{matrix}
\overset{\displaystyle O}{\overset{\|}{HOCR}} \\[6pt]
\overset{\displaystyle O}{\overset{\|}{HOCR'}} \\[6pt]
\overset{\displaystyle O}{\overset{\|}{HOCR''}}
\end{matrix}
\right.
\;\xrightarrow{-3H_2O}\;
\begin{matrix}
CH_2O_2CR \\
| \\
CHO_2CR' \\
| \\
CH_2O_2CR''
\end{matrix}
$$

Glycerol   *Fatty acids*   *A triacylglycerol,*
                                         or triglyceride

R, R', R'' = long continuous hydrocarbon chains
that may be the same or different

The difference between fats and oils is their physical state. At room temperature, fats are *solid*, while oils are *liquid*. An exception is the vegetable oil *coconut oil*, which melts at 21–25°C, near room temperature in the temperate zones but below room temperature in the tropics.

Some typical fats	Some typical oils
Beef fat, or tallow	Corn oil
Lard	Soybean oil
Butter	Cottonseed oil

With very few exceptions, naturally occurring fats and oils are not homogeneous triglycerides. That is, most triglycerides contain two or three different fatty acid residues, such as one palmitic acid, one stearic acid, and one oleic acid, as their esters. (See Table 17.1 for the structures of these compounds.) The specific fatty acid groups present in the triglyceride depend on the species and on other conditions, such as an animal's diet and the ambient temperature. Warm-blooded animals tend to biosynthesize fats that are fluid near their body temperature. When plants are grown at cooler temperatures, they synthesize a greater proportion of low-melting triglycerides.

A. Structures and Properties of Fatty Acids

Table 17.1 lists some fatty acids commonly found esterified with glycerol in fats and oils. The common fatty acid components of animal and plant triglycerides contain *even numbers* of carbon atoms in unbranched hydrocarbon chains. (Odd-numbered and branched-chain fatty acids, however, are common in bacteria.) The long continuous hydrocarbon chain of the fatty acids may be saturated or may contain one or more carbon–carbon double bonds. The double bonds in the unsaturated

TABLE 17.1   SOME COMMON FATTY ACIDS

Name	Number of carbons	Formula	Melting point (°C)
Saturated			
Lauric	12	$CH_3(CH_2)_{10}CO_2H$	44
Myristic	14	$CH_3(CH_2)_{12}CO_2H$	58
Palmitic	16	$CH_3(CH_2)_{14}CO_2H$	63
Stearic	18	$CH_3(CH_2)_{16}CO_2H$	70
Arachidic	20	$CH_3(CH_2)_{18}CO_2H$	75
Unsaturated[a]			
Palmitoleic	16	$CH_3(CH_2)_5CH{=}CH(CH_2)_7CO_2H$	32
Oleic	18	$CH_3(CH_2)_7CH{=}CH(CH_2)_7CO_2H$	7
Linoleic	18	$CH_3(CH_2)_4CH{=}CHCH_2CH{=}CH(CH_2)_7CO_2H$	−5
Linolenic	18	$CH_3(CH_2CH{=}CH)_3(CH_2)_7CO_2H$	−11
Arachidonic	20	$CH_3(CH_2)_4(CH{=}CHCH_2)_4CH_2CH_2CO_2H$	−50

[a]All carbon–carbon double bonds are cis.

fatty acids are cis and are homoallylic (separated by CH_2) rather than conjugated.

One CH_2 group

$-CH{=}CH-CH_2-CH{=}CH-$

Homoallylic double bonds,
as in fatty acids

No CH_2 group

$-CH_2-CH{=}CH-CH{=}CH-$

Conjugated double bonds

Effect of Unsaturation on Melting Point   Vegetable oils are generally liquid and animal fats are generally solid because vegetable oils usually contain a larger number of carbon–carbon double bonds in their fatty acid hydrocarbon chains. For this reason, vegetable oils are often referred to as **polyunsaturated triglycerides**. Table 17.2 shows the fatty acid composition of some fats and oils. The human fat triglycerides, for example, contain primarily two saturated fatty acid residues, palmitic (25%) and stearic (8%), and two unsaturated fatty acid residues, oleic (46%) and linoleic (10%).

Saturated fatty acids, by themselves or as components of a triglyceride, can fit into an orderly crystal lattice in which the carbon and hydrogen atoms of different hydrocarbon chains are stacked relatively closely in zigzag conformations. Because of the close stacking, the hydrocarbon chains develop relatively strong attractions for one another; additional energy is needed to overcome the stability of the crystal lattice when the triglyceride is melted. The cis double bonds of an unsaturated fatty acid alter the shape of the hydrocarbon chain so that its atoms cannot lie close together. Thus, the presence of a double bond decreases the attractive forces holding the hydrocarbon chains together. The looser packing means that comparatively less energy is needed to melt the triglyceride. Therefore, the melting points of the

TABLE 17.2 APPROXIMATE FATTY ACID COMPOSITION OF SOME FATS AND OILS

Fat or oil	Composition (%)[a]			
	Palmitic	Stearic	Oleic	Linoleic
Animal fats				
Lard	30	—	45	5
Butter	25	8	35	5
Human fat	25	8	46	10
Vegetable oils				
Coconut[b]	8	2	6	1
Corn	10	5	45	38
Soy	10	—	25	55
Olive	5	5	80	7

[a] Other fatty acids are present in lesser or greater amounts.

[b] Coconut oil (mp 21–25°C) contains 50% lauric acid and 18% myristic acid—a very high saturated fatty acid content compared with that of other vegetable oils.

unsaturated fatty acids and their triglycerides are lower than those of the saturated fatty acids and their triglycerides.

Figure 17.1 shows how one carbon–carbon double bond changes the shape of a triglyceride. The presence of more double bonds alters the shape to even a greater extent. From the figure, it can be seen why triglycerides with carbon–carbon double bonds cannot fit into an orderly zigzag crystal lattice. For example, trioleylglycerol, with three carbon–carbon double bonds, melts at −17°C, while tristearylglycerol, with no carbon–carbon double bonds, melts at +55°C. The greater the number of double bonds, the lower is the melting point.

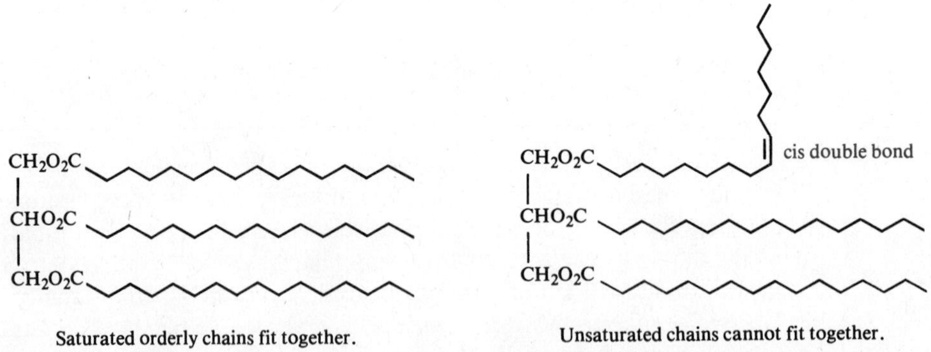

CH_2O_2C
|
CHO_2C
|
CH_2O_2C

Saturated orderly chains fit together.

CH_2O_2C cis double bond
|
CHO_2C
|
CH_2O_2C

Unsaturated chains cannot fit together.

Figure 17.1 A triglyceride with a carbon–carbon double bond cannot fit into a crystal lattice as well as a saturated triglyceride can. A fat with only one carbon–carbon double bond is shown here.

Unsaturation Lowers the Melting Point:

$$CH_3(CH_2)_5 \quad\quad (CH_2)_7CO_2H$$

$$CH_3(CH_2)_{14}CO_2H$$

$$C=C$$

$$H \quad\quad H$$

Palmitic acid
(mp 63°C)

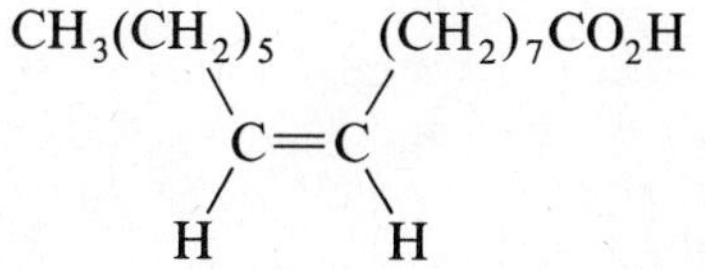

Palmitoleic acid
(mp 32°C)

Essential Fatty Acids   For mammals, linoleic acid and linolenic acid are essential dietary requirements. These animals have enzymes that can transform stearic acid to oleic acid, but do not have the enzymes needed for the introduction of additional double bonds toward the methyl end of a fatty acid chain. Mammals do have the enzymes necessary for the conversion of linoleic acid to arachidonic acid, a necessary precursor of the prostaglandins (see Section 17.5) by the pathway shown in Figure 17.2.

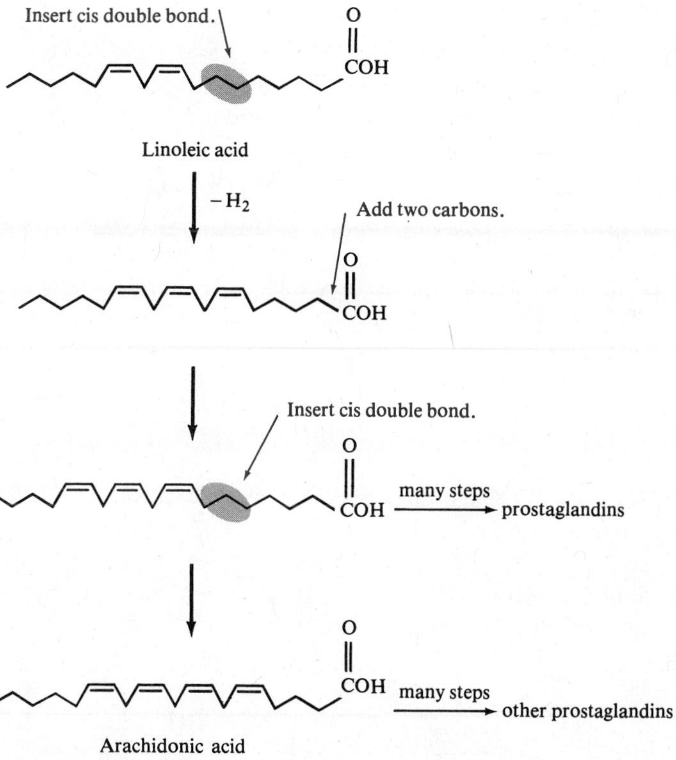

Figure 17.2 Biosynthetic pathway from linoleic acid to arachidonic acid and the prostaglandins. In the actual path, the carboxylic acids are found as derivatives of acetyl-coenzyme A.

B. Reactions of Triglycerides

Hydrogenation   The triglycerides of a typical vegetable oil contain about 20% saturated fatty acids and 80% unsaturated fatty acids. The triglycerides of a typical animal fat, however, contain more saturated fatty acids—about 50% saturated and 50% unsaturated.

Animal fats (for example, butter and lard) are more expensive than vegetable oils, but their texture is preferred for food spreads, for use in baked goods, and for cooking. Consequently, many of the solid cooking shortenings used today are prepared by the *partial hydrogenation of vegetable oils*, such as peanut oil, corn oil, or soybean oil. Partial hydrogenation is called **hardening** because the melting point of the hydrogenated product is higher.

The preparation of peanut butter usually involves adding partially hydrogenated peanut oil to crushed peanuts to make the characteristic smooth creamlike spread. Oleomargarine is prepared by emulsifying water with a partially hydrogenated vegetable oil, such as corn oil. Milk, flavoring, and food coloring are often added to make the final product more palatable. For these food products, only some of the double bonds are hydrogenated to raise the melting point of the oil so that it has a creamy consistency at or near room temperature.

Hydrogenation of an Unsaturated Triglyceride:

$$\begin{array}{c}
CH_2O_2C(CH_2)_7CH{=}CH(CH_2)_7CH_3 \\
| \\
CHO_2C(CH_2)_7CH{=}CH(CH_2)_7CH_3 \\
| \\
CH_2O_2C(CH_2)_7CH{=}CH(CH_2)_7CH_3
\end{array}
\xrightarrow[\text{moderate } T \text{ and } P]{\substack{3H_2 \\ \text{catalyst}}}
\begin{array}{c}
CH_2O_2C(CH_2)_{16}CH_3 \\
| \\
CHO_2C(CH_2)_{16}CH_3 \\
| \\
CH_2O_2C(CH_2)_{16}CH_3
\end{array}$$

Trioleylglycerol
mp −5°C

Tristearylglycerol
mp 55°C (then solidifies
and melts again at 72°)

Saponification   We have discussed the saponification (alkaline hydrolysis) of the ester groups in fats in Point of Interest 12. This reaction is used to prepare soaps, which are the sodium or potassium carboxylates of long-chain fatty acids. We also use a saponification reaction to clear fat-clogged drainpipes by pouring in a sodium hydroxide drain cleaner (lye, Drano).

$$\begin{array}{c}
CH_2O_2CR \\
| \\
CHO_2CR \\
| \\
CH_2O_2CR
\end{array}
+ \ 3Na^+ \ ^-OH \xrightarrow{\text{heat}}
\begin{array}{c}
CH_2OH \\
| \\
CHOH \\
| \\
CH_2OH
\end{array}
+ \ \underset{\textit{A soap}}{3RCO_2^- \ Na^+}$$

Water-insoluble fat   *Soluble in water*

> **PROBLEM 17.1.** Predict the products when trilinoleylglycerol is **(a)** hydrogenated at moderate temperature and pressure and **(b)** saponified with aqueous sodium hydroxide.

17.2 WAXES

Waxes are low-melting solids that have a waxy feel. The melting points of waxes are higher than those of triglycerides; therefore, waxes do not melt at body temperature. However, they are generally moldable near body temperature and hard at lower temperatures.

Waxes are simple monoesters formed from a long-chain fatty acid and a long-chain alcohol. Unlike triglycerides, waxes are resistant to saponification, probably because of their extensive hydrocarbon content.

Three typical waxes are *beeswax*, which is the wax bees use to build their honeycombs; *carnauba wax*, a relatively high-melting wax that occurs as a coating on the leaves of the Brazilian palm tree *Copernicia prunifera* and is used is automobile polishes; and *spermaceti*, a wax that separates from the oil obtained from a sperm whale's head. Paraffin wax is a mixture of alkanes obtained from petroleum and not a true ester type of wax.

Some Esters Found in Waxes:

In beeswax (mp 60–80°C):

$$CH_3(CH_2)_{24}CO_2(CH_2)_{29}CH_3 \qquad CH_3(CH_2)_{24}CO_2(CH_2)_{31}CH_3$$
$$CH_3(CH_2)_{26}CO_2(CH_2)_{29}CH_3 \qquad CH_3(CH_2)_{26}CO_2(CH_2)_{31}CH_3$$

In carnauba wax (mp 80–87°C):

$$CH_3(CH_2)_{22}CO_2(CH_2)_{31}CH_3 \qquad CH_3(CH_2)_{22}CO_2(CH_2)_{33}CH_3$$
$$CH_3(CH_2)_{26}CO_2(CH_2)_{31}CH_3 \qquad CH_3(CH_2)_{26}CO_2(CH_2)_{33}CH_3$$

In spermaceti (mp 42–47°C):

$$CH_3(CH_2)_{14}CO_2(CH_2)_{15}CH_3, \quad \text{cetyl palmitate}$$

In nature, waxes function as protective coatings on leaves, fruit, seeds, insect exoskeletons, and feathers. The secretion from the preen glands of waterfowl, for example, consists of a wax formed from esters of 1-octadecanol and branched-chain fatty acids, such as 2,4,6,8-tetramethyloctadecanoic acid:

$$CH_3(CH_2)_9\underset{\underset{CH_3}{|}}{C}HCH_2\underset{\underset{CH_3}{|}}{C}HCH_2\underset{\underset{CH_3}{|}}{C}HCH_2\underset{\underset{CH_3}{|}}{C}H\overset{\overset{O}{\|}}{C}OCH_2(CH_2)_{16}CH_3$$

A wax secreted by the preen glands of ducks and geese

In consumer products, natural waxes have been largely replaced by synthetic polymers such as silicones:

$$\left\{-\!\!\!\begin{array}{c} R \\ | \\ OSi \\ | \\ R \end{array}\!\!\!-\!\!\!\begin{array}{c} R \\ | \\ OSi \\ | \\ R \end{array}\!\!\!-\!\!\!\begin{array}{c} R \\ | \\ OSi \\ | \\ R \end{array}\!\!\!-\!\!\!\right\}$$

A silicone

17.3 PHOSPHOLIPIDS

Compounds containing phosphate ester groups are key intermediates in the metabolism of organic compounds in all living systems.

Phosphate group

$$R\!-\!O\!\!\underset{\underset{O}{\|}}{\overset{\overset{O^-}{|}}{P}}O\!-\!R'$$

From ROH   From a substituted phosphoric acid, $HOPO_3R'$

$$R\!-\!O\!-\!\underset{\underset{O}{\|}}{\overset{\overset{O^-}{|}}{P}}O^-$$

From ROH   From H_3PO_4

Lipids containing phosphate groups are called *phospholipids*. Phospholipids that are diacylesters of glycerol are called **phosphoglycerides**. These compounds have two fatty acid ester groups (RCO_2-)—generally esters of palmitic, stearic, or oleic acid. A phosphate ester group is bonded to the third hydroxyl group. Figure 17.3 shows the general structure of the phosphoglycerides.

Two common subclasses of phosphoglycerides are the **cephalins** and the **lecithins**. A cephalin is the phosphoglyceride containing a protonated ethanolamine, $HOCH_2CH_2\overset{+}{N}H_3$, bonded to the phosphate group. A lecithin contains a choline cation, $HOCH_2CH_2\overset{+}{N}(CH_3)_3$. The amine substituent and the phosphate oxygen form the dipolar ionic head of the phosphoglyceride.

Dipolar ion   From ethanolamine

$$\left\{-O\underset{\underset{O}{\|}}{\overset{\overset{O^-}{|}}{P}}OCH_2CH_2\overset{+}{N}H_3\right.$$

From choline

$$\left\{-O\underset{\underset{O}{\|}}{\overset{\overset{O^-}{|}}{P}}OCH_2CH_2\overset{+}{N}(CH_3)_3\right.$$

Dipolar end of a cephalin, or phosphatidylethanolamine

Dipolar end of a lecithin, or phosphatidylcholine

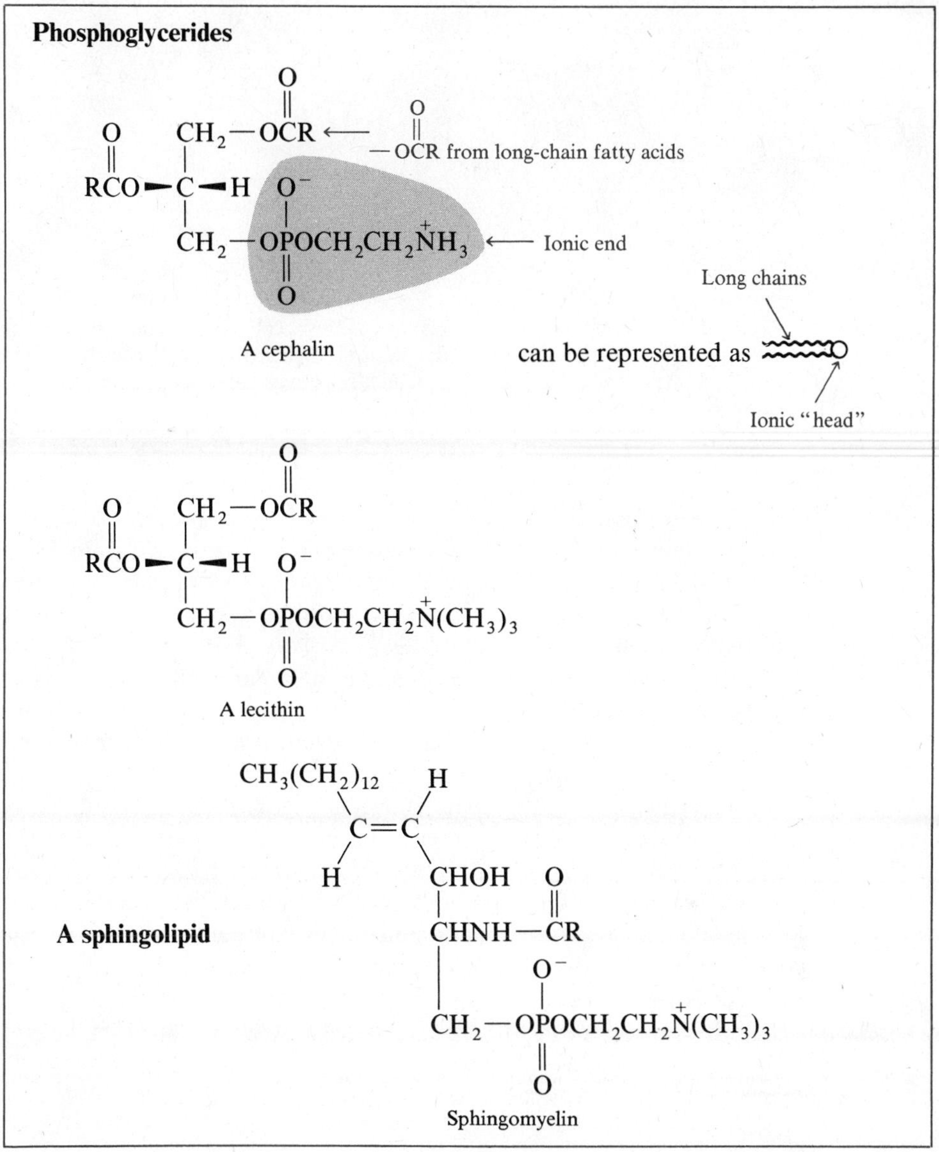

Phosphoglycerides

$$CH_2-OCR \longleftarrow \overset{O}{\underset{\|}{-}}OCR \text{ from long-chain fatty acids}$$

$$RCO \blacktriangleright C \blacktriangleleft H$$

$$CH_2-OPOCH_2CH_2\overset{+}{N}H_3 \longleftarrow \text{Ionic end}$$

A cephalin

can be represented as $\approx\approx\approx\approx$O

Long chains

Ionic "head"

$$\overset{O}{\underset{\|}{}}$$
$$CH_2-OCR$$
$$RCO \blacktriangleright C \blacktriangleleft H \quad O^-$$
$$CH_2-OPOCH_2CH_2\overset{+}{N}(CH_3)_3$$
$$\underset{O}{\overset{\|}{}}$$

A lecithin

A sphingolipid

$$CH_3(CH_2)_{12} \quad H$$
$$C=C$$
$$H \quad CHOH \quad O$$
$$CHNH-CR$$
$$O^-$$
$$CH_2-OPOCH_2CH_2\overset{+}{N}(CH_3)_3$$
$$\underset{O}{\overset{\|}{}}$$

Sphingomyelin

Figure 17.3 Structures of some phospholipids, showing the long hydrocarbon "tails" and the ionic "heads."

Cell membranes, the membranes that enclose animal cells, are composed of a double layer, or *bilayer*, of phosphoglyceride molecules. The two long hydrocarbon tails from the fatty acids of each phosphoglyceride molecule intertwine with other hydrocarbon tails, as shown in Figure 17.4. These intertwining tails form the inner membrane, while the ionic heads of the molecules face outward toward the aqueous fluids inside and outside the cell.

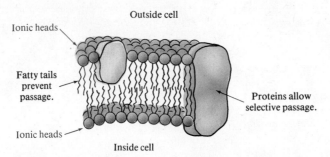

Figure 17.4 The bilayer of phospholipids in a cell membrane, showing the intertwined hydrophobic tails pointing inward; the ionic heads pointing outward toward the aqueous media; and the embedded proteins, which selectively pass water and water-soluble materials through the membrane.

The hydrophobic hydrocarbon inner membrane prevents water, ions, and other water-soluble material from passing into or out of the cell through the bilayer by osmosis. However, the bilayer contains embedded protein molecules that act as pumps, selectively passing water, ions, and other water-soluble material into and out of the cell as needed.

The fluidity of the cell membrane in different environments is regulated by the fatty acids in the phosphoglycerides. For example, bacteria grown at low temperatures contain more lower-melting unsaturated fatty acids in the membrane than do bacteria grown at higher temperatures.

Sphingolipids, such as sphingomyelin (Figure 17.3), are particularly abundant in brain and nerve cells, where they are a major constituent of the *myelin sheaths*, or the coatings around nerve fibers. The intertwining of the long hydrophobic tails strengthens the cell and insulates the nerve's electrical impulses from surrounding fluids. Sphingolipids are not glycerides, but are derivatives of sphingosine, a dihydroxy amine.

$$CH_3(CH_2)_{12}$$

$$
\begin{array}{c}
CH_3(CH_2)_{12}\qquad H \\
\diagdown\qquad\diagup \\
C = C \\
\diagup\qquad\diagdown \\
H\qquad CHOH \\
|\\
CHNH_2 \\
|\\
CH_2OH \quad\longleftarrow \text{ This portion is}\\
\text{similar to glycerol}
\end{array}
$$

Sphingosine

PROBLEM 17.2. Predict the saponification products of **(a)** the phosphoglycerides in Figure 17.3 and **(b)** sphingomyelin.

17.4 STEROIDS

Steroids are compounds with a distinctive ring system containing three six-membered rings and a five-membered ring. The rings in the steroids are said to be *fused*, meaning that they share carbon atoms. Most steroids contain methyl groups at two of the ring junctures (positions 10 and 13) of the steroid ring system. We show these two methyl groups in the following formulas so that you can see their stereochemical relationships to the rings. Figure 17.5 shows structures of some representative examples of steroids.

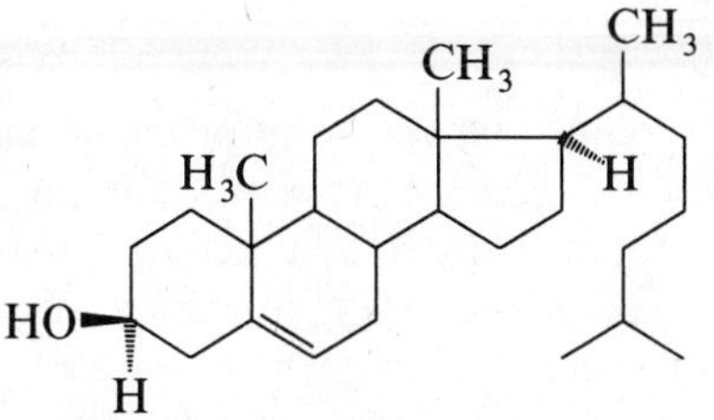

Cholesterol

found in eggs, animal tissue, and gallstones; implicated in arteriosclerosis

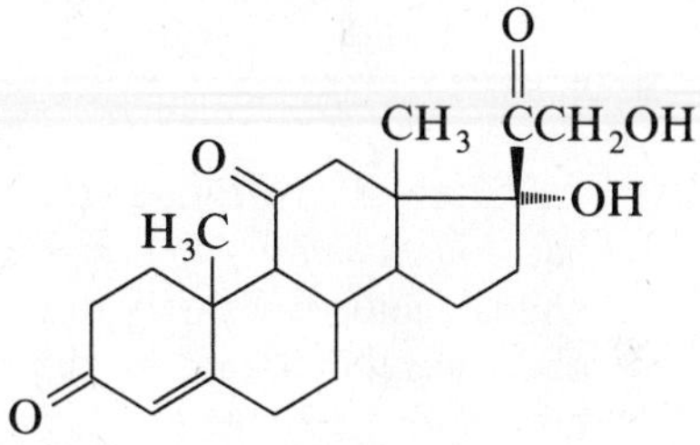

Cortisone

an adrenal hormone and anti-inflammatory agent

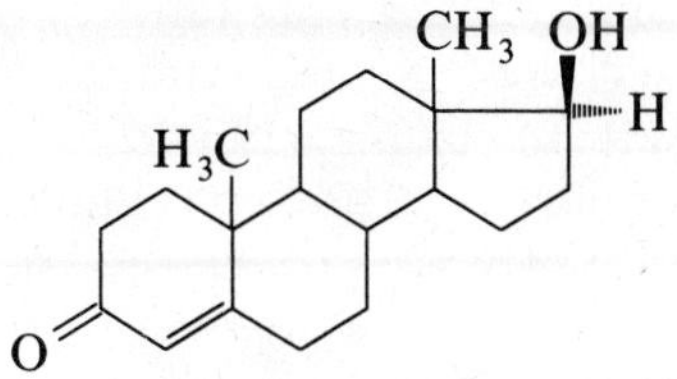

Testosterone

an androgen, or male hormone

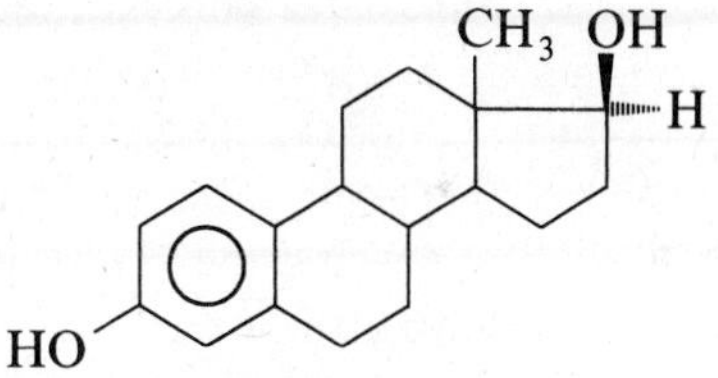

Estradiol

an estrogen, or female hormone

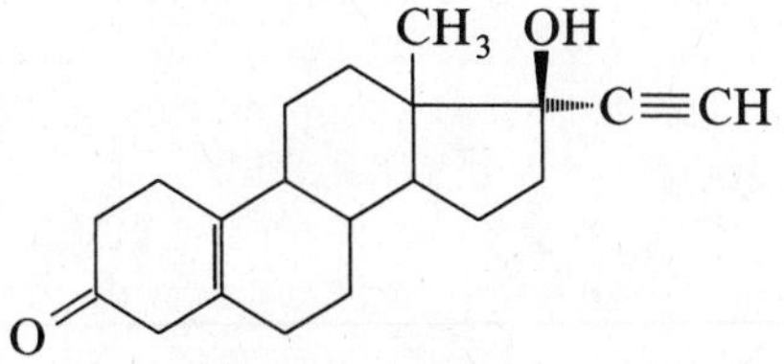

Norethynodrel

a synthetic progestin; an oral contraceptive

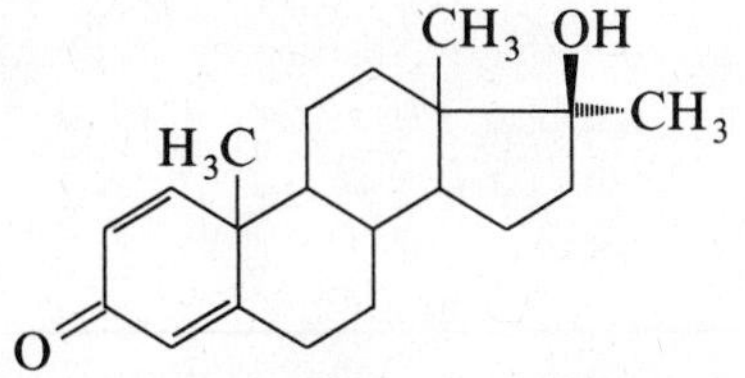

Methandrostenolone

an anabolic (tissue-building) androgen

Figure 17.5 Structures of some important steroids.

Steroid Ring System:

Chair-form shape a typical
steroid assumes

Steroids are found in almost all living systems—the principal exceptions are bacteria. *Cholesterol* (see Figure 17.5), a common steroid, is found primarily in animals and very rarely in plants. Cholesterol is a human dietary component and is also biosynthesized in the body. It is a necessary precursor of the steroidal hormones and the bile acids, the fat-emulsifying agents released into the small intestines.

The level of cholesterol in the bloodstream is a function of many factors, including diet and individual metabolism. Cholesterol is not a minor component of the blood, however. At age 55, the average cholesterol level is about 2 g per liter of blood.

Cholesterol contributes to hardening of the arteries (arteriosclerosis), a condition in which this steroid and other lipids coat the inner surface of the arteries and thus restrict the blood flow. Because arteriosclerosis can lead to heart attacks and strokes, people with this condition are generally advised to restrict cholesterol in the diet by limiting the consumption of eggs, milk fat, shrimp, and fatty meat.

Among the more important steroidal hormones are the **sex hormones**. Female hormones, called **estrogens** and **progestins** (pregnancy hormones), regulate the menstrual cycle, pregnancy, and the female secondary sex characteristics. Male hormones, called **androgens**, regulate the male primary and secondary sex characteristics. Testosterone is the principal male hormone.

In Figure 17.5, note the similarities in structure of testosterone, methandrostenolone, estradiol, and norethynodrel. Each structure has a hydroxyl group at carbon 17. Three of these structures have a keto group at carbon 3. Despite the similarities in structure, these compounds have widely different functions in the body.

Vitamin D, essential for the proper utilization of calcium in the body, is closely related to steroids and can be biosynthesized by the action of sunlight on 7-dehydrocholesterol in the skin. (The subscript in the name vitamin D_3 in the following equation refers to one of several closely related structures that have vitamin D activity.)

Biosynthesis of a Vitamin D:

Bond breaks here.

CH$_3$

H$_3$C

CH$_3$

HO

7-Dehydrocholesterol

a steroid in the skin

sunlight

Double bond
forms here

CH$_3$

H$_2$C

CH$_3$

HO

Vitamin D$_3$

PROBLEM 17.3. The following formula is an amide of glycine and the bile acid cholic acid. This amide is one agent that keeps fats emulsified in the intestines. Explain in terms of its structure how this compound can act as an emulsifying agent.

HO   CH$_3$ CH$_3$

H$_3$C

HO   OH

C

O

O

NHCH$_2$CO$^-$

From glycine,   H$_3$$\overset{+}{\text{N}}CH_2$$\overset{\displaystyle O}{\overset{\|}{\text{C}}}O^-$

From cholic acid

17.5 PROSTAGLANDINS

In the 1930s, the Swedish scientist Ulf von Euler extracted compounds now called *prostaglandins* from human seminal fluid. He was able to show that, when injected into animals, these compounds cause uterine contractions and lower blood pressure. Based on the belief that the biologically active compound originated in the prostate gland, the name "prostaglandin" was coined. It is now known that prostaglandins are a large class of compounds widespread throughout both the male and female body.

In the 1950s and 1960s, the Swedish chemists Sune Bergstrom and Bengt Samuelsson determined the structures of many of the prostaglandins. In 1982 they were awarded the Nobel Prize for this work.

Prostaglandins are 20-carbon cyclopentanediols and hydroxycyclopentanones that contain a carboxylic acid side chain. They are biosynthesized from 20-carbon fatty acids, such as arachidonic acid (see Figure 17.6).

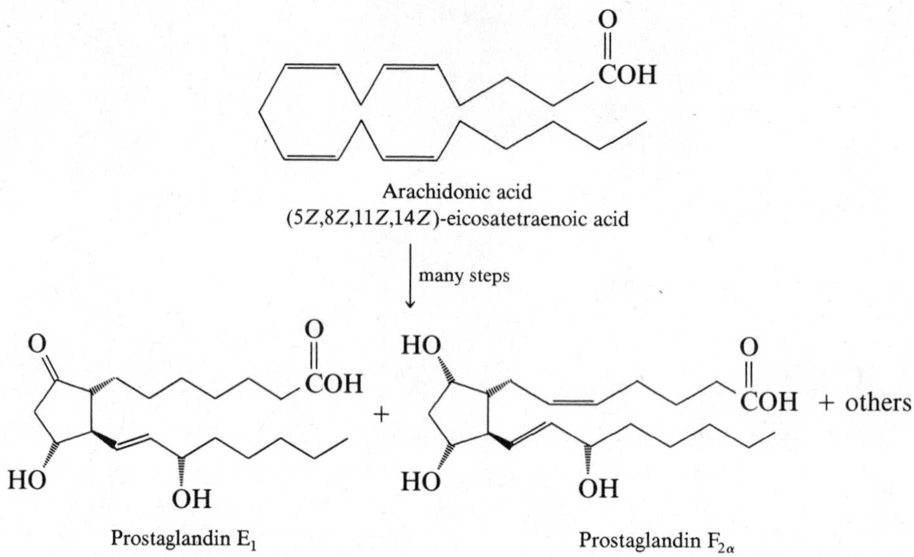

Arachidonic acid
(5Z,8Z,11Z,14Z)-eicosatetraenoic acid

many steps

Prostaglandin E₁ + Prostaglandin F₂ₐ + others

Figure 17.6 The prostaglandins are biosynthesized from 20-carbon unsaturated fatty acids.

In the body, prostaglandins act to moderate the actions of hormones and thus have many wide-reaching effects, such as control of blood pressure, involvement in blood-clotting mechanism, and in kidney function. Prostaglandins can cause dilation of arteries and bronchial tubes, inhibition of gastric secretions, induction of uterine contractions, abortion, inflammation, and pain. Aspirin acts to relieve pain and inflammation by interfering with the biosynthesis of prostaglandins. The potential use of prostaglandins themselves, as well as anti-inflammatory drugs that block their action, is a new and exciting field of research.

SUMMARY

Lipids are biological compounds that are insoluble in water but soluble in nonpolar organic solvents. **Triglycerides** (triacylglycerols) are triesters of glycerol and **fatty acids**. The common fatty acids are carboxylic acids with long unbranched hydrocarbon chains that may or may not contain carbon–carbon double bonds. The fatty acids contain even numbers of carbons and their carbon–carbon double bonds are cis.

Fats (solid at room temperature) are triglycerides containing few carbon–carbon double bonds. **Oils** (liquid at room temperature) contain a larger number of double bonds. Many vegetable oils are considered to be *polyunsaturated triglycerides*.

Triglycerides containing carbon–carbon double bonds can be *hydrogenated*. All triglycerides can be *saponified*.

Waxes are simple esters of long-chain alcohols and fatty acids.

Phosphoglycerides, such as *cephalins* and *lecithins*, are esters of glycerol with two fatty acid residues and a phosphate-alkylamine. **Sphingolipids** are derivatives of the long-chain amine-diol *sphingosine*.

Steroids are lipids with a distinctive ring system of three six-membered rings and a five-membered ring. Important steroids are cholesterol and the sex hormones.

Prostaglandins are 20-carbon cyclopentanol derivatives that moderate hormone activity in the body.

STUDY PROBLEMS

17.4. Which of the following compounds would be considered lipids? Explain your answers.

(a) CH_2CHCH_2 with OH, OH (top) and OH (bottom)

(b) CH_3COH (with O double bond)

(c) $CH_3(CH_2)_{16}COCH_2CHCH_2OH$ (with O double bond and OH)

17.5. Which compound would be higher melting? (Refer to Table 17.1 for the fatty acid structures.)
(a) trioleylglycerol   (b) trilinoleylglycerol

17.6. Write a structural formula for *trimyristylglycerol* (also called *trimyristin*), the principal triglyceride of nutmeg.

17.7. Write formulas for the pair of enantiomers of a triglyceride that contains two stearic acid residues and one oleic acid residue.

17.8. Beginning with glycerol and 16-carbon alcohols or carboxylic acids, write formulas for (a) a saturated fat, (b) a polyunsaturated vegetable oil, and (c) a wax.

17.9. Predict the product of the complete hydrogenation of trilinoleylglycerol at moderate temperature and pressure.

17.10. Predict the products when trilinoleylglycerol is heated with aqueous sodium hydroxide.

17.11. Complete the following equations:

(a) $CH_3(CH_2)_{11}CO^- Na^+ + HCl \longrightarrow$ (with O double bond on C)

(b) $CH_3(CH_2)_{11}CO^- Na^+ + Ca^{2+} \xrightarrow{H_2O}$ (with O double bond on C)

17.12. Label each of the following compounds as a cephalin, a lecithin, a sphingomyelin, or a ceramide (see Point of Interest 17).

(a)
$CH_3(CH_2)_{12}$
  $\backslash$
  $CH{=}CH$
      $\backslash$
      $CHOH$
      |
      $CHNHC(CH_2)_{14}CH_3$ (with O double bond)
      |
      CH_2OH

(b)

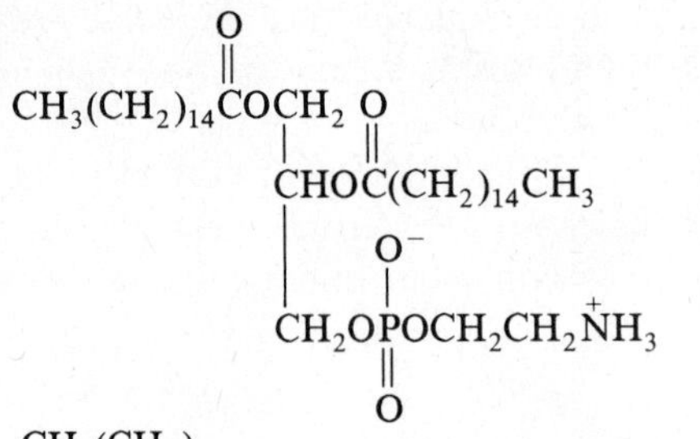

$$CH_3(CH_2)_{14}\overset{\overset{\displaystyle O}{\|}}{C}OCH_2$$

$$CHO\overset{\overset{\displaystyle O}{\|}}{C}(CH_2)_{14}CH_3$$

$$CH_2O\overset{\overset{\displaystyle O^-}{|}}{\underset{\underset{\displaystyle O}{\|}}{P}}OCH_2CH_2\overset{+}{N}H_3$$

(c)

$$CH_3(CH_2)_{12}$$
$$CH{=}CH$$
$$CHOH$$
$$CHNH\overset{\overset{\displaystyle O}{\|}}{C}(CH_2)_{14}CH_3$$
$$CH_2O\overset{\overset{\displaystyle O^-}{|}}{\underset{\underset{\displaystyle O}{\|}}{P}}OCH_2CH_2\overset{+}{N}(CH_3)_3$$

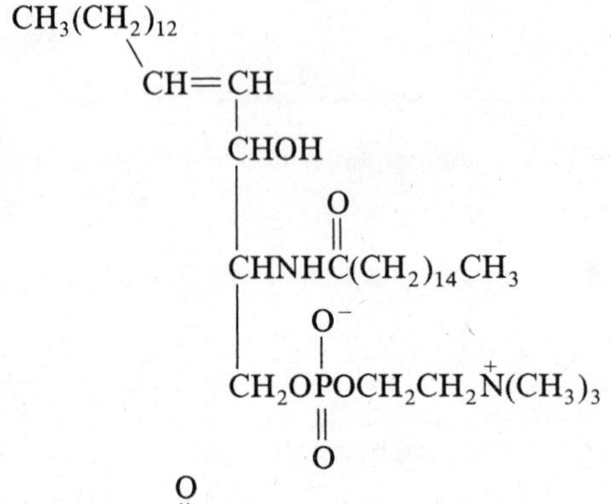

(d)

$$CH_3(CH_2)_{14}\overset{\overset{\displaystyle O}{\|}}{C}OCH_2$$
$$CHO\overset{\overset{\displaystyle O}{\|}}{C}(CH_2)_{14}CH_3$$
$$CH_2O\overset{\overset{\displaystyle O^-}{|}}{\underset{\underset{\displaystyle O}{\|}}{P}}OCH_2CH_2\overset{+}{N}(CH_3)_3$$

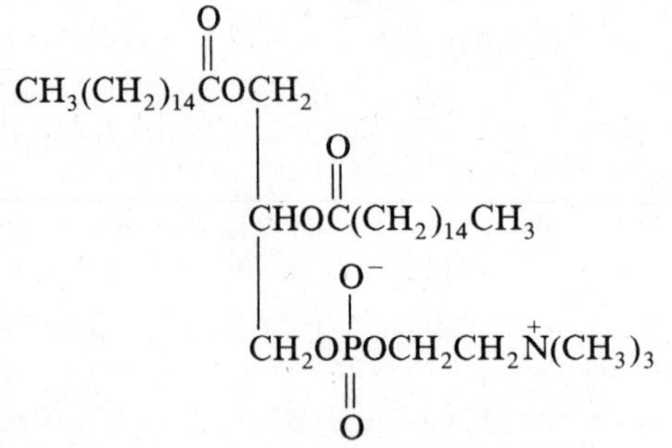

17.13. Predict the structures of the products of the following reactions. (Refer to Figure 17.5 for the structures of the starting steroids.)

(a) testosterone + 2H$_2$ $\xrightarrow[\text{catalyst}]{\text{heat, pressure}}$

(b) cholesterol + H$_2$ $\xrightarrow{\text{catalyst}}$

(c) cholesterol + CH$_3\overset{\overset{\displaystyle O}{\|}}{C}$Cl $\longrightarrow$

(d) estradiol + Na$^+$ $^-$OH $\xrightarrow{\text{H}_2\text{O}}$

POINT OF INTEREST 17

The Lipid Storage Diseases

At least 10 genetically linked lipid storage diseases (also called *sphingolipodystro-phies* and *sphingolipidoses*) are associated with the metabolism of **ceramides**, amides formed from sphingosine and a fatty acid.

A ceramide

Cerebrosides and the **gangliosides** are two major derivatives of ceramides. Both are substituted with a carbohydrate group at the terminal hydroxyl group of the ceramide—glucose or galactose in the cerebrosides and an oligosaccharide (three to five monosaccharide units) in the gangliosides.

A cerebroside   *A ganglioside*

The oligosaccharide of one ganglioside (called *Tay–Sachs ganglioside*) ends in the N-acetyl derivative of a galactosamine and contains a branch of another amino sugar called N-acetylneuramic acid.

Lipid storage diseases arise from errors in the biosynthesis and subsequent degradation of the cerebrosides and the gangliosides. In one of these lipid storage diseases, **Gaucher's disease**, one cerebroside, a glucose ceramide, is synthesized by the body at a normal rate, but the enzyme needed to cleave the glucose residue from the ceramide is missing. Consequently, the cerebroside cannot be removed metabolically and is accumulated in the body. In the adult form of the disease, the cerebroside accumulates in the liver and spleen, causing enormous enlargements of these organs (up to five to seven times their normal size). When the disease occurs in children, the cerebroside accumulates in the brain and causes mental retardation.

Cleavage of this bond is blocked in
Gaucher's disease.

Ceramide — Glucose ⟶ ⫽ ⟶ Ceramide + Glucose

In **Tay–Sachs disease**, the enzyme hexosaminidase A needed to cleave the N-acetylgalactosamine from the end of the oligosaccharide chain of the Tay–Sachs ganglioside is missing. As in Gaucher's disease, the result is accumulation of the ganglioside, primarily in the brain cells. The accumulation of the ganglioside is irreversible and results in blindness, paralysis, and finally mental deterioration. Death usually results by the age of three.

In Tay–Sachs disease, the enzyme
necessary for this hydrolysis of
this bond is missing.

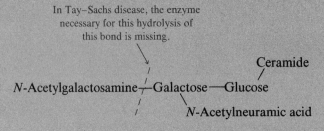

Ceramide

N-Acetylgalactosamine — Galactose — Glucose

N-Acetylneuramic acid

Chapter 18

Amino Acids and Proteins

The term protein, first used in 1838, is derived from the Greek word *proteios*, which means "primary." In living systems, proteins serve a number of important functions. All plant and animal enzymes are proteins. Proteins, together with lipids and bones, form an animal's structural components. Proteins also form muscles, antibodies, and many hormones.

During the late 1800s, the monomeric units of proteins were identified as α-amino acids. Today, we know that 20 different α-amino acids occur in proteins as a direct result of genetic coding.

The α carbon

$$H_2N-\underset{\underset{R}{|}}{CH}-\overset{\overset{O}{\|}}{C}OH$$

The amino acids differ in their R groups.

An α-amino acid,
the building block of proteins

The structures of the 20 amino acids and their names are listed in Table 18.1. Each amino acid can also be represented by a three-letter abbreviation. These abbreviations are useful in describing peptides and complex proteins.

In addition to the 20 amino acids listed in Table 18.1, other amino acids are found in proteins. These amino acids are biosynthesized either from one of the 20 amino acids or from other components. The structures of some of these amino acids and the types of proteins in which they are found are listed in Table 18.2.

In 1902, Emil Fischer and Franz Hofmeister independently showed that α-amino acids are linked together in proteins by amide bonds. These amide bonds, called

TABLE 18.1 THE COMMON AMINO ACIDS FOUND IN PROTEINS

Name	Abbreviation	Structure
Alanine	Ala	CH_3CHCO_2H $\quad\ \ |$ $\quad NH_2$
Arginine[a]	Arg	$\quad\quad\quad NH$ $\quad\quad\quad \|\|$ $H_2NCNHCH_2CH_2CH_2CHCO_2H$ $\quad\quad\quad\quad\quad\quad\quad\quad\quad |$ $\quad\quad\quad\quad\quad\quad\quad\quad\ NH_2$
Asparagine	Asn	$\quad\ \ O$ $\quad\ \ \|\|$ $H_2NCCH_2CHCO_2H$ $\quad\quad\quad\quad |$ $\quad\quad\quad NH_2$
Aspartic acid	Asp	$HO_2CCH_2CHCO_2H$ $\quad\quad\quad\quad |$ $\quad\quad\quad NH_2$
Cysteine	Cys	$HSCH_2CHCO_2H$ $\quad\quad\quad |$ $\quad\quad NH_2$
Glutamic acid	Glu	$HO_2CCH_2CH_2CHCO_2H$ $\quad\quad\quad\quad\quad |$ $\quad\quad\quad\quad NH_2$
Glutamine	Gln	$\quad\ \ O$ $\quad\ \ \|\|$ $H_2NCCH_2CH_2CHCO_2H$ $\quad\quad\quad\quad\quad |$ $\quad\quad\quad\quad NH_2$
Glycine	Gly	CH_2CO_2H $|$ NH_2
Histidine[a]	His	imidazole ring CH_2CHCO_2H $\quad\quad\quad\quad\quad |$ $\quad\quad\quad\quad NH_2$
Isoleucine[a]	Ile	$\quad\quad\ CH_3$ $\quad\quad\ \ |$ $CH_3CH_2CHCHCO_2H$ $\quad\quad\quad\quad\ |$ $\quad\quad\quad NH_2$
Leucine[a]	Leu	$\quad\quad CH_3$ $\quad\quad\ |$ $CH_3CHCH_2CHCO_2H$ $\quad\quad\quad\quad\ |$ $\quad\quad\quad NH_2$
Lysine[a]	Lys	$H_2NCH_2CH_2CH_2CH_2CHCO_2H$ $\quad\quad\quad\quad\quad\quad\quad\quad\ |$ $\quad\quad\quad\quad\quad\quad\quad NH_2$

TABLE 18.1 *continued*

Name	Abbreviation	Structure
Methionine[a]	Met	$CH_3SCH_2CH_2CHCO_2H$ $\mid$ NH_2
Phenylalanine[a]	Phe	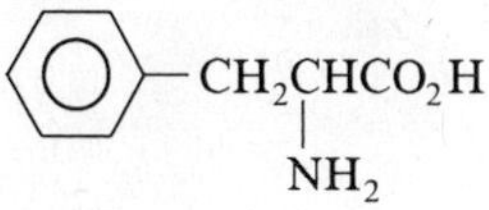 C_6H_5—CH_2CHCO_2H $\mid$ NH_2
Proline	Pro	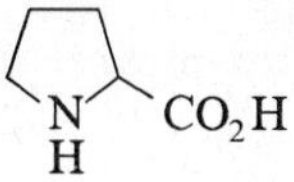 (ring) $\overset{N}{\underset{H}{}}$ CO_2H
Serine	Ser	$HOCH_2CHCO_2H$ $\mid$ NH_2
Threonine[a]	Thr	OH $\mid$ $CH_3CHCHCO_2H$ $\mid$ NH_2
Tryptophan[a]	Trp	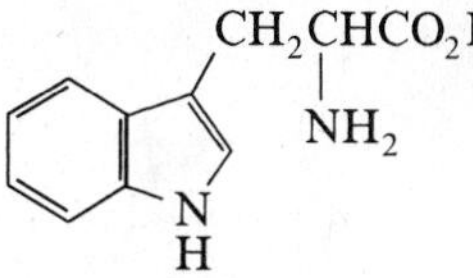 (indole)—CH_2CHCO_2H $\mid$ NH_2
Tyrosine	Tyr	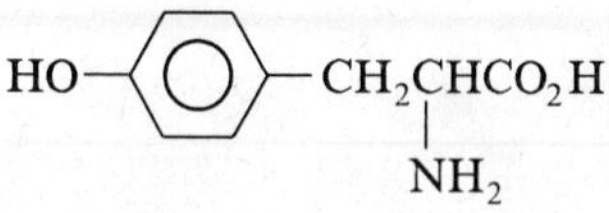 HO—C_6H_4—CH_2CHCO_2H $\mid$ NH_2
Valine[a]	Val	CH_3 $\mid$ $CH_3CHCHCO_2H$ $\mid$ NH_2

[a] Essential amino acid.

peptide bonds, are formed by the α-amino group of one amino acid and the carboxyl group of a second amino acid.

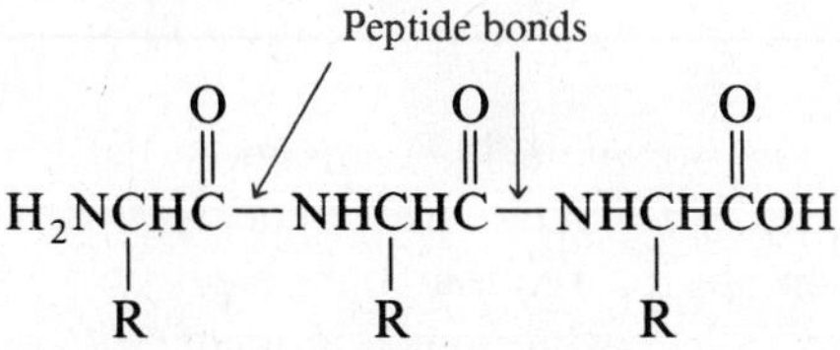

Peptide bonds

$$H_2NCHC\overset{O}{\overset{\|}{}}-NHCHC\overset{O}{\overset{\|}{}}-NHCHC\overset{O}{\overset{\|}{}}OH$$
$$\quad \underset{R}{|} \qquad \underset{R}{|} \qquad \underset{R}{|}$$

A tripeptide

TABLE 18.2   DERIVED AMINO ACIDS FOUND IN PROTEINS

Name	Formula	Protein source
Cystine	$\begin{array}{cc} CO_2H & CO_2H \\ | & | \\ H_2NCH & H_2NCH \\ | & | \\ CH_2-S-S-CH_2 \end{array}$	Keratin (hair, nails, feathers, etc.)
Hydroxyproline	HO on pyrrolidine ring, $\underset{H}{N}$, CO_2H	Collagen, elastin
Hydroxylysine	$\underset{OH}{H_2NCH_2\overset{|}{C}HCH_2CH_2\underset{NH_2}{\overset{|}{C}HCO_2H}}$	Collagen, elastin
Thyroxine	$HO-\overset{I}{\underset{I}{\bigcirc}}-O-\overset{I}{\underset{I}{\bigcirc}}-CH_2\underset{NH_2}{\overset{|}{C}HCO_2H}$	Thyroglobulin

A **peptide** is a compound formed from α-amino acids bonded together by peptide bonds. The individual amino acids within a peptide are referred to as **peptide units** or as **amino acid residues**. A peptide formed from two amino acid residues is called a **dipeptide**. A peptide with three amino acid residues is called a **tripeptide**, and so forth. A **polypeptide** is a peptide with many amino acid residues. The distinction between a polypeptide and a **protein** is arbitrary, but, in general, a polypeptide with more than about 50 amino acid residues is referred to as a protein.

18.1 AMINO ACIDS

A. Amino Acids as Dipolar Ions

An amino acid has an acidic carboxylic acid group and a basic amino group in the same molecule. Consequently, an amino acid undergoes an internal acid-base reaction to form a **dipolar ion**, an ion that has both a positive and a negative ionic charge. Dipolar ions are also called **zwitterions** (German *zwitter*, "hybrid").

Formation of a Dipolar Ion:

Acidic: loses a proton

$$\underset{\substack{\nearrow \\ \text{Basic: gains a proton}}}{\text{H}_2\overset{\displaystyle ..}{\text{N}}-\underset{\displaystyle \underset{\text{R}}{|}}{\text{CH}}\overset{\displaystyle \overset{\text{O}}{\|}}{\text{C}}\overset{..}{\text{O}}-\text{H}} \longrightarrow \underset{\substack{\\ \textit{A dipolar ion}}}{\text{H}_3\overset{+}{\text{N}}-\underset{\underset{\text{R}}{|}}{\text{CH}}\overset{\overset{\text{O}}{\|}}{\text{C}}\overset{..}{\underset{..}{\text{O}}}:^-}$$

A dipolar ion has one positive charge and one negative charge; thus, it is electrically neutral. Even though they are neutral, dipolar ions are still ionic compounds. Their physical properties reflect this fact. For example, they have high melting points, and they are soluble in water but nearly insoluble in organic solvents. These properties would not be expected if dipolar ions did not have ionic charges.

Dipolar ions are amphoteric—that is, they can react with either an acid or a base. This important chemical property is a result of the presence of the positive and the negative ionic charges. If an α-amino acid dipolar ion is treated with acid, the carboxylate group is protonated and the dipolar ion is converted to a cation. If the dipolar ion is treated with base, the amino acid loses a proton and forms an anion.

Reaction with Acid:

$$\underset{\substack{\nearrow \\ \text{Gains a proton}}}{\text{H}_3\overset{+}{\text{N}}\underset{\underset{\text{R}}{|}}{\text{CH}}\overset{\overset{\text{O}}{\|}}{\text{C}}\text{O}^-} + \text{H}^+ \longrightarrow \underset{\substack{\\ \textit{A cation}}}{\text{H}_3\overset{+}{\text{N}}\underset{\underset{\text{R}}{|}}{\text{CH}}\overset{\overset{\text{O}}{\|}}{\text{C}}\text{OH}}$$

Reaction with Base:

$$^-\text{OH} + \underset{\substack{\nearrow \\ \text{Loses a proton}}}{\text{H}_3\overset{+}{\text{N}}\underset{\underset{\text{R}}{|}}{\text{CH}}\overset{\overset{\text{O}}{\|}}{\text{C}}\text{O}^-} \longrightarrow \underset{\substack{\\ \textit{An anion}}}{\text{H}_2\text{N}\underset{\underset{\text{R}}{|}}{\text{CH}}\overset{\overset{\text{O}}{\|}}{\text{C}}\text{O}^-} + \text{H}_2\text{O}$$

PROBLEM 18.1. Write equations for **(a)** the reaction of 1.0 mol of the dipolar ion of alanine with 1.0 mol of HCl and **(b)** the reaction of 1.0 mol of the dipolar ion of serine with 1.0 mol of NaOH. (See Table 18.1 for the structures of these two amino acids.)

B. Stereochemistry

In 1851, Louis Pasteur discovered that amino acids are optically active. It was later determined that this optical activity is present because the alpha carbon of each α-amino acid except glycine is chiral.

Chiral carbons

$$
\begin{array}{ccc}
\text{CO}_2\text{H} & \text{CO}_2\text{H} & \text{CO}_2\text{H} \\
| & | & | \\
\text{H}_2\text{N}-\text{C}-\text{H} & \text{H}_2\text{N}-\text{C}-\text{H} \quad \text{H}_2\text{N}-\text{C}-\text{H} \\
| & | & | \\
\text{H} & \text{CH}_3 & \text{R}
\end{array}
$$

Glycine   Alanine   *Other α-amino acids*

The α carbon of the genetically coded amino acids (see Chapter 19) has the same configuration as L-glyceraldehyde, a compound whose configuration is (S).

—NH$_2$ projected in the same direction as the —OH group in L-glyceraldehyde

$$
\begin{array}{ccc}
\text{CHO} & \text{CHO} & \text{CO}_2\text{H} & \text{CO}_2\text{H} \\
\text{HO}-\text{C}-\text{H} \quad \text{or} \quad \text{HO}-\!\!\!-\!\!\!-\text{H} & & \text{H}_2\text{N}-\text{C}-\text{H} \quad \text{or} \quad \text{H}_2\text{N}-\!\!\!-\!\!\!-\text{H} \\
\text{CH}_2\text{OH} & \text{CH}_2\text{OH} & \text{R} & \text{R}
\end{array}
$$

L-Glyceraldehyde   L-Amino acids

Two of the 20 amino acids, isoleucine and threonine, have two chiral carbons. The configuration of the alpha carbon of these two amino acids is the same as that found in the other amino acids and in L-glyceraldehyde. In isoleucine, the second chiral carbon is (S), while in threonine it is (R).

$$
\begin{array}{cc}
\text{CO}_2\text{H} & \text{CO}_2\text{H} \\
\text{H}_2\text{N}-\!\!\!-\!\!\!-\text{H} & \text{H}_2\text{N}-\!\!\!-\!\!\!-\text{H} \\
\text{H}_3\text{C}-\!\!\!-\!\!\!-\text{H} & \text{H}-\!\!\!-\!\!\!-\text{OH} \\
\text{CH}_2\text{CH}_3 & \text{CH}_3
\end{array}
$$

Isoleucine   Threonine

All chiral amino acids found in proteins have the same configuration at carbon 2; however, amino acids with the opposite configuration, D-α-amino acids, are also found in some natural products. For example, the carbon skeleton of D-valine is found in the antibiotics penicillin and actinomycin.

C. Classification Based on Side Chain

On the basis of the structure of its side chain (R), each amino acid listed in Table 18.1 belongs to one of three classes of amino acids: (1) those with neutral side

TABLE 18.3   NEUTRAL, BASIC, AND ACIDIC AMINO ACIDS[a]

Neutral amino acids:

  Nonpolar: alanine, glycine, isoleucine, leucine, methionine, phenylalanine, proline,

    tryptophan, and valine

  Polar: asparagine, cysteine, glutamine, serine, threonine, and tyrosine

Basic amino acids: arginine, lysine, and histidine

Acidic amino acids: aspartic acid and glutamic acid

---

[a]See Table 18.1 for the structures of the amino acids.

chains, (2) those with basic side chains, and (3) those with acidic side chains. Table 18.3 lists the amino acids in these three categories.

Neutral Amino Acids   This class includes the amino acids that have neither a carboxyl group nor a basic functional group in their side chain. Fifteen of the 20 amino acids belong to this class. The neutral amino acids can be subdivided into *nonpolar* and *polar* amino acids.

Nine of the amino acids are generally considered to be nonpolar. Of the nine, four (alanine, valine, leucine, and isoleucine) have hydrocarbon side chains. One (proline) is unique in that its α amino group is bound into a ring. Two of the nonpolar amino acids (phenylalanine and tryptophan) have aromatic rings in their side chains. One of the nonpolar amino acids (methionine) has a side chain containing a methylthio group ($-SCH_3$).

Some Neutral Amino Acids with Nonpolar Side Chains:

Alanine   Proline   Methionine

The neutral, nonpolar amino acids are, in general, the least water soluble of the entire group of 20 amino acids. At pH 6–7 they exist as neutral dipolar ions. None of these amino acids have functional groups in their side chains that can hydrogen bond with water. (The heterocyclic nitrogen of tryptophan cannot hydrogen bond with water because its pair of electrons are part of the aromatic pi cloud; see Section 5.6B.) The sulfide group in methionine is nonpolar and does not form hydrogen bonds with water.

Six of the neutral amino acids are considered to be polar because their side chains contain polar groups such as $-OH$ (see Table 18.3). These amino acids are more water soluble than the neutral amino acids. Except for cysteine, the amino acids in this subgroup have a group in their side chain that can hydrogen bond with water.

Some Neutral Amino Acids with Polar Side Chains:

$$
\begin{array}{ccc}
\text{CO}_2\text{H} & \text{CO}_2\text{H} & \text{CO}_2\text{H} \\
| & | & | \\
\text{H}_2\text{N}-\text{C}-\text{H} & \text{H}_2\text{N}-\text{C}-\text{H} & \text{H}_2\text{N}-\text{C}-\text{H} \quad \overset{\text{O}}{\underset{||}{}} \\
| & | & | \\
\text{CH}_2\text{OH} & \text{CH}_2\text{SH} & \text{CH}_2\text{CH}_2\text{CNH}_2 \\
\text{Serine} & \text{Cysteine} & \text{Glutamine}
\end{array}
$$

Basic Amino Acids   The three basic (alkaline) amino acids are arginine, histidine, and lysine. Each of these amino acids has a side-chain functional group that reacts with a proton at pH 6–7 to form a positively charged ionic group. Thus, at pH 6–7 a basic amino acid has two positive charges and one negative charge, or a net positive charge.

Reaction of Basic Amino Acids with Acid:

$$
\text{H}_2\text{NCH}_2\text{CH}_2\text{CH}_2\text{CH}_2\underset{\underset{\text{+NH}_3}{|}}{\text{CH}}\overset{\overset{\text{O}}{||}}{\text{C}}\text{O}^- + \text{H}^+ \xrightarrow{\text{pH 6-7}} \text{H}_3\overset{+}{\text{N}}\text{CH}_2\text{CH}_2\text{CH}_2\text{CH}_2\underset{\underset{\text{+NH}_3}{|}}{\text{CH}}\overset{\overset{\text{O}}{||}}{\text{C}}\text{O}^-
$$

Lysine   *A cation*

The basic N ↗

$$
\underset{\underset{\text{+NH}_3}{|}}{\underset{\underset{\text{H}_2\text{N}\overset{\overset{\text{NH}}{||}}{\text{C}}\text{NHCH}_2\text{CH}_2\text{CH}_2}{}}{\text{CH}}}\overset{\overset{\text{O}}{||}}{\text{C}}\text{O}^- + \text{H}^+ \xrightarrow{\text{pH 6-7}} \text{H}_2\text{N}\overset{\overset{+\text{NH}_2}{||}}{\text{C}}\text{NHCH}_2\text{CH}_2\text{CH}_2\underset{\underset{\text{+NH}_3}{|}}{\text{CH}}\overset{\overset{\text{O}}{||}}{\text{C}}\text{O}^-
$$

Arginine   *A cation*

Histidine is a borderline case in that its reaction with acid at pH 6–7 is only about 50% complete.

The basic N ↓

$$
\underset{\text{H}}{\overset{\text{N}}{N}} \diagdown \underset{\underset{\text{+NH}_3}{|}}{\text{CH}_2\text{CH}}\overset{\overset{\text{O}}{||}}{\text{C}}\text{O}^- + \text{H}^+ \rightleftharpoons^{\text{pH 6-7}} \underset{\text{H}}{\overset{\text{H}\overset{+}{\text{N}}}{N}} \diagdown \underset{\underset{\text{+NH}_3}{|}}{\text{CH}_2\text{CH}}\overset{\overset{\text{O}}{||}}{\text{C}}\text{O}^-
$$

Histidine   *A cation*
a basic amino acid

Acidic Amino Acids   Two of the 20 amino acids are classified as acidic because they have carboxyl groups in their side chains. At pH 6–7, these side-chain carboxyl groups lose a proton to water to form a structure with two negative ionic charges and one positive ionic charge. Thus, at pH 6–7 an acidic amino acid has a net negative charge.

Reaction of an Acidic Amino Acid with Water:

$$\underset{\overset{|}{^+NH_3}}{HOCCH_2CHCO^-} + H_2O \xrightarrow{\text{pH } 6\text{--}7} \underset{\overset{|}{^+NH_3}}{^-OCCH_2CHCO^-} + H_3O^+$$

Aspartic acid   Aspartate ion
                              an anion

EXAMPLE

Complete the following equations for acid-base reactions of α-amino acids:

(a) $\underset{\overset{|}{^+NH_3}}{HO_2CCH_2CHCO_2^-} + H^+ \longrightarrow$

(b) $\underset{\overset{|}{^+NH_3}}{HO_2CCH_2CHCO_2^-} + \text{excess } {^-}OH \longrightarrow$

(c) $\underset{\overset{|}{^+NH_3}}{\text{⟨O⟩}-CH_2CHCO_2^-} + {^-}OH \longrightarrow$

Solution:

(a) $\underset{\overset{|}{^+NH_3}}{HO_2CCH_2CHCO_2H}$

(b) $\underset{\overset{|}{NH_2}}{^-O_2CCH_2CHCO_2^-} + 2H_2O$ (Both acidic protons are removed.)

(c) $\underset{\overset{|}{NH_2}}{\text{⟨O⟩}-CH_2CHCO_2^-} + H_2O$ (Only acidic proton is removed.)

PROBLEM 18.2.   Complete the following equations:

(a) $H_3\overset{+}{N}CH_2CH_2CH_2CH_2CHCO_2^-$ + excess ^-OH $\longrightarrow$
$$\underset{\overset{|}{{}^+NH_3}}{}$$

(b) $HO_2CCH_2CH_2CHCO_2^-$ Na^+ + H_2O $\rightleftharpoons$
$$\underset{\overset{|}{{}^+NH_3}}{}$$

Monosodium glutamate

a flavor enhancer

D. Isoelectric Points and Electrophoresis

The net ionic charge of an amino acid varies with changes in the pH of the solution. For example, when alanine is dissolved in acidic solution (low pH), it is protonated and forms a cation. As the pH of the solution is increased (increasing alkalinity), the alanine cation is transformed first to a neutral dipolar ion and finally to an anion.

Increasing pH →

$$H_3\overset{+}{N}\underset{\overset{|}{CH_3}}{CH}\overset{O}{\overset{\|}{C}}OH \rightleftharpoons H_3\overset{+}{N}\underset{\overset{|}{CH_3}}{CH}\overset{O}{\overset{\|}{C}}O^- \rightleftharpoons H_2N\underset{\overset{|}{CH_3}}{CH}\overset{O}{\overset{\|}{C}}O^-$$

At pH 2.3,
[cation] = [dipolar ion].
This is pK_1 for alanine.

At pH 6.0, alanine
is a neutral dipolar ion.

At pH 9.7,
[dipolar ion] = [anion].
This is pK_2 for alanine.

The pH at which an amino acid is electrically neutral is called the **isoelectric point** of that amino acid. For alanine, the isoelectric point is pH 6.0. The isoelectric points for other neutral amino acids are about the same.

The isoelectric points for acidic amino acids are about pH 3. A more acidic solution than this is required to protonate the second carboxylate group.

$$HO_2CCH_2\underset{\overset{|}{\overset{|}{{}^+NH_3}}}{CH}CO_2H \rightleftharpoons HO_2CCH_2\underset{\overset{|}{\overset{|}{{}^+NH_3}}}{CH}CO_2^- \rightleftharpoons {}^-O_2CCH_2\underset{\overset{|}{\overset{|}{{}^+NH_3}}}{CH}CO_2^-$$

$pK_1 = 2.0$

Isoelectric point,
pH 2.8

$pK_2 = 3.9$

The isoelectric points for basic amino acids are around pH 9–10. A more alkaline solution is required to remove the proton from the second ammonium group.

$$\overset{+}{N}H_3 \qquad \qquad \overset{+}{N}H_3$$
$$H_3\overset{+}{N}(CH_2)_4CHCO_2H \rightleftharpoons H_3\overset{+}{N}(CH_2)_4CHCO_2^- \rightleftharpoons$$
$$pK_1 = 2.2 \qquad \qquad pK_2 = 8.95$$

$$\overset{+}{N}H_3 \qquad \qquad NH_2$$
$$H_2N(CH_2)_4CHCO_2^- \rightleftharpoons H_2N(CH_2)_4CHCO_2^-$$

Isoelectric point, pH 9.7 $\qquad pK_3 = 10.5$

Electrophoresis At a given pH, some amino acids in a mixture will be anions, some will be cations, and some will be neutral dipolar ions. Mixtures of amino acids can be separated because of their migratory tendencies toward electrodes in solutions of different acidities. **Electrophoresis** is the process of separation of ions and neutral particles based on the attraction or lack of attraction of these particles toward electrodes at a specific pH. When a pair of electrodes is placed in an aqueous solution of amino acids, the amino acid anions (negatively charged) migrate toward the positively charged anode. Cations (positively charged) migrate toward the negatively charged cathode. Dipolar ions at their isoelectric points are neutral and are not attracted toward either electrode. Figure 18.1 illustrates how electrophoresis can be used to separate mixtures of amino acids.

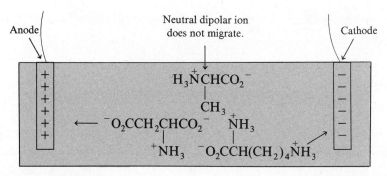

Figure 18.1 Electrophoresis is the process of separating charged and neutral particles by the attraction of the charged particles by electrodes in the solution.

PROBLEM 18.3. Predict whether the isoelectric point of each of the following amino acids would be near pH 3, pH 6, or pH 10. (Refer to Table 18.1 for the structures.) Explain your answers.

(a) glutamic acid   (b) glutamine   (c) arginine

E. Essential Amino Acids

Some amino acids cannot be synthesized by animals and must be supplied as part of the diet. These amino acids are called **essential amino acids**. The amino acids generally considered to be essential to humans are identified in Table 18.1.

The need for a specific amino acid varies with species, age, and the ingredients in the diet. Histidine, for example, is essential for human infants and for all dogs, but not essential for adult humans. The indispensability of cysteine is an example of the importance of diet. The need for cysteine depends, in part, on methionine. If methionine is in short dietary supply, cysteine becomes an essential amino acid. When methionine is abundant, the excess can be converted to cysteine—in this situation, cysteine is not an essential dietary component.

$$\underset{\text{Methionine}}{CH_3SCH_2CH_2\overset{\overset{+}{N}H_3}{\underset{|}{C}}HCO_2^-} \xrightarrow{\text{many steps}} \underset{\text{Cysteine}}{HSCH_2\overset{\overset{+}{N}H_3}{\underset{|}{C}}HCO_2^-}$$

A similar relationship is found between tyrosine and phenylalanine. Tyrosine is essential if phenylalanine is excluded from the diet because tyrosine can be biosynthesized from phenylalanine. If tyrosine is excluded from the diet, then sufficient phenylalanine must be ingested to supply both the phenylalanine and tyrosine requirements for the system.

Phenylalanine
an essential amino acid

Tyrosine
*essential only if phenylalanine
is in short supply*

Unfortunately, not all humans can synthesize tyrosine from phenylalanine. One in 10,000 humans has an inherited disease called **phenylketonuria**, or **PKU** (meaning "a phenylketone in the urine"). These people lack the enzyme *phenylalanine hydroxylase* needed for the conversion of phenylalanine to tyrosine. Excess phenylalanine builds up in the body, and some is converted to phenylpyruvic acid. When the levels of these two compounds are sufficiently high, they spill over into the urine for excretion. High levels of these compounds in the blood can cause severe mental retardation in infants. This tragedy can be avoided by control of the diet. Therefore, it is imperative that the PKU syndrome be diagnosed as early as possible after birth.

The importance of a balanced diet of amino acids cannot be overemphasized. The animal system has no storage capacity for amino acids. Thus, the biosynthesis of proteins is dependent on the availability of the proper amino acids at the time the protein is being synthesized. If the diet lacks a key amino acid, proteins requiring that amino acid are not synthesized. The other amino acids that would have been used in the protein biosynthesis are broken down and eliminated from the body.

18.2 REACTIONS OF AMINO ACIDS

A. Esterification and Acylation

Amino acids undergo reactions typical of carboxylic acids and of amines. In addition to their acid-base reactions, discussed previously in this chapter, the carboxyl group of an amino acid can be esterified by reaction with an alcohol using an acidic catalyst. The amino group can be acylated with active reagents such as acid halides or acid anhydrides.

Esterification of the Carboxyl Group:

Acylation of the Amino Group:

B. Reaction with Ninhydrin

Reactions in which amino acids form colored products are important in analytical separations. The amino acids themselves are colorless and cannot be detected in chromatography or other analytical methods by visual inspection. By converting an amino acid to a colored product, we can visually detect the location of that product.

An important color-forming reaction of amino acids is their reaction with ninhydrin. Because the intensity of the color produced in the ninhydrin reaction is proportional to the concentration of the amino acid, this reaction can be used for quantitative analysis. For example, the ninhydrin reaction is used in the **automatic amino acid analyzer**, a device that separates amino acids using an ion-exchange column and determines their relative concentrations.

$$H_3\overset{+}{N}CHCO^- + 2 \text{ Ninhydrin} \longrightarrow \text{Ruhemann's purple}$$

| An amino acid | Ninhydrin | Ruhemann's purple (blue-violet) |

$$+ RCH + CO_2 + 3H_2O + H^+$$

C. Oxidation of Cysteine to Cystine

The sulfhydryl group ($-SH$) of cysteine is easily oxidized to a disulfide group ($-SS-$), a reaction that links two cysteine residues to form the amino acid cystine. (The spelling of the names of the two amino acids differs only in the e of cysteine.) This reaction can provide a covalent bond that links two different polypeptide chains together in a protein. For example, bovine insulin has three disulfide bonds—two of these hold the two peptide chains together, and one is internal within one of the chains (see Figure 18.3).

$$2HSCH_2CHCO^- \underset{[H]}{\overset{[O]}{\rightleftharpoons}} {}^-OCCHCH_2S-SCH_2CHCO^-$$

| Cysteine | Cystine |

The protein of hair, α-keratin, contains 11–14% cystine. The chemistry of permanent waving of hair is based on breaking and reforming the disulfide bonds of cystine. (See Section 8.12.) Depilatories (hair removers) work on the same general principle. These preparations contain a reducing agent (a thioglycolate, $HSCH_2CO_2^- Na^+$) that breaks the disulfide bonds. The formulation also contains an alkali, which reacts with the resulting $-SH$ groups ($RSH + {}^-OH \longrightarrow RS^- + H_2O$), solubilizing the hair peptides so that they can be washed away.

PROBLEM 18.4. Complete the following equations for amino acid reactions:

(a)
$$\text{C}_6\text{H}_5\!-\!\text{CH}_2\!-\!\overset{\overset{+}{\text{N}}\text{H}_3}{\underset{|}{\text{CH}}}\text{CO}_2{}^- + \text{CH}_3\text{CH}_2\overset{\text{O}}{\overset{\|}{\text{C}}}\text{Cl} \longrightarrow$$

(b) $\text{CH}_3\text{OH} + \text{H}_3\overset{+}{\text{N}}\text{CH}_2\text{CO}_2{}^- \xrightarrow[\text{heat}]{\text{H}^+}$

(c) $\text{H}_3\overset{+}{\text{N}}\text{CH}_2\overset{\text{O}}{\overset{\|}{\text{C}}}\text{NH}\overset{\underset{|}{\text{CH}_2\text{SH}}}{\text{CH}}\text{CO}_2{}^- + \text{HSCH}_2\overset{\overset{+}{\text{N}}\text{H}_3}{\underset{|}{\text{CH}}}\text{CO}_2{}^- \xrightarrow{\text{oxidizing agent}}$

18.3 PEPTIDES

A. The Peptide Bond

The *peptide bond* has been defined previously in this chapter as the amide bond formed from the alpha amino group of one amino acid and the carboxyl group of a second amino acid. Recall from Section 12.4 that the nitrogen of an amide is not basic because its unshared electrons are delocalized by the carbonyl group and, thus, not readily available for reaction with a proton.

Resonance Structures for an Amide:

Unshared valence electrons
are delocalized.

Because of the double-bond character of the amide carbon–nitrogen bond, this bond is shorter (1.32 Å) than a carbon–nitrogen single bond ($\text{C}\!-\!\text{N}$; 1.48 Å), but is longer than a carbon–nitrogen double bond ($\text{C}\!=\!\text{N}$; 1.27 Å).

Bonding in an Amide Group:

Bond lengths   Bond angles

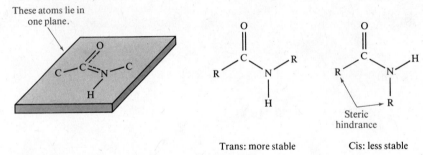

Figure 18.2 Stereochemistry around a peptide bond.

The double-bond character of the amide carbon–nitrogen bond prevents free rotation around this bond. Consequently, cis and trans stereoisomerism is possible (see Figure 18.2). Because the trans stereoisomer contains less steric hindrance than the cis isomer, the trans configuration is the more stable of the two and is the one found in peptides and proteins.

B. Amino Acid Sequence in Peptides

Two different dipeptides can be formed from two different amino acids. For example, glycine and alanine can form two dipeptides, abbreviated Gly-Ala and Ala-Gly.

The Two Dipeptides from Glycine and Alanine:

From glycine   From alanine

$$\text{H}_2\text{NCH}_2\overset{\overset{\displaystyle O}{\|}}{\text{C}}-\text{NHCHCOH}$$
$$\underset{\text{CH}_3}{|}$$

Glycylalanine (Gly-Ala)

From alanine   From glycine

$$\text{H}_2\text{NCHC}-\text{NHCH}_2\text{COH}$$
$$\underset{\text{CH}_3}{|}$$

Alanylglycine (Ala-Gly)

By convention, the amino acid residue with the free carboxyl group is written to the right in the formula and is called the **C-terminal amino acid**. The amino acid residue with the free α-amino group is written to the left and is called the **N-terminal amino acid**. These conventions apply regardless of the length of the peptide, provided the peptide is linear and not cyclic.

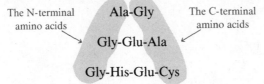

The N-terminal amino acids   Ala-Gly   The C-terminal amino acids

Gly-Glu-Ala

Gly-His-Glu-Cys

In naming peptides, the C-terminal amino acid is considered the parent. The other amino acids, including the N-terminal amino acid, are substituents on the parent and named with the ending *yl*. The amino acid substituents are listed in order starting with the N-terminal amino acid.

Ser-Tyr-Ala-Gly-Leu
Seryltyrosylalanylglycylleucine

18.4 ISOLATION AND CHARACTERIZATION OF PROTEINS

Unlike polysaccharides, such as glycogen, which are mixtures with a range of formula weights, proteins have unique structures and specific formula weights. However, proteins are often difficult to purify because they occur as complexes with lipids and carbohydrates and as mixtures with other proteins.

An additional factor that makes proteins difficult to purify is that their natural shape is easily disrupted by heat, acid, alkali, and organic solvents. When a protein loses its natural shape, it is said to have been **denatured**. A denatured protein retains its original sequence of amino acids, but loses its unique three-dimensional structure, on which its biological activity often depends. Some proteins can be **renatured** (returned to their native conformation) when they are returned to their natural environment; others cannot.

In the 1940s, the development of chromatographic techniques allowed the isolation and purification of previously inseparable proteins. With the ability to separate and purify proteins, chemists could undertake studies aimed at determining the amino acid sequence of proteins.

The first reported protein structure was that of bovine (beef) insulin, reported by the English chemist Frederick Sanger and co-workers in 1955. Sanger received a Nobel Prize for this work in 1958. (He also received a Nobel Prize in 1980 for his work on sequencing, or determining the order of, nucleic acids.) Bovine insulin contains two peptide chains, one of which contains 21 amino acids and the other, 30. The two peptide chains are bonded together by two disulfide bonds. (The third disulfide bond joins two cysteine residues within the 21-amino-acid peptide chain;

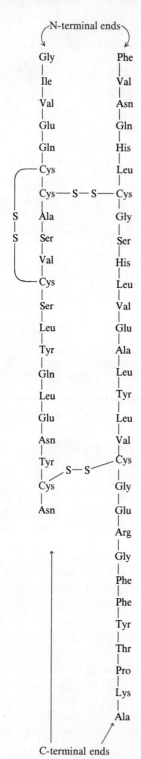

Figure 18.3 Structure of bovine insulin.

see Figure 18.3.) Today, many proteins have been sequenced, and tremendous strides have been made in understanding their three-dimensional structures.

A. Classification of Proteins

A number of methods for classifying proteins have been devised. The first classification scheme, created around 1900, was based on physical and chemical properties. We will present a version of this classical classification scheme primarily because the terms used are part of the vocabulary of clinical chemistry and biochemistry.

According to a modified form of the original classification scheme, proteins can be grouped into three classes: *fibrous proteins*, *globular proteins*, and *conjugated proteins*.

Fibrous proteins are the insoluble structural proteins found in skin, hair, connective tissue, and bone. Fibrous proteins can be subdivided into **collagens**, the principal proteins of the connective tissue, bones, teeth, and tendons; and **keratins**, the principal proteins of skin, fingernails, feathers, and hair.

Globular proteins are spherical or ellipsoid in shape. They are usually (but not always) soluble in water. Globular proteins are those most easily classified by function (such as enzymes or hormones) by the more modern classification schemes. By the classical scheme, globular proteins are subdivided into a number of subgroups, four of which are:

Albumins are identified by their solubility in water and in salt solution. Typical albumins are found in blood (serum protein) and egg whites (egg albumin).

Globulins are identified as being insoluble in water, but soluble in dilute salt solution. γ-Globulin, a typical globulin, is a mixture of proteins that can be isolated from blood serum and contains antibodies.

Histones and **protamines** are identified as water-soluble alkaline proteins. In comparison with other proteins, the histones and protamines yield unusually high concentrations of the basic amino acids. The protamines contain a large amount of arginine, which may account for up to 70–80 percent of their total amino acid content. Histones are differentiated from the protamines by their source and by the greater variety of amino acids they contain. Histones and protamines are commonly found associated with nucleic acids (see Chapter 19).

Conjugated proteins are those associated with a nonprotein group. For example, the protein in hemoglobin is associated with the iron-containing nonprotein heme. The nonprotein part of conjugated proteins, such as heme in the hemoglobin, is called a **prosthetic group**. Other typical conjugated proteins are the nucleoproteins (proteins associated with nucleic acids), mucoproteins (proteins associated with polysaccharides in mucus), and lipoproteins (proteins associated with lipids, such as cholesterol).

We are still learning more and more about proteins. This new knowledge has allowed us to devise other classification schemes. For example, many proteins can now be classified by function. A portion of a more modern classification scheme is given in Table 18.4, along with a brief explanation of each class.

TABLE 18.4   CLASSIFICATION OF SOME PROTEINS BY FUNCTION

Class	General function
Contractile proteins	Convert chemical energy into mechanical energy
Enzymes	Biochemical catalysts
Hormones	Help regulate metabolism
Protective proteins	Recognize and neutralize invading molecules; help repair cells
Storage proteins	Store amino acids in eggs and seeds
Structural proteins	Help provide the structural form of the organism

B. Levels of Protein Structure

A protein in its native state has three and sometimes four levels of structure. The first level, called the **primary structure**, is the sequence, or order, of the amino acids in the protein chain.

The higher levels of structure relate to the conformation or shape of the protein. The **secondary structure** is the shape of the long protein chain as it is held together by hydrogen bonding between the amide protons and the amide carbonyl groups.

*Hydrogen bond between amide proton
and amide carbonyl oxygen*

The secondary structure of many proteins is an **α-helix**—a right-handed spiral with the chain turning clockwise as we view it down the axis. (D-Amino acids form a left-handed helix, or a counterclockwise spiral.) An α-helix has 3.6 amino acid residues per turn, a value that results from hydrogen bonding between two amide groups that are four amino acid residues apart. Figure 18.4 shows the structure of an α-helix.

The helical structure gives rise to a flexible, elastic molecule. *Keratin*, a protein in hair, skin, nails, and so forth, is an example of a protein with molecules in an α-helical shape. *Collagen*, the principal protein of connective tissue, is also composed of helical molecules. Collagen is unusual in that its helices are left-handed instead of right-handed. The left-handed helical structure is possible because every third amino acid in the chain is an achiral glycine residue. In collagen, three left-handed helices intertwine to yield a right-handed "superhelix," called *tropocollagen* (see Figure 18.5). The intertwining of three helices in this fashion gives extra strength to tissues.

Another secondary structure is the **β-pleated sheet**. This term is derived from the appearance of the protein structure, which resembles a pleated sheet or a corrugated metal roof (see Figure 18.6). The proteins in a pleated sheet structure are

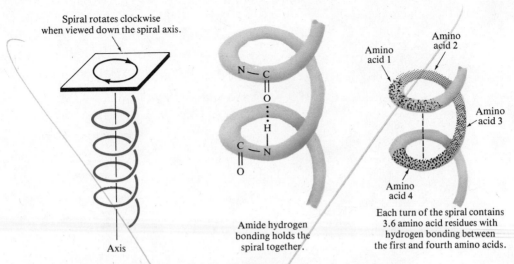

Spiral rotates clockwise
when viewed down the spiral axis.

Axis

Amide hydrogen
bonding holds the
spiral together.

Amino
acid 1

Amino
acid 2

Amino
acid 3

Amino
acid 4

Each turn of the spiral contains
3.6 amino acid residues with
hydrogen bonding between
the first and fourth amino acids.

Figure 18.4 The α-helix is a typical secondary structure of a protein.

maintained by hydrogen bonding between aligned chains. Silk fibroin is an example
of a protein with a pleated sheet structure.

   In globular proteins, the helical or pleated sheet structure is occasionally
interrupted and the chain folds back onto itself. The folded shape of a globular
protein is called its **tertiary structure**.

   The **quaternary structure** of a protein refers to the association of two or more
different chains of a multichain protein. Hemoglobin, for example, contains four
separate folded protein molecules. The association of the four protein subunits in
hemoglobin is referred to as its quaternary structure. Figure 18.7 illustrates the
tertiary and quaternary structures of hemoglobin.

   Table 18.5 summarizes the different levels of protein structure.

18.5 ENZYMES

Enzymes are biological catalysts. These are proteins whose activity depends on the
maintenance of the secondary, tertiary, and quaternary structures. A typical enzyme
is a folded globular protein with polar amino acid residues on its outer surfaces,
which ensure solubility in the aqueous body fluids. An enzyme often contains a
prosthetic group (Section 18.4A) within a pocketlike cavity in the protein structure.
The size and shape of the pocket allow recognition of a specific biomolecule.

Figure 18.5 Intertwining of three helices in tropocollagen.

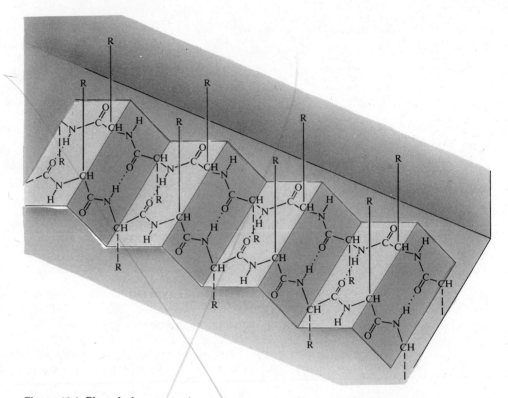

Figure 18.6 Pleated sheet secondary structure of a protein in silk fibroin. Adapted from Karlson, P.: Introduction to Modern Biochemistry, 3rd edition, 1968, Georg Thieme Verlag, Stuttgart.

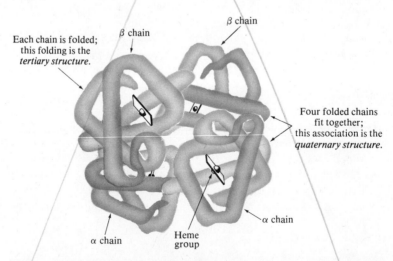

Figure 18.7 Models of hemoglobin illustrating the tertiary and quaternary structures of a protein. *Source:* R. E. Dickerson and I. Geis, The Structure and Action of Proteins, published by W. A. Benjamin. Inc., Menlo Park, California. Copyright 1969 by Dickerson and Geis.

TABLE 18.5 SUMMARY OF THE DIFFERENT LEVELS OF PROTEIN STRUCTURE

Level of Structure	Description
Primary	Sequence, or order, of amino acid residues in a protein chain
Secondary	Shape, or conformation, of a protein backbone chain
Tertiary	Folding of a protein chain back onto itself
Quaternary	Folding together of two or more discrete protein chains

Figure 18.8 illustrates how an enzyme can catalyze a biological reaction, such as hydrolysis, isomerization, or oxidation.

18.6 DETERMINATION OF A PROTEIN'S PRIMARY STRUCTURE

The first step in the complete characterization of a protein or peptide is the determination of its primary structure, the sequence of its amino acids. If the isolated peptide is homogeneous and not cyclic, the following general procedure can be used to determine its amino acid sequence:

1. cleavage of disulfide cross-links (if any) to —SH groups and blocking of all these groups;
2. end-group analysis;

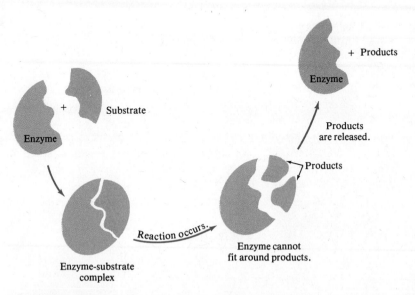

Figure 18.8 Enzyme action. The enzyme (a protein) enfolds or is induced to fit the substrate (the molecule to be acted on) and holds the substrate in place allowing (or causing) the reaction to occur. The products are released, and the enzyme assumes its original form.

3. partial hydrolysis of the peptide followed by identification of the fragments; and finally
4. elucidation of the sequence.

A. Blocking the Sulfur Groups of Cysteine and Cystine

The first step in amino acid sequencing is the blocking of the disulfide and —SH groups. If these groups are not blocked, they will interfere with subsequent sequencing reactions.

Blocking can be accomplished by one of two ways. One method is to oxidize the disulfide groups and —SH groups to sulfonic acid groups. The other method is to reduce the disulfide groups to —SH groups and then treat the mixture with a reagent, such as alkaline iodoacetate, that reacts with —SH groups.

B. End-Group Analysis

End-group analysis is used to identify the sequence of amino acids at the N-terminal and C-terminal ends of the peptide chain.

N-Terminal Amino Acid   A number of different reagents can be used to identify the N-terminal amino acid. One such reagent is phenyl isothiocyanate, called the **Edman reagent** (after the Swedish chemist Pehr Edman). Phenyl isothiocyanate reacts with the amino group of the N-terminal amino acid to form an intermediate that can be hydrolyzed to a phenylthiohydantoin. In the hydrolysis reaction, a new peptide, one amino acid shorter than the original peptide, is also released. Since the phenylthiohydantoin is a derivative of the N-terminal amino acid, its isolation and identification provide the identity of the N-terminal amino acid.

The phenylthiohydantoin of the N-terminal amino acid

The residual peptide resulting from phenylthiohydantoin hydrolysis can be recycled through the reaction sequence to identify the newly formed N-terminal amino acid. This amino acid is the next-to-last amino acid residue in the original peptide chain. In theory, the entire amino acid sequence of the peptide could be determined by repeatedly recycling the residual peptide through the procedure. In practice, the hydrolysis used to remove the phenylthiohydantoin also causes some cleavage of the peptide chain itself. After about 20 passes, the peptide chain has been sufficiently fragmented that it is no longer possible to distinguish the residual main chain from the hydrolytic fragments.

The phenyl isothiocyanate method has been incorporated into analyzers called **automated amino acid sequenators.** In these instruments, the peptide to be sequenced is bonded by its C-terminal acid to a solid support, a technique that aids in the removal of the partial hydrolysis products and allows more than 20 amino acids to be sequenced. Sequenators have been used to degrade and analyze peptides of up to 60 amino acid residues.

C-Terminal Amino Acid   Fewer methods are available for the identification of the C-terminal amino acid. One procedure employs *carboxypeptidase*, an enzyme that preferentially catalyzes the hydrolysis of the C-terminal amino acid from a peptide chain.

$$\text{Rest of peptide}-\underset{\underset{R}{|}}{\text{NHCHCO}}^- + H_2O \xrightarrow{\text{carboxypeptidase}} \text{rest of peptide} + H_3\overset{+}{N}\underset{\underset{R}{|}}{\text{CHCO}}^-$$

On cleavage of the original C-terminal amino acid, a new peptide, one amino acid shorter than the original, is formed in the reaction mixture. The C-terminal amino acid of the new peptide can also be cleaved by the carboxypeptidase. Consequently, a mixture of amino acids is released during the enzymatic hydrolysis. This fact limits the number of C-terminal acids that can be sequenced at the C-terminal end of the chain.

C. Partial Hydrolysis and Final Sequence Determination

After the end groups have been identified, the next goal is the determination of the entire sequence of amino acids in the chain. This is accomplished by partial hydrolysis of the peptide followed by matching the fragments to determine the complete sequence.

Partial hydrolysis can be carried out by mild acid hydrolysis or by enzymes. The acid hydrolysis technique is generally not satisfactory because the cleavage of amide bonds is random and a complex mixture of amino acids and small peptides is obtained. For example, a tetrapeptide will yield four amino acids, three dipeptides, and two tripeptides. If a large peptide is acid hydrolyzed, there may be so many components in the mixture that separation is not possible.

The preferred technique is to fragment the peptide using enzymes that act at specific amide bonds. For example, *trypsin* is a proteolytic enzyme (an enzyme that catalyzes the hydrolysis of proteins) commonly used in structure determination.

Trypsin catalyzes the hydrolysis of an amide bond that is on the carboxyl side of either lysine and arginine.

$$\{-NHCHC \overset{O}{\underset{|}{\|}} \overset{/}{\not{}} NH - \text{rest of peptide}$$

$$\underbrace{\qquad\qquad \underset{(CH_2)_4NH_2}{\qquad} \qquad\qquad}_{\text{Lysine residue}}$$

Trypsin attacks here

$$\{-NHCHC \overset{O}{\underset{|}{\|}} \overset{/}{\not{}} NH - \text{rest of peptide}$$

$$\underbrace{\qquad\qquad (CH_2)_3NHC\overset{\|}{\underset{NH}{}}NH_2 \qquad\qquad}_{\text{Arginine residue}}$$

A number of different proteolytic enzymes are available. A few are listed in Table 18.6 along with the types of amide bonds whose hydrolysis they catalyze.

By the use of two or more of these enzymes, fragments with overlapping segments of amino acids can be obtained. After these fragments themselves are sequenced (by the Edman method, for example), the overlapping portions of the fragments are compared and the complete sequence of the protein is deduced. To illustrate the logic used in this final step, let us assume that hydrolysis of a pentapeptide yields the following tripeptides:

Gly-Glu-Phe   Glu-Phe-Gly   Phe-Gly-Arg

TABLE 18.6 SOME PROTEOLYTIC ENZYMES AND REAGENTS USED IN THE SEQUENCE DETERMINATION OF PROTEINS

$$\{-NHCH\overset{O}{\overset{\|}{C}}\underset{\uparrow}{-}NHCH\overset{O}{\overset{\|}{C}}-\}$$
$$\quad\quad\underset{R}{|}\qquad\quad\underset{R'}{|}$$

Site of cleavage

Enzyme or reagent	Site of cleavage
Chymotrypsin	R = Tyr, Phe, Trp
Pepsin	R' = Tyr, Phe, Trp
Trypsin	R = Lys, Arg
Cyanogen bromide (BrCN)	R = Met
Thermolysin	R' = Ile, Leu, Val

By arranging the tripeptides so that the common amino acids overlap, we can determine the sequence of amino acids in the pentapeptide.

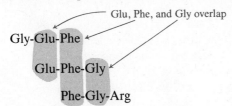

Glu, Phe, and Gly overlap

Gly-Glu-Phe

Glu-Phe-Gly

Phe-Gly-Arg

Amino acid sequence of the pentapeptide: Gly-Glu-Phe-Gly-Arg

PROBLEM 18.5.   *Bradykinin* is a pain-causing nonapeptide released by globulins in the blood as a response to toxins in wasp-sting venoms. Partial hydrolysis of bradykinin yields results in the following tripeptides:

Ser-Pro-Phe	Pro-Phe-Arg
Gly-Phe-Ser	Arg-Pro-Pro
Pro-Gly-Phe	Pro-Pro-Gly
Phe-Ser-Pro	

What is the amino acid sequence in bradykinin?

SUMMARY

Proteins are polyamides of α-amino acids. The amide bonds in proteins are called **peptide bonds**. Peptides (small proteins) can be classed as dipeptides, tripeptides, and so forth up to polypeptides.

Amino acids exist as **dipolar ions**, which are amphoteric. The alpha carbon of amino acids found in proteins has the same configuration as that of L-glyceraldehyde. Amino acids can be classed as *neutral* (polar or nonpolar), *basic*, or *acidic*, depending on their side-chain functional groups. The **isoelectric point** for an amino acid is the pH at which the dipolar ion carries no net charge and thus does not migrate toward an electrode.

Essential amino acids are those that cannot be synthesized by an organism and thus must be supplied in the diet.

Amino acids can undergo *esterification* and *acylation*. Amino acids also undergo a *ninhydrin reaction* to yield Ruhemann's purple, a blue-violet product. Cysteine can be *oxidized* to a disulfide amino acid called cystine.

Peptides are usually shown with the N-terminal amino acid on the left and the C-terminal amino acid on the right. Peptides are named starting with the N-terminal amino acid and ending with the C-terminal amino acid.

Proteins may be classed as *fibrous*, *globular* (such as albumins, globulins, histones, and protamines), or *conjugated*. They may also be classed by function, such as enzymes or hormones.

The **primary structure** of a protein is its amino acid sequence. The **secondary structure** is how the backbone arranges itself (α-helix, β-pleated sheet). The **tertiary structure** is the result of folding of a helix or other structure on itself, and the **quaternary structure** is the association among different chains, as in hemoglobin.

Denaturation of a peptide or a protein is the disruption of its hydrogen bonds and other forces that hold the molecule in the secondary, tertiary, and quaternary structural levels.

A protein's primary structure can be determined by cleaving —S—S— groups, blocking the —SH groups, determination of the end groups, and partial enzymatic hydrolysis.

STUDY PROBLEMS

18.6. Define the following terms and give an example where appropriate.
   (a) α-amino acid
   (b) dipolar ion
   (c) L-α-amino acid
   (d) neutral nonpolar amino acid
   (e) neutral polar amino acid
   (f) basic amino acid
   (g) acidic amino acid
   (h) isoelectric point
   (i) electrophoresis
   (j) PKU
   (k) peptide bond
   (l) tripeptide
   (m) C-terminal amino acid
   (n) N-terminal amino acid
   (o) enzyme
   (p) primary structure of proteins
   (q) higher structures of proteins

18.7. Write equations for the reactions showing dipolar-ion formation for the following amino acids:

(a) $H_2NCNH(CH_2)_3CHCOH$ with NH double bond and O double bond, and NH_2 group

(b) 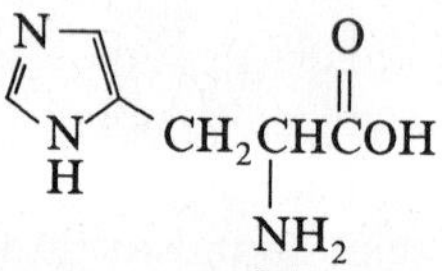 imidazole ring CH_2CHCOH with O double bond and NH_2 group

(c) pyrrolidine ring COH with O double bond

(d) $HOCH_2CHCOH$ with O double bond and NH_2 group

18.8. Write formulas showing how water can form hydrogen bonds with the polar groups in the side chains of the following amino acids:
   (a) aspartic acid   (b) threonine   (c) asparagine

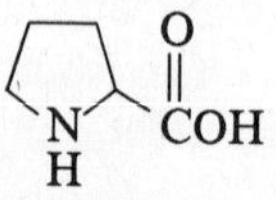

18.9. Indicate the most acidic proton in each of the following structures.

(a) $HOCH_2CHCOH$ with O double bond and $^+NH_3$ group

(b) 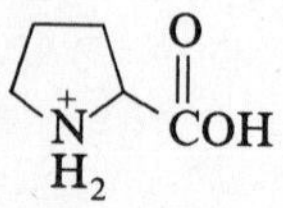 pyrrolidine ring with ^+N, H_2 and COH with O double bond

18.10. Write equations showing the reactions of 1.0 mol of the amino acids in the preceding problem with 1.0 mol of base.

18.11. Do the following amino acids require reaction with acid or with base to reach their isoelectric points? Explain.

(a) $\overset{\overset{+}{N}H_2}{\overset{\|}{C}}$ $\overset{O}{\overset{\|}{C}}$

H$_2$NCNH(CH$_2$)$_3$CHCOH
                    |
                $\overset{+}{N}H_3$

(b) $^-$O—⟨O⟩—CH$_2$CHCO$^-$
                      |
                    NH$_2$

(c) [indole ring structure] CH$_2$CHCOH
                                    |
                                $\overset{+}{N}H_3$

with O double bond

18.12. Write equations showing the reactions of the amino acids in the preceding problem going to their isoelectric points.

18.13. (a) Which of the essential amino acids have branched alkyl groups as their side chains?

(b) Do any of the nonessential amino acids have branched hydrocarbon chains in their side chains?

18.14. Write equations for (a) an esterification reaction, (b) an acylation reaction, and (c) the reaction with ninhydrin for valine. (*Hint:* In (c), refer to Section 18.2B.)

18.15. Write equations for the reactions of serine with:

(a) ⟨O⟩—CCl (with O double bond)   (b) CH$_3$OH, H$^+$, heat

(c) ninhydrin

18.16. Write condensed structural formulas for the following tripeptides:
(a) glycylalanylphenylalanine
(b) phenylalanylalanylglycine
(c) alanylphenylalanylglycine

18.17. The amino acids alanine (Ala), histidine (His), and leucine (Leu) can be combined in six different ways in a tripeptide. Write formulas for these six tripeptides using the abbreviated names.

18.18. For each of the following peptides, (1) draw a condensed structural formula, (2) identify the peptide bond(s), (3) identify the N-terminal amino acid, and (4) identify the C-terminal amino acid. For (a), draw the dipeptide in the trans conformation around the peptide bond.

(a) Asp-Phe (the C-terminal as the methyl ester): aspartame, a synthetic sweetener (Nutrasweet)

(b) Tyr-Gly-Gly-Phe-Met: an enkephalin, a brain peptide with opiate activity

18.19. Write *abbreviated* names for each of the following peptides:
(a) serylglycyltyrosylalanylleucine
(b) alanylleucylmethionylarginine

18.20. Complete the following reaction sequence:

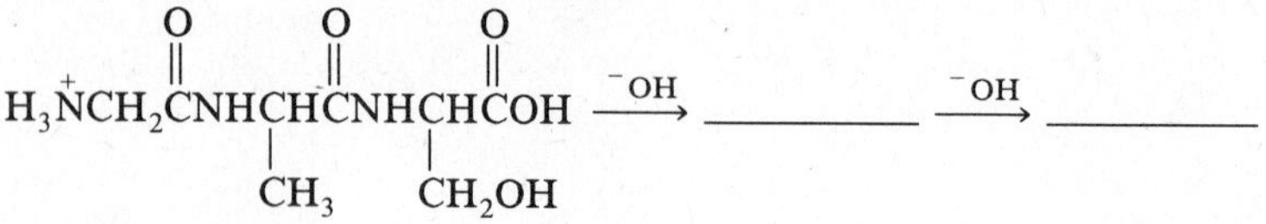

$$\overset{+}{H_3}NCH_2\overset{O}{\overset{||}{C}}NH\underset{\underset{CH_3}{|}}{CH}\overset{O}{\overset{||}{C}}NH\underset{\underset{CH_2OH}{|}}{CH}\overset{O}{\overset{||}{C}}OH \xrightarrow{\ ^-OH\ } \underline{\hspace{2cm}} \xrightarrow{\ ^-OH\ } \underline{\hspace{2cm}}$$

18.21. At pH 9 in an electrophoresis cell, would glycine migrate toward the cathode, the anode, or neither? Explain.

18.22. Which of the following features would contribute to the water solubility of a peptide?
(a) globular shape   **(b)** association with glucose
(c) rich in isoleucine residues
(d) rich in aspartic acid residues

18.23. Explain why the carboxyl group near the amino group is more acidic than the side-chain carboxyl group.

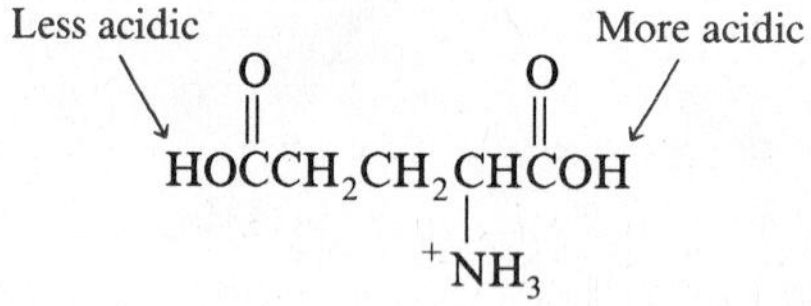

$$\overset{\text{Less acidic}}{\searrow}\ \ \ HO\overset{O}{\overset{||}{C}}CH_2CH_2\underset{\underset{^+NH_3}{|}}{CH}\overset{O}{\overset{||}{C}}OH\ \ \ \overset{\text{More acidic}}{\swarrow}$$

18.24. For each of the following amino acids, draw a formula for the structure that you would expect to be present at (1) pH 2, (2) pH 7, and (3) pH 12. Explain your answers.
(a) tyrosine   **(b)** glutamic acid
(c) isoleucine   **(d)** proline

18.25 (a) Write an equation to show how Ser-Phe-Met reacts with phenyl isothiocyanate to yield a phenylthiohydantoin.
(b) What is the structure of the dipeptide released by the reaction?

18.26. A pentapeptide obtained from a protein after treatment with trypsin contains arginine, aspartic acid, leucine, valine, and phenylalanine. The pentapeptide is treated with Edman reagent (phenyl isothiocyanate) three times. The composition of the peptide remaining after each treatment was: (1) arginine, aspartic acid, leucine, valine; (2) arginine, aspartic acid, valine; (3) arginine, valine. What is the structure of the pentapeptide?

18.27. What fragments will be formed when each of the following peptides is treated with trypsin?
(a) Lys-Gly-Ala-Ala-Glu
(b) Cys-Trp-Ala-Arg-Ala-Lys-Glu
(c) Ala-His-Arg-Glu-Lys-Phe-Ile-Glu

POINT OF INTEREST 18

Antibodies — Proteins with a Memory

When foreign substances such as bacteria invade the body, the immune system responds by producing **antibodies**—globular proteins that react with specific organic molecules to form insoluble complexes and can thus deactivate invaders. The foreign substance that reacts with an antibody is called an **antigen.** An antigen, therefore, is a substance that can induce an **immune response**, the synthesis of specific antibodies by the body.

The immune response is specific—each type of antigen reacts with only one type of antibody. How does an antibody recognize its specific antigen? To answer this question, we must look at the antigen-antibody reaction.

Antigens can be proteins, glycoproteins (carbohydrate-protein complexes), polysaccharides, lipids, or nucleoproteins (nucleic acid-protein complexes). A large molecule by itself does not act as the antigen. Rather, there are areas on the surface of the molecule, called **epitopes**, that the antibody recognizes. A large molecule may contain several epitopes on its surface. In general, about one epitope exists for each 5000 units of formula weight.

Although small molecules cannot produce an immune response, they may react with larger molecules to form new surfaces. These new surfaces may become new epitopes. Small molecules that act in this manner are called **haptens** (Greek *haptein*, "to grasp or fasten").

Allergic reactions to penicillin and to poison ivy are examples of hapten-antibody interactions. Penicillin itself is not able to elicit an immune response. Some individuals, however, can convert penicillin into other molecules, such as penicilloic acid, that bond to blood proteins. A bonded fragment can then be recognized as an epitope and evoke an immune response.

A penicillin   A penicilloic acid

The same type of process is involved in the allergic reaction to poison ivy. The active ingredients in poison ivy are a group of phenols called *urushiols*. These phenols bind to the skin proteins, the resulting phenol-protein complex is recognized as foreign, and antibodies are produced. The final result is a skin rash.

The interaction of an epitope and an antibody involves attractions between specific areas of their two surfaces. The most important antigen-antibody bond is a *hydrophobic bond* formed by the close approach of the two surfaces. Water is pushed away from the area of contact. Other noncovalent antigen-antibody attractions of importance are hydrogen bonds, electrostatic attractions between oppositely charged amino acids, and attractions resulting from induced dipoles.

The closer the two surfaces can approach, the stronger will be the binding. Therefore, the shape of the epitope and the shape of the antigen-binding site of the antibody conform to a lock-and-key type of relationship.

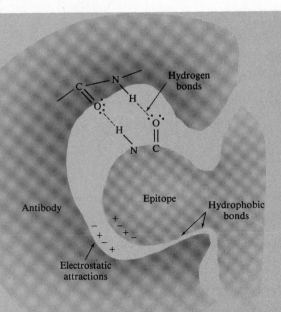

All antibodies have the same fundamental structure; however, each type of antibody has a region within its structure that is different from that in other antibodies. This region, called the **variable region**, is responsible for the antibody's ability to bind with a specific antigen. When a new antigen is introduced into the body, new antibodies with an altered variable region are produced. This altered variable region allows the antibodies to react with the epitopes of the new antigen.

Types of Antibodies and Their Production   Serum protein antibodies are called **immunoglobulins** (abbreviated **Ig**). When a new antigen encounters antibody-producing cells in the lymph system, these cells produce antibody-producing cells that are active in the bloodstream. These cells produce an immunoglobulin called **immunoglobulin M (IgM)** for about 10 days. The variable region in these IgM molecules is specific for the new antigen. After the 10-day period, the cells produce another immunoglobulin **IgG** (also called *gamma globulin*) for longer-term protection. The variable region of this IgG has the same binding site as did the IgM produced by that cell.

After the antigen has been destroyed, synthesis of the immunoglobulins ceases, but the ability to reproduce the structures is retained in *memory cells*. Should that antigen reinvade the system, the memory cells cause a rapid synthesis of antibodies. We say that the body has become *immune* to that antigen.

The body produces at least three other types of immunoglobulins. For example, immunoglobulin molecules called **IgA** are found in external secretions, such as tears.

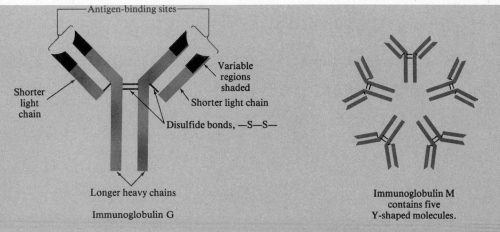

Figure 18.9 Diagrams of the structures of immunoglobulins G and M. The heavy and light chains are polypeptide chains containing both constant regions and variable regions.

Structure of Immunoglobulins   Immunoglobulin G consists of four polypeptide chains bonded together by disulfide bonds in a Y shape as shown in Figure 18.9. Two of the peptide chains, called the **heavy chains**, contain just under 500 amino acids each. The other two peptides, the **light chains**, contain about 215 amino acids.

Two identical antigen-recognition sites, located in the upper ends of the arms of the Y, consist of approximately 100 amino acids each. These sites are the variable regions of the immunoglobulin structure and differ in each type of antibody.

Immunoglobulin M is composed of an association of five Y-shaped molecules similar to the IgG molecule. This type of association is also shown in Figure 18.9.

Chapter 19

Nucleic Acids

Nucleic acids are long-chain polymers that carry the genetic code. The biosynthesis of the different proteins in an organism is accomplished on the basis of this genetic information. In recent years, great advances have been made in our understanding of nucleic acids. Ongoing research is unlocking the secrets of heredity, birth defects, the immune system, cancer, and other areas related to nucleic acid chemistry. The knowledge gained from this research is now being used to modify the genetic makeup of some living organisms, such as bacteria, so that they produce proteins useful for humans; this modification of genetic material is called *genetic engineering*. In this chapter, we discuss briefly the structures of the nucleic acids, how the genetic code is retained in their structures, and how this information is used to synthesize proteins.

19.1 GENERAL STRUCTURE OF NUCLEIC ACIDS

The two general classes of nucleic acids are **DNA**, found in the nuclei of plant and animal cells, and **RNA**, a group of related nucleic acids found in the cytoplasm, the cellular material outside the nucleus. All nucleic acids have similar structures. Their primary structure consists of monosaccharide units linked together by phosphate groups. Each monosaccharide in the chain is bonded to a nitrogen heterocyclic compound called a **base**. The structures of DNA and the different types of RNA differ primarily in the monosaccharide contained in their backbone chains and in the structure of the bases bonded to these monosaccharides.

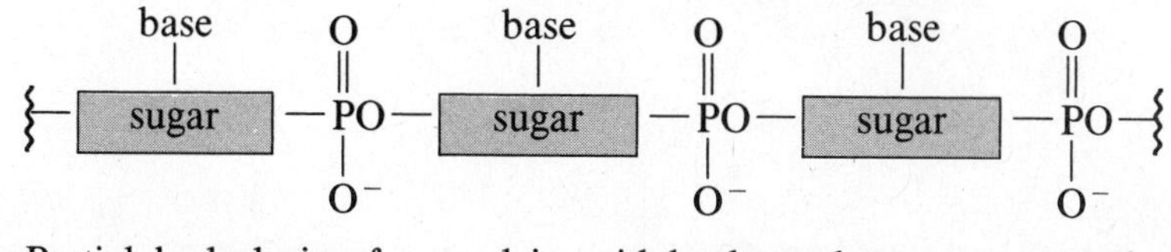

Partial hydrolysis of a nucleic acid leads to base—sugar—phosphate fragments of the main chain, called **nucleotides**. Nucleotides are the monomers of the nucleic acid polymers. Continued hydrolysis yields hydrogen phosphate ions ($HOPO_3^{2-}$) and base—sugar units called **nucleosides**. Nucleosides, in turn, can be hydrolyzed to the sugar and the nitrogen heterocyclic base.

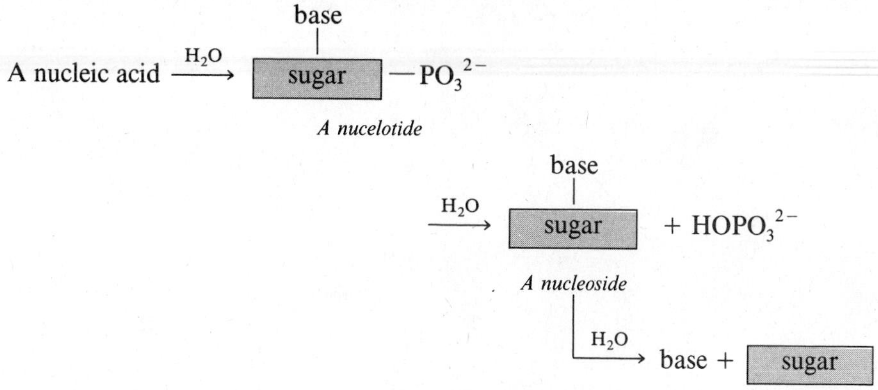

A nucelotide

A nucleoside

In acidic solution, the oxygens of the phosphate groups are protonated. As the pH of the solution is brought toward the physiological pH, near pH 7, these protonated phosphate groups donate their protons to water; that is, they act as Brønsted acids. This is the reason for the name *nucleic acids* ("acids isolated from the cell nucleus"). However, at pH 7 the phosphate groups are ionized and the oxygens carry negative charges (see Figure 19.1). We will show the structures of the nucleic acids in their anionic forms rather than in their protonated acidic forms.

Acidity of the Nucleic Acids:

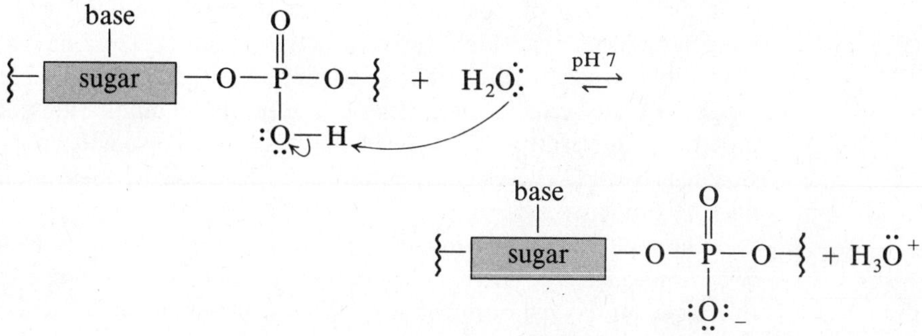

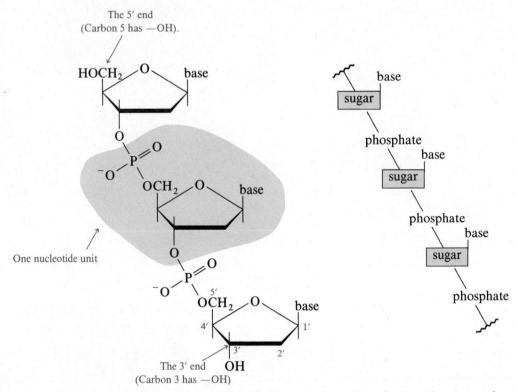

Figure 19.1 Primary structure of DNA. Each base is one of four heterocycles. A complete chain may contain more than a million nucleotide units.

19.2 THE STRUCTURE OF DNA

A. The DNA Chain

The carrier of the genetic code, found in the chromosomes in the nuclear material of cells, is a series of compounds called **deoxyribonucleic acids**, or **DNA**. DNA molecules have very high formula weights, in the millions and billions. It is, however, extraordinarily difficult to obtain an accurate measure of the formula weights of DNA molecules. Simple laboratory manipulations, such as stirring or pipetting, can result in cleavage of the huge molecules and produce lower-formula-weight mixtures. Therefore, the number of nucleotides these molecules contain can only be estimated.

The monosaccharide found in DNA is 2-deoxyribose, hence the name *deoxyribonucleic acids*. The "2-deoxy" in the name means that carbon 2 of the monosaccharide lacks an oxygen atom; that is, this carbon is not bonded to a hydroxyl group. Deoxyribose, in solution, can form either five- or six-membered rings, but only the five-membered monosaccharide ring is found in DNA.

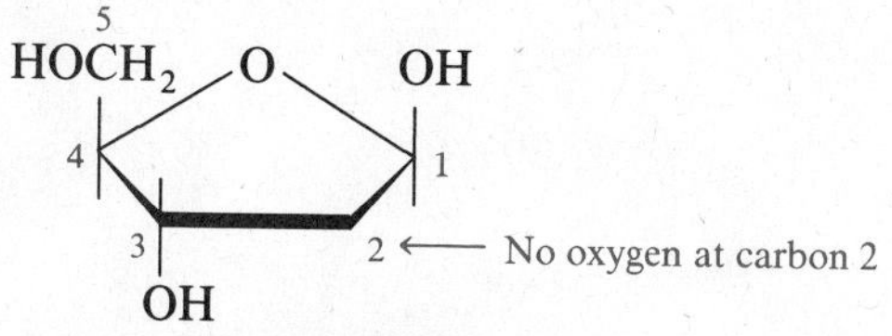

2-Deoxy-β-D-ribofuranose

the monosaccharide in DNA

In DNA, the deoxyribose units are joined by phosphate groups from the oxygen at position 3 of one deoxyribose to the oxygen at position 5 of the next deoxyribose. The heterocyclic bases are bonded to carbon 1, the anomeric carbon, of the deoxyribose units in the β configuration. Figure 19.1 shows the structure of the DNA chain. In the figure, the numbering system of the deoxyribose units is designated with a prime (′). This symbol (such as in 3′) differentiates the deoxyribose numbering from the numbering of the heterocyclic rings (no prime).

B. The Bases in DNA

Four principal bases are found in DNA. These are named thymine, cytosine, adenine, and guanine (abbreviated T, C, A, and G). Thymine and cytosine have the same parent ring skeleton, pyrimidine, while adenine and guanine have purine as their parent ring skeleton. The structures of these compounds are shown in Figure 19.2.

In DNA, the bases are bonded by a nitrogen-carbon β link to the anomeric carbon of deoxyribose.

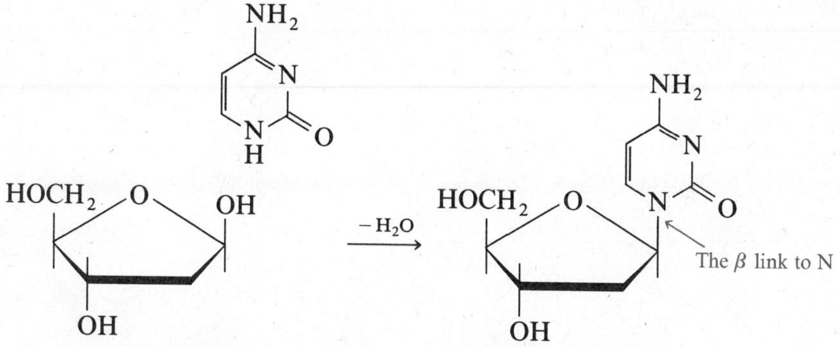

The pyrimidine bases are joined to the sugar from position 1 of the pyrimidine ring, while the purine bases are joined from position 9 of the ring.

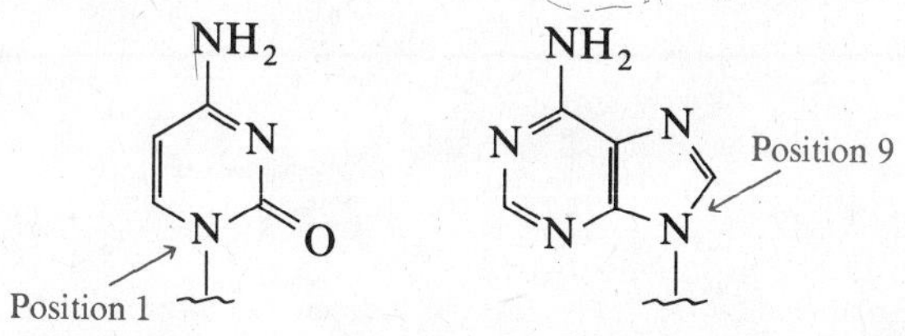

Pyrimidine bases

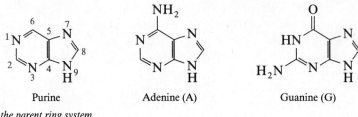

Pyrimidine

*the parent ring system
of thymine and cytosine*

Thymine (T)

Cytosine (C)

Purine bases

Purine

*the parent ring system
of adenine and guanine*

Adenine (A)

Guanine (G)

Figure 19.2 The four principal heterocyclic bases found in DNA.

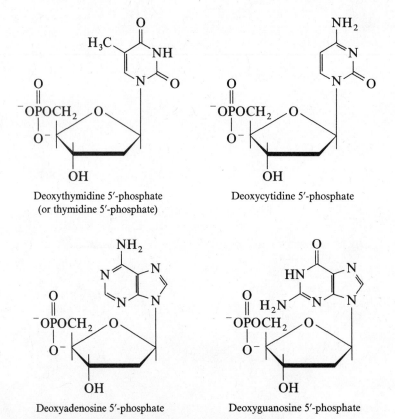

Deoxythymidine 5'-phosphate
(or thymidine 5'-phosphate)

Deoxycytidine 5'-phosphate

Deoxyadenosine 5'-phosphate

Deoxyguanosine 5'-phosphate

Figure 19.3 The four nucleotides in the DNA polymer.

In addition to these four principal bases, a few minor bases are found in DNA molecules of higher plants and animals. In general, these bases are formed from one of the four principal bases. The most abundant of these minor DNA bases is 5-methylcytosine, which forms up to 7% of the bases of some higher plants. Another minor DNA base is 6-N-methyladenine.

Minor Bases Found in Some DNA:

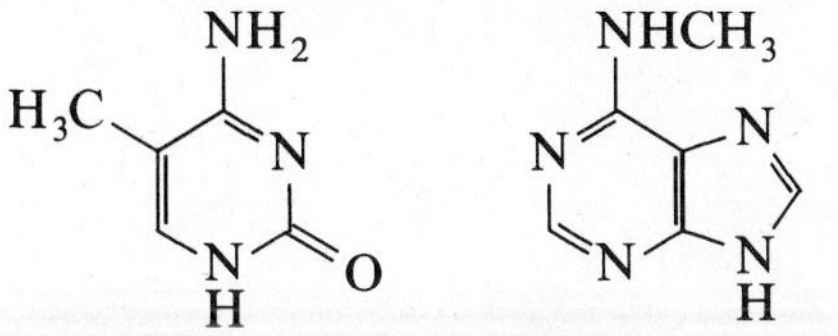

5-Methylcytosine   6-N-methyladenine

Because only four principal bases are found in DNA, the DNA polymer is composed mainly of only four types of nucleotides: the phosphate derivatives of deoxythymidine (also called simply *thymidine* because thymine is not found in RNA as the other bases are), deoxycytidine, deoxyadenosine, and deoxyguanosine. The structures are shown in Figure 19.3.

EXAMPLE

Write chemical equations that show the alkaline hydrolysis of deoxyadenosine 5'-phosphate to yield (a) a nucleoside and (b) a monosaccharide and a nitrogen heterocyclic compound.

Solution:

(a)

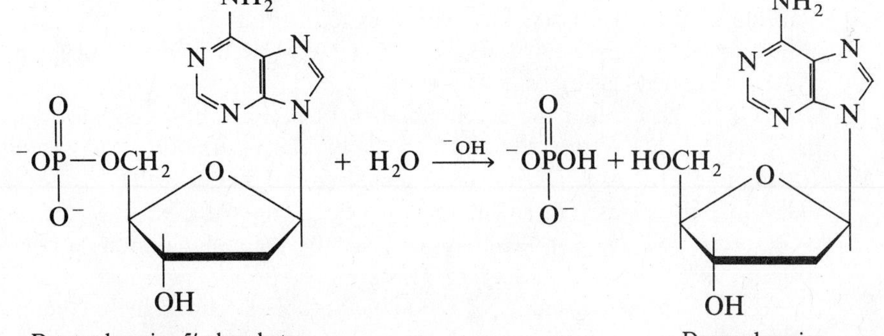

Deoxyadenosine 5'-phosphate   Deoxyadenosine

a nucleotide   *a nucleoside*

(b)

$$^-OP-OCH_2 \quad \text{(adenine-deoxyribose phosphate)} \quad + 2H_2O \xrightarrow{\ ^-OH\ }$$

$$^-OPOH + \quad \text{2-Deoxyribose} \quad + \quad \text{adenine}$$

2-Deoxyribose   adenine

> **PROBLEM 19.1.** Write equations for the alkaline hydrolysis of 2-deoxythymidine 5′-phosphate, as we have just done for the adenosine phosphate in the example.

C. The Double Helix of DNA

The DNA from different tissue samples of a species of plant or animal has the same proportions of the four principal bases. The DNA from a different species, however, contains different proportions of the same four bases. For example, the base mixture obtained from human DNA (any tissue, any individual) contains approximately 30% thymine, 30% adenine, 20% guanine, and 20% cytosine. (These values are **mole percents**, in which the percentages are proportional to the numbers of moles, not the numbers of grams.) The DNA molecules from other organisms contain different percentages of these bases, but, within experimental error, the percent of thymine always equals the percent of adenine (T = A), while the percent of guanine always equals the percent of cytosine (G = C).

In 1950, to explain the equal amounts of pairs of bases, it was suggested that the bases in DNA are actually physically paired. In 1953, biochemists James Watson and Francis Crick proposed a base-paired model for DNA that we still use today in only slightly modified form. They, along with M. Wilkins, received the Nobel Prize in 1962 for their studies and for their elucidation of the secondary structure of DNA.

The Watson-Crick model of DNA is a double-stranded helix composed of two DNA molecules coiled about each other. The strands are *antiparallel*, which means that they run in opposite directions. One strand starts with a free 5′ hydroxyl group or phosphate group and ends with a free 3′ hydroxyl group or phosphate group. The other strand starts with a 3′ end and goes to the 5′ end.

Parallel and Antiparallel Strands:

DNA strands are antiparallel.

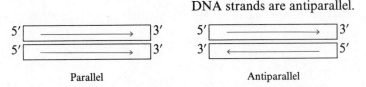

Parallel   Antiparallel

The pair of strands are held together by hydrogen bonds between bases, as shown in Figure 19.4.

The hydrogen bonds in DNA are not random, but join specific pairs of bases—adenine to thymine and guanine to cytosine. For example, any time an adenine occurs on one strand, a thymine appears on the opposite strand. Other

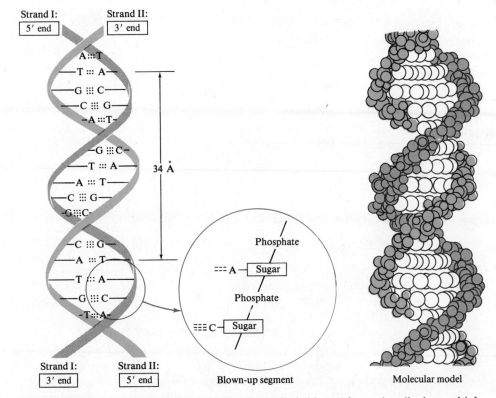

Figure 19.4 The double-stranded helix of DNA is held together primarily by multiple hydrogen bonds between pairs of complementary bases. (In the left-hand diagram, the ribbon represents the sugar-phosphate backbone of the DNA polymeric chain.)

combinations of the four bases do not form the strong multiple hydrogen bonds that the adenine–thymine pair or the guanine-cytosine pair do. This pairing is the reason that equal amounts of the base pairs are found in DNA.

Hydrogen Bonds between Thymine and Adenine (T≡≡≡A):

Hydrogen Bonds between Cytosine and Guanine (C≡≡≡G):

PROBLEM 19.2. Draw formulas for thymine and guanine to show that these bases could form only one hydrogen bond between them, instead of two or three, when they are in DNA chains.

19.3 REPLICATION OF DNA

Later in this chapter, we discuss the sequence of bases in DNA, a sequence that is the basis of the genetic code. A **gene** is a specific sequence of bases in DNA that is *the code for a specific sequence of amino acids to be incorporated into one specific protein* (hair, an enzyme, or other protein). In order that the information contained in the genes be passed from a mother cell to a daughter cell, DNA must be duplicated exactly, a process called **replication**.

In cells of higher plants and animals (eukaryotic cells), DNA is localized in the chromosomes in the cell nucleus. When a cell divides, the double helix of DNA is enzymatically unraveled. Each single strand serves as a template, or pattern, for the synthesis of a new, but complementary, chain. When the syntheses are complete, two two-stranded helices exist where only one did originally. Thus, after complete cell division, both cells will contain identical DNA.

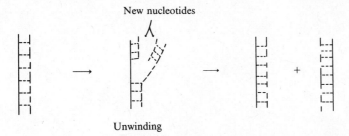

Let us look at replication in somewhat more detail. As the original helix unwinds, new nucleotides (triphosphates in this case) align themselves with their complementary bases in the intact strand. If an adenine is present on the intact DNA chain, thymine will align itself opposite the adenine. As the nucleotides become aligned, they are enzymatically polymerized to form the new and complementary chain, as shown in Figure 19.5.

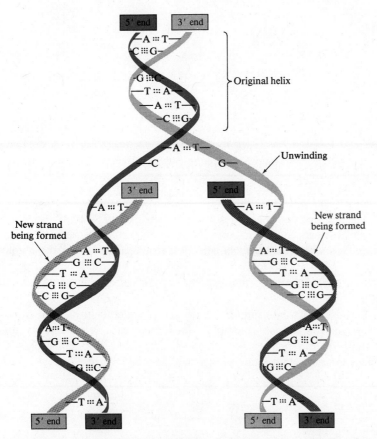

Figure 19.5 In the replication of DNA, complementary nucleotides become aligned and are polymerized into a new complementary strand.

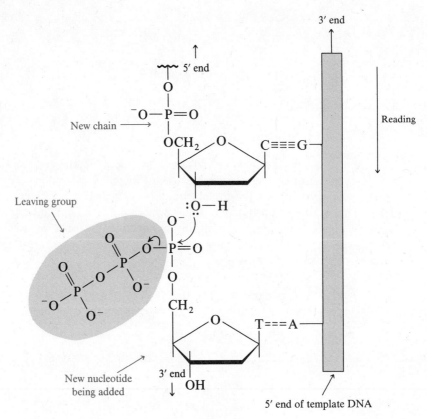

Figure 19.6 Addition of a new nucleotide to a growing DNA chain. In the substitution reaction, the leaving group is a diphosphate ion. The reaction is catalyzed by the enzyme *DNA polymerase*.

The nucleotides used to form the new strand are not polymerized simultaneously. Instead, each nucleotide undergoes a discrete reaction with the growing chain. These steps occur by substitution reactions involving the triphosphate group of an incoming nucleotide with a hydroxyl group at the 3′ carbon of the deoxyribose of the growing chain. A diphosphate ion serves as the leaving group (see Figure 19.6).

Because the two original strands are complements of each other and because each original strand is the template for its new complementary strand, the two new resultant double helices are identical. Thus, replication ensures that each cell nucleus in an organism carries identical genetic information.

19.4 STRUCTURE AND BIOSYNTHESIS OF RNA

The **ribonucleic acids** (RNAs) are the workhorses in protein synthesis. Three principal types of RNA are involved in these key biosynthetic reactions.

Messenger RNA (mRNA), a polymer of 75–3000 nucleotides, is synthesized by a portion (a single gene) of a single strand of DNA as its pattern. The mRNA of eukaryotic cells leaves the cell nucleus after it is synthesized and acts as a template from which specific amino acids are identified and incorporated in a growing peptide chain.

Transfer RNA (tRNA) is a polymer containing 75–95 nucleotides. tRNA transports a specific amino acid, as needed, to the site of protein synthesis.

Ribosomal RNA (rRNA), which contains 100–3100 nucleotides, forms part of the ribosomes, RNA-protein granules that provide the site of protein biosynthesis.

A. Structures of RNA

RNA polymers have the same general structure as DNA—a phosphate-sugar backbone with a base bonded to each monosaccharide. In ribonucleic acids, the monosaccharide is D-ribose instead of 2-deoxy-D-ribose, the monosaccharide of DNA. The four bases in RNA are two purines (adenine and guanine) and two pyrimidines (cytosine and uracil). The abbreviations for these four bases are A, G, C, and U, respectively. The first three bases are the same as those found in DNA. Uracil, found only in RNA, corresponds to the thymine found in DNA. Uracil has the same hydrogen-bonding characteristics as thymine and is also paired with adenine.

β-D-Ribofuranose   Uracil

A Portion of an RNA Chain:

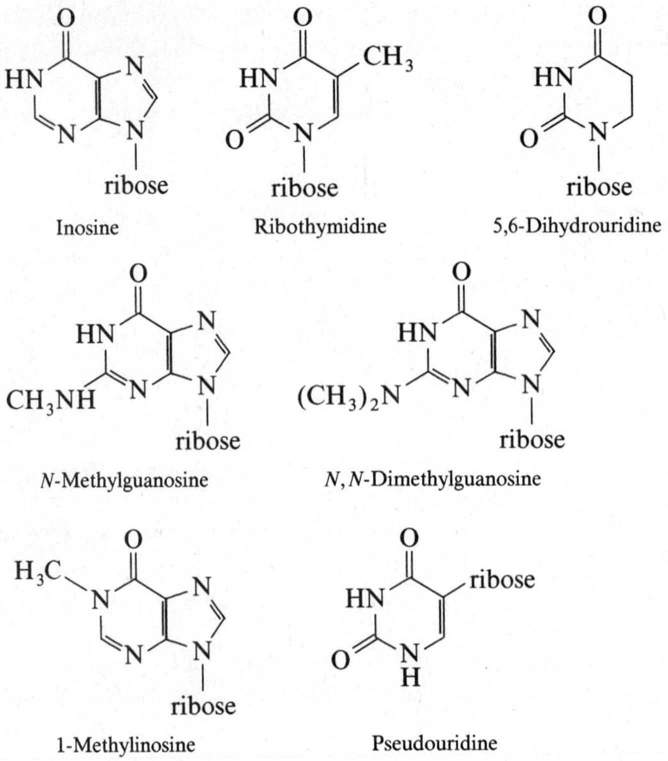

Inosine   Ribothymidine   5,6-Dihydrouridine

N-Methylguanosine   N,N-Dimethylguanosine

1-Methylinosine   Pseudouridine

Figure 19.7 Some minor nucleosides found in ribonucleic acids.

Unlike DNA, different RNA molecules, especially tRNA, have numerous minor base components. The structures of some of these minor bases are listed in Figure 19.7.

PROBLEM 19.3. Draw formulas for uracil and adenine to show the hydrogen bonds between them when they are paired.

Although the primary structures of DNA and RNA are similar, their chain lengths and secondary structures are not. *RNA chains are much shorter than DNA chains*—75–3100 nucleotides in RNA compared to millions or more in DNA. *RNA chains do not form double helices as do DNA chains, but exist as single strands.* Each type of RNA, once synthesized, assumes its own unique secondary structure. For example, one type of tRNA molecule assumes the secondary structure shown in Figure 19.9 (Section 19.5B).

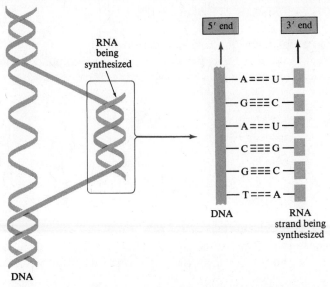

Figure 19.8 Transcription of the genetic code: the biosynthesis of an mRNA molecule from a single gene of the long DNA chain. Complementary nucleotides become aligned and polymerized.

B. Biosynthesis of RNA (Transcription)

Messenger RNA is biosynthesized using DNA as its template in a process called **transcription of the genetic code**. One mRNA molecule is specific for the biosynthesis of one type of protein molecule and thus represents the transcription of a single gene from the DNA strand.

In transcription, only a portion of the DNA double helix (the portion corresponding to one gene) unwinds, as shown in Figure 19.8. As in replication, nucleotides (ribonucleotides in transcription) are incorporated by complementary base pairings. The enzyme *RNA polymerase* catalyzes the polymerization of the ribonucleotides into the new RNA molecule. The result of transcription is an mRNA molecule that is exactly complementary to one gene of the DNA parent.

The newly synthesized RNA molecule does not remain hydrogen bonded to the DNA, but is released and leaves the cell nucleus. The DNA then returns to its stable double-helical shape.

PROBLEM 19.4. If a base sequence in DNA is T-G-A-C-G-T, what is the sequence of bases in the complementary RNA?

19.5 PROTEIN BIOSYNTHESIS (TRANSLATION)

The function of mRNA is to provide the information (as an array of bases) at the site of protein synthesis so that the amino acids are incorporated into the peptide chain in the correct sequence. This biosynthetic process is called **translation of the genetic code**.

$$\text{DNA} \xrightarrow[\text{nucleotides}]{\text{transcription}} \text{mRNA} \xrightarrow[\text{tRNA}]{\text{translation}} \text{peptide} \quad \text{or} \quad \text{protein}$$

amino acids

A. The Genetic Code

When mRNA leaves the nucleus of the cell, it becomes associated with ribosomes, which are composed of approximately 60% rRNA and 40% protein. The ribosomes attach themselves to the 5′ end of the mRNA and form the actual site of protein synthesis.

The sequence of bases in the mRNA molecule determines the sequence of amino acids to be inserted into a protein molecule. Once the mRNA molecule is bonded to ribosomes, the bases are read *three at a time* as the ribosomes progress toward the 3′ end of the mRNA molecule. Each set of three bases is called a **codon** and signals the incorporation of a single amino acid into the growing peptide chain.

Bases are read three at a time

U-G-U – C-U-A – A-U-A

To illustrate this decoding, let us consider the following hypothetical base sequence as it might be encountered in an mRNA chain:

– A-C-U – G-C-A – U-U-U –

The amino acid sequence that would be incorporated into the growing peptide chains would be threonine (signaled by the ACU codon)–alanine (from the GCA codon)–phenylalanine (from the UUU codon).

The codons for the amino acids, which are the same in all known life forms, have been determined experimentally. They are listed in Table 19.1.

In Table 19.1, note that more than one codon can signal a particular amino acid to be incorporated into a protein. In addition, some codons serve special functions. For example, the codon AUG serves two functions: (1) as an initiator codon signaling for the start of synthesis of a peptide, and (2) for the incorporation of methionine into the growing chain of a peptide. Other special-purpose codons are UAA, UAG, and UGA, all of which signal STOP. When the ribosomal synthesis site encounters one of these stop codons, the peptide chain is released and assumes its secondary and tertiary structures.

TABLE 19.1 CODONS FOR THE AMINO ACIDS

Amino acid	Codons[a]
Phenylalanine (Phe)	UUU, UUC
Leucine (Leu)	UUA, UUG
Serine (Ser)	UCU, UCC, UCA, UCG, AGU, AGC
Tyrosine (Tyr)	UAU, UAC
Cysteine (Cys)	UGU, UGC
Tryptophan (Trp)	UGG
Leucine (Leu)	CUU, CUC, CUA, CUG
Proline (Pro)	CCU, CCC, CCA, CCG
Histidine (His)	CAU, CAC
Glutamine (Gln)	CAA, CAG
Arginine (Arg)	CGU, CGC, CGA, CGG, AGA, AGG
Lysine (Lys)	AAA, AAG
Asparagine (Asn)	AAU, AAC
Isoleucine (Ile)	AUU, AUC, AUA
Methionine (Met) or N-formylmethionine (fMet)	AUG
Threonine (Thr)	ACU, ACC, ACA, ACG
Valine (Val)	GUU, GUC, GUA, GUG
Alanine (Ala)	GCU, GCC, GCA, GCG
Aspartic acid (Asp)	GAU, GAC
Glutamic acid (Glu)	GAA, GAG
Glycine (Gly)	GGU, GGC, GGA, GGG

[a] When preceded by an initiator region, the codon AUG signals: "Start a new peptide molecule beginning with N-formylmethionine, or fMet." The codons UAA, UAG, and UGA signal termination of the synthesis.

PROBLEM 19.5.   If an mRNA molecule carries the following codons, what is the amino acid sequence in the protein?

$$\{\!-CAC-AUC-UAC-GGU-\!\}$$

PROBLEM 19.6.   Write letters for all possible base sequences that would produce the following dipeptide: Glu-Arg.

B. Transport of the Amino Acids

The amino acids are carried to the mRNA-ribosome association by tRNA molecules. An organism contains many different types of tRNA—at least one for each amino acid.

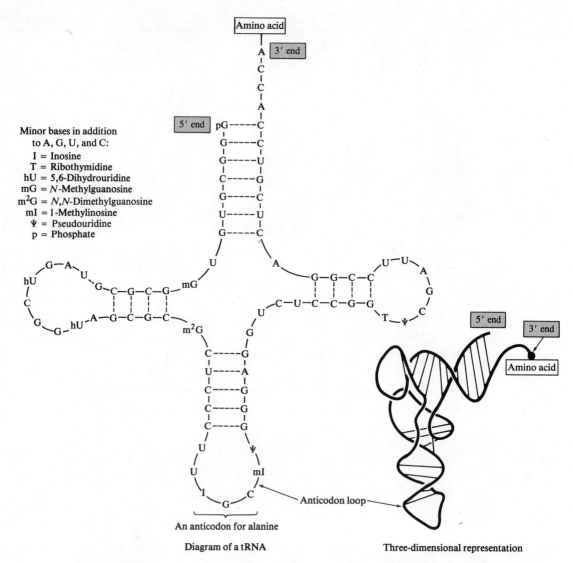

Figure 19.9 Structure of a yeast tRNA for the amino acid alanine. (See Figure 19.7 for the structures of the minor bases.) [Three-dimensional representation adapted from S. H. Kim and colleagues, *Science*, 185:436 (1974).]

Figure 19.9 shows a model of a tRNA molecule. A key feature of the tRNA structure is its **anticodon**, a series of three bases that is complementary to a particular codon. For example, if mRNA contains the codon CUA (for the amino acid leucine), the tRNA that carries leucine would contain the anticodon GAU. The anticodon allows a tRNA to recognize its rightful location on the mRNA molecule.

The tRNA molecule transports its amino acid as an ester formed with a hydroxyl group on the last ribose unit of its chain.

3' end of tRNA

Amino acid unit

C. Steps in Protein Biosynthesis

Figure 19.10 shows the overall process of protein biosynthesis. The three principal steps in this process are *initiation*, *chain elongation*, and *termination*.

Initiation   In the presence of certain protein initiator factors, mRNA becomes bound to a ribosome by the first several codons of the mRNA's 5' end. In bacterial cells, the START codon (AUG) calls for *N*-formylmethionine (fMet) as the first amino acid of the peptide to be synthesized. Depending on the protein's structure, the formyl group or even the methionine may be hydrolyzed from the final protein. However, the majority of proteins in a bacterium like *Escherichia coli*, a well-studied laboratory subject, have methionine as their N-terminal amino acid.

Formyl group

N-Formylmethionine

The codon AUG for "start with *N*-formylmethionine" is the same as the codon for the incorporation of methionine itself in the interior of the peptide chain. However, the tRNA that brings the *N*-formylmethionine to start a peptide chain differs from the tRNA that brings methionine itself. Initiation factors on the mRNA chain also help differentiate *N*-formylmethionine from methionine.

Chain Elongation   (1) The tRNA holding the second amino acid becomes hydrogen bonded to the mRNA codon and positions its amino acid at the **amino acid binding site**, or **A site**, on the ribosome. (2) A peptide bond is formed between the *N*-formylmethionine and the incoming amino acid. (3) The dipeptide is translocated to the **peptide binding site**, or **P site**, of the ribosome. (4) The ribosome moves along the mRNA chain so that the next tRNA with its amino acid can move into the A site, and the process is repeated for the next amino acid.

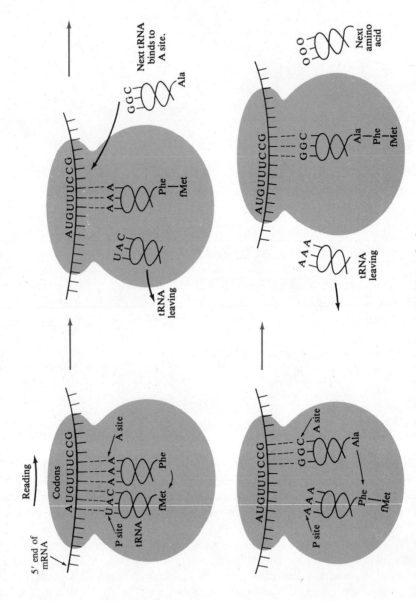

Figure 19.10 Diagram of the principal steps in protein biosynthesis.

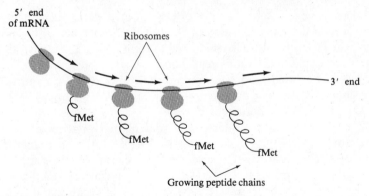

Figure 19.11 The ribosomes and growing peptides move along the mRNA molecule in assembly-line fashion.

Note that peptide synthesis starts at the 5′ end of the mRNA with the N-terminal amino acid of the peptide and proceeds to the 3′ end of the mRNA and the C-terminal amino acid of the peptide.

Termination The peptide is complete when the ribosome encounters one of the STOP codons. (These codons terminate the sequence because no tRNA molecules correspond to them.) Protein release factors catalyze the hydrolysis of the peptide so that it and the ribosome are released from each other and from the mRNA.

One mRNA molecule directs the synthesis of one specific protein, but it can direct the synthesis of many molecules of this protein. As one ribosome, holding a growing peptide, moves along the mRNA chain, other ribosomes become successively attached to the 5′ end of the mRNA to follow along behind the first tRNA. Thus, several peptide molecules can be synthesized one after the other in assembly-line fashion. Figure 19.11 illustrates this assembly-line biosynthetic process.

19.6 VARIATIONS IN THE CHEMISTRY OF HEREDITY

Hereditary diseases are those in which defective DNA, DNA containing an error in the base sequence, is passed from one generation to the next. The errors in the base sequence are caused by mutations, which may be spontaneous or induced (by exposure to radiation or mutagenic chemicals). For example, in sickle-cell anemia, a codon error signals valine instead of glutamic acid in the amino acid sequence of one of the hemoglobin proteins.

While almost all mutations are lethal, the immune system depends in part on mutation of antibodies to produce immunity to disease-causing organisms so that the body can fight new invaders.

Bacterial DNA: Not all DNA is the linear DNA of the higher plants and animals. Prokaryotic cells (bacteria and blue-green algae, for example) have a single chromosome that contains information needed to code for some 3000 proteins. In addition, some bacteria contain bits of DNA, called *plasmids*, containing only a few genes, outside the nuclear region of the cell. In both types of bacterial DNA, the ends of the DNA double helix are joined to form a double-stranded ring. Both types of DNA are capable of normal replication.

Viruses are small particles composed of either a DNA molecule or an RNA molecule surrounded by a protein coating. Viral nucleic acids may be single-stranded, double-stranded, linear, or circular.

A DNA virus is not capable of reproducing itself, but forces other cells (plant, animal, or bacterial) to replicate its viral DNA. A higher animal's DNA can replicate itself to form new DNA or the DNA can direct the synthesis of RNA. In these higher life forms, RNA *cannot* direct the synthesis of the animal's DNA. However, some RNA viruses are capable of **reverse transcription**—the viral RNA forcing the host cell to synthesize viral DNA.

Genetic engineering, frequently in the news today, is the artificial splicing or other manipulation of DNA. By genetic engineering, scientists can force common bacteria to synthesize desired human proteins, such as insulin or human growth factor, which may be expensive or in very short supply when obtained by ordinary means. Genetic engineering is discussed in Point of Interest 19.

SUMMARY

DNA (deoxyribonucleic acid) is the carrier of the genetic code in cells. These compounds are very high-formula-weight polymers of deoxyribose linked by phosphate groups. Each monosaccharide unit is also bonded to a **base**, which is a nitrogen heterocyclic compound. In the chromosomes, two strands of DNA are coiled about each other and held together by hydrogen bonds between specific base pairs—adenine to thymine and cytosine to guanosine.

DNA can be hydrolyzed to **nucleotides** (base-sugar-phosphate). In turn, nucleotides can be hydrolyzed to **nucleosides** (base-sugar) and phosphate ions.

DNA undergoes **replication** by unwinding, with each strand dictating a complementary sequence of nucleotides in the opposing new strand. The result is that two identical helices are produced from one.

RNA (ribonucleic acid) is a polymer of ribose linked by phosphate groups. Each ribose unit is also bonded to a base: cytosine, guanosine, adenine, or uracil. Messenger RNA is biosynthesized from one gene on one strand of DNA by the polymerization of complementary ribonucleotides. The biosynthesis of mRNA is called **transcription**.

Messenger RNA chains move to the ribosomes at the site of protein biosynthesis. The ribosomes become associated with one end of the mRNA, from which they move along the chain reading the mRNA base sequence. Ribosomes read the bases three at a time, and each three-base sequence is called a **codon**. Each codon signals the incorporation of one specific amino acid into a growing peptide chain. The amino acid is brought to the site by tRNA, a relatively small nucleic acid that contains an **anticodon** complementary to the codon.

Protein biosynthesis is called **translation**. The steps in translation, called *initiation*, *elongation*, and *termination*, are directed by the codons and aided by the ribosomes, tRNA, enzymes, proteins, and other chemical species.

Mutations arise from errors in the base sequence in DNA, errors that lead to changes in the amino acid sequence of proteins.

STUDY PROBLEMS

19.7. Draw the structures of **(a)** 2-deoxy-β-D-ribose 3-phosphate and **(b)** 2-deoxy-β-D-ribose 5-phosphate.

19.8. Draw the structures of **(a)** pyrimidine and **(b)** thymine.

19.9. Draw the structure of a tautomer of thymine that shows the close relationship between this compound and pyrimidine. (For a discussion of tautomers, see Section 10.10.)

19.10. Draw the structures of **(a)** purine, **(b)** guanine, and **(c)** a tautomer of guanine that shows its relationship to purine.

19.11. What is the difference in structure between a nucleotide and a nucleoside?

19.12. Using adenine, β-D-ribose, and a phosphate group, draw the structures of **(a)** a nucleoside and **(b)** a nucleotide.

19.13. Write equations for the hydrolysis of deoxythymidine 5'-phosphate to **(a)** a nucleoside and **(b)** a monosaccharide.

19.14. Adenosine 5'-triphosphate (ATP) is a very important compound in many biological schemes. What is its structure?

19.15. **(a)** If the following base sequence occurs in a DNA chain, what is the base sequence in the complementary strand?

$$5' \longleftarrow \big\{ C-A-C-T-C-T-G-T-A \big\} \longrightarrow 3'$$

(b) What would be the base sequence in an mRNA chain that is complementary to the preceding DNA strand?

(c) If the sequence in the mRNA chain in **(b)** represents three codons, what amino acid sequence would be coded for? (Refer to Table 19.1 for the codon assignments.)

19.16. Define the following terms as they apply to nucleic acids:
(a) replication   **(b)** transcription   **(c)** translation

19.17. List the following compounds in order of increasing formula weight, lowest first:
(a) mRNA   **(b)** tRNA   **(c)** DNA   **(d)** rRNA

19.18. If the sequence of a portion of DNA is

$$5' \longleftarrow \big\{ AAT-AGG-CAA \big\} \longrightarrow 3'$$

(a) What is the complementary sequence in mRNA?
(b) What tripeptide would be formed if the codons were read from the 5' end to the 3' end in mRNA?
(c) What tripeptide would be formed if the codons were read from the 3' end to the 5' end in mRNA?
(d) Which tripeptide is actually coded for?

19.19. One type of genetic mutation is that a wrong base becomes incorporated into a DNA chain, and thus a wrong base is incorporated into mRNA. What tripeptides would be formed by the following sequence shown as **(a)** in mRNA and by the mutated sequence in **(b)**?

(a) $5'$ ⟵ ⎰ UUU–UAU–AGU ⎱ ⟶ $3'$

(b) $5'$ ⟵ ⎰ UUU–UAC–AGU ⎱ ⟶ $3'$

19.20. In another type of mutation, a nucleotide is omitted from the base sequence in DNA and thus in mRNA. If the first base U is omitted from the mRNA in part (a) of the preceding problem, what would the amino acid sequence be? (Indicate the tripeptide if you can.)

19.21. Which type of mutation would more likely be lethal, the type in which an erroneous base is incorporated into DNA or the type in which an entire nucleotide is deleted? Explain your answer.

19.22. Is the purine ring flat like the benzene ring or puckered like the cyclohexane ring? Explain.

19.23. Is the uracil ring flat or puckered? Explain.

19.24. The compound *theophylline*, which occurs in tea and is also synthesized industrially, is used as a cardiac stimulant, a diuretic (a stimulant to urine secretion), and a muscle relaxant. This compound has a purine ring structure with carbonyl groups at positions 2 and 6 and methyl substituents on the nitrogens at positions 1 and 3. What is the structure of theophylline?

19.25. The stimulant caffeine has the same structure as theophylline with an additional *N*-methyl group at position 7. What is the structure of caffeine?

POINT OF INTEREST 19

Genetic Engineering

Genetic engineering is the process of creating new DNA that does not occur in nature and then cloning, or replicating, it. For example, the human gene that codes for the hormone insulin has been spliced into plasmid DNA of *E. coli*. Reproduction of the *E. coli* also reproduces the insulin gene. The gene can then be isolated and used to synthesize human insulin. Genetically engineered insulin is marketed today in Europe and the United States. Previously, all insulin available to diabetics was animal insulin isolated from the pancreases of slaughtered livestock. Other successful genetically engineered products are human growth hormone, used to treat dwarfism in children and previously available only from the pituitary glands of cadavers, and tissue plasminogen activator (t-PA), used to dissolve artery-clogging blood clots in heart attack victims. In the future, we will undoubtedly see the production of other rare human hormones and the development of disease-resistant plants.

Genetic engineering was made possible by the discovery of two components of bacteria: (1) selective cutting enzymes called **restriction endonucleases** or simply **restriction enzymes** and (2) bits of bacterial DNA called **plasmids**, which are easier to work with than chromosomes because they are much smaller.

Restriction enzymes have been evolved by bacteria to destroy invading viruses and other foreign nucleic acids. These enzymes are capable of cleaving DNA by recognition of specific base sequences. The bacteria's own DNA does not succumb to the effects of the restriction enzymes because bacteria contain enzymes that modify its own DNA by methylating key positions.

For an example of the action of a restriction enzyme, consider an enzyme that catalyzes the hydrolysis of DNA after TAA in the base sequence CTTAAG. This enzyme would cleave the following double-stranded DNA at two slightly different positions.

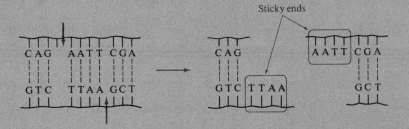

In this example, each of the DNA strands contains nucleotides (circled) that can hydrogen bond with other DNA strands. These ends, called "sticky ends," can be used to splice DNA molecules, such as human genes and bacterial plasmids, together.

In the laboratory, restriction enzymes are used first to excise the desired gene from human DNA or other DNA. The restriction enzyme is also used to cleave a plasmid, so that the sticky ends of the DNA and the plasmid are complementary. Then the desired gene can be spliced into the plasmid enzymatically to yield **recombinant DNA**, a DNA molecule formed from DNA pieces from different sources.

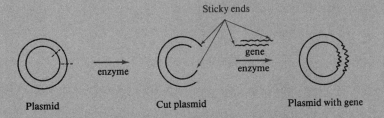

The plasmids are returned to the bacteria to be cloned, or replicated. Finally, the cloned DNA is isolated and used as a template in the synthesis of the desired protein.

Answers to In-Chapter Study Problems

CHAPTER 1

1.1. Electronegativity increases as we go up a column or from left to right. Therefore, in **(a)**, N > P > As. In **(b)**, As < Se < Br.

1.2. **(a)** $:\ddot{C}l: + \overset{\cdot}{C}a + :\ddot{C}l: \longrightarrow Ca^{2+} + 2:\ddot{C}l:^{-}$

(b) $Mg: + :\ddot{O}: \longrightarrow Mg^{2+} + :\ddot{O}:^{2-}$

1.3. **(a)** $H:\overset{..}{\underset{..}{O}}:\overset{..}{N}::\overset{..}{O}:$
$\qquad\qquad$ 1 $\quad$ 2

$\qquad$ Oxygen 1: $\quad 6 - \frac{1}{2}(4) - 4 = 0$

$\qquad$ Nitrogen : $\quad 5 - \frac{1}{2}(6) - 2 = 0$

$\qquad$ Oxygen 2: $\quad 6 - \frac{1}{2}(4) - 4 = 0$

(b) $H:\overset{\overset{\textstyle H}{..}}{\underset{..}{C}}::\underset{1}{N}::\underset{2}{\overset{..}{N}}:$

$\qquad$ Carbon: $\quad 4 - \frac{1}{2}(8) - 0 = 0$

$\qquad$ Nitrogen 1: $\quad 5 - \frac{1}{2}(8) - 0 = +1$

$\qquad$ Nitrogen 2: $\quad 5 - \frac{1}{2}(4) - 4 = -1$

(c) $H:\underset{\underset{\textstyle H}{..}}{C}:\overset{\overset{\textstyle :\ddot{O}:^{1}}{}}{\underset{..}{N}}\overset{}{\underset{\ddot{O}:}{}}_{2}$

$\qquad$ Carbon: $\quad 4 - \frac{1}{2}(8) - 0 = 0$

$\qquad$ Nitrogen: $\quad 5 - \frac{1}{2}(8) - 0 = +1$

Oxygen 1: $6 - \frac{1}{2}(2) - 6 = -1$

Oxygen 2: $6 - \frac{1}{2}(4) - 4 = 0$

1.4. **(a)** $H_2C{=}O$; **(b)** $Cl_2C{=}CHCH_3$; **(c)** $CH_3CH(CH_3)_2$, $H_3CCH(CH_3)_2$, or $(CH_3)_3CH$. In **(c)**, there are other variations.

1.5. **(a)** 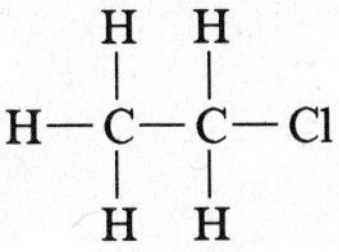 [structure] **(b)** 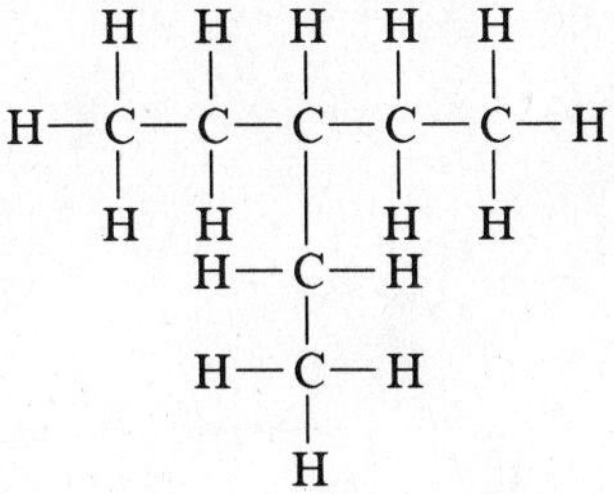 [structure]

(c) 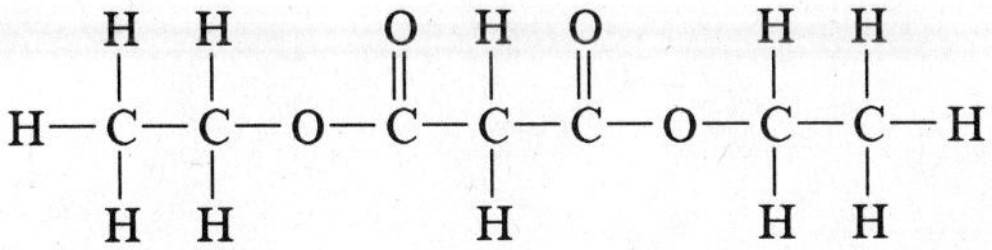 [structure]

1.6. **(a)** 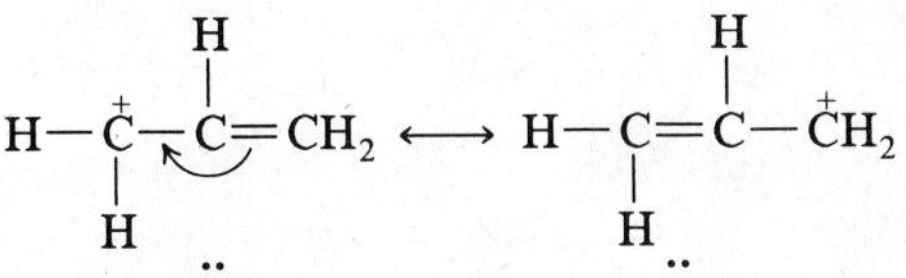 [structure]

(b) 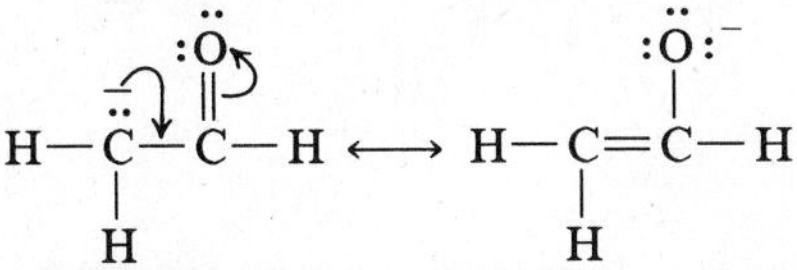 [structure]

(c) 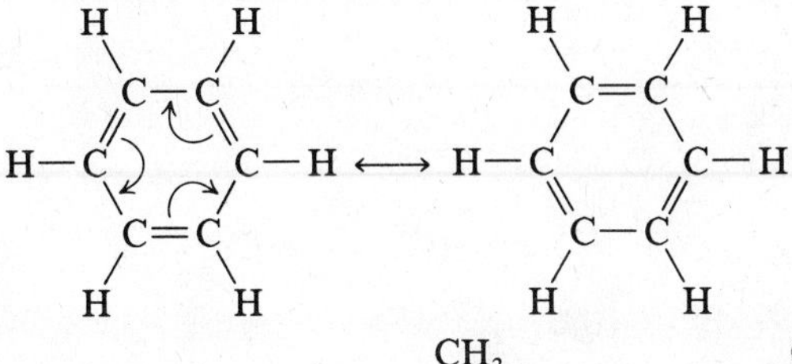 [structure]

1.7. $CH_3CH_2CH_2CH_2CH_3$ $CH_3CHCH_2CH_3$ CH_3CCH_3

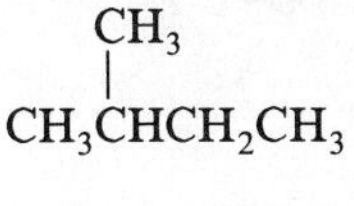

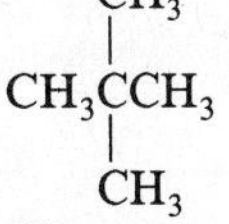

1.8. **(a)** 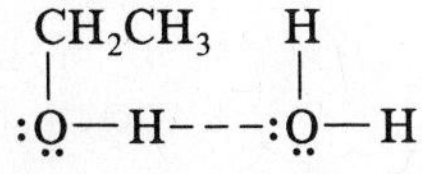 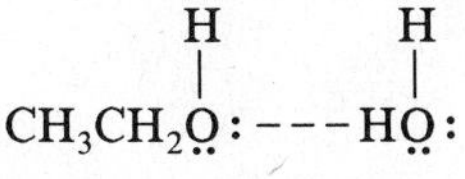 [structure]   [structure]

(b) 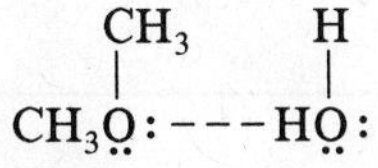 [structure]

(c) None

(d) 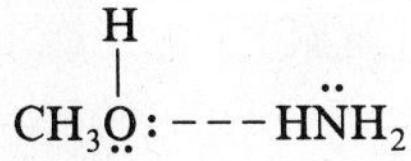 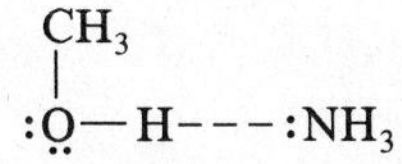 [structure]   [structure]

1.9. Compound **(a)** acts as a base (a proton acceptor) because it has an unshared pair of valence electrons that can be donated. Ion **(b)** also acts as a base. Compound **(c)** contains a carboxyl group, $-CO_2H$, and thus acts as an acid, or a proton donor. Ion **(d)** acts as an acid, or proton donor.

1.10. (a)

(b) $^-OC-CO^- + 2H_2O$ (with two C=O double bonds)

(c) $HOCH_3 + {}^-OH$

(d) $\underbrace{HO}CH_2CO^- + H_3O^+$ (with C=O double bond)

not part of a
carboxyl group

1.11. (a)

$FeBr_3 + :\ddot{Br}-\ddot{Br}: \rightleftharpoons {}^-FeBr_4 + {}^+\ddot{Br}:$

Lewis acid: Lewis base:
electron acceptor electron donor

(b)

$CH_3CH_2\ddot{O}H + H^+ \rightleftharpoons CH_3CH_2\overset{+}{\underset{\,}{\ddot{O}}}\!H$ (with H on the oxygen)

Lewis base: Lewis acid:
electron donor   electron acceptor

(c) $(CH_3)_3C^+ + :\ddot{C}l:^- \longrightarrow (CH_3)_3C-\ddot{C}l:$

Lewis acid   Lewis base

CHAPTER 2

2.1. **(a)** C_5H_{12}; C_nH_{2n+2} is the general formula. **(b)** C_8H_{16}; C_nH_{2n} is the general formula.

2.2. **(a)** 2-methylbutane; **(b)** 3-methylpentane; **(c)** 3-methylpentane; **(d)** 3-ethylpentane.

2.3. **(a)** 3,5-dimethylheptane; **(b)** 3,3-diethylpentane; **(c)** 2,3,4,6,7-pentamethyloctane; **(d)** 4,5-diisopropyloctane.

2.4. **(a)**

$$CH_3CH_2\underset{\underset{CH_3}{|}}{C}HCH_2\underset{\underset{CH_3}{|}}{C}HCH_3$$

(b)

$$CH_3CH_2CH_2CH_2\underset{\underset{C(CH_3)_3}{|}}{\overset{\overset{C(CH_3)_3}{|}}{C}}CH_2CH_2CH_2CH_3$$

(c) 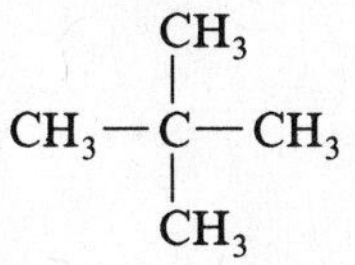
$$CH_3-\overset{\displaystyle CH_3}{\underset{\displaystyle CH_3}{C}}-CH_3$$

(d) 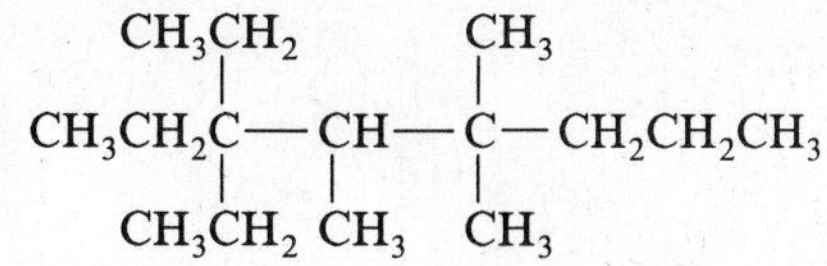
$$CH_3CH_2\overset{\displaystyle CH_3CH_2}{\underset{\displaystyle CH_3CH_2}{C}}-\overset{}{\underset{\displaystyle CH_3}{CH}}-\overset{\displaystyle CH_3}{\underset{\displaystyle CH_3}{C}}-CH_2CH_2CH_3$$

2.5.

Anti   Eclipsed   Staggered (gauche)

Eclipsed Br and CH₃   Staggered (gauche)   Eclipsed

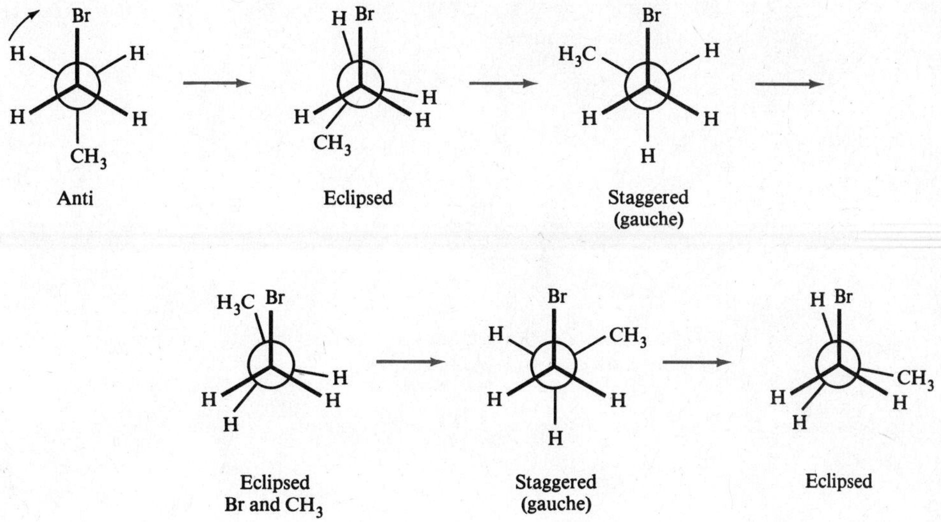

2.6. **(a)** $CH_2Cl_2 + \cdot Cl \longrightarrow \cdot CHCl_2 + HCl$

$\cdot CHCl_2 + Cl_2 \longrightarrow ClCHCl_2 + \cdot Cl$

(b) $CHCl_3 + \cdot Cl \longrightarrow \cdot CCl_3 + HCl$

$\cdot CCl_3 + Cl_2 \longrightarrow CCl_4 + \cdot Cl$

2.7. **(a)** 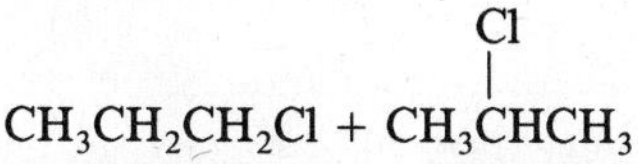 $CH_3CH_2CH_2Cl + CH_3\overset{\displaystyle Cl}{\overset{|}{C}}HCH_3$

(b) 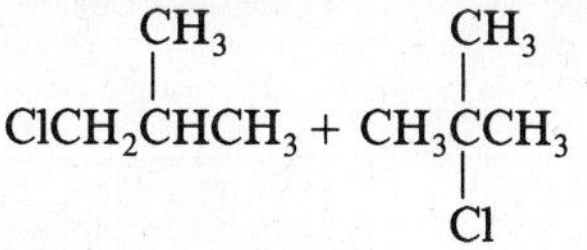 $ClCH_2\overset{\displaystyle CH_3}{\overset{|}{C}}HCH_3 + CH_3\overset{\displaystyle CH_3}{\underset{\displaystyle Cl}{C}}CH_3$

(c)

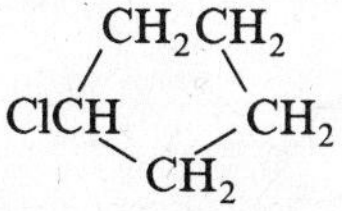

2.8.
$$Br_2 \longrightarrow 2:\ddot{Br}\cdot \qquad\qquad \Delta H = +46 \text{ kcal/mol}$$
$$CH_3CH_2-H \longrightarrow CH_3\dot{C}H_2 + H\cdot \qquad \Delta H = +98$$
$$CH_3\dot{C}H_2 + :\ddot{Br}\cdot \longrightarrow CH_3CH_2\ddot{Br}: \qquad \Delta H = -68$$
$$H\cdot + :\ddot{Br}\cdot \longrightarrow H\ddot{Br}: \qquad\qquad \underline{\Delta H = -87}$$
$$\Delta H = -11 \text{ kcal/mol}$$

2.9. **(a)** 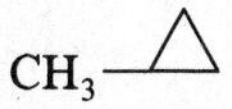   **(b)** 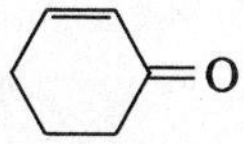   **(c)**

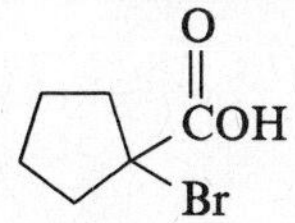

2.10. (a)

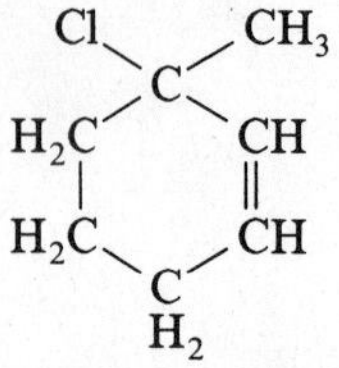

(b)

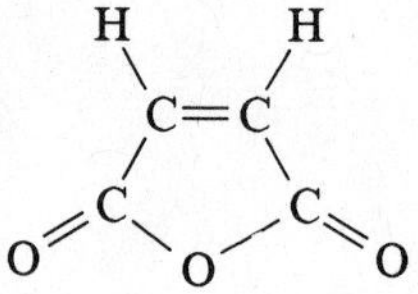

(c)

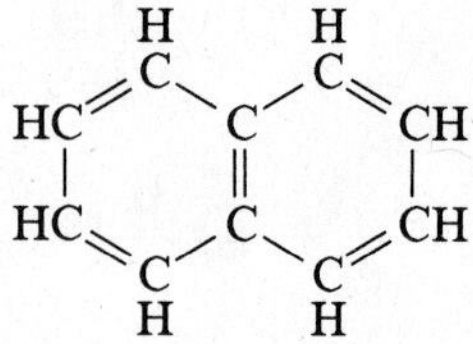

2.11. (a) *t*-butylcycloheptane;   **(b)** 1-cyclopropyl-2-ethylcyclooctane;   **(c)** 1,4-diisopropyl-2-methylcyclopentane.

2.12. (a)

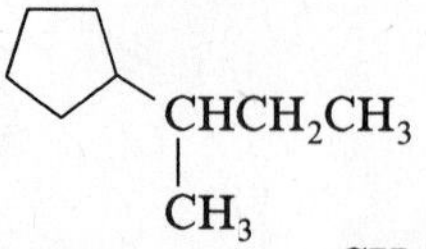

(b)

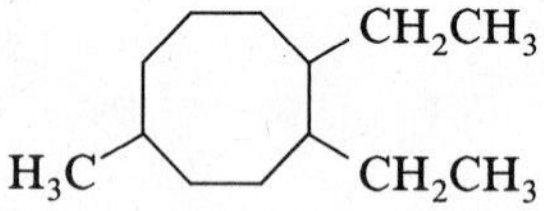

(c)

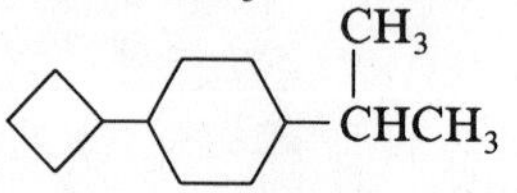

2.13. (a) *trans*-1,2-diethylcyclopropane;   **(b)** *cis*-1-ethyl-3-methylcyclopentane;   **(c)** *trans*-1-isobutyl-4-isopropylcyclohexane.

2.14. (a)

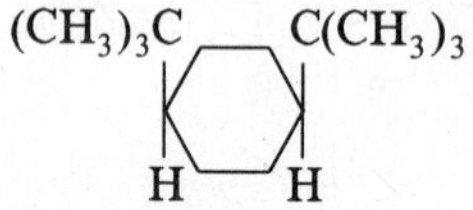

(b)

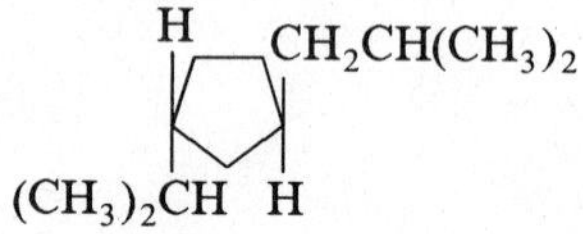

2.15. (a) equatorial;   **(b)** axial;   **(c)** equatorial.

2.16.

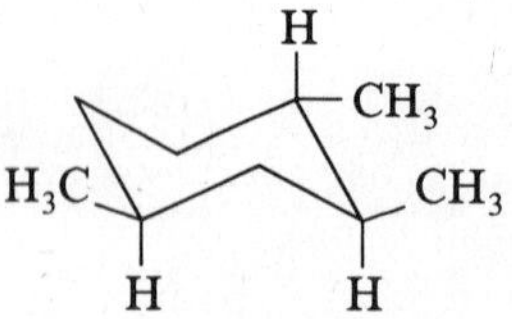

2.17. Formula **(b)** shows the predominant conformation because the larger group (the *t*-butyl group) is in the less hindered equatorial position.

CHAPTER 3

3.1. Molecular formula: C_4H_4; general formula: C_nH_{2n-4}

3.2. (a) C_5H_{10} has the general formula C_nH_{2n}; therefore, it contains one ring or one double bond.

(b) C_5H_8 has the general formula C_nH_{2n-2}; therefore, it contains two double bonds, one ring and one double bond, two rings, or one triple bond.

(c) C_6H_{10} also has the general formula C_nH_{2n-2}; therefore, this formula represents the same structural features as in **(b)**.

3.3. (a) cis:

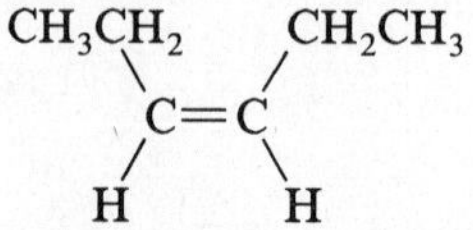

trans:

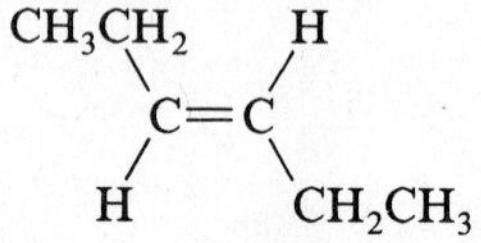

(b) cis:

$$\begin{array}{cc} Cl & CH_3 \\ \backslash & / \\ C=C \\ / & \backslash \\ H & H \end{array}$$

trans:

$$\begin{array}{cc} Cl & H \\ \backslash & / \\ C=C \\ / & \backslash \\ H & CH_3 \end{array}$$

(c) cis:

$$\begin{array}{cc} H & H \\ \backslash & / \\ C=C \\ / & \backslash \\ CH_2=CH & CH_3 \end{array}$$

trans:

$$\begin{array}{cc} H & CH_3 \\ \backslash & / \\ C=C \\ / & \backslash \\ CH_2=CH & H \end{array}$$

no cis-trans isomerism
for this double bond

(d) No cis-trans isomerism possible because one of the double-bond carbons is bonded to two hydrogens.

3.4. **(a)** 5-methyl-1-heptene;   **(b)** *cis*-3-methyl-4-octene;   **(c)** 1-isopropylcyclooctene or 1-isopropyl-1-cyclooctene;   **(d)** cyclohexylethyne.

3.5. **(a)**

$$\begin{array}{cc} CH_3 & CH_3 \\ | & | \\ CH_3CHCH_2C=CH_2 \end{array}$$

(b)

$$\begin{array}{cc} & CH_3 \\ & | \\ CH_3 & CH_2CHCH_3 \\ \backslash & / \\ C=C \\ / & \backslash \\ H & H \end{array}$$

(c)

CH₂CH₃ attached to ring, CH(CH₃)₂ attached to ring

3.6. **(a)** $CH_3CH=CHCH_3 \xrightarrow{H^+} CH_3\overset{+}{C}HCH_2CH_3 \xrightarrow{Br^-} CH_3\overset{\displaystyle Br}{\underset{|}{C}}HCH_2CH_3$

(b) cyclohexene $\xrightarrow{H^+}$ cyclohexyl cation $+ \xrightarrow{I^-}$ cyclohexyl–I

3.7. **(a)**

$$\begin{array}{c} CH_3 \\ | \\ CH_3C-CH_2CH_3 \\ | \\ Br \end{array}$$

(b) cyclopentane with Cl and CH₂CH₃

3.8. **(a)** $CH_3\overset{CH_3}{\underset{|}{C}}=CHCH_3 \xrightarrow{H^+} CH_3\overset{CH_3}{\underset{|}{\overset{+}{C}}}-CH_2CH_3,$

More stable
3° carbocation

not $CH_3CH-\overset{CH_3}{\underset{|}{\overset{+}{C}}}HCH_3 \xrightarrow{Br^-}$ product

Less stable
2° carbocation

(b)

More stable
3° carbocation

not

Less stable
2° carbocation

$\xrightarrow{Cl^-}$ product

3.9. (a) $CH_3CH_2\overset{\underset{\displaystyle OH}{|}}{C}HCH_3$ **(b)**

3.10. (a) $CH_2{=}CHCH_2CH_2CH_3 + H_2O \xrightarrow{H^+}$ product

(b)

or

or

or $\xrightarrow[H^+]{H_2O}$

3.11. (a) $CH_3CH{=}CHCH_2CH_3 + Br_2 \longrightarrow$ product

(b)

$+ Br_2 \xrightarrow{\text{trans addition}}$

3.12. $CH_3CH{=}CH_2 + Br_2 \longrightarrow$

$CH_3\overset{\overset{\displaystyle +Br}{\diagup \ \diagdown}}{CH}{-}CH_2$

$Br^- \diagup \qquad \diagdown \; Cl^- \text{ attacks position 2}$

$CH_3\overset{\underset{\displaystyle Br}{|}}{C}H\overset{\underset{\displaystyle Br}{|}}{C}H_2$ $CH_3\overset{\underset{\displaystyle Br}{|}}{C}H\overset{\underset{\displaystyle Cl}{|}}{C}H_2$

The dichloro compound is not formed because Br^+ is the attacking electrophile.

3.13. (a) $\xrightarrow{-Br^-} CH_3\overset{\overset{\displaystyle +Br}{\diagup \ \diagdown}}{CH}{-}C(CH_3)_2 \xrightarrow{+Br^-} CH_3\overset{\underset{\displaystyle Br}{|}}{C}H{-}\overset{\overset{\displaystyle Br}{|}}{\underset{\underset{\displaystyle Br}{|}}{C}}(CH_3)_2$

(b) $\xrightarrow{-Br^-} CH_3\overset{\overset{\displaystyle +Br}{\diagup \ \diagdown}}{CH}{-}C(CH_3)_2 \xrightarrow{H_2O} CH_3\overset{\underset{\displaystyle Br}{|}}{C}H{-}\overset{\overset{\displaystyle Br}{|}}{\underset{\underset{\displaystyle +OH_2}{|}}{C}}(CH_3)_2 \xrightarrow{-H^+} CH_3\overset{\underset{\displaystyle Br}{|}}{C}HC\overset{\underset{\displaystyle OH}{|}}{}(CH_3)_2$

(c) 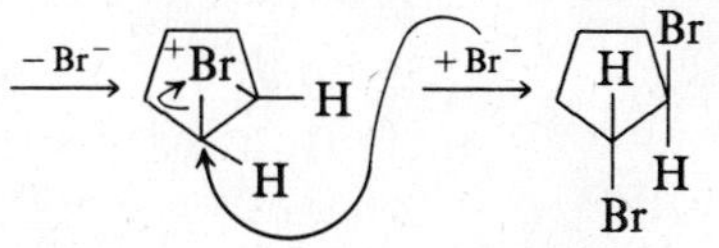 (trans)

(d) 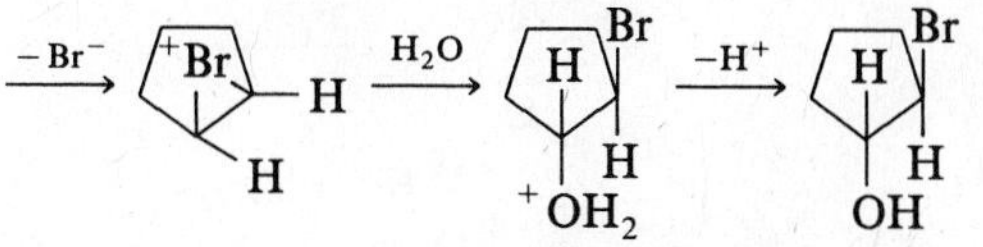 (trans)

3.14. (a) $CH_2{=}CHCH_2CH_3 \xrightarrow{H_2O,\ H^+} CH_3\overset{\displaystyle OH}{\overset{|}{C}}HCH_2CH_3$

(b) $CH_3CH_2CH{=}CH_2 \xrightarrow[(2)\ H_2O_2,\ ^-OH]{(1)\ BH_3} CH_3CH_2CH_2CH_2OH$

3.15. (a) gain of four oxygens: oxidation

(b) loss of six hydrogens: oxidation

(c) This reaction is neither an oxidation nor a reduction because the reactant and product differ only by a molecule of water. You might say that one carbon of the double bond has been oxidized and the other reduced; thus, the oxidation and reduction cancel each other for the entire molecule.

(d) gain of two hydrogens: reduction

3.16. (a) $CH_3CH_2CH_2CH_2CH_2CH_3$ **(b)** 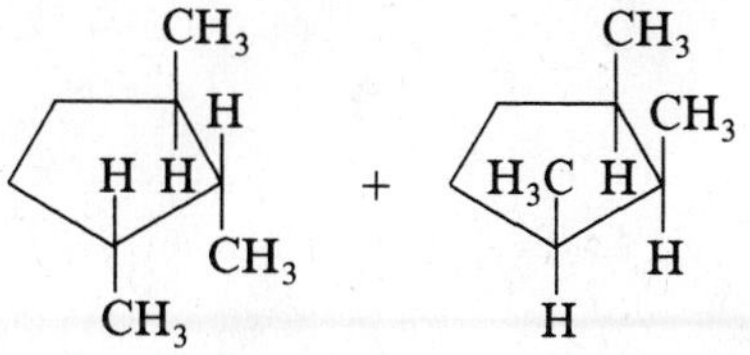

3.17. (c) (b) (a) ⟶
Increasing substitution; increasing stability

3.18. (a) $CH_3\overset{\displaystyle OH}{\overset{|}{C}}H{-}\overset{\displaystyle OH}{\overset{|}{C}}H_2$ **(b)** 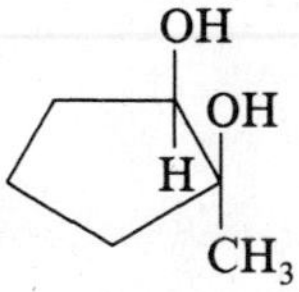

(c) 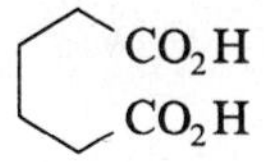 or $HO_2CCH_2CH_2CH_2CH_2CO_2H$

(d) 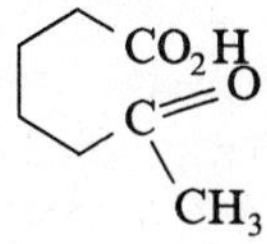 or $CH_3\overset{\displaystyle O}{\overset{\|}{C}}CH_2CH_2CH_2CH_2CO_2H$

3.19. (a) 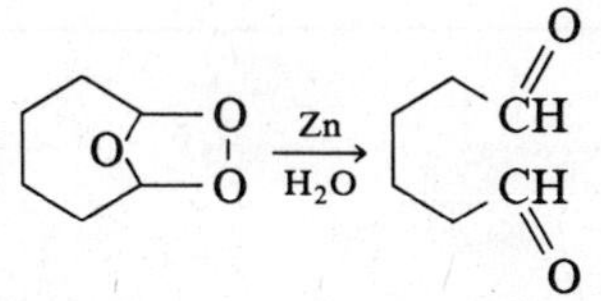 or $H\overset{\displaystyle O}{\overset{\|}{C}}CH_2CH_2CH_2CH_2\overset{\displaystyle O}{\overset{\|}{C}}H$

(b) 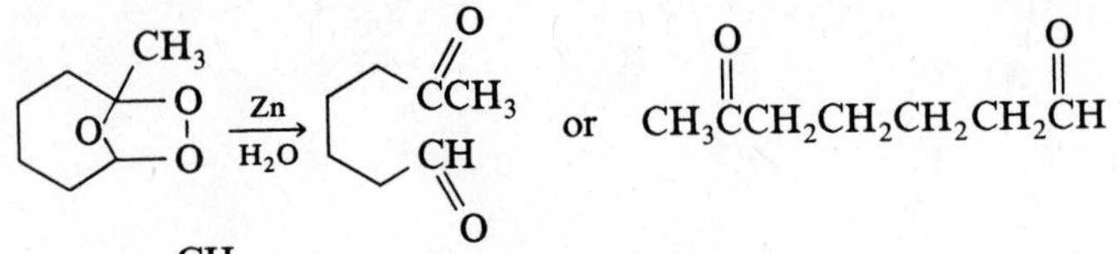 $\xrightarrow[H_2O]{Zn}$ or $CH_3\overset{O}{\overset{\|}{C}}CH_2CH_2CH_2CH_2\overset{O}{\overset{\|}{C}}H$

(c) $\xrightarrow[H_2O]{Zn}$ $+\ H\overset{O}{\overset{\|}{C}}H$

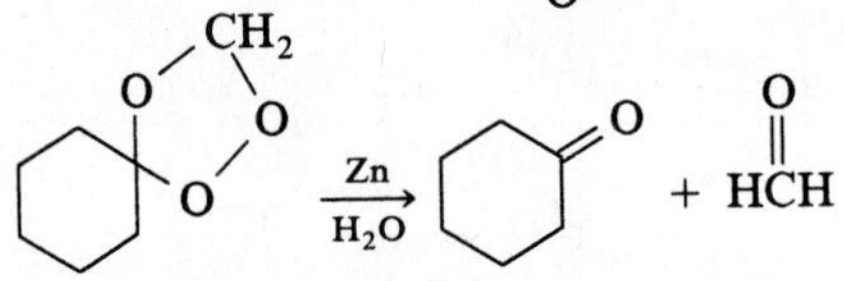

3.20. (a) 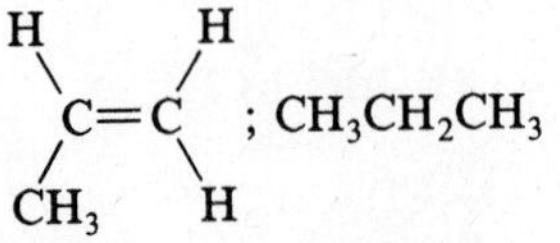 ; $CH_3CH_2CH_3$

(b) 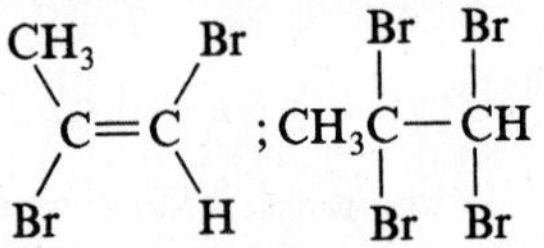 ; $CH_3\overset{Br\ Br}{\underset{Br\ Br}{\overset{|\ \ |}{\underset{|\ \ |}{C-CH}}}}$

trans Br's

(c) 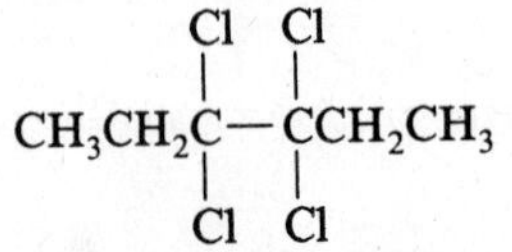 $CH_3CH_2\overset{Cl\ \ Cl}{\underset{Cl\ \ Cl}{\overset{|\ \ \ |}{\underset{|\ \ \ |}{C-CCH_2CH_3}}}}$

3.21. (a) 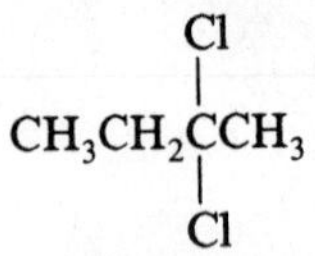 $CH_3CH_2\overset{Cl}{\underset{Cl}{\overset{|}{\underset{|}{C}CH_3}}}$ **(b)** 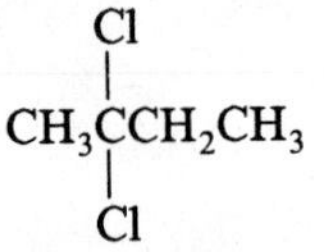 $CH_3\overset{Cl}{\underset{Cl}{\overset{|}{\underset{|}{C}CH_2CH_3}}}$

3.22. (a) 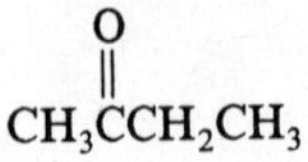 $CH_3\overset{O}{\overset{\|}{C}}CH_2CH_3$ **(b)**

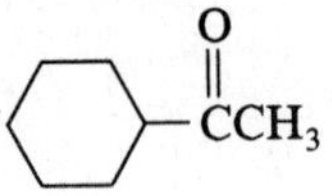

3.23. (a) $CH_3CH_2C{\equiv}C\mathord{:}^-\ Na^+ + NH_3$
(b)–(d) no reaction

CHAPTER 4

4.1. (a) 1,3,5-heptatriene; **(b)** 1,3,5,7-cyclooctatetraene; **(c)** 1-methyl-1,4-cyclohexadiene.

4.2. (a) conjugated (double bonds from adjacent carbons); **(b)** isolated (double bonds *not* from adjacent carbons); **(c)** isolated; **(d)** and **(e)** conjugated.

4.3. (a) 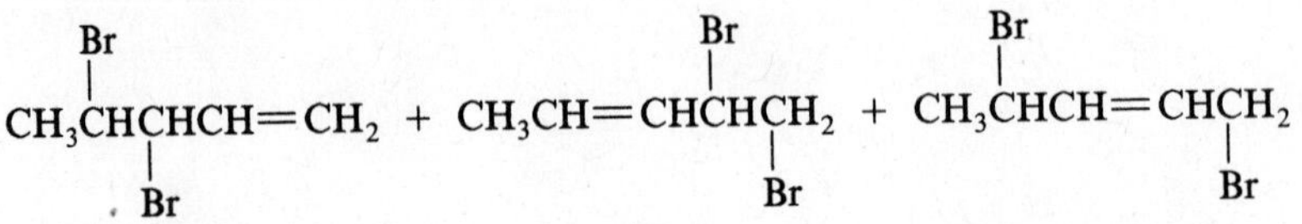 $CH_3\overset{Br}{\overset{|}{C}H}\underset{Br}{\overset{|}{C}H}CH=CH_2$ + $CH_3CH=CH\overset{Br}{\underset{Br}{\overset{|}{\underset{|}{C}H}}}CH_2$ + $CH_3\overset{Br}{\overset{|}{C}H}CH=CH\underset{Br}{\overset{|}{C}H_2}$

1,2 1,2 1,4

(b) 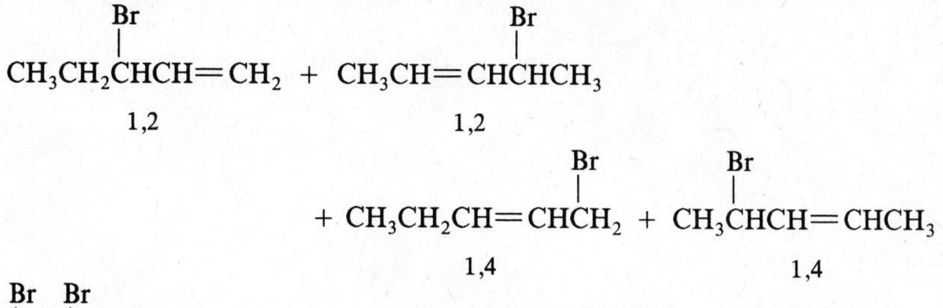 $\overset{\overset{\displaystyle Br}{|}}{CH_3CH_2CHCH}=CH_2$ + $CH_3CH=CH\overset{\overset{\displaystyle Br}{|}}{CHCH_3}$

     1,2   1,2

+ $CH_3CH_2CH=CH\overset{\overset{\displaystyle Br}{|}}{CH_2}$ + $CH_3\overset{\overset{\displaystyle Br}{|}}{CHCH}=CHCH_3$

          1,4   1,4

(c) 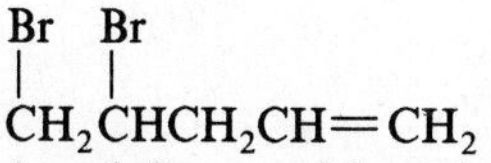 $\overset{\overset{\displaystyle Br\ \ Br}{|\ \ \ |}}{CH_2CHCH_2CH}=CH_2$ is the only product from the reaction of equimolar quantities of diene and bromine. The double bonds are not conjugated; therefore, no 1,4 addition is possible.

(d)

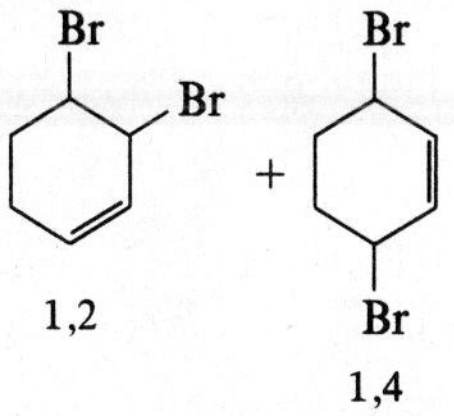

    1,2   1,4

(e) The 1,2- and 1,4-addition products are the same because both reactions proceed by way of an allylic cation intermediate.

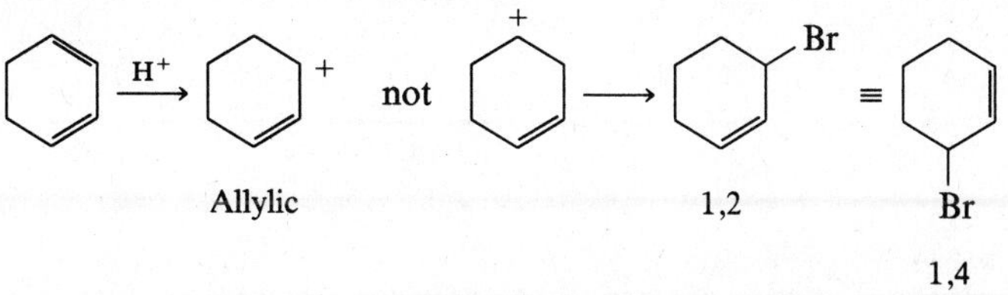

4.4. (a) $CH_3CH_2CH\overset{\frown}{=}CH\!-\!\overset{+}{C}HCH_3 \longleftrightarrow CH_3CH_2\overset{+}{C}H\!-\!CH=CHCH_3$

(b) $CH_2=CH\!-\!CH\overset{\frown}{=}CH\!-\!\overset{+}{C}HCH_3 \longleftrightarrow CH_2\overset{\frown}{=}CH\!-\!\overset{+}{C}H\!-\!CH=CHCH_3$

$\longleftrightarrow \overset{+}{C}H_2\!-\!CH=CH\!-\!CH=CHCH_3$

(c)

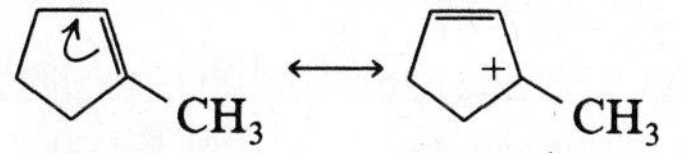

(d)

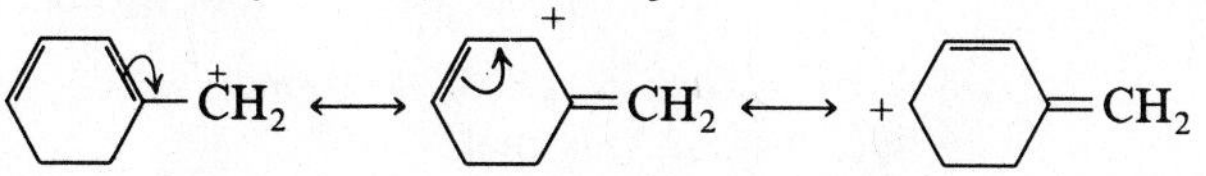

4.5. (a) 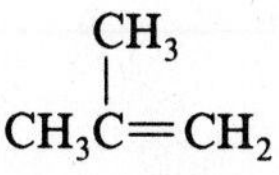 $\overset{\overset{\displaystyle CH_3}{|}}{CH_3C}=CH_2$ (b)

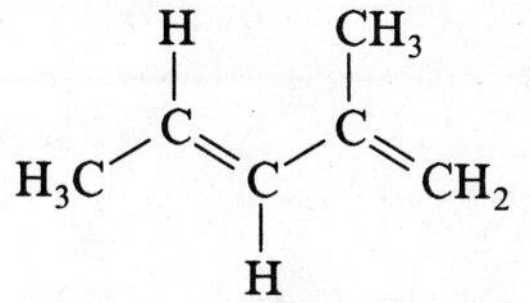

   (c)

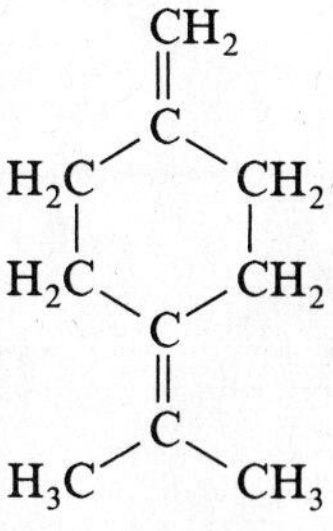

4.6. (a)

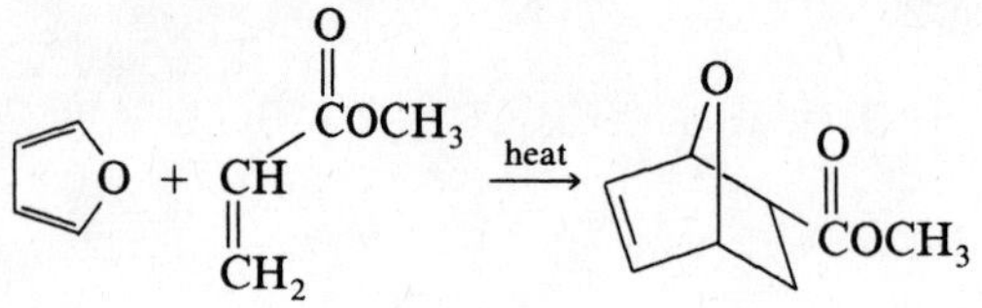

(b)

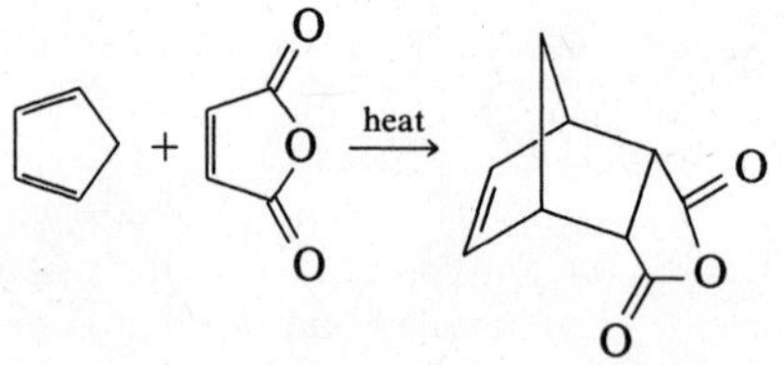

4.7. (a)

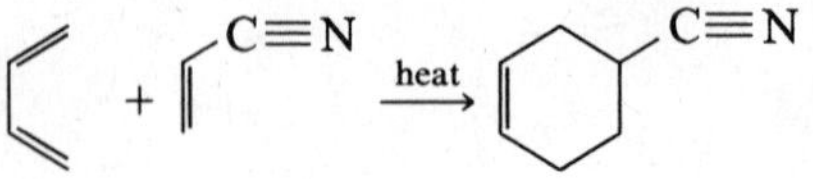

(b)

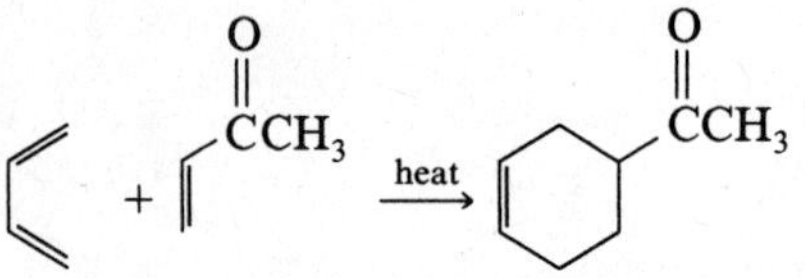

(c)

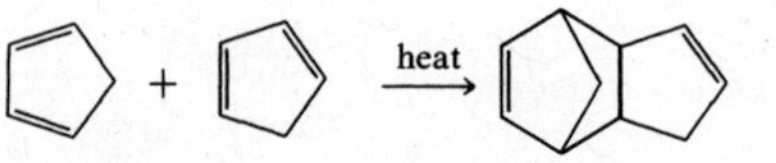

CHAPTER 5

5.1. (a)

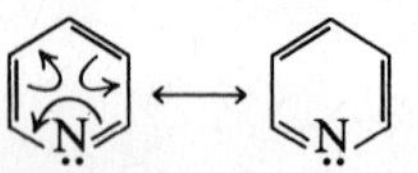

(b)

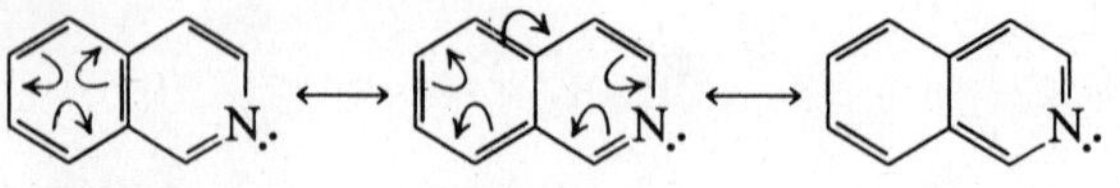

(c)

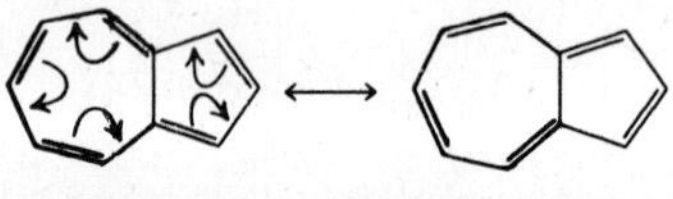

5.2. The heat liberated per double bond should be 28.6 kcal/mol. Anthracene contains seven double bonds; therefore, the calculated heat of hydrogenation is 200.2 kcal/mol.

Calculated ΔH: $7 \times 28.6 = 200.2$ kcal/mol
Actual ΔH: 116.2 kcal/mol
Resonance energy $= (200.2 - 116.2)$ kcal/mol $= 84.0$ kcal/mol

5.3. (a) *m*-methylbenzoic acid or 3-methylbenzoic acid; **(b)** *p*-dichlorobenzene or 1,4-dichlorobenzene; **(c)** *o*-nitrostyrene or 2-nitrostyrene.

5.4. (a) 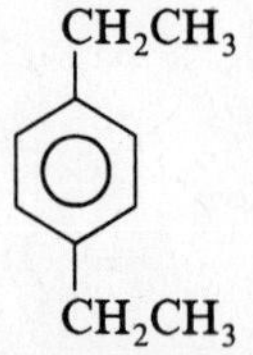 **(b)** 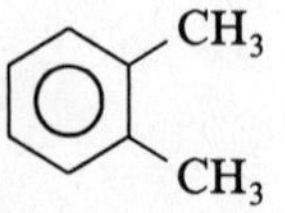 **(c)**

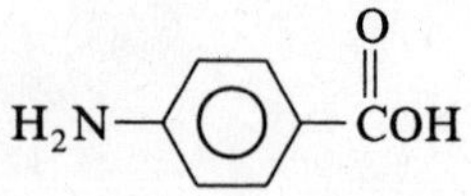

5.5. **(a)** 1,3,5-triiodobenzene; **(b)** 2,4,6-tribromoaniline; **(c)** 1-phenyldecane.

5.6. **(a)** **(b)**

(c) **(d)**

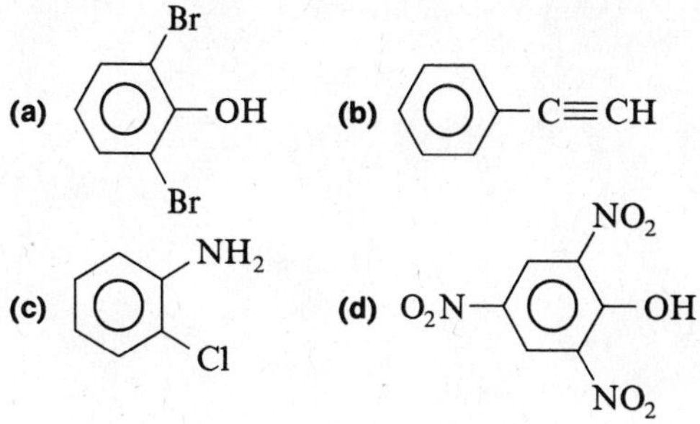

5.7.

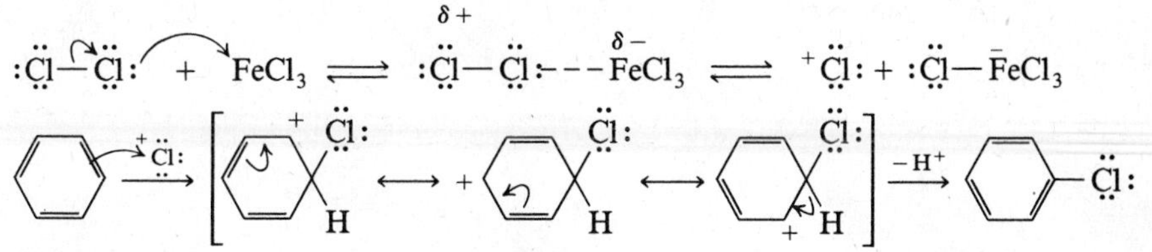

5.8.

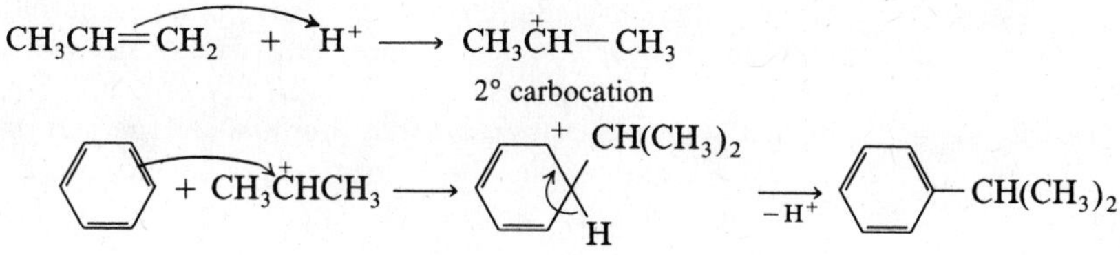

5.9. **(a)**

(b) **(c)**

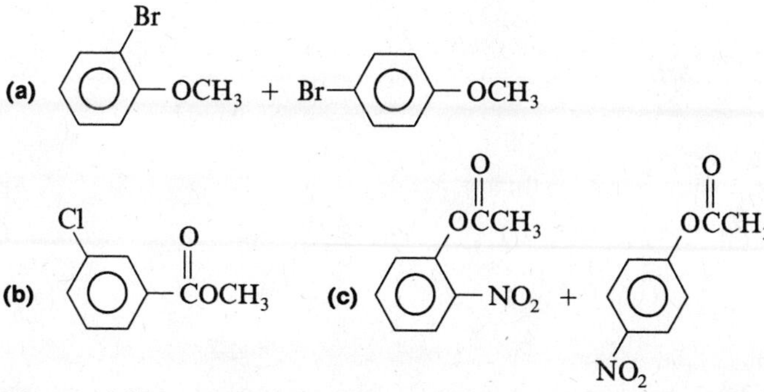

5.10. Many stabilizing resonance structures can be drawn for the intermediate.

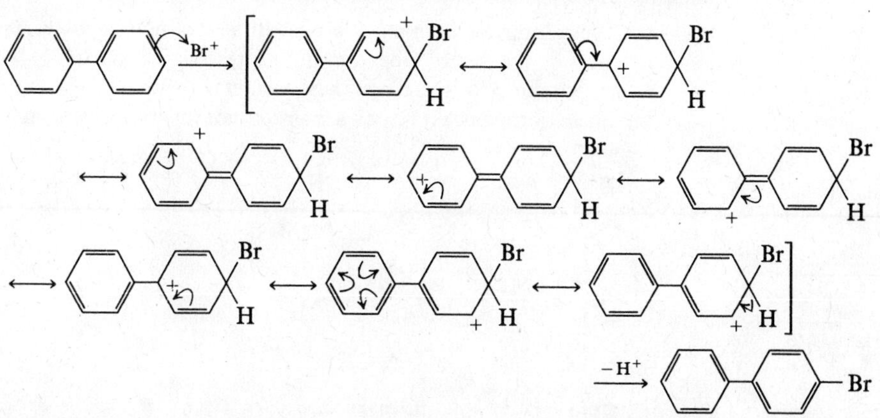

5.11. When aniline is treated with a strong acid, the amino group is protonated. The aromatic ring is then substituted with a *meta* director ($-\overset{+}{N}H_3$).

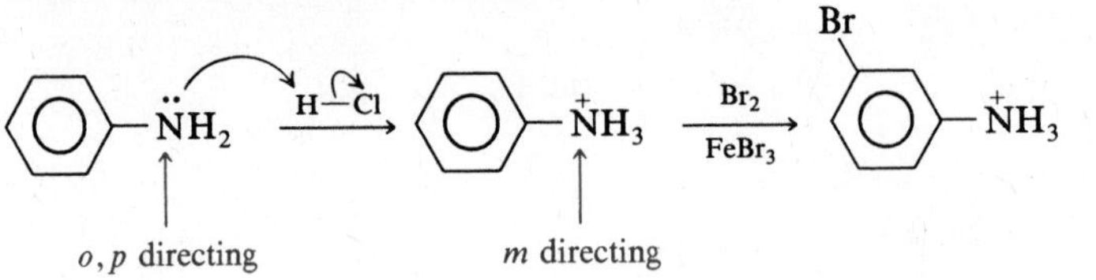

o,p directing   *m* directing

5.12. (a)

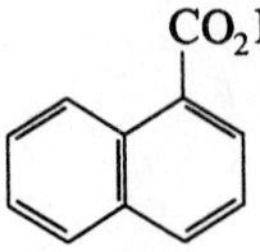

(b)

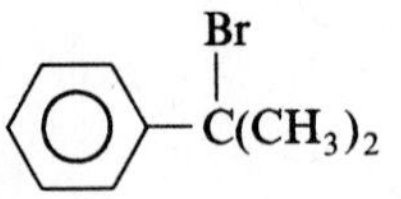

(c)

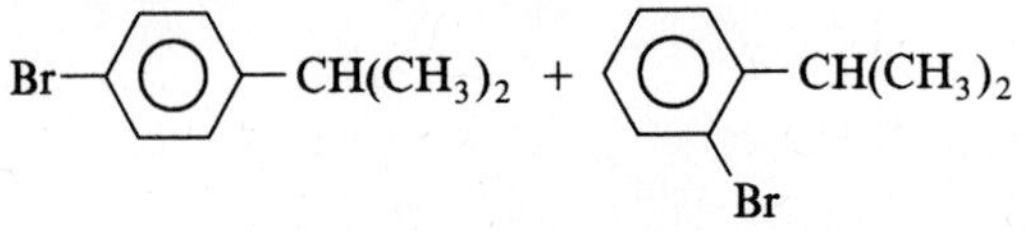

5.13. Nitrogens (a), (b), (d), and (e) are basic. Each has a pair of unshared valence electrons either in an sp^3-hybrid orbital [nitrogen (e)] or in an sp^2-hybrid orbital [nitrogens (a), (b), and (d)]. These electrons can be donated to a proton. Nitrogen (c) is not basic. Its "unshared" valence electrons are in a *p* orbital and are part of the aromatic pi cloud. These two *p* electrons allow both rings to be aromatic with a total of 10 pi electrons ($4n + 2$, where $n = 2$).

CHAPTER 6

6.1. (a)

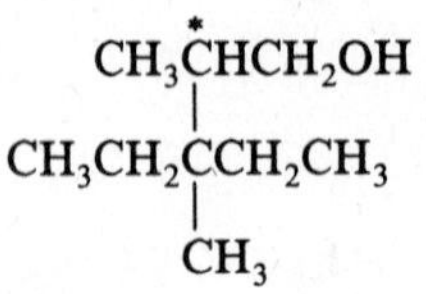

(b)

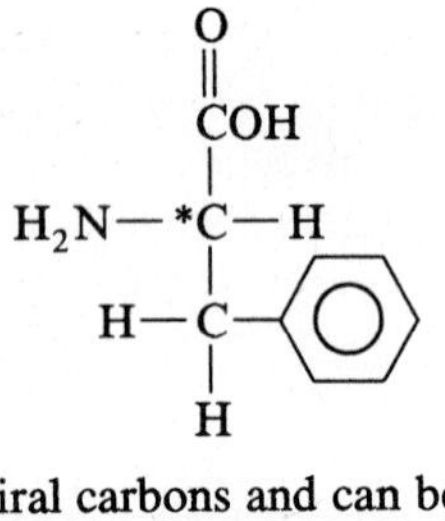

(c)

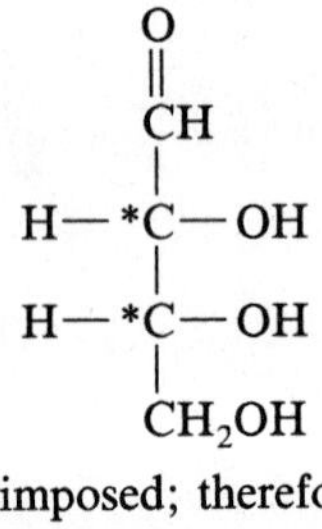

6.2. (a) These formulas contain no chiral carbons and can be superimposed; therefore, they represent the same compound.
   (b) These formulas each contain one chiral carbon, but the groups around this carbon are not the same. One formula represents an aldehyde, while one represents a carboxylic acid; therefore, these formulas represent different compounds.
   (c) These formulas represent a pair of nonsuperimposable mirror images, or enantiomers.

6.3.

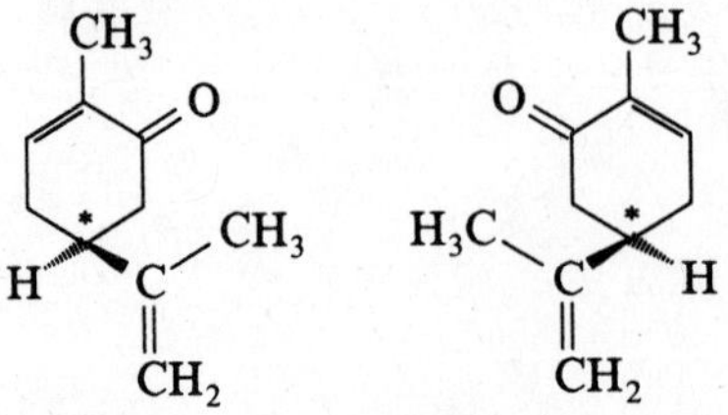

Carvone   Its enantiomer

6.4. Because the nose can detect differences in odor between enantiomers, the receptor sites must be chiral.

6.5. **(a)** $\alpha = +23°$

$$l = 5.00 \text{ cm} \times \frac{1 \text{ dm}}{10 \text{ cm}} = 0.500 \text{ dm}$$

$$c = 1.00 \text{ g}/5.00 \text{ mL} = 0.20 \text{ g/mL}$$

$$[\alpha]_D^{20} = \frac{+23°}{(0.500)(0.20)} = 230° \text{ (ethanol)}$$

(b) $\alpha = -3.5°$

$$l = 5.00 \text{ cm} \times \frac{1 \text{ dm}}{10 \text{ cm}} = 0.500 \text{ dm}$$

$$c = 0.500 \text{ g}/6.00 \text{ mL} = 0.0833 \text{ g/mL}$$

$$[\alpha]_D^{20} = \frac{-3.5°}{(0.500)(0.0833)} = -84° \text{ (ethanol)}$$

6.6.

By atomic number of the bonded atom

By atomic number of the next atom in line (H versus C)

(a)   (b)   (c)   (d)
_____→
Increasing priority

6.7.

(c) (b) (d) (a)
_____→
Increasing priority

(a) is the highest priority because the attaching atom has the highest atomic number.

(c) is the lowest priority (despite the presence of iodine atoms) because the atom bonded directly to the carbonyl carbon is C, not O.

6.8. **(a)** With the lowest priority group ($-CH_3$) in the rear, proceed from the highest priority group ($-OH$) to the next highest ($-CH_2CH_2CH_3$). The direction is counterclockwise; the compound is (S).

(b) (S);   **(c)** (R);   **(d)** (S).

6.9. **(a)** 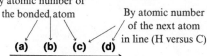 $\begin{array}{c} 1 \\ \diagdown \\ C=C \\ \diagup \\ 2 \end{array} \begin{array}{c} 1 \\ \diagup \\ \diagdown \\ 2 \end{array} (Z)$ **(b)** $\begin{array}{c} 1 \\ \diagdown \\ C=C \\ \diagup \\ 2 \end{array} \begin{array}{c} 2 \\ \diagup \\ \diagdown \\ 1 \end{array} (E)$

6.10. **(a)** diastereomers;   **(b)** enantiomers;   **(c)** enantiomers;   **(d)** enantiomers.

6.11. **(a)**

$$\begin{array}{c} CH_2OH \\ | \\ H{-}C{-}Br \\ | \\ Br{-}C{-}H \\ | \\ CH_2OH \end{array} \begin{array}{c} \\ \\ (2S) \\ (3S) \end{array}$$

$$\begin{array}{c} O \\ \| \\ CH \\ | \\ HO{-}C{-}H \\ | \\ CH_2 \\ | \\ H{-}C{-}OH \\ | \\ CH_2OH \end{array} \begin{array}{c} \\ \\ \\ (2S) \\ (4S) \end{array}$$

6.12. (a) achiral:

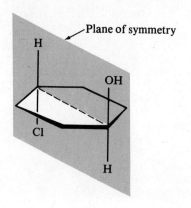

Plane of symmetry

(b) chiral, with two chiral carbons
(c) chiral, with two chiral carbons

6.13. (a) meso:

Plane of symmetry

HO—OH

H   H

(b) has no chiral carbons; **(c)** meso.

6.14.

$$\left.\begin{array}{l}(R)\text{-RNH}_2 \\ (S)\text{-RNH}_2\end{array}\right\} \text{ racemic mixture } + (+)\text{-RCO}_2\text{H}$$

$(R)\text{-R}\overset{+}{\text{N}}\text{H}_3 \; (+)\text{-RCO}_2^- \quad + \quad (S)\text{-R}\overset{+}{\text{N}}\text{H}_3 \; (+)\text{-RCO}_2^-$

mixture of diastereomeric salts

Separate by fractional
crystallization

$(R)\text{-R}\overset{+}{\text{N}}\text{H}_3 \; (+)\text{-RCO}_2^- \qquad\qquad (S)\text{-R}\overset{+}{\text{N}}\text{H}_3 \; (+)\text{-RCO}_2^-$

aqueous NaOH   aqueous NaOH

$(R)\text{-RNH}_2 + (+)\text{-RCO}_2^- \qquad\qquad (S)\text{-RNH}_2 + (+)\text{-RCO}_2^-$

Extract amine from   Extract amine from
aqueous solution.   aqueous solution.

$(R)\text{-RNH}_2 \qquad\qquad\qquad\qquad (S)\text{-RNH}_2$

CHAPTER 7

7.1. **(a)** benzylic **(b)** alkyl; **(c)** alkyl; **(d)** vinylic; **(e)** aryl; **(f)** allylic; **(g)** alkyl.

7.2. **(a)** 3-chloropentane;   **(b)** 3-chloro-1-butene;   **(c)** 1,2-dichloro-4-methylcyclohexane.

7.3. **(a)**
$$\overset{\text{Cl}}{\underset{\text{Cl}}{\text{HCCH}_2\text{CH}_2\text{CH}_2\text{CH}_2}}\overset{}{\underset{\text{CH}_3}{\text{CHCH}_3}}$$

(b) 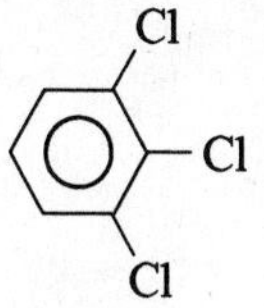   **(c)**

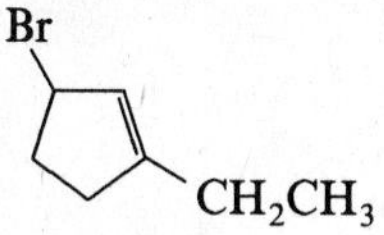

(d) 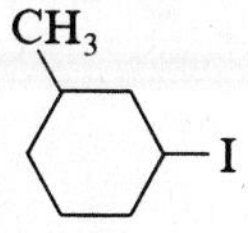   **(e)** $\text{CH}_3\overset{\text{Br}}{\text{CHCH}_3}$

7.4. **(a)** and **(b)** primary;   **(c)** secondary;   **(d)** tertiary.

7.5. **(a)** $\text{CH}_3\text{CH}=\text{CH}_2 + \text{HBr} \longrightarrow \text{CH}_3\overset{\text{Br}}{\text{CHCH}_3}$

(b) $\text{CH}_3\overset{\text{OH}}{\text{CHCH}_2\text{CH}_3} + \text{HI} \longrightarrow \text{CH}_3\overset{\text{I}}{\text{CHCH}_2\text{CH}_3}$

(c) $\text{CH}_3\text{CH}=\text{CHCH}_3 + \text{Cl}_2 \longrightarrow \text{CH}_3\overset{\text{Cl}}{\text{CH}}-\overset{\text{Cl}}{\text{CHCH}_3}$

(d) 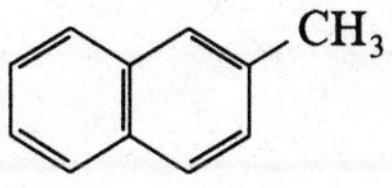 $+ \text{Br}_2 \xrightarrow{\text{light}}$ product

7.6. **(a)** $(\text{CH}_3)_2\text{CHCN} + \text{K}^+\ \text{I}^-$
(b) $\text{CH}_3\text{SCH}_2\text{CH}_3 + \text{Na}^+\ \text{I}^-$
(c) $(\text{CH}_3)_2\text{CHCH}_2\text{CH}_2\text{OCH}_2\text{CH}_3 + \text{Na}^+\ \text{Cl}^-$

(d) 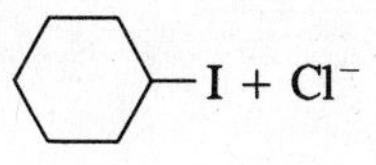 $-\text{I} + \text{Cl}^-$

7.7.

Transition state

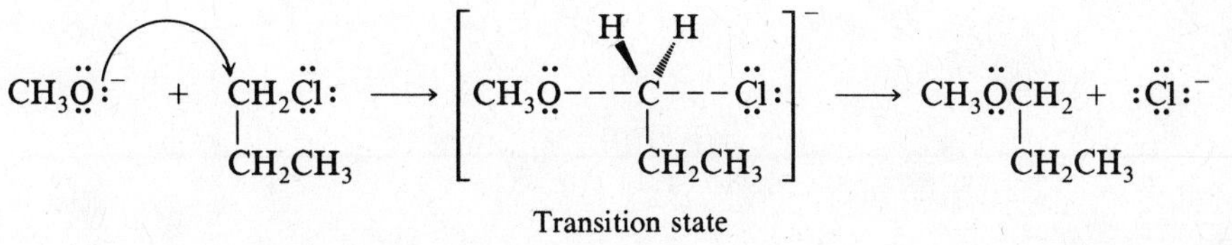

7.8. The rate of reaction would decrease by a factor of 10, or be one-tenth of the original rate.

7.9. $\underbrace{\textbf{(a)} \quad \textbf{(d)} \quad \textbf{(e)}}_{\text{No } S_N2} \ll \underset{2°}{\textbf{(c)}} < \underset{1°}{\textbf{(b)}} < \underset{\text{Methyl}}{\textbf{(f)}}$

7.10. (a) $CH_3CH_2CH_2 - C \equiv CH + Na^+ Br^-$
(b) $CH_3CH_2CH_2 - C \equiv N + K^+ Cl^-$

(c) $\overset{\displaystyle OCH_3}{\overset{|}{CH_3CHCH_3}} + Na^+ I^-$

(d) $\overset{\displaystyle OCH_3}{\overset{|}{CH_3CHCH_3}} + K^+ I^-$

7.11. In each reaction, the halogen could be Cl or I, as well as Br.

(a) $CH_3CH_2Br + {}^-SCH_2CH_3 \longrightarrow$ product

(b) ⬡$-CH_2Br + {}^-OH \longrightarrow$ product

(c) $CH_2 = CHCH_2Br + {}^-CN \longrightarrow$ product
(d) two possible methods:

⬡$-CH_2Br + CH_3O^-$ or   ⬡$-CH_2O^- + CH_3I$

7.12. (a) $(CH_3)_3C - \ddot{\underset{\cdot\cdot}{Br}} : \xrightarrow{-\,:\ddot{Br}:^-} (CH_3)_3\overset{+}{C} \longrightarrow$

$(CH_3)_3C - \overset{+}{\underset{|}{\underset{H}{\ddot{O}}}} - ⬡ \xrightarrow{-H^+} (CH_3)_3C\ddot{O} - ⬡$

(b) (cyclohexane with CH_3 and $\ddot{\underset{\cdot\cdot}{I}}:$) $\xrightarrow{-:\ddot{I}:^-}$ (ring with $\overset{+}{} - CH_3$) $\xrightarrow{H\ddot{O}CH_2CH_3}$

(ring with $\overset{H}{\overset{|}{\overset{+}{\underset{\cdot\cdot}{O}}CH_2CH_3}}$ and CH_3) $\xrightarrow{-H^+}$ (ring with $\ddot{O}CH_2CH_3$ and CH_3)

7.13. (a) ⬡$-Cl$ < **(c)** ⬡$-Cl$ < **(b)** ⬡$-Cl$

Vinylic, no reaction   2°   Allylic

Increasing reactivity →

7.14.

$$CH_3CH_2-\overset{\overset{\displaystyle :\overset{..}{Br}:}{|}}{\underset{\underset{\displaystyle :\overset{..}{O}CH_3}{}}{\underset{\displaystyle CH_3CH_2}{C}}}-CHCH_3 \quad\longrightarrow\quad \left[CH_3CH_2-\overset{\overset{\displaystyle :\overset{..\,\delta-}{Br}:}{|}}{\underset{\underset{\displaystyle H--\underset{\displaystyle \delta-}{\overset{..}{O}CH_3}}{|}}{\underset{\displaystyle CH_3CH_2}{C}}}\cdots\cdots CHCH_3 \right]$$

Transition state

$$\longrightarrow \quad \overset{\displaystyle CH_3CH_2}{\underset{\displaystyle CH_3CH_2}{>}}C=CHCH_3 \;+\; CH_3\overset{..}{O}H \;+\; :\overset{..}{Br}:^-$$

7.15. **(a)** $CH_3CH_2\overset{\overset{\displaystyle I}{|}}{\underset{\underset{\displaystyle CH_3}{|}}{C}}CH_2CH_2CH_3$ **(b)**

$$H_2C\overset{CH_2-CH_2CH_3}{\underset{\underset{\displaystyle H_2}{C}}{>}}\overset{\displaystyle C}{\underset{\displaystyle Br}{<}}$$

7.16. and **7.17.**

(a) $CH_3CH_2\overset{}{\underset{\underset{\displaystyle CH_3}{|}}{C}}=CHCH_2CH_3 \;+\; CH_3CH_2\overset{}{\underset{\underset{\displaystyle CH_2}{||}}{C}}CH_2CH_2CH_3$

$$+\; CH_3CH=\overset{}{\underset{\underset{\displaystyle CH_3}{|}}{C}}CH_2CH_2CH_3$$

(b) ⬠$\,CH_3$ + ⬠$=CH_2$

7.18. **(a)** $CH_3CH_2CH_2OH$, S_N2. The starting halide is a 1° alkyl halide, and ^-OH is a good nucleophile.

(b) $(CH_3CH_2CH_2)_2C=CHCH_2CH_3$. The reaction of a strong base and a 3° alkyl halide would proceed by an E2 path.

(c) $(CH_3CH_2CH_2)_3COH + (CH_3CH_2CH_2)_2C=CHCH_2CH_3$. The alcohol is formed by the S_N1 mechanism and the alkene by the E1 mechanism. A tertiary alkyl halide with a weak base yields the solvolysis product plus concurrent elimination.

(d) $(CH_3CH_2CH_2)_2CHOH$. The alcohol is formed by an S_N2 pathway because the reaction involves a 2° alkyl halide, a strong nucleophile, and little heat. If the reaction had been heated, the following alkene would have been expected to form (E2 pathway):

$$trans\text{-}\,CH_3CH_2CH=CHCH_2CH_2CH_3$$

(e) $CH_2=CHCH_2OH$. A reactive allylic primary halide and a strong nucleophile would react by the S_N2 mechanism.

(f) $CH_2=CHCH_2OH$. A reactive allylic halide with a weak base would react by the S_N1 mechanism.

CHAPTER 8

8.1. $\xrightarrow[\text{Increasing solubility}]{a<c<b<d}$

Compound **(a)** is the least soluble because it has a continuous-chain hydrocarbon group of seven carbons and only one hydroxyl group. Compound **(c)** is more soluble than **(a)** because the hydrocarbon portion of the structure is branched. Compound **(c)** is less soluble than **(b)** because it has one hydroxyl group and **(b)** has *two* hydroxyl groups. Compound **(d)** is the most soluble because it contains four hydroxyl groups.

8.2. **(a)** 2-chloro-1-butanol; **(b)** 3-methyl-4-nitro-2-pentanol;
(c) *trans*-3-isopropylcyclopentanol or *trans*-3-isopropyl-1-cyclopentanol.

8.3. **(a)** $\underset{\underset{Br\ \ OH}{|\ \ \ \ |}}{CH_2CHCH_2CH_2CH_2CH_2CH_2CH_3}$ **(b)** $\underset{\underset{CH_3}{|}}{\overset{\overset{OH}{|}}{CH_3CH_2CCH_2CH_3}}$

(c)

$\begin{array}{c} Cl \\ | \\ \langle \text{H} \\ \text{OH} \end{array}$ (ring structure with Cl, H, H, OH)

8.4. **(a)** primary; **(b)** tertiary; **(c)** secondary.

8.5. **(a)** *p*-methylphenol; **(b)** 2,4-dichlorophenol; **(c)** 4-methyl-2-nitrophenol.

8.6. **(a)** $O_2N-\langle O \rangle-OH$ **(b)** $Cl-\overset{\overset{Cl\ \ \ \ \ Cl}{}}{\underset{\underset{Cl\ \ \ \ \ Cl}{}}{\langle O \rangle}}-OH$ **(c)** $CH_3-\overset{\overset{C(CH_3)_3}{}}{\underset{\underset{C(CH_3)_3}{}}{\langle O \rangle}}-OH$

8.7. **(a)** $CH_3CH_2CH_2CH_2Cl \xrightarrow{^-OH} CH_3CH_2CH_2CH_2OH$

(b) $CH_3CH_2CH=CH_2 \xrightarrow[\text{(2) } H_2O_2,\ ^-OH]{\text{(1) } BH_3} CH_3CH_2CH_2CH_2OH$

(c) $CH_3CH_2CH=CH_2 + H_2O \xrightarrow{H^+} \underset{\underset{OH}{|}}{CH_3CH_2CHCH_3}$

8.8. **(a)** $CH_3CH_2O^-\ Na^+ + NH_3 \uparrow$ **(b)** $CH_3CH_2\overset{+}{O}H_2 + Cl^-$

(c) $\langle O \rangle-O^-\ K^+ + \frac{1}{2}H_2$ **(d)** $\underset{\underset{OH_2}{\overset{+}{|}}}{CH_3CH_2CHCH_3} + I^-$

8.9. **(a)** $CH_2=CHCH_2Br + H_2O$ **(b)** $\langle O \rangle-I + H_2O$

(c) $\langle O \rangle-CH_2Br + H_2O$ **(d)** $(CH_3)_2CHI + H_2O$

8.10. (a) [cyclohexene with I substituent] + H_2O

(b) $CH_3CH_2CH_2Cl + H_2O$ (only if $ZnCl_2$ is used as a catalyst)
(c) $CH_3CH_2CH_2Br + H_2O$
(d) $(CH_3)_3CBr + H_2O$

(b)	<	**(c)**	<	**(d)**	<	**(a)**
1°ROH + HCl		1°ROH + HBr		3°ROH + HBr		allylic + HI

→ Increasing rate of reaction

8.11. (a) [cyclohexane]—Cl   **(b)** [cyclohexane]—CH_2Br

8.12. (a) $CH_3\overset{\underset{\displaystyle |}{OH}}{C}HCH_2CH_3$ **(b)** $CH_3CH_2\overset{\underset{\displaystyle |}{\overset{\displaystyle OH}{|}}}{\underset{\underset{\displaystyle CH_2CH_3}{|}}{C}}CH_2CH_3$

(c)
$$\begin{array}{c} H_2C-CH_2 \\ H_2C \qquad\qquad C {\overset{CH_3}{\underset{OH}{<}}} \\ H_2C-CH_2 \end{array}$$

8.13. (a) $CH_3\overset{\underset{\displaystyle |}{\overset{\displaystyle OH}{|}}}{C}HCH_2CH_3 \xrightarrow[\text{heat}]{H_2SO_4} CH_2{=}CHCH_2CH_3 \;+\; \underset{H}{\overset{CH_3}{C}}{=}\underset{H}{\overset{CH_3}{C}}$

$+\; \underset{H}{\overset{CH_3}{C}}{=}\underset{CH_3}{\overset{H}{C}} + H_2O$

(b) $(CH_3CH_2)_2\overset{\underset{\displaystyle |}{\overset{\displaystyle OH}{|}}}{C}CH_2CH_3 \xrightarrow{H_2SO_4} (CH_3CH_2)_2C{=}CHCH_3 + H_2O$

only alkene

(c) [cyclohexane with CH_3 and OH] $\xrightarrow{H_2SO_4}$ [cyclohexene]—CH_3 + [cyclohexane]$={CH_2} + H_2O$

8.14. (a) $CH_3\overset{\underset{\displaystyle |}{\overset{\displaystyle CH_3}{|}}}{C}HCO_2H$ **(b)** [benzene ring]—CO_2H

(c) [benzene ring]—$\overset{\overset{\textstyle O}{\|}}{C}H$ **(d)** [cyclohexanone ring with CH_3 and $CH(CH_3)_2$ substituents, C=O]

8.15. (a)

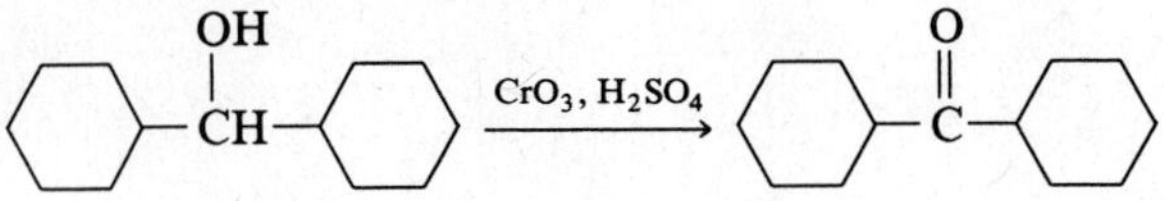

(b) $(CH_3)_2CHCH_2OH$ + PCC $\xrightarrow{CH_2Cl_2}$ $(CH_3)_2CHCH$ (with C=O)

(c)

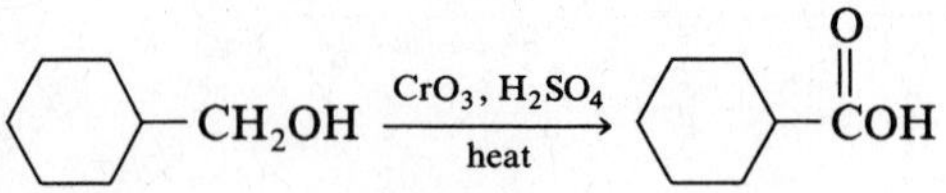

(d) $CH_3CH_2\overset{OH}{\underset{|}{C}HCH_2CH_2OH}$ $\xrightarrow[\text{heat}]{CrO_3,\,H_2SO_4}$ $CH_3CH_2CCH_2COH$ (with two C=O)

8.16.

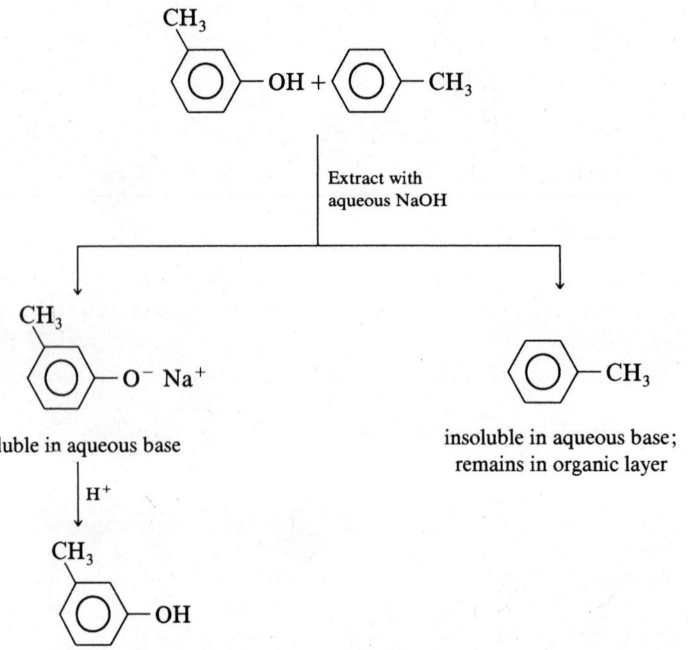

8.17. (a) 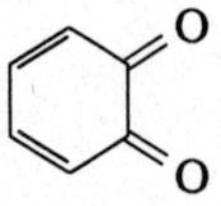   **(b)**

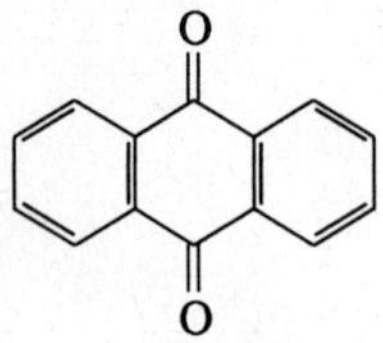

8.18.

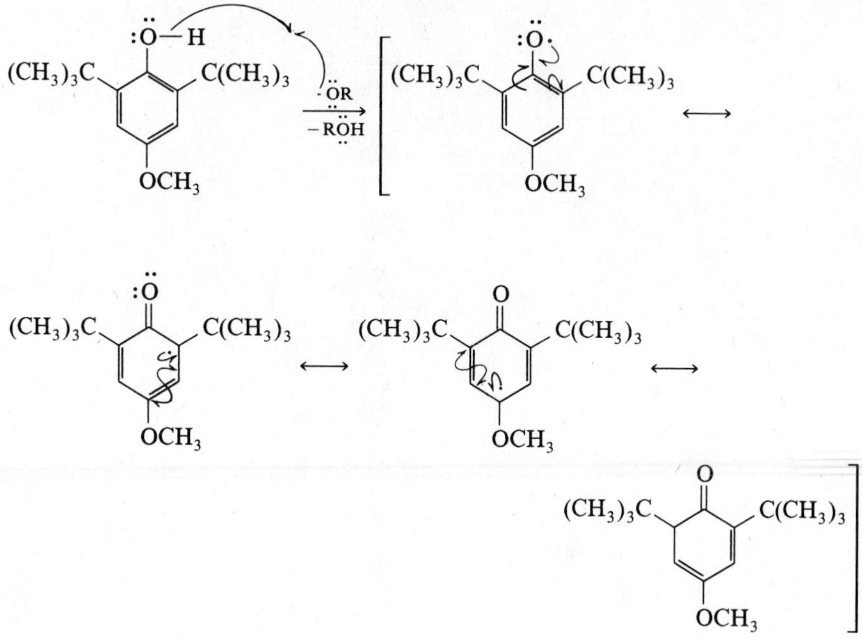

CHAPTER 9

9.1. (a) CH$_3$CH—O—CHCH$_3$ **(b)** CH$_3$OCH$_2$CH$_2$CH$_3$
     | |
    CH$_3$ CH$_3$

(c) 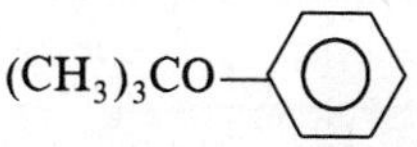 (CH$_3$)$_3$CO—⟨O⟩

9.2. (a) 1,4-dimethoxybenzene or *p*-dimethoxybenzene; **(b)** 1-methoxypropane; **(c)** 2-iso-propoxybutane; **(d)** 3-methoxy-1-propanol.

9.3. (a) CH$_3$I + CH$_3$CH$_2$O$^-$ ⟶ CH$_3$OCH$_2$CH$_3$ + I$^-$
CH$_3$O$^-$ + CH$_3$CH$_2$Br ⟶ CH$_3$OCH$_2$CH$_3$ + Br$^-$

(b) 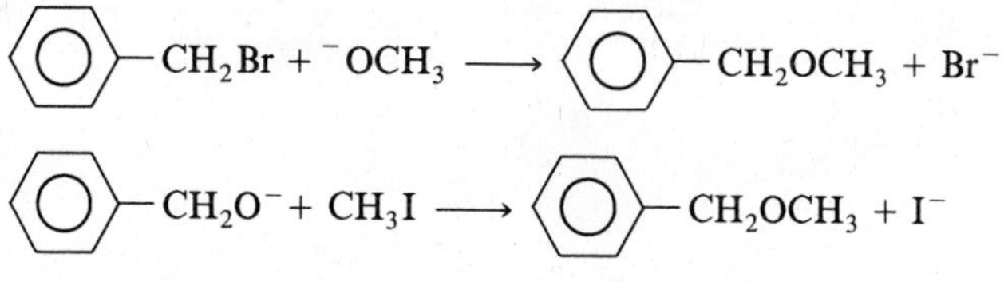 ⟨O⟩—CH$_2$Br + $^-$OCH$_3$ ⟶ ⟨O⟩—CH$_2$OCH$_3$ + Br$^-$

⟨O⟩—CH$_2$O$^-$ + CH$_3$I ⟶ ⟨O⟩—CH$_2$OCH$_3$ + I$^-$

9.4. (a) 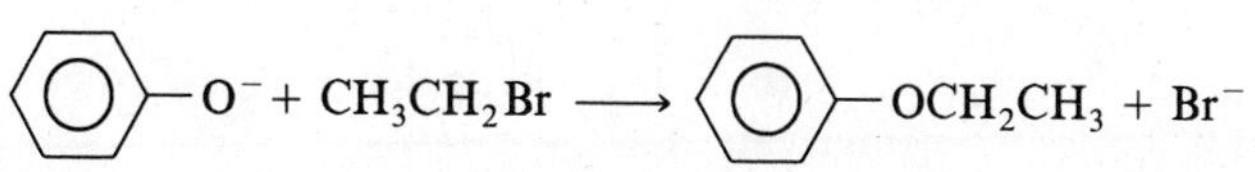 ⟨O⟩—O$^-$ + CH$_3$CH$_2$Br ⟶ ⟨O⟩—OCH$_2$CH$_3$ + Br$^-$

The other route would require bromobenzene as the halide and CH$_3$CH$_2$O$^-$ as the nucleophile. Halobenzenes do not undergo S$_N$2 reactions; therefore, no reaction would occur.

(b) $(CH_3)_3CO^- + CH_3I \longrightarrow (CH_3)_3COCH_3$

The other route would require $(CH_3)_3CBr$, which is a tertiary alkyl halide and would undergo elimination to yield an alkene.

9.5. (a) $CH_3CH_2I + CH_3CH_2CH_2I + H_2O$

(b) $CH_3CH_2Br + (CH_3)_2CHBr + H_2O$

(c) $(CH_3)_2CHOC(CH_3)_2$
$\qquad\qquad\qquad |$
$\qquad\qquad\quad OOH$

9.6. (a)
$\quad$ OH
$\quad\ |$
$\quad CH_2CH_2OCH_3$ **(b)**

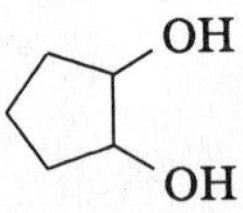

9.7. (a) $CH_2{=}CHMgI$; $CH_2{=}CH{-}CH_2CH_2OH$

(b) 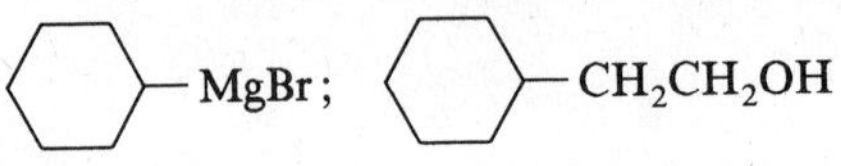 $-MgBr$; $-CH_2CH_2OH$

(c) 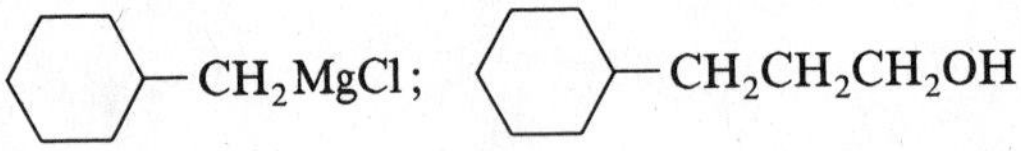 $-CH_2MgCl$; $-CH_2CH_2CH_2OH$

9.8. (a) $CH_3CH_2Br \xrightarrow[\text{ether}]{Mg} CH_3CH_2MgBr \xrightarrow[\text{(2) } H_2O,\ H^+]{\text{(1) } CH_2{-}CH_2 \ (O)} CH_3CH_2{-}CH_2CH_2OH$

(b) 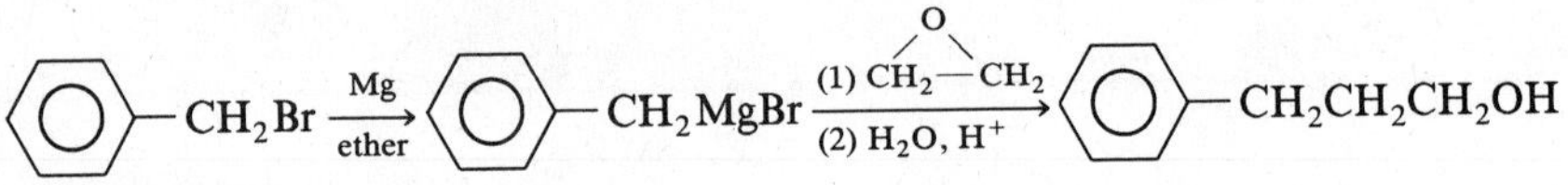 $-CH_2Br \xrightarrow[\text{ether}]{Mg}$ $-CH_2MgBr \xrightarrow[\text{(2) } H_2O,\ H^+]{\text{(1) } CH_2{-}CH_2 \ (O)}$ $-CH_2CH_2CH_2OH$

(c)

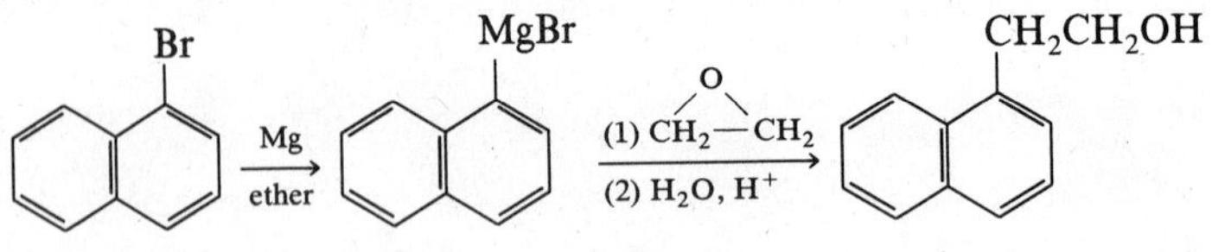

CHAPTER 10

10.1.
$\qquad\qquad \overset{\delta+}{H}{-}OH$
$\qquad\qquad\ \ |$
$CH_3\overset{..}{C}{=}\overset{\delta-}{\underset{..}{O}}$
$\qquad |$
$\qquad CH_3$

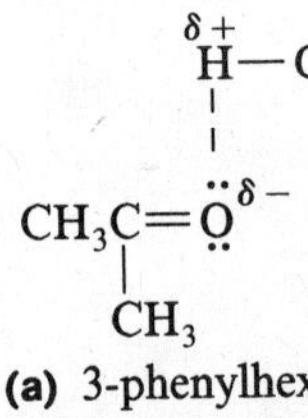

10.2. (a) 3-phenylhexanal; **(b)** *p*-bromobenzaldehyde or 4-bromobenzaldehyde; **(c)** 5-chloro-3-methylpentanal.

10.3. (a)
$\qquad CH_3CH_2 \quad O$
$\qquad\qquad |\qquad\quad \|$
$\quad CH_3CH_2CCH_2CH$
$\qquad\qquad |$
$\qquad\qquad OH$

(b)
$\qquad\qquad\qquad\qquad\qquad O$
$\qquad\qquad\qquad\qquad\qquad \|$
$\quad CH_3CH_2CH_2CHCH_2CH$
$\qquad\qquad\qquad\ \ |$
$\qquad\qquad\qquad\ \ Cl$

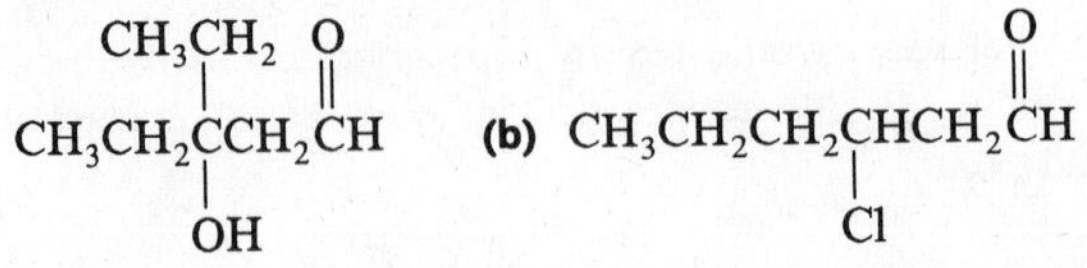

(c) $\underset{\displaystyle ||}{\overset{\displaystyle O}{}}$ $ClCH_2CH$ (d) $Cl-\langle\bigcirc\rangle-\underset{\displaystyle ||}{\overset{\displaystyle O}{}}CH$ (with Cl substituent)

10.4. **(a)** 4-methyl-3-hexanone; **(b)** 2,4-dimethyl-3-pentanone;
(c) 1-phenyl-1-propanone.

10.5. **(a)** $CH_3CHCH_2CCH_3$ with OH below first C and O (double bond) above the C **(b)** cyclohexane ring with Cl, Cl and $=O$ **(c)** $CH_2CH_2CCHCH_3$ with Br, O, Br

10.6. **(a)** $CH_3CH_2CH_2CH_2CHCH_3$ (with OH) $\xrightarrow[H_2SO_4]{CrO_3}$ $CH_3CH_2CH_2CH_2CCH_3$ (with O)

(b) $CH_3CH_2CH_2CH_2CH_2CH_2OH$ $\xrightarrow[CH_2Cl_2]{PCC}$ $CH_3CH_2CH_2CH_2CH_2CH$ (with O)

(c) $\langle\bigcirc\rangle$ + $ClCCHCH_3$ (with O and CH_3) $\xrightarrow{AlCl_3}$ $\langle\bigcirc\rangle-CCHCH_3$ (with O and CH_3)

10.7. $\underset{\text{Increasing reactivity}}{\overline{\text{(d) < (b) < (a) < (c)}}}$

10.8. The three chloro atoms withdraw electron density from the carbonyl carbon, making it more positive and thus susceptible to attack by water. The electron withdrawal also stabilizes the product hydrate and thus draws the equilibrium to the hydrate side of the equation.

$$Cl \leftarrow \underset{\displaystyle\downarrow Cl}{\overset{\displaystyle\uparrow Cl}{C}} \leftarrow \underset{\delta+}{\overset{\displaystyle O\ ||}{C}} - H$$

10.9. **(a)** CH_3CH (with O) $+ H_2O$ $\overset{H^+}{\rightleftharpoons}$ CH_3CH (with OH, OH)

(b) $CH_3CCH_2CH_3$ (with O) $+ H_2O$ $\overset{H^+}{\rightleftharpoons}$ $CH_3CCH_2CH_3$ (with OH, OH)

(c) Cl_3CCCCl_3 (with O) $+ H_2O$ $\overset{H^+}{\rightleftharpoons}$ Cl_3CCCCl_3 (with OH, OH)

10.10. (a) $CH_3\overset{\displaystyle O}{\overset{\displaystyle \|}{C}}H + CH_3CH_2OH \underset{\overset{\displaystyle\rightleftharpoons}{}}{\overset{H^+}{}} CH_3\overset{\displaystyle OH}{\underset{\displaystyle OCH_2CH_3}{\overset{|}{\underset{|}{CH}}}}$

$\underset{\overset{\displaystyle\rightleftharpoons}{}}{\overset{CH_3CH_2OH, H^+}{}} CH_3\overset{\displaystyle OCH_2CH_3}{\underset{\displaystyle OCH_2CH_3}{\overset{|}{\underset{|}{CH}}}} + H_2O$

(b) $\bigcirc\!\!-\overset{\displaystyle O}{\overset{\displaystyle \|}{C}}H + HOCH_2CH_2OH \underset{\overset{\displaystyle\rightleftharpoons}{}}{\overset{H^+}{}} \bigcirc\!\!-\overset{\displaystyle OH}{\underset{\displaystyle OCH_2CH_2OH}{\overset{|}{\underset{|}{CH}}}}$

$\underset{\overset{\displaystyle\rightleftharpoons}{}}{\overset{H^+}{}} \bigcirc\!\!<\!\!\begin{array}{c} O \\ O \end{array}\!\!> + H_2O$

(c) $CH_3\overset{\displaystyle O}{\overset{\displaystyle \|}{C}}CH_2CH_2CH_2OH \underset{\overset{\displaystyle\rightleftharpoons}{}}{\overset{H^+}{}} \overset{HO}{\underset{H_3C}{}}\!\!<\!\!\begin{array}{c} O \\ \\ \end{array}$

10.11. $CH_3CH_2\overset{\displaystyle O}{\overset{\displaystyle \|}{C}}H + HOCH_2CH_2CH_2OH \underset{\overset{\displaystyle\rightleftharpoons}{}}{\overset{H^+}{}}$

10.12. $\bigcirc\!\!-\overset{\displaystyle O}{\overset{\displaystyle \|}{C}}CH_3 + HCN \overset{^-CN}{\longrightarrow} \bigcirc\!\!-\overset{\displaystyle OH}{\underset{\displaystyle CH_3}{\overset{|}{\underset{|}{C}}}}-C\equiv N$

10.13. (a) $CH_3CH_2\overset{\displaystyle OMgBr}{\underset{\displaystyle CH_2CH_3}{\overset{|}{\underset{|}{CH}}}}$ **(b)** $\bigcirc\!\!<\!\!\overset{\displaystyle OMgI}{\underset{\displaystyle CH_3}{}}$

(c) $\bigcirc\!\!-\overset{\displaystyle OMgBr}{\underset{\displaystyle CH_3}{\overset{|}{\underset{|}{C}}}}CH_3$

10.14. (a) $CH_3\overset{\displaystyle}{\underset{\displaystyle CH_3}{\overset{|}{\underset{|}{CH}}}}CHCH_2MgBr \xrightarrow[\text{(2) } H_2O, H^+]{\text{(1) } H\overset{\displaystyle O}{\overset{\displaystyle \|}{C}}CH_3} CH_3\overset{\displaystyle}{\underset{\displaystyle CH_3}{\overset{|}{\underset{|}{CH}}}}CHCH_2\overset{\displaystyle OH}{\overset{|}{CH}}CH_3$

or $CH_3\overset{\displaystyle}{\underset{\displaystyle CH_3}{\overset{|}{\underset{|}{CH}}}}CHCH_2\overset{\displaystyle O}{\overset{\displaystyle \|}{C}}H \xrightarrow[\text{(2) } H_2O, H^+]{\text{(1) } CH_3MgI} CH_3\overset{\displaystyle}{\underset{\displaystyle CH_3}{\overset{|}{\underset{|}{CH}}}}CHCH_2\overset{\displaystyle OH}{\overset{|}{CH}}CH_3$

(b) $CH_3CH_2MgBr \xrightarrow[\text{(2) } H_2O, H^+]{\text{(1) } CH_3\overset{\displaystyle O}{\overset{\displaystyle \|}{C}}CH_3} CH_3CH_2\overset{\displaystyle OH}{\underset{\displaystyle CH_3}{\overset{|}{\underset{|}{C}}}}CH_3$

$$
\text{or} \quad CH_3CH_2\overset{\overset{\displaystyle O}{\|}}{C}CH_3 \xrightarrow[\text{(2) H}_2\text{O, H}^+]{\text{(1) CH}_3\text{MgI}} CH_3CH_2\overset{\overset{\displaystyle OH}{|}}{\underset{\underset{\displaystyle CH_3}{|}}{C}}CH_3
$$

(c) $CH_3CH_2CH_2CH_2CH_2CH_2\,MgBr \xrightarrow[\text{(2) H}_2\text{O, H}^+]{\overset{\overset{\displaystyle O}{\|}}{\text{(1) HCH}}}$

$$CH_3CH_2CH_2CH_2CH_2CH_2CH_2OH$$

10.15. (a) ⬡=NOH   **(b)** ⬡=N–⬡

(c) ⬡=NNH–(benzene ring with NO$_2$ at ortho and NO$_2$)

10.16. (a) $CH_3CH_2\overset{\overset{\displaystyle O}{\|}}{C}H + NH_2NH_2 \xrightarrow{\text{H}^+}$

(b) $CH_3CH_2\overset{\overset{\displaystyle O}{\|}}{C}H + NH_2OH \xrightarrow{\text{H}^+}$

(c) $CH_3CH_2\overset{\overset{\displaystyle O}{\|}}{C}H + NH_2NH–⬡ \xrightarrow{\text{H}^+}$

10.17. (a) $CH_3CH_2\overset{\overset{\displaystyle O}{\|}}{C}O^-$ **(b)** ⬡–$CO_2^{\,-}$

(c) no reaction

10.18. (a) ⬡–$CH_2\overset{\overset{\displaystyle O}{\|}}{C}CH_3 \xrightarrow[\text{(2) H}_2\text{O, H}^+]{\text{(1) NaBH}_4}$ ⬡–$CH_2\overset{\overset{\displaystyle OH}{|}}{C}HCH_3$

(b) (cyclohexane ring with CH$_3$ at top, =O at right, CH(CH$_3$) with H$_3$C and CH$_3$ at bottom) $\xrightarrow[\text{(2) H}_2\text{O, H}^+]{\text{(1) NaBH}_4}$

no α hydrogen

10.19. (a) $CH_3\overset{\overset{\displaystyle O}{\|}}{C}C(CH_3)_3 \rightleftharpoons CH_2{=}\overset{\overset{\displaystyle OH}{|}}{C}C(CH_3)_3$ only possible tautomer

(b) $CH_3C{\overset{\overset{\displaystyle H\cdots\ddot{O}}{\underset{\displaystyle O}{|}}}{\underset{\displaystyle CH}{\diagdown}}}COCH_2CH_3 \rightleftharpoons CH_3\overset{O}{\overset{||}{C}}CH_2\overset{O}{\overset{||}{C}}OCH_2CH_3 \rightleftharpoons$

$$CH_2{=}C{\overset{\overset{\displaystyle H\cdots\ddot{O}}{\underset{\displaystyle O}{|}}}{\underset{\displaystyle CH_2}{\diagdown}}}COCH_2CH_3$$

(c)

CHAPTER 11

11.1. Using $C_6H_5{-}$ to represent the phenyl group:

$$C_6H_5\overset{O}{\overset{||}{C}}O{-}H{-}{-}{-}{:}\underset{H}{\ddot{O}}{-}\overset{O}{\overset{||}{C}}C_6H_5 \qquad C_6H_5C{\overset{\overset{\displaystyle \ddot{O}{:}{-}{-}H{-}O}{}}{\underset{\underset{\displaystyle O{-}H{-}{-}{-}{:}\ddot{O}{.}}{}}{\diagup\diagdown}}}CC_6H_5$$

$$C_6H_5\overset{O}{\overset{||}{C}}O{-}H{-}{-}{-}{:}\overset{.}{\underset{\underset{\displaystyle HO\overset{||}{C}C_6H_5}{}}{\overset{||}{O}}} \qquad C_6H_5\overset{O}{\overset{||}{C}}O{-}H{-}{-}{-}{:}\underset{H}{\overset{..}{O}}CH_2CH_3$$

$$C_6H_5C{\overset{\overset{\displaystyle \ddot{O}{:}{-}{-}HOCH_2CH_3}{}}{\underset{\displaystyle OH}{\diagup\diagdown}}} \qquad C_6H_5\overset{O}{\overset{||}{C}}\underset{H}{\overset{..}{O}}{:}{-}{-}{-}HOCH_2CH_3$$

$$CH_3CH_2OH{-}{-}{-}{:}\underset{H}{\overset{..}{O}}CH_2CH_3$$

11.2. (a) $CH_3\underset{\underset{\displaystyle CH_3}{|}}{C}HCH_2CH_2\overset{O}{\overset{||}{C}}OH$

(b)

(c) $ClCH_2\overset{O}{\overset{||}{C}}OH$

(d)

11.3. (a) 3,3-dichloropropanoic acid
(b) 2-ethyl-2-methylpentanoic acid
(c) 2-propylpropanedioic acid (or 2-*n*-propyl)
(d) *cis*-1,2-cyclooctanedicarboxylic acid

(e) *cis*-3-nitrocyclohexanecarboxylic acid

(f) *trans*-2-chlorocyclopentanecarboxylic acid

11.4. (a) $\underset{\substack{\|\\O}}{CH_3C}OH + \underset{\substack{\|\\O}}{HOC}CH_2CH_3$

(b) $\underset{\substack{\|\\O}}{-C}OH + CO_2$

(c) and (d) $CH_3CH_2CH_2\underset{\substack{\|\\O}}{C}OH$

11.5. (a) Strong oxidation of any *p*-dialkylbenzene with a benzylic hydrogen, such as *p*-xylene, will yield the dicarboxylic acid

$$CH_3-\text{⟨O⟩}-CH_3 \xrightarrow[\text{heat}]{CrO_3,\ H_2SO_4} \text{product}$$

(b) Strong oxidation of any 1-alkylnaphthalene with a benzylic hydrogen, such as 1-methylnaphthalene, will yield the carboxylic acid.

$\xrightarrow[\text{heat}]{CrO_3,\ H_2SO_4}$ product

11.6. (a) $CH_3CH_2CH_2Br \xrightarrow[\text{ether}]{Mg} CH_3CH_2CH_2MgBr \xrightarrow[(2)\ H_2O,\ H^+]{(1)\ CO_2} CH_3CH_2CH_2\underset{\substack{\|\\O}}{C}OH$

(b) $\xrightarrow[\text{ether}]{Mg}$ $\xrightarrow[(2)\ H_2O,\ H^+]{(1)\ CO_2}$

11.7. (a) $2CH_3\underset{\substack{\|\\O}}{C}OH$ (b) and (c) $CH_3\underset{\substack{\|\\O}}{C}OH$

(d)

11.8. (a) There are two methods.

(1) $-CH_2Br \xrightarrow[\text{ether}]{Mg}$ $-CH_2MgBr \xrightarrow[(2)\ H_2O,\ H^+]{(1)\ CO_2}$ product

(2) $-CH_2Br \xrightarrow{^-CN}$ $-CH_2CN \xrightarrow[\text{heat}]{H_2O,\ H^+}$ product

(b) $\underset{\substack{|\\Cl}}{CH_3CH}CO_2H \xrightarrow{^-OH} \underset{\substack{|\\OH}}{CH_3CH}CO_2^- \xrightarrow{H^+} \underset{\substack{|\\OH}}{CH_3CH}CO_2H$

(c)

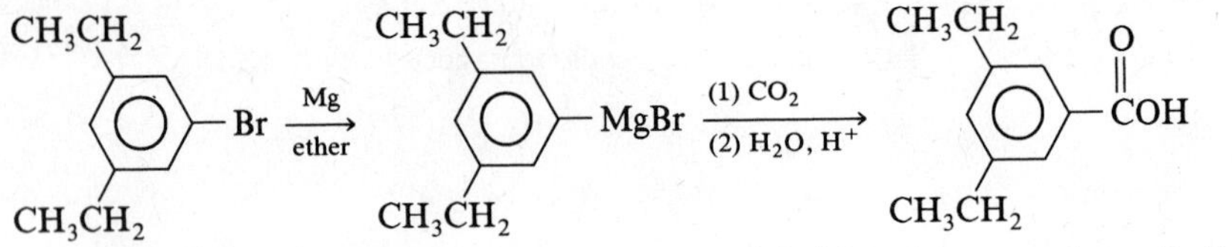

A nitrile synthesis cannot be used because aryl halides do not undergo S_N2 substitution reactions with ^-CN.

11.9.

$$\underrightarrow{\text{(c)} < \text{(b)} < \text{(a)}}$$
Increasing K_a; increasing acidity

11.10.

$$\underrightarrow{\text{(c)} < \text{(b)} < \text{(a)}}$$
Decreasing pK_a; increasing acidity

11.11. (a) $K_a = 1 \times 10^{-4}$;   **(b)** $K_a = 1 \times 10^4$;   **(c)** $K_a = 1 \times 10^{-9}$.

No Cl   Cl at position 4   Cl at position 2

11.12.

$$\underrightarrow{\text{(b)} < \text{(a)} < \text{(c)}}$$
Increasing acidity

11.13. In each case, the stronger acid loses a proton, and the stronger base gains (or retains) the proton.

(a)

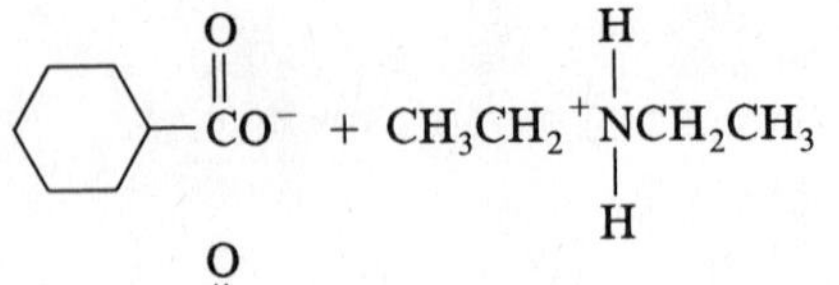

(b)

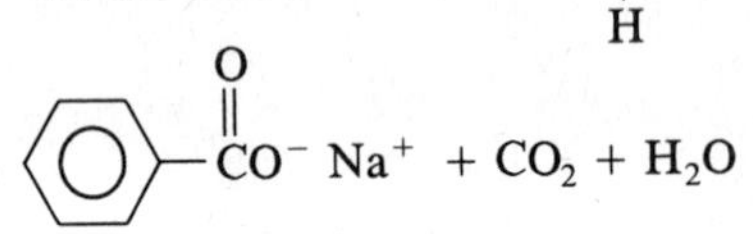

(c) and **(d)** no reaction

(e)

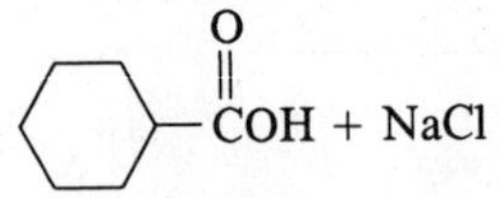

(f)

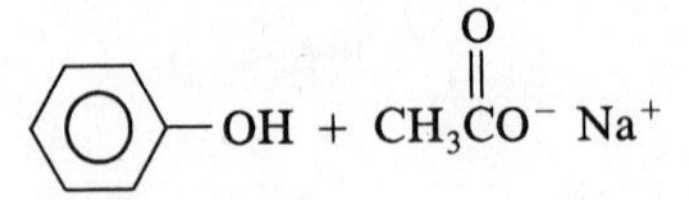

(g) no reaction

11.14. Alanine forms a dipolar ion by an acid-base reaction with itself. The dipolar ion is ionic; therefore, it is soluble in water, but insoluble in ether, a relatively nonpolar solvent.

$$CH_3CHCOH \longrightarrow CH_3CHCO^-$$

with the NH_2 and O groups as shown:

$$\underset{NH_2}{CH_3CH\overset{O}{\overset{\|}{C}}OH} \longrightarrow \underset{^+NH_3}{CH_3CH\overset{O}{\overset{\|}{C}}O^-}$$

Dipolar ion

an ionic compound

11.15. (a) C₆H₅—COH $\xrightarrow[\text{(2) H}_2\text{O, H}^+]{\text{(1) LiAlH}_4}$ C₆H₅—CH₂OH

(b) $CH_3\overset{O}{\overset{\|}{C}}CH_2\overset{O}{\overset{\|}{C}}OH$ $\xrightarrow[\text{(2) H}_2\text{O, H}^+]{\text{(1) LiAlH}_4}$ $CH_3\overset{OH}{\overset{|}{C}}HCH_2CH_2OH$

or $CH_3\overset{OH}{\overset{|}{C}}HCH_2\overset{O}{\overset{\|}{C}}OH$ $\xrightarrow[\text{(2) H}_2\text{O, H}^+]{\text{(1) LiAlH}_4}$ $CH_3\overset{OH}{\overset{|}{C}}HCH_2CH_2OH$

11.16. $CH_3\overset{\overset{\ddot{O}:}{\|}}{C}OH$ $\underset{H^+}{\rightleftharpoons}$ $CH_3\overset{\overset{+}{\overset{\ddot{O}H}}}{\overset{\|}{C}}{-}OH$ $\underset{CH_3CH_2\ddot{O}H}{\rightleftharpoons}$ $CH_3\overset{\overset{:\ddot{O}H}{|}}{C}OH$ $\underset{-H^+}{\rightleftharpoons}$
$$\underset{\underset{H}{\overset{+}{\overset{|}{\ddot{O}CH_2CH_3}}}}{}$$

$CH_3\overset{\overset{:\ddot{O}H}{|}}{C}{-}\ddot{O}H$ $\underset{H^+}{\rightleftharpoons}$ $CH_3\overset{\overset{:\ddot{O}H}{|}}{C}{-}\overset{+}{\ddot{O}H_2}$ $\underset{-H_2\ddot{O}:}{\rightleftharpoons}$ $CH_3\overset{\overset{:\overset{+}{\ddot{O}}{-}H}{\|}}{C}{-}OCH_2CH_3$ $\underset{-H^+}{\rightleftharpoons}$
$:\ddot{O}CH_2CH_3$ $\ddot{O}CH_2CH_3$

$$CH_3\overset{\overset{\ddot{O}:}{\|}}{C}{-}OCH_2CH_3$$

One of these oxygens
is lost in H₂O

11.17. C₆H₅—COH + CH₃¹⁸OH $\underset{}{\overset{H^+}{\rightleftharpoons}}$ C₆H₅—$\overset{OH}{\overset{|}{\underset{^{18}OCH_3}{C}}}$—OH $\rightleftharpoons$ product

└———— This oxygen is retained.

Key intermediate

CHAPTER 12

12.1. (a) benzoyl bromide; (b) pentanoyl chloride.

12.2. (a) $CH_3\overset{\overset{O}{\|}}{C}Cl + \tfrac{1}{3}H_3PO_3$

(b) $\overset{\overset{O}{\|}}{C}Cl + SO_2 + HCl$

(c) $Br\overset{\overset{O}{\|}}{C} - \overset{\overset{O}{\|}}{C}Br + \tfrac{2}{3}H_3PO_3$

12.3.

12.4. (a) $CH_3CH_2\overset{\overset{O}{\|}}{C}O-$$+ HCl$

(b) $CH_3CH_2O\overset{\overset{O}{\|}}{C}CH_2\overset{\overset{O}{\|}}{C}OCH_2CH_3 + 2HCl$

(c) $\overset{\overset{O}{\|}}{C}-NH-$$+$$-\overset{+}{N}H_3\ Br^-$

(d) $\overset{\overset{O}{\|}}{C}-N$$+$$\overset{+}{N}H_2\ Cl^-$

12.5. In each reaction sequence, PBr_3 could be used instead of $SOCl_2$.

(a) $CH_3CH_2\overset{\overset{O}{\|}}{C}OH \xrightarrow{SOCl_2} CH_3CH_2\overset{\overset{O}{\|}}{C}Cl \xrightarrow{(CH_3CH_2)_2NH} CH_3CH_2\overset{\overset{O}{\|}}{C}N(CH_2CH_3)_2$

(b) $-\overset{\overset{O}{\|}}{C}OH \xrightarrow{SOCl_2}$$-\overset{\overset{O}{\|}}{C}Cl \xrightarrow[\text{pyridine}]{(CH_3)_3COH}$$-\overset{\overset{O}{\|}}{C}OC(CH_3)_3$

(c) $(CH_3)_3C\overset{\overset{O}{\|}}{C}OH \xrightarrow{SOCl_2} (CH_3)_3C\overset{\overset{O}{\|}}{C}Cl \xrightarrow[\text{pyridine}]{\text{}-OH} (CH_3)_3C\overset{\overset{O}{\|}}{C}O-$

12.6. **(a) (1)** $2CH_3CH_2\overset{\displaystyle O}{\overset{\|}{C}}OH + CH_3\overset{\displaystyle O}{\overset{\|}{C}}O\overset{\displaystyle O}{\overset{\|}{C}}CH_3 \xrightarrow{\text{heat}}$

$CH_3CH_2\overset{\displaystyle O}{\overset{\|}{C}}O\overset{\displaystyle O}{\overset{\|}{C}}CH_2CH_3 + 2CH_3\overset{\displaystyle O}{\overset{\|}{C}}OH \uparrow$

Remove by
distillation

(2) $CH_3CH_2\overset{\displaystyle O}{\overset{\|}{C}}OH \xrightarrow{SOCl_2} CH_3CH_2\overset{\displaystyle O}{\overset{\|}{C}}Cl$

$+$

$\xrightarrow{\quad} CH_3CH_2\overset{\displaystyle O}{\overset{\|}{C}}O\overset{\displaystyle O}{\overset{\|}{C}}CH_2CH_3$

$\xrightarrow{NaOH} CH_3CH_2\overset{\displaystyle O}{\overset{\|}{C}}O^- \ Na^+$

(b) (1) $CH_3\overset{\displaystyle O}{\overset{\|}{C}}OH \xrightarrow{SOCl_2} CH_3\overset{\displaystyle O}{\overset{\|}{C}}Cl \xrightarrow{CH_3CH_2\overset{\displaystyle O}{\overset{\|}{C}}O^- \ Na^+} CH_3\overset{\displaystyle O}{\overset{\|}{C}}O\overset{\displaystyle O}{\overset{\|}{C}}CH_2CH_3$

(2) $CH_3CH_2\overset{\displaystyle O}{\overset{\|}{C}}OH \xrightarrow{SOCl_2} CH_3CH_2\overset{\displaystyle O}{\overset{\|}{C}}Cl \xrightarrow{CH_3\overset{\displaystyle O}{\overset{\|}{C}}O^- \ Na^+}$

(c) $+ CH_3\overset{\displaystyle O}{\overset{\|}{C}}O\overset{\displaystyle O}{\overset{\|}{C}}CH_3 \xrightarrow{\text{warm}}$ $+ 2CH_3\overset{\displaystyle O}{\overset{\|}{C}}OH$

12.7. **(a)** $+ Na^+ \ Cl^-$

(b) $CH_3CH_2\overset{\displaystyle O}{\overset{\|}{C}}OCH_3 + CH_3CH_2\overset{\displaystyle O}{\overset{\|}{C}}OH$

(c) $CH_3CH_2O\overset{\displaystyle O}{\overset{\|}{C}}CH_2CH_2\overset{\displaystyle O}{\overset{\|}{C}}OH$

12.8. **(a)** $CH_3CH_2O\overset{\displaystyle O}{\overset{\|}{C}}(CH_2)_7CH_3$ **(b)** $\underset{\displaystyle CH_3}{CH_3\overset{\displaystyle }{C}HCH_2}O\overset{\displaystyle O}{\overset{\|}{C}}CH_2CH_3$

(c) $\underset{\displaystyle CH_3}{CH_3\overset{\displaystyle }{C}HCH_2}CH_2O\overset{\displaystyle O}{\overset{\|}{C}}CH_3$

12.9. **(a)** *n*-propyl propanoate; **(b)** methyl propanoate; **(c)** ethyl butanoate; **(d)** methyl 3-methylbutanoate; **(e)** ethyl *p*-chlorobenzoate.

12.10. **(1)** $CH_3CH_2CH_2\overset{\displaystyle O}{\overset{\displaystyle \|}{C}}OH + CH_3CH_2OH \xrightarrow[\text{heat}]{H^+} CH_3CH_2CH_2\overset{\displaystyle O}{\overset{\displaystyle \|}{C}}OCH_2CH_3$

(2) $CH_3CH_2CH_2\overset{\displaystyle O}{\overset{\displaystyle \|}{C}}OH \xrightarrow{SOCl_2} CH_3CH_2CH_2\overset{\displaystyle O}{\overset{\displaystyle \|}{C}}Cl$

$\xrightarrow{CH_3CH_2OH} CH_3CH_2CH_2\overset{\displaystyle O}{\overset{\displaystyle \|}{C}}OCH_2CH_3$

(3) $CH_3CH_2CH_2\overset{\displaystyle O}{\overset{\displaystyle \|}{C}}OH \xrightarrow[\text{heat}]{(CH_3C)_2O} CH_3CH_2CH_2\overset{\displaystyle O}{\overset{\displaystyle \|}{C}}O\overset{\displaystyle O}{\overset{\displaystyle \|}{C}}CH_2CH_2CH_3$

$\xrightarrow{CH_3CH_2OH} CH_3CH_2CH_2\overset{\displaystyle O}{\overset{\displaystyle \|}{C}}OCH_2CH_3$

12.11. **(a)** $CH_3CH_2\overset{\displaystyle O}{\overset{\displaystyle \|}{C}}OH + (CH_3)_2CHOH$

(b) $^-O\overset{\displaystyle O}{\overset{\displaystyle \|}{C}}CH_2\overset{\displaystyle O}{\overset{\displaystyle \|}{C}}O^- + 2CH_3CH_2OH$

(c) ⬡—$\overset{\displaystyle O}{\overset{\displaystyle \|}{C}}OH$ +

$H-\overset{\displaystyle CH_2CH_3}{\underset{\displaystyle CH_3}{\overset{\displaystyle |}{\underset{\displaystyle |}{O\hat{C}\cdots H}}}}$

The (*R*) alcohol; not racemic or inverted to the (*S*) enantiomer because the configuration around the chiral carbon is not affected by the reaction

12.12. **(a)** $CH_3\overset{\displaystyle O}{\overset{\displaystyle \|}{C}}OCH_3 \xrightarrow[\text{(2) } H_2O,\ H^+]{\text{(1) } 2CH_3CH_2CH_2MgBr} CH_3\overset{\displaystyle OH}{\overset{\displaystyle |}{C}}(CH_2CH_2CH_3)_2$

(b) ⬡—$\overset{\displaystyle O}{\overset{\displaystyle \|}{C}}OCH_3 \xrightarrow[\text{(2) } H_2O,\ H^+]{\text{(1) } 2CH_3MgI}$ ⬡—$\overset{\displaystyle OH}{\overset{\displaystyle |}{C}}(CH_3)_2$

(c) $H\overset{\displaystyle O}{\overset{\displaystyle \|}{C}}OCH_3 \xrightarrow[\text{(2) } H_2O,\ H^+]{\text{(1) } 2CH_3MgI} (CH_3)_2CHOH$

12.13. **(a)** $CH_3CH_2CH_2OH + CH_3(CH_2)_4CH_2OH$
(b) $(CH_3)_2CHOH + HOCH_2CH_2CH_3$
(c) $CH_3\underset{\displaystyle OH}{\overset{\displaystyle |}{C}}HCH_2CH_2\underset{\displaystyle OH}{\overset{\displaystyle |}{C}}H_2$

12.14.

$$CH_3\overset{\overset{\displaystyle O}{\|}}{C}N\!-\!H$$

Hydrogen bond from NH to unshared electrons of O

$$H\!\cdots\!\overset{\cdot\cdot}{\underset{\cdot\cdot}{O}}$$

$$CH_3\overset{\overset{\displaystyle O}{\|}}{C}NH_2$$

12.15. (a) $H_3\overset{+}{N}CH_2\overset{\overset{\displaystyle O}{\|}}{C}NH\overset{\overset{\displaystyle O}{\|}}{\underset{\underset{\displaystyle CH_3}{|}}{C}H}CO^- + 2^-OH \xrightarrow{\text{heat}} H_2NCH_2\overset{\overset{\displaystyle O}{\|}}{C}O^- + H_2N\overset{\overset{\displaystyle O}{\|}}{\underset{\underset{\displaystyle CH_3}{|}}{C}H}CO^-$

(b) $H_3\overset{+}{N}CH_2\overset{\overset{\displaystyle O}{\|}}{C}NH\overset{\overset{\displaystyle O}{\|}}{\underset{\underset{\displaystyle CH_3}{|}}{C}H}CO^- \xrightarrow[\text{heat}]{H_2SO_4,\ H_2O} H_3\overset{+}{N}CH_2\overset{\overset{\displaystyle O}{\|}}{C}OH + H_3\overset{+}{N}\overset{\overset{\displaystyle O}{\|}}{\underset{\underset{\displaystyle CH_3}{|}}{C}H}COH$

(c) $H_3\overset{+}{N}CH_2\overset{\overset{\displaystyle O}{\|}}{C}NH\overset{\overset{\displaystyle O}{\|}}{\underset{\underset{\displaystyle CH_3}{|}}{C}H}CO^- \xrightarrow[\text{(2) H}_2\text{O, H}^+]{\text{(1) LiAlH}_4} H_3\overset{+}{N}CH_2CH_2\overset{+}{N}H_2\underset{\underset{\displaystyle CH_3}{|}}{C}HCH_2OH$

CHAPTER 13

13.1. $CH_3\underset{\underset{\displaystyle H}{\overset{\displaystyle\curvearrowleft}{|}}}{\overset{\overset{\displaystyle O}{\|}}{C}}HCH\ +\ {}^-\!:\overset{\cdot\cdot}{O}H \underset{-H_2O}{\rightleftharpoons} \left[CH_3\overset{\cdot\cdot}{\underset{\cdot\cdot}{C}}H\!-\!\overset{\overset{\displaystyle:\overset{\cdot\cdot}{O}}{\|}}{C}H \longleftrightarrow CH_3CH\!=\!\overset{\overset{\displaystyle:\overset{\cdot\cdot}{O}:^-}{|}}{C}H \right]$

$CH_3\overset{\cdot\cdot}{\underset{\cdot\cdot}{C}}H\overset{\overset{\displaystyle O}{\|}}{C}H\ +\ CH_3CH_2\overset{\overset{\displaystyle:\overset{\cdot\cdot}{O}}{\|}}{C}H \rightleftharpoons CH_3CH_2\overset{\overset{\displaystyle:\overset{\cdot\cdot}{O}:^-}{|}}{C}H\!-\!\underset{\underset{\displaystyle CH_3}{|}}{C}H\overset{\overset{\displaystyle O}{\|}}{C}H$

$CH_3CH_2\overset{\overset{\displaystyle:\overset{\cdot\cdot}{O}:^-}{|}}{C}H\!-\!\underset{\underset{\displaystyle CH_3}{|}}{C}H\overset{\overset{\displaystyle O}{\|}}{C}H\ +\ H\!-\!\overset{\cdot\cdot}{\underset{\cdot\cdot}{O}}H \rightleftharpoons CH_3CH_2\overset{\overset{\displaystyle:\overset{\cdot\cdot}{O}H}{|}}{C}H\underset{\underset{\displaystyle CH_3}{|}}{C}H\overset{\overset{\displaystyle O}{\|}}{C}H\ +\ {}^-\!:\overset{\cdot\cdot}{O}H$

13.2. Look at the product.

$$
\underset{\text{from } CH_3CH_2CH_2CH}{
CH_3CH_2CH_2\underset{|}{\overset{OH}{C}}H-\overset{O}{\overset{\|}{C}}H CH
}
$$

$$
CH_2CH_3
$$

from $CH_3CH_2CH_2\overset{O}{\overset{\|}{C}}H$

from $CH_3CH_2CH_2\overset{O}{\overset{\|}{C}}H$

The equation: $2CH_3CH_2CH_2\overset{O}{\overset{\|}{C}}H \underset{}{\overset{^-OH}{\rightleftharpoons}} CH_3CH_2CH_2\underset{|}{\overset{OH}{C}}H\overset{O}{\overset{\|}{C}}HCH$

$$CH_2CH_3$$

13.3. (a) $\langle\bigcirc\rangle-\underset{|}{\overset{OH}{C}}HCH_2\overset{O}{\overset{\|}{C}}C(CH_3)_3 \longleftarrow$ would dehydrate readily; see Section 13.1D.

(b) $(CH_3)_3C\underset{|}{\overset{OH}{C}}HCH_2\overset{O}{\overset{\|}{C}}H$

13.4. $\overset{O}{\overset{\|}{H}}CH + CH_3\overset{O}{\overset{\|}{C}}CH_3 \underset{}{\overset{^-OH}{\rightleftharpoons}} \underset{|}{\overset{OH}{C}}H_2CH_2\overset{O}{\overset{\|}{C}}CH_3$

13.5. (a) $2\langle\bigcirc\rangle-CH_2\overset{O}{\overset{\|}{C}}H \underset{}{\overset{^-OH}{\rightleftharpoons}} \langle\bigcirc\rangle-CH_2\underset{|}{\overset{OH}{C}}H-\underset{|}{C}H\overset{O}{\overset{\|}{C}}H \overset{-H_2O}{\longrightarrow}$ product

$$\bigcirc$$

(b) $\langle\bigcirc\rangle-\overset{O}{\overset{\|}{C}}H + CH_3\overset{O}{\overset{\|}{C}}CH_3 \underset{}{\overset{^-OH}{\rightleftharpoons}} \langle\bigcirc\rangle-\underset{|}{\overset{OH}{C}}H-CH_2\overset{O}{\overset{\|}{C}}CH_3 \overset{-H_2O}{\longrightarrow}$ product

13.6. $CH_3\overset{O}{\overset{\|}{C}}CH_2 \overset{:\ddot{O}H}{\underset{|}{C}}HCH_3 \underset{}{\overset{^-:\ddot{O}H}{\rightleftharpoons}} CH_3\overset{O}{\overset{\|}{C}}CH_2-\overset{:\ddot{O}:^-}{\underset{|}{C}}HCH_3 \rightleftharpoons$

$$CH_3\overset{O}{\overset{\|}{C}}\ddot{C}H_2 + \overset{O}{\overset{\|}{H}}CCH_3$$

the products

$\overset{H_2O}{\longrightarrow} CH_3\overset{O}{\overset{\|}{C}}CH_3 + {}^-OH$

13.7. **(a)** $CH_3\overset{O}{\underset{||}{C}}CH_2\overset{O}{\underset{||}{C}}OCH_2CH_2CH_3 + CH_3CH_2CH_2OH$

(b) $CH_3CH_2CH_2\overset{O}{\underset{||}{C}}\underset{\underset{CH_2CH_3}{|}}{CH}\overset{O}{\underset{||}{C}}OCH_3 + CH_3OH$

(c) $+ CH_3OH$

13.8. **(a)** $H\overset{O}{\underset{||}{C}}OCH_3 + CH_3CH_2\overset{O}{\underset{||}{C}}OCH_3 \xrightarrow[\text{(2) }H^+]{\text{(1) }CH_3O^-} \text{product}$

(b) $-\overset{O}{\underset{||}{C}}OCH_2CH_3 + CH_3CH_2CH_2\overset{O}{\underset{||}{C}}OCH_2CH_3 \xrightarrow[\text{(2) }H^+]{\text{(1) }CH_3CH_2O^-} \text{product}$

(c) $CH_3O\overset{O}{\underset{||}{C}}CH_2CH_2CH_2CH_2CH_2\overset{O}{\underset{||}{C}}OCH_3 \xrightarrow[\text{(2) }H^+]{\text{(1) }CH_3O^-}$

In **(c)**, first write the formula of the cyclic compound showing the C's and H's; then, determination of the starting material is easier.

13.9. **(a)** $\xrightarrow{-CH_3OH} CH_3CH_2\overset{O}{\underset{||}{C}}\underset{\underset{CH_3}{|}}{CH}\overset{O}{\underset{||}{C}}OH \xrightarrow{\text{heat}} CH_3CH_2\overset{O}{\underset{||}{C}}CH_2CH_3 + CO_2$

(b) $+ CO_2$ **(c)** $HO\overset{O}{\underset{||}{C}}\underset{\underset{CH_3}{|}}{CH_2} + CO_2$

13.10. **(a)** $2(CH_3)_2CH\overset{O}{\underset{||}{C}}H \underset{\xrightarrow{\hspace{0.3cm}}}{\overset{^-OH}{\rightleftharpoons}} (CH_3)_2CHCH-\underset{\underset{CH_3}{|}}{\overset{\overset{OH\ CH_3}{|\ \ \ \ |}}{C}}-\overset{O}{\underset{||}{C}}H$

(b) C_6H_5—$\overset{\overset{\displaystyle O}{\|}}{C}OCH_3$ + $(CH_3)_2CHCH_2\overset{\overset{\displaystyle O}{\|}}{C}OCH_3$ $\xrightarrow[\text{(2) H}^+]{\text{(1) CH}_3\text{O}^-}$

C_6H_5—$\overset{\overset{\displaystyle O}{\|}}{C}$—$\underset{\underset{\displaystyle CH(CH_3)_2}{|}}{CH}\overset{\overset{\displaystyle O}{\|}}{C}OCH_3$ $\xrightarrow[\text{(2) H}^+,\text{ cold}]{\text{(1) }^-\text{OH, heat}}$ C_6H_5—$\overset{\overset{\displaystyle O}{\|}}{C}\underset{\underset{\displaystyle CH(CH_3)_2}{|}}{CH}\overset{\overset{\displaystyle O}{\|}}{C}OH$

(c) C_6H_5—$\overset{\overset{\displaystyle O}{\|}}{C}OCH_3$ + $CH_3CH_2CH_2\overset{\overset{\displaystyle O}{\|}}{C}OCH_3$ $\xrightarrow[\text{(2) H}^+]{\text{(1) CH}_3\text{O}^-}$

C_6H_5—$\overset{\overset{\displaystyle O}{\|}}{C}$—$\underset{\underset{\displaystyle CH_2CH_3}{|}}{CH}\overset{\overset{\displaystyle O}{\|}}{C}OCH_3$ $\xrightarrow[-CO_2]{H^+,\text{ heat}}$ C_6H_5—$\overset{\overset{\displaystyle O}{\|}}{C}CH_2CH_2CH_3$

(d) $H\overset{\overset{\displaystyle O}{\|}}{C}CH_2CH_2CH_2CH_2CH_2\overset{\overset{\displaystyle O}{\|}}{C}H$ $\underset{}{\overset{^-\text{OH}}{\rightleftharpoons}}$ [cyclohexane ring]—$\overset{\overset{\displaystyle O}{\|}}{C}H$ with OH $\xrightarrow[-H_2O]{H^+,\text{ heat}}$ [cyclohexene ring]—$\overset{\overset{\displaystyle O}{\|}}{C}H$

13.11. (a) C_6H_5—$\overset{\overset{\displaystyle O}{\|}}{C}OCH_3$ + $CH_3\overset{\overset{\displaystyle O}{\|}}{C}OCH_3$ $\xrightarrow[\text{(2) H}^+]{\text{(1) CH}_3\text{O}^-}$

C_6H_5—$\overset{\overset{\displaystyle O}{\|}}{C}CH_2\overset{\overset{\displaystyle O}{\|}}{C}OCH_3$ $\xrightarrow[\text{(2) H}^+,\text{ cold}]{\text{(1) }^-\text{OH, heat}}$ product

(b) $2\,C_6H_5$—$\overset{\overset{\displaystyle O}{\|}}{C}CH_3$ $\overset{^-\text{OH}}{\rightleftharpoons}$ C_6H_5—$\underset{\underset{\displaystyle CH_3}{|}}{\overset{\overset{\displaystyle OH}{|}}{C}}$—$CH_2\overset{\overset{\displaystyle O}{\|}}{C}$—$C_6H_5$ $\xrightarrow[-H_2O]{H^+}$ product

(c) $H\overset{\overset{\displaystyle O}{\|}}{C}H$ + $CH_3\overset{\overset{\displaystyle O}{\|}}{C}H$ $\overset{^-\text{OH}}{\rightleftharpoons}$ $\underset{\underset{\displaystyle CH_2CH_2}{|}}{\overset{\overset{\displaystyle OH}{|}}{CH}}\overset{\overset{\displaystyle O}{\|}}{C}H$ $\xrightarrow[-H_2O]{H^+}$ product

CHAPTER 14

14.1. **(a)** tertiary; **(b)** primary; **(c)** and **(d)** secondary.

14.2. **(a)** *t*-butylethylamine; **(b)** *sec*-butylamine; **(c)** *N*-methyl-*o*-methylaniline; **(d)** methyl *m*-aminobenzoate.

14.3. The nitrogen has an unshared pair of electrons that can hydrogen bond very efficiently with the hydrogen of water.

$$(CH_3)_3N:--H-\overset{\diagup H}{O}$$

14.4. **(a)** $(CH_3)_3CNH_2$ because it has the largest K_b.

14.5. **(c)** $(CH_3)_2NH$ because it has the smallest pK_b.

14.6. **(a)** $CH_3CH_2\overset{+}{N}H_2CH_2CH_3 + Na^+ \, HPO_4^{2-}$ $CH_3CH_2NHCH_2CH_3$

(b) 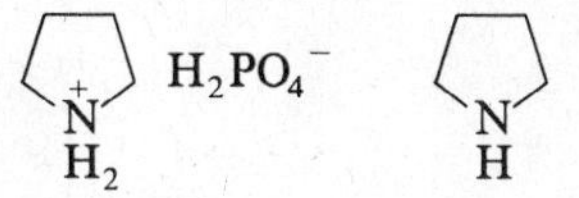 $H_2PO_4^-$

(c)

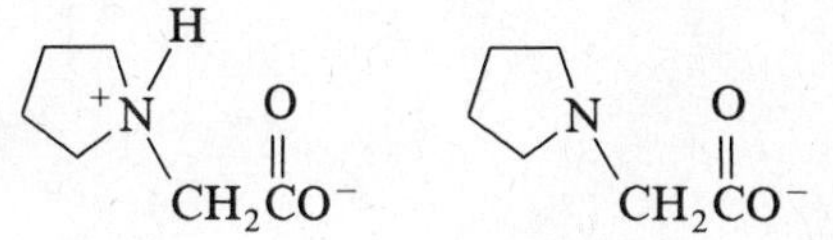

14.7. The leftover aniline will form a water-soluble salt on treatment with cold mineral acid. The acetanilide will not. Therefore, extract the solution with aqueous HCl and separate the aqueous layer from the water-insoluble material.

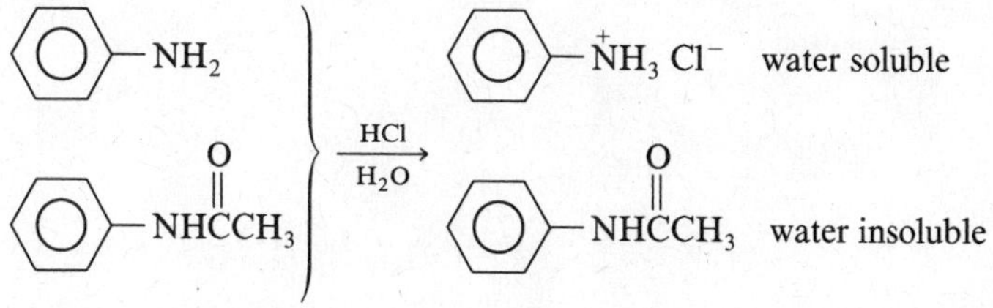

14.8. **(b)** The excess NH_3 would promote reaction with $CH_3CH_2CH_2Br$ to yield the desired product. The secondary and tertiary amines arise from the reaction of $CH_3CH_2CH_2Br$ with $CH_3CH_2CH_2NH_2$ and $(CH_3CH_2CH_2)_2NH$. An excess of $CH_3CH_2CH_2Br$ would promote these side reactions.

14.9. **(a)**

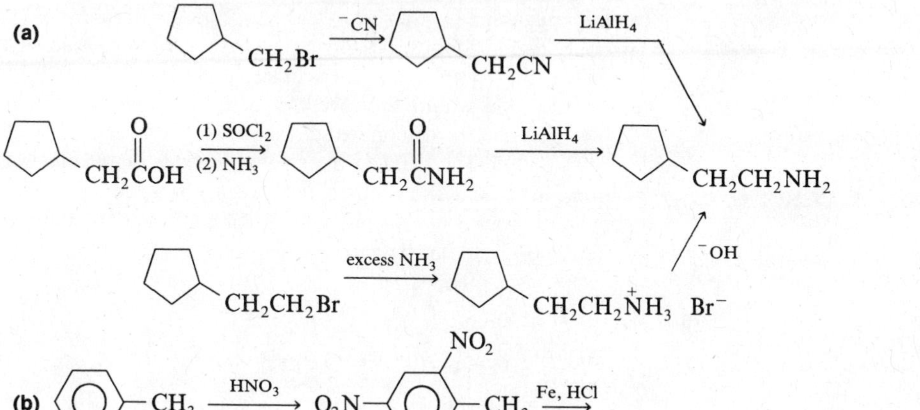

(b)

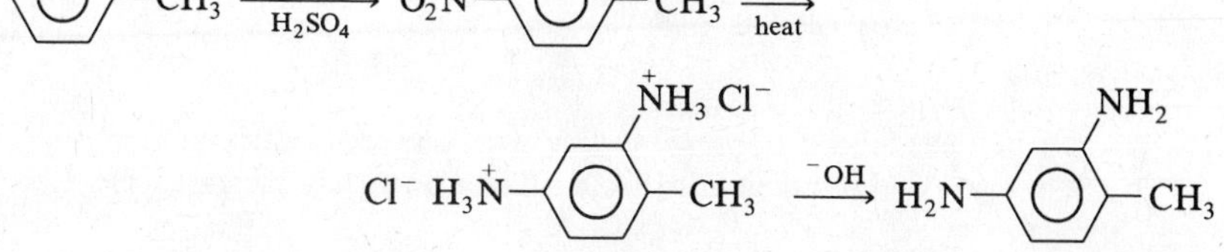

14.10. (a) and **(b)** $H_2NCH_2(CH_2)_4CH_2NH_2$

(c) $H_3\overset{+}{N}$—⟨○⟩—$\overset{\overset{\textstyle O}{\|}}{C}OH$ Cl^-

(d) $\xrightarrow[\text{(2) }^-OH]{\text{(1) heat}}$ (3-(1-methylpyrrolidin-2-yl)pyridine)

14.11. (a) ⟨○⟩—NCH_3, with $N=O$

(b) (naphthalene)—$\overset{+}{N}\equiv N$

(c) ON—⟨○⟩—N⟨piperidine⟩

(d) (naphthalene)—CN

(e) CH_3—⟨○⟩ with Br

(f) (naphthalene)—$N=N$—⟨○⟩—OH

CHAPTER 15

15.1. (a) 1-Propanol, $CH_3CH_2CH_2OH$, would show strong, broad absorption in the 3000–3500-cm^{-1} region (about 3.0–3.5 μm). *n*-Propylamine, $CH_3CH_2CH_2NH_2$, would show a weaker, sharper *pair* of overlapping peaks in the same region.

(b) *n*-Propylamine would show weak absorption with a double peak in the 3000–3500-cm^{-1} region (about 3.0–3.5 μm), while dimethylamine, $(CH_3)_2NH$, would show only a single weak peak in the same region.

(c) Benzoic acid would show very strong, broad absorption in the 3000–3500-cm^{-1} region. This absorption would slope into the aryl C—H absorption at about 3000–3300 cm^{-1} (3.0–3.3 μm). Benzaldehyde would not show this strong absorption, but would show up to four weak absorption peaks in the 2700–2900-cm^{-1} region (about 3.4–3.7 μm). Each compound would exhibit carbonyl absorption in the 1700–1750-cm^{-1} region (about 5.8 μm).

15.2. (a) (4) 217 nm
(b) (1) 270 nm
(c) (3) 312 nm   Increasing conjugation; increasing wavelength
(d) (4) 343 nm

15.3. Silicon is *electropositive* with respect to carbon. Therefore, in TMS, carbon withdraws electron density from silicon and thus increases the electron density in the C—H bonds. The increased electron density causes the protons to be shielded and to absorb upfield.

$$\overset{\delta+}{Si} \rightarrow \overset{\delta-}{C} \rightarrow \overset{\delta-}{H}$$

15.4. **(a)** $CH_3CH_2Cl + [Cl] \longrightarrow$ $\overset{\displaystyle Cl}{\underset{\displaystyle |}{CH_2CH_2Cl}}$

1,2-Dichloroethane

(b) $CH_3CH_2Cl + [Cl] \longrightarrow$ $\overset{\displaystyle Cl}{\underset{\displaystyle |}{CH_3CHCl}}$

1,1-Dichloroethane

1,1-Dichloroethane and 1,2-dichloroethane are structural isomers; therefore, the two types of protons (CH_3 and CH_2) in chloroethane are *not* equivalent. If the protons were equivalent, the two products would have yielded the same compound. For example:

$$\overset{\displaystyle H}{\underset{\displaystyle H}{CH_3CCl}} + [Cl] \longrightarrow \overset{\displaystyle Cl}{\underset{\displaystyle H}{CH_3CCl}} \quad or \quad \overset{\displaystyle H}{\underset{\displaystyle Cl}{CH_3CCl}}$$

Both formulas represent
1,1-dichloroethane;
thererfore, the two replaced
protons are equivalent.

15.5. **(a)** contains two types of protons in a 1 : 1 ratio.

Equivalent CH_2 protons;
total of 4

$ClCH_2CH_2 - O - CH_2CH_2Cl$

Equivalent CH_2 protons;
also total of 4

a b   c d

(b) $ClCH_2CH_2 - O - CH_2CHCl_2$

This formula shows four types of protons, labeled a, b, c, and d. The ratio is 2 : 2 : 2 : 1. (H_b and H_c are not equivalent because protons H_b are adjacent to CH_2Cl and protons H_c are adjacent to $CHCl_2$.)

(c) contains two types of protons in a 3 : 2 ratio.

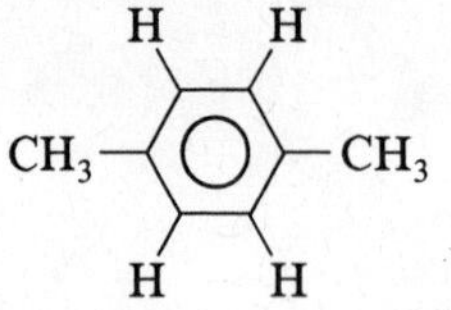

CH$_3$—⟨ring with H, H, H, H⟩—CH$_3$ 6 equivalent methyl protons
4 equivalent ring protons

(d) contains four types of protons in a 6 : 2 : 1 : 1 ratio.

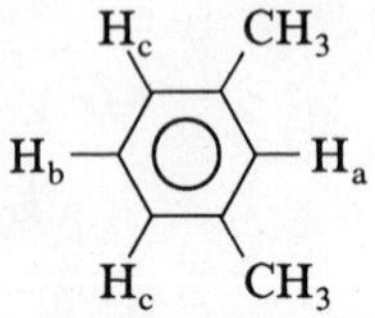

H$_c$ CH$_3$
H$_b$—⟨ring⟩—H$_a$ 6 equivalent methyl protons
H$_c$ CH$_3$ 1 H$_a$
1 H$_b$
2 equivalent H$_c$

15.6. The compound could contain eight protons or it could contain any multiple of eight (6 : 6 : 4 = 16; 9 : 9 : 6 = 24, etc.) We can tell only the ratio, not the exact numbers, of protons per molecule.

15.7. (a)

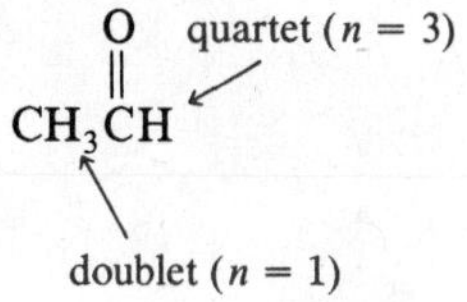

$$\underset{\text{}}{\text{CH}_3}\overset{\text{O}}{\underset{\|}{\text{CH}}}$$ quartet ($n = 3$)

doublet ($n = 1$)

(b) H$_c$—⟨ring with H$_b$, H$_a$, H$_b$, H$_a$⟩—OCH$_3$ singlet ($n = 0$)

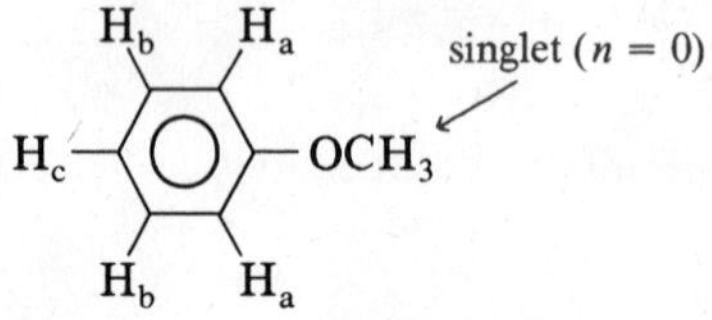

H$_a$ = doublet ($n = 1$)
H$_b$ = triplet ($n = 2$)
H$_c$ = triplet ($n = 2$)

Because H$_b$ is split by two types of protons, its signal may not be a perfect triplet.

(c) Cl—⟨ring with H$_b$, H$_a$, H$_b$, H$_a$⟩—C(=O)—H   singlet ($n = 0$)

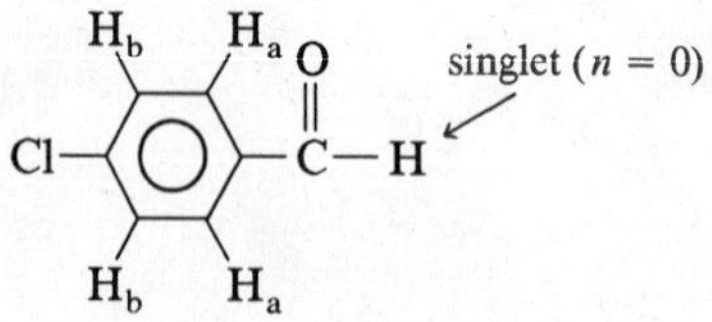

H$_a$ = doublet ($n = 1$)
H$_b$ = doublet ($n = 1$)

(d) CH$_3$CH$_2$—O—CH(CH$_3$)(CH$_3$)   ← c   d

a   b

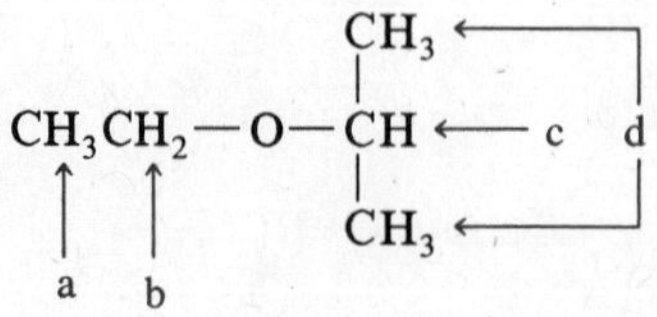

H$_a$ = triplet ($n = 2$)
H$_b$ = quartet ($n = 3$)
H$_c$ = septet (7 peaks) ($n = 6$)
H$_d$ = doublet ($n = 1$)

CHAPTER 16

16.1.

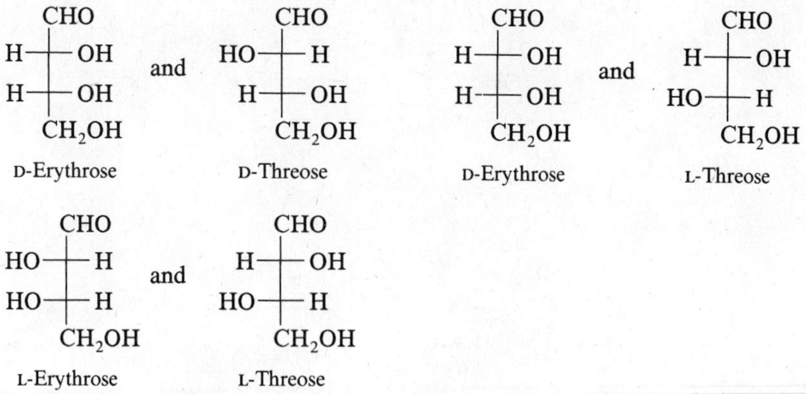

CHO
H—OH
H—OH
CH$_2$OH

D-Erythrose

and

CHO
HO—H
H—OH
CH$_2$OH

D-Threose

CHO
H—OH
H—OH
CH$_2$OH

D-Erythrose

and

CHO
H—OH
HO—H
CH$_2$OH

L-Threose

CHO
HO—H
HO—H
CH$_2$OH

L-Erythrose

and

CHO
H—OH
HO—H
CH$_2$OH

L-Threose

16.2. (a) D;   **(b)** L;   **(c)** L.

16.3. (a) furanose;   **(b)** pyranose;   **(c)** furanose.

16.4.

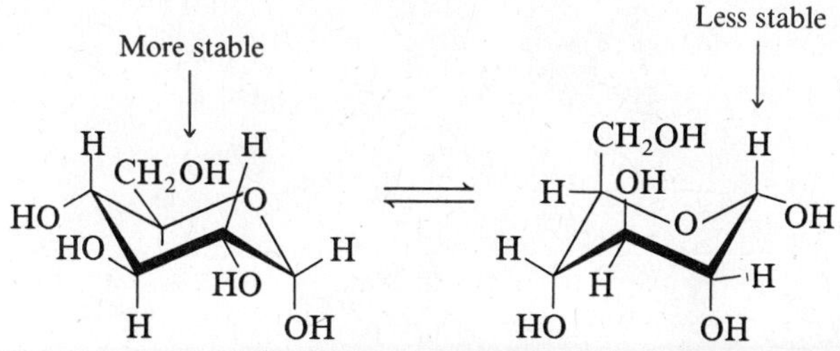

More stable

Less stable

α-D-Glucose

16.5.

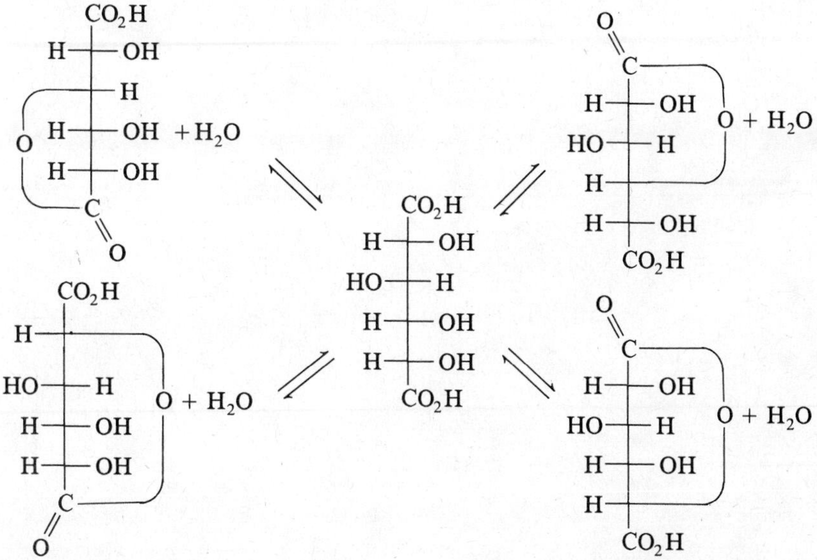

16.6. (a)

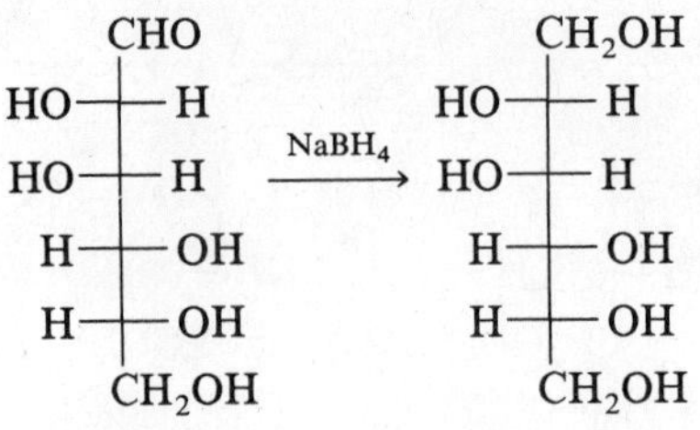

CHO
HO—H
HO—H
H—OH
H—OH
CH$_2$OH

D-Mannose

$\xrightarrow{\text{NaBH}_4}$

CH$_2$OH
HO—H
HO—H
H—OH
H—OH
CH$_2$OH

Mannitol

(b)

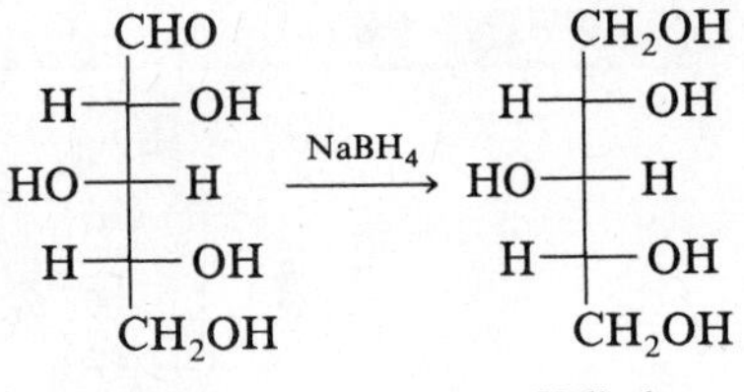

CHO
H—OH
HO—H
H—OH
CH$_2$OH

D-Xylose

$\xrightarrow{\text{NaBH}_4}$

CH$_2$OH
H—OH
HO—H
H—OH
CH$_2$OH

Xylitol

16.7. (a)

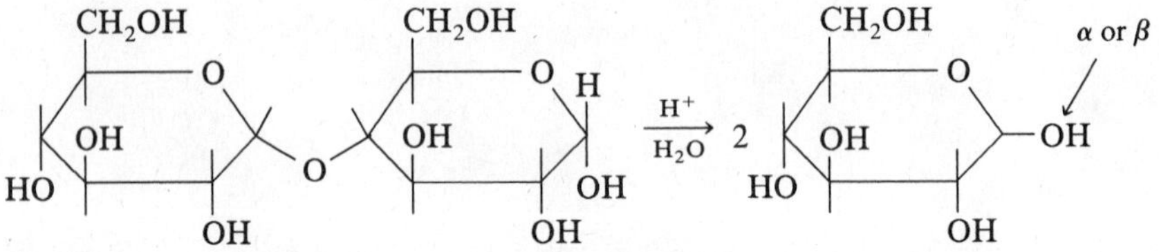

α-Maltose

$\xrightarrow[\text{H}_2\text{O}]{\text{H}^+}$ 2

α or β

D-Glucose

(b) the same: mixture of α- and β-D-glucose

16.8. (a)

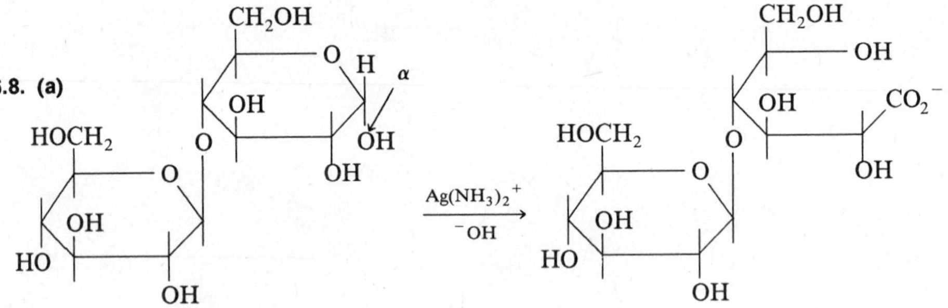

α-Cellobiose

$\xrightarrow[{}^-\text{OH}]{\text{Ag(NH}_3)_2{}^+}$

(b)

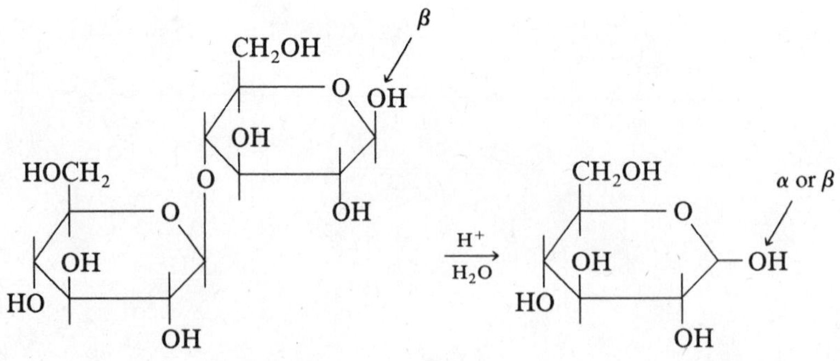

β-Cellobiose

$\xrightarrow[\text{H}_2\text{O}]{\text{H}^+}$

α or β

D-Glucose

16.9.

$$CH_2OH$$

$\alpha, 1 \rightarrow 6'$ link

α or β

Isomaltose

CHAPTER 17

$$CH_2O_2C(CH_2)_7CH=CHCH_2CH=CH(CH_2)_4CH_3$$

17.1. $$CHO_2C(CH_2)_7CH=CHCH_2CH=CH(CH_2)_4CH_3$$

$$CH_2O_2C(CH_2)_7CH=CHCH_2CH=CH(CH_2)_4CH_3$$

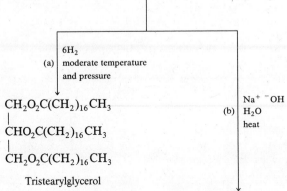

(a) $\begin{array}{c} 6H_2 \\ \text{moderate temperature} \\ \text{and pressure} \end{array}$

(b) $\begin{array}{c} Na^+\ ^-OH \\ H_2O \\ heat \end{array}$

$$CH_2O_2C(CH_2)_{16}CH_3$$

$$CHO_2C(CH_2)_{16}CH_3$$

$$CH_2O_2C(CH_2)_{16}CH_3$$

Tristearylglycerol

$$3CH_3(CH_2)_4CH=CHCH_2CH=CH(CH_2)_7\overset{\displaystyle O}{\overset{\displaystyle \|}{C}}O^-\ Na^+$$

Linoleic acid

+

$$HOCH_2CHCH_2OH$$
$$|$$
$$OH$$

Glycerol

17.2.

$$\text{RCO}\blacktriangleright\underset{\substack{|\\\text{CH}_2\text{O}-\overset{\text{O}^-}{\underset{\overset{\|}{\text{O}}}{\text{P}}}\text{OCH}_2\text{CH}_2\overset{+}{\text{N}}\text{H}_3}}{\overset{\overset{\text{O}}{\underset{\|}{\text{CH}_2\text{OCR}}}}{\text{C}\blacktriangleleft\text{H}}} \xrightarrow[\text{heat}]{\text{NaOH, H}_2\text{O}} 2\text{RCO}^-\,\text{Na}^+ + \underset{\substack{|\\\text{CH}_2\text{OH}}}{\overset{\overset{\text{CH}_2\text{OH}}{|}}{\text{CHOH}}}$$

$$+ \text{HOCH}_2\text{CH}_2\text{NH}_2 + \text{Na}_2\text{HPO}_4$$

$$\text{RCO}\blacktriangleright\underset{\substack{|\\\text{CH}_2\text{O}-\overset{\text{O}^-}{\underset{\overset{\|}{\text{O}}}{\text{P}}}\text{OCH}_2\text{CH}_2\overset{+}{\text{N}}(\text{CH}_3)_3}}{\overset{\overset{\text{O}}{\underset{\|}{\text{CH}_2\text{OCR}}}}{\text{C}\blacktriangleleft\text{H}}} \xrightarrow[\text{heat}]{\text{NaOH, H}_2\text{O}} 2\text{RCO}^-\,\text{Na}^+ + \underset{\substack{|\\\text{CH}_2\text{OH}}}{\overset{\overset{\text{CH}_2\text{OH}}{|}}{\text{CHOH}}}$$

$$+ \text{HOCH}_2\text{CH}_2\overset{+}{\text{N}}(\text{CH}_3)_3 + \text{Na}_2\text{HPO}_4$$

$$\underset{\substack{|\\\text{CHNH}-\overset{\text{O}}{\underset{\|}{\text{C}}}\text{R}\\|\\\text{CH}_2-\text{O}-\overset{\text{O}^-}{\underset{\overset{\|}{\text{O}}}{\text{P}}}\text{OCH}_2\text{CH}_2\overset{+}{\text{N}}(\text{CH}_3)_3}}{\overset{\overset{\text{CH}_3(\text{CH}_2)_{12}\quad\text{H}}{\underset{\text{H}}{\text{C}=\text{C}}}}{\text{CHOH}}} \xrightarrow{\text{NaOH}} \underset{\substack{|\\\text{CHNH}_2\\|\\\text{CH}_2-\text{OH}}}{\overset{\overset{\text{CH}_3(\text{CH}_2)_{12}\quad\text{H}}{\underset{\text{H}}{\text{C}=\text{C}}}}{\text{CHOH}}}$$

$$+ \text{RCO}_2^-\,\text{Na}^+ + \text{HOCH}_2\text{CH}_2\overset{+}{\text{N}}(\text{CH}_3)_3 + \text{Na}_2\text{HPO}_4$$

17.3. The amide of cholic acid contains an ionic head (hydrophilic) and a predominantly nonpolar portion (hydrophobic). The nonpolar portion can carry along hydrophobic fatty molecules and emulsify them.

CHAPTER 18

18.1. (a) $\overset{+}{H_3}NCH\overset{O}{\overset{\|}{C}}O^- + HCl \longrightarrow \overset{+}{H_3}NCH\overset{O}{\overset{\|}{C}}OH + Cl^-$
        $|$ $|$
       CH_3 CH_3

(b) $\overset{+}{H_3}NCH\overset{O}{\overset{\|}{C}}O^- + Na^+\ ^-OH \longrightarrow H_2NCH\overset{O}{\overset{\|}{C}}O^- + Na^+ + H_2O$
        $|$ $|$
      CH_2OH CH_2OH

18.2. (a) $H_2NCH_2CH_2CH_2CH_2CHCO_2^- + 2H_2O$
                                    $|$
                                  NH_2

(b) $^-O_2CCH_2CH_2CHC\overset{O}{\overset{\|}{C}}O^- + H_3O^+ + Na^+$
              $|$
            $^+NH_3$

18.3. (a) near pH 3; **(b)** near pH 6; **(c)** near pH 10.

18.4. (a) ⟨benzene ring⟩$-CH_2CHCO_2H + HCl$
                           $|$
                  $NH\overset{O}{\overset{\|}{C}}CH_2CH_3$

(b) $\overset{+}{H_3}NCH_2\overset{O}{\overset{\|}{C}}OCH_3 + H_2O$

(b) $\overset{+}{H_3}NCH_2\overset{O}{\overset{\|}{C}}NHCHCH_2S-SCH_2CHCO_2^- + {}^-O_2CCHCH_2S-SCH_2CHCO_2^-$
                        $|$ $|$ $|$
                      CO_2^- $\overset{+}{N}H_3$ $\overset{+}{N}H_3$

   ... $\overset{+}{N}H_3$ ...

$+ \overset{+}{H_3}NCH_2\overset{O}{\overset{\|}{C}}NHCHCH_2S-SCH_2CHNHCCH_2\overset{+}{N}H_3$
                  $|$ $|$ $\overset{O}{\overset{\|}{}}$
                CO_2^- CO_2^-

18.5. Arg-Pro-Pro
      Pro-Pro-Gly
          Pro-Gly-Phe
              Gly-Phe-Ser
                  Phe-Ser-Pro
                      Ser-Pro-Phe
                          Pro-Phe-Arg
      ─────────────────────────────────────
      Arg-Pro-Pro-Gly-Phe-Ser-Pro-Phe-Arg
                      ↖
      Amino acid sequence in bradykinin

CHAPTER 19

19.1.

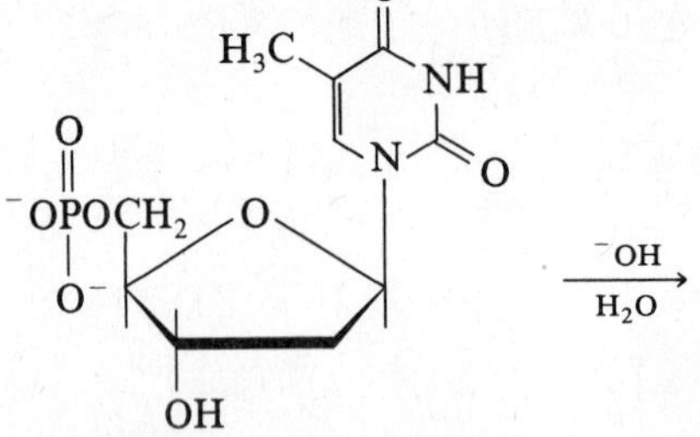

Deoxythymidine 5'-phosphate
(or thymidine 5'-phosphate)

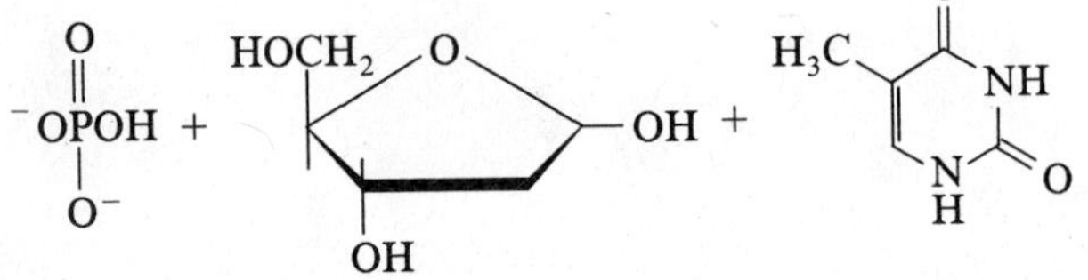

19.2.

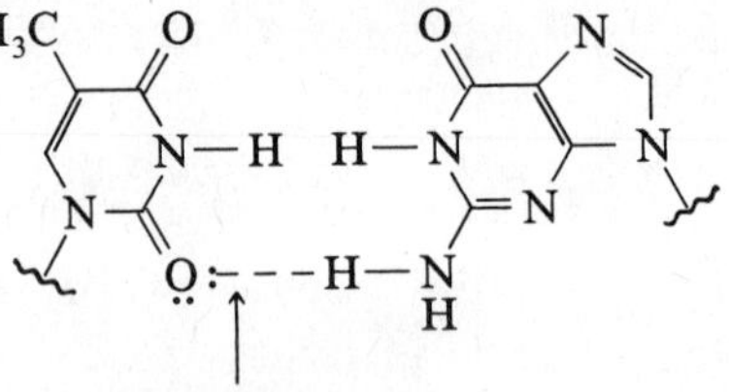

Only hydrogen bond

19.3.

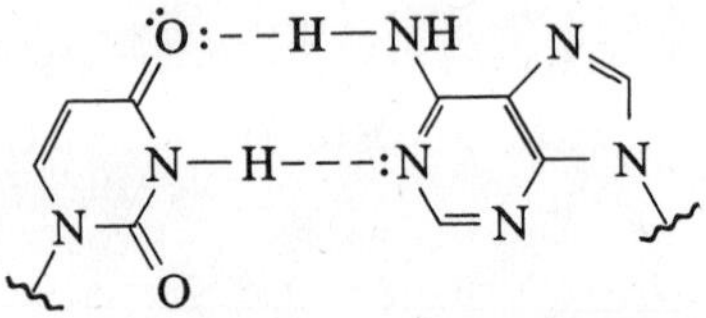

Uracil   Adenine

19.4.

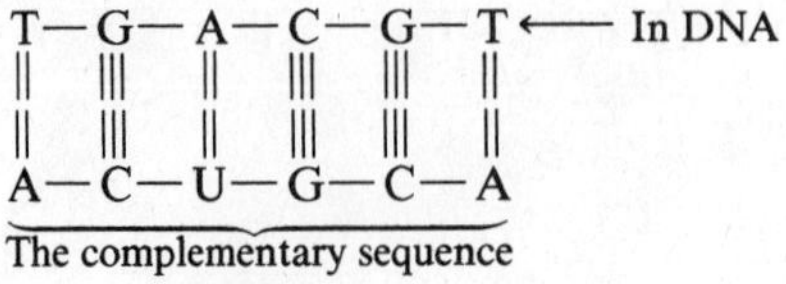

The complementary sequence
in RNA

19.5.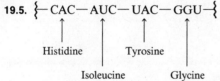

Amino acid sequence: His-Ile-Tyr-Gly

19.6.   Codons for Glu: GAA, GAG

Codons for Arg: CGU, CGC, CGA, CGG, AGA, AGG

Possible sequences: GAA-CGU, GAA-CGC, GAA-CGA, GAA-CGG,

GAA-AGA, GAA-AGG, GAG-CGU, GAG-CGC,

GAG-CGA, GAG-CGG, GAG-AGA, GAG-AGG

Index

Absorption in Infrared Spectra

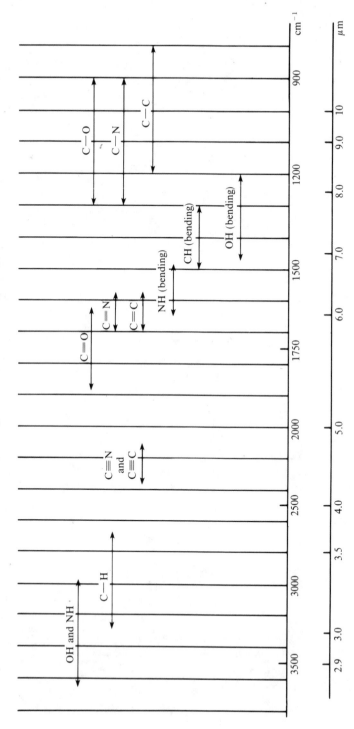